Chemistry

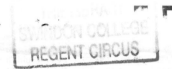

ANN AND PATRICK FULLICK

Heinemann Educational Publishers
Halley Court, Jordan Hill, Oxford, OX2 8EJ
a division of Reed Educational & Professional Publishing Ltd
Heinemann is a registered trademark of Reed Educational & Professional Publishing Ltd

OXFORD MELBOURNE AUCKLAND
JOHANNESBURG BLANTYRE GABARONE
IBADAN PORTSMOUTH NH (USA) CHICAGO

First published 2000
ISBN 0 435 57096 X

04 03 02 01 00
10 9 8 7 6 5 4 3 2 1

Edited by Anne Russell

Designed, produced and illustrated by Gecko Limited, Bicester, Oxon

Original illustrations © Heinemann Educational Publishers, 2000

Printed and bound in Spain by Mateu Cromo

Tel: 01865 888058 www.heinemann.co.uk

Acknowledgements
The authors and publishers would like to the following for permission to use photographs:

Cover photos: Stone

p2 and contents The Planetarium, Armagh; **p3 T** Archiv fü Kunst und Geschichte, Berlin, **B** Philip Parkhouse; **p4** Andrew Syred, Microscopix; **p5** SPL; **p6** Chris Coggins; **p7** Bureau International Des Poids et Mesures; **p8** Dr B Bobth/GSF Picture Library; **p9** Chris Coggins; **p10** Peter Gould; **p11** Peter Gould; **p12** SPL; **p13** The Science Museum; **p15 T** NASA, **B** SPL; **p16 (a)** SPL, **(b) and (c)** Peter Gould, **(d)**, **(e)**, **(f)**, **(g) and (h)** SPL; **p19 T** Vivien Fifield, **BR and BL** Mary Evans Picture Library; **p23** SPL; **p24** The Science Museum; **p26 both** SPL; **p32** Philip Parkhouse; **p36** NHPA; **p37** SPL; **p44** Ace Photos; **p47 T** SPL, **B** Peter Gould; **p52 T** SPL; **M** NHNM, **B** Smithsonian; **p55** Robert Harding Picture Library; **p61** Adam Hart-Davis/SPL; **p64 T** SPL; **B** Tony Stones Images; **p65** Peter Gould; **p69** SPL; **p73 T** Dr G Dobson, University of York, **B** SPL; **p74 L** SPL, **R** Peter Gould; **p75** Andrew Lambert; **p78** SPL; **p81** Peter Gould; **p82 T** Philip Parkhouse, **B** Peter Gould; **p83 M and B** Peter Gould; **p84 T** Robert Harding Picture Library, **B** Peter Gould; **p87 M and L** SPL, **TR** Chris Coggins, **BR** Robert Harding Picture Library; **p89** Philip Parkhouse; **p90** SPL; **p91 T** OSF, **B** Roger Scruton; **p93** Chris Coggins; **p96** Vivien Fifield; **p97 both** SPL; **p99** SPL; **p101 T** OSF, **B** Bruce Coleman; **p102** OSF; **p104** SPL; **p105** Peter Gould; **p107** Chris Coggins; **p108** SPL; **p109** MC Escher Foundation, Baarn, Holland; **p113** Peter Gould; **p114** Philip Parkhouse; **p119** J Allan Cash; **p126** SPL; **p127 T** Andrew Lambert, **B** SPL; **p130 (a)**, **(b) and (c)** Peter Gould, **(d)** Robert Harding Picture Library, **(e)** Chris Coggins, **(f)** SPL, **(g)** The Hulton-Deutsch Collection, **(h)** SPL; **p142 T** Ace Photo Agency, **B** SPL; **p146 T** SPL, **M** J Allan Cash Ltd; **p147** Bruce Coleman; **p149 T** Peter Gould, **B** SPL; **p153 (a)** J Allan Cash Ltd; **(b)** Peter Gould; **(c) and (d)** Chris Coggins; **(e)** Philip Parkhouse; **(f)** Chris Coggins; **(g)** Peter Gould; **p155** Hutchison Library; **p156 TL** Peter Gould; **TR** The Royal Photographic Society; **B** SPL; **ML** Chris Coggins; **MM** SPL; **MR** J Allan Cash Ltd; **p160 L** Chris Coggins; **R** SPL; **p162 T** J Allan Cash Ltd, **B** Chris Coggins; **p164** Roger Scruton; **p165 both** SPL; **p166 TL** Ancient Art and Architecture, **M** J Allan Cash Ltd, **BR** Heather Angel; **p169 TL** Peter Gould, **TR** Philip Parkhouse, **BL** Chris Coggins, **M** SPL, **BR** Ace Photo Agency; **p175** SPL; **p176** The Mansell Collection; **p180** SPL; **p182** Peter Gould; **p184 T** SPL, **B** Topham Picture Source; **p185 L** Peter Gould, **M** Chris Coggins, **R** SPL; **p187 T** Paul Brierley, **B** Chris Coggins; **p188** Paul Brierley; **p190 T** Barnabys Picture Library, **B** Peter Gould; **p191** SPL; **p193 TL** Peter Gould, **BR** SPL, **BL** Robert Harding Picture Library, **TR** Ancient Art and Architecture; **p199 T** Associated Press **B** M Perrones, University of Mexico; **p200** The Mansell Collection; **p201** Chris Coggins; **p203** Philip Parkhouse; **p206 TR** Dulux; **TL** SPL; **BL** Robert Harding Picture Library; **BM** Philip Parkhouse; **BR** Chris Coggins; **p207 L** Nicks and **M** F Chambers; **p212** Peter Gould; **p213 T** Peter Gould, **B** Ancient Art and Architecture; **p214 and contents** J Allan Cash Ltd; **p219** Philip Parkhouse; **p221 BR** Peter Gould, **M and TR** Philip Parkhouse, **TL** ICI; **p222** Peter Gould; **p223 T** Iron Age, **B** Peter Gould; **p226** Titanium Metals Corporation; **p232** OSF; **p233** Advertising Archives; **p234 T** Geoscience Features, **B** Greenpeace Communications Ltd; **p236** SPL/Mark Clarke; **p251** Peter Gould; **p255** SPL/Will McIntyre; **p256 L and R** Philip Parkhouse, **MR** Haymarket Magazines Ltd, **ML** Ace Photo Agency/Gabe Palmer; **p257** Mary Evans Picture Library; **p258** Peter Gould; **p261** Peter Gould; **p270** SPL/Simon Fraser; **p274** Allsport; **p275** Peter Gould; **p280 T** OSF, **B** Bruce Coleman; **p283** Philip Parkhouse; **p284 and contents** SPL/Jerry Schad; **p285** Peter Gould; **p287 T** Philip Parkhouse, **B** J Allan Cash Ltd; **p291** J Allan Cash Ltd; **p293 T** Mary Evans Picture Library, **B** SPL/Sam Ogden; **p295 both** Image Select; **p303** Peter Gould; **p306 T** Planet Earth Pictures, **B** Haymarket Magazines Ltd; **p307** SPL/Bill Longore; **p309 T** Peter Gould, **B** NASA; **p313** Johnson Matthey; **p314** Roger Scruton; **p315** Frank Spooner Pictures; **p317 T** Peter Gould, **B** Geoscience Features; **p318 TR** Modulus Technical Services Ltd, **B** Peter Gould; **p319 T** Bruce Coleman Ltd; **B** Peter Gould; **p326** SPL; **p337** Novo Nordisk; **p340** OSF, **B** HE; **p341 TL** SPL/Royal Observatory, Edinburgh; **TR** J Allan Cash Ltd; **BL and BR** Peter Gould; **p342** Peter Gould; **p344 TL** The Environment Picture Library, **TR** Philip Parkhouse, **B** Mark Powell; **p344 T** Mark Powell, **M** Peter Gould, **B** Philip Parkhouse; **p349** Royal Society of Chemistry; **p358** SPL; **p359 T** Peter Gould, **B** SPL; **p360** Robert Harding Picture Library; **p364 T** Peter Gould, **B** Mark Powell; **p370** A-Z Botanical Collection; **p373** Peter Gould; **p374** Peter Gould; **p385** Peter Gould; **p381 T** OSF **B** Panos Pictures; **p382** SPL/Martin Bond; **p384** Sally & Richard Greenhill; **p385 T** Valor Heating, **B** Calor Gas Ltd; **p387** Mazda Cars (UK) Ltd; **p389** Philip Parkhouse; **p394 T** SPL, **B** OSF; **p395** Philip Parkhouse; **p398** Ace Photo Agency/David Kerwin; **p403 L** SPL/Biophoto Associates, **R** Philip Parkhouse; **p404** Philip Parkhouse; **p406 L** Barnabys Picture Library, **R** Haymarket Magazines Ltd; **p409** Planet Earth Pictures; **p413 T** Mary Evans Picture Library, **B** Philip Parkhouse; **p416 TL** J Allan Cash Ltd, **TR** Image Select, **M** Royal Society of Chemistry, **B** Mary Evans Picture Library; **p424** Peter Gould; **p425** H M & S; **p430 L** Philip Parkhouse, **R** Image Select; **p432 T** Mary Evans Picture Library, **B** SPL/David Guyon, The BOC Group plc; **p438 L** The Advertising Archives, **R** Philip Parkhouse; **p439** Peter Gould; **p446** SPL/Alex Bartel; **p452 and contents** Robert Harding Picture Library; **p453** Mary Evans Pictures Library; **p454** Mark Edwards/Still Pictures; **p468** Peter Gould; **p470** Mary Evans Picture Library; **p473** Gary Moyes/BBC Good Food; **p479 T** Holt Studios International, **B** Philip Parkhouse; **p477** Norprint International Ltd; **p478** Peter Gould; **p479** Peter Gould; **p482 TL** Henning Christoph/Sylvia Katz Collection, **TR** Russell Hobbs/Sylvia Katz Collection, **B** SPL/Jack Fields; **p484** Peter Gould; **p489** Roger Scruton; **p490 T and M** Philip Parkhouse, **B** SPL/David Scharf; **p492** Peter Gould; **p493** Philip Parkhouse; **p495** SPL; **p496** The Environmental Picture Library; **p501** Courtesy of Mothercare; **p504** SPL/Alexandre/BSIP; **p507** Ace Photos, **B** Popperfoto; **p508** Frank Spooner Pictures; **p509** Allsport; **p521** SPL/Adrienne Hart-Davis; **p522 TL** and **BL** The Colour Museum, **TR** SPL/David Leah, **BR** Robert Harding Picture Library; **p523 T** SPL/Chemical Design Ltd, **B** Biophoto Associates; **p526 T** SPL/Division of Computer Research & Technology, National Institute of Health; **B** NHPA; **p528** SPL/Dr Jeremy Burgess; **p530 L** SPL/CNRI; **R** SPL/Dr Jeremy Burgess; **p531** SPL/Dr Jeremy Burgess; **p533** Philip Parkhouse; **p540 T** Peter Gould; **BL** SPL/Division of Computer Research & Technology, National Institute of Health, **BM** J Allan Cash Ltd; **BR** Plant Earth Pictures.

Picture research by Thelma Gilbert

The authors and publishers would like to thank the AQA, Edexel and OCR Examination Boards for permission to reproduce their material.

The authors and publishers would like to thank the following for permission to reproduce copyright material:

p69 John Frederick Nims for the extract from his poem; **p172** ICI Chemicals & Polymers Limited for the diagram of chlorine production; **p233** Society of Motor Manufacturers and Traders for plotting data; **p298** Maps reproduced from Ordnance Survey mapping with the permission of the Controller of Her Majesty's Stationery Office, © Crown copyright, Licence no. 398020; **p351** American Chemical Society for the graph of theoretical conditions to convert graphite to diamond.

The publishers have made very effort to trace the copyright holders, but if they have inadvertently overlooked any, they will be pleased to make the necessary arrangements at the first opportunity.

How to use this book

Heinemann Advanced Science: Chemistry has been written to accompany your AS and A2 biology courses, and contains all the core specification material you will need during your period of study. The book is divided into five sections of associated material which have been made easy to find by the use of colour coding.

Within each chapter a *Chemistry Focus* explores the real-life context of an area of the science to come. You may use this to whet your appetite for the chapter that follows, or read it when you have made some progress with the chemistry to emphasise the relevance of what you are doing. However you use it, we hope it will make you want to read on!

In addition to the main text, the chapters of the book contain two types of boxes.

The blue *information* boxes contain basic facts or techniques which you need to know. This often includes key facts you will have already met at GCSE – the boxes then carry these ideas forward to AS or A2 level. *Information* boxes are sometimes referred to in the main text, and can be read either as you meet them or when you have finished reading the chapter.

The pink headed *extension* boxes contain more advanced information which is not referred to in the main text. You do not need to address the contents of these *extension* boxes until you have really got to grips with the rest of the material in the chapter or are studying the subject at A2 level.

When you have completed the work in a chapter of the book, there are questions to help you find out how much of the material you have understood and to help you with your revision. Many chapters also have a Key Skills question. These are designed to help you develop the Key Skills needed to complete your studies successfully, particularly in the areas of Communication [C], Application of Number [A] and Use of IT [IT]. These Key Skills often refer to the Focus material as well as the chapter contents. The summaries at the end of each chapter provide further help with revision, while at the end of each section of the book there is a selection of AS and A2 level questions.

This book has been written to be an accessible, clear and exciting guide to AS and A2 chemistry. We hope that it will help to maintain your interest in the subject you have chosen to study, and that it will play a valuable role in developing your knowledge of chemistry – and with it, an increased understanding of chemistry in the world around you.

Acknowledgements

We should like to thank the team at Heinemann who helped get this new edition into print, particularly Lindsey Charles who has been stalwart in her support and Anne Russell for her calmness and editing skills.

Finally, we should like to thank our friends, family and professional colleagues for their continuing advice and contributions to all aspects of the revision. Special thanks go to our sons and Ann's mother for their support and encouragement throughout the project.

Ann and Patrick Fullick, 2000

Dedication

For William, Thomas, James and Edward

CONTENTS

1 MATTER AND ENERGY

2 THE PERIODIC TABLE

3 HOW FAR? HOW FAST?

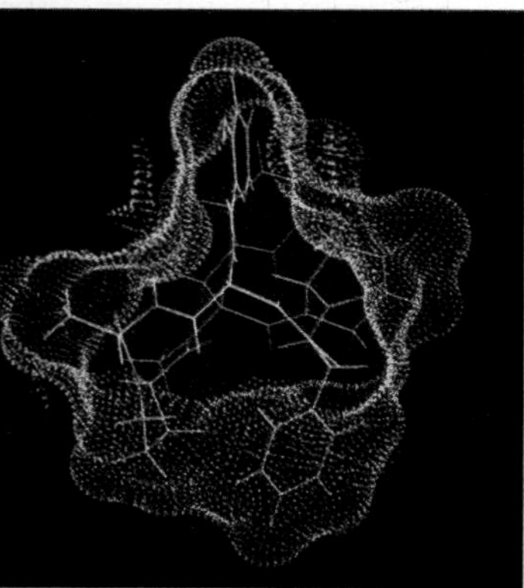

4 ORGANIC CHEMISTRY (1)

5 ORGANIC CHEMISTRY (2)

1 MATTER AND ENERGY

Introduction

Within the Earth there is an astonishing number of different chemicals which are part of both the Earth itself and the living organisms which grow and live upon it. But all these different chemicals are made up of the same type of fundamental particles, which we call atoms. As a result of the work of scientists in the early years of the twentieth century, we now have models which help us to understand the structure of atoms and their interactions with each other – this work forms part of the ideas we shall examine in the early part of this section.

Once the mysteries of the atoms themselves have been unravelled, the way that atoms form bonds and join together in molecules can be understood. This is important for understanding the world around us, and also has many implications for industrial processes in which nature-identical and artificial substances are made. Making chemical compounds which are useful to human beings is a multi-billion pound industry around the world, from the production of relatively simple compounds such as sodium hydroxide through to the development of enormously complex pharmaceutical molecules to help cure disease.

With the exception of those molecules found deep in space, molecules do not exist in isolation – they interact with other molecules around them as a result of **intermolecular forces**. These forces affect the arrangement of the molecules and how tightly they are held together. This affects the physical properties of substances, and so intermolecular forces have an enormous influence on the world around us – they play a large role in determining properties such as melting points and boiling points, the viscosity of liquids and the strength of molecular solids.

In any chemical change energy is involved. When bonds are broken energy is required, while the formation of bonds releases energy. Accounting for these energy changes helps us to predict whether a particular reaction is likely to take place, and under what conditions, making possible safe and effective chemical reactions on both laboratory and industrial scales.

Figure 1 Interactions between matter and energy can produce some spectacular results.

FOCUS THE LEGACY OF THE ALCHEMISTS

Where has chemistry come from? Throughout the history of the human race, people have struggled to make sense of the world around them. Through the branch of science we call chemistry we have gained an understanding of the matter which makes up our world and of the interactions between particles on which it depends. The ancient Greek philosophers had their own ideas of the nature of matter, proposing atoms as the smallest indivisible particles. However, although this idea seems to fit with modern models of matter, so many other Ancient Greek ideas were wrong that chemistry cannot truly be said to have started there.

Alchemy was a mixture of scientific investigation and mystical quest, with strands of philosophy from Greece, China, Egypt and Arabia mixed in. The main aims of alchemy that emerged with time were the quest for the **elixir of life** (the drinking of which would endue the alchemist with immortality), and the search for the **philosopher's stone**, which would turn base metals into gold. Improbable as these ideas might seem today, the alchemists continued their quests for around 2000 years and achieved some remarkable successes, even if the elixir of life and the philosopher's stone never appeared.

Over the centuries the alchemists discovered the first strong acids, the distillation of alcohol, the element zinc and the use of opium in pain relief. The tradition in alchemy continued into the eighteenth century, and one of the most illustrious alchemists ever was Sir Isaac Newton. Alongside his famous work in physics and mathematics, Newton was an ardent alchemist who spent much of his time in the search for the philosopher's stone. It is thought that he may eventually have gone mad due to mercury poisoning as a result of his alchemical experiments.

Towards the end of the eighteenth century, pioneering work by Antoine and Marie Lavoisier and by John Dalton on the chemistry of air and the atomic nature of matter paved the way for modern chemistry. During the nineteenth century chemists worked steadily towards an understanding of the relationships between the different chemical elements and the way they react together. A great body of work was built up from careful observation and experimentation until the relationships which we now represent as the periodic table emerged. This brought order to the chemical world, and from then on chemists have never looked back.

Modern society looks to chemists to produce, amongst many things, healing drugs, pesticides and fertilisers to ensure better crops and chemicals for the many synthetic materials produced in the twenty-first century. It also looks for an academic understanding of how matter works and how the environment might be protected from the damage humans have caused. Fortunately, chemistry holds many of the answers!

Figure 1 Alchemists – the first recognisable forerunners to the modern chemists

Figure 2 Chemistry has developed in a number of directions, from the study of fundamental particles to the understanding of the massive and complex organic molecules found both in living organisms and synthesised within the chemical industry.

1.1 Atoms: the basis of matter

Early ideas

The existence of atoms seems to us today to be one of the simplest and most fundamental of principles, but this was not always the case. The ideas behind the existence of atoms can be traced back about 2500 years, to the Greek philosopher Leucippus and his pupil Demokritos. The word 'atom' comes from the Greek word meaning 'indivisible'. Despite this long history, it was not until the turn of the nineteenth century that chemists began to describe their observations in terms of atoms.

Evidence for the existence of atoms

Although chemists began to use the idea of atoms to explain their observations in the early part of the nineteenth century, the scientific community as a whole did not generally accept their existence until much later in the century. The conclusive proof that matter is made up of atoms was provided by Albert Einstein in 1905, whose calculations showed that Brownian motion is best explained in terms of a particle theory of matter.

Figure 1.1.1 The dancing, zig-zag motion of smoke particles as they are bombarded by air molecules is known as Brownian motion. It was first observed by the botanist Robert Brown in 1827, who was looking at pollen grains suspended in water.

Even though there is powerful evidence that matter is made up of particles, our everyday observations do not confirm it – after all, a piece of steel (say) appears smooth and continuous. This appearance of the smoothness and solidity of matter is an illusion, caused by the immense number of particles packed into a small volume of matter.

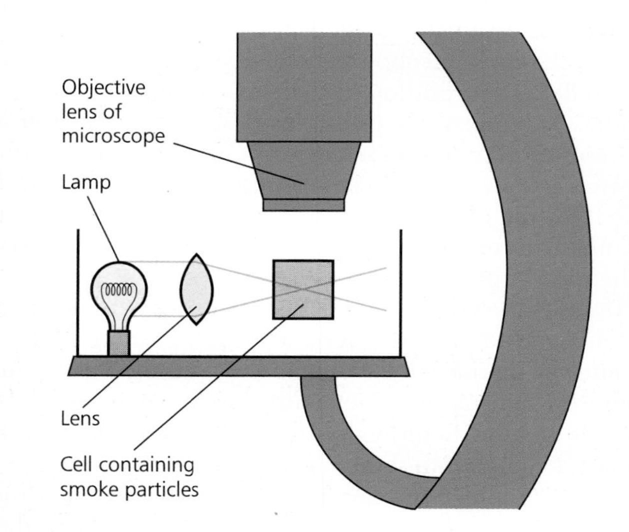

The atomic theory of John Dalton was published in 1808. In this theory, Dalton proposed the existence of indivisible particles of matter called atoms. He also proposed that 'Chemical analysis and synthesis go no further than to the separation of particles from one another and their union. No new creation or destruction of matter is within the reach of chemical agency.' Dalton's ideas can be expressed in simple terms as follows:

1 All matter consists of indivisible particles called atoms.
2 Atoms of the same element are similar in mass and shape, but differ from atoms of other elements.
3 Atoms cannot be created or destroyed.
4 Atoms combine together to form 'compound atoms' (which we would now call molecules) in simple ratios.

Although some of Dalton's ideas have subsequently been shown to be incorrect, his theory formed the basis for much of the work of chemists in the nineteenth century, including the foundation of the periodic table, as we shall see in section 2.1.

Figure 1.1.2 The work of John Dalton represented a decisive break from the work of the alchemists, yet he was held up to ridicule and scorn. 'Atoms are round bits of wood invented by Mr Dalton', wrote one of his critics.

Counting atoms

The relative atomic mass scale

Whenever we measure something, we compare the property we are measuring with a known quantity. For example, if you buy a kilogram of apples, the mass of the apples is effectively being compared with a standard 1 kg mass. Figure 1.1.4 on page 7 shows the standard kilogram, a platinum–iridium cylinder which is kept at Sèvres near Paris. This cylinder is **defined** as having a mass of exactly 1 kg. When you buy 0.5 kg of apples, you are effectively comparing their mass with the mass of this cylinder! In the same way, chemists compare the masses of atoms with the mass of a standard atom – the carbon-12 atom.

The **relative atomic mass scale** was initially based on the element hydrogen. Using this scale, hydrogen atoms were given a relative atomic mass of exactly 1, and the masses of all other atoms were then compared with them. Now we say that:

The relative atomic mass in grams of any element contains the same number of atoms as there are atoms of carbon-12 in 12 g of carbon-12.

We measure the masses of atoms using a technique called mass spectroscopy, described on pages 13–14. As we shall see, all the atoms in an element do not necessarily have the same mass. Atoms of the same element which have different masses are called **isotopes**, as will be discussed in section 1.2. Following the work of Aston and others earlier in the twentieth century, chemists realised that it was necessary to have a scale of relative atomic mass based on a single isotope. This is why they agreed to use a scale based on carbon-12 instead of hydrogen. On this scale, carbon-12 atoms are assigned a mass of 12 **atomic mass units** (this is written as 12 u). An atom which is three times heavier than an atom of carbon-12 thus has a relative atomic mass of $3 \times 12\text{ u} = 36\text{ u}$, while an atom with a mass of 55 u is ${}^{55\text{ u}}/_{12\text{ u}} \approx 4.6$ times heavier than an atom of carbon-12.

If you look up the relative atomic mass of carbon in a data book, you will find it given as 12.011, not 12.000. This is due to the fact that naturally occurring carbon contains mainly atoms of carbon-12, along with some atoms of carbon-13 and carbon-14, making the *average* relative mass of a carbon atom slightly greater than 12. Relative atomic mass is sometimes represented using the symbol A_r.

The mole

We often need to be able to count atoms, whether to mix atoms together in the correct ratio to carry out a reaction, or to calculate the ratio of numbers of different atoms in a sample of an unknown substance. The **mole** provides a way of doing this.

The relative atomic mass of iron

Mass spectrum studies of iron show that a sample of iron contains atoms in the proportions shown in table 1.1.1.

	Iron-54	Iron-56	Iron-57	Iron-58
Relative abundance /%	5.84	91.68	2.17	0.31

Table 1.1.1

To find the relative atomic mass of iron, we must calculate the average relative atomic mass of these four isotopes of iron, multiplying the relative atomic mass of each isotope by its relative abundance to calculate a **weighted mean**:

$$\begin{aligned} A_r(\text{Fe}) &= 54\,\text{u} \times \tfrac{5.84}{100} + 56\,\text{u} \times \tfrac{91.68}{100} + 57\,\text{u} \times \tfrac{2.17}{100} + 58\,\text{u} \times \tfrac{0.31}{100} \\ &= 55.9\,\text{u} \end{aligned}$$

One atom of carbon-12 is 12 times heavier than one atom of hydrogen, which is in turn 39 times lighter than one atom of potassium. Because the masses of atoms are related in this way, it follows that 1 g of hydrogen contains the same number of atoms as 12 g of carbon-12 or 39 g of potassium. As we have seen, in general, the relative atomic mass in grams of any element contains the same number of atoms as there are atoms of carbon-12 in 12 g of carbon-12. This is the basis of the definition of the **mole**, the SI unit of the amount of substance, defined as:

> **The amount of substance that contains as many particles (whether these are atoms, ions or molecules) as exactly 12 g of carbon-12.**

Just like other units, the mole has an abbreviation, mol. The number of atoms in exactly 12 g of carbon-12 is found from experiments to be 6.02×10^{23} (to three significant figures). This number is known as the **Avogadro constant**, after the nineteenth-century Italian chemist Amedeo Avogadro. The constant has units of particles mol^{-1} (particles per mole). The type of particles being counted will vary according to the substance being described – for example, it might be molecules in a mole of oxygen gas. The definition of the mole means that one mole of *any* substance contains 6.02×10^{23} particles.

Figure 1.1.3 We often count entities by weighing. Counting different numbers of sweets like these can be done quickly by weighing them, but we need to know how much each individual sweet weighs.

1

Units

A great deal of science is based on measuring physical quantities, such as length and mass. The value of a physical quantity consists of two things – a number, combined with a unit. For example, the same length may be quoted as 2.5 km or 2500 m. In order that scientists and engineers can more easily exchange ideas and data with colleagues in other countries, a common system of units is now in use in the world of science. This system is called the *Système Internationale* (SI), and consists of a set of seven **base units**, with other **derived units** obtained by combining these. The base SI units are the metre (m), kilogram (kg), second (s), ampere (A), kelvin (K), candela (cd) and mole (mol). Each of these base units relates to a standard held in a laboratory somewhere in the world, against which all other measurements are effectively being compared when they are made. Figure 1.1.4 shows the standard kilogram.

Figure 1.1.4 This platinum–iridium cylinder is the standard kilogram – it is defined as having a mass of exactly 1 kg.

The units of mass and density show how base units and derived units are related:

- Mass has units of *kilograms* in the SI system. In the laboratory, we are usually working with small amounts and so we more often measure mass in grams.
- Density measures the mass per unit volume of a substance, so the units of density are *kilograms per cubic metre*. This may be written as kilograms/metre^3, kg/m^3 or kg m^{-3}. Each of these means the same thing: kilograms ÷ metres^3. Because m^{-3} = 1/m^3, kg m^{-3} means the same as kg/m^3. We are usually concerned with smaller amounts of substance in chemistry and would more often describe density in the units g dm^{-3}.

The value of a physical quantity consists of a number multiplied by a unit, so the heading for a column in a table or the scale for the axis of a graph may be written as (for example) mass/g or density/g dm^{-3}. Since the description tells the reader that the value of the physical quantity has been divided by the unit, it is then only necessary to write down the numerical part of the physical quantity, since, for example, 5 g ÷ g = 5.

Counting by weighing

The mole enables chemists to count atoms by weighing. Since 12 g of carbon-12 contain 6.02×10^{23} atoms, 1 g of carbon-12 contains:

$$\mathbf{1\ g \times \frac{6.02 \times 10^{23}\ atoms\ mol^{-1}}{12\ g\ mol^{-1}}}$$

and 6 g of carbon-12 contain:

$$\mathbf{6\ g \times \frac{6.02 \times 10^{23}\ atoms\ mol^{-1}}{12\ g\ mol^{-1}}}$$

The relative atomic mass of magnesium is 24, so 24 g of magnesium contain 6.02×10^{23} atoms (this is one mole of magnesium atoms). 6 g of magnesium contain:

$$\mathbf{6\ g \times \frac{6.02 \times 10^{23}\ atoms\ mol^{-1}}{24\ g\ mol^{-1}}}$$

This is half the number of atoms in 6 g of carbon-12, which is only to be expected since magnesium atoms are twice as heavy as atoms of carbon-12.

Notice how the units of each physical quantity have been included in the calculations. Units behave exactly like numbers in calculations, and can be cancelled:

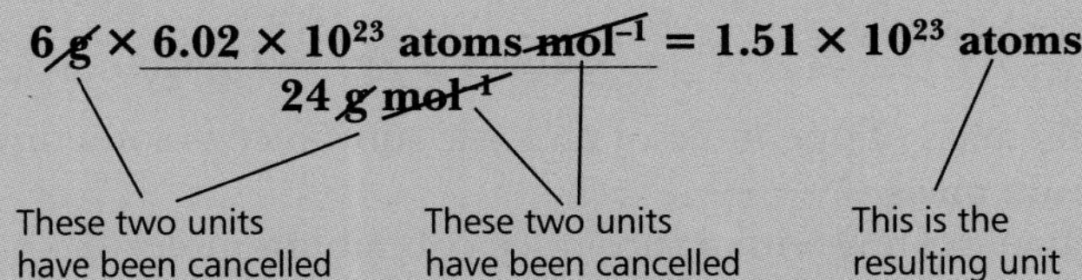
6 g × 6.02 × 10^23 atoms mol^-1 / 24 g mol^-1 = 1.51 × 10^23 atoms
These two units have been cancelled
These two units have been cancelled
This is the resulting unit

Writing equations

When representing chemical changes using equations, the symbol for an element stands for two things – an *atom* of the element and a *mole of atoms* of the element. If we consider the burning of carbon in a plentiful supply of oxygen to produce carbon dioxide, this process can be represented by the equation:

$$\mathbf{C(s) + O_2(g) \rightarrow CO_2(g)}$$

This tells us that:

1 One *atom* of carbon reacts with one *molecule* of oxygen to produce one *molecule* of carbon dioxide.
2 One *mole of atoms* of carbon reacts with one *mole of molecules* of oxygen to produce one *mole of molecules* of carbon dioxide.

The equation also tells us the **state** (solid, liquid, gas or aqueous ions) of the reactants and products.

This example shows why it is important to specify precisely what type of particle we are talking about whenever we use moles. Since oxygen is diatomic (it consists of molecules each of which contains two oxygen atoms), it would be confusing to refer simply to 'a mole of oxygen' – do we mean oxygen atoms or oxygen molecules? Using the formula for the substance or clearly stating the particles involved avoids this confusion.

For ionic compounds such as sodium chloride, one mole of the substance is calculated using the formula of the substance. One mole of sodium chloride, NaCl, thus has a mass of:

$$\mathbf{1 \times 23\ g + 1 \times 35.5\ g = 58.5\ g}$$

since it contains one mole of sodium ions (Na^+) with a mass of 23 g, and one mole of chloride ions (Cl^-) with a mass of 35.5 g. This shows that to make one mole of sodium chloride, we must react one mole of sodium atoms with one mole of chlorine atoms. Like oxygen, chlorine is diatomic, so the equation for this reaction is written:

$$\mathbf{Na(s) + \tfrac{1}{2}Cl_2(g) \rightarrow NaCl(s)}$$

To avoid having fractions in the equation, we multiply through by 2:

$$\mathbf{2Na(s) + Cl_2(g) \rightarrow 2NaCl(s)}$$

In contrast, one mole of calcium fluoride, CaF_2, has a mass of:

$$\mathbf{1 \times 40\ g + 2 \times 19\ g = 78\ g}$$

as this contains one mole of calcium ions (Ca^{2+}) and two moles of fluoride ions (F^-). The formation of calcium fluoride from calcium and fluorine can be represented by the equation:

$$\mathbf{Ca(s) + F_2(g) \rightarrow CaF_2(s)}$$

The mass of one mole of an ionic substance is sometimes referred to as its **formula mass**.

When dealing with a substance containing molecules rather than ions, the **relative molecular mass** is the sum of all the relative atomic masses of the atoms making up each molecule. In this way, the relative molecular mass of carbon dioxide, CO_2, is $12 + (2 \times 16) = 44$. By definition, the relative molecular mass of the substance in grams contains one mole of molecules of the substance, so 44 g of carbon dioxide contain 6.02×10^{23} molecules.

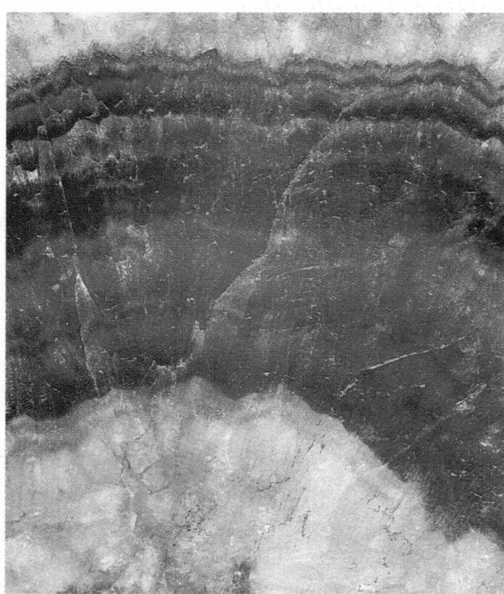

Figure 1.1.5 Calcium fluoride is found naturally as the mineral fluorspar. This variety is called blue john, and is used to make jewellery and other ornaments.

Calculating formulae

The formula of a compound shows which atoms are present in the compound, and in what ratio. The following example shows how the formula of a substance may be calculated from the masses of the different elements in it.

Analysis of a sample of aluminium chloride shows that it contains 5.8 g of aluminium and 22.9 g of chlorine. The calculation of the number of moles of atoms in this sample may be carried out using a simple table like table 1.1.2.

	Al	Cl
Mass found by analysis	5.8 g	22.9 g
Mass of one mole of atoms	27 g mol^{-1}	35.5 g mol^{-1}
Moles of atoms in sample	$\frac{5.8 \text{ g}}{27 \text{ g mol}^{-1}} = 0.215$ mol	$\frac{22.9 \text{ g}}{35.5 \text{ g mol}^{-1}} = 0.645$ mol
Ratio of moles in sample (= ratio of atoms in sample)	1	3

Table 1.1.2

This calculation shows that the ratio of the numbers of aluminium atoms to chlorine atoms in this sample of aluminium chloride is 1:3, which might indicate that the formula is $AlCl_3$. However, other formulae such as Al_2Cl_6 or Al_3Cl_9 are also possible, and the analysis does not give any guide as to which of these is correct.

The simplest formula for a chemical compound shows the ratio of numbers of atoms of each element in the compound using whole numbers. This formula is called the **empirical formula**. The formula that describes the actual numbers of atoms of each element in a molecule of a compound is called the **molecular formula**. In the case of aluminium chloride, as we shall see in section 2.5, both $AlCl_3$ and Al_2Cl_6 are correct as the molecular formula!

Giant structures of atoms and ions are described by their empirical formula, since a molecular formula has no meaning for these substances.

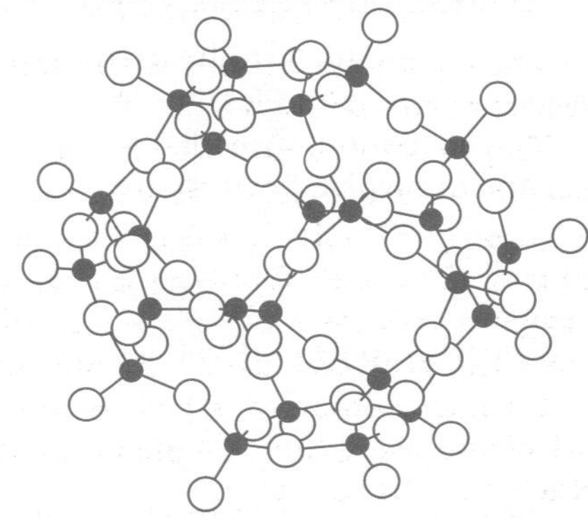

Figure 1.1.6 The water softener in this domestic dishwasher contains an **ion-exchange resin**, which belongs to a class of compounds called the **zeolites**. The basic structure of the zeolite lattice is shown in the diagram, where the large open spaces in the lattice can be seen – it is these spaces that enable the resin to exchange one type of ion for another. One zeolite used as an ion-exchange resin has the formula $Na_{12}Al_{12}Si_{12}O_{48}.27H_2O$.

Moles in solution

A great number of reactions occur in solution – that is, with a substance (solid, liquid or gas) dissolved in a solvent. It is convenient to have a measure of the amount of substance dissolved in a particular volume of solution – the **concentration**. The units used for this are **moles per cubic decimetre**, abbreviated as mol dm^{-3}.

The cubic decimetre is now used as the unit of volume in chemistry, replacing the **litre**, although the two units are equivalent as 1 dm = 10 cm:

$$\textbf{1 litre} = \textbf{1000 cm}^3 = \textbf{(10 cm)}^3 = \textbf{1 dm}^3$$

The unit mol dm^{-3} describes the concentration of a dissolved substance (called a **solute**) in terms of the number of moles of solute in 1 dm^3 of the solution. This is *not* the same as saying 'the number of moles of solute added to 1 dm^3 of solvent', as figure 1.1.7 illustrates. The symbol M is sometimes used to abbreviate mol dm^{-3} further, so that 0.5M means 0.5 mol dm^{-3}.

Calculating reacting quantities

Many chemical reactions involve gases, having gaseous reactants, products or both. The reaction of black copper(II) oxide with hydrogen is an example:

$$\textbf{Copper(II) oxide + hydrogen} \rightarrow \textbf{copper + water}$$

This **word equation** tells us very little about the reaction, other than the names of the reactants and the products. For this reason, chemists use the chemical formula of a substance in equations:

$$CuO(s) + H_2(g) \rightarrow Cu(s) + H_2O(l)$$

As we saw earlier, this gives us much more information about what happens when copper(II) oxide and hydrogen react together, although it still does not tell us how hydrogen and copper(II) oxide react together (do the hydrogen atoms in the hydrogen molecule split apart and then react with the copper oxide, for example?), nor does it tell us that the copper(II) oxide must be heated strongly for the reaction to occur. The equation tells us only about the overall change taking place, not the reaction mechanism or the reaction conditions.

So how much copper(II) oxide and hydrogen gas would be needed to produce 50 g of copper? This is just the sort of question that a chemist involved in an industrial process needs to answer, though in industry carbon would be more likely to be used than hydrogen for this reaction, since it is easier to handle and less hazardous. The equation for the process helps us to do the calculation, provided we know the relative atomic masses of the substances involved, which are:

Cu	O	H
63.5	16	1

If we wish to produce 50 g of copper, this is:

$$\frac{50\ \text{g}}{63.5\ \text{g mol}^{-1}} = 0.787\ \text{mol of copper}$$

The equation for this reaction tells us that 1 mol of copper(II) oxide is needed to produce 1 mol of copper, so 0.787 mol of copper(II) oxide will be needed to produce 0.787 mol of copper. The mass of 1 mol of copper(II) oxide is:

$$(63.5 + 16)\ \text{g mol}^{-1} = 79.5\ \text{g mol}^{-1}$$

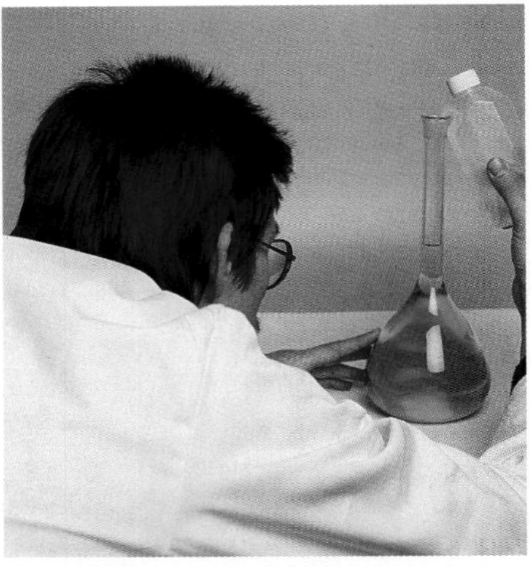

Figure 1.1.7 To make a solution of accurate concentration, you dissolve the solute in a small quantity of solvent and then add more solvent until the solution has the required volume. This technician has ensured that 0.1 mol of copper(II) sulphate is dissolved in 1 dm^3 of solution, which would not be the case if he had simply dissolved 0.1 mol of copper(II) sulphate in 1 dm^3 of water.

Calculating moles in solution

To calculate the concentration of a solution, we divide the number of moles of solute by the volume. For example, in 25.0 cm^3 of a solution containing 4.9 g of sulphuric acid,

$$\text{Amount of } H_2SO_4 = \frac{4.9\ \text{g}}{((2 \times 1)+32+(4 \times 16))\ \text{g mol}^{-1}}$$

$$= \frac{4.9\ \text{g}}{98\ \text{g mol}^{-1}}$$

$$= 0.05\ \text{mol}$$

This sulphuric acid is dissolved in 25.0 cm^3 of solution. There are 1000 cm^3 in 1 dm^3, so:

$$H_2SO_4 \text{ in 1 dm}^3 = \frac{100\ \text{cm}^3}{25\ \text{cm}^3} \times 0.05\ \text{mol}$$

$$= 40 \times 0.05\ \text{mol}$$

$$= 2.0\ \text{mol}$$

The concentration of the sulphuric acid solution is 2.0 mol dm^{-3}.

Calculating the number of moles of solute in a given volume of a solution of known strength is done in a similar way. For example, how many moles of NaOH are there in 15.0 cm^3 of 0.2 mol dm^{-3} sodium hydroxide solution?

There are 0.2 mol of NaOH in 1000 cm^3 of solution, so in 1.0 cm^3 of solution,

$$\text{Amount of NaOH} = \frac{1.0\ \text{cm}^3}{1000\ \text{cm}^3} \times 0.2\ \text{mol}$$

$$= 2.0 \times 10^{-4}\ \text{mol}$$

In 15.0 cm^3 there will be 15 times as many moles:

$$\text{NaOH in 15 cm}^3 = 15 \times 2.0 \times 10^{-4}\ \text{mol}$$

$$= 3.0 \times 10^{-3}\ \text{mol}$$

So the mass of copper(II) oxide required is:

$$0.787 \text{ mol} \times 79.5 \text{ g mol}^{-1} = 62.6 \text{ g}$$

Yield

Notice that the calculations to find the amount of copper(II) oxide required to make a certain amount of copper assume that *all* the copper oxide will be turned into copper – that is, the **yield** of the reaction will be 100%. Yields approaching 100% are very rare in practical chemical reactions for various reasons. The reactants may not be pure, or the reaction may be an equilibrium that does not go to completion – we shall find out more about this in section 3.1. Some product will be left behind on the apparatus, and the product may be difficult to purify – for example, if it is to be crystallised, some will be lost. Any calculations which are to be used on a practical basis need to take all this into account. In this example, if the practical yield is likely to be 50% (that is, half of the copper(II) oxide was converted into copper), we will need twice as much copper(II) oxide as calculated above to make 50 g of copper.

Chemical equations tell us nothing about the yield of the reaction.

The mass of hydrogen required can be calculated similarly. The equation tells us that 1 mol of hydrogen gas is needed to produce 1 mol of copper, so 0.787 mol of hydrogen gas will be needed to produce 0.787 mol of copper. The mass of 1 mol of hydrogen gas, H_2, is:

$$2 \times 1 \text{ g mol}^{-1} = 2 \text{ g mol}^{-1}$$

So the mass of hydrogen gas needed is:

$$0.787 \text{ mol} \times 2 \text{ g mol}^{-1} = 1.57 \text{ g}$$

Rather than weighing gases, it is often much more convenient to be able to measure their volume. Provided we are careful to state the conditions under which we are doing this (the density of a gas is affected by its temperature and pressure), this is quite straightforward, since:

One mole of any gas occupies 22.4 dm^3 at 0°C and 10^5 Pa pressure (these conditions are known as standard temperature and pressure, s.t.p.).

In this case, we have 0.787 mol of hydrogen gas, which at s.t.p. will have a volume of:

$$0.787 \text{ mol} \times 22.4 \text{ dm}^3 \text{ mol}^{-1} = 17.6 \text{ dm}^3$$

Figure 1.1.8 Magnesium hydroxide is present in milk of magnesia. It can be made by adding a solution of sodium hydroxide to a solution of magnesium chloride.

Ions in solution

An **ion** is an atom that has lost or gained one or more electrons. An atom that loses an electron becomes positively charged, and is called a **cation**. An atom that gains an electron becomes negatively charged, and is called an **anion**. We shall see the reasons for these names in section 3.5. Many reactions involve ions in solution. When substances react in this way, an **ionic equation** may help to show what is happening. As an example of this, consider adding a solution of sodium hydroxide to a solution of magnesium chloride.

Analysis of the precipitate formed in this reaction shows that it has the formula $Mg(OH)_2$. Careful experiment enables us to calculate the amount of sodium hydroxide solution reacting with a certain amount of magnesium chloride solution. Results indicate that 10.0 cm^3 of 0.1 mol dm^{-3} sodium hydroxide solution react exactly with 5.0 cm^3 of 0.1 mol dm^{-3} magnesium chloride solution. This shows that these substances react in the proportion

2 mol of sodium hydroxide to 1 mol of magnesium chloride. The left-hand side of the equation for this reaction is therefore:

$$2NaOH(aq) + MgCl_2(aq) \rightarrow$$

The reaction produces sodium chloride, NaCl, and magnesium hydroxide, $Mg(OH)_2$, so the balanced equation is:

$$2NaOH(aq) + MgCl_2(aq) \rightarrow 2NaCl(aq) + Mg(OH)_2(s)$$

Equations like this are sometimes called **molecular equations**, because they show the complete formulae of all the substances involved, although none of the substances involved in this example actually consists of molecules.

Since both reactants dissolve in water to form ions, we can also write an **ionic equation** to show what is going on in the reaction:

$$\begin{array}{lcl} 2Na^+(aq) + 2OH^-(aq) & \rightarrow & 2Na^+(aq) + 2Cl^-(aq) \\ + Mg^{2+}(aq) + 2Cl^-(aq) & & + Mg(OH)_2(s) \end{array}$$

From this equation we notice that the sodium ions and chloride ions appear in exactly the same way on each side of the equation. Ions like this are often called **spectator ions**, and can be eliminated from the equation:

$$\begin{array}{lcl} \cancel{2Na^+(aq)} + 2OH^-(aq) & \rightarrow & \cancel{2Na^+(aq)} + \cancel{2Cl^-(aq)} \\ + Mg^{2+}(aq) + \cancel{2Cl^-(aq)} & & + Mg(OH)_2(s) \end{array}$$

This produces the much simpler **net ionic equation**:

$$Mg^{2+}(aq) + 2OH^-(aq) \rightarrow Mg(OH)_2(s)$$

Using different types of equation

Which type of equation is most useful? This depends on what we wish to do. The molecular equation gives us a description of the reactants and the amounts required if we wish to produce magnesium hydroxide in a practical experiment. If we wish to know what is actually happening in solution when sodium hydroxide and magnesium chloride react together, the ionic equation provides us with the information. The power of the net ionic equation is that it enables us to make generalisations about the reaction, so that we can see that, for example, any solution containing magnesium ions will react with any other solution containing hydroxide ions to form a precipitate of magnesium hydroxide.

Measuring atoms

The radius of an atom

Atoms are far too small to be seen with a light microscope. However, it is possible to see atoms using more sophisticated techniques. Figure 1.1.9 shows the tip of a platinum needle seen through an **ion microscope**. This instrument uses helium ions instead of light, making it possible to obtain much higher magnifications.

The radius of an individual atom is around 10^{-10} m.

Figure 1.1.9 The tip of a platinum needle seen through an ion microscope. The pattern in the photograph shows that the needle consists of layers of platinum atoms, arranged in a highly symmetrical crystal lattice.

Powers of 10, standard form and logarithms

Powers of 10 are often used to express very large or very small numbers. For example, the approximate distance travelled by light in 1 year is 10^{16} m, or 10 000 000 000 000 000 m. In this example, 16 is called the **index** – the power to which 10 is raised, and 10 is the **base**. Similarly, the approximate size of an atom is 10^{-10} m, or 0.000 000 000 1 m.

The use of powers of 10 saves writing out all the zeros. Note that the *sign* of the index tells you which way the decimal point moves.

Numbers are very often written in **standard form**, as a number between 1 and 10 multiplied by 10 raised to the appropriate power. So, for example, the boiling point of nickel is 3.0×10^3 K, and the wavelength of red light is 6.3×10^{-7} m. Both these examples are quoted to two significant figures – the use of significant figures is dealt with in section 1.7.

A further use of powers is in **logarithms**. Two types of logarithms are in common use:

- base 10 logarithms, written as $\log_{10}$ (or sometimes just lg)
- base e logarithms or **natural logarithms**, written as $\log_e$ (or sometimes just ln).

The logarithm of a number in a given base is the power to which the base must raised in order to equal the number. So:

$$\log_{10} 10\,000 = 4, \text{ since } 10\,000 = 10^4$$

Tables of logarithms can be used to carry out calculations, although most people now prefer to use calculators. However, logarithms can still be very useful in exploring relationships between quantities.

The mass of an atom

How do we know the relative atomic masses of the atoms of different elements? Atoms are far too small for us to measure their mass directly by weighing, but an instrument called a **mass spectrometer** provides a straightforward way of comparing the masses of atoms and molecules.
Figure 1.1.10 shows the principles of this instrument.

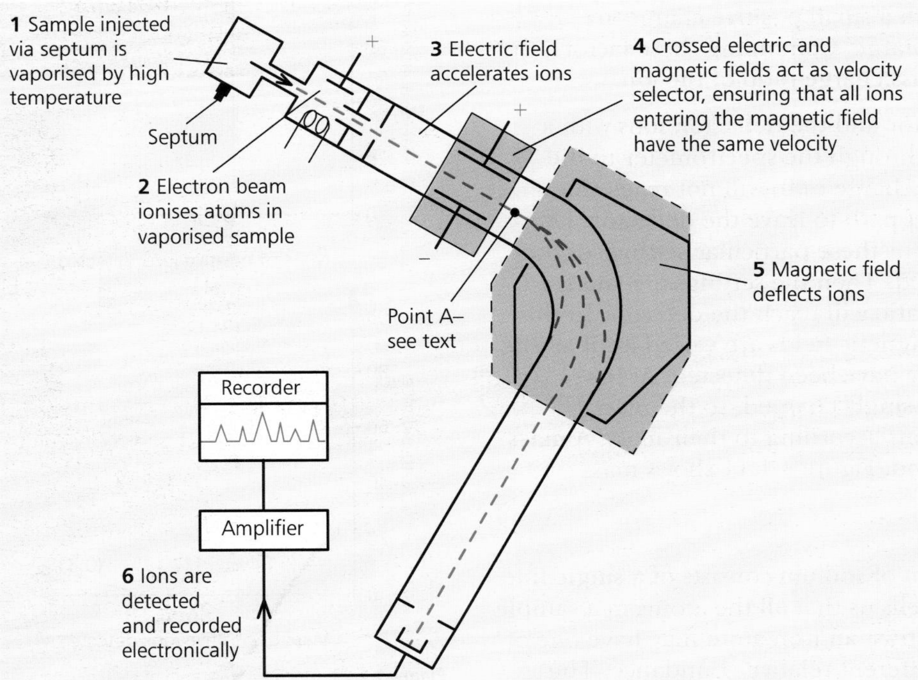
1 Sample injected via septum is vaporised by high temperature
Septum
2 Electron beam ionises atoms in vaporised sample
3 Electric field accelerates ions
4 Crossed electric and magnetic fields act as velocity selector, ensuring that all ions entering the magnetic field have the same velocity
5 Magnetic field deflects ions
Point A– see text
Recorder
Amplifier
6 Ions are detected and recorded electronically

The first mass spectrometer was built in 1918 by Francis William Aston, a student of J. J. Thomson, the discoverer of the electron. Ions of the element being studied were produced in the glass bulb. An electric field then accelerated them towards the magnetic field produced by the large coils on the right of the apparatus. The high voltages needed were produced by means of the induction coil at the bottom of the apparatus, which works in the same way as the coil in the ignition system of a car.

Figure 1.1.10

1 A sample is injected into the mass spectrometer and is vaporised by the high temperature at this point in the machine. The sample is vaporised because the machine determines the relative mass of the particles in the sample by the way they move through the spectrometer, and in order to move through the machine the particles must be in the gaseous phase. The vaporised sample enters the main part of the mass spectrometer, which is kept at a very low pressure.

2 The vaporised sample is bombarded with high-energy electrons. This produces ions, as the high-energy electrons collide with atoms of the sample, knocking one (or sometimes more) electrons out of them. If we consider a sample containing atoms of element X, then:

e^-	+	$X(g)$	$\rightarrow$	$X^+(g)$	+	$\underbrace{e^- + e^-}$
High-energy bombarding electron		**atom**		**positive ion**		**low-energy electrons (one of these is the electron knocked out of atom X, the other the bombarding electron which has lost some of its energy in the collision with X)**

3 The ions are accelerated by an electric field.

4 The rapidly moving ions enter a region where there is a magnetic field and an electric field at right angles to each other and at right angles to the direction of travel of the ions. This is called a **velocity selector**. As a result, all the ions leaving the selector at point A have exactly the same velocity.

5 The ions enter a region in which there is a uniform magnetic field. This is called the **deflector**. The magnetic field causes the ions to move in a circular path. The degree of deflection depends on two factors – the mass of the ions and their charge (+1, +2, and so on). Heavier ions are deflected less than lighter ions. Ions with a small positive charge are deflected less than ions with a greater charge. Combining these factors, the degree of deflection depends on the mass:charge ratio of the ion.

6 For any one setting of the velocity selector and the deflector, ions with a particular mass:charge ratio will travel through the spectrometer to the detector. Any ions with a different mass:charge ratio will not reach the detector as they will not have the correct path to leave the deflector. The number of ions that reach the detector for these particular settings of the velocity selector and deflector is recorded. Then the settings are changed so that ions of a different mass:charge ratio will reach the detector. In this way, the strengths of the electric and magnetic fields are varied until all the different ions produced from the sample have been detected. At low electron beam energies, the ions will be singly charged, so the mass spectrometer is effectively sorting the ions according to their mass. A **mass spectrum** can be obtained by this method. Figure 1.1.11 shows mass spectra for sodium and iron.

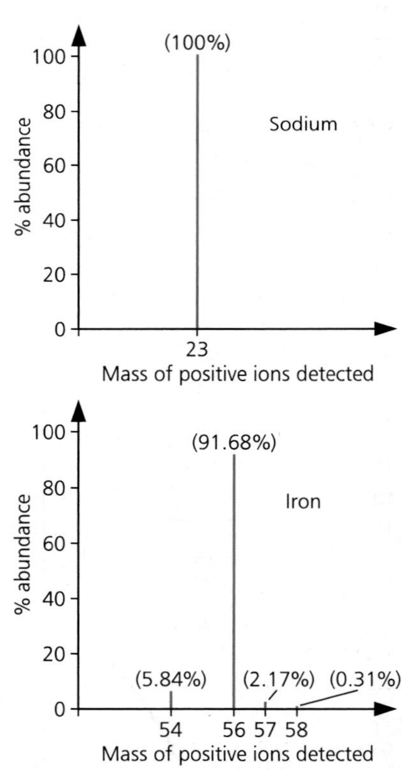

Figure 1.1.11 The mass spectrum of sodium consists of a single line, while that of iron contains four lines.

As figure 1.1.11 shows, the mass spectrum of sodium consists of a single line, while that of iron contains four lines. This tells us that all the atoms in a sample of sodium have the same mass, while in contrast an iron atom may have one of four possible masses, each of which has a different relative abundance. The reason for this is discussed in section 1.2.

Making sense of Mars

Space is still the final frontier for human exploration. We know relatively little about our nearest neighbours in space, although over the last couple of decades we have begun to find out rather more. The big question that many people ask is whether there is life on other planets. Only by discovering what the atmosphere of the planets is like, what the body of the planet is made of, what the temperature range is and what the weather is like will we have any chance of discovering if there ever has been – or ever could be – life on our neighbours in the universe.

Sending a mass spectrometer into space, which can be used to identify different elements in the atmosphere of a planet, has been a major step forward. In the 1980s, the Pioneer Venus Mission set up by NASA (the American space agency) sent a mass spectrometer into the upper atmosphere of Venus. The spectrometer identified gases which included helium, carbon dioxide and molecular nitrogen, along with carbon monoxide, atomic oxygen and atomic nitrogen. In the 1990s there was another expedition – the Galileo probe to Jupiter. Some of the images sent back by Galileo were quite amazing, and information from the mass spectrometer, combined with other instruments, gave a far clearer picture of the planet than had ever been possible before.

Jupiter has an atmosphere made up largely of hydrogen and helium, but it does contain a number of other molecules, some organic and some inorganic. These include methane, ammonia, water vapour and hydrogen sulphide. The atmosphere of Jupiter also contains a number of inert gases in addition to helium – these include neon, argon, krypton and xenon.

But the planet which seems to exert the greatest fascination for people on Earth – fuelled by science fiction tales of little green men – is Mars. In the 1990s excitement rose to fever pitch when a meteorite found in Alaska was shown to come from Mars and to contain evidence which a team of American scientists were convinced showed the possibility of life existing on the planet. The material was analysed in a mass spectrometer and was found to contain hydrocarbons, and particularly a group of chemicals known as polycyclic aromatic hydrocarbons, which the scientists felt were very similar to the results they would expect from the simple decay of organic material. In addition, the structure of the rock showed tiny structures which the American team thought might be fossilised microbacteria. Controversy still rages over this evidence. In the meantime NASA sent up another mass spectrometer on a new mission to Mars, this time in a Japanese spacecraft, the Nozomi (Planet-B). Launched in 1998, because of technical problems Planet-B will not arrive at Mars until late 2003, when hopefully the mass spectrometer will send back detailed analyses of the gases in the Martian atmosphere for several years.

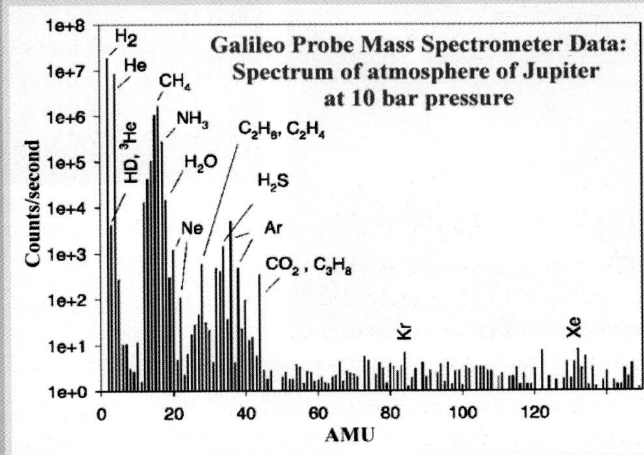

Figure 1.1.12 A sample mass spectrum from the Galileo space probe taken from the atmosphere of Jupiter.

Figure 1.1.13 Mars – just one of the planets whose secrets the spectrometer is helping us to unravel.

Spectra in chemistry

Modern chemists rely on the use of many other types of spectra as well as mass spectra to obtain information about atoms and molecules. A few of the uses of spectroscopy are shown in figure 1.1.14, and we shall meet some of them in more detail later in the book. (You can find out more about light and the other types of electromagnetic radiation used in spectroscopy in the box 'Electromagnetic waves' on pages 30–32).

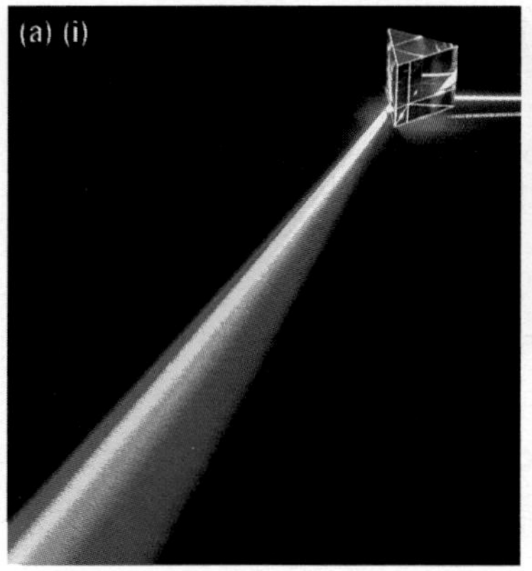

(a) (i)

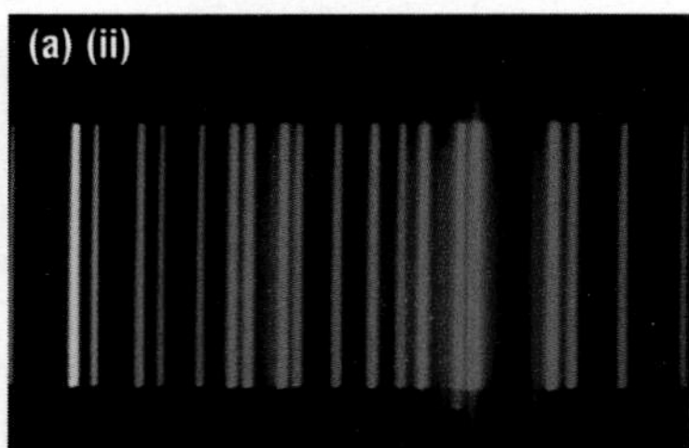

(a) (ii)

Spectrum (i) is produced by a white-hot tungsten filament, while (ii) is produced by atoms of the gas neon in an advertising sign. In section 1.3 we shall look at what spectra like this tell chemists about the electronic structure of atoms.

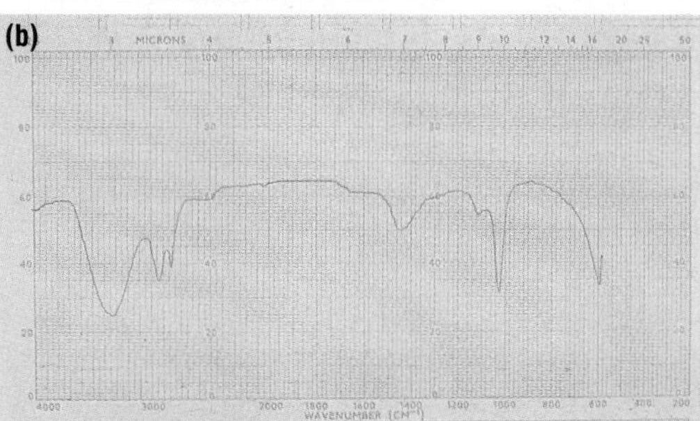

Infra-red radiation can cause the atoms in molecules to vibrate. The frequencies at which infra-red radiation is absorbed by a molecule give clues about the shape of the molecule and the way the atoms in it are joined together. We shall study this in more detail in section 3.1.

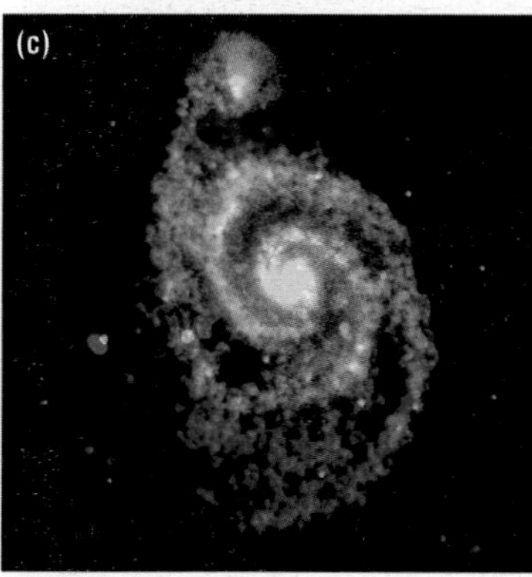

(c)

Microwave radiation is absorbed by molecules, causing them to rotate. Examination of microwave radiation coming from distant objects in outer space has provided information about the molecules present in deep space.

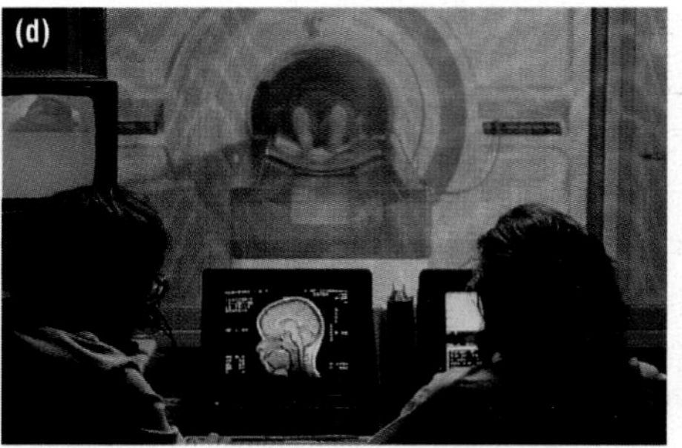

(d)

Nuclear magnetic resonance (NMR) spectroscopy examines the magnetic behaviour of protons in the nuclei of certain types of atoms. As well as being used by chemists to probe the structure of complex molecules, NMR imaging is also used for medical diagnosis.

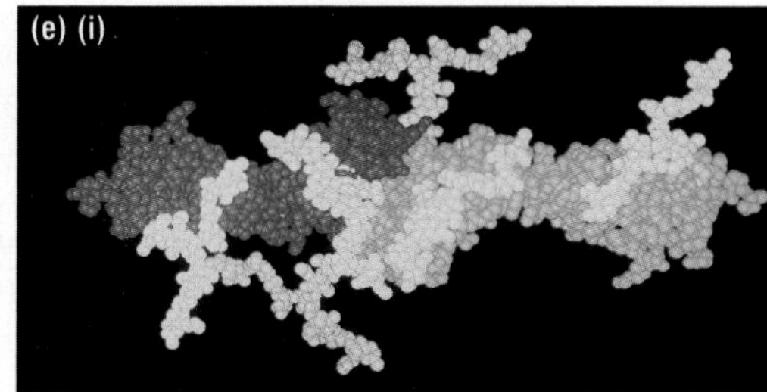

(e) (i)

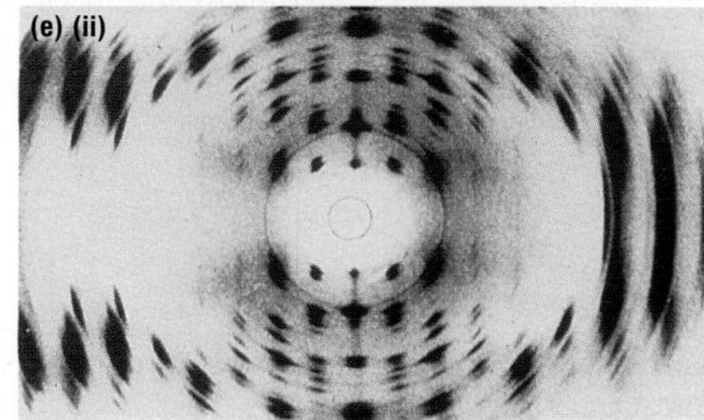

(e) (ii)

The structure of crystals, whether of a simple compound like sodium chloride or of an extremely complex molecule like the antigen glycoprotein in (e) (i), can be determined using X-rays. These short-wavelength electromagnetic waves are diffracted by the atoms in the crystal, producing a pattern on a photographic film like that in (e) (ii). X-ray crystallography is examined in more detail in section 1.6.

Figure 1.1.14 Spectra provide various invaluable tools to aid in identifying materials.

SUMMARY

- **Atoms** make up the smallest identifiable part of a chemical element.
- Atoms are counted using the **mole** as a unit. A mole is the amount of substance which contains as many particles (atoms, ions or molecules) as exactly 12 g of carbon-12.

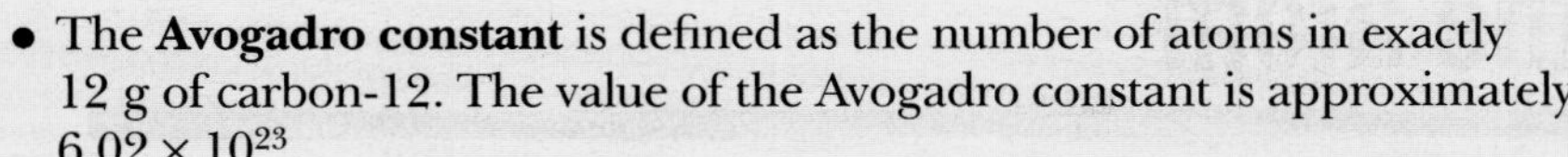

- The **Avogadro constant** is defined as the number of atoms in exactly 12 g of carbon-12. The value of the Avogadro constant is approximately 6.02×10^{23}.
- The **empirical formula** of a compound shows the ratios (in whole numbers) of the atoms of different elements that the compound contains.
- The **molecular formula** of a compound describes the *actual* number of atoms of each element contained in one molecule of the compound.
- **Concentration** is measured in **moles per cubic decimetre**.
- **Ionic equations** are used to show the reactions of ions in solution.
- The **size** of an atom may be estimated using instruments such as the ion microscope.
- The **mass** of an atom may be measured using a **mass spectrometer**. Other types of spectrometry reveal the combinations of elements within a molecule and the arrangement of the subatomic particles within an atom.

QUESTIONS

1 Calculate the average relative atomic mass of each of the following elements:
- **a** bromine (50.5% bromine-79, 49.5% bromine-81)
- **b** silver (51.3% silver-107, 48.7% silver-109)
- **c** chromium (4.3% chromium-50, 83.8% chromium-52, 9.6% chromium-53, 2.3% chromium-54).

2 Calculate the number of atoms in:
- **a** 36 g of carbon-12
- **b** 4 g of magnesium-24
- **c** 532 g of caesium-133
- **d** 57 g of lead-208.

3 Calculate the mass of one mole of the following:
- **a** As_2O_3 (the 'arsenic' beloved of mystery writers)
- **b** PbN_6 (used in explosives to 'prime' the charge)
- **c** $Ca(NO_3)_2$ (used in matches)
- **d** $Ca(C_6H_{12}NSO_3)_2$ (calcium cyclamate, an artificial sweetener).

4 Aluminium oxide, Al_2O_3, and hydrogen iodide, HI, react together as follows:

$$Al_2O_3(s) + 6HI(aq) \rightarrow 2AlI_3(aq) + 3H_2O(l)$$

- **a** How many moles of hydrogen iodide would react completely with 0.5 mol of aluminium oxide?
- **b** How many moles of aluminium iodide would this reaction form?
- **c** What mass of aluminium iodide is this?
- **d** What masses of aluminium oxide and hydrogen iodide would be required to produce 102 g of aluminium iodide?

5 The substance ATP is important as it provides a way of transferring energy in living cells. A sample of 1.6270 g of ATP was analysed and found to contain 0.3853 g of carbon, 0.05178 g of hydrogen, 0.2247 g of nitrogen and 0.2981g of phophorus. The rest was oxygen.
- **a** Calculate the number of moles of each element present in the sample of ATP.
- **b** Calculate the ratio of the number of moles of each element using the nearest whole numbers.
- **c** Write down the empirical formula of ATP.
- **d** The relative molecular mass of ATP is 507. What is its molecular formula?

6 Ammonium nitrate, NH_4NO_3, is widely used as a fertiliser. Unfortunately it can be highly explosive under certain circumstances, and there have been several examples of people being killed when large amounts of ammonium nitrate have exploded.

The equation for this reaction is

$$2NH_4NO_3(s) \rightarrow 2N_2(g) + O_2(g) + 4H_2O(g)$$

- **a** How many moles of gas are produced by the explosion of 1 mol of ammonium nitrate?
- **b** 1 tonne (1000 kg) of ammonium nitrate has a volume of about 0.6 m^3. If one mole of any gas occupies around 30 dm^3 (0.03 m^3) at 100 °C and 10^5 Pa pressure, make a rough estimate of the increase in volume when 1 tonne of ammonium nitrate explodes.
- **c** Why is your answer to **b** only approximate?

7 When solutions of potassium chloride, KCl, and silver nitrate, $AgNO_3$, are mixed together, a precipitate of silver chloride forms.
- **a** Write down a balanced equation for this reaction.
- **b** In a certain experiment, 24.0 cm^3 of 0.05 mol dm^{-3} silver nitrate solution exactly reacts with 15.0 cm^3 of potassium chloride solution. Calculate the concentration of the potassium chloride solution.
- **c** What mass of potassium chloride would be needed to produce 100 cm^3 of a solution of this concentration?

1.2 The structure of the atom

The story unfolds

Although Dalton's theory of the existence of atoms was published in 1808, there was no firm idea of what an atom might look like until the early part of the twentieth century. The last years of the nineteenth century saw a great deal of research into the structure of matter, and this work produced ideas that form the basis of the modern understanding of the atom.

Mysterious rays

In 1855 Sir William Crookes carried out a series of investigations into the behaviour of metals heated in a vacuum. The experiments of Crookes and others showed that a heated cathode produced a stream of radiation, which could cause gases at low pressure to glow and which could make other substances emit light too. The radiation emitted from the cathode was given the name **cathode rays**. By the mid-1890s it was known that these rays could be deflected by a magnetic field, and that they carried a negative charge, as shown by figure 1.2.1.

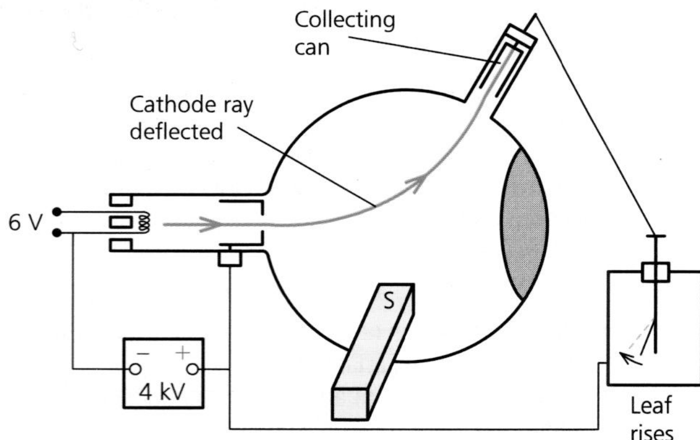

Figure 1.2.1 The negative charge on cathode rays was first demonstrated by the Frenchman Jean Baptiste Perrin in 1895, using a form of the apparatus represented here – now known as a **Perrin tube**. The beam of rays is deflected into the collecting can using a magnetic field. The electroscope leaf then deflects, showing that a charge has been collected. If the electroscope is isolated from the collecting can and then the charge tested, it is shown to be negative.

Throughout the latter part of the nineteenth century, scientists debated the nature of cathode rays, and many theories were advanced about them. Some scientists felt that the rays were waves, similar in properties to electromagnetic waves, while others were more inclined to think that the rays were particles. George Stoney even went so far as to name the particle – the **electron**. In 1897, Joseph John Thomson showed that the rays were indeed streams of electrons, since they could be deflected by an electric field. Thomson used the size of the deflection to calculate the mass:charge ratio of an electron, which he found to be about 2000 times smaller than the mass:charge ratio of a hydrogen ion. Whatever the gas used in the discharge tube, the mass:charge ratio of the electrons produced was always the same, suggesting that electrons are found in all matter and that they are a fundamental particle.

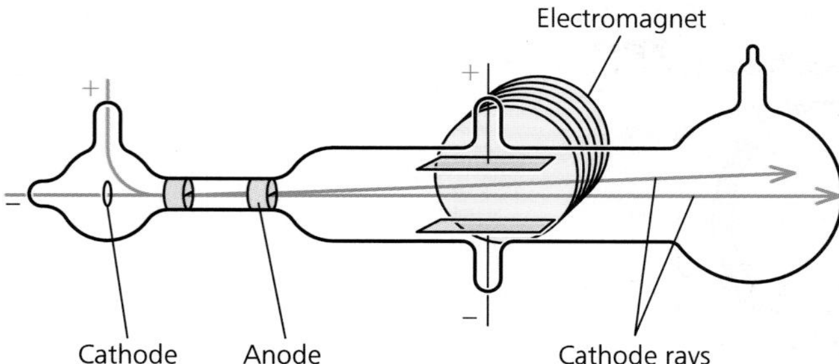

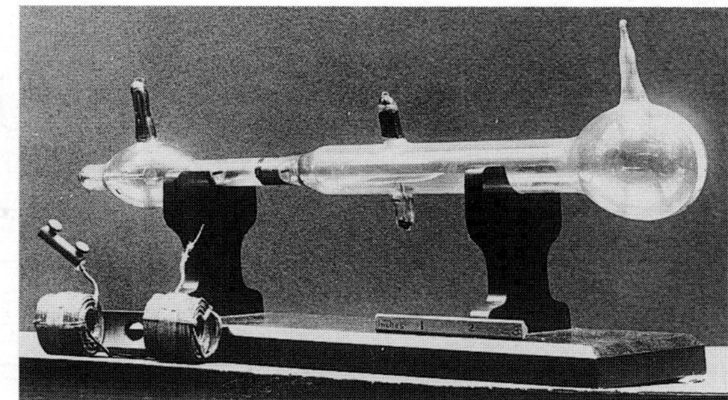

Figure 1.2.2 Sir J. J. Thomson and the apparatus he used to produce and investigate cathode rays. Thomson was awarded a Nobel prize for this work, as well as being knighted. A gifted teacher (seven of his research assistants won Nobel prizes, as did his son, George), Thomson was buried in Westminster Abbey after his death, in recognition of his outstanding contributions to science.

Models

As scientists strive to understand the world around us, it is often necessary to make intelligent guesses about cause and effect – for example, the guesses made by Thomson as he explained his observations of cathode rays. These intelligent and often imaginative guesses are usually referred to as **theories** or **models**. They represent a way of picturing the Universe and explaining the way it behaves – subject, of course, to being shown as false by observations made as part of **experiments** to test them.

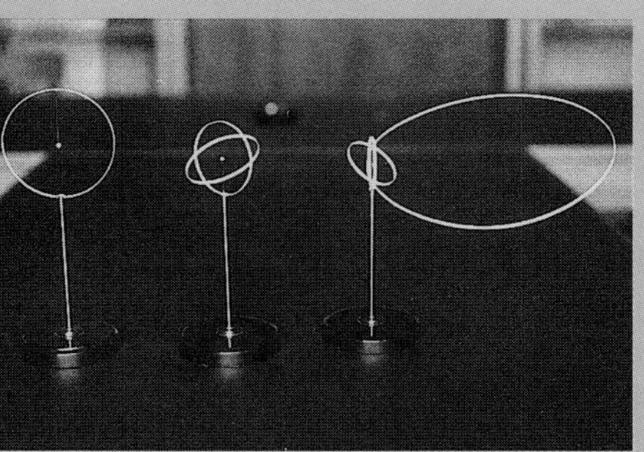

Figure 1.2.3 Early models of the atom

Models for the atom

The discovery of the electron led to a flurry of speculation about the structure of the atom. In 1898, Thomson proposed the 'plum pudding' model for the atom, in which negative electrons are embedded in a sphere of positive charge. This and other models of the period were based on the erroneous assumption that the mass of an atom was contained in its electrons.

Positive rays

Further work using a tube containing a cathode with a hole in it, as in figure 1.2.4 overleaf, showed that other rays were also present in discharge tubes. These rays travelled through the hole in the cathode, apparently in the opposite direction to the cathode rays. Investigation of the rays indicated that they consisted of a stream of positively charged particles, formed by the electrons travelling through the residual gas in the tube knocking electrons out of gas atoms, leaving positive ions behind.

In contrast to the negative rays, the mass of the particles making up the positive rays was dependent on the residual gas in the tube. The lightest particles were obtained when hydrogen was used. Other gases produced particles that had mass:charge ratios which were whole-number multiples of the mass:charge ratio of the hydrogen particle.

From this new information, the positively charged particles from the hydrogen atom appeared to be a good candidate for the positively charged

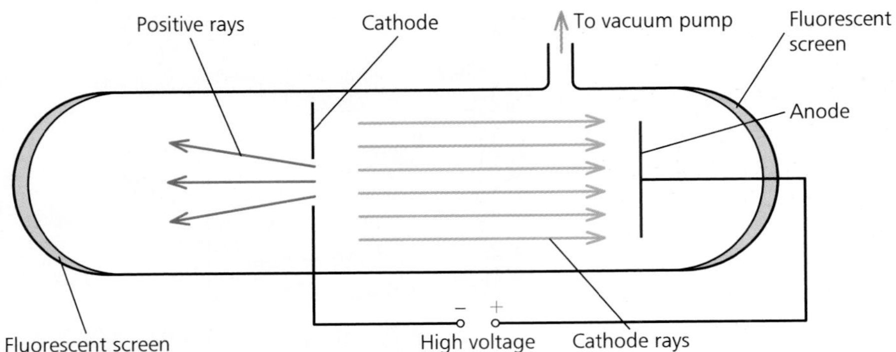

Figure 1.2.4 Positive ions formed in the tube are attracted by the cathode. Some of them travel through the hole in the cathode and strike the phosphor screen, producing a flash of light.

building block of matter, in the same way as the electron appeared to be the negatively charged building block. For this reason the hydrogen atom minus an electron was given the name **proton**, from the Greek word *proteios*, meaning 'of first importance'.

Measuring the charge on the electron – Millikan's oil-drop experiment

In 1909, the American researcher Robert Millikan designed an elegant experiment that measured the charge on the electron, and so enabled the mass of the electron to be calculated from its mass:charge ratio. (The inverse of this, the charge:mass ratio of a particle, is often referred to as the **specific charge** of the particle.) Essentially, Millikan measured the charge on a great number of tiny oil drops which had been charged by friction by spraying them out of a nozzle. He was then able to show that each drop carried a charge which was a multiple of a basic unit of charge.

A small number of the oil drops sprayed into the box above the plates pass through the hole in the top plate and are seen through the microscope. When there is no electric field between the plates, the drops fall slowly with a steady velocity. An individual drop carrying a charge may be brought to rest by applying a voltage across the plates so that its

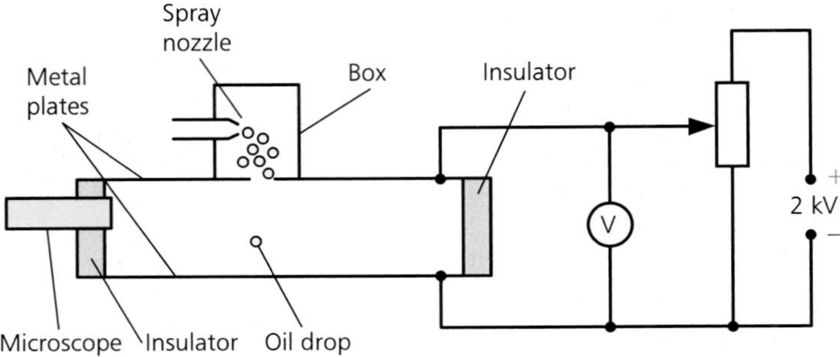

Figure 1.2.5 The principle of Millikan's experiment

weight (acting downwards) is exactly balanced by the electrostatic force on it (acting upwards). The drop's charge can be found by careful measurement of the voltage needed to bring the drop to rest and the rate at which it falls when there is no voltage between the plates.

Millikan's values for the charge on each drop were always whole-number multiples of -1.6×10^{-19} C (C is the abbreviation for coulomb, the unit of electrical charge). As a drop could only pick up whole numbers of electrons, this indicated that the charge on an individual electron e must be -1.6×10^{-19} C. Using this figure, together with Thomson's figure for the specific charge on the electron, the mass of the electron m_e could be calculated:

$$e/m_e = -1.76 \times 10^{11}\ \text{C kg}^{-1}$$

Thus:

$$m_e = \frac{e}{-1.76 \times 10^{11}\ \text{C kg}^{-1}}$$

Substituting Millikan's value of e gives:

$$m_e = \frac{-1.6 \times 10^{-19}\ \text{C}}{-1.76 \times 10^{11}\ \text{C kg}^{-1}}$$

$$= 9.1 \times 10^{-31}\ \text{kg}$$

Further experiments have enabled this mass to be measured more precisely – data for the electron and the other subatomic particles are given in table 1.2.2, page 27.

An atom that gains an electron to become an ion develops a charge of -1.6×10^{-19} C, while one losing two electrons develops a charge of $+3.2 \times 10^{-19}$ C. Normally we indicate these charges simply as –1 and +2 respectively, but it is worth remembering what this actually means.

The discovery of the nucleus

In 1911, work carried out by Ernest Rutherford and his colleagues showed that the 'plum pudding' model was false. In its place, the nuclear model of the atom was proposed. Rutherford's work depended on the use of a product of radioactive decay, namely the alpha particle, which he and Frederick Soddy had shown to be a helium ion in 1909. The helium ion is positively charged. Under the direction of Rutherford, Hans Geiger and Ernst Marsden subjected thin metal foils to a beam of alpha particles. Figure 1.2.6 shows their apparatus. The majority of alpha particles passed straight through the foils, with very little deflection, even when the thickness of the foil was as much as 10^4 atoms. In a very few cases (as few as one in 1800), alpha particles showed some deflection, while in fewer cases still the alpha particles were deflected through an angle greater than 90°.

This large angle of scattering for a very small number of particles led Rutherford to propose that the majority of the mass of the atom was concentrated in a minute positively charged region, around which the electrons in the atom circulated. When an alpha particle came very close to the nucleus, Rutherford reasoned, the electrostatic repulsion between the two would be sufficient to repel the alpha particle and so produce the large angle of scattering. Since the nucleus was small, this scattering would occur for only the few particles which approached the nucleus sufficiently closely.

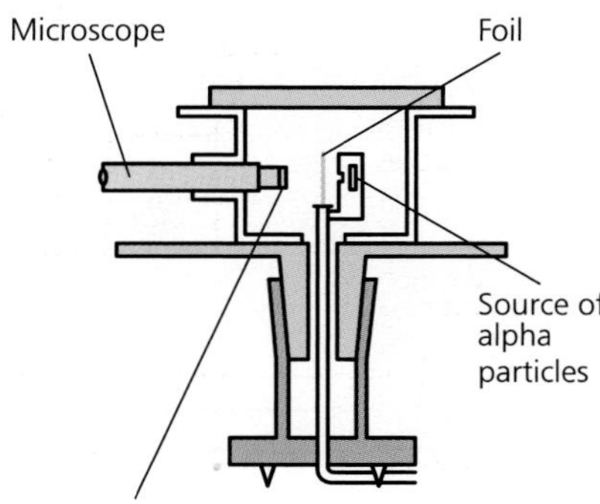

Figure 1.2.6 Geiger and Marsden's apparatus. The microscope and screen can be rotated around the foil so as to detect alpha particles deflected through different angles.

Rutherford calculated the fraction of alpha particles that should be deflected through particular angles on the basis of this model, and showed that their trajectories would be hyperbolas. Figure 1.2.7 shows some trajectories.

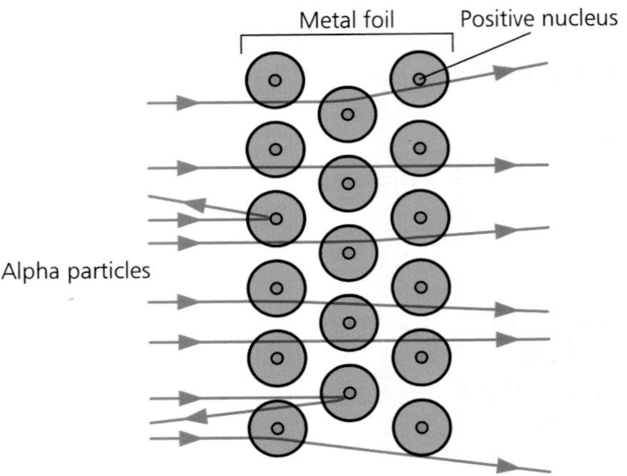

Figure 1.2.7 The trajectories of alpha particles deflected by the positively charged nuclei of the atoms in the foil. The trajectories are all hyperbolas.

Using his model, Rutherford was able to show that the nucleus of an atom is of the order of 10^{-14} m across. This compares with the size of an atom calculated in section 1.1, of the order of 10^{-10} m – that is, the diameter of the nucleus is about 1/10 000 of the diameter of the atom itself. It is therefore not surprising that most of the alpha particles in Geiger and Marsden's experiments passed through the foil with no deflection, although it may be difficult to imagine that virtually all the mass of the atom is concentrated in the tiny volume of the nucleus.

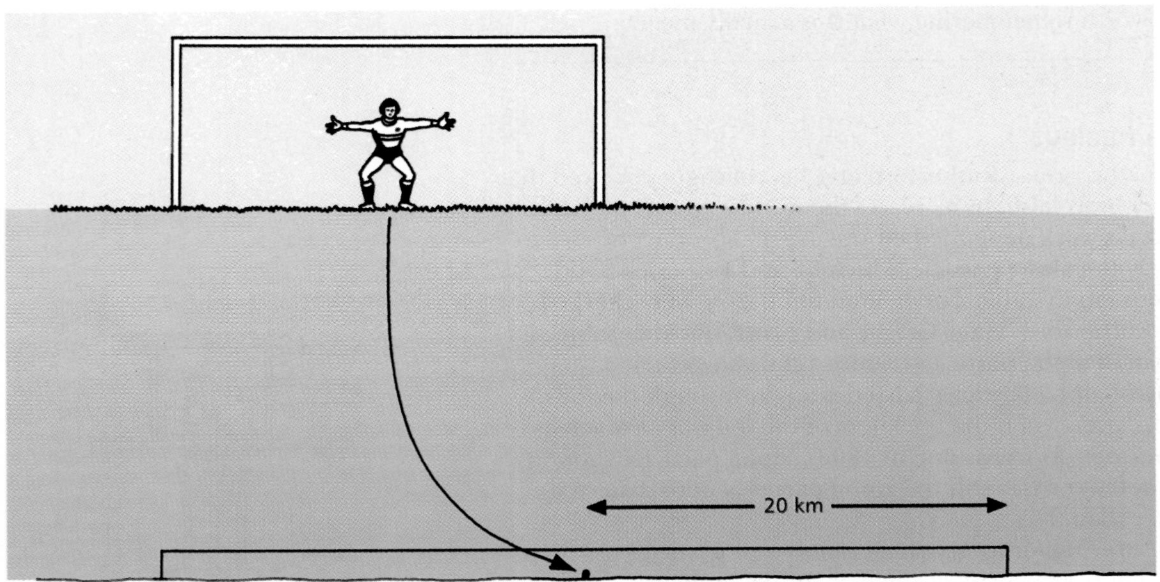

Figure 1.2.8 To get some idea of the scale of the atom, imagine a goalkeeper standing with arms outstretched in the centre of a goal. If the goalkeeper represents the nucleus of an atom, and a football kicked towards the goal represents an alpha particle, the goalposts would need to be 20 km either side of the keeper to represent the scale of the atom.

Rutherford's model pictured the atom as a miniature Solar System, in which electrons orbited the nucleus much as planets orbit the Sun. Because matter is electrically neutral, the positive charge on the nucleus of each atom must be balanced by the negative charge on the orbiting electrons. The model says no more about the distribution of the electrons outside the nucleus, other than that they move in a space that is far larger than that occupied by the nucleus. In section 1.4 we shall examine in some detail the behaviour of electrons in atoms, since it is this which determines the chemical properties of the atoms.

The significance of the number of protons in the nucleus

When a sample of an element is bombarded by very high-energy electrons, X-rays (high-frequency electromagnetic waves) are produced. Each chemical element has its own characteristic X-ray spectrum, producing X-rays of particular frequencies.

The X-ray spectra of elements were studied by Henry Moseley, a young research student at Oxford University in 1912. Moseley found that the frequency of the X-rays produced by a particular element could be expressed mathematically and related to a number which he called the element's **atomic number**, Z. The order of these numbers was exactly the same as the order of the elements in the periodic table. Scientists quickly realised that not only does the atomic number represent an element's position in the periodic table, but it does so because it represents the number of protons in the nucleus of the atom. Since atoms are electrically neutral, this is equal to the number of electrons in the atom, which is the key factor in determining an element's chemical properties.

Figure 1.2.9 Henry Moseley. In 1915, three years after his work on the X-ray spectra of elements, Moseley was killed during the British invasion of Gallipoli during the First World War. After this tragic loss it became the policy of the British government not to send research scientists into battle.

The search for the missing particle

Although Rutherford's model was enormously successful at explaining the scattering of alpha particles, there was a problem when the experimental data were used to calculate the mass of the nucleus. From the results of the scattering experiments, it was possible to calculate the charge on the nucleus, which was of course equal to the charge on the protons in the nucleus. However, the mass of these protons was only about half the overall mass of the nucleus. In 1920, William Draper Harkins, an American physicist, suggested that the missing mass could be accounted for if the nucleus contained other particles with a mass similar to that of the proton but no charge. He named this particle the **neutron**. Because it has no charge, the neutron was hard to find. The hunt for it was finally ended in 1932, when it was discovered by the British scientist James Chadwick in the experiment described in figure 1.2.10.

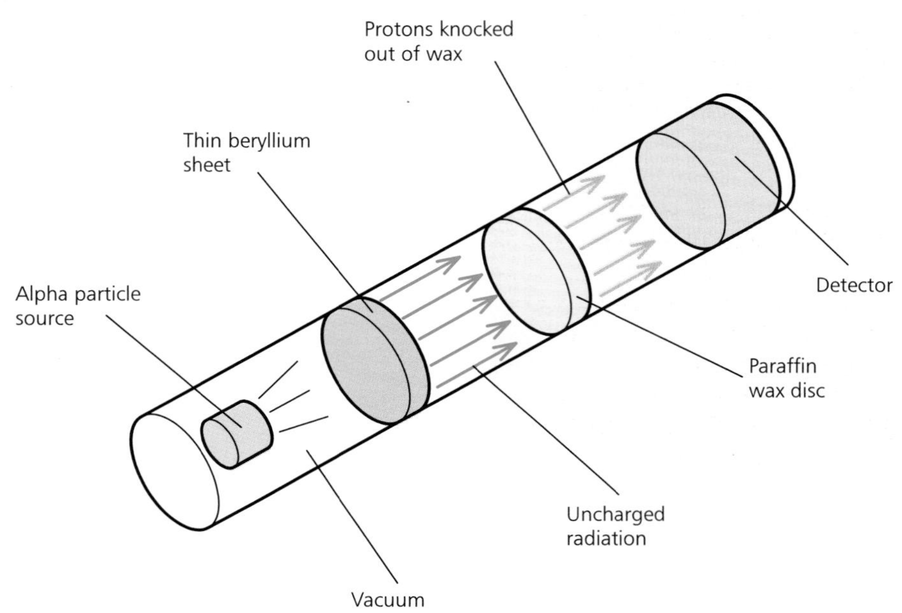

Figure 1.2.10 Chadwick's discovery of the neutron in 1932. Bombarding a beryllium plate with alpha particles produced a mysterious uncharged radiation on the opposite side of the sheet. Placing a solid material containing many hydrogen atoms (paraffin wax) in the path of this radiation caused protons to be knocked out of the wax. Chadwick showed that the unknown radiation must consist of uncharged particles with a mass close to that of the proton. The photograph shows part of Chadwick's original apparatus.

Describing the atom

Following the work of Thomson, Rutherford and many others at the beginning of this century, scientists now believe that the nuclei of atoms are made up of **protons** and **neutrons**. These two components of the nucleus are usually referred to as **nucleons**. The **electrons** occupy the space outside the nucleus. Since an atom is electrically neutral, the number of protons in the nucleus of an atom must be exactly equal to the number of electrons. This number is the **atomic number** (sometimes called the **proton number**) of the element, given the symbol **Z**. All atoms of a particular element have the same atomic number and therefore behave in the same way chemically, since it is the number of electrons in the atom that determines its chemical properties. Atoms of the same element may have different numbers of neutrons, however, giving them different masses. Atoms with the same atomic number but with different numbers of neutrons are called **isotopes**. In fact, most elements are mixtures of isotopes, as Aston first showed with his mass spectrograph.

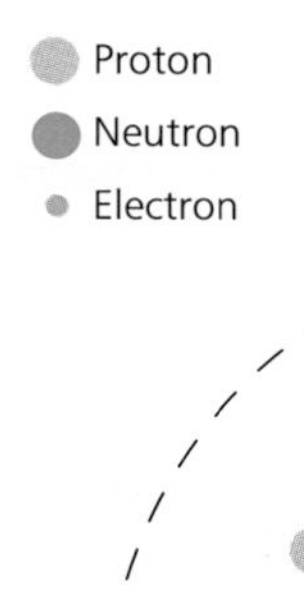

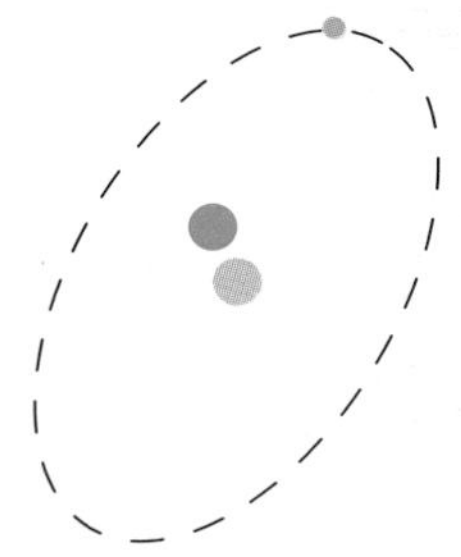

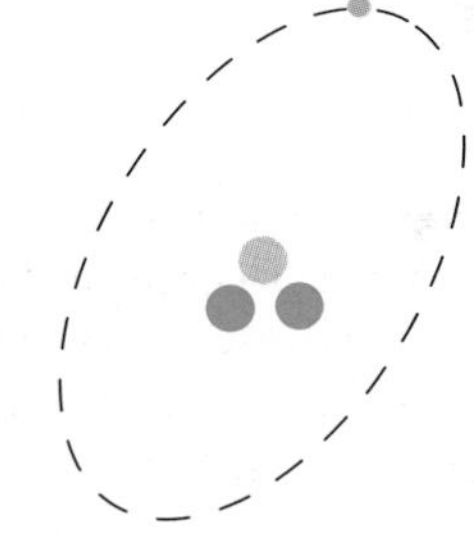

The most common isotope of hydrogen has a single proton in its nucleus, with a single electron outside the nucleus.

This is the isotope of hydrogen called **deuterium**. With a neutron in its nucleus as well as a proton, this atom is about twice as massive as the atom containing no neutron.

This isotope of hydrogen is called **tritium**. Its atom contains two neutrons, and is therefore about three times as massive as the atom containing no neutron.

Figure 1.2.11 The three isotopes of hydrogen. Around 99.985% of the atoms in a container of hydrogen will have only a proton in the nucleus, while about 0.015% will have a proton and a single neutron. The number of atoms with a proton and two neutrons in the nucleus will be variable – these nuclei are **unstable**, and undergo radioactive decay. Although their physical properties differ, all three isotopes have identical chemical properties, since each consists of atoms with only a single electron.

The number of protons plus the number of neutrons in the nucleus is called the **nucleon number** (or sometimes the **mass number**) of the element, given the symbol *A*. The average distance between the protons in the nucleus is very small, which means that there is a very large repulsive electric force between them, since they are all positively charged. At these small separations, a force called the **strong force** provides an even larger, attractive force which keeps the nucleons together. Nevertheless, for large atoms the repulsive electric force is significant, and can lead to the breaking up of the nucleus. Figure 1.2.12 shows how the data for a nucleus are written alongside its chemical symbol.

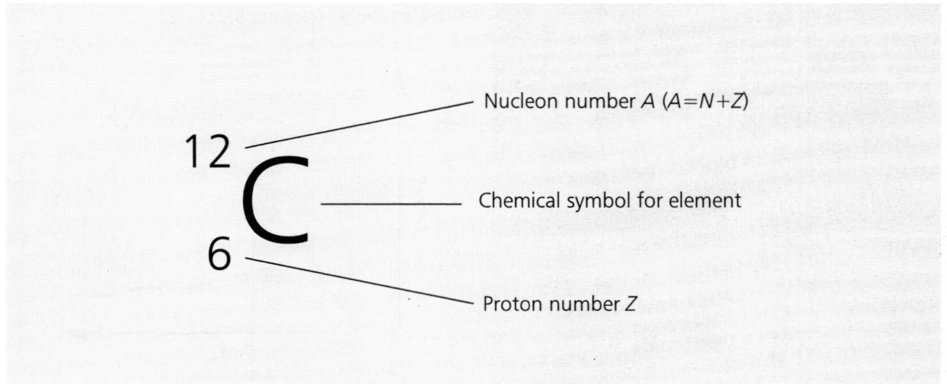

Figure 1.2.12 An atom with *Z* protons and *N* neutrons is represented like this. In text, elements may be referred to by their name followed by their mass number, for example, carbon-12 or uranium-235.

	Lithium-7	Silicon-28	Copper-65	Dysprosium-164	Uranium-238
Proton number Z	3	14	29	66	92
Nucleon number A	7	28	65	164	238
Number of neutrons N ($= A - Z$)	4	14	36	98	146
Symbol	${}^{7}_{3}Li$	${}^{28}_{14}Si$	${}^{65}_{29}Cu$	${}^{164}_{66}Dy$	${}^{238}_{92}U$

Table 1.2.1 How protons and neutrons are combined to make up the nuclei of different atoms

FOCUS ISOTOPES AND THEIR USES

Isotopes have a wide range of uses based on the fact that the isotopes of an element all have the same chemical properties. Some isotopes of some elements are unstable, however, with the result that these isotopes are radioactive – their nuclei break apart, emitting alpha particles, beta particles or gamma rays.

During the twentieth century, many advances have been made in non-invasive techniques of diagnosis, enabling doctors to obtain information about a patient's health without resorting to surgery. One such technique is to use radioactive isotopes as **tracers**. Molecules containing a radioactive atom are taken up by the organ under investigation. The radioactive substance most commonly used in tracers is an isotope of technetium, technetium-99m. The nucleus of the technetium-99m atom is unstable, and loses energy by emitting a gamma ray. This decay process has a half-life of 6 hours, making technetium-99m a good choice for tracer studies. The gamma rays emitted carry enough energy to make them reasonably simple to detect, while the half-life of 6 hours means the activity lasts long enough to be useful in clinical tests, but exposes the patient to ionising radiation for as short a time as possible. The technetium is normally used to label a molecule which is preferentially taken up by the tissue to be studied, as illustrated by figure 1.

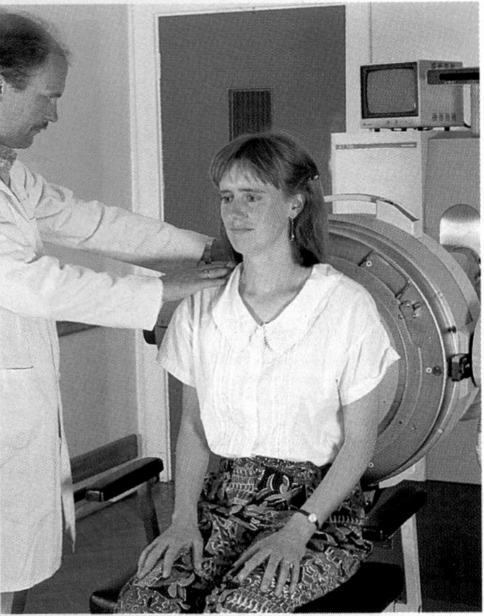

Figure 1 Images of a patient's kidneys are obtained using radioactive tracers. The graph shows the uptake of tracer by the two kidneys in one such scan. The amount of gamma radiation detected from the right kidney is much lower than that from the left kidney. This may indicate that there is a problem with the right kidney, causing this reduced activity – further tests to confirm this will probably be required.

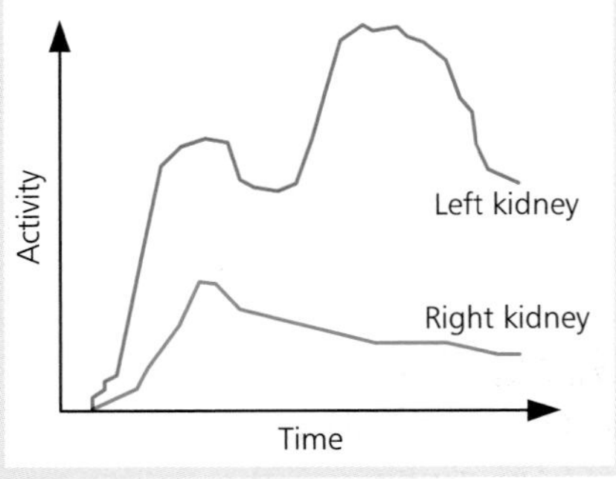

Historians and archaeologists use an isotope of carbon in dating ancient remains. This dating technique, known as **radiocarbon dating**, depends on the radioactive isotope carbon-14. Molecules of carbon dioxide containing an atom of carbon-14 are produced in the upper atmosphere by cosmic rays, which are high-energy subatomic particles from space. Plants take up carbon dioxide containing carbon-14 in exactly the same way as they take up carbon dioxide containing the more common carbon-12 atom, since carbon-12 and carbon-14 are chemically identical. As a result, the proportion of carbon-14 to carbon-12 in plant matter is the same as that in air. (This ratio is also found in animal matter, since all animals ultimately get the energy they need by eating plants.) When an organism dies, the uptake of carbon dioxide ceases. The decay of the carbon-14 in the once living matter causes the ratio of carbon-14 to carbon-12 to reduce with time, providing a way of measuring the age of the organism.

Figure 2 Part of the shroud of Turin, said to be the burial cloth in which Christ was wrapped following his crucifixion. Carbon-14 dating of the cloth of the shroud suggests that it is less than 2000 years old, and that it may therefore be a forgery.

Properties of the subatomic particles 

Table 1.2.2 shows the masses and charges for the three particles making up the atom. Since the masses are so small, it is convenient to have a unit of mass close to their mass to use for measurements. Like the mole, this unit is related to carbon-12. The atoms of this isotope of carbon have six protons, just like all atoms of carbon, and six neutrons. The **atomic mass unit** u is defined as one-twelfth of the mass of an atom of carbon-12. The experimental measurement of u shows that:

$$\mathbf{1\ u = 1.660\ 540 \times 10^{-27}\ kg}$$

This means that the masses (m) and charges (q) of the proton, neutron and electron are frequently written as:

$$m_{proton} = 1\ u \qquad q_{proton} = +1$$

$$m_{neutron} = 1\ u \qquad q_{neutron} = 0$$

$$m_{electron} = {}^{1}/_{1800}\ u \qquad q_{electron} = -1$$

and the 'u' is often omitted. The reason for this simplification becomes obvious when we look at the data for the masses and charges of the atomic particles in table 1.2.2.

	Proton	Neutron	Electron
Mass/kg	$1.672\ 623 \times 10^{-27}$	$1.674\ 929 \times 10^{-27}$	$9.109\ 390 \times 10^{-31}$
Mass/u	1.007 276	1.008 665	0.000 548 580
Charge/C	$+1.602\ 177 \times 10^{-19}$	0	$-1.602\ 177 \times 10^{-19}$

Table 1.2.2 The masses and charges of the proton, neutron and electron. These data are given to 7 significant figures, although they have in fact been measured with even greater precision than this.

SUMMARY

- The model of the **nuclear atom** was developed by Rutherford in 1911. This proposed a tiny positive **nucleus** containing most of the mass of the atom, orbited by electrons.
- The nucleus of an atom is made up of **positively charged protons** and **uncharged neutrons**, collectively known as **nucleons**. The **proton number** is the number of protons and this equals the number of **negatively charged electrons** arranged in space around the nucleus.
- The proton number of an atom is also known as the **atomic number** Z. This represents the number of protons in the nucleus. As atoms are electrically neutral, an atom contains as many electrons as it has protons.

- An element may have two or more **isotopes**. Different isotopes of the same element have atoms with the same number of protons but different numbers of neutrons.
- The **mass number** (or **nucleon number**) A of an isotope is the number of protons plus the number of neutrons in the nucleus of an atom of the isotope. Different isotopes of the same element have different mass numbers.

QUESTIONS

1 In terms of the structure of their atoms, how are isotopes of a chemical element alike? How are they different?

2 What property might two atoms of *different* elements have in common?

3 Write down the number of protons, neutrons and electrons in the atoms of the following elements (refer as necessary to the periodic table at the back of the book):

a nitrogen-15
b palladium-105
c potassium-39
d uranium-235
e tin-120
f fluorine-19.

4 Accurate values of the relative masses of some isotopes of elements are given in table 1.2.3.

Element:	$^{1}_{1}$**H**	$^{2}_{2}$**H (or D)**	$^{11}_{5}$**B**	$^{12}_{6}$**C**	$^{16}_{8}$**O**	$^{27}_{13}$**Al**
Relative atomic mass/u	1.0078	2.0141	11.0093	12.0000	15.9949	26.9815

Table 1.2.3

Using this information, calculate accurate values for the relative molecular masses of the following compounds:

a H_2O
b CO_2
c D_2O ('heavy water', used in some types of nuclear reactors)
d B_2H_6 (diborane, a gas that reacts spontaneously with the oxygen in the air)
e Al_2O_3 (one form of this compound is called corundum – it is very hard, and is used to make 'sandpaper').

5 If the radius r of a nucleus is taken to be the distance from the centre of the nucleus to a point where the density of the nuclear material has fallen to half the value at the centre, the radius in metres of a nucleus is given approximately by the relationship:

$$r = 1.4 \times 10^{-15} A^{\frac{1}{3}}$$

where A is the mass (nucleon) number of the nucleus. Starting with this relationship, calculate:

a the radius of a gold nucleus (A = 197)
b the volume of the gold nucleus.
c The density of gold is 18 880 g cm^{-3}. Calculate the approximate volume of a gold atom. Compare this with your answer to **b**.

6 Use your answers to question 5 to estimate what volume of **a** your body and **b** the Earth is filled with nuclear material. The radius of the Earth is approximately 6400 km.

7 Neutron stars are composed almost entirely of neutrons, and have a density approximately equal to that of nuclear material. The mass of the Sun is approximately 2×10^{30} kg. Use your answers to question 5 to calculate **a** the approximate density of nuclear material and **b** the diameter of a neutron star with the same mass as the Sun.

QUESTIONS

Developing Key Skills

Radiocarbon dating – death's stopwatch

In 1802 Rasmus Nyerup, a Danish antiquarian, expressed the frustration which has been felt throughout the ages by those who have wanted to study the history of human beings or indeed of the Earth itself.

'Everything which has come down to us from heathendom is wrapped in a thick fog; it belongs to a space of time we cannot measure. We know it is older than Christendom, but whether by a couple of years or a couple of centuries, or even by more than a millennium, we can do no more than guess.'

In the years immediately after the Second World War a team of scientists led by Professor Willard F. Libby of the University of California developed radicarbon dating. Suddenly, dating historical artefacts accurately to within a few years became a real possibility, an achievement for which Libby received a Nobel prize in 1960.

The C-14 isotope makes up 0.000 000 000 10% of the total carbon in the CO_2 in the atmosphere. C-14 becomes part of the tissues of living organisms and is constantly renewed in life, a process that is described in more detail in the box on page 26. After death, the amount of C-14 isotope present in the once living tissues halves in quantity every 5730 years. This makes it possible to produce figures which relate the percentage of C-14 left in a sample of biological material with the age of the sample (table 1.2.4).

% of original C-14 left in sample	Years since organism died
100	0
95	420
90	870
85	1350
80	1850
75	2400
70	2950
65	3550
60	4200
55	4950
50	5730
45	6600
40	7600
35	8700
30	9950
25	11 450
20	13 300
15	15 700
10	19 050
5	24 750
1	38 100

Table 1.2.4

a Use the data in table 1.2.4 to plot a graph showing the relationship between the percentage of C-14 left in a sample and its age.

b For remains up to 10 000 years old, the radiocarbon date is rounded to the nearest 10 years. Between 10 000 and 25 000 years old, the date is rounded to the nearest 50 years, and from 25 000 to the current limit of 55–60 000 years the date is rounded to the nearest 100 years. Why do you think this is done – and why might there be greater uncertainties when the age of older specimens is measured?

c Calculate the approximate ages of specimens found with the following percentages of their total C-14 left:
 i 3%
 ii 43%
 iii 89%

d In 1988, fragments from the Shroud of Turin were subjected to radiocarbon dating in Switzerland, the UK, the USA and France. All laboratories dated the fragments between 1260 and 1390 AD. What range of C-14 left in the fabric of the shroud would give these readings? What percentage would have been required to give a reading that dated the shroud as genuinely coming from the time of Jesus Christ, around 2000 years ago?

e Not everyone accepts the results obtained by the four laboratories for the age of the shroud. Use electronic and other sources to find out as much as you can about the *scientific* arguments over the accuracy of radiocarbon dating and the way in which readings are calibrated.

[Key Skills opportunities: A, C, IT]

1.3 The arrangement of electrons in atoms

Energy and electromagnetic waves

The work of Rutherford and his colleagues in the early part of this century resulted in a model of the atom in which electrons were arranged around a central nucleus. This model does not solve the essential questions concerning the chemical behaviour of elements, since it is only the electrons – not the nucleus – that take part in chemical reactions. Although trends in the chemical properties of the elements are caused by the effects of the nucleus, as we shall see in section 2.2, we need to understand the **electronic structure** of an atom to explain its chemical properties. The key to understanding electronic structure comes from the study of electromagnetic waves.

Electromagnetic waves

Electromagnetic waves are disturbances that can travel through space or matter. Like water waves, they are oscillations, but unlike water waves these oscillations involve not matter but electric and magnetic fields instead.

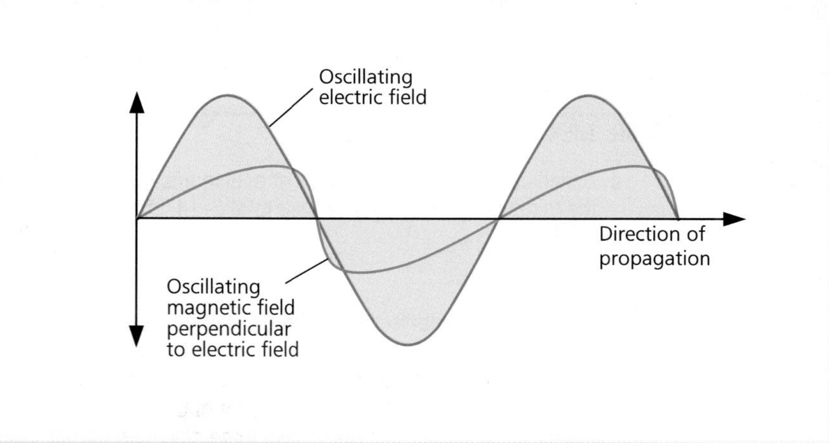

Figure 1.3.1 As an electromagnetic wave travels, the electric and magnetic fields oscillate together. In a vacuum, electromagnetic waves travel at the speed of light: 3×10^8 m s^{-1}.

As with any wave, the speed of travel c of an electromagnetic wave is related to its wavelength λ and its frequency f by the relationship:

$$c = f\lambda$$

Using SI units, speed is measured in metres per second (m s^{-1}), wavelength in metres (m) and frequency in seconds^{-1} (s^{-1}). This last unit of frequency is given the name **hertz** (Hz).

Electromagnetic waves have a wide range of wavelengths, from radio waves with wavelengths of thousands of metres to gamma rays whose wavelength is less than one thousand millionth of a millimetre. In the middle of this vast range is a narrow band that makes up the visible part of the spectrum, which our eyes perceive as light.

Figure 1.3.2 gives some details of the electromagnetic spectrum.

Frequency/Hz	Applications	Produced by	Type of radiation	Detected by	Wavelength/m
10^{24}					10^{-16}
10^{23}		Energy released from the nuclei			10^{-15}
10^{22}		of radioactive atoms			10^{-14}
10^{21}			Gamma rays	Ionisation in	10^{-13}
10^{20}	X-ray therapy			Geiger–Muller tube	10^{-12}
10^{19}		Bombarding metal			10^{-11}
10^{18}	X-ray radiography	targets with high-energy electrons	X-rays	Photographic	10^{-10}
10^{17}				film	10^{-9}
10^{16}					10^{-8}
10^{15}		High-temperature matter	Ultraviolet	Skin (sunburn) Fluorescence	10^{-7}
10^{14}	Photography	Sun	Visible light	Eye	10^{-6}
10^{13}	IR	Hot solids	Infra-red	Thermopile Skin (heat)	10^{-5}
10^{12}	spectroscopy				10^{-4}
10^{11}	Microwave	Klystron oscillators		Oscillation of	10^{-3}
10^{10}	radio links		Microwaves	molecules	10^{-2}
10^{9}	Radio				10^{-1}
10^{8}	astronomy		Short wave		1
10^{7}	TV	Electrons oscillating	Radio	Resonance in tuned electrical	10^{1}
10^{6}		in transmitting aerials	waves Medium wave	circuits	10^{2}
10^{5}	Inductive		Long wave		10^{3}
10^{4}	heating				10^{4}
10^{3}					10^{5}

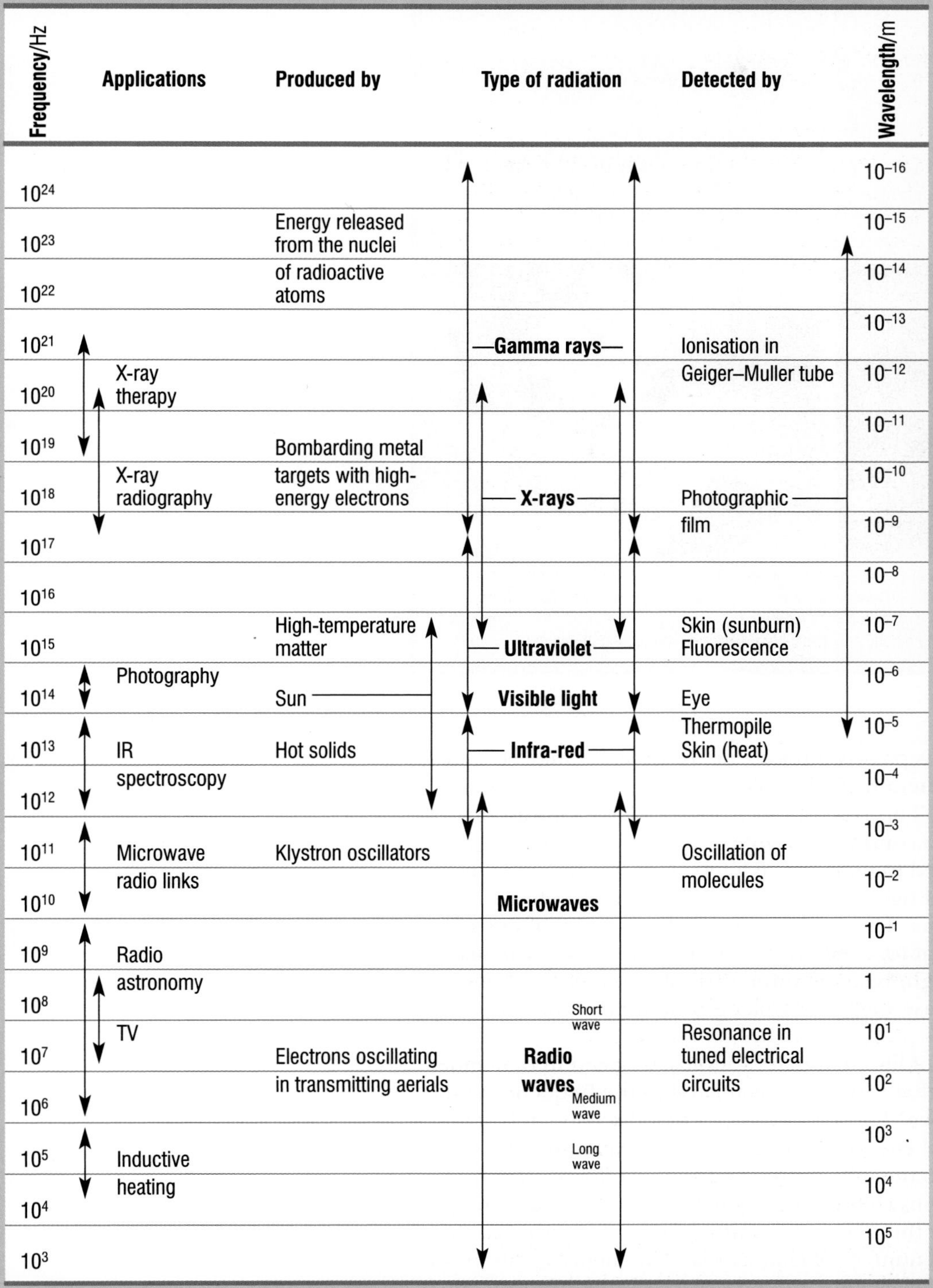

Figure 1.3.2 The electromagnetic spectrum

From work by the German physicist Max Planck, carried out in 1900, we know that under certain circumstances electromagnetic radiation may be regarded as a stream of particles, rather than as a wave. These particles are called **photons**. Their existence was confirmed by Albert Einstein, who was able to show that the energy carried by electromagnetic radiation depends on its frequency. The relationship between the frequency f of an electromagnetic wave and the energy E carried by a photon of this frequency is:

$$E = hf$$

where h is a constant called the Planck constant, which has the value 6.63×10^{-34} J s. The energy of a single photon is referred to as one **quantum** (plural quanta) of energy.

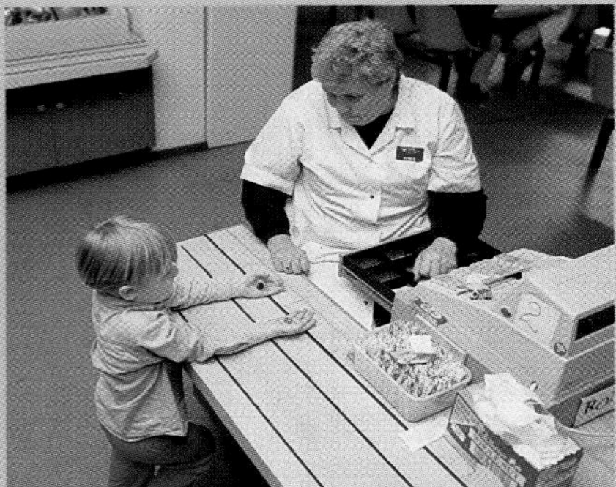

Figure 1.3.3 The smallest unit of money is unimportant when dealing in very large sums, but it is very important when the sum of money is small. Energy is like this – when we deal in large amounts, as we do in our everyday lives, the fact that it comes in quanta can be ignored. But when we deal with very small amounts, as we do when we deal with individual atoms, the energy quantum becomes very important indeed.

Atomic spectra

If light from a tungsten lamp is passed through a glass prism, it may be split into its constituent colours – literally all the colours of the rainbow, as figure 1.1.14(a)(i) on page 16 shows. This type of spectrum is called a **continuous spectrum**, since it consists of a continuous range of wavelengths. In contrast, if a potential difference is applied to neon gas in a tube (a discharge tube), it produces the **line spectrum** of figure 1.1.14(a)(ii), consisting of a characteristic number of discrete wavelengths. These spectra are both examples of **emission spectra**, when light is given out (or emitted) by a substance. (Atoms can also form **absorption spectra**. If light is shone through a tube containing atoms of an element in the gaseous phase, certain wavelengths may be absorbed, showing as dark bands.)

To understand the spectra of light emitted by atoms, we need to remember that light carries energy, and that this energy is related to the frequency of the light, as shown in the box above. The atoms of neon in a discharge tube emit light because they have gained energy in a process called **excitation**. The potential difference applied to the tube causes a current – a flow of electrons – through the gas. These electrons collide with neon atoms and transfer energy to them. The neon atoms lose this energy by emitting light. The fact that this light consists only of a limited number of characteristic frequencies tells us that only a limited number of energy changes can take place within the atom, as figure 1.3.4 illustrates.

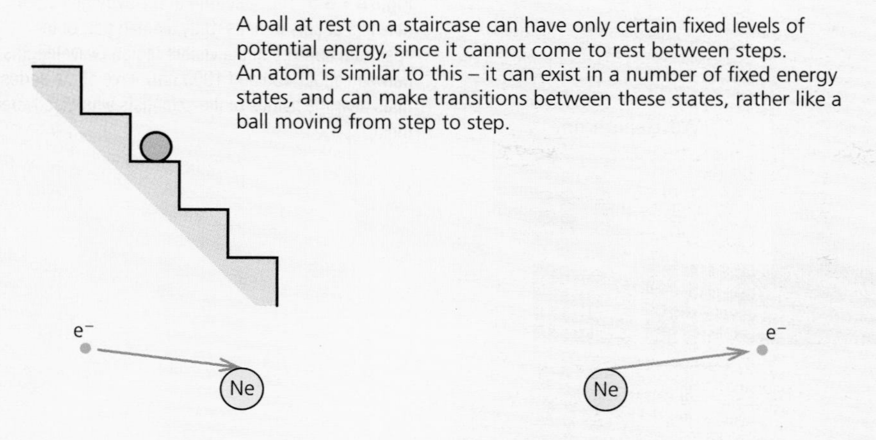

A ball at rest on a staircase can have only certain fixed levels of potential energy, since it cannot come to rest between steps.
An atom is similar to this – it can exist in a number of fixed energy states, and can make transitions between these states, rather like a ball moving from step to step.
e⁻
Ne
e⁻
Ne
Incoming electron has kinetic energy E_1
Rebounding electron has kinetic energy E_2
An electron may transfer some of its kinetic energy to a neon atom as a result of a collision.
In this case ΔE, the energy transferred from the electron to the atom, is given by $\Delta E = E_1 - E_2$.
This energy is then lost in the form of a photon of light, the frequency of the emitted light being given by $f = \Delta E/h$.

Figure 1.3.4

Hydrogen gas in a discharge tube produces a simpler line spectrum than neon, shown in figure 1.3.5 overleaf. Using the relationships $c = f\lambda$ and $E = hf$, we can convert the wavelengths in figure 1.3.5 into a series of energy differences. The box below illustrates how this conversion is carried out.

Energy lost from the hydrogen atom

One of the lines in the spectrum of the hydrogen atom has a wavelength of 6.57×10^{-7} m. What decrease in energy must the hydrogen atom undergo in order to emit light of this wavelength?

We know that:

$$c = f\lambda$$

so:

$$f = c/\lambda$$

$$= \frac{3.00 \times 10^{8} \text{ m s}^{-1}}{6.57 \times 10^{-7} \text{ m}}$$

$$= 4.57 \times 10^{14} \text{ Hz}$$

The energy transition associated with the emission of a photon of light of this frequency is given by:

$$E = hf$$

$$= 6.63 \times 10^{-34} \text{ J s} \times 4.57 \times 10^{14} \text{ Hz}$$

$$= 3.03 \times 10^{-19} \text{ J}$$

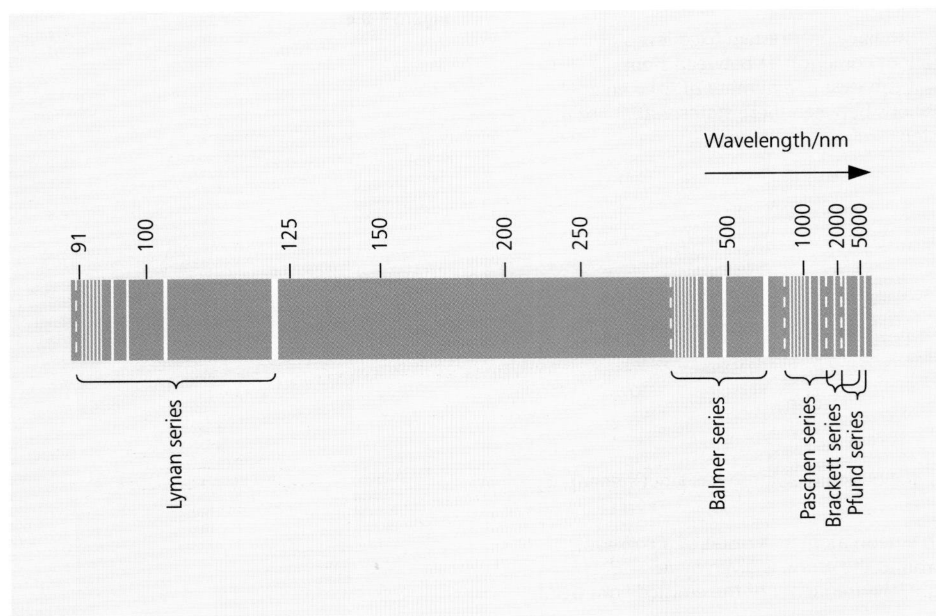

Figure 1.3.5 The spectrum of the hydrogen atom consists of sets of lines. Only a small part of the spectrum appears in the visible region (wavelengths between about 400 and 1000 nm). Five of the series of lines bear the names of the scientists who discovered them.

Using this technique, we end up with a series of values of ΔE – what do they mean? On the basis of what we know about line emission spectra and Einstein's photon theory, it seems reasonable to think of an atom emitting a photon of light when the atom moves from a state of high energy to one of low energy. For the hydrogen atom, assume that the transition with the greatest energy difference ($\Delta E = 2.18 \times 10^{-18}$ J) represents a transition from a level for which $E = 0$ to one for which $E = -2.18 \times 10^{-18}$ J. This line is the line of shortest wavelength in the Lyman series. In this transition the hydrogen atom will *lose* 2.18×10^{-18} J of energy. A calculation similar to that in the box on page 33 shows that the photon emitted has a wavelength of 9.17×10^{-8} m. This is the largest transition possible for the hydrogen atom (because there is no line in the hydrogen emission spectrum with a shorter wavelength). This line must therefore represent a transition from the highest energy level to the lowest energy level, that is, from $n = \infty$ to $n = 1$ in figure 1.3.6(a). The energy level for which $n = 1$ is the lowest possible energy state for the atom and is called its **ground state**. Other lines in the Lyman series have longer wavelengths and can be fitted onto the same diagram, with transitions between levels below $n = \infty$ down to the ground state. The line with the longest wavelength in the Lyman series is thus due to a transition between $n = 2$ and $n = 1$.

Having constructed the energy level diagram in this way using the Lyman series, calculations for the Balmer series show that the lines in this series are associated with transitions from higher levels to $n = 2$. In exactly the same way, lines in the Paschen series are associated with transitions from higher levels down to $n = 3$, and so on. Figure 1.3.6(b) shows the energy levels and transitions for these series.

Energy levels and electrons

Although the model we have just seen explains the spectrum of the hydrogen atom quite neatly, what is an energy level and what does a transition between two levels mean?

The negatively charged electron is outside the nucleus of the hydrogen atom. We might imagine that the removal of that electron is associated with an

(a) The transition between energy levels associated with the line with shortest wavelength in the spectrum of the hydrogen atom. In going from $n = \infty$ to $n = 1$, the atom loses 2.18×10^{-18} J of energy, and releases a photon with 2.18×10^{-18} J of energy.

(b) Transitions from higher energy levels to lower ones are responsible for the emission of light with wavelengths corresponding to the lines in the known series for the hydrogen atom.

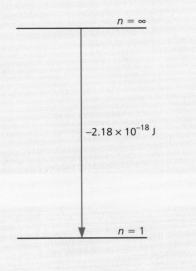

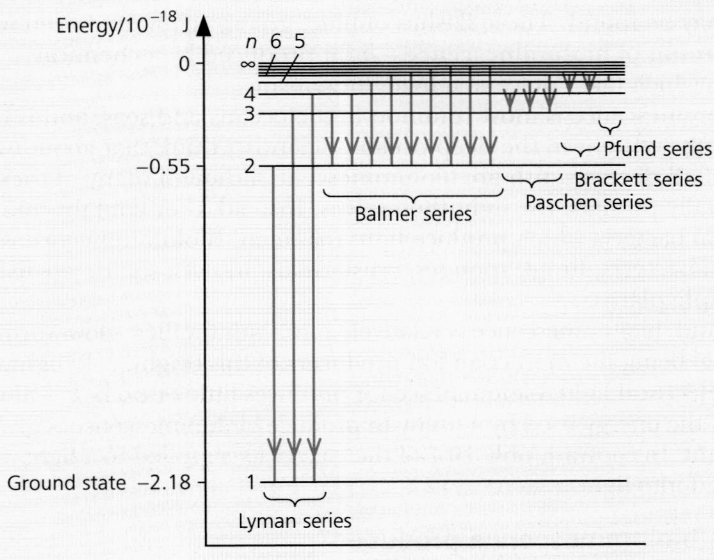

Figure 1.3.6 Energy levels in the hydrogen atom. Transitions between different levels produce the spectral lines.

input of energy to the atom – in other words, the further away the electron gets from the nucleus, the higher the potential energy of the atom. Emission and absorption of light can then be explained in terms of electrons moving between different energy levels in the atom.

As we have seen, an electron transition is called an **excitation**, when an electron moves from a lower level to a higher one. **Ionisation** is a transition in which an electron is removed from an atom completely. Ionisation of a hydrogen atom in its ground state involves a transition from $n = 1$ to $n = \infty$, which as we have seen corresponds to an energy change of 2.18×10^{-18} J. This is the **ionisation energy** or **ionisation potential** of hydrogen, which is normally quoted for 1 mole of hydrogen atoms:

$$\textbf{IE of hydrogen} = 2.18 \times 10^{-18}\ \text{J atom}^{-1}$$

$$= 2.18 \times 10^{-18}\ \text{J atom}^{-1} \times 6.02 \times 10^{23}\ \text{atom mol}^{-1}$$

$$= 1.3 \times 10^{6}\ \text{J mol}^{-1}$$

$$= 1300\ \text{kJ mol}^{-1}$$

Evidence that absorption and emission of radiation are concerned with the movement of electrons in atoms came from experiments carried out by James Franck and Gustav Hertz in 1913.

FOCUS LIGHTING UP THE DARK

In the depths of the ocean, in the branches of a tree, in the quiet of an English churchyard – in the most unexpected places – strange coloured lights may be found. These flashes of blue, green, yellow and even red are the result of **bioluminescence** – light produced by a chemical reaction which takes place in a living organism.

Bioluminescence is most common in the oceans and seas, and is the only source of light in the deep oceans. Scientists think that about two-thirds of all deep sea fish are bioluminescent, although many of them do not actually produce the light themselves. Instead, they contain colonies of special bacteria which produce light for them! Bioluminescence is also found in bacteria, fungi, sponges, crustaceans, insects, squid, jellyfish and simple plants.

On land, bioluminescence is relatively rare, with fireflies, glow-worms and fungi being the most common producers of this bright, cold light. Unlike electrical light, bioluminescence produces little or no heat – almost 100% of the energy used by a firefly to produce bioluminescence is given off as light. In contrast, only 10% of the electricity supplied to a light bulb is converted into light.

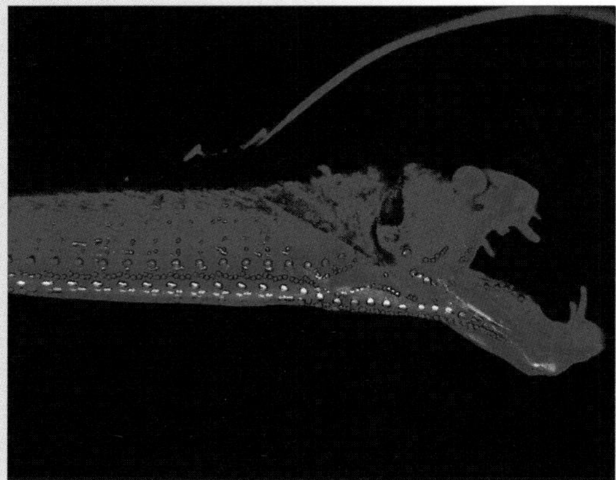

Figure 1 Bioluminescence – beautiful and efficient light

How is bioluminescence produced?

Bioluminescence is the result of an oxidation reaction which takes place in specialised light-producing cells. A chemical called **luciferin** is oxidised in a reaction which involves an enzyme called **luciferase**. In this reaction luciferin first combines with another molecule called ATP (short for **adenosine triphosphate**), to form luciferyl adenylate. This molecule then binds tightly (through intermolecular forces) to luciferase. When oxygen is present, luciferyl adenylate is oxidised to form oxyluciferin, with water and bioluminescence as the by-products. The bioluminesence is the result of the excitation of electrons in the molecule to a high energy state. As the excited molecule returns to the ground state it emits energy as visible light. If the chemicals are extracted from living organisms and simply mixed together in a jar and shaken, the same light emission can be seen.

Figure 2 Bioluminescence without the 'bio'. Even when extracted from cells, the excitation of the luciferin molecules causes the emission of light as the electrons return to their original shells.

What is bioluminescence for?

The animals and plants which have evolved bioluminescence use it for a wide variety of purposes. Firefly larvae use it to convince predators that they are not good to eat whilst the adults use their flashing lights to attract a mate. Bioluminescent fungi glow to attract insects which then spread their spores. Fish use bioluminescence to see in the dark, keep together, lure prey, and escape from predators. Perhaps one of the weirdest examples is found in Caribbean brittle stars. If caught by a crab, these creatures lose an arm. The rest of the brittle star then goes dark, but the detached arm not only bioluminesces, but flashes the light on and off to attract the crab's attention while the rest of the animal makes its escape. For these brittle stars, excited electrons save lives!

The Franck–Hertz experiment

In this experiment, electrons from the cathode were accelerated towards the grid. These electrons then continued towards the anode, from which they could flow back to the power supply, via a galvanometer. A small potential of about 2 V was applied between the grid and the anode, so the anode had a potential about 2 V lower than the grid.

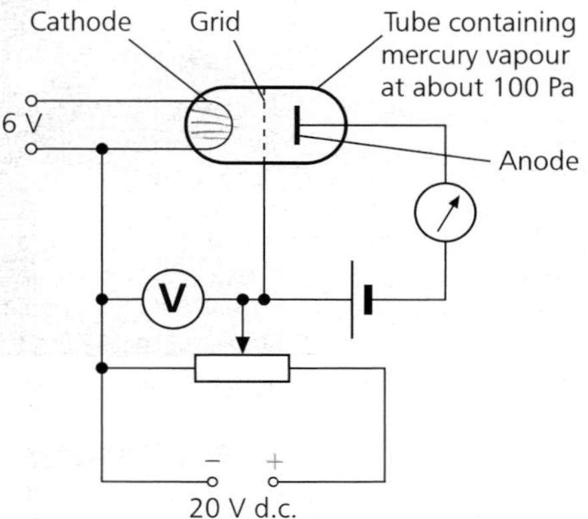

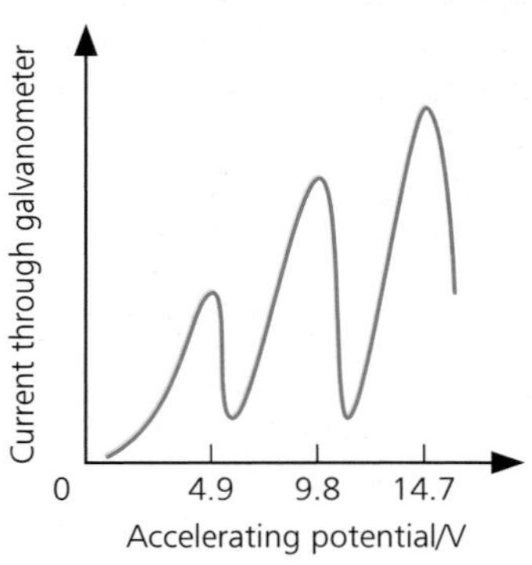

Figure 1.3.7 Franck and Hertz's experiment used a tube containing low-pressure mercury vapour. Triode valves of this type used to be commonly found in radio sets until the advent of the transistor.

The graph shows the results of the experiment. As the accelerating potential (the potential between the cathode and the grid) was increased, the current through the galvanometer increased, reaching an initial maximum at 4.9 V and then decreasing sharply. Further increasing the accelerating potential caused the current to rise again, when it peaked again at 9.8 V and then fell. Another peak at 14.7 V was also seen. At these peaks of current, the mercury vapour glowed.

Franck and Hertz explained these observations in terms of the energy levels of electrons in the mercury atom. With an accelerating potential below 4.9 V, electrons collided with mercury atoms in their path without any transfer of energy between electrons and atoms. However, electrons accelerated by a potential of 4.9 V had just enough energy to excite an electron in the mercury atom to a higher energy level, and in the collision lost all their own energy. As the accelerating potential was increased above 4.9 V, electrons had some energy remaining after a collision, so could overcome the 2 V potential between the grid and anode. The subsequent peaks at 9.8 V and 14.7 V corresponded to electrons having sufficient energy to excite two and three mercury atoms respectively.

A similar arrangement can be used to measure the ionisation energy of an atom.

Figure 1.3.8 Laser light is emitted when many atoms undergo similar energy transitions at the same time. This is achieved by promoting a large number of atoms to an energy level above the ground state. As an electron in one of the excited atoms jumps down from its higher energy level it emits a photon. As this photon travels past another atom in an excited state, it causes the electron in this atom to jump down to the lower level. The passage of light thus encourages or **stimulates** the emission of radiation from other atoms - producing the intense beam of light characteristic of the laser. The word 'laser' stands for '**l**ight **a**mplification by **s**timulated **e**mission of **r**adiation'.

The electrons in the atom – Bohr's model

Working at around the same time as Franck and Hertz, the Danish physicist Niels Bohr developed a model of the hydrogen atom, supplementing the laws of classical physics with ideas of his own, since the classical laws did not appear to provide consistent insight into the behaviour of electrons in atoms.

Using these ideas, Bohr's model was a striking success. The equations derived from it could successfully be used to calculate values for the radius of the hydrogen atom and its energy levels (particularly its ionisation energy), and they could be combined to produce a relationship resembling Balmer's formula. Yet it could not successfully explain the behaviour of atoms with more than one electron (even helium) – a new set of laws was needed to get to grips with such things. This new science is called **quantum mechanics**.

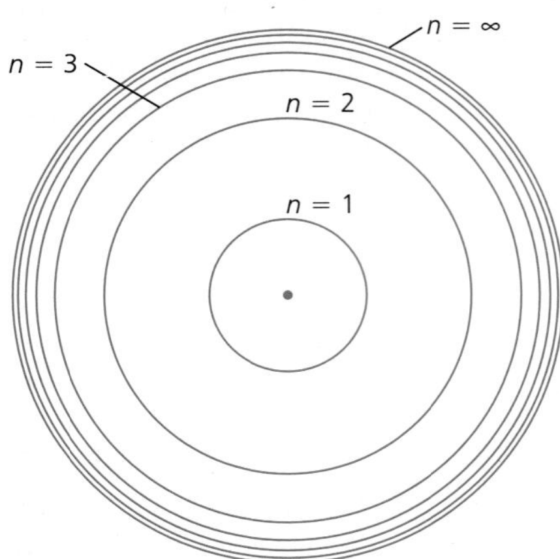

Figure 1.3.9 The energy level diagram of figure 1.3.6(b) results in a model for the hydrogen atom in which possible electron orbits are arranged like this. Moving out from the centre of the atom, the difference in radii between successive orbits decreases in the same way as the gap between energy levels decreases.

Electrons in atoms – the modern story

Quantum mechanics

We have seen that electrons can undergo transitions from one level to another, and that these transitions are associated with changes in the energy of the atom. But how are electrons arranged within atoms?

The new science of quantum mechanics uses the idea that the behaviour of electrons in atoms is best understood by considering them as waves rather than as particles. (This idea was first proposed by the Frenchman Louis Victor de Broglie in 1923, and was taken up and developed by the Austrian Erwin Schrödinger in 1926.) Although the full treatment is highly mathematical, the results of the theory are fortunately quite straightforward.

Shells, subshells and orbitals

Quantum mechanics shows that the electrons in atoms are arranged in a series of **shells**, which resemble the layers in an onion. Each shell is described by a number, known as the **principal quantum number** n, which tells us about the size of the shell. The larger the value of n, the further from the nucleus we are likely to find the electron. (Notice the use of the term 'likely to find'. This is because the shell is the region where the *probability* of finding the electron is greatest – although it does not completely rule out the possibility that it may be somewhere else altogether. This is rather like a pet cat – the most likely place to find it is around the house, but this does not rule out the possibility that it could be somewhere else – in a field, over the road or under the neighbour's bed!) The value of n ranges from $n = 1$ to $n = \infty$, and the shells are sometimes labelled with capital letters, starting with K for the first shell.

Quantum mechanics also shows that each shell may contain a number of **subshells**. These subshells are described by the letters s, p, d, f, g, and so on. Calculations show that every shell has an s subshell, all the shells except the first have a p subshell, all the shells except the first and second have a d subshell, and so on. The subshells can be represented like this:

Shell	*Subshells*
1	**1s**
2	**2s, 2p**
3	**3s, 3p, 3d**
4	**4s, 4p, 4d, 4f**

The subshells within the shell are associated with different energies, increasing like this:

s (lowest)→ p → d

Each type of subshell (s, p, d and so on) contains one or more **orbitals**. The number of orbitals in a subshell is determined by the subshell's type:

Subshell	*Number of orbitals*
s	**1**
p	**3**
d	**5**
f	**7**

Table 1.3.1 summarises how the first four shells are made up.

	First shell (n = 1)	Second shell (n = 2)		Third shell (n = 3)			Fourth shell (n = 4)			
Subshells	s	s	p	s	p	d	s	p	d	f
Number of orbitals	1	1	3	1	3	5	1	3	5	7

Table 1.3.1

Electrons in orbitals

Orbital energy

Quantum mechanics makes it possible to calculate the energy associated with each orbital in atoms containing many electrons. This is important to know when examining some of the properties of the elements, as we shall see in section 2. Figure 1.3.10 (overleaf) shows an energy level diagram for atoms with two or more electrons.

Notice that all the orbitals in a particular subshell are at the same energy level. As the principal quantum number n increases, the energy gap between successive shells gets smaller (compare the gap between the second and third shells with the gap between the sixth and seventh shells, for example). As a result of this, an orbital in an inner shell may be associated with a higher energy level than an orbital in the next shell out. This can be seen in the case of the 3d orbital for example, which has an energy level *above* that of the 4s orbital, but *below* that of the 4p orbital. This order of energy levels will be of particular interest when we come to examine the arrangement of elements in the periodic table in section 2.

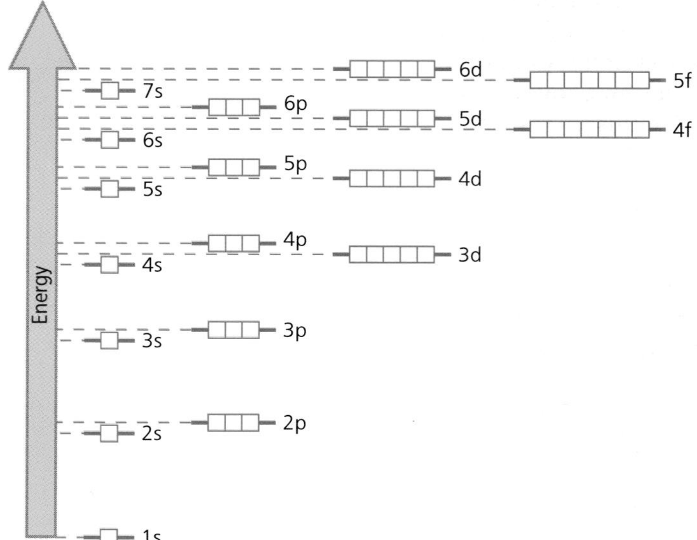

Figure 1.3.10 An energy level diagram for an atom with two or more electrons. Each orbital is indicated by a single square box.

Orbital shape

Just as quantum mechanics provides a way of calculating the energies of orbitals in atoms, so it also provides a way of visualising their shapes. There is nothing in our everyday experience which allows us to visualise the shape of a particle behaving as a wave, which seems to us very strange. (Niels Bohr is quoted as saying 'Anyone who is not shocked by quantum theory has not understood it'.) We deal with this difficulty by talking in terms of probabilities, as we saw on page 38. An electron behaves as though it is spread out around the nucleus as a cloud – we sometimes speak of an **electron cloud**. The shapes of orbitals calculated from quantum mechanics provide us with a map of the **electron density** of this electron cloud.

Figure 1.3.11(a) shows the electron cloud calculated for the 1s orbital. There are several ways of interpreting such a diagram. One of these is to say that an electron in a 1s orbital spends most of its time in a sphere close to the nucleus. Another is to say that the most likely place to find an electron in a 1s orbital is in a sphere close to the nucleus. We might also interpret the diagrams in terms of electron density, showing how much of an electron's charge is likely to be found in a given volume. Each of these interpretations is essentially similar, and is an attempt to picture the results of quantum mechanics in a way that we can deal with in terms of our own experience of the world – which is very different from the subatomic world of the electron!

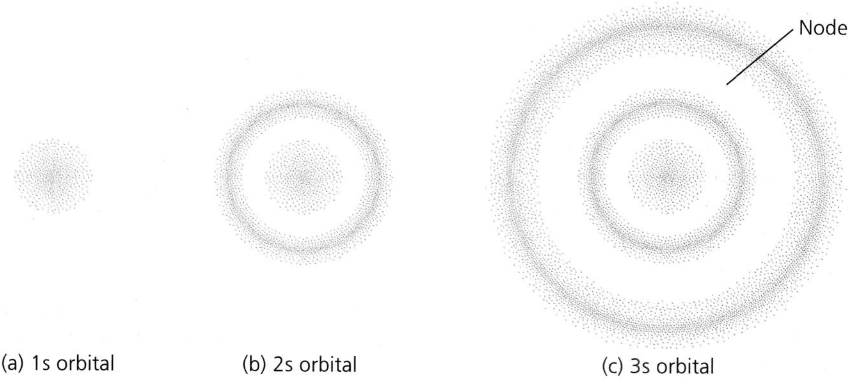

Figure 1.3.11 The electron clouds for the s orbitals in the first, second and third shells

The electron clouds for the s orbitals in the second and third shells are shown in figures 1.3.11(b) and (c). The distribution of electron density in these orbitals is similar to that in the 1s orbital (they are spherically symmetrical around the nucleus). However, there are regions called nodes where the electron wave has zero amplitude, and where the electron density is zero as a result. This is the case for the p and d orbitals too, which are shown in figure 1.3.12.

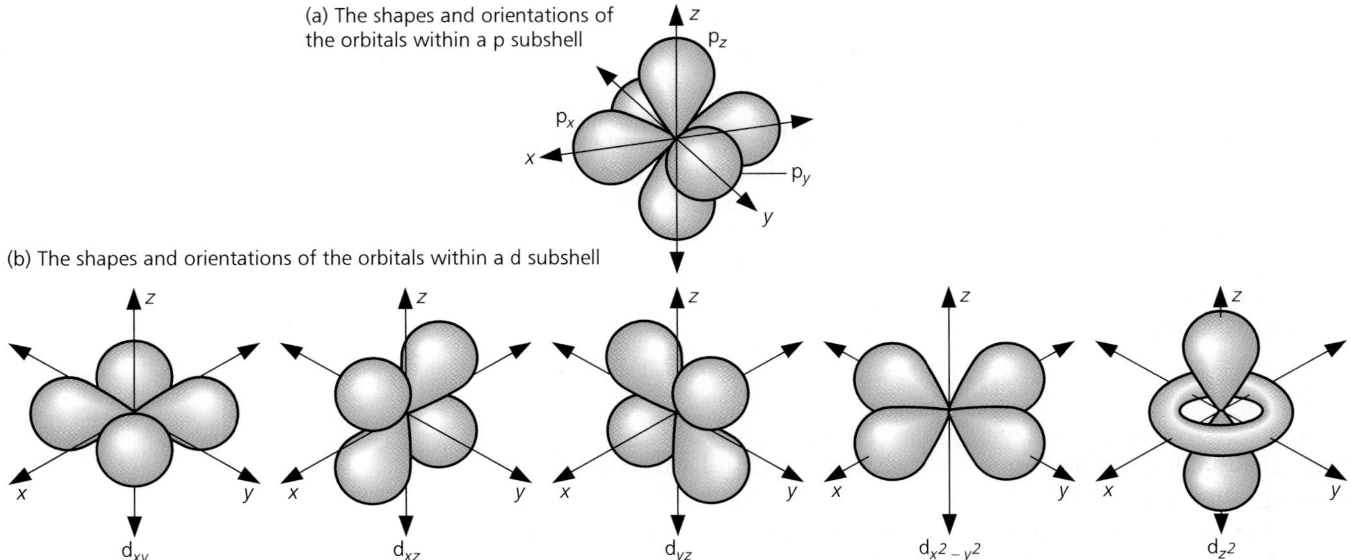

Figure 1.3.12

The p orbitals are approximately dumb-bell-shaped, and lie at 90° to each other as figure 1.3.12(a) shows. As with the s orbitals, the size of the p orbitals increases as the principal quantum number increases. Each orbital is labelled as if it lay along one of the axes of an *xyz* coordinate system. Notice that each p orbital has a node at the nucleus of the atom.

The shapes of the d orbitals are complex, and are difficult to draw on the same set of axes. They are shown in figure 1.3.12(b). Four of the five d orbitals are a similar windmill shape, while the fifth is similar in shape to a p orbital, except that it has a doughnut-shaped ring of electron density around its middle. The labels of these orbitals come from quantum mechanics. These orbitals will be important when we come to look at the properties of the transition elements, in section 2.7.

Filling the orbitals

An atom will be in its lowest state of energy (its ground state) when its electrons are arranged in the orbitals with the lowest possible energy levels. The way the electrons fill up the orbitals to make up elements in their ground states is governed by some straightforward rules. One of the factors influencing the filling of the orbitals is **electron spin**.

An electron in an atom behaves like a tiny magnet. This can be explained by imagining that an electron spins on its axis, in much the same way as the Earth does, as shown in figure 1.3.13. (In this visualisation, we are returning to the particle model of the electron rather than the wave model.)

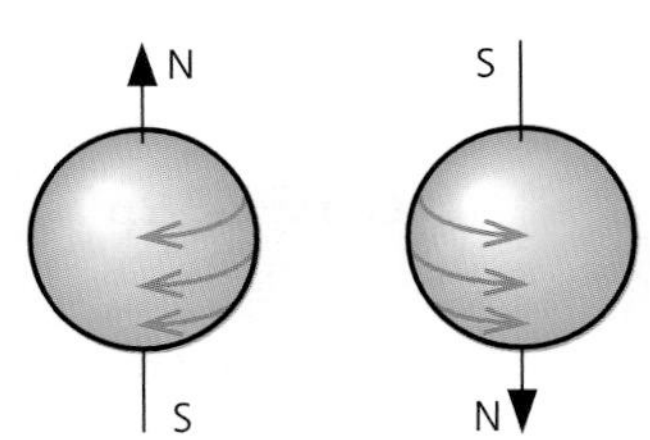

Figure 1.3.13 The magnetic properties of the electron can be explained by imagining that the electron spins on its axis.

We can visualise that an electron can spin in either direction – clockwise or anticlockwise. Because of this magnetic behaviour, we represent an electron as a small arrow, showing its spin by pointing the arrow up (↑) to represent spin in one direction or down (↓) to represent spin in the opposite direction.

Electron spin has an important influence on the way in which electrons occupy orbitals, since no two electrons in an orbital may have the same spin. As a result of electron spin, each orbital can contain a maximum of two electrons. This can be summarised as follows:

Subshell	*Maximum number of electrons in subshell*	
s	[↑↓]	= 2
p	[↑↓] [↑↓] [↑↓]	= 6
d	[↑↓] [↑↓] [↑↓] [↑↓] [↑↓]	= 10
f	[↑↓] [↑↓] [↑↓] [↑↓] [↑↓] [↑↓] [↑↓]	= 14

Let us now look at how the orbitals of an atom are filled. Beginning with the hydrogen atom in its ground state, this will have a single electron in its 1s shell. We may represent the electronic arrangement or **configuration** of hydrogen as $1s^1$. It is often useful to represent electronic configuration by means of a diagram like that above, in which each orbital is represented by a box. Each electron is represented by an arrow, pointing up or down to denote its spin. For hydrogen the diagram is:

H [↑]
1s

The atomic number *Z* of helium is 2, which means that it has two protons in its nucleus. As atoms are electrically neutral, the helium atom must also have two electrons. These are both allowed to occupy the 1s orbital, provided that they have opposite spins, so the electronic configuration of helium is $1s^2$, and its orbital diagram is:

He [↑↓]
1s

After helium, the next two elements are lithium ($Z = 3$) and beryllium ($Z = 4$) with three and four electrons respectively. The 1s shell is full, so the next two electrons go into the next lowest energy orbital, which is the 2s orbital. This gives the electronic configuration of lithium and beryllium as:

	1s	2s
Li $1s^2\ 2s^1$	[↑↓]	[↑]
Be $1s^2\ 2s^2$	[↑↓]	[↑↓]

Boron, carbon and nitrogen are the next elements, with $Z = 5$, $Z = 6$ and $Z = 7$ respectively. The 2s orbital is full, so the fifth, sixth and seventh electrons go into the next lowest energy orbital, which is the 2p orbital:

	1s	2s	2p
B $1s^2\ 2s^2\ 2p^1$	[↑↓]	[↑↓]	[↑] [ ] [ ]
C $1s^2\ 2s^2\ 2p^2$	[↑↓]	[↑↓]	[↑] [↑] [ ]
N $1s^2\ 2s^2\ 2p^3$	[↑↓]	[↑↓]	[↑] [↑] [↑]

Notice that all three of the 2p orbitals are always shown, even if only one or two of them contain an electron. Which of the 2p orbitals is filled first is unimportant, since they all have the same energy.

Filling orbitals – the Hund rule

Notice that in constructing the electron configuration for carbon, the two electrons in the 2p subshell are placed in different orbitals, rather than placing them in the same orbital with opposite spins, like this:

C $1s^2 2s^2 2p^2$ [↑↓] [↑↓] [↑↓] [ ] [ ]
1s 2s 2p

This is the result of the **Hund rule**, which says that when electrons are placed in a set of orbitals with equal energy, they spread out to maximise the number of unpaired electrons. This can be clearly seen in the case of nitrogen, where the three 2p orbitals each contain a single, unpaired electron.

The remainder of the second shell is completed by adding three more electrons, in the elements oxygen, fluorine and neon:

O $1s^2 2s^2 2p^4$ [↑↓] [↑↓] [↑↓] [↑] [↑]
1s 2s 2p

F $1s^2 2s^2 2p^5$ [↑↓] [↑↓] [↑↓] [↑↓] [↑]
1s 2s 2p

Ne $1s^2 2s^2 2p^6$ [↑↓] [↑↓] [↑↓] [↑↓] [↑↓]
1s 2s 2p

Using the energy level diagram in figure 1.3.10, further electrons can be added to build up more shells, electrons being put into subshells in order of increasing energy. The worked examples in the box below will show you how this is done.

Electronic configurations – worked examples

(1) Write down the electronic configuration of a vanadium atom ($Z = 23$), and represent this configuration diagrammatically, showing how electron spins are paired.

Vanadium has 23 protons in its nucleus, so the neutral vanadium atom will have 23 electrons. Following figure 1.3.11, we can begin filling subshells in order of increasing energy until we run out of electrons (and remembering the Hund rule too). This gives the configuration:

$$\text{V } 1s^2\,2s^2\,2p^6\,3s^2\,3p^6\,3d^3\,4s^2$$

The arrangement of electrons in orbitals can be represented diagrammatically as:

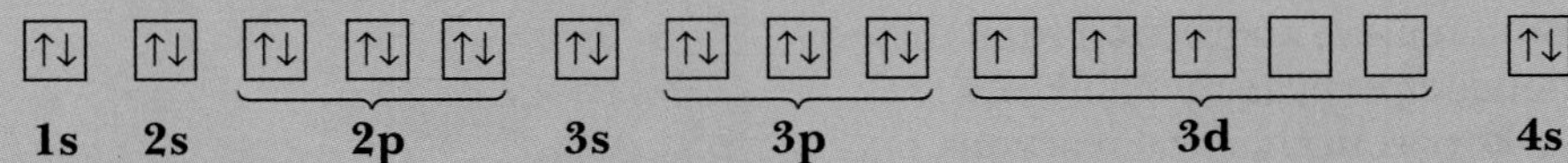

Notice that the orbitals are not written strictly in order of increasing energy, but grouped together in shells, so the 3d orbital comes before the 4s orbital, even though the 4s orbital fills first. The three electrons in the 3d subshell go into separate orbitals so that the number of unpaired electrons is as large as possible.

(2) Write down the electronic configuration of a polonium atom ($Z = 84$).

With 84 electrons, we follow the order of figure 1.3.10 again, giving:

Po $1s^2\ 2s^2\ 2p^6\ 3s^2\ 3p^6\ 3d^{10}\ 4s^2\ 4p^6\ 4d^{10}\ 4f^{14}\ 5s^2\ 5p^6\ 5d^{10}\ 6s^2\ 6p^4$

Writing electronic configurations more simply

Chemical changes concern only the outer electrons of the atom, not those in the inner shells. This is because these inner electrons simply do not come into contact with the electrons of other atoms when chemical bonds are formed. It is therefore convenient to be able to represent electronic configurations simply so that attention is focused on the outer electrons.

As an example, consider the elements potassium and calcium. The electronic configurations for these elements are:

K $1s^2\ 2s^2\ 2p^6\ 3s^2\ 3p^6\ 4s^1$

Ca $1s^2\ 2s^2\ 2p^6\ 3s^2\ 3p^6\ 4s^2$

The electrons in the first three shells of both atoms are identical – it is only the fourth shell that differs. The inner electrons have the configuration of the noble gas argon. The 'shorthand notation' of the elements can therefore be written as the symbol of the noble gas, followed by the outer electrons:

K [Ar] $4s^1$

Ca [Ar] $4s^2$

Electronic configurations and the properties of elements

The arrangements of electrons in atoms can help to explain some of the properties of the elements. In particular, chemists are interested to know the answers to questions like: 'Why are the elements known as the noble gases so chemically unreactive?' 'In general, why do elements in the same group of the periodic table have similar properties?' 'What reason is there for the patterns in properties such as ionisation energy, which are seen as we move around the periodic table?'

We shall be looking at questions like these in some detail in section 2, but for the moment we can make some general observations showing how the trends in properties, the periodic table and the electronic structures of elements are linked.

Examination of table 1.3.2, page 45, shows that the noble gases helium, neon and argon have electronic structures that involve full subshells of electrons. Measurement of ionisation energies (see figure 2.1.4 on page 133) shows that the noble gases have exceptionally high ionisation energies. The fact that a large amount of energy needs to be supplied in order to remove an electron from these atoms suggests that their electronic configuration is a particularly stable one, which also helps to explain their reluctance to take part in chemical reactions.

Other elements in the table can be compared in the same way. Doing this we see that lithium and sodium have very similar electronic structures, in which each has a single electron outside a full shell of inner electrons. Similarly, fluorine and chlorine have outer shells of electrons which are just one electron short of a full shell. The position of an element in the periodic table is determined by its electronic structure. We shall see in section 1.4 that this directly influences their chemical properties, since it determines the way in which they form chemical bonds.

Figure 1.3.14 Neon and other inert gases provide the colours in these brightly lit signs. Electrical energy causes electrons in the atoms' outer shells to become excited. As they return to their ground state they give out light energy of varying wavelengths, which is seen as different colours.

Element	Number of electrons	Arrangement of electrons in ground state 1s	2s	2p	3s	3p
Hydrogen (H)	1	1				
Helium (He)	2	2				
Lithium (Li)	3	2	1			
Beryllium (Be)	4	2	2			
Boron (B)	5	2	2	1		
Carbon (C)	6	2	2	2		
Nitrogen (N)	7	2	2	3		
Oxygen (O)	8	2	2	4		
Fluorine (F)	9	2	2	5		
Neon (Ne)	10	2	2	6		
Sodium (Na)	11	2	2	6	1	
Magnesium (Mg)	12	2	2	6	2	
Aluminium (Al)	13	2	2	6	2	1
Silicon (Si)	14	2	2	6	2	2
Phosphorus (P)	15	2	2	6	2	3
Sulphur (S)	16	2	2	6	2	4
Chlorine (Cl)	17	2	2	6	2	5
Argon (Ar)	18	2	2	6	2	6

Table 1.3.2 The electronic configurations of the first 18 elements

Predicting electronic configurations

The rules that we have looked at generally work well for predicting electronic configurations, although like most rules, there are exceptions.

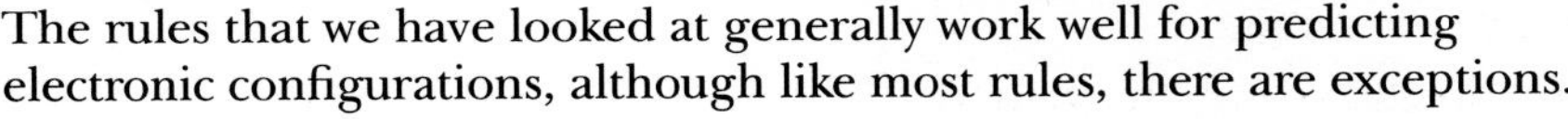

For example, the rules suggest that the elements molybdenum and silver should have the electronic configurations:

Mo [Kr] 4d⁴ 5s² or [Kr] ↑ ↑ ↑ ↑ ☐ ↑↓

Ag [Kr] 4d⁹ 5s² or [Kr] ↑↓ ↑↓ ↑↓ ↑↓ ↑ ↑↓

4d 5s

Experimental determination of their configurations, however, shows them to be:

Mo [Kr] 4d⁵ 5s¹ or [Kr] ↑ ↑ ↑ ↑ ↑ ↑

Ag [Kr] 4d¹⁰ 5s¹ or [Kr] ↑↓ ↑↓ ↑↓ ↑↓ ↑↓ ↑

4d 5s

In both cases an electron is 'borrowed' from the 5s subshell in order to produce a half-filled 4d subshell (in the case of molybdenum) or a filled 4d subshell (in the case of silver). This example illustrates a general observation that half-filled and filled subshells appear to be favourable energetically. In the case of molybdenum, this means that it is preferable to have a half-full 4d subshell and a half-full 5s subshell rather than the predicted $4d^4\ 5s^2$ configuration, while for silver it is preferable to have a full 4d subshell and half-full 5s subshell rather than the predicted configuration of $4d^9\ 5s^2$.

The extra stability that half-full and full subshells seem to give to atoms influences not only the elements in their ground state but also the ions formed by them – particularly the transition elements. We shall see this effect later when we consider the formation of chemical bonds (section 1.4), as well as the properties of elements and their compounds in the periodic table (section 2).

SUMMARY

- The **electronic structure** or electronic configuration of an atom determines its chemical properties.
- The **emission spectrum** of an element appears as a series of distinct lines (a **line spectrum**) indicating that only a limited number of energy changes are possible within the structure of the atom.
- Each element has a characteristic line spectrum which can be used to identify it.
- The modern model of the atom is based on quantum mechanics. Electrons can be thought of as a cloud of negative charge surrounding the nucleus. Electrons are found in **shells** which are identified by principal quantum numbers $n = 1, 2, 3$, etc. Each shell contains one or more **subshells**, identified by letters s, p, d, f, etc.
- Each subshell contains one or more **orbitals**. One orbital can contain a maximum of 2 electrons.
- s subshells hold a maximum of 2 electrons in a single orbital.
  p subshells hold a maximum of 6 electrons in 3 orbitals.
  d subshells hold a maximum of 10 electrons in 5 orbitals.
  f subshells hold a maximum of 14 electrons in 7 orbitals.
- Orbitals have characteristic shapes depending on the subshell type. The shapes of the orbitals provide a guide to the distribution of electron density around the nucleus.

QUESTIONS

1 **a** Why does the emission spectrum of an element consist of a series of lines?
  b How can the ionisation energy of an element be found from its spectrum?
  c One line in the hydrogen spectrum has a frequency of 7.3×10^{14} Hz. What change in energy within the hydrogen atom is this line associated with?

2 Write down the electronic configurations of the following elements:
  a Li **b** O **c** Mg **d** P **e** Cl **f** Ga.

3 Draw energy level diagrams like that in figure 1.4.10 for the following elements:
  a Be **b** N **c** S **d** Ti **e** Ge **f** Kr.

4 The gold atom, Au, has 79 electrons. On this basis, we might expect the electronic configuration of the gold atom in its ground state to be:

[Xe] $4f^{14}\ 5d^9\ 6s^2$

However, the actual configuration is:

[Xe] $4f^{14}\ 5d^{10}\ 6s^1$

Explain this.

1.4 Chemical bonding

1

Now that we know something about the structure of atoms, we are in a position to look at how atoms combine to form chemical substances. Atoms are held together in compounds by **chemical bonds**, and understanding these is a central part of chemistry, since changes in bonding underlie all chemical reactions.

There are two main types of chemical bond – **ionic** and **covalent**. We shall begin by looking at the formation of chemical compounds from ions.

Figure 1.4.1 A model of the complex molecule DNA (deoxyribonucleic acid), the substance in each cell of our body that codes for the production of proteins from individual amino acids. As this and other chemical processes in the body take place, each atom must be precisely bonded to another to construct the correct molecule. With so little room for error, complex systems have evolved to ensure the correct chemical bonds are formed in exactly the right way – it is not just *which* atoms are joined together, but also *how* they are joined, that matters.

Ionic compounds

The formation of compounds from ions

Figure 1.4.2 shows sodium metal reacting with chlorine gas, liberating a substantial amount of energy in the process. Apart from this large amount of energy, one of the most remarkable things about this reaction is the production of a compound with properties quite different from those of the two elements that formed it, as table 1.4.1 highlights.

Sodium	Chlorine	Sodium chloride
Soft metal, melting point 98 °C	Greenish gas, boiling point –35 °C	White crystalline solid, melting point 801 °C
Reacts vigorously with water	Soluble in water	Soluble in water
Good conductor of electricity in solid and liquid state	Does not conduct electricity	Good conductor of electricity in liquid state and in aqueous solution

Table 1.4.1

Figure 1.4.2 Sodium (a soft metal with a low melting point) reacts violently with chlorine (a greenish gas) to produce sodium chloride (a white solid with a very high melting point).

The explanation for these differences lies in the fact that, like many compounds formed between metals and non-metals, sodium chloride is an **ionic** compound, made up of a **lattice** of Na^+ and Cl^- ions. In contrast, sodium consists of a lattice of metal atoms, and chlorine is a gas made up of Cl_2 molecules. We shall look at the effect of the structure of a substance on its chemical and physical properties in section 1.6, but for now let us concentrate on why sodium and chlorine react together in this way. Why does sodium chloride contain Na^+ and Cl^- ions, rather than (for example) Na^- and Cl^+ ions, or even Na^{2+} and Cl^{2-} ions?

Part of the answer to this question lies in the electronic structure of the sodium atom, which is:

$$\textbf{Na} \quad 1s^2\ 2s^2\ 2p^6\ 3s^1$$

In section 1.3 we saw that full shells of electrons are particularly stable. Evidence from the ionisation energies of the elements (figure 2.1.4 on page 133) suggests that there is a particular stability associated with the electronic configuration of the noble gases. Sodium can achieve a full outer shell of electrons by losing the electron in its 3s subshell to become a Na^+ ion. This gives it the same electronic configuration as the noble gas neon – Na^+ and Ne are said to be **isoelectronic**:

$$\mathbf{Na^+} \quad \mathbf{1s^2\ 2s^2\ 2p^6}$$

$$\mathbf{Ne} \quad \mathbf{1s^2\ 2s^2\ 2p^6}$$

The electronic structure of chlorine is:

$$\mathbf{Cl} \quad \mathbf{1s^2\ 2s^2\ 2p^6\ 3s^2\ 3p^5}$$

Just as sodium achieved a full outer shell of electrons by losing an electron, chlorine can achieve a noble gas configuration by gaining an electron. The Cl^- ion so formed is isoelectronic with the argon atom:

$$\mathbf{Cl^-} \quad \mathbf{1s^2\ 2s^2\ 2p^6\ 3s^2\ 3p^6}$$

$$\mathbf{Ar} \quad \mathbf{1s^2\ 2s^2\ 2p^6\ 3s^2\ 3p^6}$$

So sodium and chlorine can each achieve the electronic configuration of a noble gas by transferring the electron in sodium's 3s subshell to chlorine's 3p subshell. We can represent this as shown in figure 1.4.3, which also shows a similar process occurring between a calcium atom and two fluorine atoms. This type of chemical bond, involving a complete transfer of electrons from one atom to another, is known as an **ionic** or **electrovalent bond**. In the process of electron transfer, each atom becomes an ion that is isoelectronic with the nearest noble gas, and the substance formed is held together by electrostatic forces between the ions.

The tendency for the ions formed by elements to have a full outer shell of electrons is expressed in the **octet rule**:

> **When atoms react, they tend to do so in such a way that they attain an outer shell containing eight electrons.**

Like most simple rules, this one has exceptions, although it is useful in many situations concerned with bonding, as we shall see shortly.

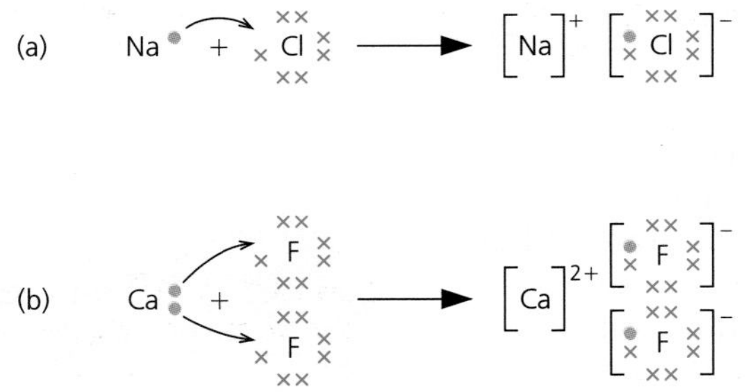

(a) Na + Cl ⟶ [Na]$^+$ [Cl]$^-$

(b) Ca + F + F ⟶ [Ca]$^{2+}$ [F]$^-$ [F]$^-$

Figure 1.4.3 The transfer of electrons involved in the formation of (a) sodium chloride and (b) calcium fluoride. Each atom forms an ion with an outer shell containing eight electrons.

Accounting for electrons – dot and cross diagrams

It is necessary to count electrons when thinking how atoms combine in order to see how the atoms may donate or receive electrons to obtain a full outer shell. This is most usually done through the use of **dot and cross diagrams**, in which the outer electrons of one element are represented by dots and those of the other element by crosses.

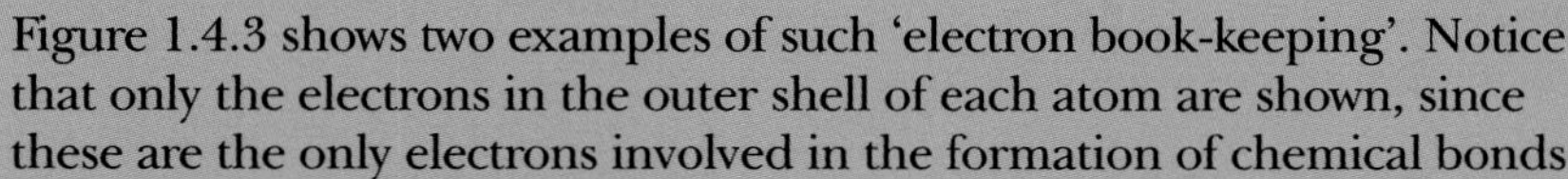

Figure 1.4.3 shows two examples of such 'electron book-keeping'. Notice that only the electrons in the outer shell of each atom are shown, since these are the only electrons involved in the formation of chemical bonds.

The other important point to note about dot and cross diagrams is that, although the diagrams distinguish between the electrons in different atoms, all electrons are in fact identical. This means that when a sodium atom loses an electron to a chlorine atom, it is impossible to tell which of the electrons surrounding the Cl^- ion comes from the sodium atom.

Factors affecting the formation of ions

This picture of ions being formed as elements attempt to attain a noble gas structure is simple and appealing. However, it rather oversimplifies the formation of a compound from its elements. What factors should we consider in this process?

Factor 1 – ionisation energy

Table 1.4.2 and figure 1.4.4 show the energy required to remove the first four electrons from atoms of sodium, magnesium and aluminium. Ionisation always involves an input of energy, since work must be done on an electron in order to overcome the attractive force between it and the nucleus, as we saw in section 1.3. The energy needed to remove one mole of electrons from one mole of gaseous atoms is the **first ionisation energy** for that element. The energy needed to remove one mole of electrons from one mole of singly charged gaseous cations (positive ions) is the **second ionisation energy** for that element. The equations for these processes are given in table 1.4.2. To calculate the total energy required to remove the second electron from an atom, we have to add the first and second ionisation energies together. The ionisation energies in the table suggest that the formation of ions with a noble

	First ionisation energy [$M(g) \rightarrow M^+(g) + e^-$]/ kJ mol^{-1}	Second ionisation energy [$M^+(g) \rightarrow M^{2+}(g) + e^-$]/ kJ mol^{-1}	Third ionisation energy [$M^{2+}(g) \rightarrow M^{3+}(g) + e^-$]/ kJ mol^{-1}	Fourth ionisation energy [$M^{3+}(g) \rightarrow M^{4+}(g) + e^-$]/ kJ mol^{-1}
Na	496	4563	6913	9544
Mg	738	1451	7733	10 451
Al	578	1817	2745	11 578

Table 1.4.2 The first four ionisation energies of sodium, magnesium and aluminium

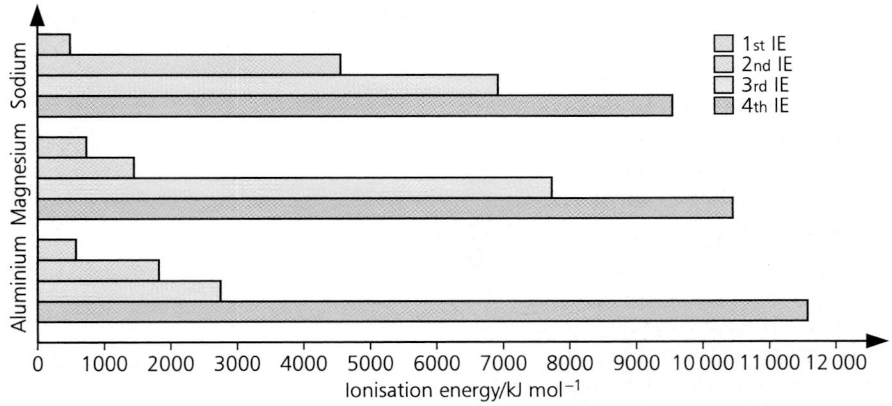

Figure 1.4.4

gas structure (Na^+, Mg^{2+} and Al^{3+}) requires considerably less energy than the formation of ions such as Na^{2+}, Mg^{3+} or Al^{4+}. Before we can decide how important this might be, we need to look at the formation of negative ions too.

Factor 2 – electron affinity

When anions (negative ions) are produced, we need to know the energy change involved in adding an electron to a neutral atom.

$$X(g) + e^- \rightarrow X^-(g)$$

This energy change is known as the **electron affinity** of the atom X. It is effectively the opposite of the ionisation energy for $X^-(g)$. The electron affinity for an element is defined as the energy change when one mole of singly charged gaseous anions is formed from one mole of gaseous atoms and one mole of electrons. In the same way as for ionisation energy, this is the *first* electron affinity. The second electron affinity for an element is the energy change when one mole of doubly charged gaseous anions is formed from one mole of singly charged gaseous anions and one mole of electrons.

Table 1.4.3 shows the electron affinities of some elements on the right-hand side of the periodic table. The electron affinities shown are all negative (energy is released), with the exception of the addition of a second electron to oxygen.

Element	Process	Electron affinity/kJ mol^{-1}
F	$F(g) + e^- \rightarrow F^-(g)$	−328
Cl	$Cl(g) + e^- \rightarrow Cl^-(g)$	−349
Br	$Br(g) + e^- \rightarrow Br^-(g)$	−325
I	$I(g) + e^- \rightarrow I^-(g)$	−295
O	$O(g) + e^- \rightarrow O^-(g)$	−141
	$O^-(g) + e^- \rightarrow O^{2-}(g)$	+798
H	$H(g) + e^- \rightarrow H^-(g)$	−73

Table 1.4.3 Electron affinities for some elements. Measurement of electron affinity can be done using spectroscopic techniques, in a way similar to that used for determining ionisation energies.

The addition of one electron to the elements shown releases energy, although the addition of further electrons (as in the case of the formation of O^{2-} from O^-) requires energy, since a negative electron must be forced into an already negative ion. Comparison of the figures in this table with those in table 1.4.2 shows that less energy is released when a Cl^- ion is formed than is required to remove an electron from a sodium atom. This means that the formation of Na^+ and Cl^- ions from sodium and chlorine atoms is a process that requires energy, and is therefore unlikely to occur spontaneously.

Factor 3 – lattice energy

Ionisation energies and electron affinities give us information about the energy changes involved in the production of ions from neutral atoms, but this is only part of what happens when an ionic compound is formed. The story is completed by knowledge of the **lattice energy** of the compound. This is the energy released when the ions in the gas phase come together to form one mole of solid. For sodium chloride, this is the process:

$$Na^+(g) + Cl^-(g) \rightarrow NaCl(s)$$

which releases 780 kJ mol^{-1} of energy.

You can find out more about ionic compounds and their properties in section 1.6. Lattice energies (more properly called lattice enthalpies) are examined in more detail in section 1.8, while periodic trends in ionisation energy and electron affinity will be explored in section 2.1.

Combining the factors

So what can we learn about the formation of ions and ionic compounds from this brief survey of ionisation energy, electron affinity and lattice energies?

1 The removal of electrons to form positive ions that have a noble gas structure involves much less energy than the removal of electrons that would leave a different electronic structure.
2 Addition of one electron to an atom may release energy, although addition of subsequent electrons requires energy.
3 Energy is released when ions come together to form an ionic compound.

Putting these ideas together, it becomes clear that these three factors are connected. The formation of Na^+ ions requires energy, which is not fully compensated for by the fact that energy is released when the electron removed from the sodium atom is used in the formation of a Cl^- ion. However, the formation of a lattice of Na^+ and Cl^- ions releases a large amount of energy, so that the formation of sodium chloride from gaseous sodium and chlorine atoms appears to be a favourable process, releasing energy. Overall, the energy required to produce ions (an energetically unfavourable process) must be balanced by the energy released when the ions are brought together to form the solid lattice. As we shall see in section 1.8, we need to do a little more calculation to decide whether in fact sodium chloride in the form NaCl really is the most likely product of the reaction of sodium and chlorine.

Covalent compounds

Sharing electrons

For many elements, compounds cannot be formed by the production of ions, since the energy released in the formation of the lattice of ions would be insufficient to overcome the energy required to form the ions in the first place. In this case, atoms use another method of achieving a noble gas configuration – electron sharing. This is especially true for elements in the middle of the periodic table, for which the loss or gain of three or four electrons would be required to achieve a noble gas configuration.

To understand the sharing of electrons between atoms, consider what happens as two hydrogen atoms approach one another. Each atom consists of a single proton as its nucleus, with a single electron orbiting it. As the atoms get closer together, each electron experiences an attraction towards the two nuclei, and the electron density shifts so that the most probable place to find the two electrons is between the two nuclei, as shown in figure 1.4.5. Effectively, each atom now has a share of both electrons. The electron density between the two nuclei exerts an attractive force on each nucleus, keeping them held tightly together in a **covalent bond**.

Figure 1.4.5 A covalent bond forming between two hydrogen atoms

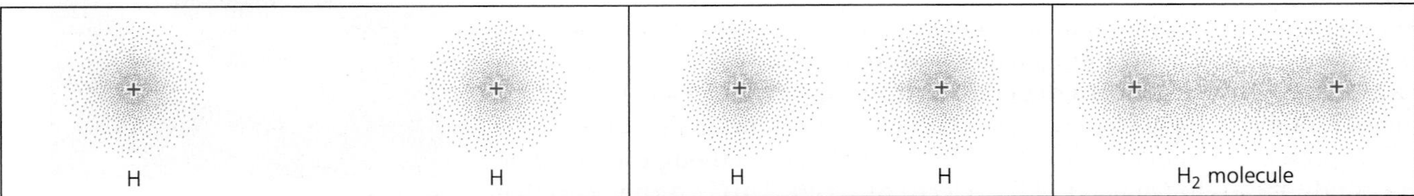

Two hydrogen atoms a large distance apart

As the atoms approach, the electron density between them increases.

Eventually the electron density becomes greater between the two atoms than anywhere else.

FOCUS DIAMONDS ARE FOREVER

Figure 1 An immensely strong lattice of covalent bonds give diamonds their strength and their beauty.

Carbon is a non-metallic element which is of major importance on the planet Earth. In compounds it forms the basis of all known life forms. As an element carbon exhibits some fascinating chemistry because of the different ways the atoms bond together within the carbon molecules. Amorphous carbon and graphite are known and understood, but probably do not excite the imagination as much as the most recently discovered forms of carbon, the buckminsterfullerenes (see page 196 for details of all the forms of carbon). But it is as diamond that carbon has been best known and most highly valued for centuries.

Pure diamond is made up of carbon atoms held together by very strong covalent bonds in an interlocking tetrahedral crystal structure. It is the hardest naturally occurring substance, is colourless and transparent, the best conductor of heat and has the highest refractive index, responsible for its brilliance and sparkle when cut. Diamonds are valued both for their rarity and their great beauty when cut and polished.

But not all diamonds are colourless – in fact many of the best known diamonds in history have been coloured diamonds. One of the most famous of these is the Hope diamond. This fabulous blue stone is reputed to be a harbinger of ill fortune, with stories of death and misfortune befalling its many owners. Much of this is exaggeration, but it is a fascinating tale. Mined in India, the Hope weighed 22 g, an enormous stone. It was the centrepiece of the French crown jewels until the revolution, when it was put on public display after the king and queen were guillotined. However, it was soon stolen. About twenty years later it reappeared in London, recut and smaller, now weighing only 8.8 g. There are only two other blue diamonds known of the same colour and quality – probably the offcuts of the Hope. In 1830 it was owned by Henry Hope, who died young. His grandson gambled heavily and sold the family diamond to pay off his debts. It was lent to an actress by a Russian nobleman – she was shot and killed on stage. Later the diamond was owned by an American, Mrs McClean. Her eldest son was run over and killed aged 9, her daughter died of a drug overdose and her husband became an alcoholic and died in an institution. There were even rumours (later disproved) that the stone was aboard the liner *Titanic* on its ill-fated voyage. But in between all these unfortunate owners the stone spent much of its time in the hands of people who lived perfectly normal and happy lives. Eventually the beautiful stone was given to the Smithsonian Institution to be the centrepiece of a gem stone collection.

It seems strange that stones with such a straightforward chemical structure can on rare occasions be found with such rich and intense colours. Where does the colour come from?

The secret lies in the inclusion of atoms of other elements within the covalent crystals of the carbon. A single nitrogen atom substituted for one carbon atom in a few places in the lattice produces a diamond with a deep yellow colour. The amazing blue of the Hope diamond is the result of boron atoms in the carbon lattice. The presence of different atoms within the crystal, even though still held by covalent bonds, means the number of free electrons varies. This in turn means that different wavelengths of light will be absorbed by the crystal, so we see different colours – a relatively simple explanation for a rare and beautiful phenomenon.

Figure 2 Coloured diamonds such as these (the Hope, the Shepard and the Deyoung Red) are so beautiful and so rare that they are virtually priceless.

The covalent bond represents a balance between the attractive force pulling the nuclei together (due to the electron density between the nuclei) and the repulsive force of the two positively charged nuclei pushing each other apart. For two hydrogen atoms, this balance of the attractive and repulsive forces occurs when the nuclei are separated by a distance of 0.074 nm, as shown in figure 1.4.6. This distance is known as the **bond length**. Energy is released as the two atoms come together to form the bond. This is of course the same as the amount of energy required to break the bond, the **bond energy** (or bond enthalpy, see section 1.8). In the case of the H_2 molecule, the bond energy is 436 kJ mol^{-1}, so 436 kJ mol^{-1} are released when one mole of gaseous hydrogen molecules are formed from gaseous hydrogen atoms.

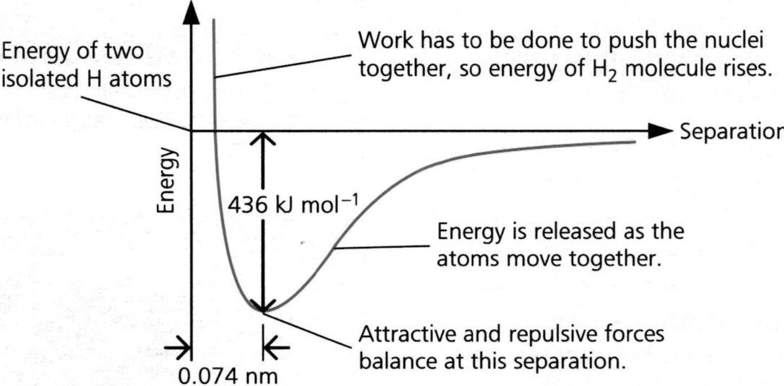

Figure 1.4.6 The energy of two hydrogen atoms at different separations

A covalent bond is thus a shared pair of electrons. Covalent bonds form between atoms of non-metal elements like Cl_2 and P_4, as well as in compounds like H_2O and CH_4. A covalent bond is usually represented as a line, so that the hydrogen molecule is represented like this:

H—H

This is called the **structural formula** of the hydrogen molecule, which as we know has the molecular formula H_2.

Covalent bonds may also be represented using dot and cross diagrams, as in the case of the formation of ions in ionic compounds. Once again, the octet rule can be applied in many cases, as figure 1.4.7 shows. Notice that it is possible for two atoms to share more than one pair of electrons – sharing two pairs results in a **double bond** (O═O), and sharing three pairs produces a **triple bond** (N≡N). However, such diagrams are only used for 'electron book-keeping' – they do not tell us anything about the shapes of molecules, and they do not represent the real positions of electrons.

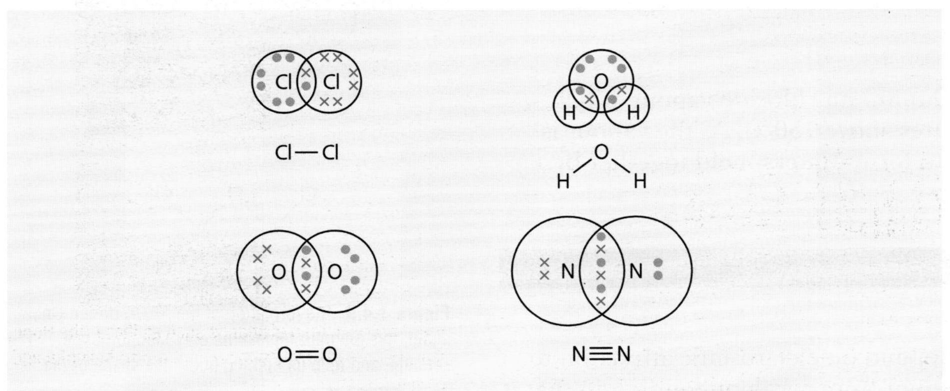

Figure 1.4.7 Dot and cross diagrams can be used to show how the atoms in a covalently bonded molecule share electrons in order to obtain a full outer shell of electrons.

Exceptions to the octet rule

The octet rule does not always work, especially for elements in the third and subsequent periods of the periodic table. For example, if you draw dot and cross diagrams for SF_6 or PCl_5, you will see that the sulphur and phosphorus atoms have more than eight electrons around them (although the fluorine and chlorine atoms all have an octet of electrons). This is possible because the elements in the third period have a principal quantum number of 3, and so they may have up to 18 electrons in their outer shell.

Elements in the second period can only have a maximum of eight electrons in the outer shell (because $n = 2$ for these elements), so the octet rule usually works for them. The most common exception to this is in compounds of boron – for example BCl_3, in which boron has six electrons in its outer shell.

Coordinate covalent bonds

Sometimes both the electrons making up a covalent bond come from the same atom. This type of covalent bond is called a **coordinate covalent bond** or sometimes a **dative covalent bond**. An example of such a bond occurs when ammonia is dissolved in a solution containing hydrogen ions, for which we can write:

$$NH_3(g) + H^+(aq) \rightarrow NH_4^+(aq)$$

Figure 1.4.8 shows a dot and cross diagram for this reaction. Notice that the NH_4^+ ion produced (called the ammonium ion) contains a nitrogen atom with a full outer shell of electrons, together with four hydrogen atoms, each of which also has a full outer shell.

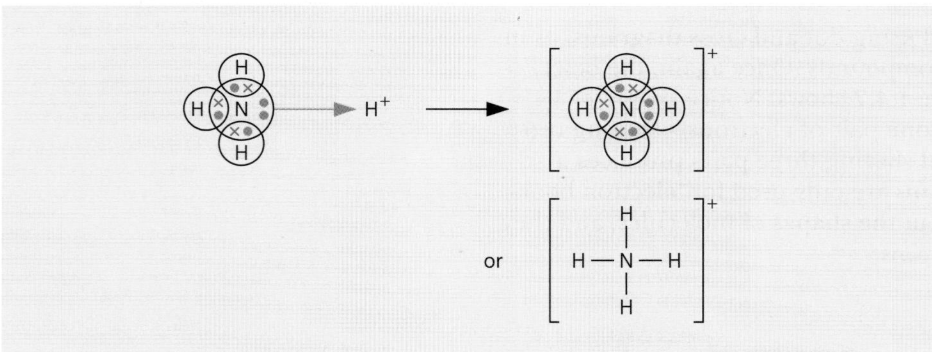

Figure 1.4.8 The formation of an ammonium ion. Notice that all the N–H bonds are equivalent, so that in practice it is impossible to distinguish between them.

Another example of such a bond is in aluminium chloride vapour, which consists of molecules of $AlCl_3$ at temperatures above 750 °C. As the vapour is cooled, pairs of molecules come together to form **dimers**, held together by coordinate bonds as shown in figure 1.4.9.

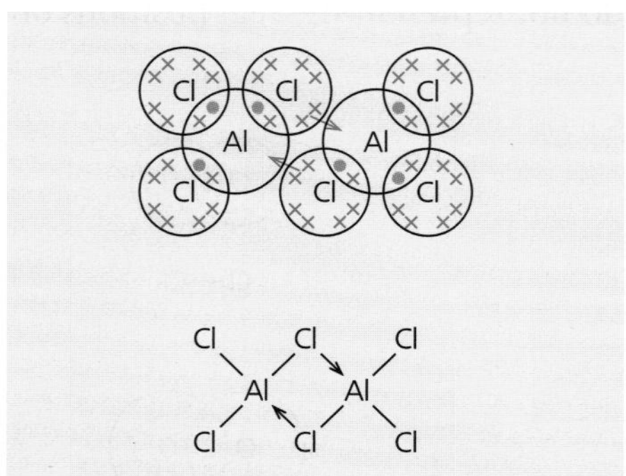

Figure 1.4.9 The formation of the Al_2Cl_6 dimer. Once again, this diagram represents only electron book-keeping, and tells us nothing about the shape of the Al_2Cl_6 molecule.

Predicting the shapes of molecules

Ionic compounds

Although both ionic and covalent bonds depend on electrostatic attraction to hold atoms together, ionic bonding is different from covalent bonding in that it is *non-directional.* The ions in an ionic compound attract ions of the opposite charge in every direction around them, and repel ions of the same charge in the same way. Because of this, the structure of ionic compounds is simply the arrangement of ions in a lattice that maximises the attractive forces between oppositely charged ions and minimises the repulsion between similarly charged ions. The analysis of the structure of ionic lattices can be quite complex – section 1.6 examines this in more detail.

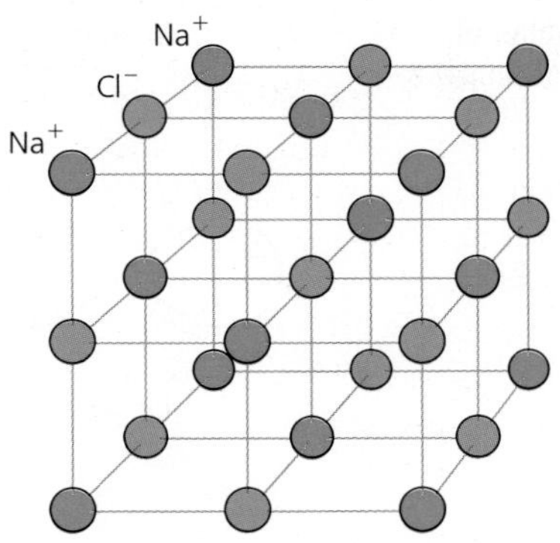

Figure 1.4.10 A model of the ionic lattice of sodium chloride. This is the arrangement with the lowest possible energy of these ions, producing the most stable compound possible.

Covalent compounds

In contrast to ionic bonds, covalent bonds are highly directional. This leads to molecules having a very definite shape (whether they are in the solid, liquid or gas phase), in which the three-dimensional relationship between the atoms is constant. This spatial relationship is important, because it governs the chemical and physical properties of molecules.

Electron pair repulsion

A very simple model of electron pair repulsion can be used to explain the shape of molecules. This is based on the idea that the outer shell electron pairs stay as far away from each other as possible, so that the repulsive forces between them are as small as possible. The simplest example of such repulsion occurs in the molecule $BeCl_2$, which has the electronic structure shown in figure 1.4.12.

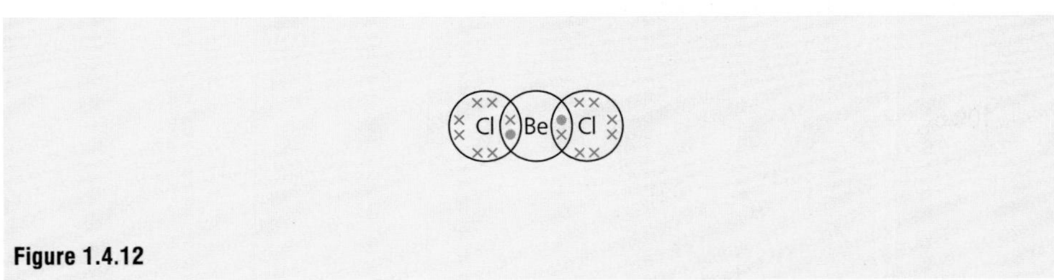

Figure 1.4.12

Figure 1.4.11 Nowhere is molecular shape more important than in the structure of enzymes, substances which control the rate of chemical reactions in biological systems. This spray contains a molecule with exactly the right shape to interfere with an enzyme controlling the nerve impulses in an insect's body, so killing it.

In the $BeCl_2$ molecule, beryllium has only two pairs of electrons around it (it is an exception to the octet rule). To minimise the repulsion between these pairs, they must be arranged so that they are on opposite sides of the beryllium atom, as shown in figure 1.4.13(a). This gives the $BeCl_2$ molecule the shape shown in figure 1.4.13(b) – it is a **linear** molecule.

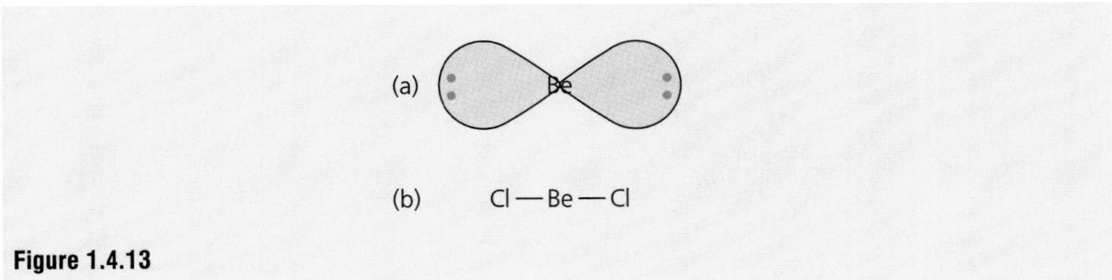

Figure 1.4.13

Extending this idea, we can predict the structure of BF_3, which has three pairs of electrons around the boron atom (again, this is an exception to the octet rule), giving the molecule the **trigonal planar** shape shown in figure 1.4.14.

Table 1.4.4 summarises the predictions we can make about the shapes of molecules with two to six pairs of bonding electrons based on this simple model of minimising repulsion between electron pairs. (Six is the greatest number of pairs around a central atom that we are likely to come across.) The table gives the predicted **bond angles** in each molecule, a convenient way of describing a molecule's shape.

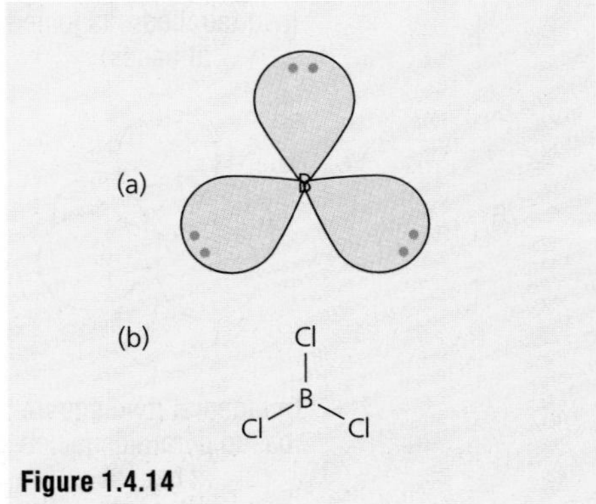

Figure 1.4.14

Number of electron pairs	Shape	Bond angle	Example
2	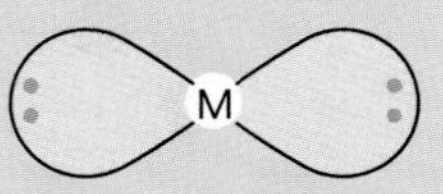 Linear	180°	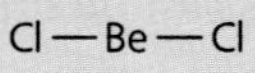 Cl—Be—Cl $BeCl_2$
3	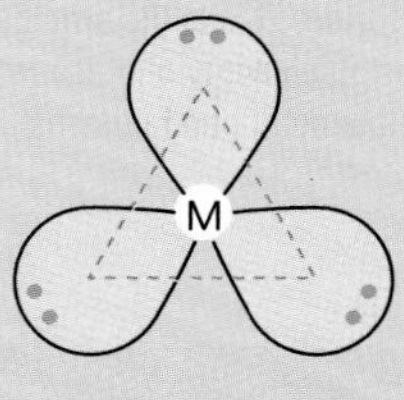 Trigonal planar	120°	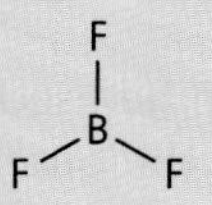 BF_3
4	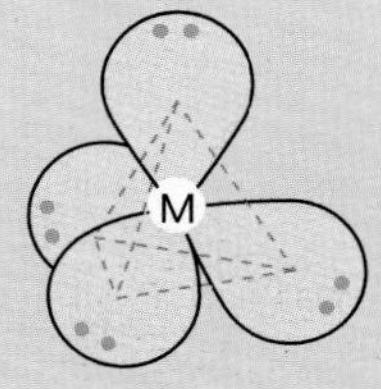 Tetrahedral (triangular-based pyramid)	109.5°	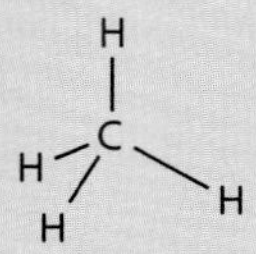 CH_4
5	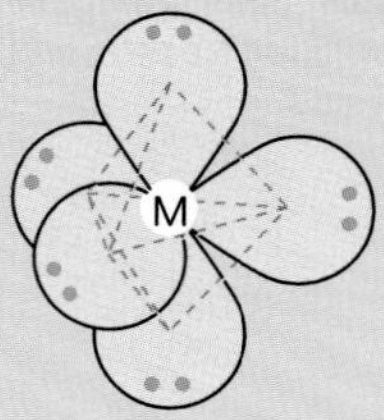 Trigonal bipyramidal (two tetrahedrons joined at bases)	120° and 90°	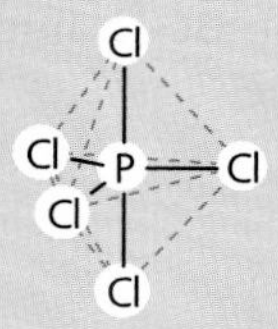 PCl_5
6	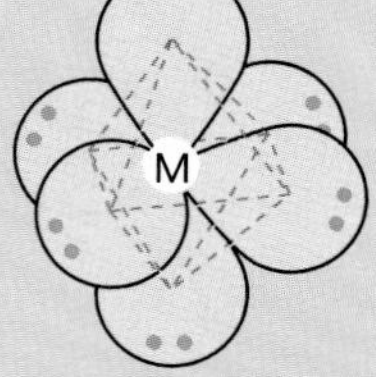 Octahedral (two square-based pyramids joined at bases)	90°	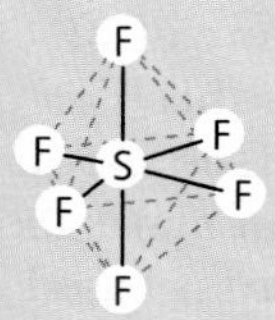 SF_6

Table 1.4.4 The shapes predicted by the electron pair repulsion model for two to six electron pairs

Lone pairs

Not all molecules contain a central atom with electron pairs in bonds around it. For example, ammonia can be represented as shown in figure 1.4.15(a). The central nitrogen atom has four electron pairs around it, but three of them make up N—H bonds – the fourth is a non-bonding pair or **lone pair**. As a result, the electron pairs take up a shape which is tetrahedral (figure 1.4.15(b)), and the shape of the molecule is **pyramidal** – a tetrahedron with the nitrogen atom at its centre, as shown in figure 1.4.15(c). The same argument shows that the water molecule is **bent linear**, as shown by figures 1.4.15(d) to (f).

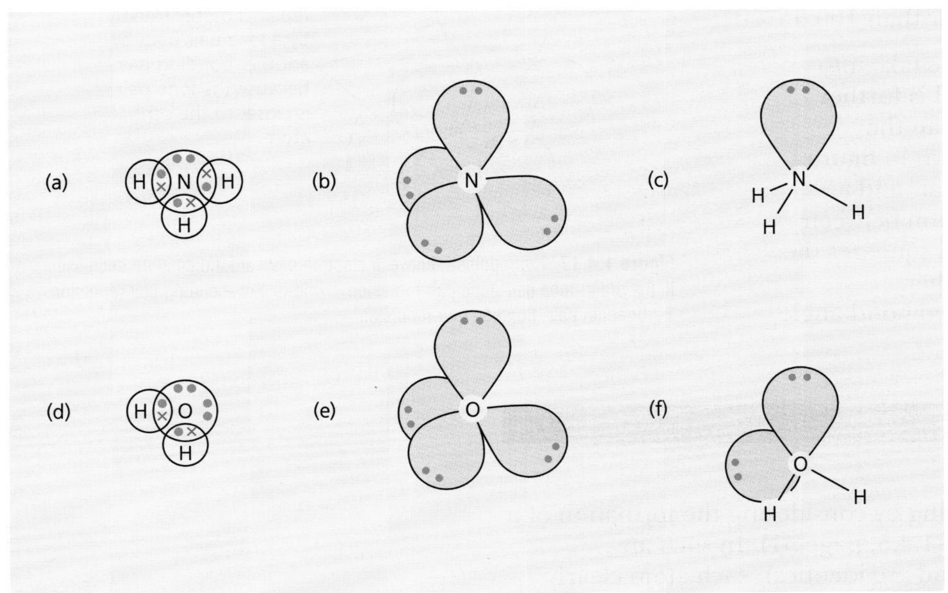

Figure 1.4.15 The electron arrangement and shapes of the ammonia molecule and the water molecule

Multiple bonds

For atoms involved in double or triple bonds, all the bonding electron pairs are found in between the two atoms they join, so molecules containing multiple bonds can be treated in the same way as those with single bonds. Figure 1.4.16 shows how this leads to a linear shape for carbon dioxide, but a bent linear shape for sulphur dioxide.

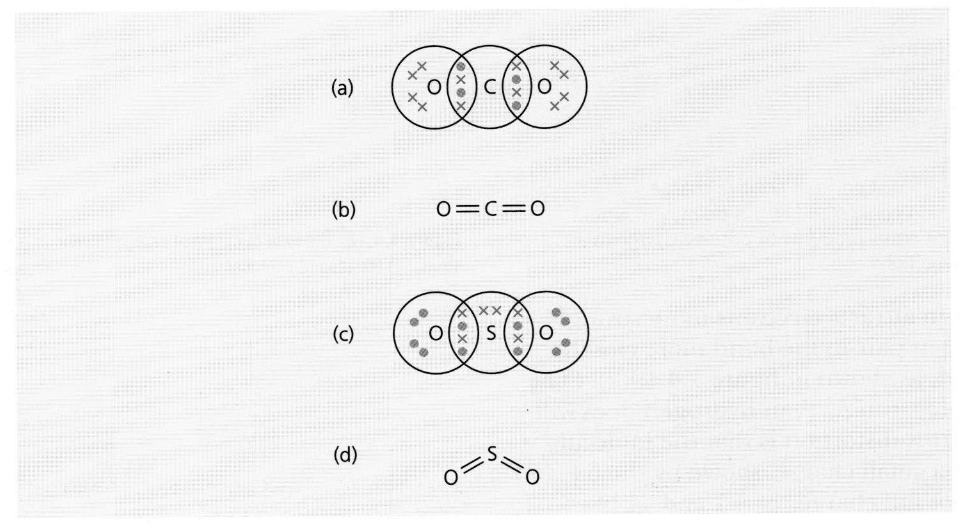

Figure 1.4.16 The electron arrangements and shapes of the carbon dioxide molecule and the sulphur dioxide molecule

Unequal electron pair repulsion

The predicted bond angles in ammonia would be 109.5°, if the electron pairs were really arranged around the nitrogen atom in a tetrahedral shape, all repelling each other equally as they do in methane. However, careful measurements show that they are 107°, 2.5° smaller than predicted. An even larger departure is shown by the water molecule, in which the H–O–H bond angle is 104.5° rather than 109.5°.

These departures can be explained if we take into account that the electron pair in a bond is further from the nucleus of the central atom than the electron pair in a lone pair. This is shown in figure 1.4.17. The repulsion between a lone pair and a bonding pair is greater than the repulsion between two bonding pairs. This extra repulsion squeezes the bonding pairs in ammonia closer together, accounting for the smaller than expected bond angle.

The effect is even greater in the water molecule, where the extra repulsion between two lone pairs causes a greater decrease in bond angle.

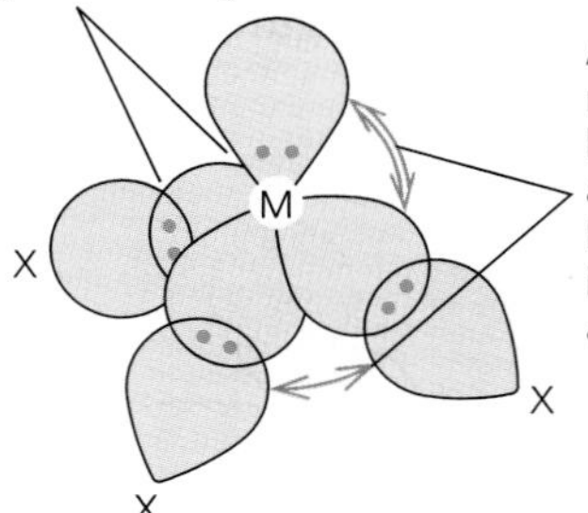

The lone pair is pulled closer to the central atom than the bonding pair since it is attracted strongly by only one centre of positive charge instead of two.
A bonding pair is repelled more strongly by a lone pair than by another bonding pair, because the lone pair is closer to the central atom.
M
X
X
X

Figure 1.4.17 The repulsion between electron pairs around an atom decreases in the order: lone pair–lone pair repulsion > lone pair–bonding pair repulsion > bonding pair–bonding pair repulsion.

Bond character

Polar covalent bonds

We began our discussion of covalent bonding by considering the formation of a bond between two hydrogen atoms (figure 1.4.5, page 51). In such an arrangement, where both atoms in the bond are identical, each atom clearly gets an equal share of the electron pair forming the bond, which is attracted equally to both nuclei. Another way of saying this is to say that the **centres of charge** in the molecule coincide – the idea of centre of charge is directly analogous to the idea of centre of mass, as shown in figure 1.4.18.

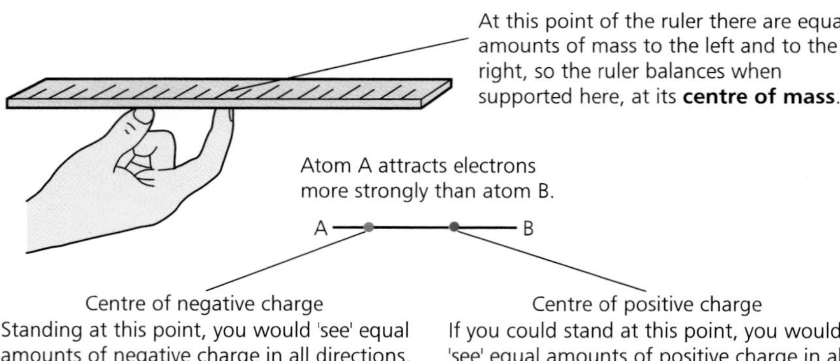

At this point of the ruler there are equal amounts of mass to the left and to the right, so the ruler balances when supported here, at its **centre of mass**.

Standing at this point, you would 'see' equal amounts of negative charge in all directions.

If you could stand at this point, you would 'see' equal amounts of positive charge in all directions.

Figure 1.4.18 The idea of centre of charge is very similar to the idea of centre of mass.

In a molecule like HCl, the chlorine atom attracts electrons more strongly than the hydrogen atom, pulling the electron pair in the bond more closely towards it and distorting the electron cloud, as shown in figure 1.4.19(b). (The reasons why chlorine attracts electrons more strongly than hydrogen does will be examined in section 2.1.) The result of this distortion is that the molecule is **polarised** – the ends of the molecule have a small charge, shown by the δ+ and δ– in figure 1.4.19(c). These charges are not full charges of +1 and –1 like those on an ion, which is why they are written using the lower case Greek letter delta. In the case of the HCl molecule, the hydrogen carries a charge of +0.17 and the chlorine a charge of –0.17.

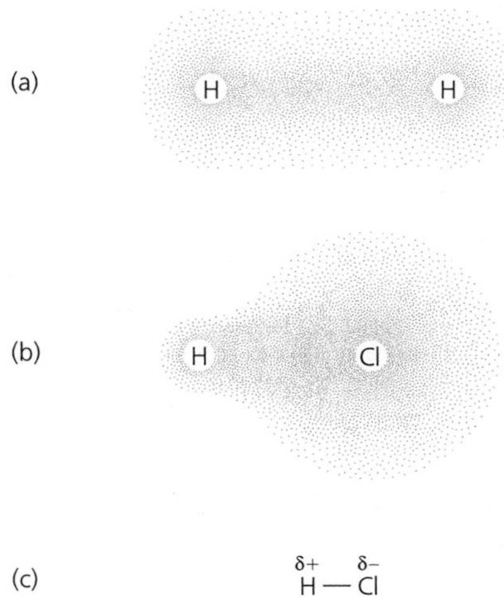

Figure 1.4.19 (a) In the H_2 molecule, the electron density is evenly distributed and the centres of positive and negative charge coincide. (b) The uneven distribution of electron density in the HCl molecule gives rise to partial charges at the ends of the molecule. These can be represented on the displayed formula as shown in (c).

Partial ionic bonds

As we have seen, covalent bonds may be polarised if one of the atoms attracts electrons more strongly than the other. In a similar way, ionic bonds may be distorted by the attraction of the cation for the outer electrons of the anion. Figure 1.4.20(a) shows a wholly ionic bond, with the attraction of the electron cloud around the anion shown in figure 1.4.20(b). If the distortion is great, it may even lead to a charge cloud which begins to resemble that of a covalent bond, as in figure 1.4.20(c).

Covalent or ionic? – The scale of electronegativity

From what we have seen, it is obvious that the terms ionic and covalent represent two extremes of a scale. While some covalent bonds do result from equal sharing of electrons between two atoms (when the atoms are identical, for instance), there are many instances where this is not so. Equally, electrons are partially shared in many ionic bonds, resulting in some covalent character. There is no sharp division between the two types of bond.

In an attempt to measure the degree to which the bond between two atoms is ionic or covalent, chemists use the concept of **electronegativity**. The electronegativity of an atom is a measure of its ability to attract electrons in a bond. The greater its electronegativity, the greater the tendency to attract electrons. The scale commonly used was first proposed by the American chemist Linus Pauling, and runs from 0 to 4.

On the Pauling scale, the most electronegative element (fluorine) has a value of 4.0, with the other elements ranging down to 0.7 (caesium). The difference in electronegativity between the atoms provides a measure of the **ionic character** of the bond. The ionic character exceeds 50% if the difference is greater than 1.7, as shown in figure 1.4.21.

We shall use the idea of electronegativity in section 1.5, and again later in the book. You can find out more about how electronegativity varies between elements in section 2.

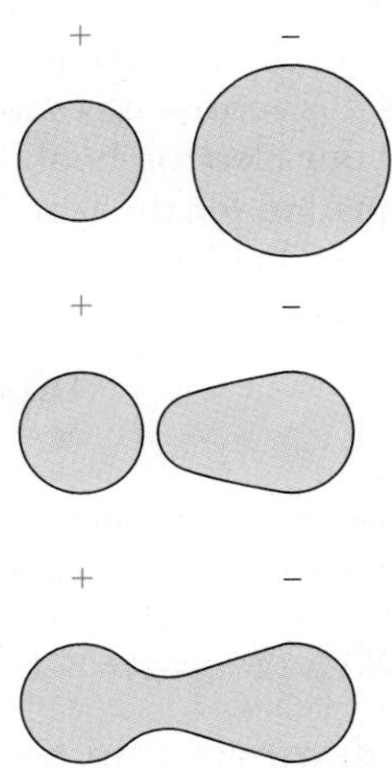

Figure 1.4.20 The distortion of an ionic bond. The distortion is favoured if the cation is small and has a large positive charge, and the anion is large with a large negative charge.

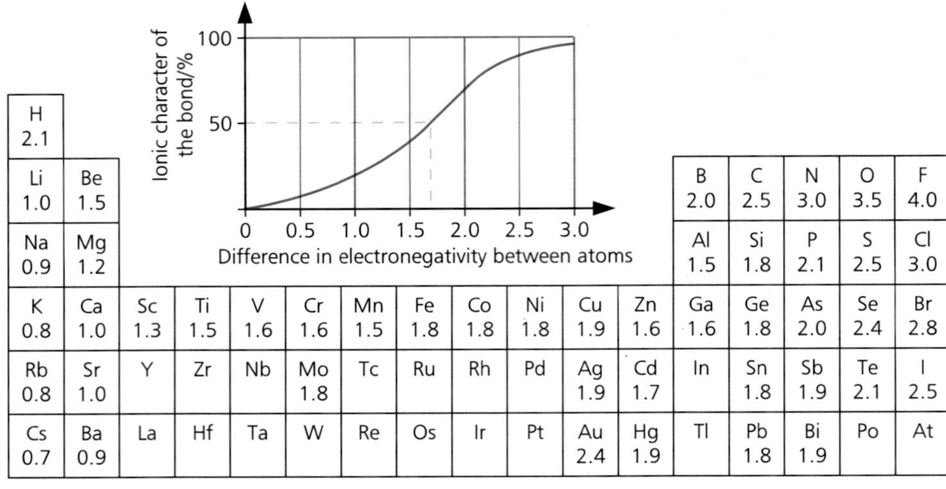

H 2.1																
Li 1.0	Be 1.5											B 2.0	C 2.5	N 3.0	O 3.5	F 4.0
Na 0.9	Mg 1.2											Al 1.5	Si 1.8	P 2.1	S 2.5	Cl 3.0
K 0.8	Ca 1.0	Sc 1.3	Ti 1.5	V 1.6	Cr 1.6	Mn 1.5	Fe 1.8	Co 1.8	Ni 1.8	Cu 1.9	Zn 1.6	Ga 1.6	Ge 1.8	As 2.0	Se 2.4	Br 2.8
Rb 0.8	Sr 1.0	Y	Zr	Nb	Mo 1.8	Tc	Ru	Rh	Pd	Ag 1.9	Cd 1.7	In	Sn 1.8	Sb 1.9	Te 2.1	I 2.5
Cs 0.7	Ba 0.9	La	Hf	Ta	W	Re	Os	Ir	Pt	Au 2.4	Hg 1.9	Tl	Pb 1.8	Bi 1.9	Po	At

Figure 1.4.21 The electronegativities of some of the elements. The graph shows how the character of a bond varies with the difference in electronegativity of the two atoms bonded.

Keeping track of electrons

More electron book-keeping – oxidation numbers

Keeping track of electrons during chemical reactions is important, since the charges on both sides of a chemical equation must balance. One way of doing this is to use **oxidation numbers**, an idea that has several applications in chemistry.

If we consider the reaction of chlorine with sodium, we know that one electron is transferred from each sodium atom to each chlorine atom. We write the equation for the reaction as:

$$\underset{0}{2\mathbf{Na(s)}} + \underset{0}{\mathbf{Cl_2(g)}} \rightarrow 2\mathbf{\underset{+1}{Na}\underset{-1}{Cl}(s)}$$

The small number underneath the symbol for each element represents its oxidation number. By convention, the oxidation number of an uncombined element is zero. The loss of an electron by sodium is represented by its oxidation number increasing from 0 to +1. In contrast, chlorine gains an electron, which is shown by its oxidation number going from 0 to –1. Notice how the sum of the oxidation numbers on each side of the equation is the same (in this case it is zero) – this shows that we have not lost or gained any electrons overall, and our electron book-keeping is correct.

In the equation for the reaction of potassium with oxygen, the oxidation numbers can be balanced by taking into account the numbers of atoms on each side of the equation:

$$\underset{0}{4\mathbf{K(s)}} + \underset{0}{\mathbf{O_2(g)}} \rightarrow 2\mathbf{\underset{+1}{K_2}\underset{-2}{O}(s)}$$

At first sight the oxidation numbers do not seem to balance. However, there are *four* potassium ions and *two* oxide ions on the right-hand side of the equation. The *total* oxidation number associated with potassium is (4 × +1) = +4, and that for oxide is (2 × –2)= –4. Now the book-keeping is correct!

Oxidation numbers and covalent compounds

The oxidation number concept can be extended to reactions involving covalent compounds – for example, the reaction of hydrogen and oxygen:

$$\mathbf{2H_2(g) + O_2(g) \rightarrow 2H_2O(l)}$$

This reaction looks very different from the reaction of potassium with oxygen, since that involves the formation of K^+ ions and O^{2-} ions while this produces a molecule held together by shared electrons. But there is a similarity between the two reactions.

If you look at the data in figure 1.4.21, you will see that the electrons in the O—H bond are not completely shared, since oxygen is more electronegative than hydrogen (it has a greater tendency to attract electrons than hydrogen in a chemical bond). This means that the electron density around an atom changes during the reaction – a hydrogen atom starts with zero charge in the H_2 molecule, and finishes with a partial positive charge in H_2O, while an oxygen atom starts with zero charge in the O_2 molecule, and finishes with a partial negative charge in H_2O. The reaction therefore involves a shift in electron density, which is not so very different from the change that happens in the reaction between potassium and oxygen.

Oxidation numbers and electron density

Oxidation numbers provide a way of following shifts in electron density during reactions. We first need to assign oxidation numbers to the elements in the molecules involved. In doing this, we continue to assign an oxidation number of zero to uncombined elements. In addition, we assign oxidation numbers to elements in compounds so that:

> **The oxidation number of an atom in a molecule is the charge it would have if the electrons in each of its bonds belonged to the more electronegative element.**

In this way, the oxidation number of hydrogen in water is +1, while that of oxygen is –2, and the reaction of hydrogen with oxygen can be written with its oxidation numbers included:

$$\underset{0}{2H_2(g)} + \underset{0}{O_2(g)} \rightarrow 2\underset{+1}{H_2}\underset{-2}{O}(l)$$

(Once again, notice how the oxidation numbers balance once the number of atoms involved on each side of the equation is taken into account.)

Why is the concept of oxidation number useful? Many chemical reactions involve a shift of electron density, and such a reaction is called a **redox reaction**, a term that is short for **red**uction and **ox**idation. The many important processes involving redox reactions include burning, respiration and rusting, as well as the generation of electricity in batteries. Originally the definition of reduction and oxidation was very narrow, being used by chemists to describe only those reactions that involve oxygen and hydrogen. Nowadays the term is applied to any process involving a transfer of electrons, in which the oxidation number of an element changes.

The rules for assigning and using oxidation numbers are as follows:

1 The oxidation number of any uncombined element is zero.
2 The oxidation number of an ion of an element is the same as its charge.
3 The sum of all the oxidation numbers in a molecule (or in a complicated ion) is equal to the charge on the particle.
4 Fluoride *always* has an oxidation number of –1.
5 Combined oxygen has an oxidation number of –2, except in peroxides (when it has an oxidation number of –1), and when it combines with fluorine.
6 Combined hydrogen has an oxidation number of +1, except in metal hydrides (when its oxidation number is –1).

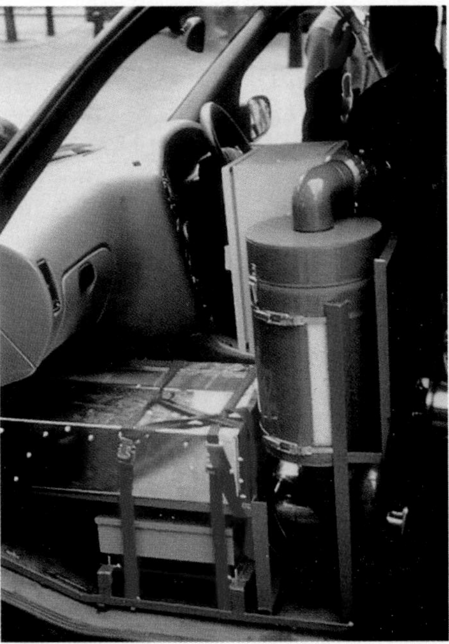

Figure 1.4.22 Fuel cells use a redox process – the reaction of hydrogen and oxygen - to produce electricity. This is a fuel cell from a spacecraft. Apart from electricity, the fuel cell produces a chemical product that is very useful in this application – water.

Using these rules, we can assign oxidation numbers to compounds like this:

NH_3 Treat as $(N^{3-})(H^+)_3$ So: oxidation number of N in NH_3 = –3
and: oxidation number of H in NH_3 = +1

PCl_5 Treat as $(P^{5+})(Cl^-)_5$ So: oxidation number of P in PCl_5 = +5
and: oxidation number of Cl in PCl_5 = –1

MnO_4^- Treat as $[(Mn^{7+})(O^{2-})_4]^-$
So: oxidation number of Mn in MnO_4^- = +7
and: oxidation number of O in MnO_4^- = –2

OF_2 Treat as $(O^{2+})(F^-)_2$ So: oxidation number of O in OF_2 = +2
and: oxidation number of F in OF_2 = –1

H_2O_2 Treat as $(H^+)_2(O_2)^{2-}$ So: oxidation number of H in H_2O_2 = +1
and: oxidation number of O in H_2O_2 = –1

We shall use the concept of oxidation number during our exploration of the periodic table in section 2, and shall look at oxidation and reduction in more detail in section 3.

Naming compounds

Some elements can exist with more than one oxidation number. Oxidation numbers are used in naming compounds containing these elements, to avoid confusion. (Chemists often refer to the **oxidation state** of an element, which is simply another way of talking about oxidation number.)

For compounds involving two elements, the oxidation number is usually quite clear, so we have:

PCl_3	phosphorus **tri**chloride
PCl_5	phosphorus **penta**chloride
CO	carbon **mon**oxide
CO_2	carbon **di**oxide
OF_2	oxygen **di**fluoride
O_2F_2	**di**oxygen **di**fluoride

In more complex cases, the name of the element with the variable oxidation number has its oxidation state in Roman numerals after it:

$KMnO_4$	potassium manganate(VII)
K_2CrO_4	potassium chromate(VI)
$FeCl_3$	iron(III) chloride

In addition to these systematic names, many substances also have common (or **trivial**) names, so iron(II) chloride is sometimes called ferrous chloride, while iron(III) chloride is called ferric chloride, copper(I) oxide is cuprous oxide, and copper(II) oxide is cupric oxide. (It is usual for trivial names to have the endings -ous and -ic to distinguish the lower and higher of two oxidation states.) Potassium manganate(VII), $KMnO_4$, has the common name potassium permanganate.

The systematic names of chemical compounds are agreed by an organisation called IUPAC (the International Union of Pure and Applied Chemistry). Generally speaking, it is preferable to use the systematic names of compounds, although of course you will often find the common names used too. The IUPAC rules for naming substances also recommend the use of some trivial names for common chemicals, where the systematic names are particularly cumbersome. These are given in table 1.4.5.

Substance	IUPAC systematic name	IUPAC recommended common name
HNO_2	Nitric(III) acid	Nitrous acid
HNO_3	Nitric(V) acid	Nitric acid
NO_2^-	Nitrate(III)	Nitrite
NO_3^-	Nitrate(V)	Nitrate
H_2SO_3	Sulphuric(IV) acid	Sulphurous acid
H_2SO_4	Sulphuric(VI) acid	Sulphuric acid
SO_3^{2-}	Sulphate(IV)	Sulphite
SO_4^{2-}	Sulphate(VI)	Sulphate

Table 1.4.5 IUPAC recommended common names

SUMMARY

- When atoms react together they may **lose**, **gain** or **share** electrons to attain an outer shell containing eight electrons – a **stable electronic configuration** like that of a noble gas.
- Many of the compounds formed when metals and non-metals react together are **ionic**. Atoms that lose electrons during a reaction form **positive ions** (**cations**) whereas those that gain electrons form **negative ions** (**anions**). These positive and negative ions attract each other strongly, forming an **ionic lattice**. The electron losses and gains in ionic bonding may be represented using **dot and cross diagrams**.
- The formula of an ionic compound is affected by:

  the **ionisation energy** of the element – the amount of energy required to remove one mole of electrons from one mole of gaseous atoms

  the **electron affinity** of the element – the energy change when one mole of gaseous atoms gain one mole of electrons

  the **lattice energy** – the energy released when gaseous ions come together to form one mole of an ionic compound.
- **Covalent bonding** occurs when the energy needed to form ions is greater than the energy that would be released with the formation of the ionic lattice. Covalent bonding involves the **sharing** of electrons between two nuclei and results in the formation of **molecules**. Dot and cross diagrams may be used to show the sharing of the electrons. Covalent bonds are directional, so covalent molecules containing more than two atoms have distinctive shapes which may be predicted.
- **Coordinate covalent bonding** involves the sharing of electrons, but both shared electrons are donated by the same atom.
- **Oxidation numbers** provide a way of understanding the shifts in electron density that occur during a chemical reaction.

QUESTIONS

1 'Elements in the middle of the periodic table tend to form bonds by sharing electrons, while elements at the ends of the periods tend to form bonds by transferring electrons.' Is this a fair summary of the behaviour of the elements? Justify your answer.

2 Draw dot and cross diagrams to illustrate the bonds in the simplest compounds formed between chlorine and:
a potassium **b** calcium **c** carbon
d phosphorus.

3 Draw a dot and cross diagram to show how a coordinate covalent bond is formed in the reaction:

$$BF_3 + F^- \rightarrow BF_4^-$$

4 Draw dot and cross diagrams and hence predict the arrangement of the atoms in the following:
a TeF_6 **b** CS_2 **c** SbH_3 **d** SO_2 **e** NO_3^-

5 Assign oxidation numbers to all the atoms in the following:
a S^{2-} **b** PH_3 **c** H_2SO_4 **d** IF_7 **e** NO^+
f KIO_3 **g** $Na_2S_2O_3$

6 Are the following reactions redox reactions? Justify your answers:
a $2NO_2 \rightarrow N_2O_4$
b $Zn + Cu^{2+} \rightarrow Zn^{2+} + Cu$
c $2CrO_4^{2-} + 2H^+ \rightarrow Cr_2O_7^{2-} + H_2O$

Developing Key Skills

Produce a PowerPoint presentation on chemical bonding for use with pupils in Key Stage 3 (11–14). Your presentation should be able to be used by two different levels of pupils – there should be a simple initial approach suitable for the younger pupils or less able older pupils, but the presentation should be extended to catch the imagination and interest of the most able year 9 pupils as well.

[Key Skills opportunities: C, IT]

FOCUS FORCES OF LIFE

The importance of the bonds between atoms which form molecules is plain to see. But there are other, far weaker forces which have major effects on the world around us. Intermolecular forces – dipole–dipole interactions, hydrogen bonds, van der Waals forces (see page 71) – affect the shape of many large molecules as they form complex structures. They also affect the properties of compounds where the individual molecules are held close together.

The role of intermolecular forces is especially clear in molecules which make up living organisms, particularly carbohydrates and proteins (see section 5.5) and in similar large synthetic molecules such as plastics (see page 404). Living organisms depend on special proteins known as enzymes (biological catalysts) for the chemical reactions necessary for life to take place. Enzymes work because within their complex structures are **active sites**. These have a very specific shape which will only bind to the appropriate substance, and the shape of the active site is maintained by intermolecular forces.

Similarly the chemical messages which control most of the events in a large organism – or even a simple cell – work by means of special receptor molecules within cell membranes. Again the shape of each receptor is very specific to the hormone or other chemical to which it must respond – and again that shape is maintained by intermolecular forces, particularly hydrogen bonds.

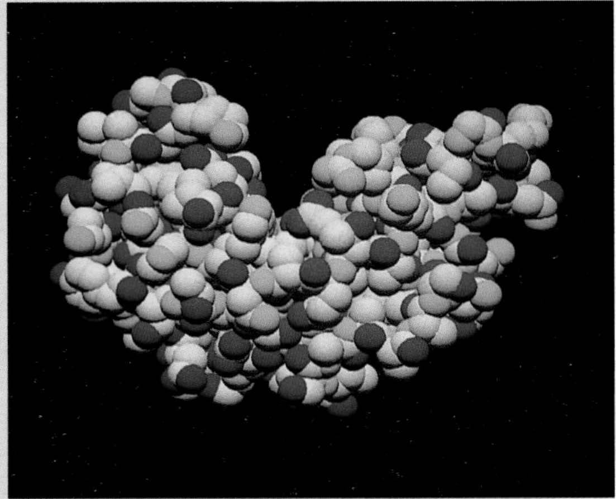

Figure 1 Enzymes are vital for everything, from digesting food to creating new cells. To work they depend on the shape of their active sites, held in place by intermolecular forces.

Bad hair day?

Human hair is made up of the protein keratin, coiled into α-helices and held in shape by hydrogen bonds. Very often people want to change the way their hair looks, and the most common way of doing this is to blow-dry it, to dry it wrapped around curlers or to use hot tongs or brushes on it. Each of these hair treatments does the same thing: they break down the hydrogen bonds already present in the hair and reform them with your hair held in a different position – curlier, straighter or just different. But the next time you wash your hair, if not before, the hair returns to its natural style as the original hydrogen bonds reform. The only way to make a permanent change to the hair is to have a perm (short for 'permanent wave'). Chemicals used in perming break covalent bonds between sulphur atoms in the polypeptide coils, and these are then reformed in a different place when the hair is restyled. (You can find out more about the structure of proteins on pages 534–9.) Covalent bonds are much stronger than hydrogen bonds, so the hair stays in its new arrangement.

The behaviour of hair demonstrates clearly the difference between intermolecular forces such as hydrogen bonds and covalent bonds like disulphide bridges – and means that bad hair days can all be blamed on chemistry!

Figure 2 Perming is an involved procedure and the effect on that piece of hair is permanent – if you don't like it you have to grow it out or cut it off, or have another perm.

1.5 Intermolecular forces

1

So far we have seen that the forces between the atoms in matter may involve transferring electrons to form ions, or sharing them to form molecules. This is not the whole story though, for to understand fully the behaviour of matter we need to know how molecules interact with each other – in other words, about **intermolecular forces**. As we shall see, intermolecular forces have at least as much influence on the world around us as the ionic and covalent bonds we studied in section 1.4.

Polar molecules

Some surprisingly different properties

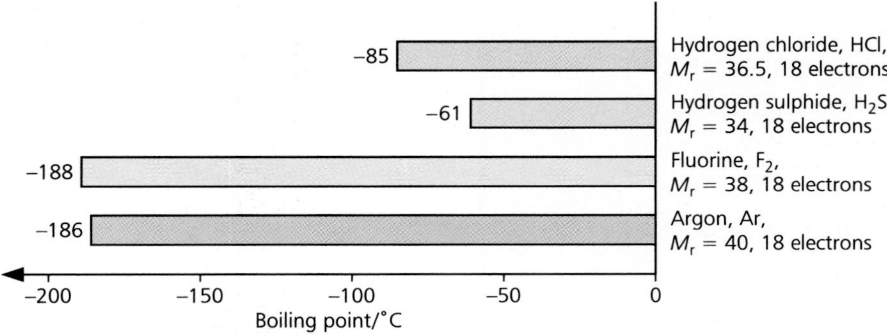

Figure 1.5.1 Two of these substances have very low boiling points, while the boiling points of the other two are much higher.

Look at the boiling points of the substances shown in figure 1.5.1. The particles of matter that make up these four substances are similar, yet the substances divide into two pairs with very different boiling points. Why is this? Before we set about answering this question, we should look at a simple demonstration.

The deflection of liquids in electric fields

If you bring a plastic comb that has been rubbed on a woollen jumper up to a thin stream of water from a tap, the water bends towards the comb dramatically, as figure 1.5.2 shows.

The explanation for this is that the comb is given an electric charge when it is rubbed on a piece of wool, and the resulting electric field

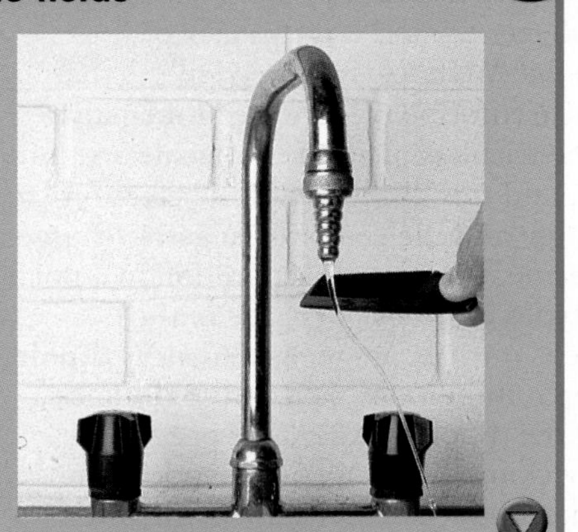

Figure 1.5.2 The deflection of a stream of water by a plastic comb

around the comb then attracts the stream of water. If the water is replaced by another liquid, the liquid stream is not always deflected, as table 1.5.1 shows.

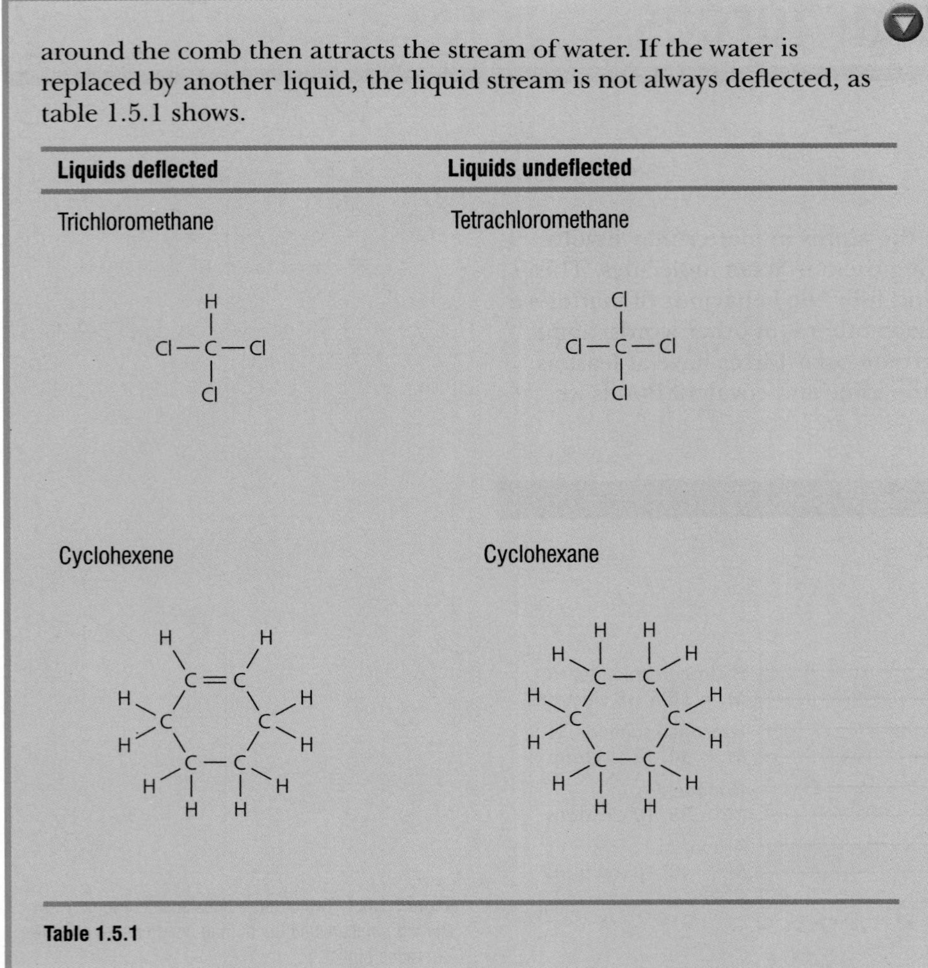

Liquids deflected	Liquids undeflected
Trichloromethane	Tetrachloromethane
Cyclohexene	Cyclohexane

Table 1.5.1

Why are some molecules polar?

The results of the demonstration in the box above show us that some liquids are deflected by an electric field, while others are not. Careful examination of the structures of the molecules that are deflected shows that they are not symmetrical – for example, trichloromethane and cyclohexene – and it is this lack of symmetry that leads to a molecule being polar. We know from section 1.4 that bonds like the C—Cl bond are polar, because chlorine is more electronegative than carbon. As a result, the electrons in the trichloromethane molecule are attracted towards one end of the molecule, and this distortion of the electron cloud means that the centre of positive charge and the centre of negative charge do not occur in the same place in the molecule. The charge separation in a polar molecule makes it a **dipole** (there are two types or **poles** of charge in the molecule). The degree to which a molecule is polarised is measured as its **dipole moment**, which is the amount of charge separation multiplied by the distance between the centres of charge. (Measuring the dipole moments of polar molecules provides an important way of checking the predictions made from theories of bonding.) The tetrachloromethane molecule is symmetrical and therefore not polar. Figure 1.5.3 compares the two molecules.

Point X

Figure 1.5.3 Chlorine is more electronegative than carbon or hydrogen. Because of this, point X in the $CHCl_3$ molecule has a net amount of positive charge to the right, and a net amount of negative charge to the left – it is a polar molecule. By contrast, the electron-attracting effects of the chlorine atoms in the CCl_4 molecule cancel out – the centres of positive and negative charge do coincide in this molecule, and it is not polar.

Differences in boiling point and polar molecules

Now look at figure 1.5.1 again. Hydrogen chloride and hydrogen sulphide are both polar molecules, while fluorine and argon are non-polar (from section 1.4 you should be able to predict the shape of H_2S and to show why it is a polar molecule). The boiling points of these two pairs of substances differ by more than 100 °C, with the non-polar molecules having much lower boiling points than the polar molecules. This tells us that the forces between the polar molecules are much greater than the forces between the non-polar molecules, since it takes more energy to separate the molecules. How do they come about?

1

Forces between molecules

Dipole–dipole interactions

Polar molecules have a permanent dipole – that is to say, a permanent separation of charge. As a result of this, polar molecules are attracted towards one another by forces called **permanent dipole–permanent dipole interactions**, in which the negative end of one molecule is attracted towards the positive end of another (figure 1.5.4). These interactions decrease quite rapidly as the distance between molecules increases, and they are about 100 times weaker than covalent bonds. This kind of interaction accounts for the forces between the molecules of trichloromethane.

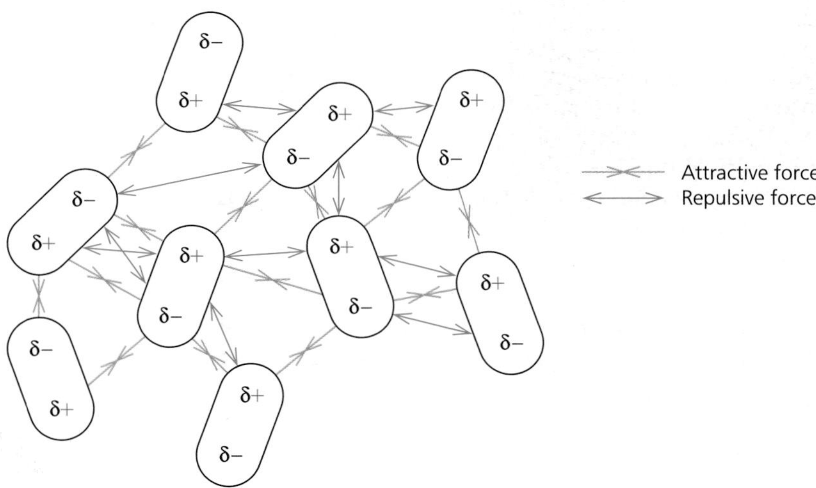

Figure 1.5.4 Permanent dipole–permanent dipole interactions between molecules. The molecules arrange themselves so that the positive end of one molecule is near to the negative ends of nearby molecules. Attractive forces between oppositely charged portions of molecules are therefore generally stronger than repulsive forces between similarly charged portions of molecules, and molecules experience a net attractive force as a result.

Hydrogen bonds

The noble gases are in group 0 of the periodic table (see section 2). The boiling points of the noble gases increase down the group, due to the increase in the size of the electron cloud. Do other groups of the periodic table behave in a similar way?

Figure 1.5.5 overleaf shows the boiling points of the hydrides formed by the elements in groups IV, V, VI and VII. While the hydrides of group IV behave in a very similar way to the noble gases, the hydrides of the other groups do not – at least, those of the lightest elements in the groups do not. This suggests that the intermolecular forces in these hydrides are much stronger than expected compared with the hydrides of the other elements in each group. The strength of these intermolecular forces can be illustrated by comparing the molar heats of vaporisation (or enthalpy changes of vaporisation, see section 1.8) of the hydrides of the elements of group VI, as shown in table 1.5.2.

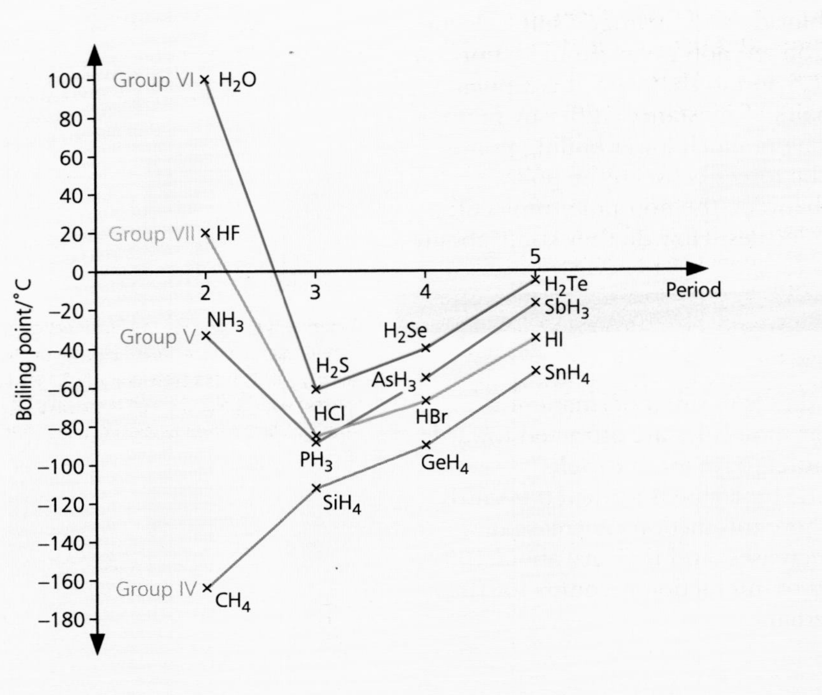

Figure 1.5.5 The variation in boiling point of the hydrides of groups IV, V, VI and VII

The molar enthalpy change of vaporisation of a liquid is the amount of energy required to turn one mole of the liquid into one mole of its gas at the boiling point. This is a direct measure of the intermolecular forces in the liquid. Notice how the energy required to turn one mole of liquid water at 100 °C into steam at 100 °C is more than twice the energy needed to turn one mole of liquid hydrogen sulphide into its gas at its boiling point.

Compound	H_2O	H_2S	H_2Se	H_2Te
Standard molar enthalpy change of vaporisation/ kJ mol^{-1}	41.1	18.7	19.9	23.8

Table 1.5.2 Standard molar enthalpy changes of vaporisation of the hydrides of group VI

These particularly strong interactions are due to a different type of intermolecular force. This is called a **hydrogen bond**, and exists between two molecules that each contain a polar bond between hydrogen and another atom. A hydrogen bond is a particularly strong dipole–dipole interaction.

Group IV	Group V	Group VI	Group VII
C 2.5	N 3.0	O 3.5	F 4.0
Si 1.8	P 2.1	S 2.5	Cl 3.0
Ge 1.8	As 2.0	Se 2.4	Br 2.8
Sn 1.8	Sb 1.9	Te 2.1	I 2.5

Table 1.5.3 The electronegativities of the elements of groups IV to VII. The electronegativity of hydrogen is 2.1.

We know that differences in electronegativity are good indicators of polar bonds. The data in table 1.5.3 show that there are large differences in electronegativity between hydrogen and oxygen, and between hydrogen and fluorine. This means that the H—F and H—O bonds are very polar, with a substantial amount of the electron density being drawn away from the hydrogen atom, leaving it with a δ+ charge.

The highly polarised bonds in HF and H_2O lead to hydrogen bonds being formed. The hydrogen atom has no inner shells of electrons. As a result, the nucleus of the hydrogen atom in these molecules is left unusually exposed by the shift in electron density within the bond, making it easily accessible for strong permanent dipole–permanent dipole interactions to occur. This happens when the lone pair of electrons of another atom (oxygen in the case of water, fluorine in the case of hydrogen fluoride) is attracted to the positive nucleus of the hydrogen, leading to the formation of a hydrogen bond, as shown in figure 1.5.6.

Figure 1.5.6 Hydrogen bonding in water, hydrogen fluoride and ammonia. Notice how the two lone pairs in the oxygen atom can each form a hydrogen bond with another water molecule, so it is possible for two hydrogen bonds to be formed per water molecule.

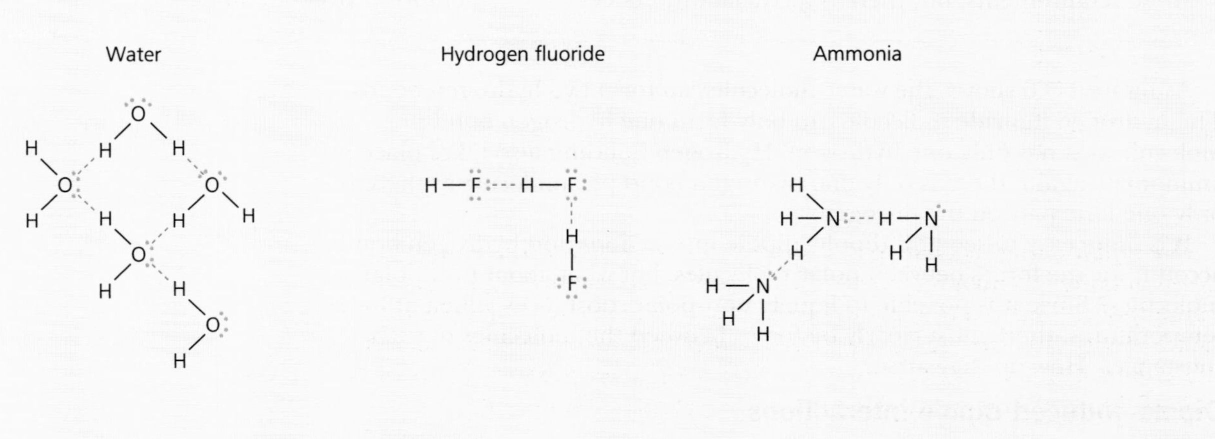

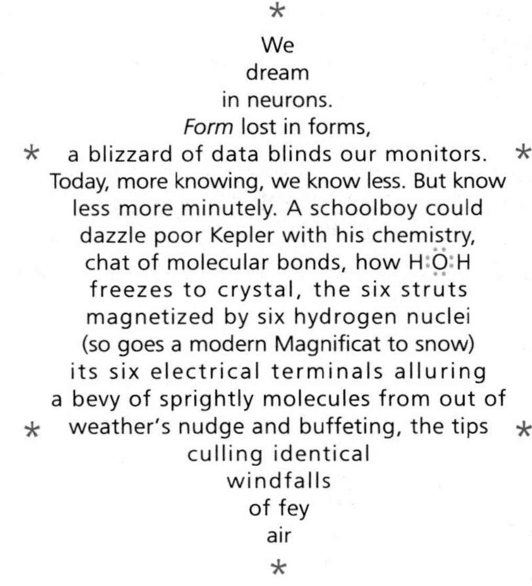
*
We
dream
in neurons.
Form lost in forms,
* a blizzard of data blinds our monitors. *
Today, more knowing, we know less. But know
less more minutely. A schoolboy could
dazzle poor Kepler with his chemistry,
chat of molecular bonds, how H:O:H
freezes to crystal, the six struts
magnetized by six hydrogen nuclei
(so goes a modern Magnificat to snow)
its six electrical terminals alluring
a bevy of sprightly molecules from out of
* weather's nudge and buffeting, the tips *
culling identical
windfalls
of fey
air
*

Figure 1.5.7 Intermolecular forces are responsible for the beautiful shapes of snowflakes. This debt is acknowledged in this extract from a beautiful poem by John Frederick Nims.

Hydrogen bonding in ammonia and hydrogen chloride

Careful examination of figure 1.5.5 shows that the boiling point of ammonia is higher than expected, while that of hydrogen chloride is approximately in line with the expected trend based on the rest of the group. However, table 1.5.3 shows that the electronegativity of chlorine is the same as that of nitrogen. It seems that hydrogen bonding happens in ammonia but not in hydrogen chloride – why?

The formation of a hydrogen bond requires the presence of a hydrogen attached to an electronegative atom, and also at least one lone pair of electrons. Both ammonia and hydrogen chloride satisfy these requirements, but there is a crucial difference – the lone pair of electrons has a principal quantum number of 2 in ammonia, whereas the hydrogen chloride lone pairs have a principal quantum number of 3.

This small difference is important, since the electron clouds of the 3s and 3p orbitals containing the lone pairs in chlorine are larger and more diffuse than the clouds containing the electron pair in ammonia, with electrons in the second quantum shell. The smaller, less diffuse electron cloud is able to form a strong interaction with the δ+ hydrogen of another molecule, whereas the larger clouds surrounding the chlorine atom are not.

As figure 1.5.6 shows, the water molecule can form two hydrogen bonds. The hydrogen fluoride molecule can only form one hydrogen bond per molecule as it has only one hydrogen. Hydrogen bonding also takes place in ammonia – again, there is only one hydrogen bond per molecule as there is only one lone pair on the nitrogen.

It is quite easy to see how dipole–dipole interactions and hydrogen bonds account for the forces between polar molecules, but what about non-polar molecules? Since it is possible to liquefy non-polar substances, albeit at lower temperatures, there must clearly be forces between the molecules of such substances. How do they arise?

Dipole–induced-dipole interactions

The forces of attraction that exist between two non-polar molecules also arise due to an uneven charge distribution. As we have seen, electrons are not fixed in space. If we consider a neutral atom, at any particular instant the centres of positive and negative charge may not coincide, due to an instantaneous asymmetry in the electron distribution around the nucleus. Thus there is an **instantaneous dipole** in the molecule. (Over time, of course, these dipoles average out, producing a net dipole of zero.) Any other atom next to an atom with an instantaneous dipole will experience an electric field due to the dipole, and so will itself develop an **induced dipole** – figure 1.5.8(a) shows this. The **instantaneous dipole–induced dipole interactions** between neighbouring molecules enable non-polar molecules to come together and form a liquid when the temperature is sufficiently low. The forces are small, however, since the

Figure 1.5.8 The existence of an instantaneous dipole in one atom gives rise to an induced dipole in a nearby atom, thus causing a net attractive force between the two atoms. The instantaneous dipole varies over time, sometimes having a value of zero. As a result, the attractive force between the atoms fluctuates, and is sometimes zero.

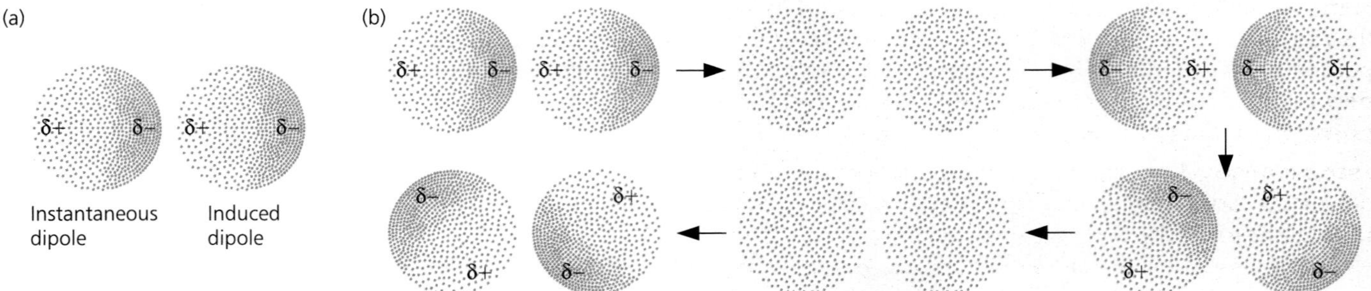

attraction between two molecules lasts only a very short time, as the instantaneous dipole is 'turned on' and 'turned off' as the electron density fluctuates (figure 1.5.8(b)). These interactions are the source of the forces between the molecules of tetrachloromethane, and between the atoms of the inert gases.

The size of the dipole–induced-dipole forces depends on the size of the electron cloud of the interacting particles. Atoms with large electron clouds tend not to hold on to their outer electrons so tightly, so large electron clouds tend to be easily deformed. This favours the existence of an instantaneous dipole (and also favours the creation of an induced dipole too), so dipole–induced-dipole forces tend to increase with the size of the electron cloud. This can be clearly seen in the boiling points of the noble gases, shown in table 1.5.4. Dipole–induced-dipole forces between molecules are affected by the size of the molecules for the same reason, so ethane, C_2H_6, is a gas at room temperature, while hexane, C_6H_{14}, is a liquid.

	He	Ne	Ar	Kr	Xe	Rn
Boiling point/°C	–269	–246	–186	–152	–107	–62

Table 1.5.4 The boiling points of the noble gases

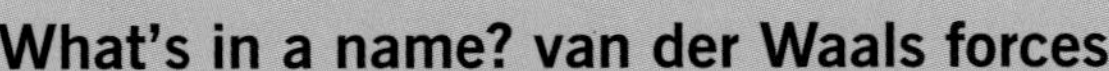

What's in a name? van der Waals forces

The term 'van der Waals forces' is used sometimes to mean just dipole–induced-dipole interactions, but usually includes dipole–dipole interactions as well. Johannes van der Waals first suggested that there were forces between molecules in 1873, and it was later realised that these two different types of attractive force were possible. Hydrogen bonds are not generally thought of as van der Waals forces.

Comparing intermolecular forces

Covalent bond strengths are typically between 200 and 500 kJ mol^{-1} (although they can vary outside this range). Hydrogen bonds are weak in comparison – they vary between about 5 and 40 kJ mol^{-1}. (See section 1.8 for more about the strength of bonds.) Van der Waals forces are weaker still, having about 10% of the strength of the average hydrogen bond, that is, around 2 kJ mol^{-1}. Hydrogen bonds and van der Waals forces are not strong enough to influence the chemical behaviour of most substances, although they may both affect the physical properties of substances. The importance of the hydrogen bond in determining the properties of water is described on page 147.

SUMMARY

- Some covalent molecules contain bonds with an unequal charge distribution – they have some polar nature. **Polar molecules** have permanent dipoles and the opposite ends attract each other, forming **permanent dipole–permanent dipole interactions**. These may result in raised melting and boiling points in a compound.
- **Instantaneous dipoles** appear and disappear continually in atoms and non-polar molecules, setting up **induced dipoles**. **Instantaneous dipole–induced-dipole interactions** bring atoms and non-polar molecules together to form liquids at low temperatures.

- When two molecules each contain a polar bond between a hydrogen atom and a very electronegative atom such as fluorine, oxygen or nitrogen, intermolecular forces known as **hydrogen bonds** occur. Hydrogen bonding is responsible for many of the unexpected properties of water, for example, high melting and boiling point, and surface tension.

QUESTIONS

1 Which of the following molecules would you expect to be polar and why?
a HI **b** SO_2 **c** SO_3 **d** CH_2Cl_2 **e** CO
f BCl_3 **g** $POCl_3$

2 Predict the main type of intermolecular force present in each of the following substances. Justify your answers.
a CH_4 **b** CH_3OH **c** CH_3OCH_3 **d** PCl_3
e NH_2OH

3 Make a list of all the forces that exist in the solid KIO_3.

4 Ethanol and methoxymethane both have the molecular formula C_2H_6O. Their structural formulae are as follows:

CH_3CH_2OH Ethanol CH_3OCH_3 Methoxymethane

At atmospheric pressure, ethanol boils at 79 °C, while methoxymethane boils at –25 °C. Ethanol is soluble in water, while methoxymethane is not. Explain the difference in the properties of these two substances in terms of their molecules.

5 **a** Plot a graph of standard molar enthalpy change of vaporisation against relative molecular mass for the data given in table 1.6.2 on page 78.
b Extrapolate the line which passes through the points for H_2Te, H_2Se and H_2S. Use this to obtain an estimate for the standard molar enthalpy change of vaporisation of water if there were only van der Waals forces between the molecules in water. (*Hint*: There are no hydrogen bonds between the molecules in H_2Te, H_2Se and H_2S, only van der Waals forces.)
c From your value obtained in **b** and the actual standard molar enthalpy change of vaporisation of water, calculate the approximate strength of the hydrogen bonds in water.

Developing Key Skills

A top hairdresser wants to give customers more information about their hair and what they are doing to it when they choose different styles. You are commissioned to produce a leaflet for people who heat-treat their hair – blowdrying, hot brushes, crimping and curlers – and for people having a perm. Some of the information (on the structure of hair itself, etc.) will be common to both leaflets, some will be specific to the particular treatment they have chosen. The leaflets need to be eyecatching, interesting and, most of all, informative with plenty of easy-to-understand diagrams.

[Key Skills opportunities: C, IT]

FOCUS SEEING INTO SOLIDS

In the early twentieth century a new science emerged which was a complex combination of mathematics, physics and chemistry. It was known as **X-ray crystallography**. If a beam of X-rays is shone into a solid crystal the atoms within the crystal diffract the X-rays. An image of the diffracted rays can be obtained using a piece of photographic film, which reacts to X-rays in the same way as to light. The diffracted rays then *interfere* with each other. If this interference is *constructive*, a bright spot on the photographic film is formed, but if the interference is *destructive* then the waves cancel out and a dark spot appears.

Whilst crystallography was emerging as a science, a young girl called Dorothy Crowfoot was developing a growing fascination for the world of crystals, organising a small laboratory where she could grow crystals and analyse chemicals to her heart's content. By 1932, at 22 years old, Dorothy had a degree in chemistry from Oxford University. She married Thomas Hodgkin and continued with her chemistry, working with Desmond Bernal, the foremost proponent of X-ray crystallography at that time.

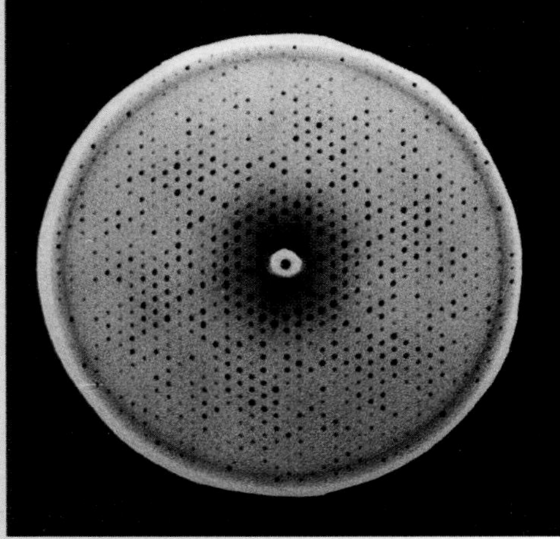

Figure 1 Dorothy Hodgkin did not choose simple molecules to analyse – this crystallogram is of insulin, one of the compounds she is most famous for working on.

Deciphering an X-ray crystallogram of even the simplest solid in the days before computers was no easy matter, but Dorothy Hodgkin had a rare talent for it. The skill is a mixture of mathematical vision, imagination and rigorous measuring and observation, and success depends both on preparing a suitable crystal to X-ray and on the ability to interpret the results.

Over a period of years Dorothy Hodgkin completed the detailed analysis of a large number of compounds, including cholesterol and over 100 steroids. Throughout this time she was interested in an analysis of insulin, and in 1934 she finally managed to grow crystals of it and to photograph this complex and vital molecule, although it was to be years before she finally teased out the details of the structure. Yet in spite of her outstanding work, she was barred at Oxford University from the meetings of the chemistry faculty club because she was a woman!

Dorothy Hodgkin worked on a structural analysis of the penicillin molecule, finally determining its structure in 1945. She then began her work on vitamin B_{12}. For her work on this molecule, four times bigger than penicillin, Hodgkin introduced new techniques using cobalt, against the advice of fellow workers. However, she was proved right, and by uncovering the atomic arrangement of the molecule paved the way to its use as a successful treatment for pernicious anaemia.

In 1964 Dorothy Hodgkin was awarded the Nobel Prize for Chemistry for her work on the structure of vitamin B_{12} and her other unprecedented discoveries which had extended the boundaries of science.

Still she continued working, and 34 years after she first became interested in the molecule she managed to discover the structure of insulin using X-ray crystallography. By this time she was helped by computers, which have radically changed the science, making it faster and easier for all.

As well as her work in crystallography, Dorothy Hodgkin was a tireless worker for world peace. She had three children, nine grandchildren and, at the time of her death in 1994, three great-grandchildren. By all accounts she was not only a genius in her field but also a warm, affectionate and caring human being – a substantial legacy for anyone to leave.

Figure 2 Dorothy Hodgkin, a woman who pushed back the boundaries of science with her magnificent work in the field of X-ray crystallography. Yet when she was awarded the Nobel Prize for chemistry the headline in the *Daily Mail* ran: 'Nobel Prize for British Wife'.

1.6 Solids

Chemistry is obviously the study of chemical properties and changes, although in our everyday lives the physical properties of materials are likely to concern us rather more than their chemical properties. For example, the fashion designer producing next season's fashions will select fabrics with properties that best suit the designs in terms of texture, drape, and so on, while the engineer choosing a material to manufacture a car body might be more concerned with the density of the material and its ability to be formed into complex shapes. All these properties are ultimately determined by the interactions between the particles of the materials, and it has been one of the great successes of chemistry that we can design materials with particular physical properties, based on an understanding of these interactions.

In this section we shall explore the physical properties of solids. We shall go on in section 1.7 to look in detail at the other two states of matter – liquids and gases.

Figure 1.6.1 The outside of the space shuttle is covered in heat-resistant tiles like this, to prevent the enormous heat of re-entry damaging the craft. These tiles have an extremely low thermal conductivity, resulting from the careful design of the material of which they are made.

Crystal structure

The solid state

Solids form an important part of the world around us, providing materials with a definite shape and predictable properties. From the work of chemists during this century, we know that the properties of solids are a result of the particular way the particles within them are arranged – this arrangement is clear even on a brief examination.

Figure 1.6.2 shows some crystals of copper(II) sulphate. The regular shape of these crystals seems quite remarkable – why did the manufacturers go to so much trouble to ensure that each crystal has the same shape? The truth of the matter is that they did not! Whenever crystals of copper(II) sulphate form, they have the same basic shape, so that all these crystals have the same shape quite naturally.

Figure 1.6.2 Crystals of the familiar substance copper(II) sulphate

The shapes of crystals vary from substance to substance, although any particular substance has crystals with a common shape. This common structure results from the regular arrangement of particles within the crystal, and reflects a high degree of order within it, whether the particles are atoms, ions or molecules.

Figure 1.6.3 The regular shape of this chrome alum crystal is a result of the arrangement of the ions in the crystal lattice which is highly regular.

Crystal lattices

Despite the millions of different chemical substances, there is a very limited range of crystal structures that exist. The arrangement of particles in a crystal is called the **crystal lattice**. Crystal lattices have a regular structure, made up of a number of repeating parts called **unit cells**.

Close packing

The problem of packing atoms together in a lattice is not so very different from the problem a fruit packer has in packing apples into a box. How should the first layer be packed in order to use the space most effectively? A brief experiment (either with apples or with polystyrene spheres) should convince you that the arrangement in figure 1.6.4 makes the best use of space.

Figure 1.6.4 shows an arrangement of spheres in which any given sphere is surrounded by six other spheres at the corners of a hexagon. The spheres in this arrangement are packed as closely together as possible – they are referred to as being **close packed**.

Having decided on the packing for the first layer, how should the second layer be packed? To use space effectively (to pack as many spheres into a given space as possible), we should maintain the close packed arrangement. This means putting the next layer into the hollows between the spheres in the first layer, as shown by figure 1.6.5.

Figure 1.6.4 The most efficient way of packing spheres within a single layer

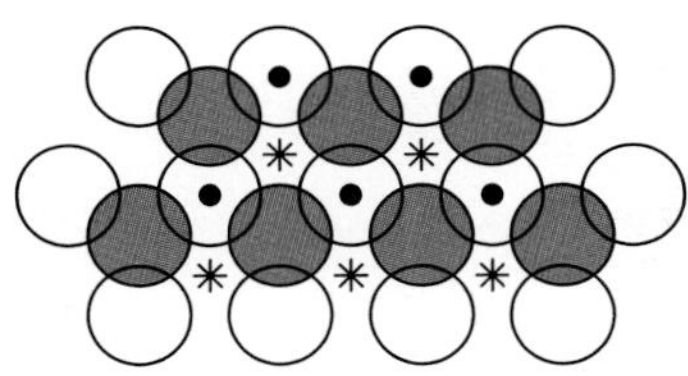

Figure 1.6.5 The second layer is put in the depressions between the spheres in the first layer.

The third layer of spheres can now go on top of the second layer in one of two ways:

1 The third layer can be placed so that the spheres in this layer rest in the depressions in the second layer that are directly above the spheres in the first layer. (These depressions are labelled • in figure 1.6.5.) This arrangement of layers is described as **ABAB**, reflecting how atoms in alternate layers are placed one above the other.

2 The third layer can be placed so that the spheres in this layer rest in the depressions in the second layer that are *not* directly above the spheres in the first layer. (These depressions are labelled * in figure 1.6.5.) This arrangement of layers is described as **ABCABC**, reflecting the repeating pattern of three layers.

These two arrangements are shown in figure 1.6.6.

Arrangement 1 – hexagonal close packing

Figure 1.6.7 shows the ABAB arrangement from a different angle, showing how the arrangement can be made up from a repeating unit. The hexagonal nature

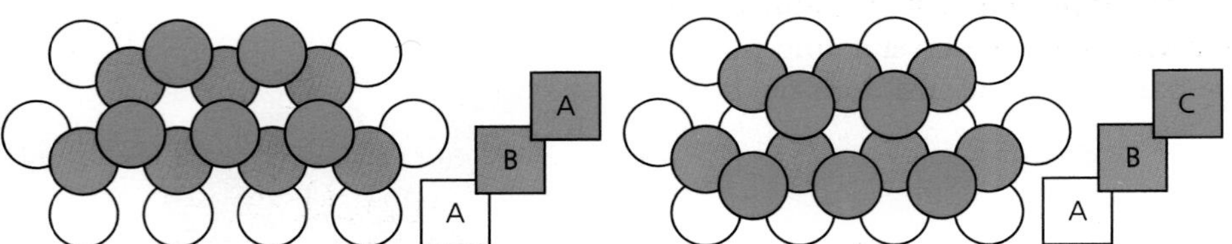

Figure 1.6.6 The two close-packed arrangements of layers

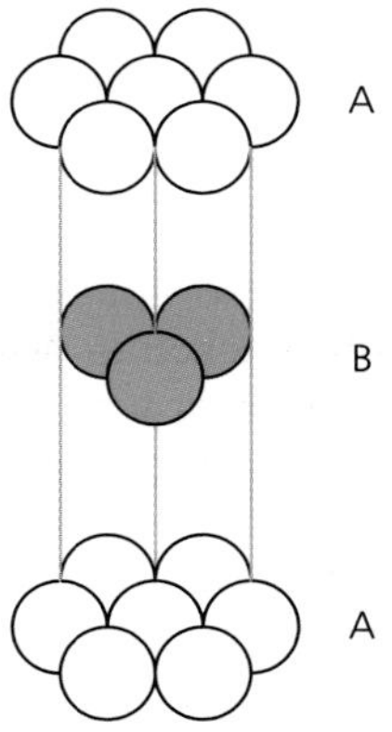

of this structure can clearly be seen, and as a result this type of packing is called **hexagonal close packed** (h.c.p.). The spheres shown here comprise the unit cell of the h.c.p. structure – by repeating this unit cell over and over again a complete ABAB structure can be built up.

Figure 1.6.7 This drawing of the ABAB arrangement of spheres shows its hexagonal structure .

Arrangement 2 – cubic close packing

The ABCABC structure is shown again in figure 1.6.8(a). This arrangement can be built up from a unit cell with a cubic structure, and as a result the arrangement is called **cubic close packed** (c.c.p.). The existence of this cubic structure is not easy to see – figure 1.6.8(b) shows its arrangement. The shading shows the spheres in layer B (shaded red), the spheres in layer A (white) and the spheres in layer C (blue). (You may need to look quite hard at this diagram to convince yourself! A little time spent with some polystyrene spheres modelling this arrangement is well worth the effort.) Because the unit cell is a cube with spheres at the centre of each face, the structure is also known as a **face-centred cubic** (f.c.c.) structure.

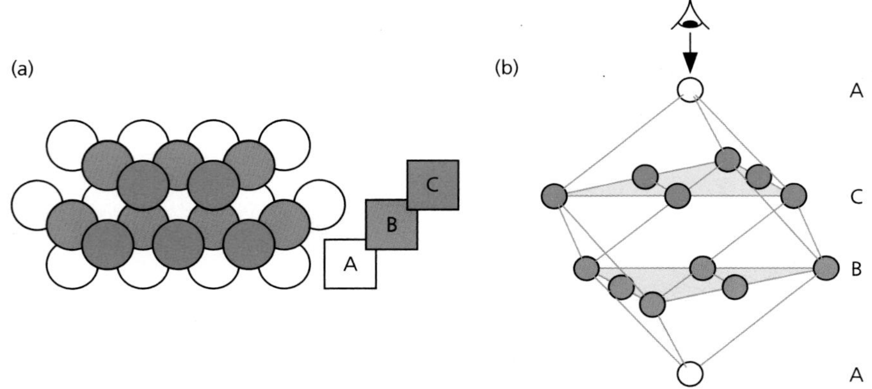

Figure 1.6.8 The face-centred cubic unit cell of the ABCABC arrangement is not so clear as the unit cell of the h.c.p. arrangement. Looking down through the layers in diagram (a) in the order ACBA, your eye is looking along the line shown in diagram (b).

Coordination number

Both the h.c.p. and f.c.c. structures contain spheres that are in contact with 12 other spheres around them – six within their own layer, and three in each of the layers above and below them. This number is known as the **coordination number** of the sphere.

Body-centred cubic packing

A different arrangement of spheres is possible if we do not close pack each layer of spheres, but instead arrange them in rows, but not touching each other, as shown in figure 1.6.9(a). A second layer can be formed by placing a layer of spheres in the depressions of the first layer, and a third layer formed in a similar way. This produces an arrangement with the unit cell shown in figure 1.6.9(b). This diagram shows why this arrangement is called **body-centred**

cubic (b.c.c.). Notice that the coordination number of this arrangement is 8 – each sphere is in contact with four spheres in the layer above it and four in the layer below it, but in contact with no spheres within its own layer.

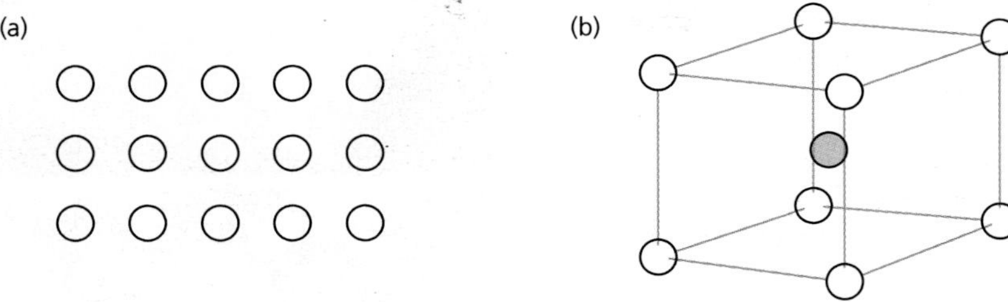

Figure 1.6.9 The body-centred cubic arrangement

These three crystal structures form the basis of a great many of the solid substances we meet in our everyday lives. We shall now go on to look at some examples of these lattice structures in chemical substances.

Crystal structures in chemical substances

Metals

The h.c.p., f.c.c., and b.c.c. arrangements are all represented in the structures of the metals. Table 1.6.1 gives examples of how the atoms are arranged in some metals. Comparison of this with the periodic table at the back of the book shows that there is not always an immediate relationship between a metal's position in the periodic table and its crystal structure.

Hexagonal close packed	Face-centred cubic	Body-centred cubic
Mg, Ti, Zn	Ag, Al, Pb	Fe, Mo, Na, K

Table 1.6.1 The crystal structures of some metals

Metals are good conductors of heat and electricity. A simple model of the metal crystal lattice helps to explain these properties.

The simple model of metals views a metal crystal as containing positive metal ions surrounded by a 'sea' of mobile electrons (the term **delocalised** is sometimes applied to this mobile cloud of electrons). Metals conduct electricity well because the electrons in the cloud are free to move through the lattice under the influence of an electric field. This motion of electrons also explains the high thermal conductivity of metals, since moving electrons may transfer their kinetic energy rapidly through the lattice.

The high melting and boiling points of many metals suggest that the forces between metal atoms must be large. This simple model seems to provide a reasonable explanation for this, in terms of the lattice of positive ions being held tightly together by the negatively charged electron 'glue'. We should not try to push this model too far, however – the metals have a wide range of melting points, and a simple model is not sufficient to produce an explanation for this.

The crystal lattice of a metal is not usually uniform throughout the metal. Most metals consist of a large number of small crystals called **grains**, separated by distinct **grain boundaries**. Within each grain, the metal atoms form a regular crystal structure. This polycrystalline nature of metals is shown clearly in figure 1.6.11.

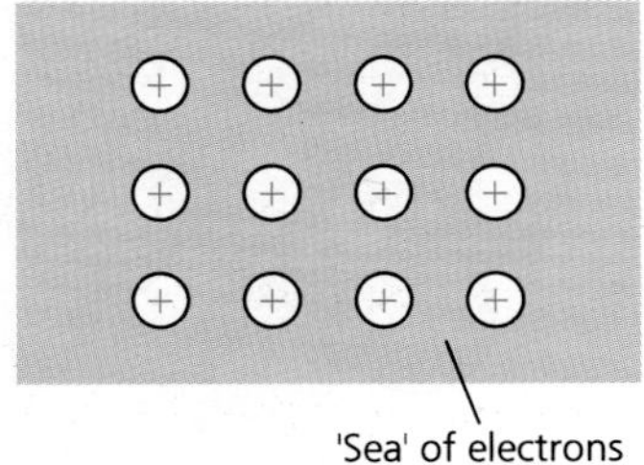

Figure 1.6.10 The model of a metal as a lattice of positive ions surrounded by delocalised electrons

Refining the model – the strength of metallic bonds

What determines the strength of the metallic bonds in a metal lattice? The amount of electrical charge in a given volume is called the charge density. Increasing the charge density of the delocalised electron cloud and the charge density of the ions in the metal lattice will lead to greater electrostatic forces of attraction between the electron cloud and the ions in it.

On this basis, we would expect potassium (large, 1+ ions) to have weaker metallic bonds than iron (smaller, 2+ ions), since the ions and the electron cloud in the potassium lattice will have a smaller charge density than the ions and the electron cloud in the iron lattice. The physical properties of potassium (soft, melting point 63 °C) and iron (hard, melting point 1535 °C) suggest that this is a fair interpretation.

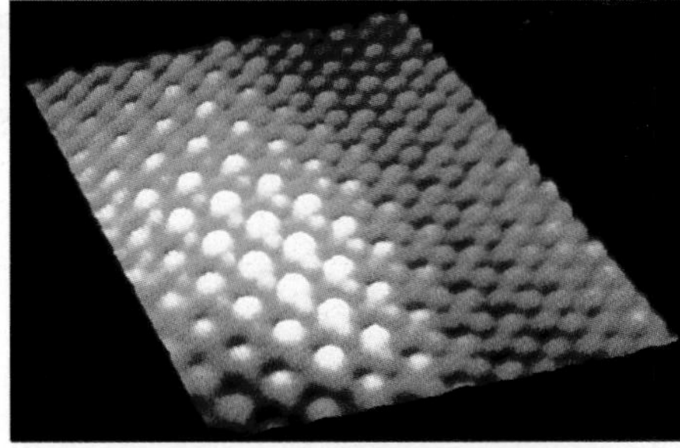

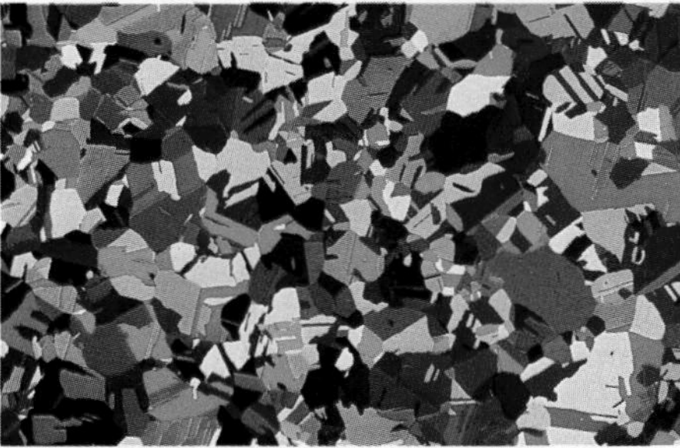

Figure 1.6.11 The scanning tunnelling electron micrograph shows the regular arrangement of palladium atoms in the crystal lattice of metallic palladium (top). On a larger scale, the grain boundaries between crystals can quite clearly be seen in the photograph of an etched piece of copper (bottom). (The surface film of oxide has been removed from this piece of copper using acid, and the surface then polished to reveal the crystals.)

Ionic solids

Ionic solids or salts are substances that contain positive and negative ions – for example, sodium chloride, NaCl. These substances are similar to metals in that they have high melting and boiling points and are quite hard. They differ from metals in the important respect that they are **brittle** – they shatter when hit, whereas metals are generally both **malleable** (they can be hammered into sheets) and **ductile** (they can be drawn out into wires).

The model for an ionic solid consists of a lattice containing both positive and negative ions. The properties of ionic solids reflect the strong interactions that exist between ions with opposite charges in this lattice. Such a structure also accounts for the observation that ionic solids conduct electricity when molten, but not in the solid state. This is reasonable, as mobile ions will exist in the molten state, and these will be able to carry charge through the liquid. In the solid state, the ions will be held immobile in the lattice, and so will be unable to conduct electricity.

The exact arrangement of ions in an ionic lattice varies, according to the ions in the solid. In an ionic lattice the positive and negative ions may be quite different in size, unlike the situation in metal lattices. This, as well as the charges on the ions, will affect the crystal structure. Sodium chloride has a unit

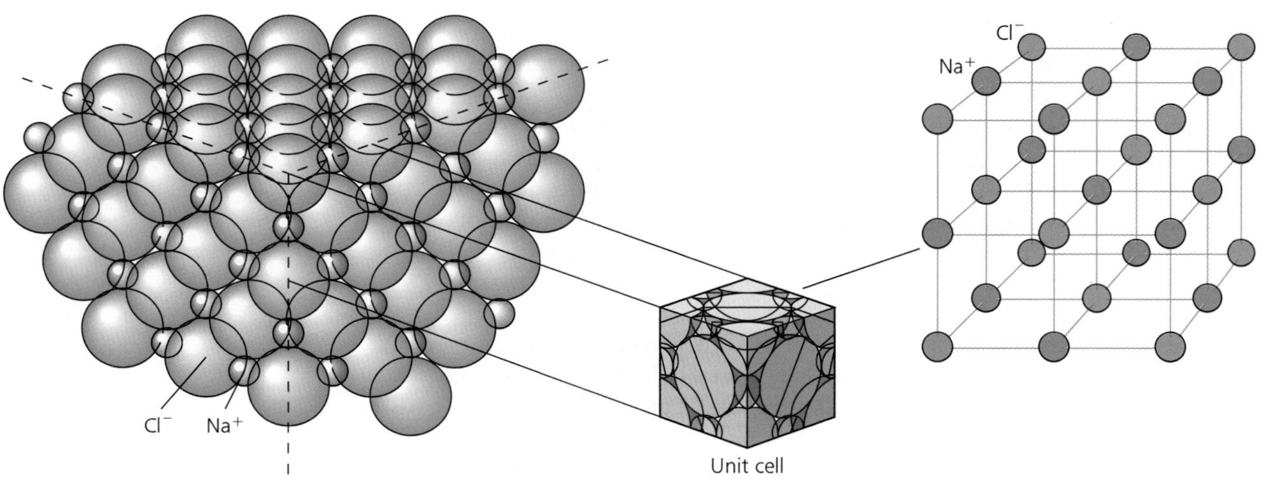

Figure 1.6.12 The rock salt structure. The inset diagram gives an alternative way of showing the unit cell.

cell similar to the face-centred cubic arrangement, as shown in figure 1.6.12. Each cation is surrounded by six anions, and vice versa. This is known as 6:6 coordination. The packing of ions in the sodium chloride lattice is commonly found in ionic compounds, and is known as the **rock salt structure**. The structure of caesium chloride is similar to the body-centred cubic structure, as shown in figure 1.6.13. The caesium chloride structure shows 8:8 coordination.

1

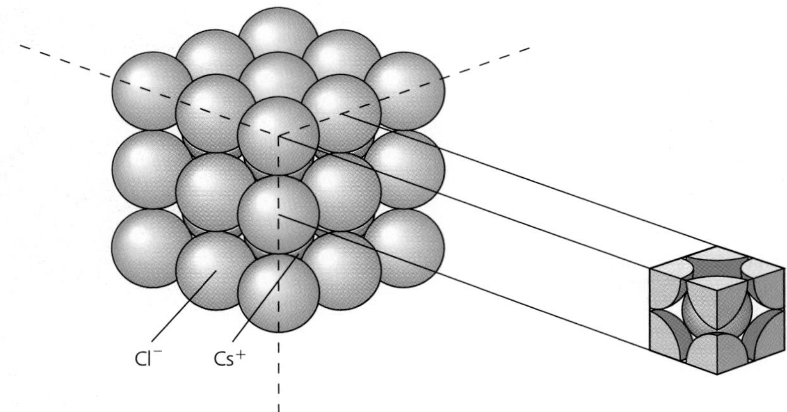

Figure 1.6.13 The caesium chloride structure

Calculating the formula of caesium chloride

Using the unit cell of caesium chloride, we can work out its formula.

Figure 1.6.13 shows that each unit cell contains one caesium atom at its centre. At each corner of the cell is a chloride ion. Each of these ions is shared between seven other unit cells, so effectively one-eighth of each chloride ion 'belongs' to the unit cell. Calculation of the formula is then quite simple:

$$1 \times Cs^+ \text{ ion} = 1 \times Cs^+$$

$$8 \times \tfrac{1}{8}\ Cl^- \text{ ion} = 1 \times Cl^-$$

The formula is CsCl

This is what we would expect on the basis of the octet rule. A similar calculation can be carried out for sodium chloride.

Why are ionic solids brittle?

Ionic solids shear when a force is applied to them. To understand why this is, we shall look at how the force between two particles in a solid varies as the distance between them changes. In figure 1.4.6 on page 53, we saw a curve of potential energy as two hydrogen atoms approach each other to form a H—H covalent bond. Figure 1.6.14 shows a generalised graph of force against separation for two particles forming a chemical bond. This graph is quite similar in shape to the graph in figure 1.4.6 – note that the force between the two particles is exactly zero at their equilibrium separation, the point at which the attractive and repulsive forces between them exactly balance.

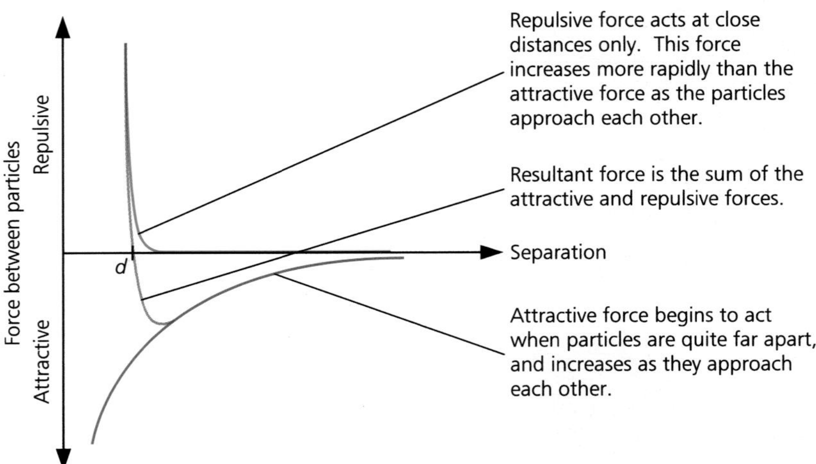

Force between particles
Repulsive
Attractive
Repulsive force acts at close distances only. This force increases more rapidly than the attractive force as the particles approach each other.
Resultant force is the sum of the attractive and repulsive forces.
Separation
d
Attractive force begins to act when particles are quite far apart, and increases as they approach each other.

Figure 1.6.14 The force between two particles in a solid has a repulsive element and an attractive element. These two elements exactly cancel out at *d*, the equilibrium separation. At separations less than *d* the repulsive element outweighs the attractive element, and so the resultant force acts to push the particles apart. The reverse is true for separations greater than *d*, where the resultant force acts to push them together.

This graph can help us to explain several things about the behaviour of solids. To begin with, notice how rapidly the repulsive force between two particles increases as they approach each other more closely than the equilibrium separation. This explains why solids are incompressible, and so we can treat the particles of a solid as hard, incompressible spheres, a bit like billiard balls. As the separation between two particles becomes greater than their equilibrium separation, an attractive force pulls them together. As a result of this, solids have **tensile strength** – they are able to resist forces that pull them apart. In this respect, ionic solids and metallic solids are quite similar in their behaviour.

The reason why ionic solids are brittle and metallic solids are not lies in the behaviour of the crystal lattice when it is deformed. As the force on a metal increases, a point is reached where the attractive force between layers of atoms is no longer able to hold them together, and two layers slip over one another. This leads to a permanent shift in the position of the two layers relative to one another, and the metal is permanently deformed. Engineers refer to this kind of deformation as **plastic**, which means 'flowing'. This is the sort of deformation that occurs when a metal is drawn into a wire or hammered out into a new shape, as shown in figure 1.6.15(a).

This slipping of layers cannot occur in an ionic lattice in the same way, since a small shift in the position of one layer of ions relative to another causes ions of the same charge to be next to each other. The repulsive force that results literally forces the two layers apart, so ionic solids do not show plastic behaviour, and simply snap.

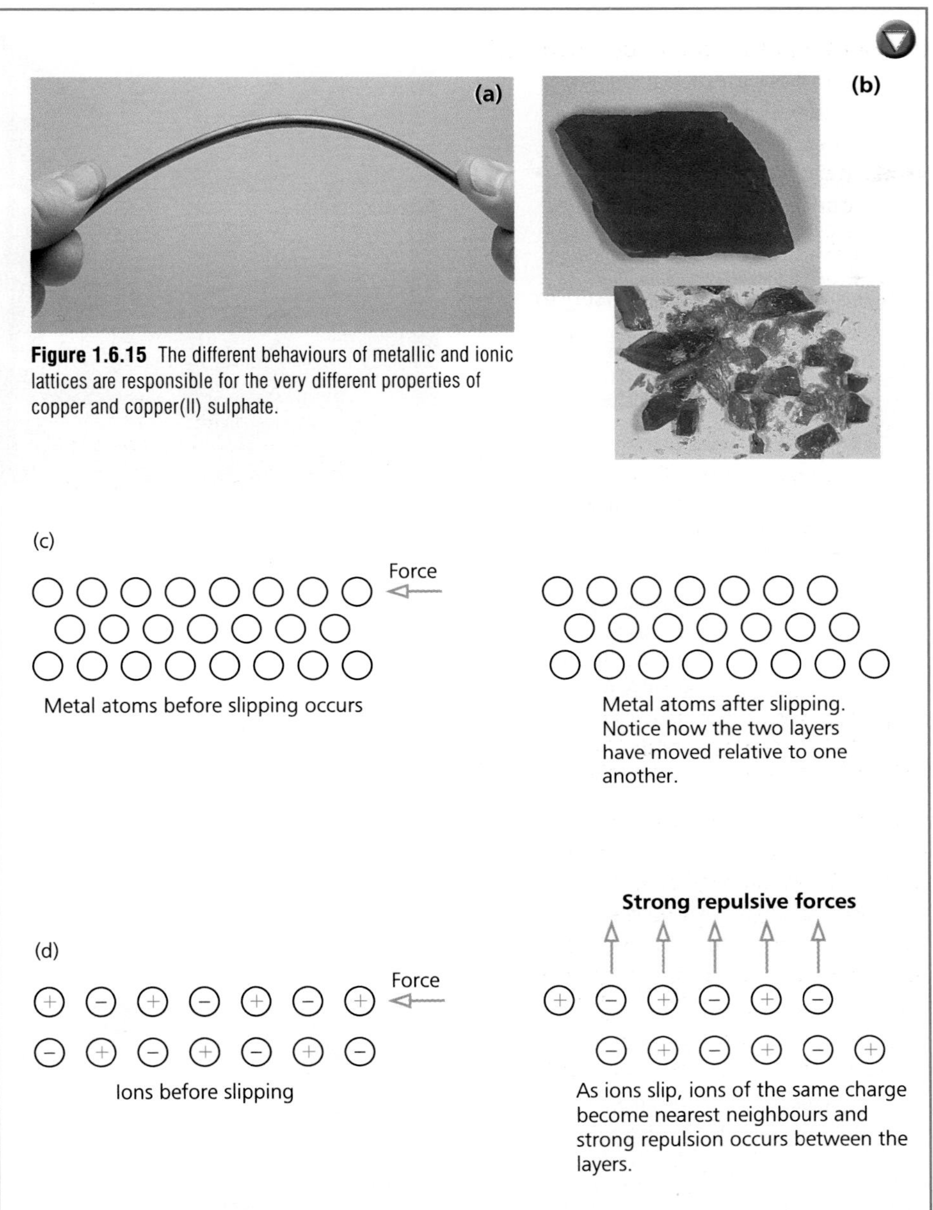
(a)
(b)
(c)
Force
Metal atoms before slipping occurs
Metal atoms after slipping. Notice how the two layers have moved relative to one another.
(d)
Force
Ions before slipping
Strong repulsive forces
As ions slip, ions of the same charge become nearest neighbours and strong repulsion occurs between the layers.

Figure 1.6.15 The different behaviours of metallic and ionic lattices are responsible for the very different properties of copper and copper(II) sulphate.

Molecular and covalent crystal structures

We have looked at ionic and metallic crystal structures. Molecular substances also form crystals in the solid state. In **molecular crystals**, the lattice positions are occupied by individual molecules (or by atoms in the case of the noble gases in their solid state). These solids tend to be soft and have low melting points, because the forces between the particles are small. These forces are van der Waals forces in the case of the noble gases, permanent dipole–permanent dipole interactions in the case of a substance like sulphur dioxide, and hydrogen bonds in the case of ice.

Some solids are made up of **covalent crystals**, in which the lattice positions contain atoms held together by covalent bonds. The lattice can be thought of as a giant molecule, and such solids are often known as **macromolecular** or

giant molecular structures. One of the best known of these structures is diamond, shown in figure 2.6.3 on page 197. Covalent crystals are usually hard, with high melting points, due to the great strength of covalent bonds. An excellent example of such properties is found in silicon carbide, SiC, an abrasive commonly found in 'sandpaper', and shown in figure 1.6.16.

Both molecular crystals and covalent crystals are non-conductors of electricity, since the particles making up the substance have no net charge.

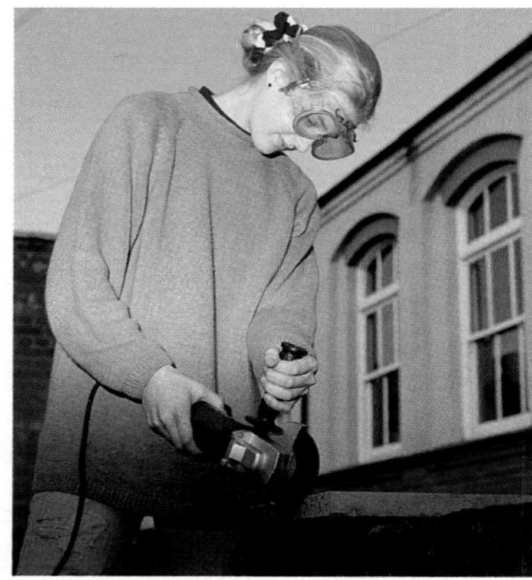

Figure 1.6.16 The hardness of silicon carbide, coupled with its high melting point, make it ideal for use in cutting wheels.

Crystal structures and physical properties

The properties of the different types of crystal structures are summarised in table 1.6.2.

Type of crystal	Particles in lattice	Forces between particles	Examples	Properties
Metallic	Positive ions	Attractions between positive ions and electron cloud	Na, Cu, Fe, Mg, Al	Wide range of hardness and melting points. Good electrical and thermal conductivity. Can be deformed plastically.
Ionic	Positive and negative ions	Attractions between oppositely charged ions	NaCl, CaF_2, KNO_3	Hard, high melting points. Non-conductors of electricity in solid state, but conduct when molten. Brittle.
Molecular	Atoms or molecules	Permanent dipole–permanent dipole interactions, van der Waals forces, hydrogen bonds	HCl, Ar, H_2O, sugar	Soft, low melting points. Non-conductors of electricity in solid and liquid states.
Giant molecular	Atoms	Covalent bonds between atoms	Diamond, SiC, SiO_2 (sand)	Very hard, high melting points. Non-conductors of electricity.

Table 1.6.2 Summary of the types of crystal structure and their properties

Non-crystalline solids

There are two types of solid structure we commonly meet that do not have a crystal structure. These are the **amorphous solids** and the **composites**.

Amorphous solids

If you break a piece of glass, the edges of the pieces have a wide range of shapes and forms – some are flat, others are curved, some may be ridged. If you compare what happens when a crystal of copper(II) sulphate is broken, you can see that the behaviour of the salt is very different.

This difference in behaviour of glass and copper(II) sulphate is due to the fact that glass is an **amorphous solid**. The word 'amorphous' means 'without

Figure 1.6.17 Glass and copper(II) sulphate both shatter when hit hard, but the edges of the resulting pieces look quite different.

form'. Amorphous solids do not have the kind of long range order that crystalline solids have, and their particles are jumbled up. In fact, amorphous solids are more like liquids than solids in terms of their structures, and are sometimes referred to as 'supercooled liquids', that is, liquids cooled below their melting points.

Many types of polymers are examples of amorphous solids. This type of solid has no definite melting point, but simply softens as it is heated, which is why you can shape both glass and plastics by heating them.

Crystalline plastics

By clever chemistry, it is possible to obtain plastics with a crystalline rather than an amorphous structure. A very strong material can be produced by arranging the molecules of a plastic to produce a structure with high order. (This material is sometimes called an **oriented polymer**, describing the way the molecules line up together.)

A recent example of such polymer engineering is a substance called 'Spectra', produced by an American chemical company. Spectra fibres have enormous strength, and yet are very flexible. These fibres can be mixed with other materials to make strong, rigid articles like helmets for cycling and other sporting activities, or they can be woven into fabric.

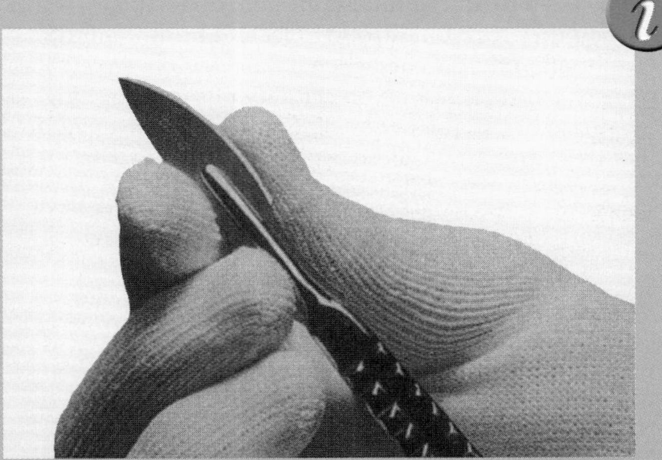

Figure 1.6.18 Gloves made of fabric woven from oriented polymer fibres can help to protect surgeons' hands from scalpel cuts.

Composites

The final class of solids we shall look at is the **composite materials**. These materials consist of one type of material combined with another in order to obtain a material that combines the advantages of the two chosen materials without their weaknesses.

One of the best known composite materials is **reinforced concrete**. This is made by pouring liquid concrete over a network of steel rods. Concrete is very strong when compressed, but it has little tensile strength. On the other hand, steel has good tensile strength. This combination produces a building material that is inexpensive, but has much better ability to withstand tensile forces than concrete alone.

Biological systems also make use of composite materials. Wood (a composite of cellulose fibres joined together by lignin) was used by nature as a construction material long before humans were around! Bone is another example of a natural composite, made up of living cells in a matrix of collagen fibres and calcium salts. Such a composite has enormous compressive strength, with the ability to withstand shear deformation too. At the same time, bone is light – particularly important for flying animals.

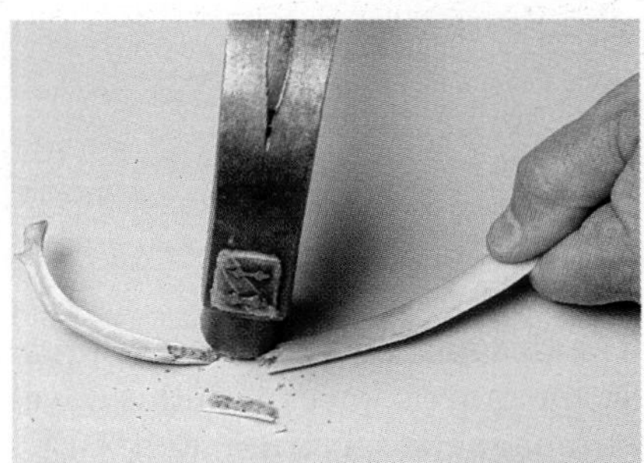

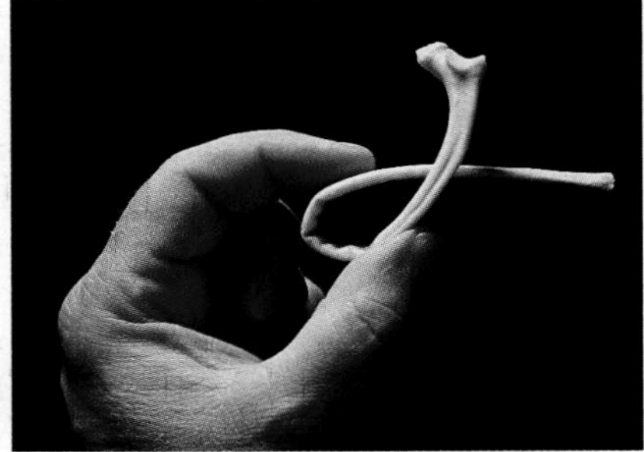

Figure 1.6.19 The combined properties of collagen and calcium salts make bone an extremely strong material weight for weight. Removal of the collagen fibres leaves a material which is very brittle. On the other hand, removal of the calcium salts leaves a material with little rigidity.

Finding out about solid structures

Using X-rays

We have seen that the properties of solids such as melting point and hardness are determined by the arrangement of particles in them. In the same way, the behaviour of a solid when X-rays are passed through it also depends on its structure.

The use of X-rays to probe the structure of solids goes back to 1912, when the German scientist Max von Laue obtained photographs of X-ray diffraction patterns caused by crystals. His work was taken up by the father-and-son team of William and Lawrence Bragg, who were able to determine the structure of a number of crystals, including sodium chloride.

Diffraction and interference

The term **diffraction** describes the deflection of waves when they travel past solid objects. The amount of diffraction that occurs when a wave travels through a narrow gap or past an object depends on two things – the size of the gap or the object, and the wavelength of the wave. When these two quantities are close together in size, diffraction occurs.

Figure 1.6.20 Diffraction occurs only when the wavelength of the wave and the size of the object or gap are close together. Light waves are not appreciably diffracted by a large object, hence the crisp shadow formed on a sunny day. They are appreciably diffracted by the point of a pin, however.

As well as diffraction, waves may **interfere** with each other. At its simplest, this can be explained by saying that if two waves coincide in phase as in figure 1.6.21(a), they produce another wave with an amplitude that is the sum of the amplitudes of the two waves. If, on the other hand, the two waves coincide antiphase as in figure 1.6.21(b), then they cancel each other out, and both waves are destroyed.

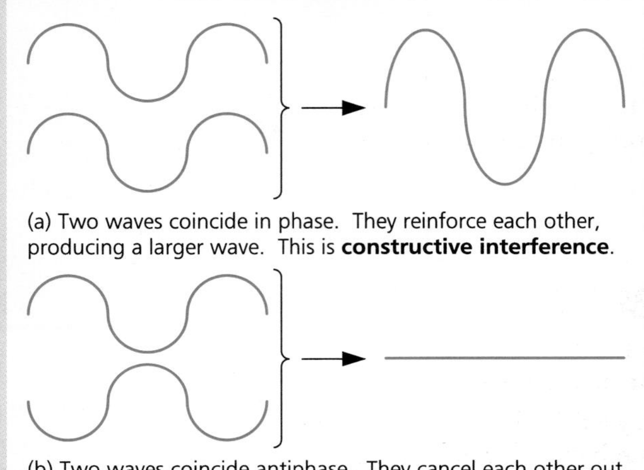

(a) Two waves coincide in phase. They reinforce each other, producing a larger wave. This is **constructive interference**.
(b) Two waves coincide antiphase. They cancel each other out, leaving no wave at all. This is **destructive interference**.

Figure 1.6.21 The interference of two waves

The distance between layers of particles in crystals is around 10^{-10} m. Electromagnetic waves with this wavelength fall in the X-ray region of the electromagnetic spectrum, and X-rays are diffracted strongly when they pass through a crystal lattice. As a result, beams of X-rays emerge from the crystal at different angles. These beams interfere with each other constructively only at certain angles, with destructive interference occurring at all other angles. The structure of the crystal can be determined from photographs of these patterns of constructive and destructive interference.

An example of the diffraction patterns obtained by passing X-rays through crystals was shown in figure 1.1.14(e) on page 16. X-ray diffraction studies of crystals are now used widely by biochemists to help find out about the shapes of complex biological molecules. The diffraction patterns produced by such crystals are complex, and require the use of computer programs to interpret them. These programs, together with other calculations, produce a set of **electron density maps**, which can then be interpreted by chemists in order to understand the shape of the molecule. Figure 1.6.22 shows an electron density map of urea (carbamide), a substance present in urine, together with its structural formula.

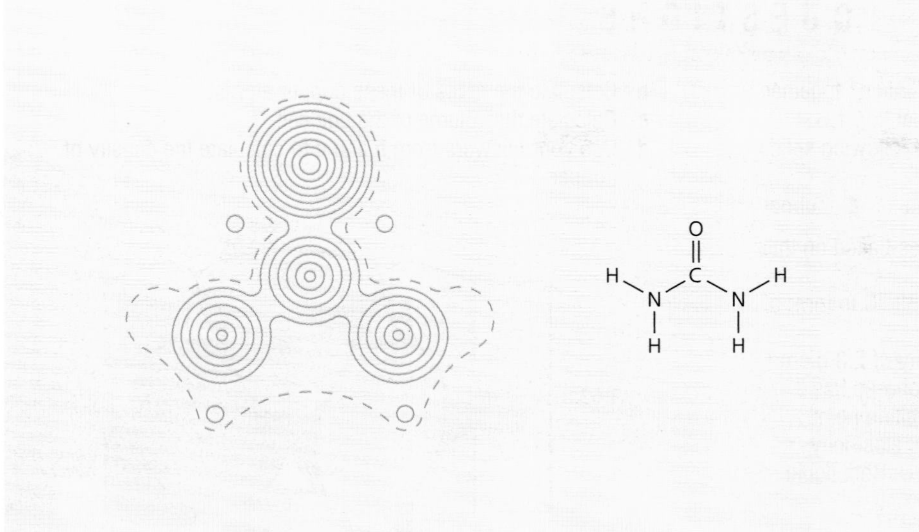

Figure 1.6.22 An electron density map of urea (carbamide). The closer the contour lines, the higher the electron density. The oxygen atom has the highest electron density, while the hydrogen atoms have a very low electron density.

In this section we have examined some important ideas about the solid state. We now go on to look at the two other states of matter – liquids and gases.

SUMMARY

- Solids have a definite shape and predictable properties which depend on the arrangement of the particles within them.
- **Crystal lattices** involve the arrangement of solid particles in a small number of regularly repeating patterns. **Close-packed** and **face-centred cubic** structures have a coordination number of 12, **body-centred cubic** structures have a coordination number of 8.
- A simple model of the structure of a **metal** consists of a 'crystal' of metal nuclei in a 'sea' of mobile or **delocalised** electrons.
- **Ionic solids** consist of a lattice containing negative and positive ions held together by strong interactions between oppositely charged particles. Common ionic crystal lattices are the **rock salt structure** with 6:6 coordination, and the **caesium chloride structure** with 8:8 coordination.
- **Molecular crystals** have the lattice positions occupied by covalent molecules held together by relatively weak intermolecular forces.
- **Covalent crystals** occur where the lattice positions are filled by atoms and the lattice is held together by covalent bonds. They are also known as **macromolecular** or **giant molecular** structures.
- **Amorphous solids** such as glass have no crystal structure – there is a jumbled arrangement of particles. These structures may be referred to as **supercooled liquids**.
- **Composite solids** combine two different types of solid to take maximum advantage of the properties of both.
- Information about the crystalline structure of materials may be obtained by **X-ray crystallography**.

QUESTIONS

1 Using the terms 'atom', 'molecule', 'ion' and 'lattice', together with the ideas about intermolecular forces in section 1.6, explain as fully as you can the structure of the following solids and how this affects their properties:
a sugar **b** sodium chloride **c** diamond **d** copper

2 Suggest structures for the following substances based on their properties:
a $TiBr_4$ – soft orange crystals which melt at 39 °C to form a liquid which does not conduct electricity.
b Phosphorus – red phosphorus has a density of 2.3 g cm^{-3} and a high melting point, while white phosphorus has a density of 1.8 g cm^{-3} and a much lower melting point. Neither red nor white phosphorus conducts electricity.
c Gallium – shiny crystals which melt at 30 °C. Both liquid and solid gallium conduct electricity.

3 a What does the term 'coordination number' mean?
b Draw diagrams to show the coordination number of an atom in a structure which is **i** close-packed and **ii** body-centred cubic.
c How would you expect the physical properties of solids with these two structures to differ?

4 High-density polythene has about three times the tensile strength of low-density polythene. Suggest a reason for this.

5 Figure 1.6.23 shows a face-centred cubic unit cell of copper. The edges of the cell are 3.62×10^{-10} m long, and the average mass of a copper atom is $1.055\,2 \times 10^{-25}$ kg.
a Calculate the number of copper atoms which 'belong' to the unit cell.
b Calculate the mass of these copper atoms.
c Calculate the volume of the unit cell.
d Use your answers from **b** and **c** to calculate the density of copper.

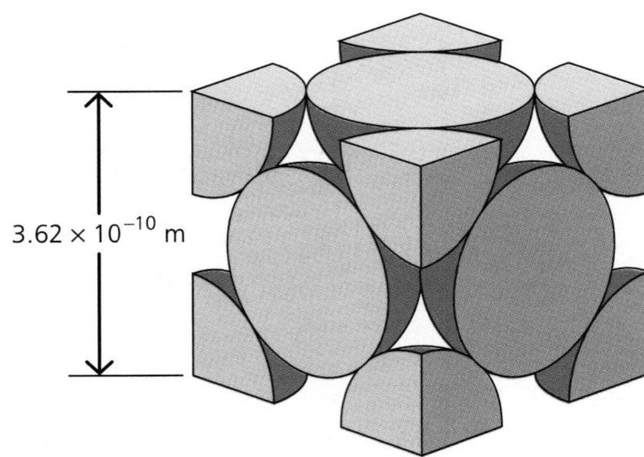

Figure 1.6.23

Developing Key Skills

Rosalind Franklin was an outstanding scientist of the last century who did not win a Nobel prize. Unfortunately she died of cancer at a very young age before her part in the unravelling of the structure of DNA could be recognised. Her specialist field was X-ray crystallography. Using library and Internet resources, put together a brief biography of Rosalind Franklin, including clear explanations of her work and its importance.

[Key Skills opportunities: C, IT]

1.7 Liquids and gases

Having examined the solid state in some detail, we shall now explore the liquid and gaseous states. Before doing so, it is useful to think very briefly about changes of state, basing our thoughts on the kinetic theory.

Changing state

The kinetic theory

The **kinetic theory of matter** is based on the idea that all matter is made up of particles which are indivisible. As well as the word 'atom', the Ancient Greeks also gave us the word 'kinetic' (from the Greek word meaning 'moving'). As we saw in section 1.1, Brownian motion provides some direct evidence that matter can be modelled in this way, and the fact that the kinetic theory works as a model provides powerful evidence for its being true.

As energy flows into a piece of matter, its temperature may rise or it may change its state. How can this behaviour be understood in terms of the kinetic theory?

As energy flows into the matter, the total amount of energy in it increases – we say that its heat content or **enthalpy** H increases. (We shall look at energy changes like this in section 1.8.) The heat content is simply the total amount of energy – both kinetic and potential – possessed by all the individual particles in the matter. In visualising this increase in heat content, we usually think of the kinetic energy of the molecules increasing as the matter is heated, with a mental picture of the particles moving more rapidly. If the piece of matter is a solid, as more and more energy is supplied to it and as its temperature increases, it will reach a point at which it melts, assuming that it does not undergo some sort of chemical reaction before it reaches its melting point.

Applying the kinetic theory

As with any model, the model we use to understand matter must do more than give a general description of the behaviour of matter – it must also have the power to *predict*. The development of the kinetic theory of matter is one of the great success stories of science – the vast majority of people have heard of atoms and molecules. It is a theory that is simple to understand in principle, and yet can provide an insight into the inner workings of the world around us.

Figure 1.7.1 Matter in all states. Not every substance can exist in all states – for example, wood reacts with the oxygen in the air long before it reaches a temperature at which it would melt. The stars in the sky represent a fourth state of matter – plasma – in which the electrons are stripped away from atoms, leaving a 'soup' of positive nuclei and electrons.

This model helps to explain the properties of solids, which have a definite shape because of the strong forces holding their particles together. These particles are fixed in one place in the solid lattice, although they are able to vibrate about their fixed positions. As a piece of solid matter is heated, its temperature increases, and its particles vibrate more vigorously, taking up more room and so causing the solid to expand. Eventually, once enough energy has been supplied to raise the temperature of the solid to its melting point, the vibrations are vigorous enough to give the particles sufficient energy to break free of the forces holding them in the lattice, and the solid melts. The energy required to turn a solid into a liquid at its melting point is called the **enthalpy change of fusion** or **heat of fusion** of that substance. This is explained in more detail in section 1.8.

The particles in a liquid move around relatively freely, although there is still sufficient force between them to keep them together in one place. Liquids therefore flow, but they have a **free surface**, taking up the shape of their container. Once the heat content of a liquid is such that many of its particles are able to overcome completely the forces between them, the liquid boils. The particles in the gas that results move around completely independently of each other, except when they collide – so a gas completely fills its container. As mentioned in section 1.5, the energy required to turn a liquid into a gas at its boiling point is called the **enthalpy change of vaporisation** or **heat of vaporisation** of that substance. Figure 1.7.2 outlines the arrangements of particles in a solid, a liquid and a gas.

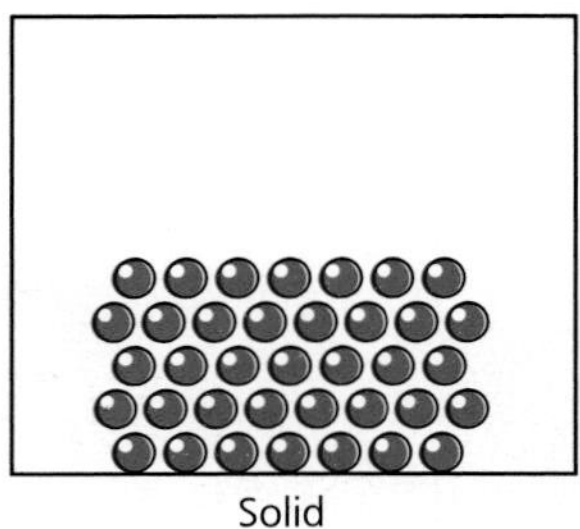

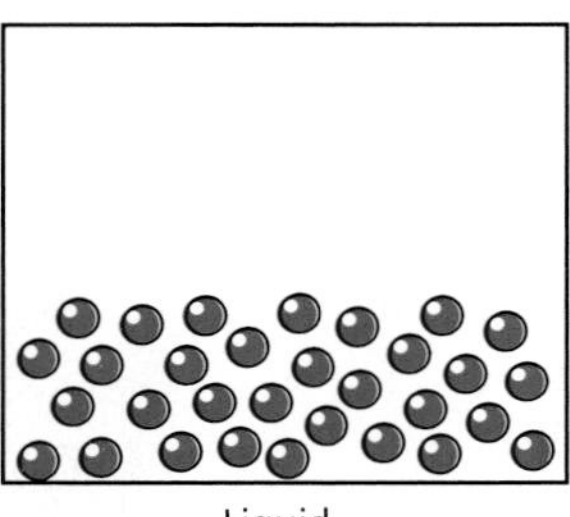

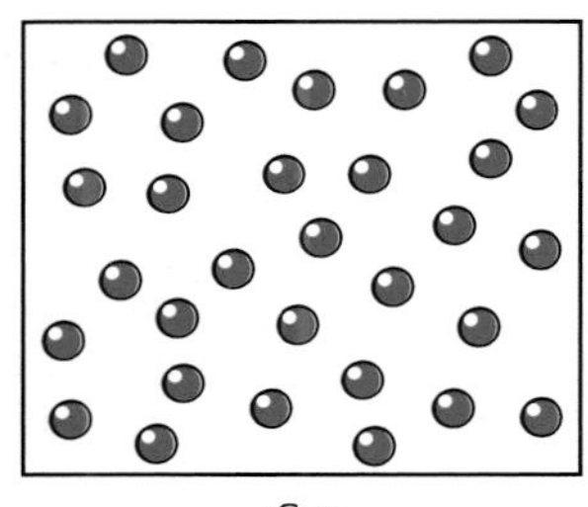

Figure 1.7.2 A simple interpretation of the particulate nature of matter

Temperature, heat and the kinetic theory

The particles of matter, although in motion, do not all move at the same speed – some molecules move slowly, while others move much faster (we shall see evidence for this in section 3.6). A sample of matter, for example, a volume of gas, thus contains particles with a range of kinetic energies.

Work carried out by physicists shows that the temperature of an object is proportional to the *average* kinetic energy of its particles. This temperature is measured on the **absolute scale** of temperature. Using this scale (in which the units of temperature are **kelvins**, K), when the average kinetic energy of the particles of a piece of matter is zero, the temperature is 0 K, **absolute zero**. No temperature lower than absolute zero can exist, since a particle cannot have a negative kinetic energy. As the temperature of an object increases, the average kinetic energy of its particles increases – we measure this increase as a change in temperature. Figure 1.7.3 compares the absolute scale of temperature with the Celsius scale.

We often say that a hot object *contains heat,* and that it can transfer this heat to a colder object if we bring the two into contact. Using this kind of language,

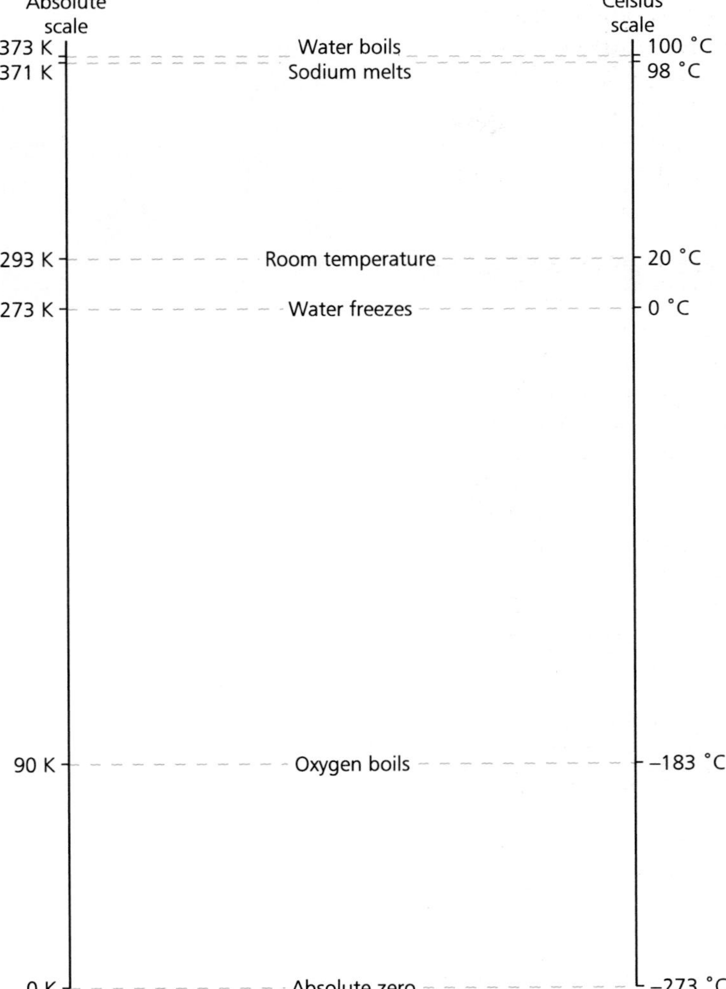

Figure 1.7.3 The absolute scale of temperature and the Celsius scale. To convert from one scale to the other, use the relationship:
Absolute temperature = Celsius temperature + 273

it is easy to think that heat is a 'thing', and that this 'thing' flows from the hot object to the colder one. The kinetic theory shows us that this is not so. Instead of thinking about heat as a substance, we should think about the transfer of heat at a molecular level, visualising the more rapid random motion of particles in the hot object 'jostling' the slower-moving particles in the cold object, and so transferring some of their kinetic energy to them. In this way, the average kinetic energy of the particles in the hot object falls – it cools down – and the average kinetic energy of the particles in the cold object rises – it warms up.

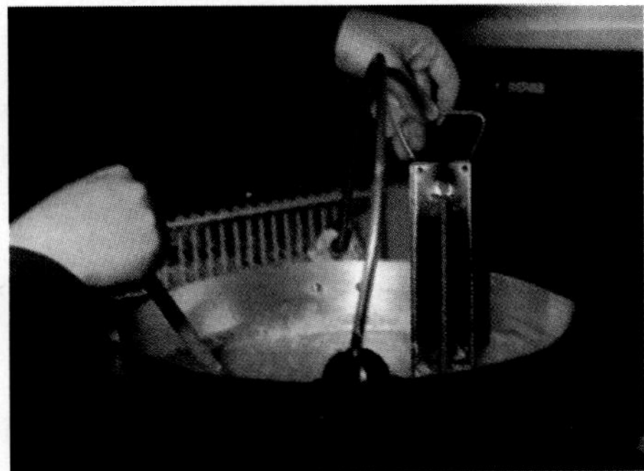

Figure 1.7.4 The distinction between temperature and heat is an important one. The amount of heat in an object is the sum of all the kinetic energies of its individual particles, while its temperature is proportional to the average of all these kinetic energies.

Liquids

A particle model for liquids

Liquids are difficult to model well, as their physical properties lie between those of a gas and a solid. Since solids do not expand appreciably when they melt, it seems reasonable to conclude that the separation of particles in a liquid is similar to that in a solid. However, the particles in a liquid must be freer and more easily able to move than those in a solid, since a liquid has the ability to flow (it is a **fluid**). On the other hand, the free surface of a liquid does not need to be restrained by a container in order to keep the liquid together, which implies that there must be forces keeping the particles in a liquid together.

This balance of freedom to move versus forces keeping the particles together leads to a model of a liquid in which there is considerable short-term and short-range structure, although in the long term any one particle does not remain in contact with the same other particles for any length of time. It is this 'neither one thing nor the other' behaviour that makes a liquid much more difficult to model successfully than a solid or a gas.

Figure 1.7.5 The particles of a liquid have sufficiently strong forces to hold them together in a drop, yet they are sufficiently mobile to allow it to flow.

Gases

The gas laws

The interdependence of the pressure, temperature and volume of a gas were investigated in the seventeenth and eighteenth centuries, when the relationships known as **Boyle's law**, **Charles's law** and the **pressure law** were established. These relationships form part of the basis of the kinetic theory of gases.

The units of pressure

The SI unit of pressure is the **pascal** (Pa), where 1 Pa = 1 N m^{-2}. Two other units are in common use when dealing with pressures in gases. The first of these measures pressure in **atmospheres** (atm), comparing the pressure in the gas with the pressure exerted by the Earth's atmosphere. (Atmospheric pressure is 101.325 kPa, or about 1.0×10^5 Pa.) The other unit measures pressure in terms of the height of a column of mercury that can be supported by the gas. Using this unit, 1 atm = 760 mm of mercury (written as 760 mmHg).

Boyle's law states that:

> **For a constant amount (number of moles) of gas, the pressure is inversely proportional to the volume if the temperature remains constant.**

or:

$$p \propto \frac{1}{V} \text{ for constant } T$$

Charles's law states that:

> **For a constant amount of gas, the volume is proportional to the absolute temperature if the pressure remains constant.**

or:

$$V \propto T \text{ for constant } p$$

The **pressure law** states that:

> **For a constant amount of gas, the pressure is proportional to the absolute temperature if the volume remains constant.**

or:

$$p \propto T \text{ for constant } V$$

FOCUS DIVING DANGERS

Scuba diving is a popular hobby, and is also part of the job for some people – the salvage industry, the police and the oil industry are just three of the organisations who have to call routinely on the services of divers. But of course, human beings are not adapted to breathe underwater. While tanks of compressed air have made it possible for us to survive (and even enjoy) life in an alien environment, diving is obviously not without its dangers.

The gas laws allow us to understand the behaviour of gases under different conditions. Thus at sea level the overall air pressure is 1 atmosphere, made up of the partial pressures of oxygen (159.6 mmHg), nitrogen (592.8 mmHg) and other gases. When we breathe air into our lungs some of the gases dissolve in our blood. The body uses oxygen for respiration, producing carbon dioxide as a waste product.

Figure 1 Scuba diving has opened up the beauties of the undersea world to human beings, animals who have evolved to breathe air at atmospheric pressure.

As depth under water increases, the pressure on a diver's body increases because of the weight of the water pushing down. As a result, more of the gases breathed in by the diver dissolve in the blood and tissues.

Nitrogen, an inert gas which is not metabolised by the body, is the one which causes the most problems for divers. Most of the extra oxygen which is inhaled is used by the body (although if divers go too deep they may die of oxygen poisoning) but the nitrogen simply accumulates, dissolved in the blood and tissues.

Nitrogen narcosis is the result of these raised nitrogen levels. At depths greater than about 30 m dissolved nitrogen may have an anaesthetic effect on divers, making them feel and act as if they are drunk. As soon as they begin to ascend the problem is resolved, since the reduced pressure allows the nitrogen to come out of solution and it is exhaled. However, the confusion and intoxication of nitrogen narcosis can lead divers to remove their masks and drown, or to go deeper rather than ascending, when they may run out of air.

Decompression sickness is another, different, problem. This may occur as a diver ascends from a deep dive, or from a very long dive, when a lot of nitrogen has become dissolved in the body. If the ascent is too rapid some of the excess nitrogen in the tissues and blood comes out of solution as the pressure falls and forms bubbles – just like taking the top off a bottle of fizzy drink. If these bubbles are very small there should be no problem. Larger bubbles can travel in veins to the lungs, where they become trapped in the tiny lung capillaries. They can also become trapped in capillaries elsewhere in the body, causing symptoms which range from slight skin mottling and tingling in the hands and feet to shock and death. Blockage of the blood flow to the joints by bubbles is a classic problem which causes great pain, often in the knees and legs, and the doubling over of the diver with 'the bends'. Because nitrogen dissolves readily in the fatty sheath around the nerves, bubbles of nitrogen often form in the nerves, compressing them and causing both pain and paralysis.

For many divers the symptoms of decompression sickness are relatively mild and soon wear off, leaving the individual determined to ascend more slowly in future. But some divers are so badly affected that they have to enter a decompression chamber (a hyperbaric chamber) where they are repressurised so the nitrogen is redissolved in the body tissues. They can then be decompressed very slowly – sometimes over days – to avoid tissue damage.

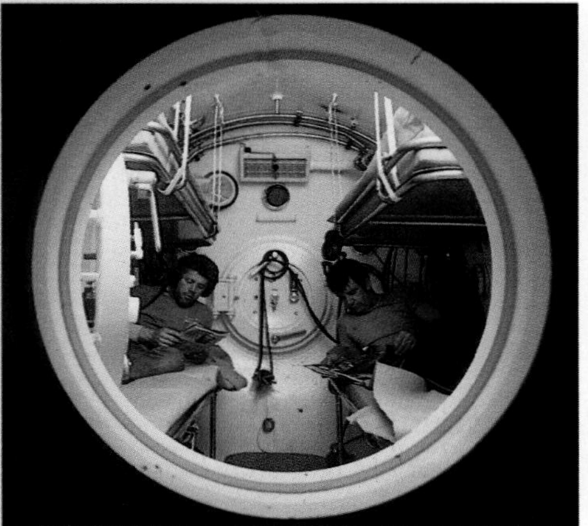

Figure 2 A hyperbaric chamber – the only hope for divers with severe decompression sickness.

Investigating the gas laws

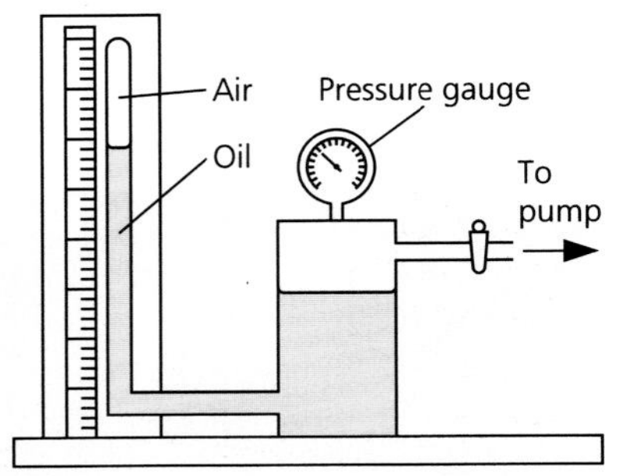

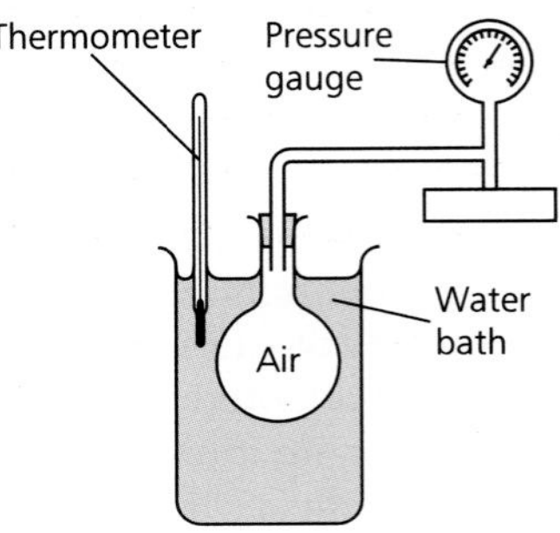

Boyle's law
Gas (usually air) is trapped in a glass tube by light oil. Pressure is exerted on the oil by a pump, which compresses the gas. Pressure is measured with a gauge and volume by the length of the air column.

Charles's law
Gas (usually air) is trapped in a capillary tube by a bead of concentrated sulphuric acid (this ensures that the air column is dry). The gas is heated in a water bath, and the volume measured by the length of the column.

Pressure law
Gas (usually air) is trapped in a flask and heated in a water bath, or an oil bath if temperatures above 100 °C are required. Pressure is measured with a pressure gauge.

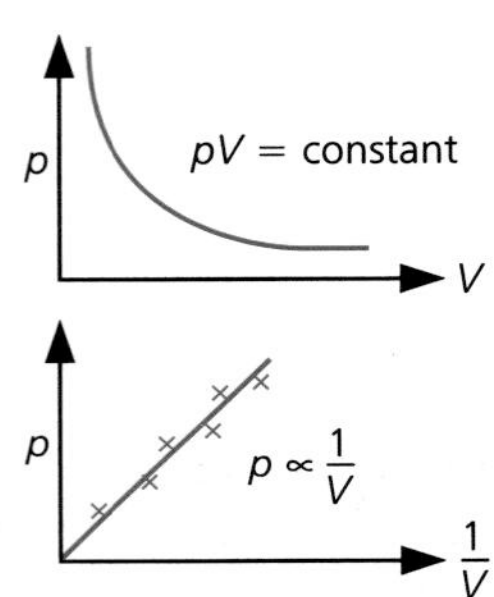

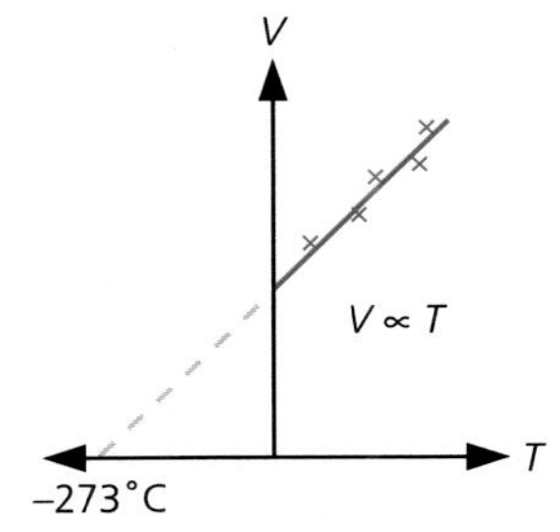

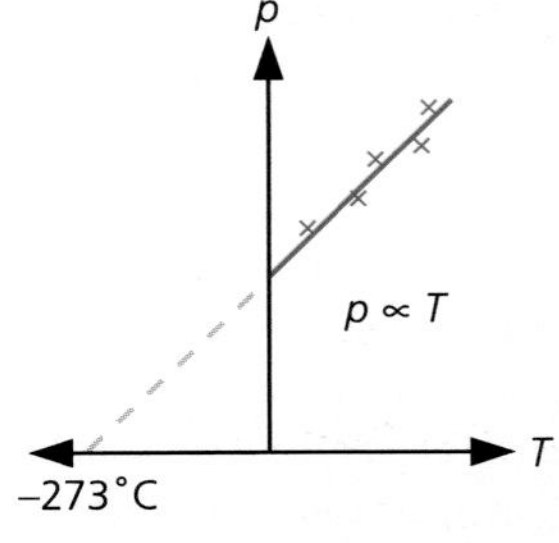

Figure 1.7.6

In the seventeenth century, Robert Boyle investigated the relationship between the pressure and the temperature of a fixed amount of gas, and found that:

$$\mathbf{pV = constant}$$

if the temperature of the gas was kept constant. A plot of p against V results in a curve called an **isotherm**, shown in figure 1.7.6. A straight line graph results if p is plotted against $1/V$, since $p \propto 1/V$.

The investigation of the variation of the volume of a gas with its temperature shows that there is a simple relationship between these two quantities, as the graph in figure 1.7.6 shows. The plot of volume versus temperature intercepts the x-axis at –273 °C, which suggests that if a gas did not eventually liquefy as it was cooled, it would occupy zero volume at this temperature, and we could write $V \propto T$, where T is the absolute temperature measured in kelvins. This behaviour we have described applies only to **ideal gases**. **Real gases** often behave somewhat differently.

Investigation of the variation of the pressure of a gas with its temperature shows a similar relationship to that found between volume and temperature.

Combining the gas laws

The three gas laws may be summarised as follows:

$$pV = \text{constant (if } T \text{ is constant)}$$
$$p/T = \text{constant (if } V \text{ is constant)}$$
$$V/T = \text{constant (if } p \text{ is constant)}$$

Combining them gives:

$$\frac{pV}{T} = \textbf{constant}$$

The constant depends on the amount of gas involved. This is an amount of matter, which is measured in moles. Investigation shows that for a given pressure and temperature, doubling the amount of a gas doubles its volume, so the constant may be written as nR, where n is the number of moles of gas. R is a constant known as the **molar gas constant**, which has the approximate value 8.3 J mol^{-1} K^{-1}.

We can now write down the **equation of state** for a gas:

$$\frac{pV}{T} = nR \quad \text{or} \quad pV = nRT$$

This equation describes the way a fixed amount of gas behaves as its pressure, volume and temperature change. Experimental evidence shows that this mathematical model for the behaviour of a gas is only satisfactory when the gas is at relatively low pressures and high temperatures, that is, when it is far from liquefaction. A gas that does obey the equation of state is an **ideal gas**. The behaviour of real and ideal gases is discussed in the box 'Gases – ideal and real' on page 98.

If a gas undergoes a process that involves a change in conditions (a change in pressure, $p_1 \rightarrow p_2$, volume, $V_1 \rightarrow V_2$, and temperature, $T_1 \rightarrow T_2$), we can write:

$$\frac{p_1V_1}{T_1} = \frac{p_2V_2}{T_2}$$

Using the gas equation

Example 1

An electric light bulb with a volume of 180 cm^3 contains argon gas at a pressure of 300 mmHg when the temperature of the gas is 27 °C. When the bulb is switched on, it is found that the pressure of the gas in the bulb reaches a steady value of 550 mmHg. What is the temperature of the gas in the bulb?

To solve this question, we need to use the gas equation in the form:

$$\frac{p_1V_1}{T_1} = \frac{p_2V_2}{T_2}$$

We can assume that (to a very good approximation) the volume of the bulb remains constant in this process, so that $V_1 = V_2$. Therefore:

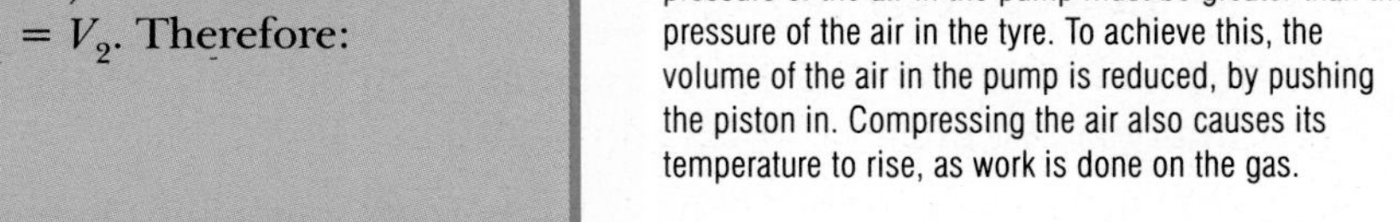

$$\frac{p_1}{T_1} = \frac{p_2}{T_2}$$

When solving a problem like this, it is always good practice to write down the data that we know first:

$$p_1 = 300\,\text{mmHg} \qquad p_2 = 550\,\text{mmHg}$$

$$T_1 = 27\,^{\circ}\text{C} = (273 + 27)\,\text{K} \qquad T_2 = ?$$

$$= 300\,\text{K}$$

Figure 1.7.7 In order to get air into a bicycle tyre, the pressure of the air in the pump must be greater than the pressure of the air in the tyre. To achieve this, the volume of the air in the pump is reduced, by pushing the piston in. Compressing the air also causes its temperature to rise, as work is done on the gas.

(Notice that in using the gas equation in this form, the units of pressure – and volume too – are not important, although they must *always* be the same on both sides of the equation. However, temperature must *always* be converted to absolute temperature.)

Substituting into the equation:

$$\frac{300\ \text{mmHg}}{300\ \text{K}} = \frac{550\ \text{mmHg}}{T_2}$$

So:

$$T_2 = \frac{550\ \text{mmHg} \times 300\ \text{K}}{300\ \text{mmHg}}$$

$$= 550\ \text{K (or 277 °C)}$$

The argon in the light bulb is at a temperature of 277 °C.

Example 2

A student wishes to produce ammonia by heating ammonium chloride with calcium hydroxide. The equation for the reaction is:

$$Ca(OH)_2(s) + 2NH_4Cl(s) \rightarrow CaCl_2(s) + 2NH_3(g) + 2H_2O(g)$$

After drying the gas, the student needs to fill a 500 cm^3 glass bulb with ammonia at a pressure of 1.5 atm and a temperature of 20 °C. What is the minimum mass of ammonium chloride that will be needed?

We need to use the equation of state for a gas. We must first convert *all* these figures into SI units – pressure in pascals, volume in cubic metres and temperature in kelvins.

p: $1.5\ \text{atm} = 1.5 \times 1.0 \times 10^5\ \text{Pa} = 1.5 \times 10^5\ \text{Pa}$

V: $500\ \text{cm}^3 = \dfrac{500\ \text{cm}^3}{1.0 \times 10^6\ \text{cm}^3\ \text{m}^{-3}} = 5.0 \times 10^{-4}\ \text{m}^3$

T: $20\ °\text{C} = (20 + 273)\ \text{K} = 293\ \text{K}$

Substituting these into the equation of state:

$$pV = nRT$$

$$1.5 \times 10^5\ \text{Pa} \times 5.0 \times 10^{-4}\ \text{m}^3 = n \times 8.31\ \text{J mol}^{-1}\ \text{K}^{-1} \times 293\ \text{K}$$

$$\text{So: } n = \frac{1.5 \times 10^5\ \text{Pa} \times 5.0 \times 10^{-4}\ \text{m}^3}{8.31\ \text{J mol}^{-1}\ \text{K}^{-1} \times 293\ \text{K}}$$

$$= 3.1 \times 10^{-2}\ \text{mol}$$

The student needs 0.031 moles of ammonia to fill the 500 cm^3 glass bulb with ammonia at a pressure of 1.5 atm at 20 °C.

The equation for the reaction tells us that 1 mole of ammonium chloride is needed to produce 1 mole of ammonia, so the student will need 0.031 moles of ammonium chloride to produce 0.031 moles of ammonia. The formula mass of ammonium chloride is 53.5 g mol^{-1}, so 0.031 moles has a mass of:

$$0.031\ \text{mol} \times 53.5\ \text{g mol}^{-1}$$

$$= 1.66\ \text{g}$$

The minimum mass of ammonium chloride needed is 1.7 g.

Precision

What is a reasonable answer to the question about the amount of ammonium chloride needed? A calculator gives the answer as 1.6585 g, which is 5 significant figures. Notice that the data in the question are to 2 significant figures at most.

In this case it seems reasonable to give the answer to 2 significant figures (1.7 g) since this is the greatest precision that the data warrant. In solving numerical problems it is generally bad practice to give a solution to a greater number of significant figures than the most precise item of data given.

Avogadro and the gas laws

At the end of the eighteenth century, the English scientist Henry Cavendish established that hydrogen and oxygen always react to form water in the ratio of 2:1 by volume. This investigation was repeated and extended by the French chemist Joseph-Louis Gay-Lussac. In 1805, Gay-Lussac published his findings of these investigations. He discovered that the volumes of all reacting gases and the volumes of the gaseous products of these reactions are in simple ratios of whole numbers, provided that the volumes are measured under the same conditions. This point is illustrated in figure 1.7.8.

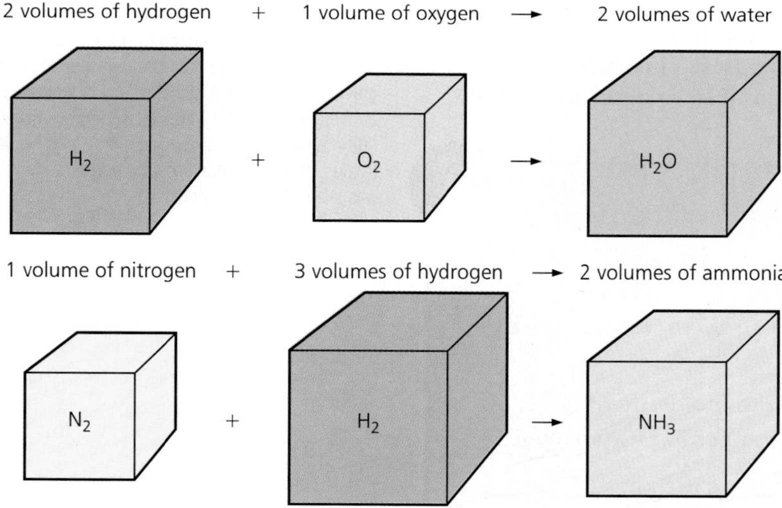

Figure 1.7.8 The volumes of gases involved in reactions are always simple ratios of whole numbers.

The Italian chemist Amedeo Avogadro seized on Gay-Lussac's work and in 1811 published his law (now named for him as **Avogadro's law**) stating that equal volumes of all gases contain equal numbers of particles, provided that they are at the same temperatures and pressures. By interpreting Gay-Lussac's work in this way, Avogadro not only showed how Dalton's atomic theory could be understood in practical terms, but also provided chemists with a very powerful practical tool for making measurements of gases. Unfortunately, like many great thinkers, Avogadro was ahead of his time, and it was not until the middle of the nineteenth century that his ideas were accepted by the majority of chemists.

How did Avogadro's law explain Gay-Lussac's observations? If we take as an example the synthesis of ammonia from nitrogen and hydrogen, it was known from experiment that:

1 volume of nitrogen + 3 volumes of hydrogen → 2 volumes of ammonia

Using Avogadro's ideas, it follows from this that if one volume of each gas contains *n* molecules, then:

1*n* molecules of nitrogen + 3*n* molecules of hydrogen → 2*n* molecules of ammonia

from which:

1 molecule of nitrogen + 3 molecules of hydrogen → 2 molecules of ammonia

or:

$\frac{1}{2}$ molecule of nitrogen + $1\frac{1}{2}$ molecules of hydrogen → 1 molecule of ammonia

Now one molecule of ammonia must contain at least one nitrogen atom. Therefore half a nitrogen molecule is at least one atom of nitrogen. This means that the simplest formula for nitrogen is N_2. Similarly, as one molecule of ammonia is produced from $1\frac{1}{2}$ molecules of hydrogen, ammonia must contain at least three atoms of hydrogen, while the hydrogen molecule contains two. Hence the chemical equation for the reaction can be written as:

$$N_2(g) + 3H_2(g) \rightarrow 2NH_3(g)$$

The power of Avogadro's law lay in the fact that it provided chemists with a way of comparing the relative masses of the molecules in two gases simply by comparing the masses of equal volumes of the two gases. This enabled chemists to measure the relative molecular masses of gases, and so to determine their formulae.

Figure 1.7.9 The English scientist Henry Cavendish (1731–1810). Cavendish left Cambridge University without a degree and lived frugally in London, even after an inheritance made him enormously wealthy. He explored the properties of gases, and measured the heats of fusion and vaporisation of many substances, as well as publishing a version of Ohm's law. His last great scientific work was to measure the value of Newton's gravitational constant, the accuracy of which was not bettered for over 100 years. Exceedingly shy and retiring, Cavendish was sociable only with his scientific friends. This is the only existing portrait of him, which was sketched surreptitiously.

Applying Avogadro's ideas

Imagine that we have x dm³ of an unknown gas, X. Avogadro's law tells us that this volume of X contains the same number of molecules as x dm³ of hydrogen. If we weigh x dm³ of X and compare its mass with that of x dm³ of hydrogen, we then know that:

$$\frac{\text{Mass of } x \text{ dm}^3 \text{ of X}}{\text{mass of } x \text{ dm}^3 \text{ of hydrogen}} = \frac{\text{mass of one molecule of X}}{\text{mass of one molecule of hydrogen}}$$

If we then have y dm³ of another gas Y, we can also say that:

$$\frac{\text{Mass of } y \text{ dm}^3 \text{ of Y}}{\text{mass of } y \text{ dm}^3 \text{ of hydrogen}} = \frac{\text{mass of one molecule of Y}}{\text{mass of one molecule of hydrogen}}$$

We can therefore compare the masses of the molecules of X and Y, by comparison with the mass of the hydrogen molecule. Provided other chemists agree to use hydrogen as a 'standard measure', we have a way of comparing the masses of gases found by different chemists.

We now use carbon-12 as the standard for comparing the mass of molecules, but the principle of comparing masses remains the same as it did in Avogadro's time.

Molar volume

One of the major implications of Avogadro's law is that one mole of any gas must occupy the same volume under the same conditions. This is the **molar volume** of the gas. In order to have a standard set of conditions under which to compare molar volumes, scientists have agreed to use **standard temperature and pressure** (s.t.p.). Standard temperature and pressure is 273 K and 10^5 Pa. Under these conditions, the volume of one mole of a gas is 22.4 dm^3, or very close to it, as shown in table 1.7.1.

Gas	Molecular formula	Molar volume/dm^3
Helium	He	22.398
Hydrogen	H_2	22.410
Oxygen	O_2	22.414
Carbon dioxide	CO_2	22.414

Table 1.7.1 The volumes of some real gases at s.t.p.

Figure 1.7.10 The French chemist Joseph-Louis Gay-Lussac (1778–1850). A highly respected academic, Gay-Lussac was the first person to isolate the element boron, as well as carrying out meticulous research in chemical analysis which led him to his conclusion about the combining volumes of gases. In 1825, he and his fellow countryman Michel Eugène Chevreul patented a way of making candles from fatty acids. Candles made in this way were an enormous improvement on those made from tallow (animal fat), and the two chemists received enormous acclaim for their invention. It is interesting to note that although Gay-Lussac is now remembered for his academic work, it was the practical invention of candles for which he received most acknowledgement in his own time.

Finding the relative molecular mass of a gas

By using the equation of state for a gas, we can find the relative molecular mass of a gas in a container, provided we can measure its pressure, temperature, mass and volume.

A chemist collected 0.299 g of a gas in a 400 cm^3 container. The pressure exerted by the gas was 316 mmHg, and its temperature was 25 °C.

Once again, we must first convert these figures into SI units.

$$p\text{:}\quad 316\ \text{mmHg} = \frac{316\ \text{mmHg}}{760\ \text{mmHg}} \times 1.0 \times 10^5\ \text{Pa} = 4.2 \times 10^4\ \text{Pa}$$

$$V\text{:}\quad 400\ \text{cm}^3 = \frac{400\ \text{cm}^3}{1.0 \times 10^6\ \text{cm}^3\ \text{m}^{-3}} = 4.0 \times 10^{-4}\ \text{m}^3$$

$$T\text{:}\quad 25\,°\text{C} = (25 + 273)\ \text{K} = 298\ \text{K}$$

Substituting these into the equation of state:

$$pV = nRT$$

$$4.2 \times 10^4\ \text{Pa} \times 4.0 \times 10^{-4}\ \text{m}^3 = n \times 8.31\ \text{J mol}^{-1}\ \text{K}^{-1} \times 298\ \text{K}$$

So:

$$n = \frac{4.2 \times 10^4\ \text{Pa} \times 4.0 \times 10^{-4}\ \text{m}^3}{8.31\ \text{J mol}^{-1}\ \text{K}^{-1} \times 298\ \text{K}}$$

$$= 6.78 \times 10^{-3}\ \text{mol}$$

This is the number of moles in 0.299 g of the gas. Therefore the mass of *one* mole of the gas (its relative molecular mass, M_r) is given by:

$$M_r = \frac{0.299\ \text{g}}{6.78 \times 10^{-3}\ \text{mol}}$$

$$= 44.1\ \text{g mol}^{-1}$$

Figure 1.7.11 Lorenzo Romano Amedeo Carlo Avogadro, Count of Quaregna and Cerreto (1776–1856). Beginning his career as a lawyer, Avogadro quickly turned his sharp mind to matters of chemistry, using a mathematical approach to understanding such matters as the combining volumes of gases. He is widely regarded as one of the founders of physical chemistry, the branch of chemistry concerned with the physical properties of matter.

This value of M_r on its own is unlikely to be sufficient to enable the chemist to determine the chemical composition of the gas. Coupled with other information, however, the calculation of relative molecular masses like this provides chemists with a straightforward tool for exploring the nature of chemical substances.

Gases – ideal and real

The kinetic theory and the gas equations describe the behaviour of **ideal** gases. Such gases have negligible forces between their particles, which take up negligible volume. Their volume and pressure decrease linearly as the temperature decreases, until both reach zero at the absolute zero of temperature. Clearly such gases cannot exist! **Real** gases do have attractive forces between their particles, since it is possible to liquefy them. Similarly, the particles of real gases do occupy a significant volume, since they liquefy to form incompressible liquids.

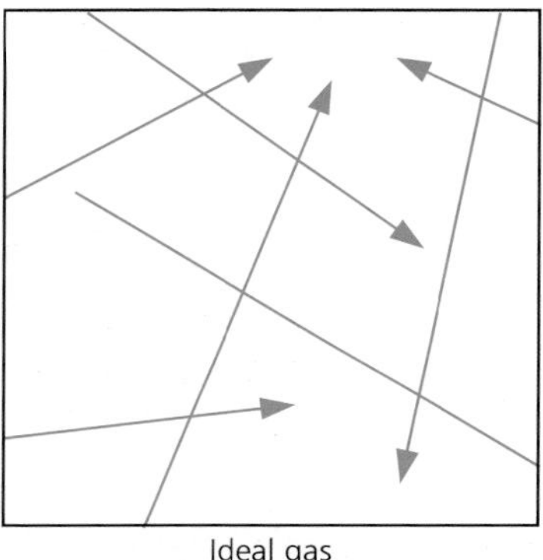

Ideal gas

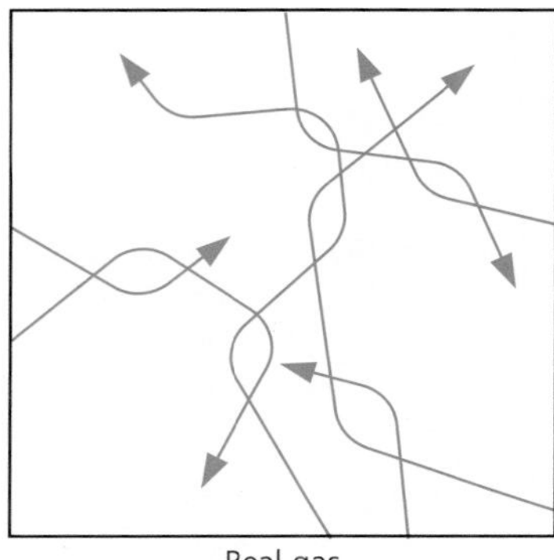

Real gas

Figure 1.7.12 In an ideal gas, there are no attractive forces between the molecules – as a result, they all travel in straight lines except when they collide with another molecule or with the wall of the container. In a real gas, there are attractive forces between the molecules, so when molecules pass close to one another, their paths bend.

However, under the right circumstances, real gases *can* behave ideally. In order for this to be so, a gas must be well away from the conditions at which it would liquefy. This ensures that the particles have sufficient energy to overcome the attractive forces between them, and that the size of the particles of the gas is negligible compared with the distance between them.

These conditions are achieved by making sure that the temperature of the gas is well above its **critical temperature**, which is the temperature at which it may be liquefied by pressure alone.

The gas must also be at a pressure well below its **critical pressure**, which is the pressure at which it becomes a liquid when at its critical temperature.

One piece of circumstantial evidence that suggests that the model for a real gas can be quite simple is the remarkably constant value of the molar volume of different gases seen in table 1.7.1.

Dalton's law

In many situations, we come across mixtures of gases – the air around us is one of many such mixtures. How do the pressures exerted by the individual gases compare with the total pressure exerted by the mixture?

The answer to this question was first given by John Dalton. Imagine that we have a mixture of three gases, A, B and C. Dalton discovered that the contribution to the total pressure made by A is in direct proportion to the fraction by volume of it present in the mixture. In other words, if the mixture consists of a dm^3 of A, b dm^3 of B and c dm^3 of C, then the contribution that A makes to the total pressure p is:

$$p_A = \frac{a}{a + b + c} \times p$$

p_A, the contribution of a gas to the total pressure, is called the **partial pressure** of the gas. It is the pressure that the gas would exert if it alone were filling the container. The law relating the partial pressures to the total pressure is called **Dalton's law of partial pressures**. This can be written as:

The total pressure of a mixture of ideal gases is the sum of their individual partial pressures.

or:

$$p = p_A + p_B + p_C + \ldots$$

The partial pressure of, say, oxygen in a mixture of gases is written as $p(O_2)$.

	Inhaled air	Exhaled air
$p(O_2)$/Pa	1.185	0.870
$p(CO_2)$/Pa	0.002	0.240

Table 1.7.2

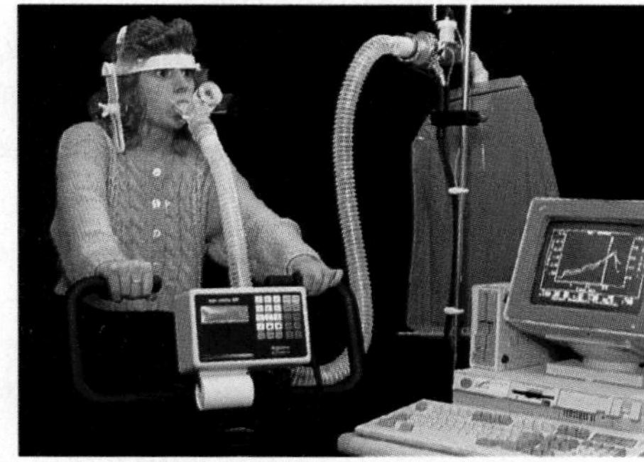

Figure 1.7.13 By measuring the partial pressures of gases in inhaled and exhaled air, it is possible to find out about a person's metabolic rate. Table 1.7.2 shows typical values for someone at rest. The partial pressures change according to exercise, and are also influenced by the individual's metabolism and diet.

Applying Dalton's law

Scuba divers need to know a great deal about the constituents of the air they are breathing. The water above a diver exerts considerable pressure on the diver's body. At depths below about 30 m, the partial pressure of nitrogen increases to a value that may lead to **nitrogen narcosis**. This results from the anaesthetic effect of nitrogen under pressure, making a diver behave as if intoxicated by alcohol. What is the partial pressure of nitrogen at 30 m depth?

The pressure due to the weight of water at this depth is about 300 kPa above atmospheric pressure, so the total pressure on the diver is around 300 kPa + 100 kPa = 400 kPa. The partial pressure of nitrogen at this depth is given by Dalton's law. If $f_{nitrogen}$ is the mole fraction of nitrogen present and p is the total pressure, then:

$$p_{nitrogen} = f_{nitrogen} \times p$$

We know that p = 400 kPa, and that $f_{nitrogen}$ = 0.8 (as air consists of 80% nitrogen by volume), so:

$$p_{nitrogen} = 0.8 \times 400 \text{ kPa}$$
$$= 320 \text{ kPa}$$

In other words, nitrogen narcosis is likely to result when the partial pressure of nitrogen exceeds 320 kPa.

SUMMARY

- The **kinetic theory of matter** provides a model of matter as being made up of moving particles.
- An increase in the kinetic energy of the particles of a solid (for example on heating) causes the particles to vibrate more rapidly until the particles can break free of the forces holding them together – the solid **melts** and becomes a **liquid**.

- The particles in a liquid move around relatively freely but there are sufficient forces between them to maintain a loose association. Thus liquids flow, but have a **free surface**, taking up the shape of their container.
- The kinetic energy of the particles in a substance is related to the temperature of the substance. This is measured using the **absolute temperature scale**, in units **kelvins**, K. At **absolute zero** (0 K) the particles of matter have no kinetic energy and no lower temperature can exist.
- **Absolute temperature = Celsius temperature + 273.**
- The particles of a gas move around quite independently of each other unless they collide. The particles of a gas move randomly and fill any container into which the gas is put.
- The kinetic model of gases and the interdependence of pressure, temperature and volume of a gas have given rise to the gas laws:

  Boyle's law: For a constant amount of gas, the pressure is inversely proportional to the volume if the temperature remains constant.

  Charles's law: For a constant amount of gas, the volume is proportional to the absolute temperature if the pressure remains constant.

  The **pressure law**: For a constant amount of gas, the pressure is proportional to the absolute temperature if the volume remains constant.
- Combining the gas laws gives the **equation of state for an ideal gas**:

$$pV = nRT$$

  where p is the pressure of the gas, V its volume, n the number of moles of gas, T the absolute temperature and R the **molar gas constant**, 8.3 J mol^{-1} K^{-1}. This equation describes the way a fixed amount of an ideal gas behaves as the temperature, pressure and volume are changed.
- **Avogadro's law** states that equal volumes of all gases contain equal numbers of particles if they are at the same temperatures and pressures.
- At standard temperature and pressure, one mole of any gas occupies 22.4 dm^3.
- **Dalton's law of partial pressures** states that the total pressure of a mixture of ideal gases is the sum of their individual partial pressures.

QUESTIONS

1 A friend, sceptical about the existence of atoms and molecules, suggests that Brownian motion provides no evidence for the existence of these tiny particles – the movement of smoke particles can equally well be explained in terms of the movement of the air as a whole surrounding them. How would you argue against this interpretation of the experimental observations?

2 Why is it possible to compress a gas, but almost impossible to compress a liquid or a solid?

3 Explain the following in terms of simple kinetic theory:
- **a** A gas fills any container into which it is put and exerts a pressure on the walls of the container.
- **b** The pressure of a gas rises if it is heated in a vessel of fixed volume.
- **c** The pressure in an oxygen cylinder falls as oxygen is withdrawn from it, whereas the pressure in a chlorine cylinder remains constant until almost all of the chlorine has been withdrawn.
  (The critical temperature of chlorine is 146 °C, while that of oxygen is –118 °C.)

4 The volume of a helium atom is around 10^{-30} m^3. What fraction of any volume of helium gas at s.t.p. is free space?

FOCUS RESPIRATION AND PHOTOSYNTHESIS

When energy is lost or given out during a chemical reaction it is said to be **exothermic**. If, on the other hand, energy is gained during a reaction then it is said to be **endothermic**. For many of the endothermic and exothermic reactions we meet in the laboratory it is fairly obvious whether heat energy is given out or taken in. But around us and within us, all of the time, two reactions fundamental to life are taking place. To determine whether they involve an output or an input of energy needs careful consideration.

Figure 1 When the chemicals in trees react with the oxygen in the air a spectacular amount of heat and light energy may be given out.

Respiration

Cells are the basic unit of all living organisms, and within almost every cell the same chemical reaction takes place. Cellular respiration involves a complex series of reactions, many of which are redox reactions. It occurs in animals and plants alike. The process involves a form of controlled combustion, where glucose from digested food is oxidised in a reaction with oxygen from the air to produce carbon dioxide and water and energy in the form of the chemical ATP. Heat energy is produced because the reaction is exothermic.

$$C_6H_{12}O_6 + 6O_2 \rightarrow 6CO_2 + 6H_2O \qquad \Delta H = -2880\,kJ$$

Warm-blooded animals such as mammals (e.g. dogs, cats, people) have a far faster rate of cellular respiration than cold-blooded animals (e.g. fish, tortoises, insects). It is the extra heat produced by the exothermic process of this respiration that allows warm-blooded animals to control their own body temperature almost regardless of what happens to the external temperature. The presence of this exothermic reaction makes itself clearly felt when we take exercise. A workout in the gym, or a game of tennis, football or hockey means our muscles are working hard and much glucose is being 'burnt up'. As a result of the rapid rate of respiration we produce lots of extra heat, so our bodies have to cool down using familiar automatic reactions such as sweating and going red.

Photosynthesis

The other reaction which is fundamental to life takes place only in plants. Photosynthesis is the process by which plants capture light energy and use it to synthesise glucose from carbon dioxide and water. The light is captured by molecules of the green pigment chlorophyll and converted into chemical energy for use in the synthetic process. Without this input of energy photosynthesis cannot take place – it is an endothermic reaction.

$$6CO_2 + 6H_2O \rightarrow C_6H_{12}O_6 + 6O_2 \qquad \Delta H = +2880\,kJ$$

Photosynthesis is a reaction of prime importance for three reasons. Plants are the producers of food for the whole planet – they make glucose which they turn into new plant material. This feeds plant-eating animals, which then become the prey of carnivores, but ultimately they all rely on the plants at the bottom of the food chain. Secondly, photosynthesis uses up carbon dioxide, helping to prevent the build-up of greenhouse gases which threaten to change the climate of the world. And thirdly, the waste product of photosynthesis is oxygen, the gas vital for almost all living organisms to enable them to break down glucose in the exothermic reaction of respiration …

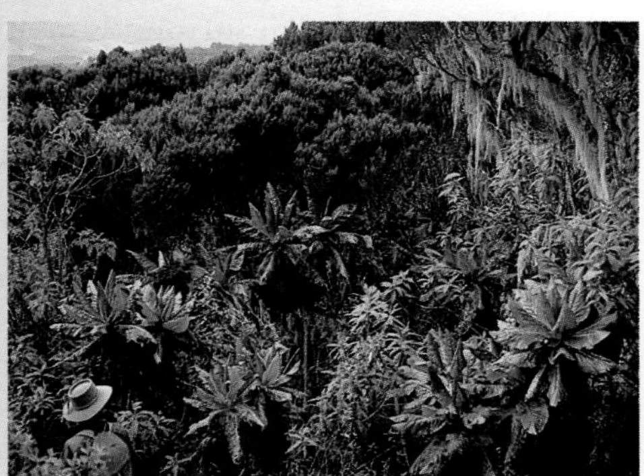

Figure 2 Plants throughout the world are photosynthesising all day long, using an input of light energy to produce glucose and oxygen.

1.8 Energy changes

So far we have looked at writing balanced equations, at the structure of the atom, and at the way the interactions between the particles of matter affect its physical properties. In doing this we have not focused closely on one of the most important aspects of chemical reactions – that of **energy changes**. In this section we shall examine energy changes in reactions, and look at how these energy changes reflect changes at a molecular level.

Figure 1.8.1 Energy from chemical reactions enables the cheetah to move its limbs to chase the gazelle and pounce on it, to move its eyes to follow the prey as it tries to escape and to convert the sound waves made by the terrified animal into electrical impulses which are interpreted by the brain.

Energy

Transferring and measuring energy

We commonly think of energy as coming in a number of forms, such as chemical, electrical, kinetic, and so on. We talk about chemical energy being transferred to kinetic energy by a petrol engine. This is a quick way of saying that the petrol – because of the way the bonds in its molecules are arranged – may be burned in an engine to produce movement, and that in doing this there is a relationship between the amount of petrol used and the movement produced. We do *not* mean that the engine literally takes 'something' out of the petrol and uses it to turn the wheels.

In chemical systems, the most common way of transferring energy is by the transfer of heat. (Remember this means that the jostling of molecules causes other molecules to vibrate more. It is not the movement of something called 'heat' – heat is a form of energy, and energy is not a thing!). For example, if you add water to some anhydrous copper(II) sulphate in a test tube, the test tube becomes hot and energy is transferred from the copper(II) sulphate to your hand holding the test tube. It is this kind of change that we shall be concerned with in the rest of this section. Before we examine this in any more detail, we need to think about how chemists look at energy changes like this.

Introducing thermochemistry

Looking at chemical systems

Thermochemistry is the study of energy transfers between reacting chemicals and the Universe that surrounds them. In the language of thermochemistry, the particular part of the Universe that we wish to study is referred to as the **system**, around which there is a **boundary**. This separates the system from the rest of the Universe – the **surroundings**. This is illustrated in figure 1.8.2. If the boundary prevents heat entering or leaving the system, the system is said to be **insulated** from its surroundings. If it also prevents matter entering or leaving the system, we have an **isolated** system.

The reaction of anhydrous copper sulphate with water produces heat. This is because the product of the reaction (hydrated copper(II) sulphate) has less chemical energy than the reactants (anhydrous copper(II) sulphate and water), and this energy imbalance is released in the form of heat, sometimes known as the **heat of reaction**. In this example the system *loses* energy in the reaction – the reaction is said to be **exothermic**, because heat leaves the system. For reactions in which the system *gains* energy, heat enters the system – this is called an **endothermic** process.

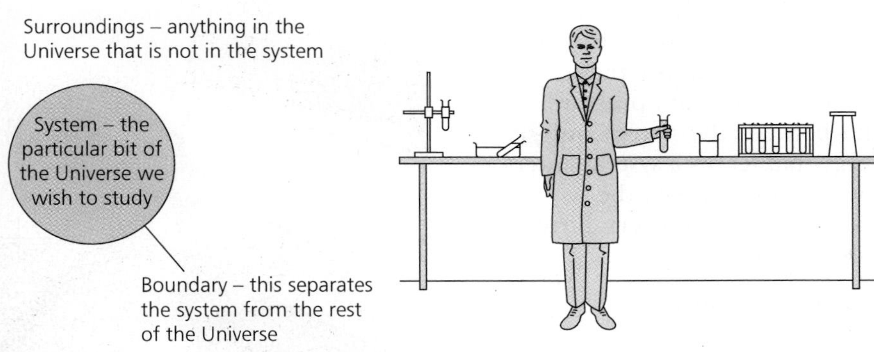

Figure 1.8.2 In the example of the reaction of anhydrous copper(II) sulphate with water, the water and the anhydrous copper(II) sulphate are the system. The test tube containing them, the person holding the test tube and the entire rest of the Universe are the surroundings.

In order to show the direction of these heat exchanges, the heat of reaction is given a sign. In the case of exothermic reactions, the heat of reaction is negative, showing that the heat content of the system has decreased as some heat has left the system and entered the surroundings. For endothermic reactions, the heat of reaction is positive, showing that the heat content of the system has increased.

Measuring energy changes

The amount of heat transferred in a given chemical reaction depends to some extent on the conditions under which the reaction occurs. Most chemical changes that we study in the laboratory – and a great many outside it too – take place at constant pressure. Chemists define the total energy content of a system held at constant pressure as its *enthalpy*, represented by the letter H. When a system reacts at constant pressure and gives out or takes in energy, we say that it undergoes an **enthalpy change**, represented as ΔH so that:

$$\Delta H = H_{\text{products}} - H_{\text{reactants}}$$

This definition of enthalpy change is consistent with our discussion of energy transfers in reactions. For exothermic changes:

$$H_{\text{products}} < H_{\text{reactants}}$$

so ΔH is negative, and for endothermic changes:

$$H_{\text{products}} > H_{\text{reactants}}$$

so ΔH is positive. Figure 1.8.3 shows how the enthalpy changes involved in reactions can be represented in an **enthalpy diagram**.

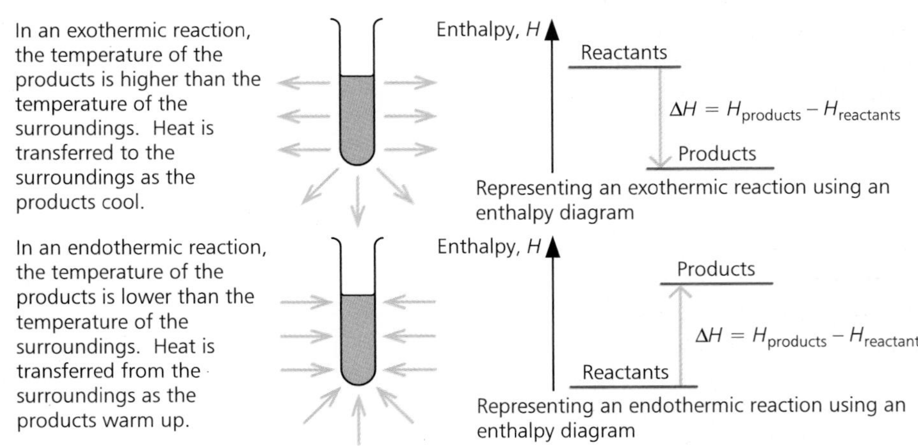

In an exothermic reaction, the temperature of the products is higher than the temperature of the surroundings. Heat is transferred to the surroundings as the products cool.
Enthalpy, H
Reactants
ΔH = H_products – H_reactants
Products
Representing an exothermic reaction using an enthalpy diagram
In an endothermic reaction, the temperature of the products is lower than the temperature of the surroundings. Heat is transferred from the surroundings as the products warm up.
Enthalpy, H
Products
ΔH = H_products – H_reactants
Reactants
Representing an endothermic reaction using an enthalpy diagram

Figure 1.8.3 Enthalpy diagrams provide a useful way of representing enthalpy changes in reactions.

We cannot *measure* the enthalpy of a system – but fortunately this does not matter! We are really only interested in the *enthalpy change* during a reaction, which is a measure of what the reaction can do for us (or even in some circumstances what it can do to us!) under conditions of constant pressure.

So far in our discussions of energy movements between the system and the surroundings, we have assumed that all the energy that leaves a system enters the surroundings (or vice versa). This is an important principle in the behaviour of the Universe, and one that scientists accept as a universal truth. It is usually called the **principle of conservation of energy**:

The total energy content of the Universe is constant.

Figure 1.8.4 When considering how much energy (how many miles) we can get from burning petrol in a car engine, we really do not care what the total enthalpy of the petrol and exhaust gases is either before or after the reaction. All we are interested in is the *change* in enthalpy, since this is the only energy that is available to us.

Enthalpy and internal energy

Sometimes we measure changes in a quantity called the **internal energy** U of a system to find out about energy changes. In other situations we talk about enthalpy changes. What's the difference between a change in internal energy and a change in enthalpy?

Chemical changes often involve changes in volume. Such changes are particularly important when gases are involved, for example, when petrol burns to form carbon dioxide and water:

$$2C_6H_{14}(g) + 19O_2(g) \rightarrow 12CO_2(g) + 14H_2O(g)$$

In this reaction, the carbon dioxide and steam produced have to do work against the Earth's atmosphere as they expand. The energy to do this work comes from the burning petrol itself. Because of this expansion, there is less energy available to do useful work than there would be if the change took place without expansion – in other words, at constant volume.

We can express this difference in the following way. If ΔU represents the change in energy content of the system at constant volume, and ΔH represents the change in energy content at constant pressure, then:

$$\Delta H = \Delta U - p\Delta V$$

where ΔV is the change in volume of the system when the reaction occurs at constant pressure p.

If we wished to study systems under a wide range of conditions, including situations where the pressure is not constant, we would be interested in ΔU – the change in the system's **internal energy**. In contrast, when we investigate changes under conditions of constant pressure, our attention is focused on ΔH, the change in the system's enthalpy.

Conditions for thermochemical measurements

The enthalpy change associated with a reaction is affected by the physical conditions of temperature and pressure. Chemists need therefore to specify a set of **standard states** under which enthalpy changes are measured. The standard conditions for thermochemical measurements (standard temperature and pressure, s.t.p.) are agreed as 298.15 K and 10^5 Pa.

An enthalpy change measured for a reaction carried out under these conditions is referred to as the **standard enthalpy change** of the reaction, and is represented by the symbol $\Delta H^{\ominus}$ or sometimes $\Delta H^{\ominus}(298\ \text{K})$. The enthalpy change for a reaction obviously also depends on the quantities of materials reacting. Enthalpy changes are expressed per mole of reaction. It is very important that the reaction is written alongside the enthalpy change, so it is clear what quantities of reactants or products are meant. Enthalpy changes are then called **standard molar enthalpy changes**, $\Delta H^{\ominus}_{m}$.

Measuring enthalpies of reaction – calorimetry

Measuring the enthalpies of reaction for different chemical processes has exercised the ingenuity of chemists for over two centuries. Essentially, the process involves carrying out the chemical change in an insulated container called a **calorimeter**, and measuring the temperature rise that results. This temperature rise may then be used in one of two ways:

1 It may be multiplied by the **heat capacity** of the calorimeter plus contents, giving the energy change in the reaction.

2 An electrical method of heating the calorimeter may be used in order to bring about the same temperature change. The electrical energy needed to do this is then the same as the energy produced by the chemical change. (This is sometimes called **electrical compensation calorimetry**.)

Heat capacities

It is common experience that different objects require different amounts of heat to change their temperatures by similar amounts. The **heat capacity** C of an object is the amount of heat required to raise its temperature by a given amount. The SI unit of heat capacity is J K^{-1}.

Whilst heat capacity is useful in certain circumstances, it applies only to a single object, such as a particular beaker of water. **Specific heat capacity** c gives the heat capacity per unit mass of a particular substance, and is obviously more generally useful. Its SI unit is $\text{J kg}^{-1}\ \text{K}^{-1}$. If we represent the amount of heat transferred as E and the temperature change as ΔT, then from these definitions:

$$E = C\Delta T \quad \textbf{for an object with a heat capacity } C$$

and:

$$E = mc\Delta T \quad \textbf{for a mass } m \textbf{ of a substance with a specific heat capacity } c$$

In practice, the first method provides a convenient way of measuring enthalpy changes for all but the most accurate requirements. A simple form of this type of calorimetry makes use of a 'coffee-cup calorimeter', which is just two expanded polystyrene cups one inside the other with a lid. Since expanded polystyrene is an excellent insulator, any heat changes occurring in the calorimeter can easily be measured before the heat finds its way out of the calorimeter into the laboratory. Figure 1.8.5 shows such a calorimeter, and the following box shows how it is used to make measurements.

Standard states

As well as temperature, pressure and amount of substance, we must consider two other factors regarding standard states for a chemical reaction.

The first is that the reacting substances must be in their normal physical states under the standard conditions. Where two or more forms of a substance exist, the most stable one at 10^5 Pa and 298 K is used (so, for example, standard enthalpy changes involving carbon relate to graphite, not diamond).

The second factor involves solutions. Where solutions are used, these should have **unit activity**. For our purposes, this effectively means that they should have a concentration of $1\ \text{mol dm}^{-3}$. (The concept of activity allows for the fact that concentrated solutions behave in such a way that they appear to be less concentrated than they really are. For simple measurements this effect is negligible under standard conditions.)

Figure 1.8.5 A coffee-cup calorimeter

Measurements with a coffee-cup calorimeter

A student placed 25.0 cm^3 of 1.0 mol dm^{-3} hydrochloric acid in a coffee-cup calorimeter, and measured its temperature as 22.5 °C. Then 25.0 cm^3 of 1.0 mol dm^{-3} sodium hydroxide solution, also at 22.5 °C, was added quickly to the acid. The mixture was stirred and the final temperature was recorded as 29.2 °C. If the specific heat capacity of the solutions is 4.2 J g^{-1} K^{-1} and their densities can be taken as 1.00 g cm^{-3}, calculate the enthalpy change for this reaction per mole of HCl, assuming that no heat was lost either to the cup or the surroundings.

In the calorimeter, 50 cm^3 of solution increased in temperature from 22.5 °C to 29.2 °C, an increase of 6.7 K. Using the density figure, the total mass of the solutions can be taken as 50 g. The energy required for this increase in temperature is given by:

$$E = mc\Delta T$$

$$= 50 \text{ g} \times 4.2 \text{ J g}^{-1} \text{ K}^{-1} \times 6.7 \text{ K}$$

$$= 1407 \text{ J}$$

As 25 cm^3 of a 1.0 mol dm^{-3} solution of HCl was used, the number of moles of HCl present is:

$$\frac{25 \text{ cm}^3}{1000 \text{ cm}^3} \times 1.0 \text{ mol}$$

$$= 0.025 \text{ mol}$$

So the enthalpy change ΔH per mole of HCl is given by:

$$\Delta H = \frac{1407 \text{ J}}{0.025 \text{ mol}}$$

$$= 56.3 \text{ kJ mol}^{-1}$$

The molar enthalpy change (per mole of HCl used) for this reaction is 56.3 kJ mol^{-1}.

The bomb calorimeter

The calorimeter used for accurate work to determine energy changes in chemical reactions is known as a **bomb calorimeter**, so called because it consists of a sealed vessel that looks like a bomb. This apparatus is particularly useful for studying the energy changes when a fuel burns, and it is also used to find out the 'calorific value' (the energy content) of foods. Figure 1.8.6 shows a simplified diagram of such a calorimeter and explains how it works.

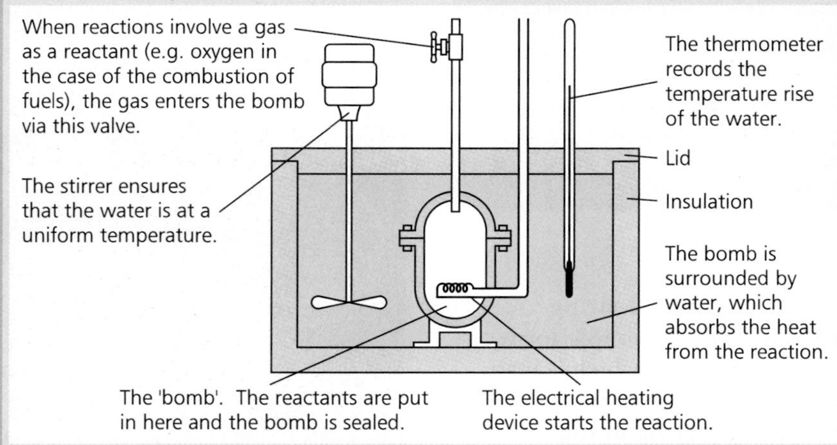

When reactions involve a gas as a reactant (e.g. oxygen in the case of the combustion of fuels), the gas enters the bomb via this valve.
The stirrer ensures that the water is at a uniform temperature.
The thermometer records the temperature rise of the water.
Lid
Insulation
The bomb is surrounded by water, which absorbs the heat from the reaction.
The 'bomb'. The reactants are put in here and the bomb is sealed.
The electrical heating device starts the reaction.

Figure 1.8.6 The energy released by the reaction in the bomb can be calculated from the temperature rise of the water, the specific heat capacity of the water and the heat capacity of the bomb and its contents.

Some of the most accurate methods of calorimetry use electrical heaters to cause the same temperature change as the chemical reaction, over the same time period. This provides a measurement of the energy change occurring in the reaction. It also has the advantage that the energy supplied by the heater not only raises the temperature of the calorimeter, but also duplicates the heat losses from it. Although the calorimeter is insulated, some heat will be lost from it, and this needs to be taken into account. This method therefore avoids the need for a heat-loss correction to be made. In an alternative technique, the water surrounding the calorimeter is kept at a steady temperature close to that of the laboratory by heating or cooling it. The energy used for heating or cooling is carefully measured. This leads to extremely small heat losses from the calorimeter, and the enthalpies of reaction measured in this way can be extremely accurate.

Figure 1.8.7 A chocolate bar like this contains 294 kJ. The energy contents of foods are calculated by burning them in a bomb calorimeter. The process of respiration in the body is a slower version of the same reaction.

Some important enthalpy changes

Enthalpy change of combustion

The bomb calorimeter is commonly used to measure energy changes when a substance burns. Chemists define the **standard molar enthalpy change of combustion** of a substance as the enthalpy change which occurs when one mole of the substance is completely burnt in oxygen under standard conditions. The term 'completely burnt' is important, since many elements form more than one oxide. For example, carbon may burn to form two oxides:

$$\text{C(graphite)} + \tfrac{1}{2}\text{O}_2\text{(g)} \rightarrow \text{CO(g)}$$

$$\text{C(graphite)} + \text{O}_2\text{(g)} \rightarrow \text{CO}_2\text{(g)}$$

Complete combustion of carbon produces carbon dioxide, so it is the enthalpy change in the second process that is the standard enthalpy change of combustion of carbon. The standard molar enthalpy change of combustion of the compound concerned can be conveniently written in 'chemist's shorthand':

$$\Delta H^{\ominus}_{c,m}\ [\text{C(graphite)}] = -393.5\ \text{kJ mol}^{-1}$$

The standard molar enthalpy change of combustion may sometimes be called the **heat of combustion**.

Fractions of moles

Notice that when we write thermochemical equations, we may need to use fractions like '$\frac{1}{2}O_2$'. This is so that we have the correct number of moles involved in the change. In this case, the standard molar enthalpy change refers to one mole of carbon, so to get a balanced equation for the reaction forming carbon monoxide, we need half a mole of oxygen molecules.

Enthalpy change of formation

Figure 1.8.8 Combustion is usually a strongly exothermic process!

Another important enthalpy change is the **standard molar enthalpy change of formation** of a compound, $\Delta H^{\ominus}_{f,m}$. This is the enthalpy change when one mole of the compound is formed from its elements under standard conditions. For example, the standard molar enthalpy change of formation of methane, CH_4, refers to the change:

$$\mathbf{C(graphite) + 2H_2(g) \rightarrow CH_4(g)}$$

The accepted value for the enthalpy change of this reaction is –74.8 kJ mol^{-1}. Once again, we can use 'chemist's shorthand':

$$\Delta H^{\ominus}_{f,m}\ [CH_4(g)] = -74.8\ \text{kJ mol}^{-1}$$

The units are kJ mol^{-1}, as we would expect. The 'per mole' refers to the formation of *one mole* of the compound, not to the quantity of elements reacting. The statement 'the standard molar enthalpy change of formation of sodium carbonate is –1131 kJ mol^{-1}' means that for every mole of sodium carbonate formed from its elements, the change in enthalpy is –1131 kJ. By definition, the enthalpy change of formation of an element is zero, since this is the change in enthalpy for:

$$\mathbf{Na(s) \rightarrow Na(s)}$$

or:

$$\mathbf{Cl_2(g) \rightarrow Cl_2(g)}$$

which are obviously not changes at all!

Enthalpy changes of formation may also be called **heats of formation**. They may be calculated from measurements made using a bomb calorimeter, although in many cases this is not possible. We shall see examples of finding enthalpy changes of formation by calculation later on in this section.

Enthalpy change of atomisation

The **standard molar enthalpy change of atomisation** of an element is the enthalpy change when one mole of its atoms in the gaseous state is formed from the element under standard conditions. The standard molar enthalpy change of atomisation of carbon (as graphite) is +716.7 kJ mol^{-1}. This is the enthalpy change for:

$$\mathbf{C(graphite) \rightarrow C(g)}$$

Once again, this can be shortened:

$$\Delta H^{\ominus}_{at,m}\ [C(graphite)] = +716.7\ \text{kJ mol}^{-1}$$

Note that the molar enthalpy change of atomisation *always* refers to the *formation* of one mole of the atoms of the element. It is *not* the enthalpy change when one mole of the element is atomised. Atomisation is always endothermic, since it involves increasing the separation between atoms, which requires energy.

Enthalpy changes (or heats) of atomisation may be calculated from other measurements (in the case of a solid, from the enthalpy changes of fusion and vaporisation together with its specific heat capacity), or they may be measured by spectroscopic means in the case of gases like oxygen. (If we know that electromagnetic radiation of frequency f has sufficient energy to cause a molecule to split into atoms, then we can use the relationship $E = hf$ to calculate the energy associated with this change.)

Enthalpy change of fusion and enthalpy change of vaporisation

In section 1.7 we mentioned enthalpy change of fusion and enthalpy change of vaporisation. The **standard molar enthalpy change of fusion** $\Delta H^{\ominus}_{fus,m}$ is the enthalpy change when one mole of a solid is converted to one mole of the liquid at its melting point at standard pressure. Similarly, the **standard molar enthalpy change of vaporisation** $\Delta H^{\ominus}_{vap,m}$ is the enthalpy change when one mole of a liquid is converted to one mole of gas at its boiling point at standard pressure. For example:

$$\Delta H^{\ominus}_{fus,m}\ (H_2O) = 6.01\ kJ\ mol^{-1}$$

$$\Delta H^{\ominus}_{vap,m}\ (H_2O) = 41.09\ kJ\ mol^{-1}$$

Working with enthalpy changes of reaction

Enthalpy changes for 'difficult' reactions

We know that the standard molar enthalpy change of combustion of carbon is –393.5 kJ mol^{-1}, which tells us that for:

$$C(graphite) + O_2(g) \rightarrow CO_2(g) \qquad \Delta H^{\ominus}_{c,m} = -393.5\ kJ\ mol^{-1}$$

What happens if we wish to know the enthalpy change for the decomposition of carbon dioxide to its elements?

$$CO_2(g) \rightarrow C(graphite) + O_2(g) \qquad \Delta H^{\ominus}_{m} = ?$$

This reaction would be very difficult to carry out, but fortunately the law of conservation of energy comes to our rescue, telling us that the enthalpy change for the reaction must be +393.5 kJ mol^{-1}. If it were not so, then we could use the reactions to build a perpetual motion machine, going from carbon and oxygen to carbon dioxide and back again, producing energy out of nothing each time. On the basis of the law of conservation of energy, the enthalpy change for a reverse reaction has the same numerical value as the enthalpy change for the forward reaction, but the opposite sign.

Reactions and perpetual motion

For the sake of argument, assume that the enthalpy change for the decomposition of carbon dioxide into its elements was only +350 kJ mol^{-1}, not +393.5 kJ mol^{-1}. If this was so, we should be able to take one mole of carbon atoms and one mole of oxygen molecules, turn them into one mole of carbon dioxide molecules and then decompose these to form carbon and oxygen again – and also make a 'profit' of 43.5 kJ of energy in the process! We could then repeat this over and over, using the energy produced to do all sorts of things such as run a car or heat a house.

Of course, we know that nature doesn't function like this, and that we cannot get energy out of nowhere. Therefore the enthalpy changes for forward and reverse reactions must add together to give zero.

Figure 1.8.9 Look closely at this picture – does the scene obey the law of conservation of energy?

Combining reactions to calculate enthalpy changes

We can take the conservation of energy idea a little bit further. Think about the reaction of carbon with oxygen. We know that two oxides are possible, so we can envisage a sort of 'triangle of reactions' in which carbon reacts with oxygen to form carbon monoxide, which reacts with more oxygen to form carbon dioxide, which then decomposes to form carbon and oxygen again. This is shown in figure 1.8.10(a).

Can we work out the energy changes involved here? The standard molar enthalpy change of formation of carbon dioxide is $-393.5\ kJ\ mol^{-1}$, while the standard molar enthalpy change of reaction for carbon monoxide and oxygen forming carbon dioxide is $-283.0\ kJ\ mol^{-1}$. On the basis of these figures, the law of conservation of energy requires that the standard molar enthalpy change of formation of carbon monoxide is $(-393.5\ kJ\ mol^{-1} - (-283.0\ kJ\ mol^{-1})) = -110.5\ kJ\ mol^{-1}$. The enthalpy diagram in figure 1.8.10(b) makes this clearer. This is exactly the figure we get if we measure the enthalpy change for this process experimentally.

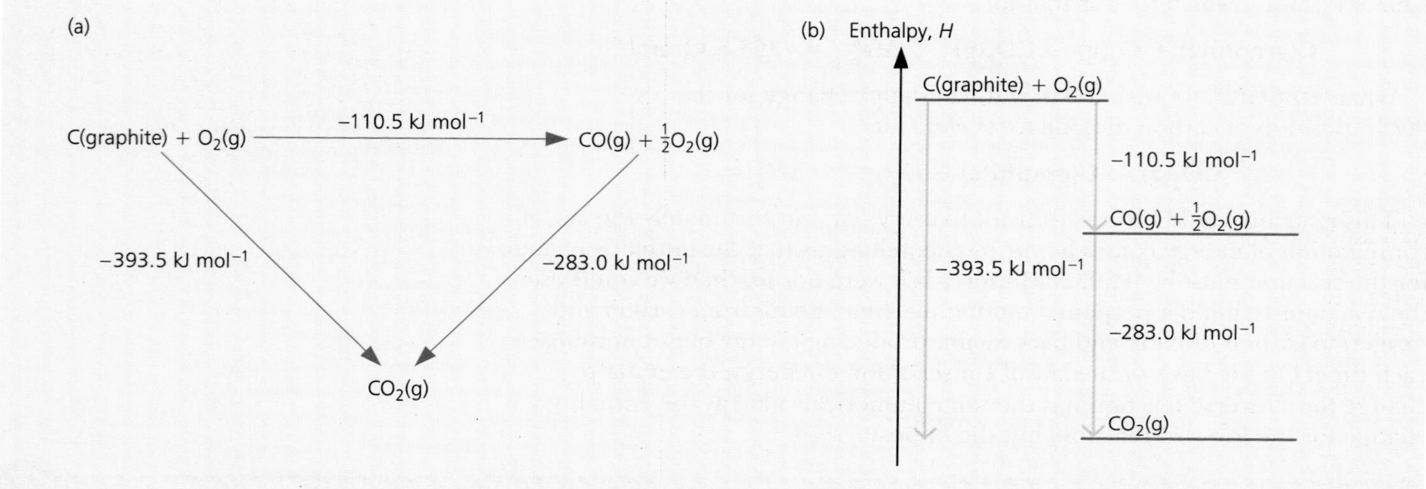

Figure 1.8.10 The energy changes in this cyclical process are governed by the law of conservation of energy, just like all other energy changes. Notice how the overall enthalpy change is zero if we go round a closed loop in this cycle, for example,

$$C + O_2 \rightarrow CO + \tfrac{1}{2}O_2 \rightarrow CO_2 \rightarrow C + O_2.$$

Hess's law

There are many cases where the law of conservation of energy can be used to combine enthalpy changes like this. Because of its usefulness, the law of conservation of energy has been rewritten in chemical terms. The chemical version of the law is called **Hess's law**, named after the chemist Germain Henri Hess. Hess's law states that:

> **For any reaction that can be written in a series of steps, the standard enthalpy change for the reaction is the same as the sum of the standard enthalpy changes of all the steps.**

In other words, if we have a reaction:

$$\mathbf{A + B \rightarrow C + D}$$

it does not matter whether we go directly from A + B to C + D, or whether we go via some intermediate compound X – the standard enthalpy change for the overall reaction will always be the same.

Using Hess's law to measure enthalpy changes of formation

Hess's law enables us to calculate standard molar enthalpy changes of formation for substances that might otherwise be difficult to measure. One such example of this is the formation of sucrose (sugar) from carbon, hydrogen and oxygen:

$$\mathbf{12C(graphite) + 11H_2(g) + 5\tfrac{1}{2}O_2(g) \rightarrow C_{12}H_{22}O_{11}(s)}$$

No chemist has ever been able to carry out this reaction, so direct measurement of the enthalpy change is impossible. However, we *can* measure the standard molar enthalpy change of combustion of sucrose, and this can lead us to the standard molar enthalpy change of formation for sucrose, as figure 1.8.11 shows.

1

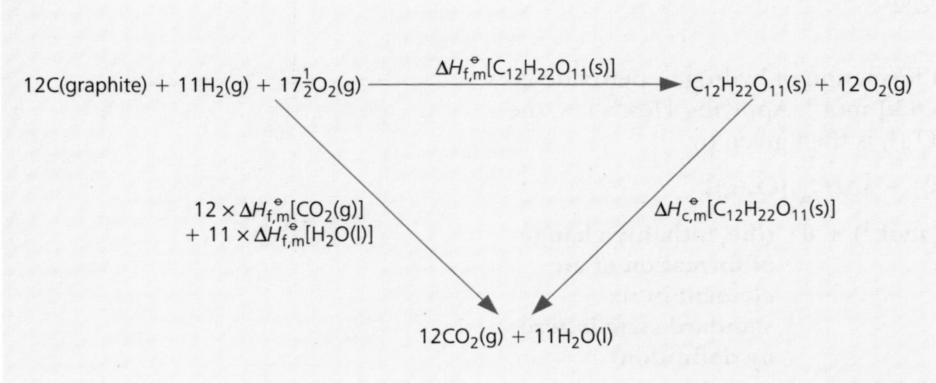

Figure 1.8.11 Whilst it is impossible to measure the standard molar enthalpy change of formation of sucrose directly, we can measure the standard molar enthalpy changes of formation of carbon dioxide and water, and the standard molar enthalpy change of combustion of sucrose. Between them, these can lead us to the enthalpy change we want to know.

Looking at the cycle of reactions in figure 1.8.11, we can see that the standard molar enthalpy change of formation of sucrose can be calculated from the standard molar enthalpy changes of formation of carbon dioxide and water, and the standard molar enthalpy change of combustion of sucrose. Applying Hess's law:

$$\Delta H^{\circ}_{f,m}\,[C_{12}H_{22}O_{11}(s)] = \begin{array}{l} 12 \times \Delta H^{\circ}_{f,m}\,[CO_2(g)] + 11 \times \Delta H^{\circ}_{f,m}\,[H_2O(l)] \\ -\,\Delta H^{\circ}_{c,m}\,[C_{12}H_{22}O_{11}(s)] \end{array}$$

The enthalpy changes on the right-hand side of this equation are:

$$\Delta H^{\circ}_{f,m}\,[CO_2(g)] = -393.5\text{ kJ mol}^{-1}$$

$$\Delta H^{\circ}_{f,m}\,[H_2O(l)] = -285.8\text{ kJ mol}^{-1}$$

$$\Delta H^{\circ}_{c,m}\,[C_{12}H_{22}O_{11}(s)] = -5639.7\text{ kJ mol}^{-1}$$

Substituting these into the equation we get:

$$\begin{aligned} \Delta H^{\circ}_{f,m}\,[C_{12}H_{22}O_{11}(s)] &= \begin{array}{l} 12 \times (-393.5\text{ kJ mol}^{-1}) + 11 \times (-285.8\text{ kJ mol}^{-1}) \\ -\,(-5639.7\text{ kJ mol}^{-1}) \end{array} \\ &= -4722\text{ kJ mol}^{-1} + -3143.8\text{ kJ mol}^{-1} + 5639.7\text{ kJ mol}^{-1} \\ &= -2226.1\text{ kJ mol}^{-1} \end{aligned}$$

So the standard molar enthalpy change of formation of sucrose (to four significant figures) is –2226 kJ mol^{-1}.

Hess's law can also be applied to the decomposition of hydrogen peroxide, a strong oxidising agent used to bleach paper and cloth, and also to bleach hair. Hydrogen peroxide decomposes slowly to form water and oxygen:

$$\mathbf{H_2O_2(l) \rightarrow H_2O(l) + \tfrac{1}{2}O_2(g)}$$

Figure 1.8.12 shows a cycle of reactions that can be used to calculate ΔH°_{m} for this reaction.

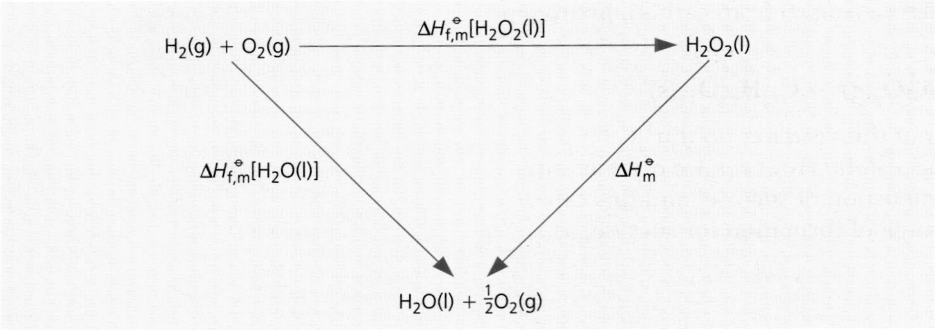

$H_2(g) + O_2(g) \xrightarrow{\Delta H^\ominus_{f,m}[H_2O_2(l)]} H_2O_2(l)$

$\Delta H^\ominus_{f,m}[H_2O(l)]$

$\Delta H^\ominus_m$

$H_2O(l) + \frac{1}{2}O_2(g)$

Figure 1.8.12 The reactions involved in calculating $\Delta H^\ominus_m$ for the decomposition of hydrogen peroxide

The standard molar enthalpy change of formation of hydrogen peroxide is –187.8 kJ mol^{-1}, and that for water is –285.8 kJ mol^{-1}. Applying Hess's law, the value of $\Delta H^\ominus_m$ for the decomposition of H_2O_2(l) is then given by:

$$\Delta H^\ominus_m = -\Delta H^\ominus_{f,m}[H_2O_2(l)] + \Delta H^\ominus_{f,m}[H_2O(l)] + \tfrac{1}{2}\Delta H^\ominus_{f,m}[O_2(g)]$$

$$= -(-187.8 \text{ kJ mol}^{-1}) + (-285.8 \text{ kJ mol}^{-1}) + 0$$

(the enthalpy change of formation of an element in its standard state is zero, by definition)

$$= -98.0 \text{ kJ mol}^{-1}$$

So the standard molar enthalpy change for the decomposition of hydrogen peroxide to water and oxygen is –98 kJ mol^{-1}.

This calculation based on Hess's law can be generalised, as shown in figure 1.8.13. We can find $\Delta H^\ominus_m$ for any process using standard molar enthalpy changes of formation. In general, for any chemical process for which the standard enthalpy change is $\Delta H^\ominus_m$, then:

$$\Delta H^\ominus_m = \textbf{sum of standard molar enthalpy changes of formation of products} - \textbf{sum of standard molar enthalpy changes of formation of reactants}$$

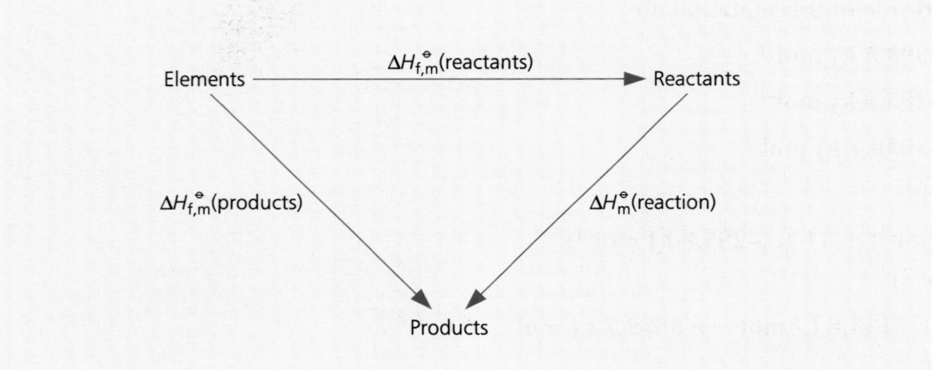

Figure 1.8.13 The application of Hess's law – using enthalpy changes of formation to find enthalpy changes of reaction

Using Hess's law

If the standard molar enthalpy changes of formation of nitrogen oxide, NO, and nitrogen dioxide, NO_2, are +90.2 kJ mol^{-1} and +33.2 kJ mol^{-1} respectively, calculate $\Delta H^\ominus_m$ for the change:

$$NO(g) + \tfrac{1}{2}O_2(g) \rightarrow NO_2(g)$$

The standard enthalpy change for this process will be given by:

$$\Delta H^{\circ}_{m} = \Delta H^{\circ}_{f,m}\,[NO_2(g)] - \Delta H^{\circ}_{f,m}\,[NO(g)]$$

(Note that the standard molar enthalpy change of formation of oxygen is zero, by definition.)

Substituting the data into this relationship gives:

$$\Delta H^{\circ}_{m} = +33.2\ kJ\ mol^{-1} - +90.2\ kJ\ mol^{-1}$$

$$= -57.0\ kJ\ mol^{-1}$$

For the reaction shown, ΔH°_{m} is –57.0 kJ mol^{-1}.

1

What can enthalpy changes tell us?

Stability

Chemists often talk of the **stability** of a compound. What does this mean?

Taking hydrogen peroxide as an example again, we saw that it had a standard molar enthalpy change of formation of –187.8 kJ mol^{-1}. This tells us that one mole of hydrogen peroxide contains 187.8 kJ mol^{-1} less energy than one mole of hydrogen and one mole of oxygen. Our everyday experience tells us that a decrease in energy is associated with an increase in stability (see figure 1.8.14), and we can say that:

Hydrogen peroxide is more stable than its elements.

or:

Hydrogen peroxide is stable with respect to its elements.

It is important to compare the stability of hydrogen peroxide with something (in this case, with its elements). Hydrogen peroxide decomposes to water and oxygen, as we have just seen, which suggests that it is **unstable** with respect to these two substances. The decomposition of hydrogen peroxide to water and oxygen is an exothermic change, which confirms this suggestion, and so we can say:

Hydrogen peroxide is stable with respect to its elements but unstable with respect to water and oxygen.

We must also be careful when talking about stability of other substances, for example nitrogen oxide, NO. The figures in the box on page 112 show that nitrogen oxide is unstable with respect to its elements, since its standard molar enthalpy change of formation is positive. Yet nitrogen oxide can be stored for long periods at room temperature, as can ethyne, C_2H_2 ($\Delta H^{\circ}_{f,m}$ = +228 kJ mol^{-1}). The reason for this apparent discrepancy is that we are considering a particular type of stability called **energetic stability** or **thermodynamic stability**.

Although nitrogen oxide and ethyne are energetically unstable, they both have a kind of stability called **kinetic stability**. If an energetically unstable substance has kinetic stability, its situation is very similar to the ruler standing on end in figure 1.8.14. The ruler is unstable, but it requires a small push to make it fall over, so it can stand on end indefinitely – at least, until someone bumps into the table! It is just the same with nitrogen oxide and ethyne – although they are energetically unstable, they require a small 'nudge' to push them to decompose. We shall see in section 3 exactly what this nudge is.

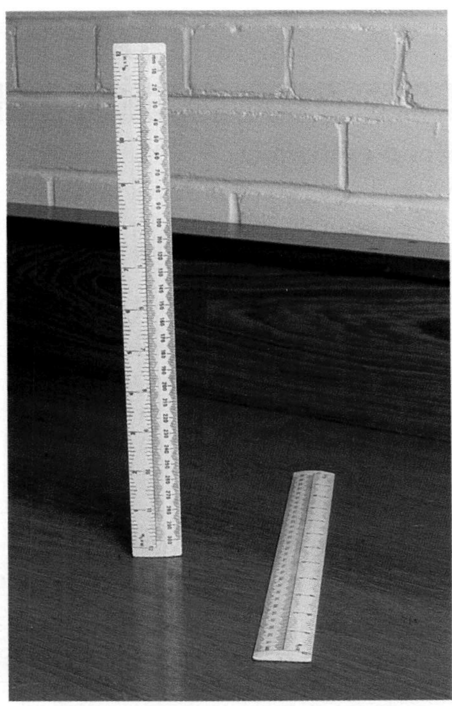

Figure 1.8.14 A ruler is less stable on end than lying flat – the ruler has *more* potential energy standing on end than lying flat.

Will it go or won't it?

The example of nitrogen oxide makes it clear that we must be very careful when talking about the stability of substances. Although nitrogen oxide is kinetically stable with respect to its elements, in the presence of oxygen it is kinetically and energetically unstable with respect to nitrogen dioxide. Exposed to the air, colourless nitrogen oxide therefore rapidly forms brown fumes of nitrogen dioxide.

Chemists often use the value of $\Delta H^{\ominus}_{m}$ for a reaction to predict whether the reaction is likely to happen or not. Exothermic reactions produce products that are energetically more stable than the reactants from which they are formed. This suggests that reactions with a large negative value of $\Delta H^{\ominus}_{m}$ are very likely to happen. Although many spontaneous reactions *are* strongly exothermic, the value of $\Delta H^{\ominus}_{m}$ alone is not sufficient to predict whether a particular reaction is likely to 'go' or not. There are several reasons for this:

1 $\Delta H^{\ominus}_{m}$ tells us about the energetic changes that occur in a reaction, but tells us nothing about the *kinetic* stability of the reactants. The enthalpy change of combustion of petrol is enormous, yet the relative safety of our system of road transport relies on the fact that a mixture of petrol and air is kinetically stable.

2 $\Delta H^{\ominus}_{m}$ tells us about enthalpy changes under *standard conditions*. Actual enthalpy changes under different conditions of temperature, pressure and concentrations are often very different.

3 Other factors apart from enthalpy changes are often concerned with chemical changes. These factors are concerned with the way the system and its surroundings are organised. We shall look at these in section 3.7.

Figure 1.8.15 $\Delta H^{\ominus}_{m}$ for the reaction C(diamond) $\rightarrow$ C(graphite) is -2 kJ mol^{-1}, so energetically these diamonds are destined to become graphite. However, the kinetic stability of diamond means that this is unlikely to happen, at least at room temperature.

Enthalpy changes and bonding

Bond enthalpies

We saw in section 1.4 that bringing atoms or oppositely charged ions together releases energy, and that the separation of atoms or oppositely charged ions requires energy – in other words:

Bond making releases energy while bond breaking requires energy.

When dealing with covalent compounds, the idea of bond energy which we met in section 1.4 is a useful one. The **standard molar enthalpy change of bond dissociation** $\Delta H^{\ominus}_{d,m}$ is the enthalpy change when one mole of bonds of a particular type are broken under standard conditions.

As an example, consider methane, CH_4, which contains four C—H bonds as shown in figure 1.8.16. The bond enthalpy of the C—H bond in methane can be found by considering the change:

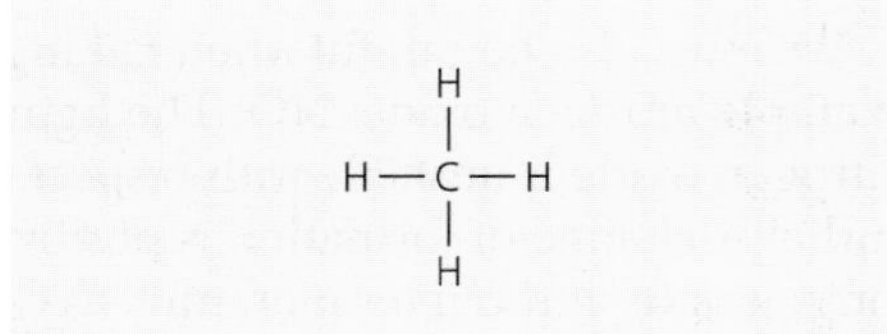

Figure 1.8.16

$$\mathbf{CH_4(g) \rightarrow C(g) + 4H(g)}$$

which is the atomisation of methane. The enthalpy change $\Delta H^{\ominus}_{at,m}$ for this process can be found by applying Hess's law to the series of changes shown in figure 1.8.17.

Figure 1.8.17 shows that $\Delta H^{\ominus}_{at,m}$ for methane is

$$\mathbf{-(-75\ kJ\ mol^{-1}) + 717\ kJ\ mol^{-1} + 4 \times 218\ kJ\ mol^{-1} = 1664\ kJ\ mol^{-1}}$$

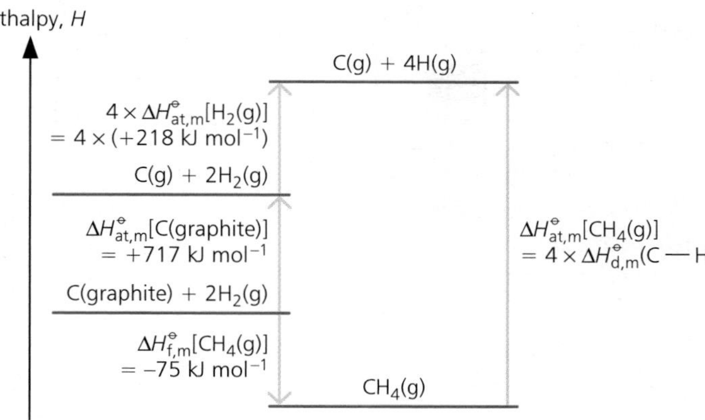

Figure 1.8.17 Calculating the bond enthalpy of the C—H bond in methane

1

Since this process involves the breaking of four C—H bonds, it follows that the C—H bond enthalpy in methane, written as $\Delta H^{\ominus}_{d,m}$ (C—H) is obtained from:

$$\Delta H^{\ominus}_{d,m}(\text{C—H}) = \frac{1664 \text{ kJ mol}^{-1}}{4} = 416 \text{ kJ mol}^{-1}$$

Notice that this is the energy required to break one mole of C—H bonds and form one mole of carbon and one mole of hydrogen. Of course, the calculation assumes that all four C—H bonds in methane are the same, so that we have calculated the *average* C—H bond enthalpy. Evidence from the chemical behaviour of methane supports the idea that all four C—H bonds are equivalent, so this assumption is justified.

This value for the C—H bond enthalpy is very close to the value in table 1.8.1, which is the C—H bond enthalpy average over a number of compounds. (The other bond enthalpies in the table are also averages for a number of compounds.)

Bond	**Bond enthalpy** $\Delta H^{\ominus}_{d,m}$/kJ mol^{-1}	**Bond**	**Bond enthalpy** $\Delta H^{\ominus}_{d,m}$/kJ mol^{-1}
C—H	413	C—N	286
C—O	358	C—F	467
C=O	743	C—Cl	346
C—C	347	C—Br	290
C=C	612	H—O	463
C≡C	838	H—N	388

Table 1.8.1 These bond enthalpies are the average values calculated for a number of polyatomic (many atom) molecules.

The bond enthalpy of a given bond is found to be similar in a wide range of compounds. This suggests that the C—C bond is very similar in (say) ethane, C_2H_6, and in butane, C_4H_{10}. This is quite in order with observations of the chemical properties of substances, which show that compounds with similar bonds behave in similar ways. If the bonds were not very similar, this would not be so.

Because bond enthalpies are so constant, they can be used to estimate the enthalpy changes of formation of substances, as illustrated in the following box. Bond enthalpies also enable us to make predictions about the behaviour of molecules containing different bonds, and also to understand the structure of molecules – we shall see examples of both of these later on in the book.

Enthalpy changes of formation from bond enthalpies

Using the bond enthalpies in table 1.8.1, estimate the molar enthalpy change of formation of methanol vapour from its elements.

To carry out this calculation we also need to know the standard molar enthalpy changes of atomisation of the elements in methanol. These are:

$$\Delta H^{\circ}_{at,m}\,[\mathrm{C(graphite)}] = +717\ \mathrm{kJ\ mol^{-1}}$$

$$\Delta H^{\circ}_{at,m}\,[\mathrm{H_2(g)}] = +218\ \mathrm{kJ\ mol^{-1}}$$

$$\Delta H^{\circ}_{at,m}\,[\mathrm{O_2(g)}] = +249\ \mathrm{kJ\ mol^{-1}}$$

The enthalpy diagram shown in figure 1.8.18 sets out the enthalpy changes in the formation of methanol.

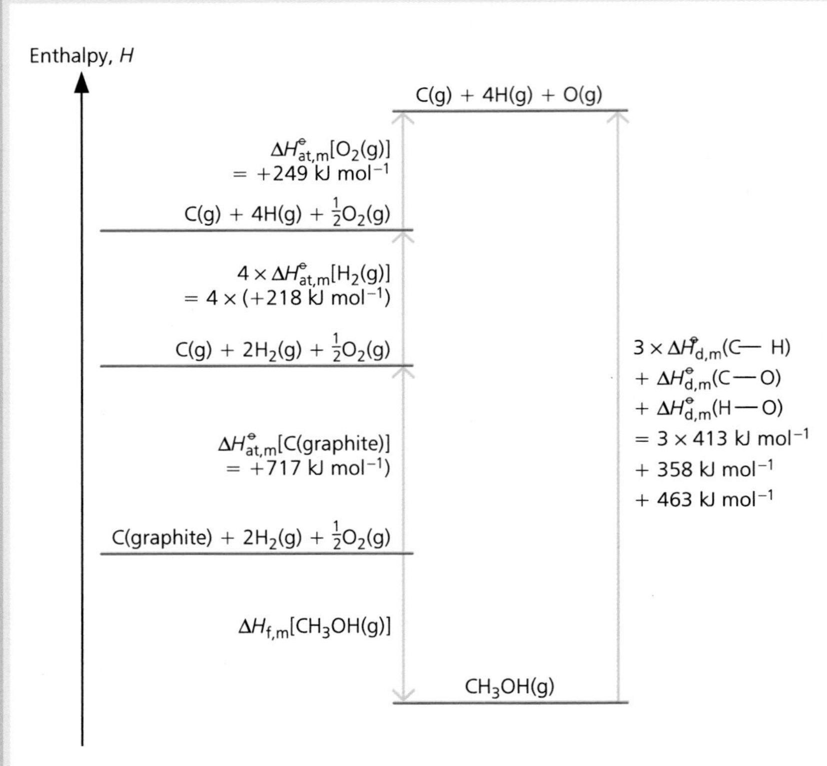

Figure 1.8.18 Enthalpy diagram showing formation of methanol

From the enthalpy diagram, the molar enthalpy change of formation of methanol can be calculated from:

$$\Delta H_{f,m} = \begin{array}{l} +717\ \mathrm{kJ\ mol^{-1}} + 4 \times 218\ \mathrm{kJ\ mol^{-1}} + 249\ \mathrm{kJ\ mol^{-1}} \\ -\ (3 \times 413\ \mathrm{kJ\ mol^{-1}} + 358\ \mathrm{kJ\ mol^{-1}} + 463\ \mathrm{kJ\ mol^{-1}}) \end{array}$$

$$= -222\ \mathrm{kJ\ mol^{-1}}$$

The molar enthalpy change of formation of methanol vapour from its elements is –222 kJ mol^{-1}. (Note that this is not a *standard* enthalpy change, since we are forming methanol in the vapour state, whereas methanol is actually a liquid under standard conditions.)

Forming ionic bonds

We saw in section 1.4 how the formation of ionic bonds involves the transfer of electrons from one atom to another to form oppositely charged ions, and the coming together of these ions to form a lattice. The formation of ions in the gaseous state from elements in the standard state is endothermic, while the formation of the lattice involves a release of energy – the lattice enthalpy. The **standard molar enthalpy change of lattice formation**, $\Delta H^{\circ}_{lat,m}$, is the enthalpy change when one mole of ionic compound is formed from its gaseous ions under standard conditions.

The lattice enthalpies of substances can be calculated from a special type of enthalpy cycle called a **Born–Haber cycle**. All the enthalpy changes in this cycle can be measured, which enables the lattice enthalpy of the compound to be calculated. Figure 1.8.19 shows the Born–Haber cycle for sodium chloride.

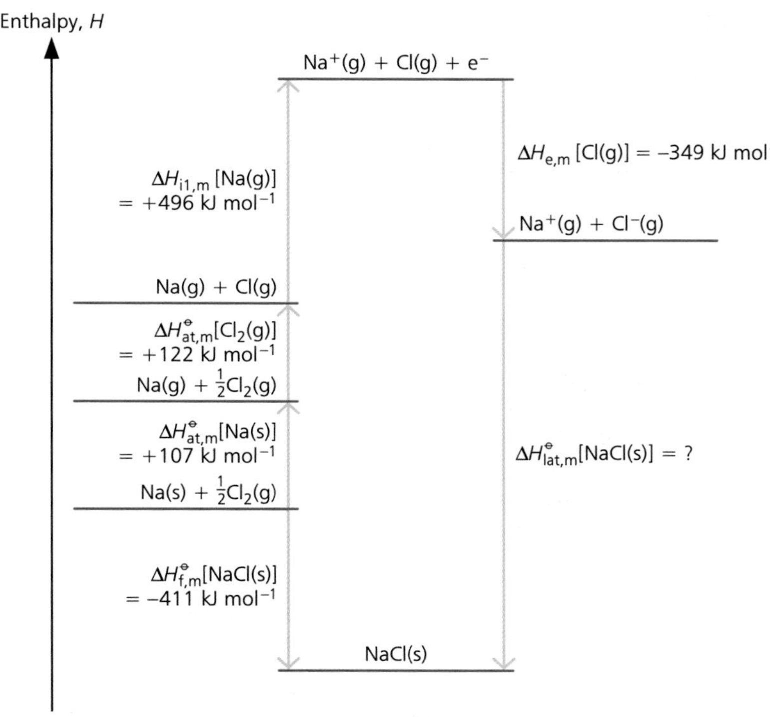

Figure 1.8.19 The Born–Haber cycle for sodium chloride

The terms in the Born–Haber cycle are as follows:

$\Delta H^{\circ}_{f,m}$ **[NaCl(s)]**	**Enthalpy change of formation of sodium chloride, from measurements obtained using a bomb calorimeter**
$\Delta H^{\circ}_{at,m}$ **[Na(s)]**	**Enthalpy change of atomisation of sodium, calculated from enthalpy changes of fusion and vaporisation and the specific heat capacity of sodium**
$\Delta H^{\circ}_{at,m}$ **[Cl_2(g)]**	**Enthalpy change of atomisation of chlorine, found by spectroscopic measurements**
$\Delta H_{i1,m}$ **[Na(g)]**	**First ionisation energy of sodium, found by spectroscopic measurements**
$\Delta H_{e,m}$ **[Cl(g)]**	**First electron affinity of chlorine, found by methods similar to those used by Franck and Hertz to measure ionisation and excitation energies (see page 37)**

Given measured quantities for all these terms, the lattice enthalpy of sodium chloride can be calculated:

$$\Delta H^\ominus_{\text{lat,m}}\,[\text{NaCl(s)}] = \begin{aligned}&-(-349\text{ kJ mol}^{-1}) - 496\text{ kJ mol}^{-1} - 122\text{ kJ mol}^{-1}\\ &- 107\text{ kJ mol}^{-1} - 411\text{ kJ mol}^{-1}\end{aligned}$$

$$= -787\text{ kJ mol}^{-1}$$

This figure for the lattice enthalpy of sodium chloride gives us some idea of the size of the attractive forces between the ions in sodium chloride. Notice that it is of the same order of magnitude as the bond enthalpies in table 1.8.1. Another example of a Born-Haber cycle is shown here, this time for the formation of magnesium chloride. Table 1.8.2 shows some lattice enthalpies for other compounds, also calculated from Born–Haber cycles.

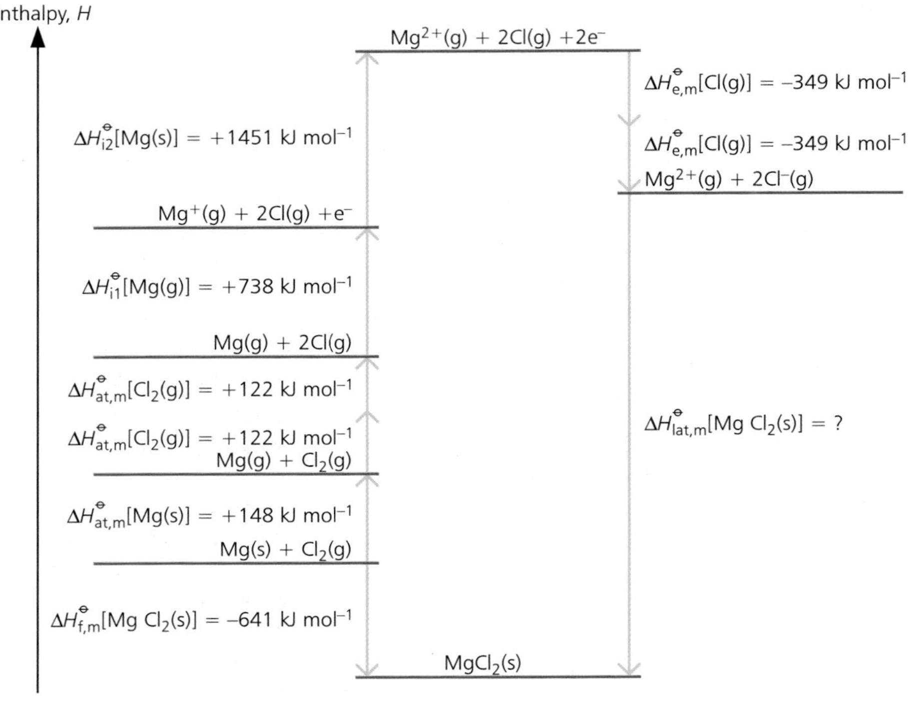

Figure 1.8.20 The Born–Haber cycle for magnesium chloride gives a value for its lattice enthalpy as 2524 kJ mol^{-1}.

Compound	Lattice enthalpy $\Delta H^\ominus_{\text{lat,m}}$/kJ mol^{-1}	Compound	Lattice enthalpy $\Delta H^\ominus_{\text{lat,m}}$/kJ mol^{-1}
NaF	918	MgF_2	2957
NaCl	780	CaF_2	2630
NaBr	742	MgO	3791
NaI	705	MgS	3299

Table 1.8.2 Lattice enthalpies for a range of compounds

What affects the size of the lattice enthalpy?

Looking at the lattice enthalpies in table 1.8.2, we can see that the lattice enthalpy of a compound is affected by both the charge and the size of the ions. The effect of the size of the ions is shown clearly by the trend in the sodium halides, and in the case of the fluorides of magnesium and calcium, and the oxide and sulphide of magnesium. In all these cases the lattice enthalpy decreases as the size of the ions increases. Notice too that ions with greater charge (Mg^{2+}, Ca^{2+}) have a larger lattice enthalpy.

Why is lattice enthalpy affected in this way? The size of the force of attraction F of one ion on another is given by Coulomb's law:

$$F = k\frac{q_1 q_2}{r^2}$$ **where q_1 and q_2 are the charges on the ions and r is the distance between them.**

This relationship shows us that increasing the charge on either ion increases the attractive forces between two oppositely charged ions in the lattice. Decreasing the size of one or both ions decreases the distance between them and so increases the attractive force.

A closer look at lattice enthalpies

By using Coulomb's law in conjunction with knowledge about the structure of ionic lattices (see section 1.6), it is possible to calculate theoretical lattice enthalpies for ionic compounds. When we do this and compare these with the lattice enthalpies obtained from Born–Haber cycles, we get the results shown in table 1.8.3.

In the case of the sodium halides, notice the similarity between the lattice enthalpies obtained in the different ways, agreeing within 3% or better. In the case of the silver halides, agreement is not so good – why?

The theoretical lattice enthalpies of the silver halides are considerably less than those calculated from Born–Haber cycles. This suggests that the silver halides are more stable than a purely ionic model indicates – there

Compound	Lattice enthalpy/kJ mol^{-1}		Compound	Lattice enthalpy/kJ mol^{-1}	
	Born–Haber	Theoretical		Born–Haber	Theoretical
NaF	918	912	AgF	958	920
NaCl	780	770	AgCl	905	833
NaBr	742	735	AgBr	891	816
NaI	705	687	AgI	889	778

Table 1.8.3 Lattice enthalpies calculated from Born–Haber cycles and from theoretical models of attraction and repulsion between ions in crystal lattices

must be a considerable degree of covalent character in them. Our knowledge of the effect of relative electronegativities on the nature of bonds supports this idea. (Electronegativities are shown in figure 1.4.21, page 60.) When the difference in electronegativity between the different ions in a crystal is high, the ionic model works well, and there is good agreement between two values of lattice enthalpy. When the difference in electronegativity is smaller, the bonding in the crystal has a considerable degree of covalent character. The melting points of the silver halides are about 20% lower than those of the sodium halides, in support of the idea that they have a greater covalent character.

Using magnesium oxide

Magnesium oxide is a compound with very high lattice energy. As a result it has a very high melting point. Because of this property magnesium oxide is often used as a refractory lining. This means it is used as the lining material in kilns, ovens and furnaces when pottery and china are fired (baked). The magnesium oxide withstands the high temperatures needed to fire the china, pots and the glazes used to decorate them without either melting or undergoing chemical reactions.

Figure 1.8.21 The high lattice energy of magnesium oxide makes it ideal for use as a refractory lining in kilns like this.

SUMMARY

- The transfer of energy is a vital factor in chemical reactions.
- When a reacting system loses energy (heat leaves the system) the reaction is said to be **exothermic**. When a reacting system gains energy (heat enters the system) the reaction is said to be **endothermic**.
- **Enthalpy** H is the total energy content of a system at constant pressure. ΔH is the **enthalpy change** when a system at constant pressure gives out or takes in energy.
- The **standard molar enthalpy change of reaction** $\Delta H^{\ominus}_{m}$ is the enthalpy change when molar quantities of reactants react together under standard conditions. For an exothermic reaction $\Delta H^{\ominus}_{m}$ is **negative**, for an endothermic reaction it is **positive**.
- The standard conditions for thermochemical measurements are **10^5 Pa pressure** and **298 K (25 °C)**.
- **Calorimeters** of various types and levels of sophistication are used to measure energy changes during reactions.
- The **standard molar enthalpy change of combustion** $\Delta H^{\ominus}_{c,m}$ is the enthalpy change when one mole of a substance is completely burned in oxygen under standard conditions.
- The **standard molar enthalpy change of formation** $\Delta H^{\ominus}_{f,m}$ is the enthalpy change when one mole of the compound is formed from its elements under standard conditions.
- The **standard molar enthalpy change of atomisation** $\Delta H^{\ominus}_{at,m}$ of an element is the enthalpy change when one mole of its atoms in the gaseous state is formed from the element under standard conditions.
- The **standard molar enthalpy change of fusion** $\Delta H^{\ominus}_{fus,m}$ of a substance is the enthalpy change when one mole of the solid is converted to one mole of the liquid at its melting point and at standard pressure.
- The **standard molar enthalpy change of vaporisation** $\Delta H^{\ominus}_{vap,m}$ of a substance is the enthalpy change when one mole of the liquid is converted to one mole of the gas at its boiling point and at standard pressure.
- **Hess's law of heat summation** states that for any reaction that may be written as a series of steps, the standard molar enthalpy change for the reaction is the same as the sum of the standard enthalpy changes of all the steps, regardless of the route by which the reaction occurs.
- A substance is stable if it tends neither to react nor to decompose spontaneously. There are different types of stability. The enthalpy change of formation of a compound gives a measure of its **energetic stability** (**thermodynamic stability**) relative to its elements. If $\Delta H^{\ominus}_{f,m}$ is a negative number, energy is given out in the formation of the compound and it is likely to be more stable than its elements. However, even if a substance is stable with respect to its elements, it may be unstable with respect to some other substance, to which it may decompose.
- Some compounds are energetically unstable but do not change under normal conditions because they are **kinetically stable** – they decompose extremely slowly.
- Chemical reactions involve **making and breaking bonds**. Breaking chemical bonds is an endothermic process, whilst bond-making is exothermic.

- Each type of covalent bond has a particular amount of energy associated with it, known as the **standard molar enthalpy change of bond dissociation** $\Delta H^{\ominus}_{d,m}$. This is the enthalpy change when one mole of a particular bond is broken to form uncharged particles under standard conditions. The equivalent quantity for ionic substances is the **standard molar enthalpy change of lattice formation** $\Delta H^{\ominus}_{lat,m}$, which is the enthalpy change when one mole of ionic compound is formed from its gaseous ions under standard conditions.
- Born–Haber cycles may be used to consider the energy relationships when ionic compounds are formed.

QUESTIONS

1 To lose 0.5 kg of fatty tissue, an average person needs to consume about 3.2×10^6 kJ of food energy less than needed. You decide to lose weight simply by cutting out sugar, keeping your exercise level and other aspects of your diet exactly the same. How much less sugar must you eat in order to lose 2 kg of fatty tissue, if 100 g of sugar provides 1.68×10^6 kJ of food energy?

2 A coffee-cup calorimeter contains 55.0 cm³ of a dilute solution of copper(II) sulphate at a temperature of 22.8 °C. A small amount of zinc powder also at 22.8 °C is added to the solution. Copper metal is formed, and the temperature of the solution rises to 32.3 °C. The copper is collected, dried and weighed, when it is found to have a mass of 0.324 g.

a Calculate the total amount of energy released in this reaction, ignoring the heat capacity of the zinc and of the calorimeter. (Take the specific heat capacity of the solution as 4.2 J g^{-1} K^{-1}.)

b Calculate the enthalpy change for this reaction per mole of copper formed.

3 Hydrogen chloride, HCl, can be made by heating potassium chloride with concentrated sulphuric acid:

$$H_2SO_4(l) + 2KCl(s) \rightarrow 2HCl(g) + K_2SO_4(s)$$

Given the information:

$$H_2SO_4(l) + 2KOH(s) \rightarrow K_2SO_4(s) + 2H_2O(l) \quad \Delta H^{\ominus}_m = -342 \text{ kJ}$$

$$HCl(g) + KOH(s) \rightarrow KCl(s) + H_2O(l) \quad \Delta H^{\ominus}_m = -204 \text{ kJ}$$

calculate the standard molar enthalpy change for the formation of hydrogen chloride from potassium chloride and concentrated sulphuric acid.

4 The standard molar enthalpy change of combustion of propanoic acid is –1527.2 kJ mol^{-1}. Given that the standard molar enthalpy change of formation of water is –285.8 kJ mol^{-1} and that of carbon dioxide is –393.5 kJ mol^{-1}, calculate the standard molar enthalpy change of formation of propanoic acid.

5 Use the following data to construct Born–Haber cycles to show why calcium and chlorine react together to form $CaCl_2$ rather than CaCl or $CaCl_3$:

$\Delta H^{\ominus}_{at}$ [Ca(s)] = +178.2 kJ mol^{-1}

Ionisation energies of Ca(g)/kJ mol^{-1}: 1st = 590, 2nd = 1145, 3rd = 4912

$\Delta H^{\ominus}_{at}$ [$\frac{1}{2}Cl_2(g)$] = +121.7 kJ mol^{-1}

Electron affinity of Cl(g)/kJ mol^{-1} = –349 kJ mol^{-1}

Theoretical lattice enthalpy changes:

$Ca^+(g) + Cl^-(g) \rightarrow CaCl(s)$	–719 kJ mol^{-1}
$Ca^{2+}(g) + 2Cl^-(g) \rightarrow CaCl_2(s)$	–2218 kJ mol^{-1}
$Ca^{3+}(g) + 3Cl^-(g) \rightarrow CaCl_3(s)$	–4650 kJ mol^{-1}

Developing Key Skills

Plan a four minute slot for a science programme on children's television. You have to use chemistry to explain the release of energy in animals, both warm blooded and cold blooded, and the way people react to exercise. You need to plan for lots of visual impact and interest, yet at the same time ensure that plenty of chemistry comes across.

[Key Skills opportunities: C]

1 Questions

1 a The atomic masses of some elements are shown in the table below:

Element	Relative atomic mass
hydrogen, ^{1}H	1.0078
carbon, ^{12}C	12.0000
nitrogen, ^{14}N	14.0031
oxygen, ^{16}O	15.9949

i Using the data in the table, explain why nitrogen monoxide (NO) and ethene (C_2H_4) can be distinguished with high resolution mass spectrometry.

ii High-resolution mass spectrometers are included on planetary space probes. Such a probe recorded gases with masses of 27.0109 and 31.0421.
Identify these two gases, which contain elements from the group shown in the table:
Gas **A**, mass 27.0109
Gas **B**, mass 31.0421 **(4 marks)**

b It is possible to identify that a given compound contains either chlorine or bromine from the presence of M and (M+2) peaks in the mass spectrum of the compound.
State what species causes the (M+2) peaks in each of the following compounds:
i C_3H_7Cl
ii C_2H_5Br **(2 marks)**

c Describe the differences in relative heights of M and (M+2) peaks in the mass spectra of chlorine- and bromine-containing compounds. **(2 marks)**

(Total 8 marks)

(OCR specimen)

2 When an electrical discharge passes through gaseous hydrogen at low pressure, electromagnetic radiation is emitted.

a Explain what processes within a hydrogen atom cause radiation to be emitted. **(3 marks)**

b If the radiation in **a** is passed through a spectrometer, several series of converging lines are observed.
i Explain why there are several series of lines.
ii Why does each series of lines converge? **(2 marks)**

c The convergence limit of the Lyman series of lines occurs at a wavelength of 1.00×10^{-7} m
i What does the limit represent?
ii Calculate the energy, in kJ mol^{-1}, of the convergence limit.
($c = 3.00 \times 10^{8}$ m s^{-1}; $L = 6.02 \times 10^{23}$, $h = 6.34 \times 10^{-34}$ J s) **(4 marks)**

d State one use of *flame* emission spectroscopy. **(1 mark)**

(Total 10 marks)

(OCR specimen)

3 Bond enthalpies can provide information about the energy changes that accompany a chemical reaction.

a What do you understand by the term *bond enthalpy*? **(2 marks)**

b The table below shows some average bond enthalpies.

Bond	Average bond enthalpy/kJ mol^{-1}
C–C	350
C=C	610
H–H	436
C–H	410

i Use this information to calculate the enthalpy change for the process:

$$\mathrm{H_2C{=}CH_2} + \mathrm{H{-}H} \longrightarrow \mathrm{H_3C{-}CH_3}$$

(3 marks)

ii The enthalpy change of this reaction can be found by experiment to be –136 kJ mol^{-1}. Explain why this value is different from that determined above from average bond enthalpies. **(1 mark)**

c Sketch a fully labelled enthalpy profile diagram for this reaction. **(2 marks)**

(Total 8 marks)

(OCR specimen)

4 The hydrocarbon heptane, C_7H_{16}, is one of the hydrocarbons present in petrol. Its combustion reaction with oxygen provides some of the energy to propel a vehicle.

a i Define the term *standard enthalpy change of combustion*.
ii State the temperature and pressure that are conventionally chosen for quoting standard enthalpy changes. **(3 marks)**

b Use the data below to calculate the standard enthalpy change of combustion of heptane.

Compound	$\Delta H^{\ominus}_{f}$ / kJ mol^{-1}
$C_7H_{16}(l)$	–224.4
$CO_2(g)$	–393.5
$H_2O(l)$	–285.9

$$C_7H_{16}(l) + 11O_2(g) \rightarrow 7CO_2(g) + 8H_2O(l)$$

(3 marks)

c Combustion in a car engine also produces polluting gases, such as carbon monoxide, unburnt hydrocarbons and nitrogen monoxide, NO.

Explain how CO and NO are produced in a car engine. **(2 marks)**

d The catalytic converter removes much of this pollution in a series of reactions.

i Balance the equation below showing the reaction to remove carbon monoxide and nitrogen monoxide.

$$CO(g) + NO(g) \rightarrow CO_2(g) + N_2(g)$$

ii The equation above is an example of a redox reaction. Identify the element being reduced and deduce the change in its oxidation number. **(3 marks)**

(Total 11 marks)

(OCR specimen)

5 The lattice enthalpy of rubidium chloride, RbCl, can be determined indirectly using a Born–Haber cycle.

a Use the data in the table below to construct the cycle and to determine a value for the lattice enthalpy of rubidium chloride.

Enthalpy change	Energy/kJ mol^{-1}
formation of rubidium chloride	–435
atomisation of rubidium	+81
atomisation of chlorine	+122
1st ionisation energy of rubidium	+403
1st electron affinity of chlorine	–349

(6 marks)

b Explain why the lattice enthalpy of lithium chloride, LiCl, is more exothermic than that of rubidium chloride. **(2 marks)**

(Total 8 marks)

(OCR specimen)

6 Potassium was discovered and named in 1807 by the British chemist Sir Humphrey Davy. The mass spectrum of a sample of potassium is shown below:

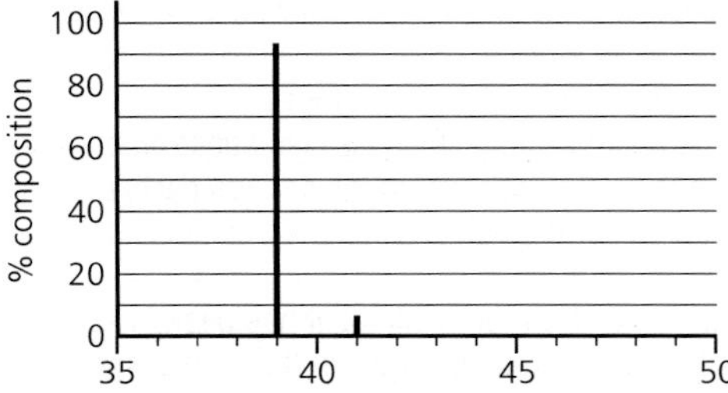

a i Use this mass spectrum to copy and complete the table below to show the percentage composition and atomic structure of each potassium isotope in the sample.

Isotope	Percentage composition	Protons	Neutrons	Electrons
^{39}K				
^{41}K				

ii The relative atomic mass of the potassium sample can be determined from its mass spectrum. Explain what you understand by the term *relative atomic mass*.

iii Calculate the relative atomic mass of the potassium sample. **(8 marks)**

b Complete the orbital electronic configuration of a potassium atom. **(1 mark)**

c The first and second ionisation energies of potassium are shown in the table below:

Ionisation	1st	2nd
Ionisation energy/kJmol^{-1}	419	3051

i Explain what you understand by the term *first ionisation energy* of potassium.

ii Why is there a **large** difference between the values for the first and the second ionisation energies of potassium? **(5 marks)**

(Total 14 marks)

(OCR specimen)

7 Lead compounds are extensively used to provide the colour in paints and pigments.

a 'White lead', used for over 2000 years as a white pigment, is based upon lead carbonate. Analysis shows that lead carbonate has the following percentage composition by mass:
Pb, 77.5%; C, 4.5%; O, 18.0%.
Calculate the empirical formula of lead carbonate.
[Ar: C, 12.0; O, 16.0; Pb, 207.0] **(3 marks)**

b 'Red lead' is the pigment in paint used as a protective coating for structural iron and steel. It is based upon the lead oxide Pb_3O_4, a scarlet powder formed by oxidising lead(II) oxide with oxygen.

i Balance the equation for the oxidation of PbO.

$$PbO(s) + O_2(g) \rightarrow Pb_3O_4(s)$$

ii What is the molar mass of Pb_3O_4? [Ar: O, 16.0; Pb, 207.0]

iii Calculate the mass of Pb_3O_4 that could be formed from 0.300 mol of PbO. **(4 marks)**

(Total 7 marks)

(OCR specimen)

8 a Showing outer electron shells only, draw 'dot-and-cross' diagrams to show the bonding in ammonia and water. **(2 marks)**

b Draw diagrams to illustrate the shape of a molecule of each of these compounds. State the size of the bond angles on each diagram and name each shape. **(6 marks)**

c On mixing with water, ammonia forms an alkaline solution containing the ammonium ion, NH_4^+:

$$NH_3(g) + H_2O(l) \rightarrow NH_4^+(aq) + H_2O(l)$$

i The ammonium ion shows *dative covalent (co-ordinate)* bonding. Explain what is meant by this term.

ii Draw a 'dot-and-cross' diagram of the ammonium ion. Label on your diagram a dative covalent bond. **(6 marks)**

(Total 14 marks)

(OCR specimen)

9 Wines often contain a small amount of sulphur dioxide that is added as a preservative. The amount of sulphur dioxide added needs to be carefully calculated; too little and the wine readily goes bad; too much and the wine tastes of sulphur dioxide.
The sulphur dioxide content of a wine can be found using its reaction with aqueous iodine.

$$SO_2(aq) + I_2(aq) + 2H_2O(l) \rightarrow SO_4^{2-}(aq) + 21^-(aq) + 4H^+(aq)$$

a i State the oxidation number of sulphur in SO_2 and in SO_4^{2-}.

ii State, with a reason, whether sulphur is oxidised or reduced in the conversion of SO_2 into SO_4^{2-}. **(3 marks)**

b The sulphur dioxide content of a wine can be found by titration. An analyst found that the sulphur dioxide in 50.0 cm^3 of white wine reacted with exactly 16.4 cm^3 of 0.0100 mol dm^{-3} aqueous iodine.

i How many moles of iodine, I_2, did the analyst use in the titration?

ii How many moles of sulphur dioxide were in the 50.0 cm^3 of wine? **(2 marks)**

iii What was the concentration of sulphur dioxide in the wine in mol dm^{-3} and in g dm^{-3}? **(3 marks)**

c The generally accepted maximum concentration of sulphur dioxide in wine is 0.25 g dm^{-3}. A concentration of less than 0.01 g dm^{-3} is insufficient to preserve the wine.
Comment on the effectiveness of the sulphur dioxide in the wine analysed in **b**. **(1 mark)**

(Total 9 marks)

(OCR specimen)

10 a What is a covalent bond? **(1 mark)**

b Draw diagrams to show the shapes of the following molecules and in each case show the value of the bond angle on the diagram.

i $BeCl_2$ **ii** BF_3 **iii** CCl_4 **iv** SF_6 **(8 marks)**

c Explain why the shape of NF_3 is not the same as the shape of BF_3. **(3 marks)**

(Total 12 marks)

(AQA specimen)

11 a Define the term *standard enthalpy of combustion*. **(2 marks)**

b i Write an equation for the complete combustion of ethane, C_2H_6.

ii Use the standard enthalpies of formation given below to calculate the standard enthalpy of combustion of ethane.

Formula and state of compound	$C_2H_6(g)$	$CO_2(g)$	$H_2O(l)$
Standard enthalpy of formation (at 298 K) /kJ mol^{-1}	–85	–394	–286

(4 marks)

c A vessel and its contents of total heat capacity 120 J K^{-1} were heated using a methane burner. Calculate the maximum theoretical temperature rise when 0.10 g of methane was completely burned. The standard enthalpy of combustion of methane is –890 kJ mol^{-1}. **(4 marks)**

(Total 10 marks)

(AQA specimen)

12 a Define the term *relative molecular mass*. **(2 marks)**

b The mass of one atom of ^{12}C is 1.993×10^{-23}g. Use this mass to calculate a value for the Avogadro constant (*L*) showing your working. **(1 mark)**

c A 153 kg sample of ammonia gas, NH_3, was compressed at 800 K into a cylinder of volume 3.00 m^3.

i Calculate the pressure in the cylinder assuming that the ammonia remained as a gas.

ii Calculate the pressure in the cylinder when the temperature is raised to 1000 K. **(5 marks)**

(Total 8 marks)

(AQA 1999)

13 The graph below shows the trend in first ionisation energy from oxygen to magnesium.

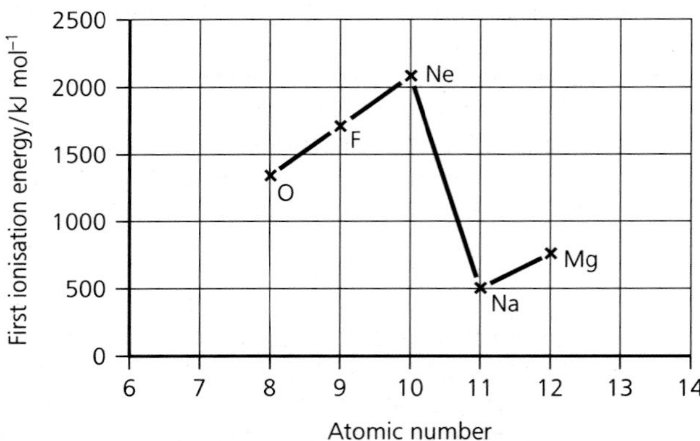

a Copy the graph and, using crosses, mark on it the first ionisation energies of nitrogen and of aluminium. Label each of your crosses with the symbol for the element. **(2 marks)**

b Explain why the first ionisation energy of neon is greater than that of sodium. **(2 marks)**

c Of the elements neon, sodium and magnesium, predict which one has the largest second ionisation energy. Explain your answer. **(3 marks)**

d Published values of electronegativity are available for oxygen, fluorine, sodium and magnesium but not for neon.

i Explain why a value of electronegativity is not available for neon.

ii Of the elements oxygen, fluorine, sodium and magnesium, predict which one has the smallest electronegativity value. **(2 marks)**

(Total 9 marks)

(AQA 1999)

14 a Describe, in terms of charge and mass, the properties of protons, neutrons and electrons. Explain fully how these particles are arranged in an atom of ^{14}N. **(6 marks)**

b Account for the existence of isotopes. **(2 marks)**

c Isotopes can be separated in a mass spectrometer. Show how this is possible by describing the various parts of a mass spectrometer and by discussing the principles of operation of each part. **(14 marks)**

d The mass spectrum of an element has peaks with relative intensity and *m/z* values shown in the table below.

m/z	80	82	83	84	86
Relative intensity	1	5	5	25	8

Identify this element and calculate its accurate relative atomic mass. **(4 marks)**

e The mass spectrum of a compound has a molecular ion peak at $m/z = 168$. Elemental analysis shows it to contain 42.9% carbon, 2.4% hydrogen and 16.7% nitrogen by mass. The remainder is oxygen.
Calculate the empirical and molecular formulae of this compound. **(4 marks)**

(Total 30 marks)

(AQA 1999)

15 a State the type of bonding in a crystal of potassium bromide. Write an equation to show what happens when potassium bromide is dissolved in water and predict the pH of the resulting solution. **(3 marks)**

b When iodine reacts directly with fluorine, a compound containing 57.2% by mass of iodine is formed.

i Determine the empirical formula of this compound.

ii The empirical formula of this compound is the same as the molecular formula. Write a balanced equation for the formation of this compound. **(4 marks)**

c i Sketch a diagram to show the shape of a BrF_3 molecule. Show on your sketch any lone pairs of electrons in the outermost shell of bromine and name the shape.

ii BrF_3 reacts with an equimolar amount of potassium fluoride to form an ionic compound which contains potassium ions. Give the formula of the negative ion produced, sketch its shape, show any lone pair(s) and indicate the value of the bond angle. **(6 marks)**

(Total 13 marks)

(AQA 1999)

16 Below are some standard enthalpy changes including the standard enthalpy of combustion of nitroglycerine, $C_3H_5N_3O_9$

$\frac{1}{2}N_2(g) + O_2(g) \rightarrow NO_2(g)$ $\Delta H^\circ_f = +34\,kJ\,mol^{-1}$

$C(s) + O_2(g) \rightarrow CO_2(g)$ $\Delta H^\circ_f = -394\,kJ\,mol^{-1}$

$H_2(g) + \frac{1}{2}O_2(g) \rightarrow H_2O(g)$ $\Delta H^\circ_f = -242\,kJ\,mol^{-1}$

$C_3H_5N_3O_9(l) + \frac{11}{4}O_2(g) \rightarrow 3CO_2(g) + \frac{5}{2}H_2O(g) + 3NO_2(g)$ $\Delta H^\circ_c = -1540\,kJ\,mol^{-1}$

a Standard enthalpy of formation is defined using the term *standard state*. What does the term *standard state* mean? **(2 marks)**

b Use the standard enthalpy changes given above to calculate the standard enthalpy of formation of nitroglycerine. **(4 marks)**

c Calculate the enthalpy change for the following decomposition of nitroglycerine.

$C_3H_5N_3O_9(l) \rightarrow 3CO_2(g) + \frac{5}{2}H_2O(g) + \frac{3}{2}N_2(g) + \frac{1}{4}O_2(g)$

(3 marks)

d Suggest one reason why the reaction in part **c** occurs rather than combustion when a bomb containing nitroglycerine explodes on impact. **(1 mark)**

e An alternative reaction for the combustion of hydrogen, leading to liquid water, is given below.

$H_2(g) + \frac{1}{2}O_2(g) \rightarrow H_2O(l)$ $\Delta H^\circ = -286\,kJ\,mol^{-1}$

Calculate the enthalpy change for the process $H_2O(l) \rightarrow H_2O(g)$ and explain the sign of ΔH in your answer. **(2 marks)**

(Total 12 marks)

(AQA 1999)

17 The boiling temperatures, T_b, of some Group IV and Group V hydrides are given below.

Compound	CH_4	SiH_4	NH_3	PH_3
T_b/K	112	161	240	185

a The polarity of a carbon–hydrogen bond can be shown as $\overset{\delta-}{C}$—$\overset{\delta+}{H}$

i What does the symbol $\delta+$, above the hydrogen atom, signify?

ii Explain briefly, in terms of its shape, why a CH_4 molecule has no overall polarity. **(3 marks)**

b Name the type of intermolecular forces which exist between CH_4 molecules in liquid methane. **(1 mark)**

c Explain why the boiling temperature of PH_3 is greater than that of CH_4. **(3 marks)**

d Explain why the boiling temperature of NH_3 is greater than that of PH_3. **(2 marks)**

e Sketch a diagram to show the shape of a molecule of NH_3 and indicate on your diagram how this molecule is attracted to another NH_3 molecule in liquid ammonia. **(3 marks)**

f Suggest why the strength of the C—H bond in CH_4 is greater than that of the Si—H bond in SiH_4. State the relationship, if any, between the strength of the covalent bond in CH_4 and the boiling temperature of CH_4. **(2 marks)**

(Total 14 marks)

(AQA 1999)

18 a Give the symbol, including mass number and atomic number, for the isotope which has a mass number of 34 and which has 18 neutrons in each nucleus. **(2 marks)**

b Give the electronic configuration of the F^- ion in terms of levels and sub-levels. **(1 mark)**

c Give a reason why it is unlikely that an F^- ion would reach the detector in a mass spectrometer. **(1 mark)**

d Some data obtained from the mass spectrum of a sample of carbon are given below.

Ion	$^{12}C^+$	$^{13}C^+$
Absolute mass of one ion/g	1.993×10^{-23}	2.158×10^{-23}
Relative abundance/%	98.9	1.1

Use these data to calculate a value for the mass of one neutron, the relative atomic mass of ^{13}C and the relative atomic mass of carbon in the sample. You may neglect the mass of an electron. **(6 marks)**

(Total 10 marks)

(AQA 1999)

19 a Sodium chloride is a compound which is almost completely ionic.

i Draw a Born–Haber cycle and use it to explain why the formation of solid sodium chloride is energetically favourable.

ii Explain why sodium chloride, unlike some metal chlorides, is predominantly ionic. **(10 marks)**

b Saturated solutions of potassium iodate(VII), which is not very soluble in water, contain the following equilibrium when the solution is in contact with excess solid:

$$KIO_4(s) \rightleftharpoons K^+(aq) + IO_4^-(aq)$$

Potassium iodate(VII) is an oxidising agent, which will oxidise iodide ions to iodine in acidic solution, being reduced to iodine in the process.

i Derive the equation for the reaction between iodate(VII) ions and iodide ions in acidic solution. **(2 marks)**

ii A saturated solution of KIO_4 at 10 °C was analysed by placing a 25.0 cm^3 sample into a conical flask, acidifying with dilute sulphuric acid, and adding an excess of potassium iodide. The liberated iodine was titrated with 0.100 $mol\,dm^{-3}$ sodium thiosulphate solution, 30.0 cm^3 being required.
Determine the concentration in $mol\,dm^{-3}$ of the potassium iodate(VII) solution. **(4 marks)**

iii If a potassium salt is added to a saturated solution of potassium iodate(VII) which is then treated as above, the titre with sodium thiosulphate solution is much lower.
Explain why this is so, and suggest what you would see as the potassium salt is added. **(3 marks)**

(Total 19 marks)

(Edexcel 1999)

2 THE PERIODIC TABLE

Introduction

Chemistry is a potentially explosive science. When chemicals are mixed together the results may range from no apparent reaction at all to a massively exothermic reaction in which energy is released as light, heat and sound.

Over the years, many chemists have suffered from their inability to predict the outcome of mixing chemicals. Almost two centuries ago the young Michael Faraday got his first chance to work at the Royal Institution when Sir Humphrey Davy injured his eyes whilst carrying out an experiment on the dangerously unstable substance nitrogen trichloride, and could neither read nor write. He called on the young Faraday to make notes of his work for him. Years later Faraday himself was to complain in his journal of eye injuries from explosions, feeling fortunate that none had caused permanent damage.

Modern chemists are much more aware of safety, with goggles, screens and gauntlets to protect eyes and hands and fume hoods to prevent the inhalation of dangerous substances. But more than this, chemists now have the tools to predict with great accuracy how different chemicals will react with one another. Central to this understanding is the periodic table.

The chemical behaviour of an element is determined by a combination of factors, one of the most influential of these being the electronic configuration of its atoms. It is only during the twentieth century that scientists developed an understanding of the electronic structure of the atom, but earlier chemists struggled to make sense of the chemical properties of substances long before this knowledge was available.

There are clear patterns in the chemical properties of the elements that enabled chemists to arrange them in the groups which made up the early versions of the periodic table. As knowledge of the electronic structure of atoms grew and the number of known elements increased, the version of the periodic table we know today emerged.

The modern periodic table is arranged in vertical **groups** and horizontal **periods**, in which the elements appear in order of their **atomic number**. The elements in each group or period have particular characteristics and display trends in their chemical and physical behaviours. These characteristics and trends can be explained in terms of the atomic numbers of the elements concerned. The periodic table provides chemists with valuable information about what is likely to happen when particular chemicals react – which makes chemistry an altogether more predicable, and safer, science than it was in the days of Davy, Faraday and their fellow workers.

Figure 1 Most chemical plants work day and night. This is not only to increase the amount of product manufactured, but is also a practical move to maintain reacting mixtures and vessels at constant temperatures and pressures.

FOCUS WHOSE IDEA WAS IT ANYWAY?

The history of the development of the periodic table is full of good ideas which were ignored. The basis of the work started as far back as 1799 when Joseph Proust showed that the proportions by mass of the elements in the compounds he investigated were constant for each different substance. A year later John Dalton proposed that elements were made up of atoms, and that each element is made up of one particular type of atom which differs from all the others by its mass. Unfortunately, Dalton wasn't very good at actually working out what these different atomic masses were, so other scientists were not entirely convinced by his work. This was a shame, because in 1928 Jöns Jacob Berzelius published a list of atomic masses which were quite accurate, but after Dalton's earlier efforts no-one was interested.

In 1829 Johann Döbereiner showed that many of the known elements could be arranged in groups of three, with all the members of the group showing similar properties. He called these groups **triads**. This was the first time elements had been grouped in this way. Döbereiner's ideas were picked up by a number of other scientists who continued working to group the elements. In 1862 Alexandre-Emile Beguyer de Chancourtois developed an early version of the periodic table, showing similarities between every eighth element. Unfortunately, when his article was published, the diagram was not printed. Without this visual element the periodicity of the elements was not clear. As a result his ground-breaking work was largely ignored!

One year later John Newlands announced his **law of octaves**. This involved arranging elements in order of their atomic masses in groups of eight. Again here was someone who had produced a basic periodic table, but Newlands' work was dismissed because it contained some major flaws. He assumed all the elements had been discovered, even though new ones were turning up each year. He put two elements in the same place several times to make things fit and even put dissimilar elements like copper, lithium and sodium together. As a result of these errors the scientific community was unimpressed with the whole idea. It was even suggested that arranging the elements alphabetically would probably produce as many similarities as Newton's carefully arranged octaves.

Finally, in 1869/70 Dmitri Mendeleev and Julius Meyer published clear representations of periodicity in the elements. Meyer plotted various physical properties of the known elements against their atomic masses and produced curves that demonstrated the periodic relationships. However, Mendeleev published first, and scored a major publicity coup by publishing a table with gaps in it for as yet undiscovered elements. Based on his table he predicted the properties of these missing elements. Then in 1895 and 1896 gallium and germanium were discovered, two of the elements predicted by Mendeleev. They showed exactly the properties he had deduced from his periodic table. This evidence convinced any remaining sceptics of the periodicity of the chemical elements based on their atomic mass and made sure it was the name of Mendeleev which became associated with the modern periodic table.

Figure 1 One example of Döbereiner's triads is the group of chlorine, iodine and bromine – elements we still recognise as related today.

Figure 2 Dmitri Mendeleev – the father of the modern periodic table.

2.1 Patterns of matter

The modern periodic table

From Mendeleev to the present day

Mendeleev arranged the known elements of his time in a table in accordance with his hypothesised **periodic law**. This stated that:

> **The properties of the elements are a periodic function of their relative atomic masses.**

This law, and the table that resulted from it, were valuable to scientists in two ways. First, they summarised the properties of the elements and arranged them into groups with similar properties. Second, they enabled predictions to be made about the properties of both known and unknown elements. Table 2.1.1 shows Mendeleev's predictions about the properties of 'eka-aluminium' and 'eka-silicon', which were subsequently discovered and given the names gallium and germanium. ('Eka' means 'one' in Sanskrit, an ancient Indian language historically used for scientific literature.)

There were some discrepancies in Mendeleev's table, but these were the inevitable result of the limited knowledge of chemistry available at the time – remember that the structure of the atom was still completely unknown. As we saw in section 1.3, the behaviour of an element is determined by the number of electrons in its atoms (shown by the element's atomic number) rather than by its atomic mass. Modern periodic law states that:

> **The properties of elements are a function of their atomic numbers**

and in modern periodic tables such as that in figure 2.1.1, all the elements are arranged in strict order of their atomic numbers.

Property	Predicted	Found experimentally
	Eka-aluminium	**Gallium**
Relative atomic mass	68	69.9
Density g cm^{-3}	6.0	5.90
Atomic volume	11.5	11.7
	Eka-silicon	**Germanium**
Relative atomic mass	72	72.3
Oxide	EsO_2 density 4.7 g cm^{-3}	GeO_2 density 4.23 g cm^{-3}
Chloride	$EsCl_4$ boiling point < 100 °C density 1.9 g cm^{-3}	$GeCl_4$ boiling point 84 °C density 1.84 g cm^{-3}

Table 2.1.1 Mendeleev's predictions about as yet undiscovered elements were later found to be surprisingly accurate.

The organisation of the periodic table

The **wide form** of the periodic table shown in figure 2.1.1 is the most commonly used, enabling all the known elements and their relationships with one another to be seen at the same time. The horizontal sets of elements are known as **periods**, the vertical sets as **groups** and the elements between groups II and III as **transition metals**. There are 91 naturally occurring elements, but more are shown on the periodic table. This is because a number of unstable, radioactive elements have been made in recent years by scientists from a variety of countries. These elements are included in the periodic classification for completeness.

The basis of the periodic table

When Mendeleev first proposed his periodic arrangement of elements by their atomic masses, he had no idea why the different elements had different atomic masses. There was no theoretical basis for the similar or dissimilar behaviours of elements either, only observations. We now understand why elements are arranged in the patterns of the periodic table, because we have a clear

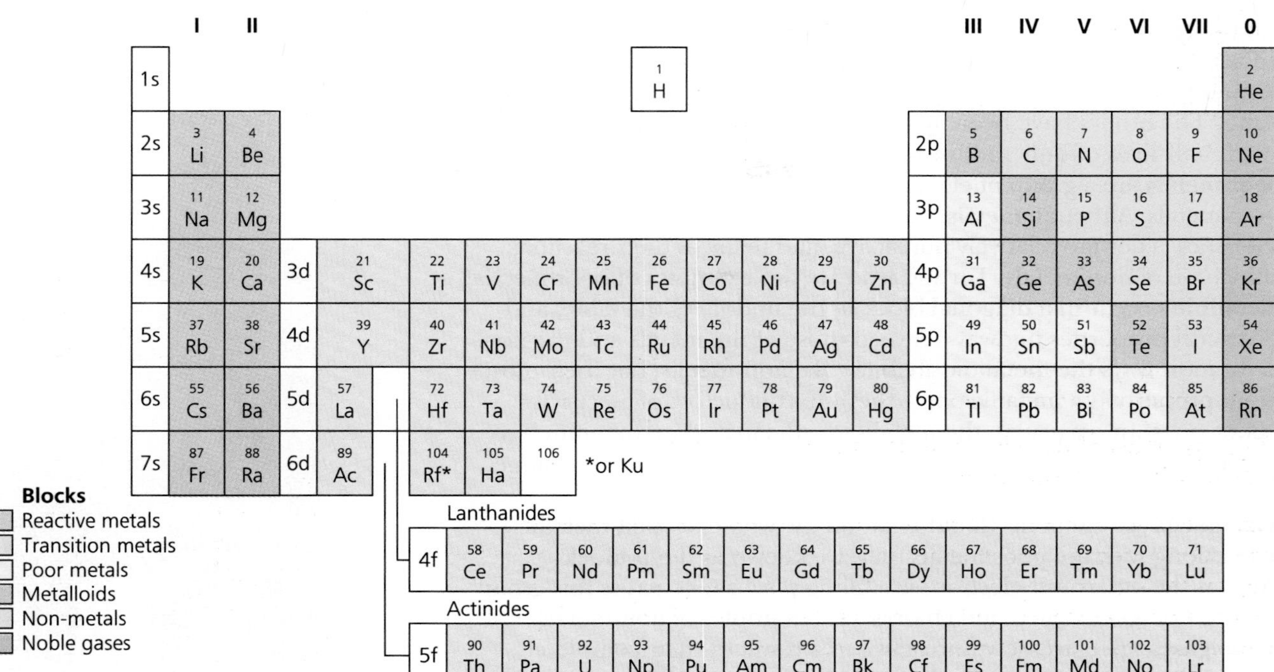

Figure 2.1.1 The modern periodic table, shown here in a relatively simple wide form. The names of the individual elements and their relative atomic masses may also be shown.

understanding of the structure of individual atoms, and how the arrangement of the electrons in different energy levels affects the reactivity of the elements. A new period is started each time electrons begin to enter a new energy level. Groups are made up of elements whose atoms have a similar electronic configuration. The effects of these relationships will be considered in more detail later in this section.

Blocks of elements

Some of the patterns of the periodic table are immediately obvious, namely the groups and the periods. Others are less easily recognised and require a knowledge of the properties of the elements as well as their electronic configuration. There are six main **blocks** of elements which are grouped together. These are indicated on figure 2.1.1 by different colours and examples are shown in figure 2.1.2.

- The **reactive metals** are the elements of groups I and II. They include sodium, potassium, calcium and magnesium and are, as the name suggests, very reactive, forming stable ionic compounds. These elements have lower melting points and boiling points and lower densities than most other metals. They are also known as the **s-block elements**, because the outermost electrons are in the s subshells.
- The **transition metals** are found in blocks between groups II and III. They include familiar everyday metals such as copper, iron, silver and chromium, and are much less reactive than the elements in groups I and II. The metals have similar properties across as well as down the block. They are also known as the **d-block elements**, as electrons are added to the d subshells across this group of elements.

  Within the transition metals is a subgroup known as the **f-block elements**, in which electrons are being added to the f subshells. The **lanthanides** are the 14 elements after lanthanum, from cerium (Ce) to lutetium (Lu). They show remarkable similarity to each other, to the extent that they are very difficult to separate. The lanthanides are also known as the **rare earths**. The **actinides** are the 14 elements after actinium, from thorium (Th) to lawrencium (Lr), although only three of them – up to uranium – are naturally occurring. The others have been artificially produced and are very unstable, often with half-lives of only milliseconds.

- The **poor metals** are a group of elements positioned to the right of the transition metals. Although metals, they do not exhibit strong metallic characteristics. They are relatively unreactive, and many of their reactions resemble those of non-metals. Tin and lead are two examples of poor metals.
- The **metalloids** occur in a diagonal block in the middle of the table, and have chemical properties midway between those of the metals and the non-metals. In most ways the metalloids behave like non-metals, but their most important property is a metallic one – they are conductors of electricity, albeit poor ones. In particular, the metalloids silicon and germanium have been responsible for the microchip revolution, which continues to change our lifestyle so dramatically.
- The **non-metals** are elements that do not, on the whole, exhibit metallic character. Along with the poor metals they are known as the **p-block elements**, as the outermost electrons are filling p subshells across this group of elements. Carbon, oxygen and chlorine are examples of non-metals.
- The **noble gases** are extremely unreactive as a result of their completely filled shells. It was originally thought that they were completely unreactive or **inert** (older books may refer to them as the **inert gases**), but several compounds of noble gas elements have been made in recent years.

The poor metals, including lead, show both metallic and non-metallic characteristics. The ability to be rolled into sheets is an example of metallic behaviour.

s-block metals are very reactive, as the behaviour of this piece of potassium with cold water demonstrates.

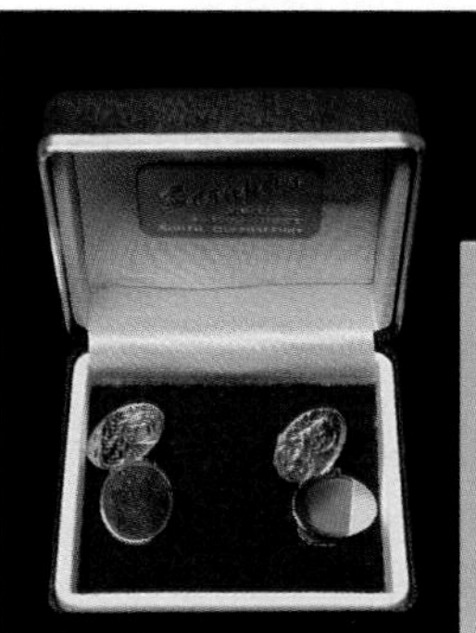

d-block elements – the transition elements – are much less reactive than the s-block elements. This enables them to be used in elemental form.

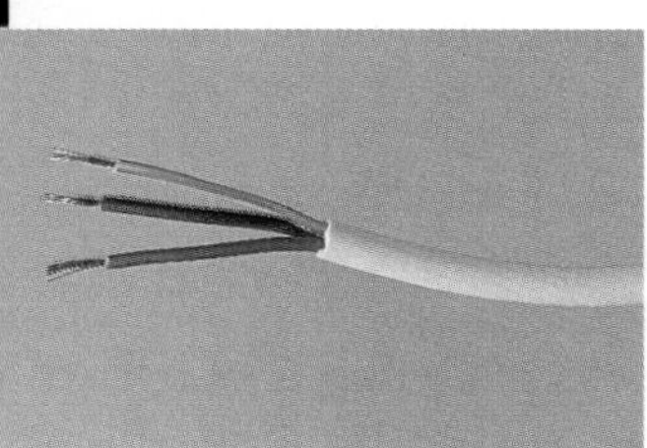

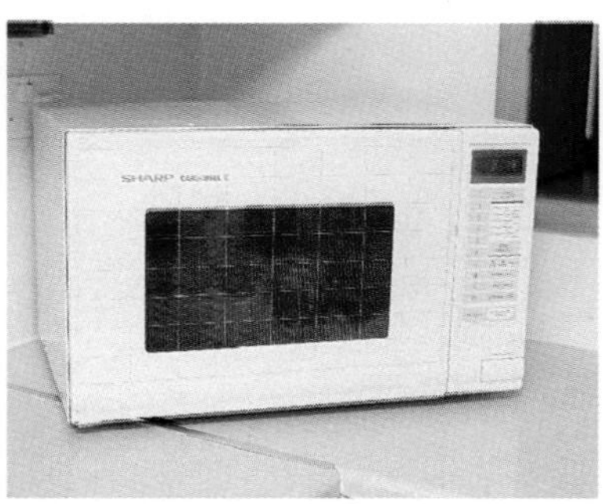

Many modern domestic appliances are controlled by microchips, made from the metalloid silicon.

Although a few non-metals, like the sulphur shown here, are solids, the majority, like chlorine (below), are gases.

The noble gases, used in these illuminated tubes, are so unreactive that for many years they were considered completely inert.

Figure 2.1.2 These photographs illustrate typical members of the six blocks of the periodic table.

Periodicity – the properties of the elements

Within the periodic table, the elements show predictable patterns of properties. Some of these properties are the direct result of the atomic structure of the element. They are known as the **atomic properties**, and we shall consider these first. Other properties depend not simply on the structure of the atoms, but also on how the atoms are linked together to give the structure of the element – these are known as the **bulk properties**.

Atomic properties

Atomic radius

As we move through the periodic table, the atomic radii of the elements show distinct trends which can be explained by considering their electronic structures.

> **What is meant by atomic radius?**
>
> Before considering trends in atomic radius, we must be clear what is meant by the term. As we saw in section 1.3, the electron cloud of an atom has no definite dimensions, since in theory the electrons could be found anywhere.
>
> Because of this difficulty, the atomic radius is frequently considered to be the distance of closest approach to another identical atom. Even this does not solve the problem, as it is not clear from this definition whether the atoms are chemically bonded or not. To be more precise, half the equilibrium distance between two covalently bonded atoms is called the **covalent radius**. For metals, the **metallic radius** is used instead of the covalent radius, which is half the distance between the nuclei of neighbouring ions in the crystalline metal. In addition, half the distance of closest approach between two atoms which are not chemically bonded is called the **van der Waals radius**.
>
> For atoms with nearly full shells, the electron cloud may be quite diffuse, and the difference between the covalent and van der Waals radii can be quite marked. For example, in chlorine the covalent radius is 0.099 nm while the van der Waals radius is 0.180 nm.
>
> When chemists refer to atomic radius they usually mean the covalent radius, although other measures of atomic size may also be used.

- *The atomic radius generally decreases across a period*. Moving across a period, the nuclear charge becomes increasingly positive as the number of protons in the nucleus increases. Although the number of electrons also increases, the outer electrons are all in the same shell. The electrons are attracted more strongly to the increasingly positive nucleus, thus reducing the total atomic radius, as shown by figure 2.1.3 overleaf.
- *The atomic radius generally increases down a group*. The outer electrons enter new energy levels as we pass down a group, so although the nucleus gains protons, the electrons are not only further away but also screened by more electron shells. As a result they are not held so tightly, and the atomic radius increases, as shown in figure 2.1.3.

The atomic radius of an atom of an element is different from the radius of an ion of that element. The term **ionic radius** is used to describe the size of an ion. A positive ion has a smaller ionic radius than the original atom, as the loss of one or more electrons means that the remaining electrons each have a larger share of the positive charge and so are more tightly bound to the positive nucleus. Similarly, a negative ion has a larger radius than that of the parent atom, as the addition of extra negative charge introduces more electron–electron repulsion. The electrons are less tightly bound to the nucleus and so the radius is larger.

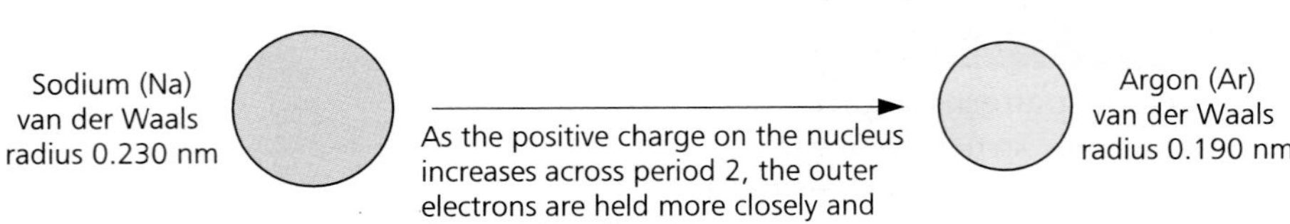
Sodium (Na)
van der Waals
radius 0.230 nm
As the positive charge on the nucleus
increases across period 2, the outer
electrons are held more closely and
the atomic radius *decreases*.
Argon (Ar)
van der Waals
radius 0.190 nm

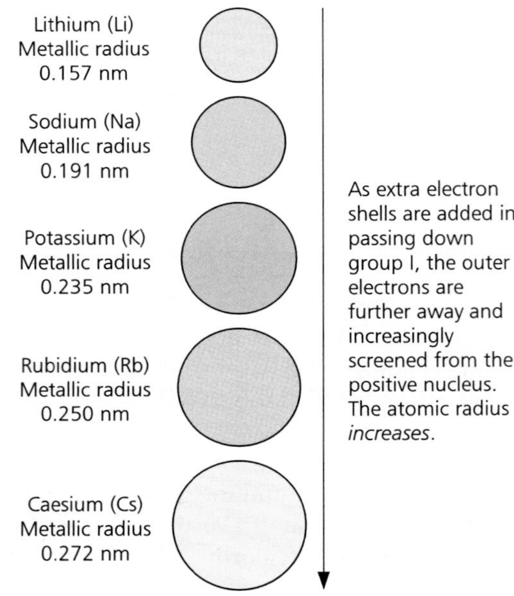
Lithium (Li)
Metallic radius
0.157 nm
Sodium (Na)
Metallic radius
0.191 nm
Potassium (K)
Metallic radius
0.235 nm
Rubidium (Rb)
Metallic radius
0.250 nm
Caesium (Cs)
Metallic radius
0.272 nm
As extra electron shells are added in passing down group I, the outer electrons are further away and increasingly screened from the positive nucleus. The atomic radius *increases*.

Figure 2.1.3 The trends in the atomic radii of the elements are related to their electronic structures, as this diagram illustrates.

Ionisation energy

As we saw in section 1.4, the first ionisation energy of an element is directly related to the attraction of the nucleus for the most loosely bound of the outer electrons. The more tightly held the outer electrons, the higher the first ionisation energy. There are three main factors that affect the ionisation energy of an atom:

1. As the distance from the outermost electron to the positive nucleus increases, the attraction of the nucleus for the electron decreases. The ionisation energy gets smaller as the atomic radius increases.
2. The magnitude of the positive nuclear charge also has an effect – a more positive nucleus will have a greater attraction for the outer electrons and so the ionisation energy of the atom will be greater.
3. The inner shells of electrons repel the outermost electrons, thus screening or shielding them from the attractive positive nucleus. The more electron shells there are between the outer electrons and the nucleus, the lower will be the ionisation energy of the atom, as the outer electrons will be less firmly held.

These factors produce two striking patterns in ionisation energies within the periodic table:

- *Ionisation energy generally increases across a period* – it becomes harder to remove an electron. This is the result of the increasing positive nuclear charge across the period without the addition of any extra electron shells to screen the outer electrons. The atomic radius gets smaller and the electrons are held more firmly. It therefore requires more energy to bring about ionisation. Figure 2.1.4 shows the first ionisation energies of the first 54 elements, in which the repeating patterns can clearly be seen. Notice that the end of each

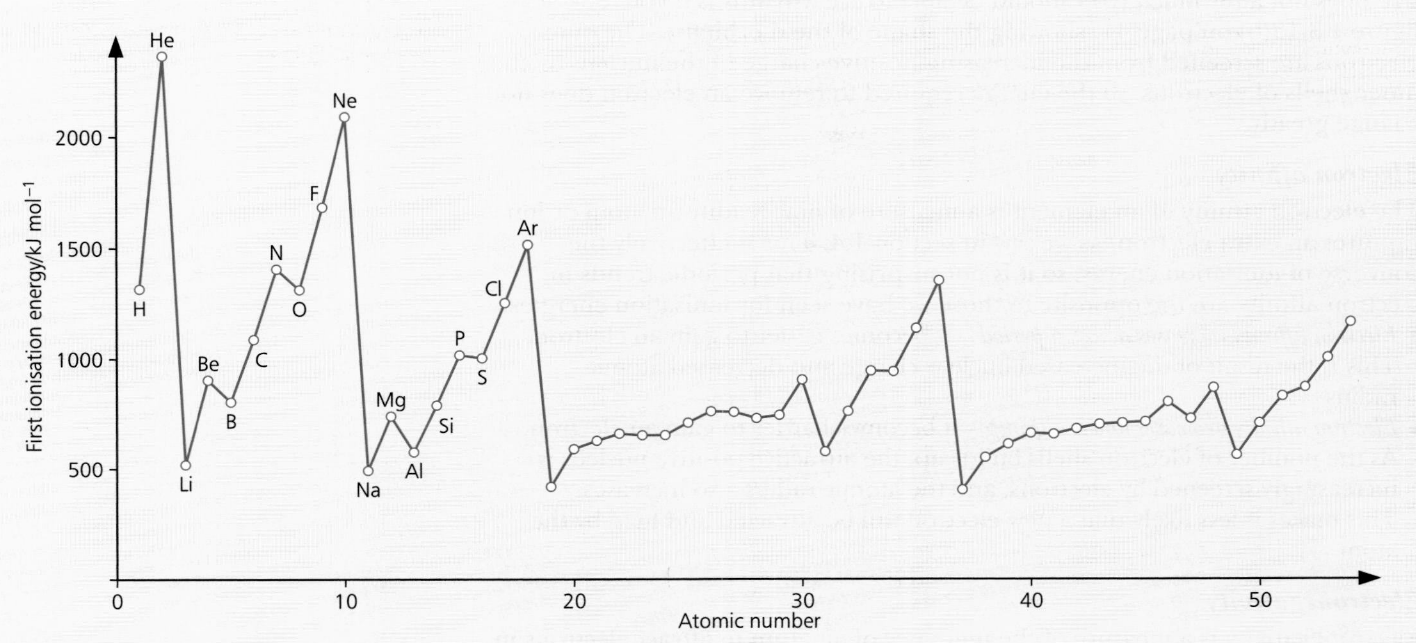

Figure 2.1.4 The first ionisation energies of the first 54 elements of the periodic table. Clear patterns are immediately obvious, and a closer look reveals more subtle fluctuations.

period is marked by the high ionisation energy of a noble gas. This distinctively high ionisation energy is a result of the stable electronic structure of the noble gases and is indicative of their unreactive natures.

As figure 2.1.4 shows, first ionisation energies do not increase smoothly across a period. This is due to the presence of subshells within each shell. Notice that in the period lithium to neon, the first ionisation energy of beryllium is larger than that of boron. The same relationship is seen for magnesium and aluminium in the next period. The higher first ionisation energies of beryllium and magnesium are associated with the full s subshells in the outer shells of these elements. Removing one electron from an atom of boron or aluminium removes the single electron in the p subshell. However, for beryllium or magnesium an electron is removed from a full s subshell. Full subshells are particularly stable, so this requires more energy than removing the single p electron from boron or aluminium.

In the same way, the unexpectedly high first ionisation energies of nitrogen and phosphorus can be explained by the fact that both these elements contain half-full p subshells in their outermost shell. Half-full subshells also appear to be associated with greater stability. Removing an electron from an atom of oxygen or sulphur removes the fourth electron in the p subshell, leaving a half-full p subshell. On the other hand, removing one electron from a nitrogen or phosphorus atom requires breaking into a half-full subshell, a process that requires more energy.

- *Ionisation energy generally decreases down a group* – it becomes easier to lose an electron. Passing down a group, the nuclear charge increases, but more importantly in this context, extra shells of electrons are added. This has the effect of increasing the distance of the outer electrons from the positive attraction of the nucleus, and screening them from its positive charge. They are thus held less tightly and the energy required to bring about ionisation is lower.

The ionisation energies of the d-block or transition elements are relatively similar. This is because electrons are entering the d subshell, and the atomic size does not alter much (you should be able to see why this is if you look at figure 1.3.12(b) on page 41, showing the shape of the d orbitals). The outer electrons are screened from the increasing positive charge of the nucleus by the inner shells of electrons, so the energy required to remove an electron does not change greatly.

Electron affinity

The electron affinity of an element is a measure of how readily an atom or ion captures an extra electron, as we saw in section 1.4. This is effectively the converse of ionisation energy, so it is not surprising that periodic trends in electron affinity are the opposite to those we have seen for ionisation energies.

- *Electron affinity increases across a period* – it becomes easier to gain an electron. This is the result of the increased nuclear charge and decreased atomic radius.
- *Electron affinity decreases down a group* – it becomes harder to gain an electron. As the number of electron shells builds up, the attractive positive nucleus is increasingly screened by electrons, and the atomic radius also increases. This makes it less likely that a new electron will be attracted and held by the atom.

Electronegativity

Electronegativity is a measure of the tendency of an atom to attract electrons in a covalent bond. The main factor affecting the electronegativity of an atom is the positive nucleus – the larger the positive charge on the nucleus, the more effectively will it attract electrons. The periodic trends in electronegativity are summarised below and in figure 2.1.5.

- *Electronegativity increases across a period*. This is due to the increasing positive charge on the nucleus in combination with the falling atomic radius.
- *Electronegativity decreases down a group*. Although the positive charge on the nucleus increases, this is more than offset by the increase in the atomic radius and the additional screening effects of extra electron shells.

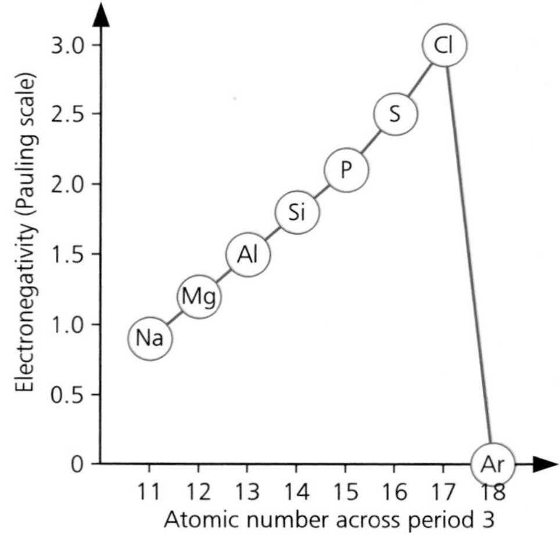

Figure 2.1.5 Electronegativity increases markedly across a period, as can be seen quite clearly in period 3. The result of this trend is that all the most strongly electronegative elements are found in the top right-hand corner of the periodic table (the non-metals), whilst the least electronegative elements are the reactive metals in the bottom left-hand corner.

Oxidation numbers

The oxidation numbers of the elements show clear patterns across the periodic table, whilst they remain the same down each group. As we saw in section 1.4, the oxidation number of an element corresponds to the number of electrons lost or gained in order to acquire a stable configuration of electrons. This is why the principal oxidation number of all the members of a group is the same

Diagonal relationships in the periodic table

So far our discussion of the patterns in the periodic table has been centred around the horizontal and vertical associations known as periods and groups. However, in certain cases there are clear diagonal similarities too.

A careful look at the periodic properties considered so far shows that in each case a trend in atomic properties going from left to right (for example, the decrease in atomic radius) will tend to be cancelled out by the effect of moving down the group (for example, the increase in atomic radius). As a result, the elements along diagonal lines running from top left to bottom right of the periodic table will tend to have similar atomic sizes and reactivities. These relationships are particularly clearly seen before the added dimension of the transition elements intervenes.

– all the members of a group have the same configuration of electrons in the outer shell, although the number of inner shells varies.

The maximum oxidation number of any element is the same as the number of the group it is in, although many elements show more than one oxidation number. For example, magnesium in group II has a maximum oxidation number of 2, whilst silicon in group IV has a maximum number of 4.

The behaviour of elements in the second period

The elements in the period lithium to fluorine often have similar chemical properties to the elements in the periods below them, although this is not always so. This is illustrated by the group VI elements oxygen and sulphur in their reactions with hydrogen and fluorine.

Similar ...

Both oxygen and sulphur react similarly with hydrogen, forming the hydrides H_2O and H_2S respectively. In these compounds the group VI elements both have an oxidation number of –2.

... but different

The elements behave differently with fluorine. Oxygen reacts with fluorine to form OF_2. In this compound, oxygen has an oxidation number of +2, since fluorine is more electronegative than oxygen. No other fluorides of oxygen are found, and oxygen never shows an oxidation number of more than +2. However, sulphur reacts with fluorine to form several fluorides. The maximum oxidation number of sulphur is +6, found in the compound SF_6. Both OF_2 and SF_6 are gases, showing that the elements in both compounds are covalently bonded.

Why are there such differences in the oxidation states of these two elements, both of which are in group VI? The answer to this question lies in the orbitals available to the atoms when they form chemical bonds.

The electronic configuration of oxygen is $1s^2 2s^2 2p^4$, and that of sulphur is $1s^2 2s^2 2p^6 3s^2 3p^4$. In forming two covalent bonds with fluorine, oxygen acquires a noble gas configuration. No further orbitals are available to the oxygen atom, and therefore no higher fluorides (OF_4, etc.) are possible. Oxygen therefore has a maximum oxidation number of +2. The electronic configuration of oxygen in OF_2 is shown below, with electrons from the fluorine atom in red.

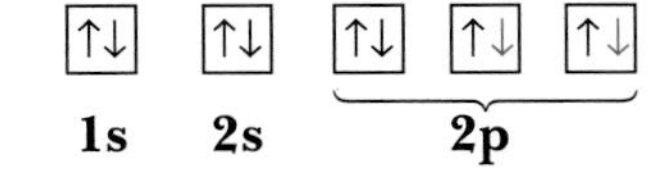

Unlike oxygen, the sulphur atom has the empty 3d orbitals available to it when it forms compounds. The availability of these empty orbitals means that each of the six electrons in the outer shell of the sulphur atom may be used to form a covalent bond, since the outer shell of the sulphur atom has sufficient orbitals. This is illustrated in the electron configuration of sulphur in SF_6, with electrons from fluorine once again shown in red.

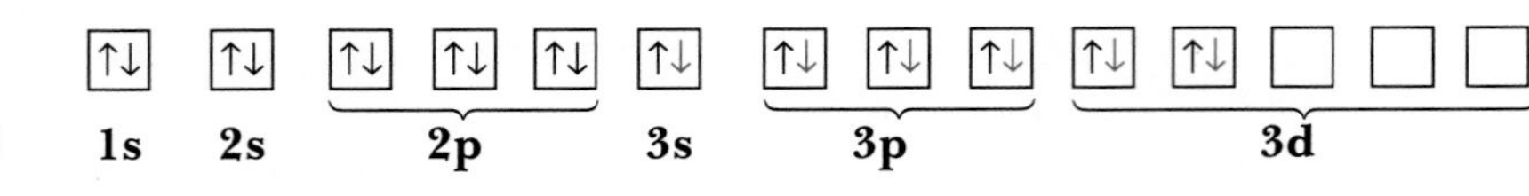

Figure 2.1.6 shows the patterns in oxidation numbers of the elements in periods 2 and 3 when combined with oxygen to form oxides, and with hydrogen to form hydrides.

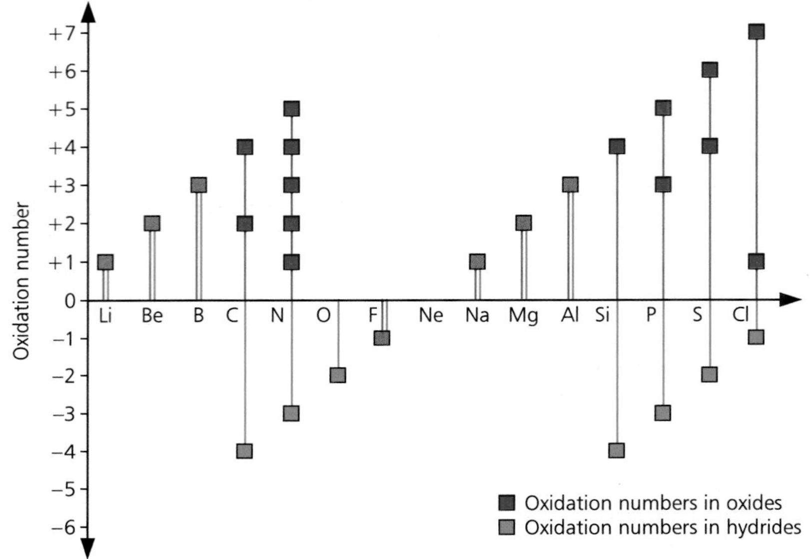

Figure 2.1.6 Although the oxidation numbers of the elements vary when combined with different elements, similar patterns are observed across the periods, as can be seen here.

Physical or bulk properties

Periodicity is seen in the physical properties of the elements (their melting points, boiling points, densities, etc.), and this is closely related to the types of elements involved and their usual structures, as shown in figure 2.1.7. The following summary looks at the periodic patterns of some of the physical properties of the elements in more detail.

Density

The densities of the elements show distinct patterns across a period. Density is related to the size of the atoms and their packing, as shown in figure 2.1.7, and there are changes in this packing across a period.

- For the metals (s- and d-block elements), the effects of atomic radius are more important than the packing, since the packing is broadly similar in the metals. As we move across the period, the electrons in each atom are held more tightly by the nucleus. This causes a decrease in the atomic radius, and so more atoms can be squeezed into a given volume. Hence the density of the metals increases across a period. It decreases down a group, due to the

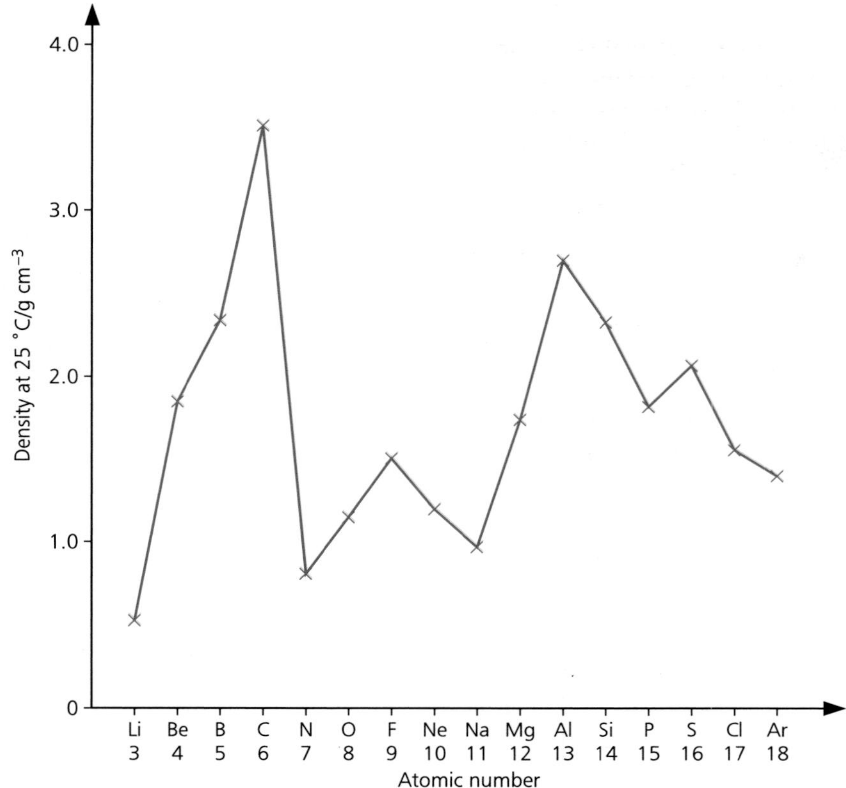

Figure 2.1.7 The periodic patterns of density depend on the size of the atoms and their packing within the elements.

increasing number of electron shells. The transition metals have the highest densities.

- Across a period, the density reaches a maximum in the middle of the periodic table. For example, in period 2 (lithium to neon), the maximum occurs at carbon, with its strongly covalently bonded lattice in the form of diamond. The maximum in period 3 occurs at aluminium. To the right of this maximum, the density of most of the elements falls, as they exist as small covalent molecules which are widely spaced, with small intermolecular forces.

Melting point

The **melting point** of a given substance is defined as the temperature at which the pure solid is in equilibrium with the pure liquid at atmospheric pressure. The melting point is affected by both packing and bonding within a substance. For example, the metallic structure of metals, with the atoms held tightly together in a sea of electrons, explains the high melting and boiling points of these elements as well as their good electrical conductivity.

Giant molecular structures such as those found in the metalloids and diamond have strong covalent bonds between the atoms, holding them tightly within a crystal structure. As a result it is very difficult to remove individual atoms and the elements have extremely high melting points. In contrast, the simple molecular structures found in most non-metals means that they exist as small discrete molecules. The atoms in the molecules are held together by strong covalent bonds, but the molecules are only held together by weak intermolecular van der Waals forces. Thus the molecules can be separated quite easily and the non-metals have low melting points as a result. Because of these structural differences, dramatic changes are seen in melting points across a period along with the changes in bonding and structure of the elements, as figure 2.1.8 overleaf illustrates.

Standard molar enthalpy change of fusion

Closely related to the melting point of an element is its **standard molar enthalpy change of fusion**. This is the amount of energy required to turn one mole of solid at its melting point into one mole of liquid at the melting point. The enthalpy change of fusion is affected by the bonding and structures of the elements in just the same way as the melting point, and so the two properties follow almost identical patterns across the periodic table, as shown in figure 2.1.8.

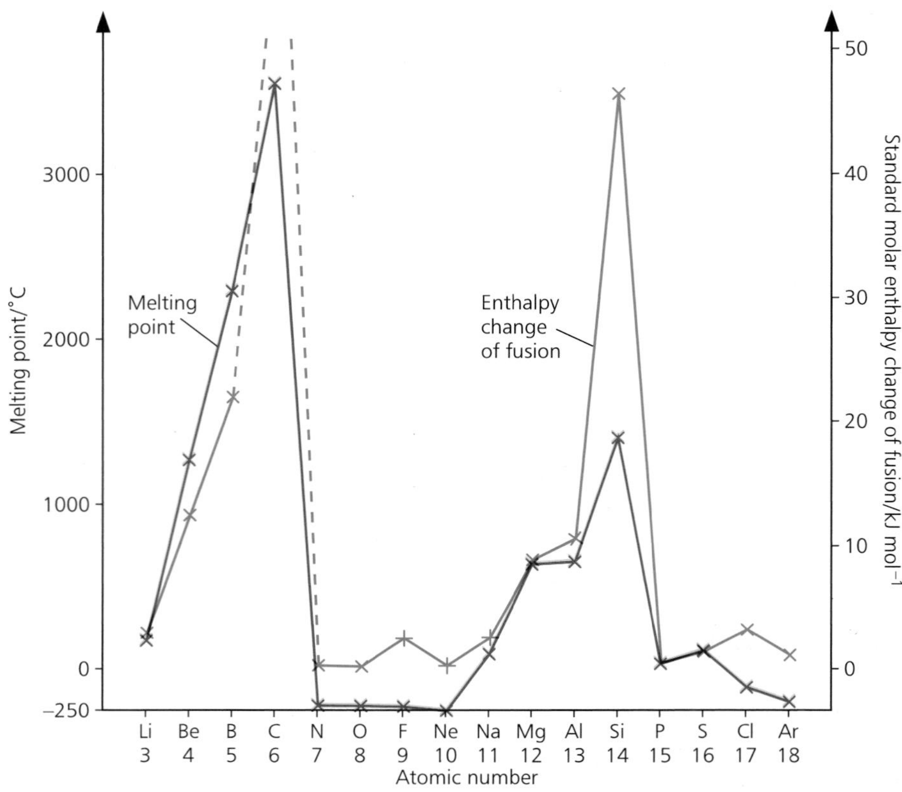

Figure 2.1.8 The bonding between atoms and the structure of the elements affects both their melting points and their enthalpy changes of fusion in a similar way, hence the clear similarity in the two graphs. The marked changes in the values occur where metal and metalloid structures give way to those of the non-metals.

Boiling point and standard molar enthalpy change of vaporisation

The **boiling point** of a substance is the temperature at which the pure liquid is in equilibrium with pure vapour at atmospheric pressure. The **standard molar enthalpy change of vaporisation** is the energy needed to convert one mole of liquid at its boiling point to one mole of vapour, also at the boiling point. Both quantities are affected by packing and bonding between the atoms of the elements in a similar way to melting point and enthalpy change of fusion. The patterns across periods 2 and 3 can be seen in figure 2.1.9, with the boiling point increasing through the metals and metalloids and then decreasing dramatically with the non-metals.

Electrical and thermal conductivities

The electrical and thermal conductivities of the elements also vary across a period with the changes in bonding and packing described above. Metals are good conductors of both electricity and heat, properties which are the result of the mobile electrons in the lattice (see section 1.6). Under normal conditions, electron movement through the lattice is relatively random, but when a potential difference is applied the electrons tend to move away from the negative electrode which repels them and towards the attractive positive

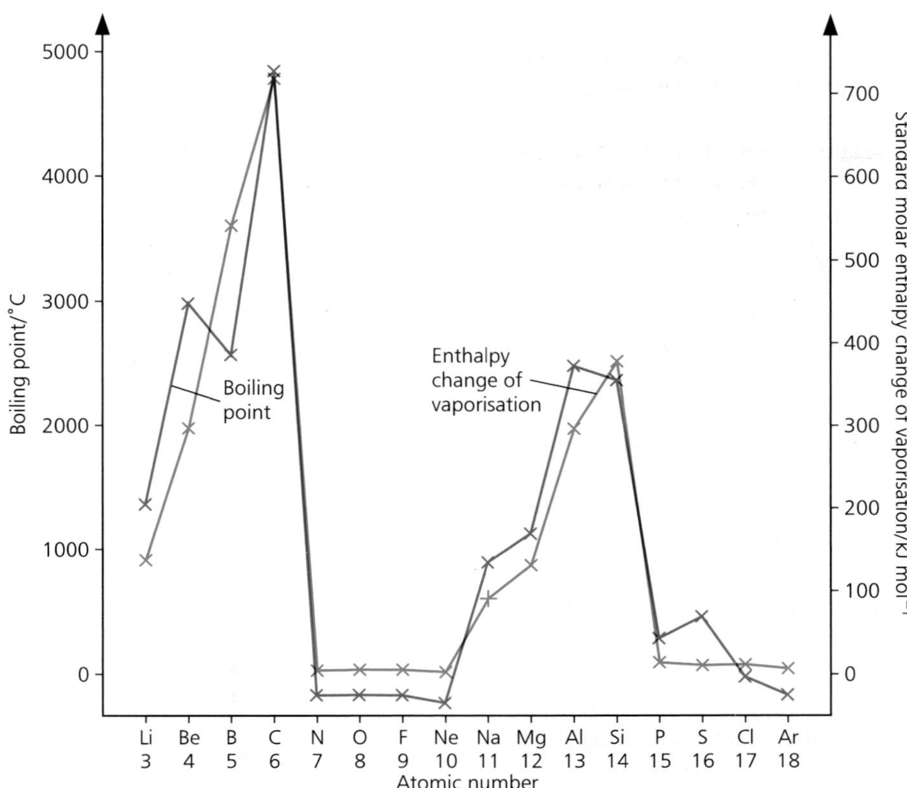

Figure 2.1.9 The boiling points and standard molar enthalpy changes of vaporisation are affected by the structure of the elements in the same way as the melting points, and so the graphs are similar.

electrode. Similarly, when a metal is heated, the electrons in the areas of high temperature have high kinetic energy and so move rapidly (and randomly) towards the cooler regions of the metal, transferring their energy to other electrons as they go.

Within the metalloids, the electrons of the covalent bonds are held much more tightly than in the metals, and so both electrical and thermal conductivities are lower than those of the metals.

Finally, the non-metals, with their discrete molecules, have no free electrons within their crystal structure – indeed many have no crystal structure under normal conditions of temperature and pressure, since they are liquids or gases. As a result, their electrical and thermal conductivities are very low indeed.

We have seen that the elements of the periodic table show clear patterns that can be related to the structure of the atoms themselves. In section 2.2 we shall go on to consider patterns in the compounds formed when these elements react together.

SUMMARY

- The chemical elements are arranged in the **periodic table** in which the horizontal sets of elements are known as **periods**, and the vertical sets as **groups**.
- Mendeleev hypothesised a periodic law stating that the properties of the elements are a periodic function of their relative atomic masses.
- The modern periodic table is based on the law that the properties of the elements are a function of their atomic numbers.

- The patterns of behaviour of the elements within the periodic table are the result of their atomic structure. A new period begins each time a new electron shell is started.
- The elements may be broadly classified as **metals** (good conductors of electricity), **metalloids** (poor conductors of electricity) and **non-metals** (non-conductors). Major areas of the periodic table include:

  the reactive metals of groups I and II (the s block)
  the transition metals – the elements between groups II and III (the d block)
  the poor metals – the relatively unreactive metallic elements of the p block
  the metalloids – a diagonal block in the middle of the p block
  the non-metals of the p block
  the noble gases with their completely filled outer electron shells.
- Elements within the periodic table show predictable patterns and trends of behaviour. **Atomic properties** are the direct result of the atomic structure of the atom. **Bulk properties** are the result of the way the atoms are linked together within the element.
- **Atomic radius** generally decreases across a period, as the electrons are held increasingly tightly by the more strongly positive nucleus. Atomic radius generally increases down a group, with the addition of extra electron shells.
- **Ionisation energy** generally increases across a period – it becomes harder to remove an electron. Ionisation energy generally decreases passing down a group – it becomes easier to lose an electron.
- **Electron affinity** generally increases across a period and decreases down a group.
- **Electronegativity** increases across a period and decreases down a group.
- The **oxidation numbers** of the elements show clear patterns across the periodic table. The maximum oxidation number shown by all the elements in a given group is the same.
- The **density** of the elements shows periodic patterns depending on the way the atoms or ions are packed. Density tends to reach a maximum in the middle of the periodic table.
- **Melting points** change across a period and down groups as a result of trends in both packing and bonding. **Molar enthalpy changes of fusion** follow almost identical patterns to those of the melting points.
- **Boiling points** and **molar enthalpy changes of vaporisation** of the elements are also affected by both packing and bonding.
- **Electrical** and **thermal conductivities** depend on the bonding between the atoms or ions. Metals are good conductors of both heat and electricity, whilst non-metals are very poor conductors.

QUESTIONS

1 Using the periodic table of the elements at the back of this book, answer the following questions:

- **a** Strontium-90 is an isotope of strontium formed when atomic weapons are exploded. Why does it replace calcium in newly formed bones?
- **b** Why does potassium chloride taste very similar to sodium chloride?
- **c** Why were copper, silver and gold very commonly used to make old coins?
- **d** Where are the densest elements found in the periodic table?
- **e** Where are the gaseous elements found in the periodic table?
- **f** Which non-metals occur as monatomic gases?

2 a Using the periodic table, write down the symbols for the ions formed when the following elements attain a noble gas configuration:

Mg K Al Sr S Br N Cl O

- **b** Using your answers to part **a**, write down the formulae of the following compounds:
  - **i** potassium bromide
  - **ii** aluminium chloride
  - **iii** magnesium nitride
  - **iv** magnesium sulphide
  - **v** strontium oxide
  - **vi** potassium oxide.

3 a Using the periodic table, write down the most likely formulae of the hydrides of the following elements:

C N Cl S P O

- **b** Which of the compounds in your answer to **a** is likely to have the most polar bonds? Justify your answer.

4 Look at figure 2.1.4. Why does ionisation energy:

- **a** increase across a period
- **b** decrease down a group?
- **c** Explain why the first ionisation energies of sulphur and aluminium are lower than those of phosphorus and magnesium respectively.

5 Explain why the electronegativities of elements:

- **a** decrease down a group
- **b** increase across a period.

6 Look at the graphs in figures 2.1.7, 2.1.8 and 2.1.9.

- **a** What elements appear on the peaks of the graphs?
- **b** What elements appear in the troughs of the graphs?
- **c** What type of structure do elements occurring at the peaks have?
- **d** What type of structure do elements appearing in the troughs have?
- **e** Explain the general similarity of the graphs.

Developing Key Skills

Design a timeline to show the development of the periodic table. This should be large, clear and colourful, with information about both the people and the ideas involved. It should be designed to be used on the laboratory wall in association with the modern periodic table.

[Key Skills opportunities: C]

2.2 The periodicity of compounds

Section 2.1 considered the periodic patterns in the properties of the elements. These patterns are related to the atomic structures of the elements, and the arrangement of the electrons in the atom affects a wide range of properties. It should not surprise us to find that these periodic patterns can also be seen in the compounds formed when elements react together. For example, clear trends in properties such as melting point, electrical conductivity and solubility in water are apparent in the chlorides of the elements across a period, and trends of this sort are repeated throughout the periodic table. In this section we shall concern ourselves with the compounds formed by members of periods 2 and 3, as these are not complicated by the d-block elements. However, similar patterns are observable in any of the periods of the table.

Patterns in the chlorides

Enthalpy changes of formation and physical properties

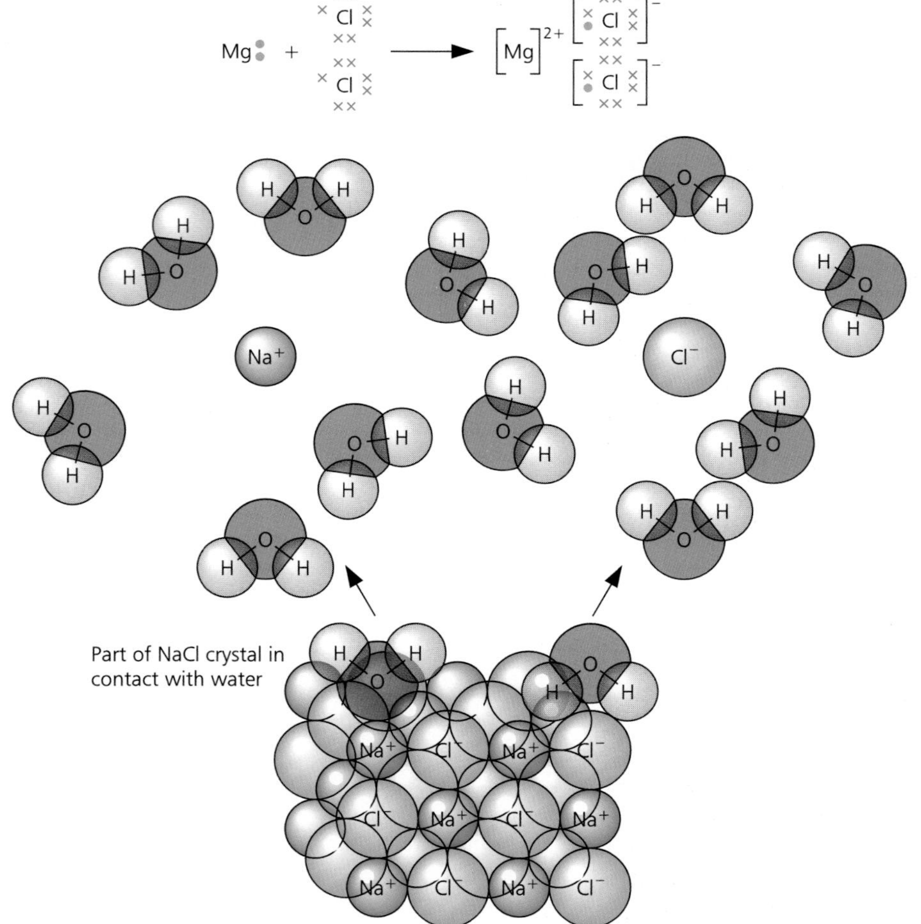

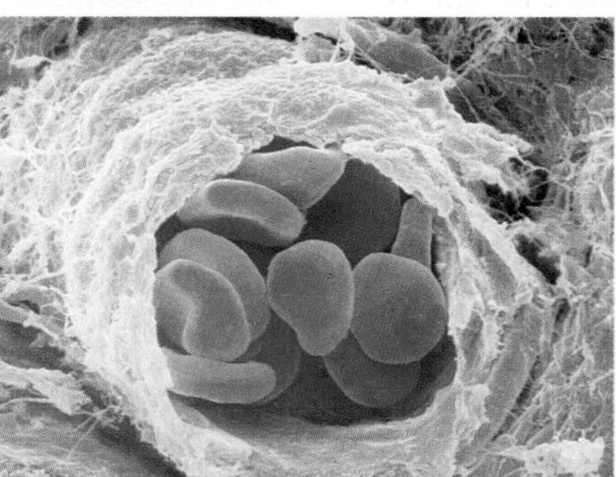

Figure 2.2.1 The electropositive metals such as sodium react with electronegative chlorine to form ionic chlorides. Sodium chloride dissolves easily in water, with each ion becoming surrounded by polar water molecules (left). Sodium chloride in solution is of vital importance in the natural world, providing the medium of life not only for all marine animals and plants but also within the individual cells of all living things.

The elements sodium to argon demonstrate clearly the patterns in the properties of the chlorides across a period. Chlorine is a very electronegative element – it receives electrons readily and is reactive. The metals on the left-hand side of the periodic table are very electropositive – they lose electrons easily. As a result, these metals react vigorously or even violently with chlorine to form ionic chlorides. Considerable energy is produced in these reactions – for example, the standard molar enthalpy change of formation of sodium chloride is -411 kJ mol^{-1}. The chlorides of the electropositive elements have giant ionic structures that are extremely stable, having properties seen in section 1.6 – very high melting and boiling points, conducting electricity in the molten state, and dissolving readily in water to form neutral solutions (see figure 2.2.1). In period 3, sodium and magnesium form chlorides of this sort.

The remainder of the elements in the period become increasingly electronegative and so tend to form covalently bonded molecular chlorides. The enthalpy change when these chlorides are formed is low compared with the formation of chlorides of the more electropositive elements ($\Delta H^{\ominus}_{f,m}$ for the formation of sulphur(II) chloride, SCl_2, is -19.7 kJ mol^{-1}). These molecules are held together only by relatively weak van der Waals forces, and as a result these chlorides have low melting points and boiling points and do not conduct electricity, even when molten.

2

Reactions with water

Sodium chloride dissolves in water to produce a neutral solution, while magnesium chloride produces a solution which is slightly acidic. Aluminium chloride produces a solution which may be very acidic, with a pH as low as 3.

Ionic substances like sodium chloride dissolve in water because water molecules are able to surround the ions and solvate them, holding them in solution as figure 2.2.1 shows. In an ionic compound with smaller positive ions, the O—H bonds in the water molecules surrounding the positive ions become more strongly polarised, with less electron density around the hydrogen atom. This causes some of the water molecules surrounding these ions to give up H^+ ions, producing an acidic solution. (This effect is discussed in more detail in section 3.4, on pages 288–9.) The increasing acidity of aqueous solutions of the chlorides as we move from sodium to aluminium can therefore be explained by the decreasing size and increasing charge of the positive ions present.

The chlorides of silicon, phosphorus and sulphur all react with water to form acidic products. Both silicon and phosphorus chlorides react in a straightforward way to produce hydrochloric acid:

$$\mathbf{SiCl_4(l) + 4H_2O(l) \rightarrow Si(OH)_4(aq) + 4HCl(aq)}$$

$$\mathbf{PCl_3(l) + 3H_2O(l) \rightarrow H_3PO_3(aq) + 3HCl(aq)}$$

The hydrolysis of the chlorides of sulphur is more complex, forming sulphur, hydrogen sulphide and sulphite ions as well as hydrochloric acid.

Summarising patterns in the chlorides

Table 2.2.1 (overleaf) provides numerical values for the trends described here, showing the patterns in the chlorides of both periods 2 and 3. The formulae of the chlorides depend on the oxidation numbers of the elements – where there are two or more compounds, the most common one is given, with the others shown in brackets. Patterns in the compounds of the elements also occur within groups. The lower elements in a group tend to be the most strongly ionic as they are less electronegative. However, the trends within groups will be considered in more detail later. The patterns in the values of enthalpy change

Property of chloride	Period 2						
Formula	LiCl	$BeCl_2$	BCl_3	CCl_4	NCl_3	Cl_2O	ClF
State at 20 °C	Solid	Solid	Gas	Liquid	Liquid	Gas	Gas
Melting point/°C	605	405	−107	−23	−40	−20	−154
Boiling point/°C	1340	520	13	77	71	5	−101
Electrical conductivity in liquid state	Good	Poor	None	None	None	None	None
Structure	Giant ionic	Partially covalent			Simple covalent molecules		
$\Delta H^{\ominus}_{f,m}$/kJ mol^{-1}	−409	−490	−404	−130	230	80	−56
$\Delta H^{\ominus}_{f,m}$ per mol of Cl atoms/kJ mol^{-1}	−409	−255	−135	−33	77	40	−56
Reaction with water	Soluble	Soluble	Soluble	Insoluble	Soluble	Soluble	Soluble
pH of aqueous solution	Weakly acidic	Acidic	Acidic	—	Acidic	Acidic	Acidic
Property of chloride	**Period 3**						
Formula	NaCl	$MgCl_2$	$AlCl_3$	$SiCl_4$	PCl_3 (PCl_5)	S_2Cl_2 (SCl_2) (SCl_4)	Cl_2
State at 20 °C	Solid	Solid	Solid	Liquid	Liquid (solid)	Liquid	Gas
Melting point/ °C	801	714	178 (sublimes)	−70	−112	−80	−101
Boiling point/°C	1413	1412	—	58	76	136	−35
Electrical conductivity in liquid state	Good	Good	Poor	None	None	None	None
Structure	Giant ionic				Simple covalent molecules		
$\Delta H^{\ominus}_{f,m}$/kJ mol^{-1}	−411	−641	−704	−687	−320	−60	0
$\Delta H^{\ominus}_{f,m}$ per mol of Cl atoms/kJ mol^{-1}	−411	−321	−235	−172	−107	−30	0
Reaction with water	Solid dissolves easily			Fumes of HCl produced			Degree of reaction with water
pH of aqueous solution	Neutral	Weakly acidic	Acidic	Acidic	Acidic	Acidic	Acidic

Table 2.2.1 The properties of the chlorides of periods 2 and 3. As we move from left to right across the periodic table, the chlorides of the elements change, from ionic chlorides with high boiling points and electrical conductivities in the liquid state, to simple molecular chlorides that are volatile and do not conduct electricity when liquid. The noble gases are not shown in the table as they do not form chlorides under normal circumstances.

of formation become clear if we compare them *per mole of chlorine atoms*, since this takes into account how many bonds are being formed. The enthalpy changes of formation of the chlorides are usually negative, and those of the most electropositive elements have very large negative values. It can be seen that the enthalpy changes of formation become less negative across a period, as the difference in electronegativity between the reacting elements becomes less marked. In general, the enthalpy change of formation of the chloride of an element becomes increasingly negative towards the left-hand side of the periodic table, as the electronegativity of the elements decreases. This results in an increase in the stability of the compounds with respect to their elements.

Patterns in the hydrides

The hydrides show similar patterns across periods 2 and 3 to those of the chlorides. With very electropositive metals, hydrogen forms ionic hydrides, in which hydrogen obtains a full shell of electrons by gaining an electron rather

Property of hydride	Period 2						
Formula	LiH	BeH_2	B_2H_6	CH_4	NH_3	H_2O	HF
State at 20 °C	Solid	Solid	Gas	Gas	Gas	Liquid	Liquid
Melting point/°C	680 (decomposes)	260 (decomposes)	–165	–182	–78	0	–83
Boiling point/°C	—	—	–92	–164	–33	100	20
Electrical conductivity in liquid state	Good	Good	None	None	None	Poor	Poor
Structure	Ionic	?Macromolecular		Simple covalent molecules			
$\Delta H^{\ominus}_{f,m}$ per mol of H atoms/kJ mol^{-1}	–91	?	12	–19	–15	–143	–271
Reaction with water	Soluble – LiOH + H_2	Complex behaviour	Complex behaviour	Insoluble	Soluble – NH_4OH	—	Soluble
pH of aqueous solution	Strongly alkaline			—	Weakly alkaline	Neutral	Weakly acidic
Property of hydride	**Period 3**						
Formula	NaH	MgH_2	AlH_3	SiH_4	PH_3	H_2S	HCl
State at 20 °C	Solid	Solid	Not known	Gas	Gas	Gas	Gas
Melting point/°C	800 (decomposes)	280 (decomposes)	150 (decomposes)	–185	–133	–85	–115
Boiling point/°C	—	—	—	–112	–88	–61	–85
Electrical conductivity in liquid state	Good	Not known	None	None	None	Poor	Poor
Structure	Ionic	?	?Macromolecular		Simple covalent molecules		
$\Delta H^{\ominus}_{f,m}$ per mol of H atoms/kJ mol^{-1}	–56	–38	–15	9	2	–10	–92
Reaction with water	Soluble – NaOH + H_2	Complex behaviour	Complex behaviour	Reacts – $Si(OH)_4$	Insoluble	Soluble	Soluble
pH of aqueous solution	Strongly alkaline			Neutral	—	Weakly acidic	Strongly acidic

Table 2.2.2 The properties of the hydrides of periods 2 and 3. As we move from left to right across the periodic table, the hydrides of the elements change from ionic metal hydrides with high boiling points and electrical conductivities in the liquid state, to simple molecular hydrides which have low boiling points and do not conduct electricity when liquid. The situation is complicated by the highly unstable nature of some hydrides, and the hydrogen bonds formed between molecules of others making them more stable than might be expected. The noble gases are not shown in the table as they do not form hydrides under normal circumstances.

than losing one – it is reduced. These hydrides are all colourless solids with low volatility and ionic lattices in which hydrogen exists as the H^- (hydride) ion. The hydride ion is a strong base, and easily removes an H^+ ion from a water molecule. As a result, ionic hydrides react with water to form hydrogen and an alkaline solution, for example, sodium hydride:

$$2NaH(s) + 2H_2O(l) \rightarrow H_2(g) + 2NaOH(aq)$$

Because hydrogen is not particularly electronegative, most hydrides are covalent molecules formed by sharing electrons. These covalent hydrides are all colourless, electrically non-conducting compounds with low boiling points. The most common hydride of oxygen, water, is one of the most important molecules for life on Earth. The enthalpy changes of formation of the hydrides are greatest where there is a large difference in electronegativity between hydrogen and the other element. Hence the enthalpy changes of formation of the hydrides in the middle of a period are lower than those at the ends of a period.

Where an element forms chains of atoms (**catenation**), then several hydrides are possible – this is particularly relevant in the case of carbon chemistry, as we shall see in sections 4 and 5. Some of the non-transition metals form hydrides

FOCUS THE CHEMISTRY OF WATER

While the patterns in the properties of compounds like the oxides and the chlorides of the elements are generally fairly straightforward, the properties of the hydrides have some remarkable exceptions. As we saw in section 1.5, the melting points and boiling points of the hydrides of some elements can vary widely from those that would be predicted on the basis of the other elements in the same group, because of the existence of strong intermolecular forces called **hydrogen bonds**. One hydride which shows particularly marked discrepancies in its properties is the hydride of oxygen, commonly known as water. It is these differences which make water the unique substance that it is – vital not only for many chemical reactions in laboratories and industry, but also as a medium for the reactions of life in biological systems.

The simple chemical formula of water is H_2O, which tells us that two atoms of hydrogen are joined to one atom of oxygen to make up each water molecule. The O—H bonds in the water molecule are highly polar, and as a result the molecule has a permanent dipole. This dipole gives the water molecule its unusual properties. Perhaps the most important result of this charge separation is the tendency of water molecules to form hydrogen bonds between molecules.

Many of the properties of water are important in both physical and biological systems. Some of the most important are given here.

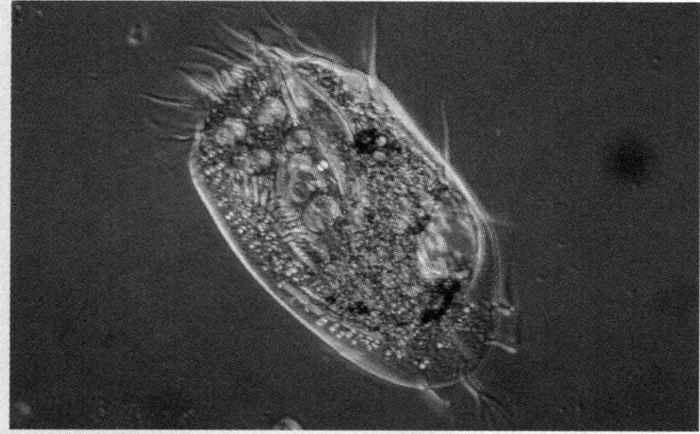

Figure 1 The structure of the water molecule and its electrical properties mean that water is an excellent solvent for a wide variety of other molecules. This behaviour has been important in the evolution of life.

(1) An excellent solvent

Many different substances will dissolve in water. The fact that the water molecule has a dipole means that polar substances will dissociate (form ions) and dissolve in it. Polar substances will generally not do this in organic solvents. Once the ions have dissolved in water, they become surrounded by water molecules, and so they remain dissolved. Water can also act as a solvent to many non-polar substances, that do not form ions. All the chemical reactions that take place within cells occur in aqueous solution.

Figure 2 If water solidified in the same way as other substances, then ice would form at the bottom of a pond rather than at the top, and the whole pond would rapidly freeze solid. As a result, pond life would be impossible.

(2) A density which changes with temperature in an unexpected way

As water cools to below 4 °C, the molecules take on an arrangement that occupies more space than the arrangement at room temperature. At room temperature, the molecules are associated by hydrogen bonds in various combinations, which are constantly changing. When freezing takes place at 0 °C, the two hydrogen atoms and two lone pairs of electrons in the oxygen atom result in fixed three-dimensional tetrahedral arrangements of hydrogen bonds. This new arrangement or lattice is rigid, and so ice is less dense than liquid water. This makes water unique. The fact that ice floats on water means that living things can survive in ponds and rivers when the temperatures fall below freezing. The ice forms at the top and acts as an insulating layer, helping to prevent the rest of the water mass from freezing. If ice formed from the bottom up, freshwater life would only be found in those areas where the water never freezes. Ice also thaws quickly because it is at the top, nearest to the warming effect of the Sun.

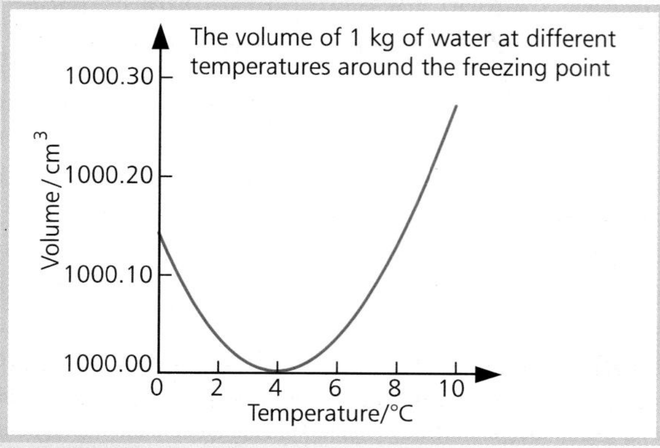

Figure 3 The density of water around its freezing point

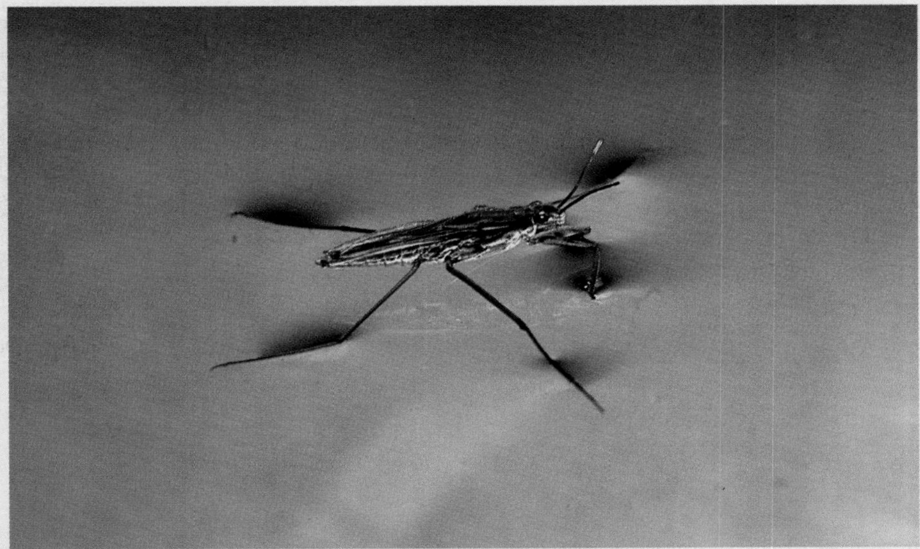

Figure 4 Surface tension is of great importance in plant transport systems, and also affects life at the surface of ponds, lakes and other water masses.

(3) Taking large amounts of energy to turn from liquid to gas and from solid to liquid

The **standard molar enthalpy change of vaporisation** is a measure of the energy needed to overcome the attractive forces between the molecules of a liquid and turn it into a gas. In the case of water, the amount of energy needed for this is very high. This is because all the hydrogen bonds holding the molecules together have to be overcome before water can become a gas, and although each hydrogen bond is in itself very weak, there are a great many of them. This is why all the water on the surface of the Earth has not vaporised into the atmosphere. The fact that the evaporation of water uses up a lot of heat energy means that water is an effective coolant, used for example in the sweating mechanism of the mammals.

(4) A high surface tension

Water has one of the highest known surface tensions. This is because of the hydrogen bonds between the water molecules, which tend to pull them down and together where water and air meet (see figure 1.7.5).

(5) An amphoteric molecule

The water molecule acts both as an acid and as a base, as the following reaction shows:

$$H_2O + H_2O \rightarrow H_3O^+ + OH^-$$

Notice the formation of the H_3O^+ ion, called the **oxonium ion**. This is the form that H^+ ions take in aqueous solution, although for convenience we often simply write H^+ rather than the more cumbersome H_3O^+. We shall return to the topic of acids and bases in section 3.

which are not ionic and which are relatively unstable – BeH_2, MgH_2 and AlH_3 are examples.

Summarising patterns in the hydrides

A glance at table 2.2.2 showing the properties of the hydrides reveals several interesting aspects of their chemistry. Unlike the chlorides, there are gaps in the table of the hydrides because some of them are so unstable that their properties cannot be measured and their structure is incompletely understood. Also unlike the chlorides, trends in the hydrides do not always follow the expected pattern. Melting points and electrical conductivities are good examples of this. The melting points of the hydrides of nitrogen, fluorine and particularly oxygen are far higher than would be predicted simply by considering the trend of the earlier elements in the period. The fact that the hydrides of oxygen and fluorine also conduct electricity, albeit poorly, in the molten state is another indication that the situation is not entirely straightforward. The discrepancies are due to attraction between the covalent molecules in the form of hydrogen bonding, which we studied in section 1.5.

Hydrides of the transition metals

The hydrides of the transition metals have an unusually complex chemistry. Some have definite formulae like nickel hydride, NiH_2, while others behave as if hydrogen is 'dissolved' in the metal, which continues to be a good conductor of electricity. The hydrogen is in fact hidden in holes or **interstices** within the metal lattice, and such compounds are called **interstitial compounds**.

Patterns in the oxides

The oxides of the elements of periods 2 and 3 show trends in their physical and chemical properties similar to those in the chlorides and hydrides. These trends in properties are closely linked to the pattern of electronegativity of the elements.

Physical properties

Just like the hydrides and the chlorides, the oxides of the elements at the left-hand end of a period have giant ionic structures, formed when oxygen removes electrons from an electropositive element. Such oxides have high melting points and conduct electricity when molten – examples are beryllium oxide, BeO, sodium oxide, Na_2O and aluminium oxide, Al_2O_3. At the other end of the period, the difference in electronegativity between oxygen and the other elements is much smaller, resulting in covalent bonds in which electrons are shared rather than transferred. This gives rise to molecular structures that have low melting and boiling points and that do not conduct electricity – carbon dioxide, CO_2 and the oxides of sulphur, SO_2 and SO_3, are examples of this. In the middle of each period, the elements boron and silicon (both metalloids) form oxides with giant covalent structures. As we saw in section 1.6, such structures lead to very high melting points and make the solids very hard – properties shared by both boron oxide, B_2O_3 and silicon(IV) oxide, SiO_2. The tightly bound electrons, together with the lack of ions in these substances, means that neither boron oxide nor silicon(IV) oxide conducts electricity, whether molten or solid. The extremes of the reactivities of the elements with oxygen are illustrated in figure 2.2.2.

Enthalpy changes of formation

The oxides are generally very stable compounds, as oxygen is one of the most electronegative elements, and the enthalpy changes of formation of the oxides are usually negative. As we move across and up the periodic table, decreasing differences in electronegativity cause enthalpy changes of formation to become increasingly less negative, reducing the thermodynamic stability of the oxides with respect to their elements. These trends are shown in table 2.2.3.

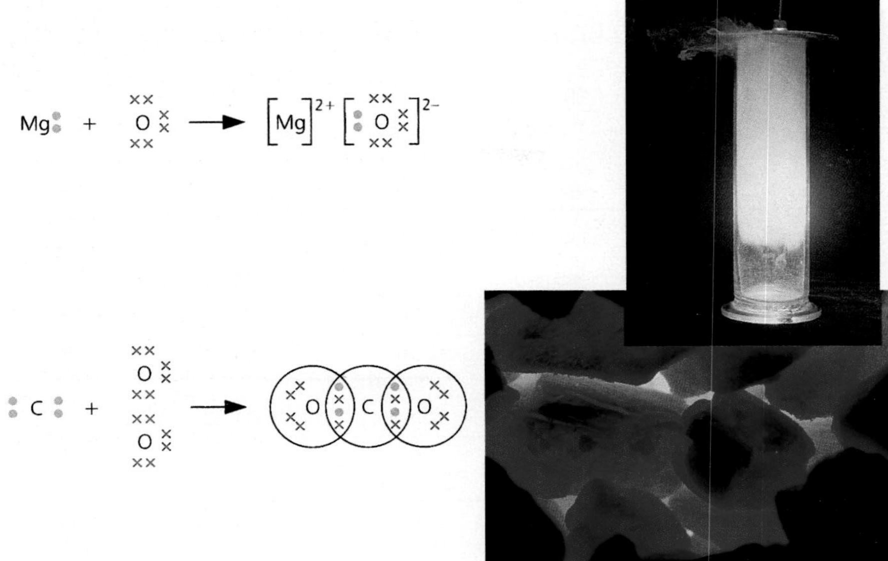

Figure 2.2.2 The reaction of magnesium and oxygen is rapid and strongly exothermic, giving off a flash of bright white light. Ionic oxides such as this are formed when electropositive elements donate two electrons to an oxygen atom, forming a positive ion along with the negative oxide ion. The more gently exothermic reaction between carbon and oxygen has been used for centuries to provide heat in the home. Covalent oxides are the result of sharing electrons between oxygen and another relatively electronegative element. Either giant or simple molecular structures are the result.

2

Reactions with water

The acid–base properties of the oxides are also linked to their structure. At the left-hand end of a period, the oxides of the electropositive elements are highly basic, and form solutions that are alkaline. The oxide ion removes H^+ ions from water, hence sodium oxide reacts with water to form sodium hydroxide:

$$\mathbf{Na_2O(s) + H_2O(l) \rightarrow 2NaOH(aq)}$$

Most of the oxides of the elements at the right-hand end of the periodic table show acidic behaviour, the exceptions to this being carbon monoxide, CO, dinitrogen oxide, N_2O and nitrogen oxide, NO. These oxides do not behave as acids or bases, and so are classified as neutral oxides. The reactions of some oxides with water are shown below.

Boron(III) oxide reacts with water forming boric acid, a very weak acid:

$$\mathbf{B_2O_3(s) + 3H_2O(l) \rightarrow 2H_3BO_3(aq)}$$

Carbon dioxide dissolves readily in water and reacts to form carbonic acid, H_2CO_3. This acid cannot be isolated, and carbon dioxide is usually shown reacting directly with water to form hydrogen ions and hydrogencarbonate ions:

$$\mathbf{CO_2(aq) + H_2O(l) \rightleftharpoons H^+(aq) + HCO_3^-(aq)}$$

The oxides of nitrogen, phosphorus, sulphur and chlorine form strong acids:

$$\mathbf{2NO_2(g) + H_2O(l) \rightarrow \underset{\text{nitric acid}}{HNO_3(aq)} + \underset{\text{nitrous acid}}{HNO_2(aq)}}$$

$$\mathbf{P_4O_{10}(s) + 6H_2O(l) \rightarrow 4H_3PO_4(aq)}$$
phosphoric(V) acid ('orthophosphoric acid')

$$\mathbf{SO_2(g) + H_2O(l) \rightarrow H_2SO_3(aq)}$$
sulphurous acid

$$\mathbf{SO_3(l) + H_2O(l) \rightarrow H_2SO_4(aq)}$$
sulphuric acid

$$\mathbf{Cl_2O_7(l) + H_2O(l) \rightarrow 2HClO_4(aq)}$$
chloric(VII) acid ('perchloric acid')

Property of oxide	Period 2						
Formula	Li_2O	BeO	B_2O_3	CO_2 (CO)	N_2O (NO) (NO_2) (N_2O_4) (N_2O_5)	O_2	OF_2
State at 20 °C	Solid		Gases except the solid N_2O_5				
Melting point/°C	>1700	2550	450	—	−91	−218	−224
Boiling point/°C	—	3900	1860	−78	−88	−183	−145
Electrical conductivity in liquid state	Good	Fair	Very poor	None	None	None	None
Structure	Giant structures – ionic and molecular			Simple covalent molecules			
$\Delta H^\ominus_{f,m}$ per mol of O atoms/kJ mol^{-1}	−598	−610	−418	−197	82	0	−22
Reaction with water	Forms $LiOH$ (aq) – an alkaline solution	Does not react with water	Forms H_3BO_3, a very weak acid	Forms H_2CO_3, a weak acid	NO_2 forms HNO_3 and HNO_2, strong acids	No reaction, small amount dissolves	Reacts slowly to form O_2 and a solution of the acid HF
Nature of oxide	Basic	Amphoteric	Acidic	Acidic	Acidic		Acidic
Property of oxide	**Period 3**						
Formula	Na_2O	MgO	Al_2O_3	SiO_2	P_4O_{10} (P_4O_6)	SO_3 (SO_2)	Cl_2O_7 (Cl_2O)
State at 20 °C	Solid	Solid	Solid	Solid	Solid (Solid)	Liquid (Gas)	Liquid (Gas)
Melting point/°C	1275	2852	2027	1610	24	17	−92
Boiling point/°C	—	3600	2980	2230	175	45	80
Electrical conductivity in liquid state	Good	Good	Good	Very poor	None	None	None
Structure	Giant structures – ionic and molecular				Simple covalent molecules		
$\Delta H^\ominus_{f,m}$ per mol of O atoms/kJ mol^{-1}	−414	−602	−559	−456	−298	−147	38
Reaction with water	Forms $NaOH$(aq), an alkaline solution	Forms $Mg(OH)_2$, weakly alkaline	Does not react	Does not react	P_4O_{10} reacts to form H_3PO_4, an acidic solution	SO_3 forms H_2SO_4, a strong acid	Cl_2O_7 forms $HClO_4$, an acidic solution
Nature of oxide	Basic	Weakly basic	Amphoteric	Acidic	Acidic	Acidic	Acidic

Table 2.2.3 The oxides of the elements in periods 2 and 3 change from being ionic, basic metal oxides with high melting and boiling points to simple molecular, acidic non-metal oxides with low melting and boiling points. In between are the giant molecular oxides of the metalloids, which have high melting and boiling points and are amphoteric in character.

The oxide of silicon, SiO_2, also behaves as a very weak acid. Although it does not react with water, it will react with a base like sodium hydroxide, forming the silicate ion, SiO_3^{2-}:

$$\mathbf{SiO_2(s) + 2OH^-(aq) \rightarrow SiO_3^{2-}(aq) + H_2O(l)}$$

The oxides of beryllium and aluminium are **amphoteric**, showing both acidic and basic character. Beryllium oxide, BeO, does not dissolve in water, but it dissolves in acids, reacting like a basic oxide to form beryllium salts:

$$\mathbf{BeO(s) + 2H^+(aq) \rightarrow Be^{2+}(aq) + H_2O}$$

The aqueous Be^{2+} ion is in fact a **complex ion**, surrounded by four water molecules, $[Be(H_2O)_4]^{2+}$. (We shall study complex ions in section 2.7.) Beryllium oxide reacts with alkalis like an acidic oxide, to form beryllates:

$$\mathbf{[Be(H_2O)_4]^{2+}(aq) + 4OH^-(aq) \rightarrow [Be(OH)_4]^{2-}(aq) + 4H_2O(l)}$$

Aluminium oxide is also amphoteric. It reacts with acidic solutions to form Al^{3+} salts and with alkalis to form aluminates:

$$\mathbf{Al_2O_3(s) + 6H^+(aq) \rightarrow 2Al^{3+}(aq) + 3H_2O(l) \quad (basic\ character)}$$

$$\mathbf{Al_2O_3(s) + 2OH^-(aq) + 3H_2O(l) \rightarrow 2Al(OH)_4^-(aq) \quad (acidic\ character)}$$

The aqueous Al^{3+} ion is also a complex ion, surrounded by six water molecules, $[Al(H_2O)_6]^{3+}$. We shall meet it again in section 2.5.

Having looked at the properties of the elements of the periodic table and some of their compounds, we have seen that they fall into regular patterns. We shall now move on to consider in more detail some of the chemistry of the periodic table, shifting our emphasis from the horizontal to the vertical groupings.

SUMMARY

- Like the elements, the compounds formed by the elements show periodic patterns in their properties.
- At the left-hand end of the second and third periods, the chlorides are ionic, dissolving in water to form neutral solutions. Moving to the right across the period, the bonding in the chlorides becomes covalent and they form acidic solutions in water.
- At the left-hand end of the second and third periods, the hydrides are ionic, reacting with water to form hydrogen and alkaline solutions. Moving right across the period, the bonding in the hydrides becomes covalent. Hydrides at the extreme right-hand end of each period display acidic character.
- The hydrides of nitrogen, oxygen and fluorine have unexpectedly high melting and boiling points due to hydrogen bonding between the molecules.
- At the left-hand end of the second and third periods, the oxides are ionic, reacting with water to form alkaline solutions. Moving right across the period, the bonding in the oxides becomes covalent and they form acidic solutions in water. In the middle of the periods the oxides display **amphoteric** character, behaving as both acids and bases.
- Water is of universal importance as a solvent and the medium for life. Its properties include:
  it is an excellent solvent, being able to solvate ions and polar molecules
  its density decreases with decreasing temperature between 4 °C and 0 °C, causing ice to float on liquid water
  it has high enthalpy changes of fusion and vaporisation and high melting and boiling points due to hydrogen bonding
  it has a high surface tension.

QUESTIONS

1 Look at the data concerning the chlorides of the period 2 elements in table 2.2.1. How are the physical properties of the following compounds explained by their structure?
a LiCl **b** $BeCl_2$ **c** NCl_3

2 Table 2.2.1 gives details of the reactions of the chlorides of period 3 elements with water. How are these reactions related to the electronegativities of the period 3 elements?

3 a Using the data given in table 2.2.2, plot a graph of the enthalpy change of formation per mole of hydrogen atoms for the hydrides of the elements of periods 2 and 3. (The enthalpy change of formation should appear on the graph's vertical axis, with the atomic number of the elements on the horizontal axis.)
b Using the electronegativity data in figure 1.4.21, page 60, calculate the difference in electronegativity between hydrogen and each of the elements in periods 2 and 3.
c Comment on the patterns in your answers to **a** and **b**.

4 Elements A and B form oxides with the formulae A_2O and BO_2. Some properties of these two oxides are given in table 2.2.4.

Table 2.2.4

Property	A_2O	BO_2
Solubility in water	Good	Good
pH of solution	Alkaline	Acidic

Suggest what you can deduce about:
a the groups to which elements A and B belong
b the structure of each oxide
c the melting and boiling points of each oxide
d the electrical conductivity of each element.

Developing Key Skills

Write an AS or A2 level question on the chemistry of water. The question should have several parts and should test the level of a student's knowledge of both the chemistry of water and the way it does – or does not – fit into the expected pattern of periodicity. If possible, include some data handling within the question.

[Key Skills opportunities: C, A]

2.3 The s-block elements

The elements of groups I and II of the periodic table are known as the **s-block elements** because their outermost electrons are found in the s subshells. The elements in group I are called the **alkali metals**, whilst those of group II are the **alkaline–earth metals**.

Group I – The alkali metals

3 **Lithium, Li** $1s^2 2s^1$
11 **Sodium, Na** $1s^2 2s^2 2p^6 3s^1$
19 **Potassium, K** (Ar) $4s^1$
37 **Rubidium, Rb** (Kr) $5s^1$
55 **Caesium, Cs** (Xe) $6s^1$
87 **Francium, Fr** (Rn) $7s^1$

Sodium lighting is a common feature on our roads.

Sodium hydroxide is the cheapest and most used industrial alkali, produced by the electrolysis of brine. It is used for the manufacture of cellulose acetate (rayon), paper and soap.

Sodium and potassium are stored under oil to prevent reaction with the air.

Sodium cyanide is used in the extraction of gold and the hardening of steel.

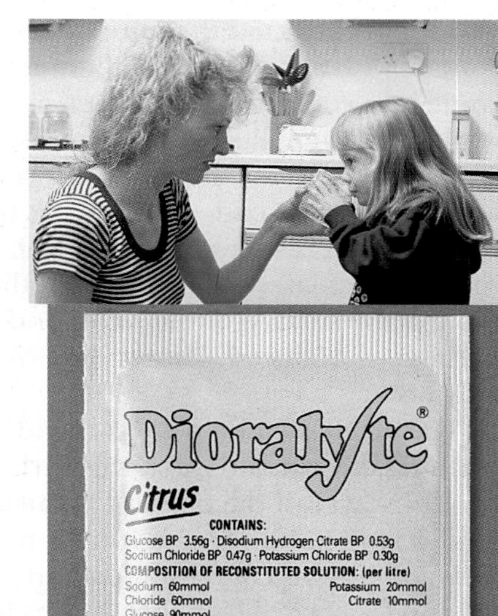

When severe sickness and diarrhoea dehydrate the body, death can follow quickly, particularly in young children. An electrolyte mixture such as this contains salts of both sodium and potassium and rapidly restores the fragile balance of the body.

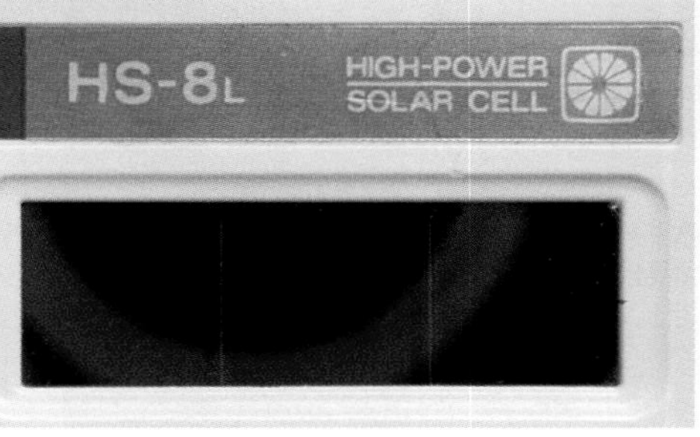

The very low first ionisation energy of caesium makes it ideal for use in photocells. When struck by light, a caesium atom emits an electron, which is attracted to a positive electrode and initiates an electric current in the circuit.

Figure 2.3.1 The alkali metals are very reactive – so much so that they are of relatively little use in their elemental state. But they make stable compounds that are useful – a few examples are shown here.

The alkali metals are a distinctive group. (Francium was discovered in 1939 by Marguerite Perey. As it is radioactive, with a half-life of less than 5 minutes, its chemical properties are difficult to determine and it will not feature further in this discussion of the alkali metals.) They are chemically the most reactive metals, and become increasingly reactive down the group. This is because the single outer electron is increasingly screened from the positive charge of the nucleus by successive electron shells, and so is lost more and more readily. Because of this great reactivity, the metals do not occur uncombined in nature, and when extracted the pure metals are usually stored in an inert liquid, out of contact with air or water. Rubidium and caesium burn spontaneously when they come into contact with air. Potassium and sodium burn when heated gently, whilst lithium melts before it burns.

All the alkali metals react with cold water, displacing hydrogen and forming an alkaline hydroxide solution – it is from this that the group takes it name. The reaction is shown on page 159. Lithium floats and reacts quietly, but the reaction increases in vigour down the group. Sodium and potassium dart around the surface of the water and the heat generated may ignite the hydrogen produced. Unless these reactions are carefully controlled, explosions may occur.

The alkali metals are all soft and white, with a characteristic lustre or sheen when freshly cut, although this surface tarnishes very rapidly with the formation of a thin layer of the oxide. The metals are all good conductors of heat and electricity. The single electron in the outer s subshell is readily lost to form ions with a noble gas configuration and a single positive charge, for example Na^+, K^+ and Cs^+. The alkali metals thus have only one oxidation state, +1.

The small size of the lithium ion means that it tends to distort the charge cloud around neighbouring anions in crystal lattices. This is particularly marked in the case of large anions like the iodide ion, I^-, with the result that lithium compounds often show quite marked covalent character.

Polarising power

We saw in section 1.4 how covalent and ionic bonds represent two extremes of bonding, and how the bonding in the majority of compounds lies somewhere in between these two extremes. The factors that affect the distortion of the electron cloud between a pair of ions are summarised by **Fajans' rules**. These give the conditions under which a pair of ions is likely to form a bond with some degree of covalent character:

Positive ion (cation)	Negative ion (anion)
High charge	High charge
Small size	Large size
⇒ large tendency to attract electrons towards ion	⇒ diffuse electron cloud which is easily distorted

The combination of a small positively charged ion (such as Li^+) and a large negative ion (like I^-) leads to a substantial distortion of the electron cloud around the negative ion, as was shown in figure 1.4.20 on page 59. Note that Fajans' rules are consistent with the predictions made about ionic and covalent character using the concept of electronegativity – they simply represent another way of looking at the ionic/covalent character of a bond.

Sodium metal and nuclear power

Liquid sodium can be used as the primary coolant in a nuclear reactor. It readily absorbs the heat produced by the nuclear reactions in the core of the reactor. The hot liquid sodium can then be used to boil water (which is the secondary coolant), producing steam. This steam then drives turbines to generate electricity in the usual way.

A major advantage of using sodium as a primary coolant is that it does not boil at atmospheric pressure until it reaches 881 °C. This enables the reactor cooling system to operate at very high temperatures, but at atmospheric pressure. If water is used as the primary coolant, it must be contained under high pressure to prevent it boiling, and even so, temperatures greater than 374 °C cannot be used, since water will boil above this temperature no matter what the pressure. Using liquid sodium, the cooling pipework in the reactor does not have to be built to withstand enormous pressures.

Figure 2.3.2 Dounreay nuclear power plant, decommissioned in 1994, used sodium as its primary coolant. Although there are now no sodium-cooled reactors in the UK, this technology is still used in other parts of the world.

The physical properties of the elements of group I are summarised in table 2.3.1. The first ionisation energy is that for the single electron in the outer s subshell. The much higher second ionisation energy represents in part the increased attraction of the nucleus to the remaining electrons once one negative charge has been lost, but primarily reflects the stability associated with a full shell of electrons.

The atoms of the elements get larger down the group as more shells of electrons are added, screening the outer electrons from the increasing nuclear charge. The addition of extra shells of electrons down the group also leads to the trend in electronegativity shown in the table. Atoms of the lower elements are more likely to have electron density transferred away from them in a chemical bond than atoms higher up the group, due to their larger clouds of electrons. This tendency is reflected in the decreasing electronegativity of the elements down the group.

Atoms of the lower elements of the group are held together less strongly than those higher up, leading to lower melting points. The decreasing strength of the metallic bond can be explained in terms of the decreasing charge density

Element	Li	Na	K	Rb	Cs
Metallic radius/nm	0.16	0.19	0.24	0.25	0.27
Ionic radius/nm	0.074	0.102	0.138	0.149	0.170
Electronegativity	1.0	0.9	0.8	0.8	0.7
First ionisation energy/kJ mol^{-1}	520	496	419	403	376
Second ionisation energy/kJ mol^{-1}	7298	4563	3051	2632	2420
Melting point/°C	181	98	63	39	29
Boiling point/°C	1342	883	760	686	669
Density/g cm^{-3}	0.53	0.97	0.86	1.53	1.88

Table 2.3.1 Some physical properties of the group I elements

Flame tests

When burned in air, all the alkali metals produce flames with a characteristic colour. This can be used to identify their presence in unknown compounds, using a **flame test**. The colours produced are shown in table 2.3.2.

Element	Flame colour
Lithium	Carmine red ▶
Sodium	Golden yellow
Potassium	Lilac
Rubidium	Red
Caesium	Blue

Table 2.3.2 The distinctive colours of the alkali metals in a flame test is a useful tool in the analysis of unknown compounds in the laboratory.

of the positive ions in the metal lattice as the ions get larger, holding the lattice together less strongly – see box on page 78.

All the alkali metals have a body-centred cubic structure at 25 °C, which together with their large atomic radius accounts for their low density. The density of the alkali metals increases as the group is descended, showing that the increase in the mass of the atoms more than offsets the increase in radius.

The intense white light emitted by magnesium when it burns in air was used for photography in dim conditions in days gone by.

Group II – The alkaline–earth metals

4 **Beryllium, Be** $1s^22s^2$
12 **Magnesium, Mg** $1s^22s^22p^63s^2$
20 **Calcium, Ca** $(Ar)4s^2$
38 **Strontium, Sr** $(Kr)5s^2$
56 **Barium, Ba** $(Xe)6s^2$
88 **Radium, Ra** $(Rn)7s^2$

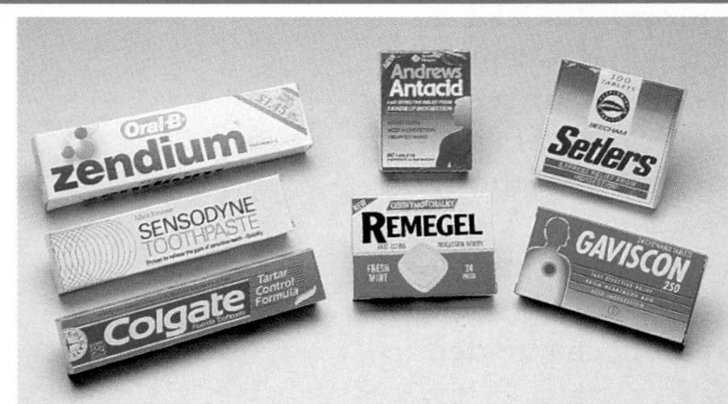

Magnesium hydroxide is a weak alkali and is used as a component of toothpastes, to neutralise acid in the mouth and prevent tooth decay. It also forms an important part of many indigestion medicines, helping to neutralise excess acid in the stomach.

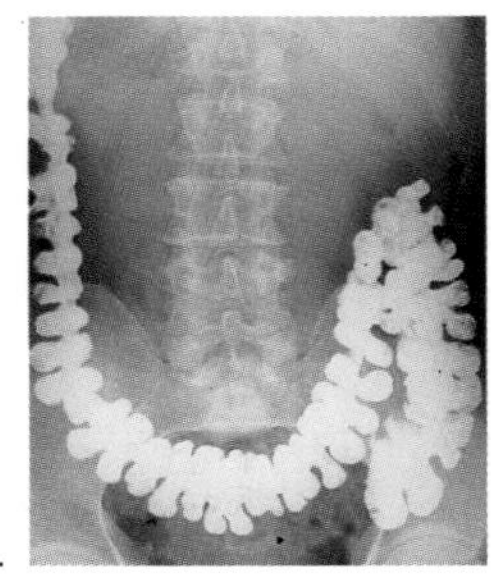

Barium sulphate is insoluble, and opaque to X-rays. Made up into a porridge-like drink and swallowed by the patient, it helps in the diagnosis of disorders of the digestive system, as it gives clear definition to the soft tissues.

The most important compounds of calcium in commercial terms are lime, CaO and limestone, $CaCO_3$. Lime is used in the manufacture of cement, mortar and plaster, and is also spread in vast quantities onto the soil to help reduce soil acidity.

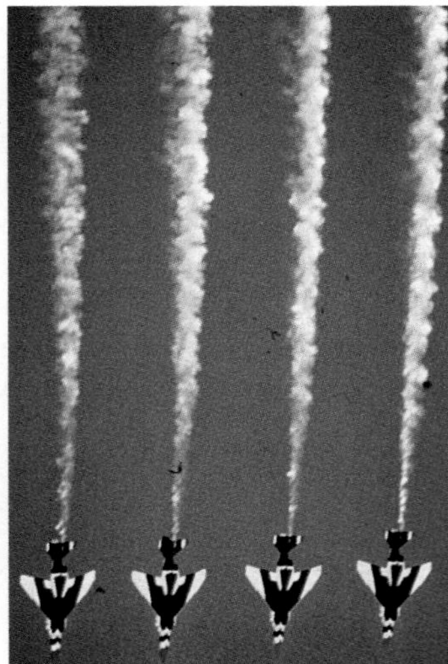

Magnesium is the member of the alkaline–earths most used in engineering, but in recent years beryllium has been used too, for example in the rudders of fighter jets.

Figure 2.3.3 Some of the alkaline–earth metals are more useful in their elemental forms than the alkali metals, but in general it is their compounds that are used in manufacturing and industry.

Group II of the periodic table is another family of very reactive metals in which the electropositive character and reactivity increases markedly on passing down the group. Radium, the last member of the group, was discovered by Marie and Pierre Curie in 1898. Like francium, it is radioactive, and whilst the properties of the known compounds of radium follow the expected trends, it will not be included in the general discussion of the group. Beryllium is also somewhat unusual. Whilst it belongs in group II, it is much more similar to the diagonally placed aluminium in group III. Its properties do not clearly follow group trends, in particular its tendency to form covalent compounds such as beryllium chloride (see table 2.2.1, page 144) and the amphoteric properties of its oxide (page 150).

The alkaline–earth metals have two electrons in their outer s subshell. Thus they tend to be less reactive than the group I elements, as they need to lose two electrons to achieve a noble gas configuration. All the alkaline–earth metals burn in air, in strongly exothermic reactions. Several of them also produce typical flame colours – calcium gives brick-red, strontium red and barium green. All the metals except beryllium react with cold water, displacing hydrogen and forming an alkaline hydroxide. Magnesium reacts only very slowly, even when powdered, but calcium, strontium and barium react with increasing vigour.

Apart from beryllium, the alkaline–earth metals are white, relatively soft metals which are good conductors of heat and electricity. The freshly cut surfaces are shiny, but they quickly oxidise and this oxide film protects the metal from further reactions. The ions formed by all the members of group II are the result of the loss of the two valency electrons in the s subshell. The ions all have an ionisation number of 2, for example Ca^{2+}, Mg^{2+} and Ba^{2+}. The ionisation energies show clearly that whilst it is slightly more difficult to remove the second electron in the outer s subshell than the first, due to the overall increased attraction of the nucleus, the main difference comes with the removal of the third electron, from the next shell in. It is also worth noting that the ionisation energies get smaller down the group – as the number of shells increases, so the attractive pull of the nucleus is felt less strongly by the outer electrons. They are screened more effectively from its positive charge, and so are removed more easily.

Element	**Be**	**Mg**	**Ca**	**Sr**	**Ba**
Metallic radius/nm	0.11	0.16	0.20	0.22	0.22
Ionic radius/nm	0.027	0.072	0.100	0.113	0.136
Electronegativity	1.5	1.2	1.0	1.0	0.9
First ionisation energy/kJ mol^{-1}	900	738	590	550	503
Second ionisation energy/kJ mol^{-1}	1757	1451	1145	1064	965
Third ionisation energy/kJ mol^{-1}	14 849	7733	4912	4210	3390
Melting point/°C	1278	649	839	769	725
Boiling point/°C	2970	1107	1484	1384	1643
Density/g cm^{-3}	1.85	1.74	1.54	2.62	3.51

Table 2.3.3 Some physical properties of the group II elements

As with the alkali metals, the electronegativity of the alkaline earth metals falls down the group, due to the addition of outer shells of electrons. The trends in the other physical properties of the group II elements are not so apparent as those of the group I metals, especially as far as melting point and density are concerned. In general the overall trends are the same as for the alkali metals, but the smooth trend is disrupted by changes in the type of packing in the metal lattices.

Patterns in the properties of groups I and II

Several important patterns can be observed in studying the s-block elements. The similarities between the elements making up groups I and II are already becoming apparent. A closer look reveals not only the differences between the elements of the two groups, but also the trends in the properties of their compounds.

Patterns in the physical properties

Some of the patterns in the physical properties of the elements have already been considered. Other trends can be picked out by careful observation. The elements of groups I and II have larger atomic radii than the other elements in their periods. A simple explanation of this is that the outer s electrons are held relatively weakly by the nucleus. Because of this increased atomic size, the attractive forces between the atoms in the metal lattices are also weakened, and so the melting and boiling points of the elements are relatively low for metals. For example, most of the transition metals melt at over 1000 °C, whereas the metals of group I all melt at less than 200 °C. This also explains why the metals are relatively soft and have low densities.

A fuller explanation of the trends in the physical properties uses the ideas of charge density we met in section 1.6. As we move from group I through group II to the transition elements, the charge density of the ions in the metal lattice and of the delocalised electron cloud surrounding them increases. This leads to increasing attraction between the electron cloud and the metal cations, increasing the strength of the metallic bond and so increasing the hardness of the metals and their melting points.

Energy changes during reactions

The differences in reactions between the elements of groups I and II can be explained in part by considering their ionisation energies. It requires less energy to remove the first electron from group II elements than from group I elements. However, group II elements need more energy than group I elements to achieve a noble gas configuration, since this requires the removal of two electrons rather than one. This means that group I elements react more readily, and that also their reaction rates tend to be faster as activation energies are lower. (The activation energy is the 'push' needed to start a reaction, explained in section 3.6.) By the same token, the elements become more reactive down both groups as the ionisation energies fall with increased screening of the nucleus as the atomic radius gets larger. This trend is reflected in the enthalpy changes of formation of the chlorides of the group II elements (table 2.3.4).

Element	$\Delta H^\ominus_{f,m}$ of chloride/kJmol^{-1}
Be	–490
Mg	–641
Ca	–796
Sr	–829
Ba	–859

Table 2.3.4 The enthalpy changes of formation for the chlorides of the group II elements

Patterns in the chemical properties

All the elements of groups I and II are high in the electrochemical series (see box on page 159). As a result, they are all good reducing agents, and most of them react with water, reducing it to hydrogen and the metal hydroxide. The equations for the reaction with water may be represented in various ways, as table 2.3.5 shows. Notice that the group II hydroxides are much less soluble than the group I hydroxides – we shall discuss the reasons for this very shortly.

	Group I	Group II
General equations	$M(s) + H_2O(l) \rightarrow \frac{1}{2}H_2(g) + MOH(aq)$	$M(s) + 2H_2O(l) \rightarrow H_2(g) + M(OH)_2(s)$ or (aq)
	$M(s) \rightarrow M^+(aq) + e^-$ $\quad H_2O(l) + e^- \rightarrow \frac{1}{2}H_2(g) + OH^-(aq)$	$M(s) \rightarrow M^{2+}(aq) + 2e^-$
Example:	$2Na(s) + 2H_2O(l) \rightarrow 2NaOH(aq) + H_2(g)$	$Mg(s) + 2H_2O(l) \rightarrow Mg(OH)_2(s) + H_2(g)$

Table 2.3.5 Representing the reactions of group I and II elements with water

The equations in table 2.3.5 do not tell us anything about reaction rates or conditions. In the examples, sodium and magnesium are shown reacting with water. Both these reactions are accompanied by the release of a large amount of energy, but the conditions needed for each reaction are very different – sodium reacts violently with cold water, while magnesium reacts at an appreciable rate only with steam. We shall find out more about making predictions about reactions in section 3.

Redox reactions and the electrochemical series

The reactions of metals with water are examples of **redox** reactions – reactions in which **red**uction and **ox**idation take place, as mentioned in section 1.4.

Probably one of the first types of redox reaction that most of us come across is the reaction of metals with oxygen in the air. If the metal is sodium, this can be represented as:

$$4Na(s) + O_2(g) \rightarrow 2Na_2O(s)$$

When this reaction happens, sodium has electrons removed from it, which are transferred to oxygen. We say that sodium is **oxidised** and oxygen is **reduced**. These processes can be shown more clearly by writing two **half-equations** for the reaction:

$$4Na \rightarrow 4Na^+ + 4e^-$$

$$O_2 + 4e^- \rightarrow 2O^{2-}$$

Notice how this way of writing the reaction emphasises the fact that the four electrons required by the two oxygen atoms in order to achieve a noble gas configuration are supplied by the four sodium atoms. In the process, the sodium atoms achieve a noble gas configuration too.

We saw in section 1.4 that redox reactions can be followed using oxidation numbers, which balance on either side of the equation. Oxidation numbers obviously do *not* balance on either side of a half-equation – both half-equations need to be considered together when balancing oxidation numbers.

In the reaction of sodium with oxygen, the sodium atom loses an electron and is oxidised to Na^+, while an oxygen atom gains electrons and is reduced to O^{2-}. Sodium acts as a **reducing agent**, while oxygen behaves as an **oxidising agent**. This illustrates the modern meaning of reduction and oxidation, in which:

Oxidation involves a *loss* of electrons, while reduction involves a *gain* of electrons.

Figure 2.3.4 Redox reactions are part of everyday life, for example, the rapid burning of a fuel or the much slower 'burning' of food in respiration.

The **electrochemical series** arranges elements in order of their power as reducing agents, with the most powerful normally placed first in the series. We shall find out how we place the elements in the series in section 3.5. A familiar form of the electrochemical series shows the relative reactivities of some of the metals. This is often used to predict whether one metal will **displace** another in a given reaction. The series is often called the **reactivity series of metals** for this reason. For example, the series:

Na Mg Al Zn Fe Sn H Cu Ag

shows that zinc is a stronger reducing agent than copper, and so zinc metal will displace copper from a solution of copper(II) sulphate, since it can 'force' its electrons into the copper ion:

$$Zn(s) + Cu^{2+}(aq) \rightarrow Zn^{2+}(aq) + Cu(s)$$

Half-equations: $Zn \rightarrow Zn^{2+} + 2e^-$

$$Cu^{2+} + 2e^- \rightarrow Cu$$

The presence of hydrogen in the series shows which metals can be obtained from their ores by reduction with hydrogen. For example, copper oxide can be reduced by hydrogen, since hydrogen is a stronger reducing agent than copper:

$$H_2(g) + Cu^{2+}(aq) \rightarrow 2H^+(aq) + Cu(s)$$

Half-equations: $H_2 \rightarrow 2H^+ + 2e^-$

$$Cu^{2+} + 2e^- \rightarrow Cu$$

On the other hand, hydrogen cannot be used to reduce zinc oxide at room temperature, since zinc is a stronger reducing agent than hydrogen.

In many circumstances, oxidation numbers can be used to help decide whether an element has been oxidised or reduced in a reaction. An obvious example of this is in the reaction of carbon with oxygen:

$$\underset{0}{C} + \underset{0}{O_2} \rightarrow \underset{+4\,-2}{CO_2}$$

Even though this reaction does not involve a complete transfer of electrons, it is clearly oxidation. The change in oxidation number of

carbon from 0 to +4 shows that it has been oxidised, while oxygen has changed from 0 to –2, so it has been reduced.

Oxidation numbers can also help to decide which element has been reduced and which oxidised in reactions involving complicated ions. As an example, consider the half-equation:

$$Cr_2O_7^{2-} + 14H^+ + 6e^- \rightarrow 2Cr^{3+} + 7H_2O$$

Which element has been reduced? Rewriting the half-equation with oxidation numbers shows us:

$$\underset{+6\ -2}{Cr_2O_7^{2-}} + \underset{+1}{14H^+} + 6e^- \rightarrow \underset{+3}{2Cr^{3+}} + \underset{+1\ -2}{7H_2O}$$

It is chromium that has been reduced, since its oxidation number has changed from +6 to +3.

We shall look in more detail at the process of oxidation and reduction in section 3.5. In particular, we shall be interested in how we can be more systematic about measuring the power of oxidising and reducing agents.

The oxides of the group I elements

At first sight the most likely formula for the oxides of the group I elements is M_2O, in which there is a lattice of M^+ and O^{2-} ions. However, when the alkali metals are heated strongly in oxygen, the following oxides are formed:

Li_2O Na_2O_2 KO_2 RbO_2 CsO_2

The reason for this is connected with the increase in the size of the alkali metal ions as the group is descended. Li^+ ions and O^{2-} ions form a stable lattice in which there are two lithium ions for each oxide ion, since the lithium ion is very small. K^+, Rb^+ and Cs^+ ions are much larger, and combine with superoxide ions, O_2^-, making a stable lattice in which there are equal numbers of metal ions and superoxide ions. Na^+ ions – which are larger than Li^+ ions but not as large as K^+ ions – combine with peroxide ions, O_2^{2-}, producing a stable lattice in which there are two sodium ions for every peroxide ion.

Thermal stability of the salts of the s-block elements

When the s-block elements react, they form salts containing positive metal ions, either M^+ or M^{2+}. As we saw in section 1.8, pages 117–19, there are several key factors in predicting the stability of an ionic compound, not the least of which is its enthalpy change of lattice formation, which is determined by the charge and size of the ions in the lattice. Because of this, the stability of ionic compounds increases (a) as ionic radius decreases and (b) as the charge on the ions increases.

In considering the stability of the salts of the s-block elements, as well as thinking about decomposing them to their constituent elements, we must also remember that other compounds may be formed. In particular, the CO_3^{2-} anion may decompose to produce CO_2 and the O^{2-} anion, while the NO_3^- anion may also decompose to give the O^{2-} anion, or the larger NO_2^- anion. The thermal stability of the carbonates and nitrates of the group I and II elements is greatly dependent on these reactions, as we shall now see.

- When heated in an ordinary Bunsen flame, the group I carbonates are stable, with the exception of lithium carbonate, Li_2CO_3. The group II carbonates, together with lithium carbonate, decompose to form stable oxides, with the formation of carbon dioxide gas, for example:

$$CaCO_3(s) \rightarrow CaO(s) + CO_2(g)$$

The stability of the carbonates to heat increases as group II is descended, as the temperatures for complete decomposition (table 2.3.6) show.

$MgCO_3$	$CaCO_3$	$SrCO_3$	$BaCO_3$
540 °C	900 °C	1290 °C	1360 °C

Table 2.3.6 The temperatures required to decompose fully the group II carbonates

- The nitrates decompose on heating in a Bunsen flame. Group I nitrates, with the exception of lithium nitrate, $LiNO_3$, form their corresponding nitrites, and these are then stable to heat. The relatively small decrease in size from

2

the NO_3^- ion to the NO_2^- ion is sufficient to achieve thermal stability, when paired with the group I cations.

$$2NaNO_3(s) \rightarrow 2NaNO_2(s) + O_2(g)$$

In contrast, lithium nitrate and all the group II nitrates decompose on heating to form the corresponding oxide, these cations needing the much smaller O^{2-} ion to give thermal stability to the compound.

$$4LiNO_3(s) \rightarrow 2Li_2O(s) + 4NO_2(g) + O_2(g)$$

$$2Mg(NO_3)_2(s) \rightarrow 2MgO(s) + 4NO_2(g) + O_2(g)$$

- The hydroxides follow the same pattern as the carbonates and nitrates. All group I hydroxides are stable up to quite high temperatures, with lithium hydroxide being the first to decompose, at around 650 °C. All the hydroxides decompose to form the corresponding oxide and water, for example:

$$2LiOH(s) \rightarrow Li_2O(s) + H_2O(g)$$

$$Ca(OH)_2(s) \rightarrow CaO(s) + H_2O(g)$$

Figure 2.3.5 The thermal decomposition of limestone (calcium carbonate), originally carried out in kilns like this one in Derbyshire, has been important in agriculture and building for years. Calcium oxide (lime) is used to treat acidic soil in farming, and forms the basis of mortar and some types of plaster. When treated with water – a process called **slaking** – calcium oxide forms calcium hydroxide, also called **slaked lime**.

Solubility of the salts of the s-block elements

When an ionic solid dissolves in water, the ionic lattice breaks apart, and the ions become surrounded by water molecules in a process called **solvation**. It is difficult to make predictions about the solubilities of salts, because the ease with which this process occurs depends on two factors.

One of these is the reverse of the lattice formation process, in which the salt forms ions. This requires an input of energy (it is endothermic), and the quantity of energy required is given by the lattice enthalpy of the solid, since we are dealing with the process:

$$M^+X^-(s) \rightarrow M^+(g) + X^-(g)$$

As we saw in section 1.8, as the size of the ions in a solid increases, so its lattice enthalpy decreases. This means that less energy will be required to separate a lattice containing large ions than one containing small ions. On the other hand, as lattice enthalpy increases with ionic charge, more energy will be required to break up a lattice containing ions with a greater charge.

The second stage of the process is the hydration of the separated ions. This is an exothermic process, so energy is given out. Predictions about the solubility of a salt are therefore difficult to make because it is not easy to know how the energy accounting in these two elements of the dissolving process will balance out.

Enthalpy changes of solution from enthalpy cycles

When one mole of an ionic solid dissolves to form an infinitely dilute solution under standard conditions, the enthalpy change is called the **standard molar enthalpy change of solution** of that substance, $\Delta H^{\ominus}_{soln,m}$. As we have seen, this process is made up of two parts, and so the enthalpy changes can be represented in a cycle. One such cycle is shown in figure 2.3.6, for sodium chloride dissolving. Notice that the enthalpy change of solution of sodium chloride is a small positive number, so the dissolving of sodium chloride is endothermic.

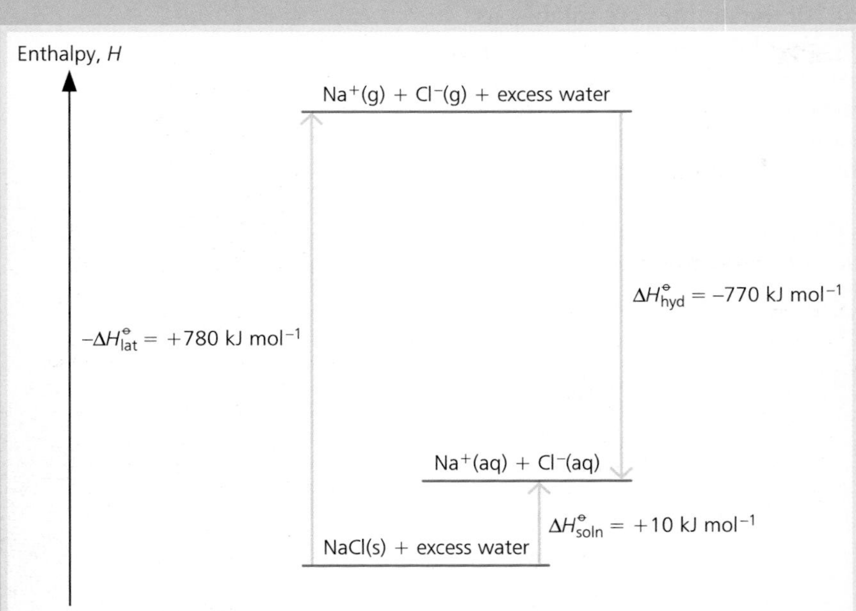

Figure 2.3.6 The enthalpy cycle shows the relationship between lattice enthalpy, enthalpy change of hydration and enthalpy change of solution of a substance.

Lattice enthalpies may be calculated theoretically, and enthalpies of solution may be measured. This provides us with a way of obtaining **standard molar enthalpy changes of hydration** for ions, which of course cannot be obtained directly since it is not possible to carry out the process:

$$M^+(g) \rightarrow M^+(aq)$$

To find the hydration enthalpy of an ion, it is necessary to measure enthalpy changes of solution and to calculate lattice enthalpies for a range of compounds containing the ion, which of course always occurs in combination with other ions of opposite charge. Enthalpy cycles can then be constructed, from which the hydration enthalpy of the ion may be deduced. Some hydration enthalpies for ions are given in table 2.3.7, although the exact figures can sometimes vary greatly between sources. Notice that ionic radius and ionic charge both affect hydration enthalpy so that small, highly charged ions have large enthalpy changes of hydration. The reason for this is that as the charge density of an ion increases, larger and larger numbers of water molecules become attracted to it by electrostatic forces. The change $M^+(g) \rightarrow M^+(aq)$ or $X^-(g) \rightarrow X^-(aq)$ thus involves the formation of greater numbers of ion–dipole interactions, and so is accompanied by a larger decrease in enthalpy.

Cations		Anions	
H^+	–1091	OH^-	–460
Li^+	–519	F^-	–506
Na^+	–406	Cl^-	–364
K^+	–322	Br^-	–335
Mg^{2+}	–1920	I^-	–293
Al^{3+}	–4690		

Table 2.3.7 Enthalpy changes of hydration $\Delta H^\ominus_{hyd,m}$ of some ions/kJ mol^{-1}

The factors that determine whether or not a substance dissolves are complex. For example, sodium chloride is readily soluble in water, even though the enthalpy change for this process is positive. Many other ionic solids behave similarly, since the energy required to break up the lattice of ions is not completely offset by the hydration of the free ions formed. We shall see in section 3.7 why such changes, which involve an increase in the enthalpy content of a system, occur spontaneously.

There are a few general trends in solubility within the compounds of the s-block elements which can be readily observed. All the nitrates are soluble, as are the chlorides. All the common group I salts are soluble in water. In group II, salts with anions having a charge of –1 are soluble, with the exception of the hydroxides. Group II salts with anions having a charge of –2 are largely insoluble, with the exceptions of a few magnesium and calcium salts. In addition to these trends, there is a tendency for the solubility of the salts to decrease down the group, as the metal ion size increases. This can be seen clearly in table 2.3.8. The table also illustrates the pattern of solubility of the hydroxides, which is opposite to the others, the salts becoming increasingly soluble down the group.

Element	**Solubility of sulphate/** mol per 100 g water	**Solubility of nitrate/** mol per 100 g water	**Solubility of carbonate/** mol per 100 g water	**Solubility of hydroxide/** mol per 100 g water
Magnesium	1830×10^{-4}	4.9×10^{-1}	1.5×10^{-4}	0.2×10^{-4}
Calcium	11×10^{-4}	6.2×10^{-1}	0.13×10^{-4}	15×10^{-4}
Strontium	0.71×10^{-4}	1.86×10^{-1}	0.07×10^{-4}	33.7×10^{-4}
Barium	0.009×10^{-4}	0.39×10^{-1}	0.09×10^{-4}	150×10^{-4}

Table 2.3.8 Solubilities of some group II metal salts

The occurrence and extraction of the s-block elements

The elements of groups I and II are too reactive to be found naturally as free elements, and always occur in combination with other elements as salts. Whilst some of these salts make up the more common minerals in the Earth's crust, others are very rare. Francium does not exist naturally. It has only been obtained fleetingly during nuclear reactions and it very rapidly decays to form other more stable elements.

Many of the salts of the alkali metals are found in sea water as they are very soluble – sodium chloride is the obvious example. They are also found combined as aluminates and silicates in clays. The alkaline–earth metals largely occur as carbonates, silicates and sulphates. Table 2.3.9 shows the relative abundance of the different elements and their major sources.

As the s-block elements are not found free in nature, the pure metals need to be extracted from their common sources. Because the elements themselves are such strong reducing agents, and because the compounds they form are so stable, it is not feasible to extract the metals by chemical reactions. The most appropriate method for extracting group I and II elements is by electrolysis of the molten (fused) salt.

$$M^+(l) + e^- \rightarrow M(l)$$

$$M^{2+}(l) + 2e^- \rightarrow M(l)$$

The two s-block elements most in demand by industry are sodium and magnesium, both of which are extracted by electrolysis.

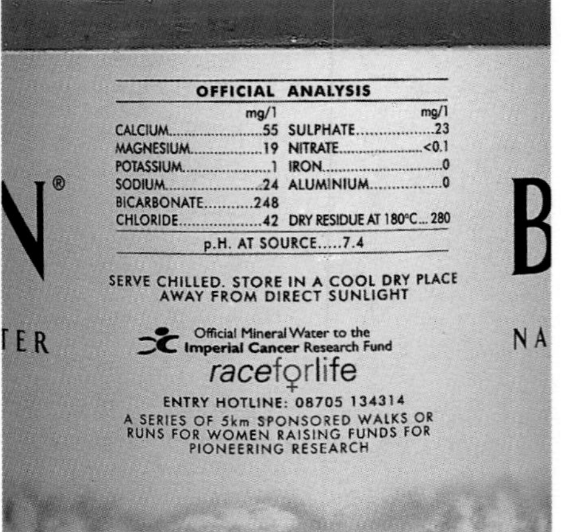

OFFICIAL ANALYSIS
mg/l
CALCIUM........55
MAGNESIUM........19
POTASSIUM........1
SODIUM........24
BICARBONATE........248
CHLORIDE........42
mg/l
SULPHATE........23
NITRATE........<0.1
IRON........0
ALUMINIUM........0
DRY RESIDUE AT 180°C... 280
p.H. AT SOURCE.....7.4
SERVE CHILLED. STORE IN A COOL DRY PLACE AWAY FROM DIRECT SUNLIGHT
Official Mineral Water to the Imperial Cancer Research Fund
raceforlife
ENTRY HOTLINE: 08705 134314
A SERIES OF 5km SPONSORED WALKS OR RUNS FOR WOMEN RAISING FUNDS FOR PIONEERING RESEARCH

Figure 2.3.7 The very low solubility of barium sulphate can be used to determine the amount of sulphate ions in a sample such as this mineral water.

FOCUS S-BLOCK ELEMENTS AND LIFE

The living world is made up of a seething mass of organic compounds consisting largely of four elements: carbon, hydrogen, oxygen and nitrogen. But other elements are also present in the tissues of living organisms, playing a wide variety of roles. The names of many of these elements are familiar to us from simple studies of chemistry and the periodic table, but their importance in the natural world is often less well known.

The *s-block* elements of groups I and II of the periodic table include sodium, potassium, magnesium and calcium. These are all reactive elements, taking part in vigorous chemical reactions with air and water. They form extremely stable ionic compounds, and so they are found in the cells and bodies of living organisms as compounds or as ions in solution.

Magnesium is of vital importance in the living world because of its role in photosynthesis, the process by which green plants and some other organisms capture light energy from the sun and transfer it to the chemical bonds of sugar molecules produced by combining carbon dioxide and water. The magnesium ion is at the centre of the chlorophyll molecule, the green pigment used by plants to capture the solar energy falling on the leaf.

Sodium chloride or common salt is a major part of the marine environment, which supports an enormous range of living organisms and is probably the medium in which the first life forms evolved. Within every living cell a fine balance of sodium and potassium ions must be maintained in order to ensure the healthy functioning of both the individual cell and the entire animal or plant. Sodium ions are also of vital importance in the functioning of nerve cells and the way co-ordinating messages are sent around the body of almost all multicellular organisms.

Calcium plays an important role in living organisms, being involved in the contraction of muscles which cause movement. Calcium compounds secreted by coral polyps form the hard external skeletons which make up the structure of the coral reef. Calcium compounds are also important in the protective shells of many invertebrates. However, perhaps the best known role of calcium is in the bones of vertebrates (animals with backbones). It is the presence of calcium in the structure of the bones which makes them rigid and thus able to perform their major function of support.

Even from a brief examination such as this, the importance of the s-block elements in the natural world can be seen. In a detailed study of the physical and chemical properties of the elements along with their methods of extraction and industrial uses, their roles in the maintenance of healthy living things can become obscured, if not completely ignored. However, the maintenance of life remains a central role for these and a large number of other elements.

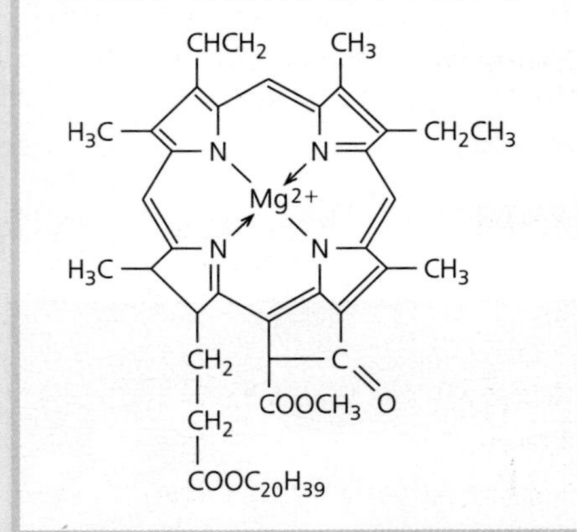

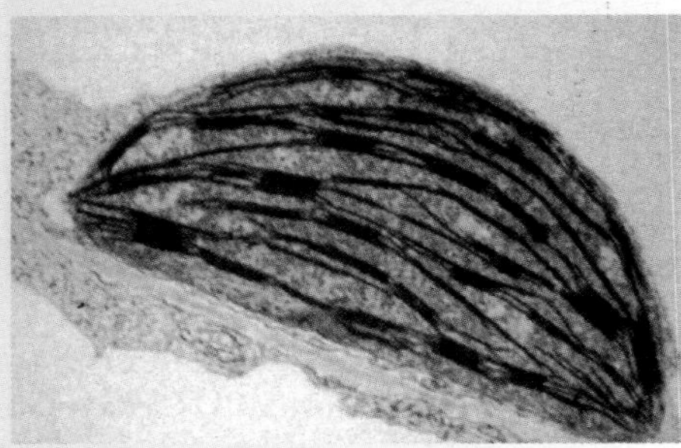

Figure 1 Chlorophyll molecules have a central magnesium ion. The molecules are contained in complex intracellular structures called **chloroplasts**, arranged to maximise the amount of light falling upon them.

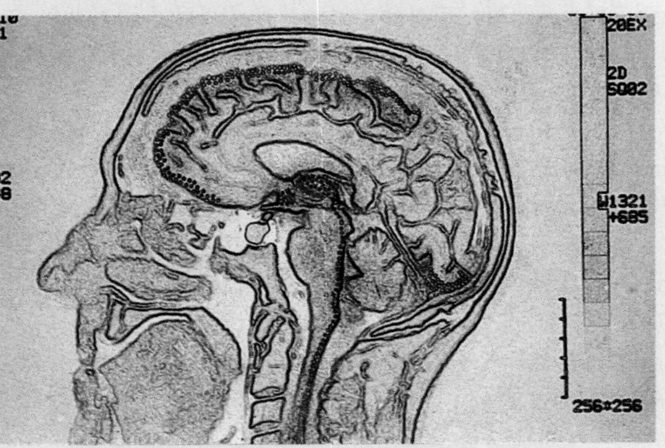

Figure 2 The human brain receives messages from all over the body, analyses information and sends out messages to control what we do and how we do it. All this is made possible by the movement of sodium ions into and out of the nerve cells, and by the action of calcium ions at the junctions between nerves.

Group I				Group II			
Element	**Proportion of the Earth's crust by mass**	**Naturally occurring compound**	**Common name**	**Element**	**Proportion of the Earth's crust by mass**	**Naturally occurring compound**	**Common name**
Lithium	Trace	$LiAl(SiO_3)_2$	Spodumene	**Beryllium**	Trace	$Be_3Al_2(SiO_3)_6$	Beryl (emerald)
Sodium	2.8%	$NaCl$ $NaNO_3$	Rock salt Chile saltpetre	**Magnesium**	2.1%	$MgCO_3$	Magnesite
Potassium	2.6%	KNO_3 $KCl.MgCl_2.6H_2O$	Saltpetre Carnallite	**Calcium**	3.6%	$CaCO_3$ $CaSO_4.2H_2O$	Chalk/limestone/marble Gypsum
Rubidium **Caesium**	Trace Trace	As chlorides, frequently with sodium and potassium compounds		**Strontium** **Barium**	Trace Trace	As carbonates and silicates, frequently with calcium and magnesium compounds	

Table 2.3.9 The group I and II elements, and their most commonly occurring compounds in the Earth's crust

Figure 2.3.8 Emeralds are a compound of beryllium, with aluminium and silicon oxide.

Figure 2.3.9 Marble has many uses in building and for ornamental stonework and statues.

Having looked at the elements in groups I and II of the periodic table, we have built up a picture of closely related metals which show many similarities in their properties due to their reactive electrons being in the s subshell. The s-block elements have many uses in industry, and are found commonly in biological systems too. The attempts of the human race to win the elements and their compounds from the Earth have a long and colourful history – this is illustrated by the use of a small marsh plant in the ancient glass industry, as shown by figure 2.3.10.

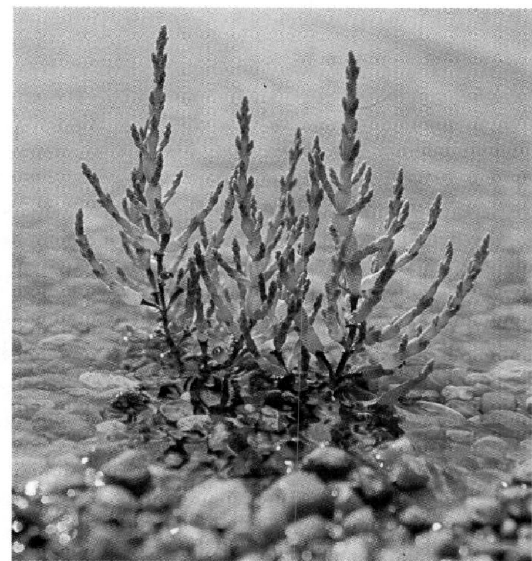

Figure 2.3.10 Samphire is a plant found growing in salt marshes in the south and east of England. It is rich in sodium, and was used for many years in the manufacture of glass, until the industrial production of cheap sodium carbonate made its use unnecessary. Its ancient use lives on today in its alternative common name – glasswort.

SUMMARY

- The elements of groups I and II of the periodic table are known as the **s-block elements** because their outermost electrons are found in the s subshells. The group I elements are known as the **alkali metals** and the group II elements are known as the **alkaline–earth metals.**
- Alkali metals have a single electron in the outer shell and they are the most reactive group of metals. They become increasingly reactive passing down the group, losing their outer electron to form ions with a single positive charge (M^+).
- Alkali metals all react with cold water to give hydrogen, forming an alkaline hydroxide solution.
- When the alkali metals are burned in air, they give characteristic flame colours and form oxides.

- The alkaline–earth metals have two electrons in their outer s subshell and are electropositive, reactive metals. They become increasingly reactive passing down the group, losing both the outer electrons to form ions with two positive charges (M^{2+}).
- The metals in groups I and II each show only one oxidation state.
- The group I and II elements are all good reducing agents.
- **Redox reactions** involve **reduction** and **oxidation** occurring simultaneously. Oxidation involves a *loss* of electrons whilst reduction involves a *gain* in electrons.
- The **electrochemical series** arranges elements in order of their power as reducing agents.
- All common group I metal salts are soluble in water. All group II salts where the anions have a charge of –1 are soluble, with the exception of the hydroxides. All group II salts where the anion has a charge of –2 are insoluble, with the exception of a few magnesium and calcium salts.
- With the exception of lithium carbonate, all group I carbonates are stable when heated in a Bunsen flame. The group II carbonates and lithium carbonate decompose on heating to form stable oxides and carbon dioxide gas.
- The nitrates of all the s-block elements decompose on heating with a Bunsen flame. Group I nitrates except lithium give stable nitrites whilst group II nitrates and lithium nitrate give oxides.
- Hydroxides show a similar pattern of stability to heat to that of the nitrates, and are stable to quite high temperatures. They decompose to give the oxide and water.
- As a result of their reactivity, group I and II elements are not found free in nature. They always occur in combination with other elements and are extracted from their ores by electrolysis.

QUESTIONS

1 Write down the electronic configurations of the ions Na^+ and Ca^{2+}.
  a Why do the alkali metals form ions with a charge of +1 but not +2?
  b Why do the alkaline–earth metals not form ions with a charge of +3?

2 Draw dot and cross diagrams for the following compounds:
  a NaBr **b** Cs_2O **c** SrO **d** MgF_2.

3 Using data from tables 2.3.1 and 2.3.3, explain the trends in:
  a atomic radius
  b ionisation energies
  as groups I and II are descended.

4 **a** How much energy is required to form one mole of Mg^+(g) ions from one mole of Mg(g) atoms?
  b How much energy is required to form one mole of Mg^{2+}(g) ions from one mole of Mg(g) atoms?
  c Why is the Mg^+ ion not generally found in magnesium compounds?

5 The elements in period 2 often show behaviour different from that of elements in periods below them. How does the behaviour of lithium and beryllium differ from the other elements of groups I and II?

QUESTIONS

6 Calcium carbonate, $CaCO_3$, is found in great quantities as limestone. When heated to around 1000 °C, calcium carbonate decomposes to form calcium oxide (quicklime), CaO.

a Write an equation for this reaction. What gas is formed when calcium carbonate decomposes in this way?

When water is added to quicklime, slaked lime is formed. Slaked lime is calcium hydroxide, $Ca(OH)_2$.

b Write an equation for the production of slaked lime from quicklime.

Slaked lime is only sparingly soluble in water. A filtered, saturated solution of slaked lime is called limewater, and is a well known test solution for carbon dioxide. When carbon dioxide is passed through limewater, a white precipitate of calcium carbonate appears at first.

c Write an equation for this reaction.

Passing further carbon dioxide through the solution eventually causes the white precipitate to dissolve, as carbonate ions are converted to hydrogencarbonate ions.

d Write an equation for this reaction.

e Why is calcium carbonate insoluble (or very nearly so) in water, while calcium hydrogencarbonate is soluble?

Developing Key Skills

You have been asked to make a brief presentation in a Sixth Form Choices evening entitled 'Why Study Chemistry?' Choose one of the following s-block elements: sodium, potassium or calcium. For the element you have chosen, find out as much as possible about it, focusing mainly on *either* its role in the human body *or* its role in industry. Prepare a brief (3–4 minutes) presentation on this one element and then use it as an example to demonstrate the enormous importance of chemistry in our lives and society.

[Key Skills opportunities: C]

2.4 The p-block elements: The halogens

The p-block elements of the periodic table make up groups III to VII. They all have electrons in p subshells and need to gain or lose varying numbers of electrons to achieve stable electronic configurations. Unlike the s-block metals, which all demonstrate similar related properties, the elements of the p block are much more disparate in character. They range from metals to non-metals and encompass much fascinating chemistry. In sections 2.4–2.6 we shall consider in some detail a few of the p-block elements that are of particular interest or importance. Group VII shows some strong group characteristics, so we shall start by looking at this group before going on to groups III and IV. (The chemistry of the more important group VI elements, oxygen and sulphur, is dealt with elsewhere in the book. Patterns in the oxides were discussed in section 2.2, and other aspects of their chemistry described throughout sections 2.3–2.7. The main industrial application of sulphur is in the manufacture of sulphuric acid, discussed in section 3.6.)

Group VII – The halogens

The elements of group VII of the periodic table are better known as the **halogens.** The term 'halogen' is derived from Greek and means 'salt producing'. The name dates back to the early years of the nineteenth century, when Jöns Jacob Berzelius used it to indicate that chlorine, bromine and iodine all occur in the sea as salts. It remains appropriate, as the halogens are very reactive and readily form salts. They require one electron to complete their outer p subshell.

About 70% of the industrial use of chlorine goes into the manufacture of products such as the plastic PVC (polyvinyl chloride).

Iodine crystals are black and shiny and produce a purple vapour. This vapour kills bacteria, and can be used to prevent infection of wounds and also to make drinking water safe on expeditions.

Non-stick coatings and fluoride toothpastes are two everyday uses of fluoride compounds.

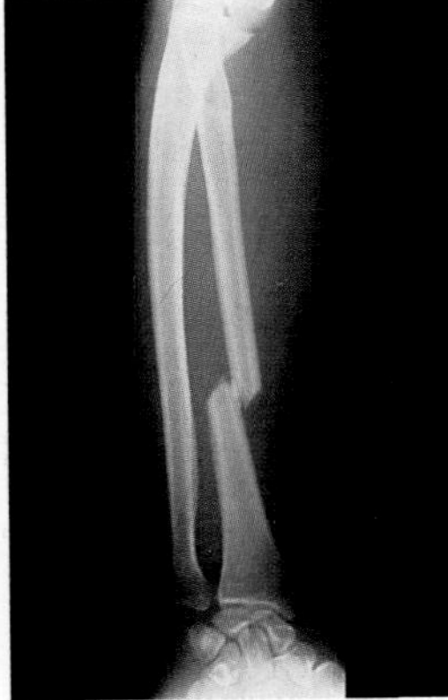

Silver compounds of bromine and iodine are light sensitive, and are used in the production of photographic film, and also for X-ray pictures.

Chlorine compounds are also used as disinfectants. Bleach (sodium chlorate(I)) is a well known example around the home, and swimming pools and water companies frequently rely on chlorine-based products to keep the water free of bacteria.

9 **Fluorine, F** $1s^22s^22p^5$
17 **Chlorine, Cl** $1s^22s^22p^63s^23p^5$
35 **Bromine, Br** (Ar) $3d^{10}4s^24p^5$
53 **Iodine, I** (Kr) $4d^{10}5s^25p^5$
85 **Astatine, At** (Xe) $5d^{10}6s^26p^5$

Figure 2.4.1 The halogens are reactive non-metals, rarely used as the elements. However, their compounds are widely used in a variety of situations.

The halogens show some very strong group characteristics, although there are also clear differences between them. The final member of the group, astatine, is intensely radioactive. The most stable isotope has a half-life of only 8.3 hours. Astatine is found only very rarely in certain uranium deposits and has no known significance outside the research laboratory. We shall therefore not consider it further in our discussions of group VII.

The halogens are a family of non-metallic elements that exist as diatomic molecules. They are all very reactive and are strong oxidising agents – we shall see the evidence for this in section 3.5. Fluorine, chlorine and bromine are all poisonous. The chemical behaviour of the halogens is the result of the seven electrons in their outer shell. There are two electrons in the s subshell and five in the p subshell. Thus the addition of only one further electron by either ionic or covalent bonding will confer a noble gas configuration to the halogen atom. The most common oxidation state for the group VII elements is –1, although other oxidation states do exist.

As the atomic number increases, the group VII elements become less reactive. They also become less volatile and darker in colour. Fluorine is a pale yellow gas at room temperature, whilst chlorine is a darker greenish yellow. Bromine (which takes its name from the Greek word *bromos* meaning 'stench') is a dark red liquid giving off a dense red vapour, whilst iodine is a shiny, greyish black crystalline solid which can be sublimed to a purple vapour.

Extraction and use of the halogens

Occurrence and extraction

Like the alkali metals on the opposite side of the periodic table, the halogens are so reactive that they are not found free in nature, but combined with other elements. Indeed fluorine, the most electronegative element, will combine with almost every other element, including some of the noble gases. Chlorine too is very electronegative and reacts directly with all other elements except carbon, nitrogen, oxygen and the noble gases. Chlorine also forms compounds with both carbon and oxygen, but not by the direct combination of the elements.

Chlorine is by far the most common of the halogens in nature, being found in the form of sodium chloride in sea water and rock salt, as well as playing a part in all living organisms. Every kilogram of sea water contains around 30 g of sodium chloride. Fluorine is the next most abundant halogen, usually occurring as fluorspar (fluorite), CaF_2, and cryolite, Na_3AlF_6. These deposits are generally quite thin and are rarely economically workable. Sometimes these fluoride-containing ores occur as semiprecious minerals which are mined, polished and used for their appearance rather than the fluoride they contain – an example is the blue john mined in Derbyshire (see figure 1.1.5, page 8).

Bromine and iodine are much rarer elements. About 70 parts per million of sea water are bromides which, surprisingly, can be extracted economically. Iodides are present as 0.05 p.p.m. in sea water, although certain seaweeds are capable of concentrating this greatly. The main source of iodine is sodium iodate(V), $NaIO_3$, and this is found only in Chile.

Some factors to be considered in industrial processes are discussed on page 184. Extracting the halogens from their sources leads to difficulties due to their powerful oxidising tendencies. They are usually obtained by oxidation of the halide ion, but in the case of fluoride there is no oxidising agent strong enough to be used in the extraction process and so the oxidation is performed by electrolysis. The electrolysis is carried out using potassium fluoride dissolved in liquid anhydrous hydrogen fluoride, as the fluorine produced would react with water. A graphite anode and steel cathode are used.

Twenty-nine million tonnes or more of chlorine are used throughout the world each year, and for the large-scale extraction needed to provide this quantity of the element a very efficient manufacturing method is needed. Chlorine is used as a cheap industrial oxidant in the manufacture of bromine, as a bleach and a germicide, but more importantly it is vital for the manufacture of many everyday materials.

The chlor-alkali industry

The production of chlorine by electrolysis of brine (a concentrated solution of rock salt in water), and the associated production of sodium hydroxide and hydrogen, is the basis of the massive **chlor-alkali industry**. A flowing mercury cathode is used, as shown in figure 2.4.2, and the sodium produced dissolves in the mercury to form an **amalgam**. After extracting the sodium, the mercury is recycled. The use of mercury in this way means that all the products of the process are useful. It also means that no energy input is required to produce the liquid metal for the electrode, as mercury is a liquid at room temperature.

During the electrolysis, chlorine is produced and liberated at the graphite anodes, and sodium is produced at and dissolves in the mercury cathode.

At the anode (+):

$$2Cl^-(aq) \rightarrow Cl_2(g) + 2e^-$$

At the cathode (–):

$$2Na^+(aq) + 2e^- \rightarrow 2Na(Hg)$$

The chlorine is collected and pressurised for storage, whilst the sodium–mercury amalgam passes on into a second 'soda cell'. Here the sodium reacts with water to form sodium hydroxide solution (caustic soda) and hydrogen:

$$2Na(Hg) + 2H_2O(l) \rightarrow 2Na^+(aq) + 2OH^-(aq) + H_2(g) + 2Hg(l)$$

Thus three products of immense use to the chemical industries are made, and the mercury cathode is ready to be used again.

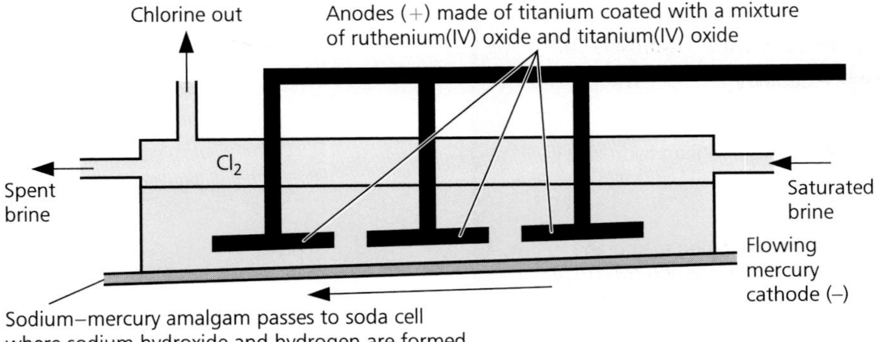

Figure 2.4.2 The flowing mercury cell is the source of chlorine, sodium hydroxide and hydrogen, all resulting from the electrolysis of brine.

Other products from the electrolysis of brine

The chlorine produced in the electrolysis of brine may be reacted with the other products of electrolysis, hydrogen and sodium hydroxide solution, to produce two further useful products.

Hydrogen and chlorine can be reacted to form hydrogen chloride gas. Dissolved in water, this produces hydrochloric acid:

$$H_2(g) + Cl_2(g) \rightarrow 2HCl(g)$$

$$HCl(g) \xrightarrow{H_2O} HCl(aq)$$

Chlorine may also be reacted with sodium hydroxide solution to produce a solution which contains sodium chlorate(I), also known as sodium hypochlorite:

$$Cl_2(g) + 2NaOH(aq) \rightarrow NaCl(aq) + NaClO(aq) + H_2O(l)$$

Sodium chlorate(I) solution is widely sold in shops as bleach. It is a powerful oxidising agent, and is used as a disinfectant, and as a domestic and industrial bleaching agent. You can find out more about halate(I) ions on page 181.

The perfect process?

The production of chlorine, sodium hydroxide and hydrogen from the electrolysis of brine in the flowing mercury cell might appear to be the perfect chemical process. It uses cheap feedstock and requires a relatively low energy input. All the products can be sold, and at first glance the potential for pollution is low. Unfortunately, this is not the

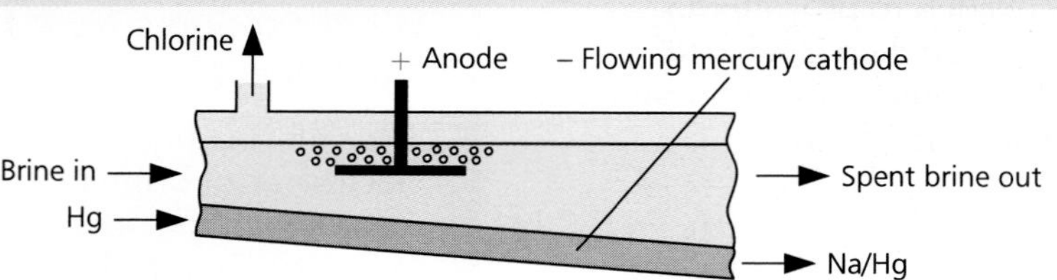

The mercury cell. The anode is made of titanium coated with a mixture of ruthenium(IV) oxide and titanium(IV) oxide. This is highly resistant to corrosion by chlorine, and also aids the production of the gas. Sodium amalgam is removed from the cell and reacted with water to form hydrogen gas and a solution containing about 50% sodium hydroxide. The mercury is then fed back into the cells.

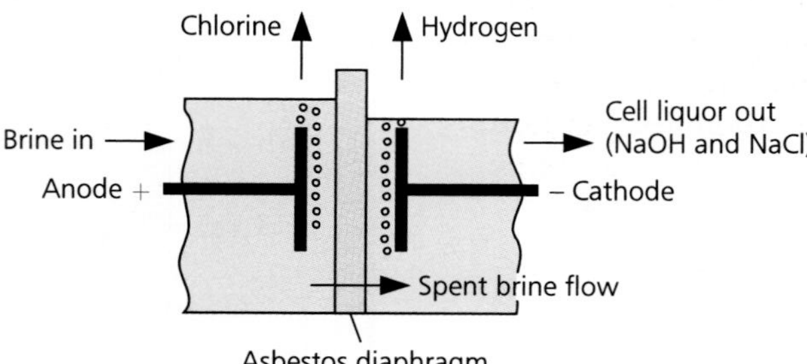

The diaphragm cell. Chlorine is produced at the anode, and hydrogen and sodium hydroxide solution are produced at the cathode. An asbestos diaphragm prevents the products mixing, which would lead to the formation of sodium hypochlorite solution (bleach):

$$2NaOH(aq) + Cl_2(aq) \rightarrow NaOCl(aq) + NaCl(aq) + H_2O(l)$$

The weak solution of sodium hydroxide which is formed also contains about 15% sodium chloride. As much of this residual salt as possible must be removed and the solution concentrated to about 50% sodium hydroxide for the solution to be marketed commercially. When the removal of the salt and the concentration of the solution are taken into account, the costs of operating mercury cells and diaphragm cells are very similar.

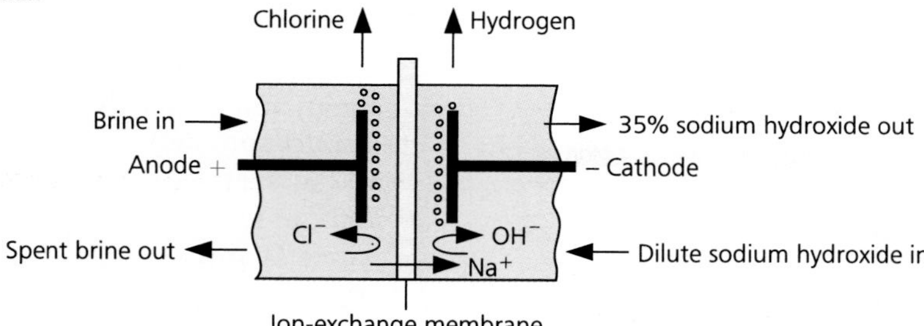

The membrane cell. There is no net flow of liquid across the separating membrane. Sodium ions are able to pass through the membrane, while chloride ions and hydroxide ions are not. In this way a current can flow through the cell, but there is no possibility of sodium hypochlorite being formed or of the sodium hydroxide solution being contaminated with sodium chloride.

The materials used for membrane construction are invariably synthetic polymers, designed and selected for their ability to transport cations rather than anions and for their resistance to the corrosive solutions and high temperatures (≈90 °C) inside the cell.

Figure 2.4.3 The three types of cell used in the production of chlorine and sodium hydroxide

case. In the 1950s, a terrible disease occurred in Minamata in Japan. People, and also domestic animals that fed on fish, developed symptoms including loss of balance, muscle wasting, paralysis and eventual death.

Babies were born suffering terrible physical deformities and mental retardation. Eventually the problem was pinpointed to mercury poisoning. One of the major sources of the mercury was the effluent released from a plant producing chlorine from brine using flowing mercury cells. By the mid-1960s, legislation was in place which placed much tighter restrictions on the amount of mercury that could be discharged from such plants, and in the ensuing years there has been a move by the chlor-alkali industry away from the flowing mercury cells to **membrane cells**. This involves the capital cost of replacing old plant, but does not carry the risk of poisoning the population.

Figure 2.4.3 shows the three types of cell used to produce chlorine and sodium hydroxide solution. Mercury and diaphragm cells are now being replaced by membrane cells.

2

Patterns in the properties of the halogens

Physical properties

Element	F	Cl	Br	I
Atomic radius/nm	0.071	0.099	0.114	0.133
Ionic radius/nm	0.133	0.180	0.195	0.215
Electronegativity	4.0	3.0	2.8	2.5
Electron affinity/kJ mol^{-1}	–328	–349	–325	–295
Melting point/°C	–220	–101	–7	114
Boiling point/°C	–188	–35	59	184
Density/g cm^{-3}	1.11*	1.56*	3.12	4.93
$\Delta H^{\ominus}_{at,m}[\frac{1}{2}X_2(g) \rightarrow X(g)]$/kJ mol^{-1}	79	122	112	107
$\Delta H^{\ominus}_{hyd,m}[X^-(g) \rightarrow X^-(aq)]$/kJ mol^{-1}	–506	–364	–335	–293
$\Delta H^{\ominus}_{vap,m}[X_2(l) \rightarrow X_2(g)]$/kJ mol^{-1}	3.3	10.2	15.0	20.9

*the density measured for the liquid at its boiling point

Table 2.4.1 Some physical properties of the group VII elements

Some of the patterns in the physical properties of the group VII elements have already been discussed in section 2.1. Others become clearer looking at the data in table 2.4.1. The atomic radius increases down the group as the number of electron shells increases. As we would expect from this, the electron affinities decrease from chlorine to bromine to iodine. However, at the top of the group, the electron affinity of fluorine is close to that of bromine. On the face of it this is surprising, since we should normally expect an extra electron added to a small atom to be attracted strongly by the positive nucleus. The explanation is that the electrons in the fluorine atom are close together, and the repulsive forces experienced by an extra electron added to the atom make the formation of the F^- ion less favourable than expected.

The decrease in volatility of the elements down the group from gaseous fluorine and chlorine through liquid bromine to the solid iodine is the result of increasingly strong van der Waals forces between the molecules as the relative molecular mass increases. These increased intermolecular forces also account for the observed increases in melting points, boiling points and enthalpy changes of vaporisation.

Because the halogens form simple, non-polar molecules, they dissolve readily in non-polar organic solvents such as tetrachloromethane. However, chlorine, bromine and iodine are sparingly soluble in water too, although chlorine also undergoes a very slow reaction with the water forming initially chloric(I) acid, HClO(aq), which decomposes to give oxygen and hydrochloric acid:

$$\mathbf{Cl_2(g) + H_2O(l) \rightleftharpoons HCl(aq) + HClO(aq)}$$

$$\mathbf{2HClO(aq) \rightarrow 2HCl(aq) + O_2(g)}$$

This second reaction is speeded up by sunlight and catalysts such as platinum. It is chloric acid that gives the familiar 'swimming pool' smell to water in which chlorine is dissolved.

Fluorine is such a powerful oxidising agent that it reacts with water, producing oxygen and hydrogen fluoride:

$$\mathbf{2F_2(g) + 2H_2O(l) \rightarrow 4HF(aq) + O_2(g)}$$

Fluorine and the noble gases

e

In 1924, the Austrian chemist Friedrich Paneth wrote:

> **The unreactivity of the noble gas elements belongs to the surest of experimental results.**

In 1962, Neil Bartlett prepared the first compound containing a noble gas, a yellow-orange solid which was stable at room temperature, with the formula $XePtF_6$.

Bartlett had previously been working with a compound of platinum and fluorine, PtF_6, which he found to be a very strong oxidising agent. Its oxidising power was so great that it proved possible to make a compound in which the O_2^+ cation appeared, $[O_2]^+[PtF_6]^-$. Bartlett's success in making a noble gas compound was due to the fact that he noticed that the first ionisation energy of the xenon atom was slightly less than that of the oxygen molecule, and reasoned that it ought to be possible to make the noble gas react in the same way as the oxygen.

Following Bartlett's success, the next few years saw the synthesis of a considerable number of noble gas compounds, ranging in oxidation state from +2 to +8. The compounds now known include XeF_2, XeF_4, $XeOF_4$ and XeO_4.

It is wise never to be too certain of anything in chemistry!

Chemical properties

While the alkali metals become increasingly reactive down the group, the trend in the halogens is the opposite, and they become increasingly unreactive down the group. This trend in reactivity can be clearly demonstrated by the tendency of the elements to displace each other from solutions of ions. Iodine is displaced from solution by the more reactive bromine, which in turn is displaced by chlorine. Chlorine would be displaced by fluorine, except that fluorine quickly reacts with the water, and so this displacement cannot be carried out.

$$\mathbf{Br_2(aq) + 2I^-(aq) \rightarrow 2Br^-(aq) + I_2(aq)}$$

$$\mathbf{Cl_2(aq) + 2Br^-(aq) \rightarrow 2Cl^-(aq) + Br_2(aq)}$$

$$\mathbf{I_2(aq) + 2Br^-(aq) - \text{no reaction}}$$

These displacement reactions can be used as laboratory preparations for bromine and iodine, and industrially, chlorine is used to produce bromine from sea water. Apart from fluorine, the halogens can also be obtained in the laboratory by oxidation reactions using manganese(IV) oxide, MnO_2 or potassium manganate(VII), $KMnO_4$. Chlorine is produced, for example, by the action of concentrated hydrochloric acid on potassium manganate(VII):

$$2MnO_4^-(aq) + 16H^+(aq) + 10Cl^-(aq) \rightarrow 5Cl_2(g) + 2Mn^{2+}(aq) + 8H_2O(l)$$

The halogens react with a wide range of both metallic and non-metallic elements, and these reactions are outlined below. The reactions of chlorine and bromine show great similarity, whilst those of fluorine and iodine vary to some extent.

The halogens as oxidising agents

The halogens are all oxidising agents, fluorine being the most powerful and iodine the least. The reasons for this trend are explored in the box on page 179. Chlorine is used in industry as a bleach, its oxidation of complex organic molecules resulting in the formation of colourless compounds. In the laboratory, its bleaching properties mean that it decolorises damp universal indicator paper.

2

The oxidising uses of chlorine

The great nappy dilemma

For many years chlorine has been used as a bleaching agent by the paper industry to produce very white paper pulp. The same chlorine bleaches were used regardless of whether the pulp was destined for the production of paper for writing or printing purposes, or whether it was to be used as absorbent padding in disposable nappies and similar products. The chlorine-bleaching process results in the production of **dioxins**, poisonous chemicals that persist and accumulate in the bodies of living organisms. Another process, ozone bleaching, can do the same job as chlorine bleaching, although some of the resulting pulp is cream in colour. The industry claimed that people would not want to use nappies that were less than snow-white on their babies.

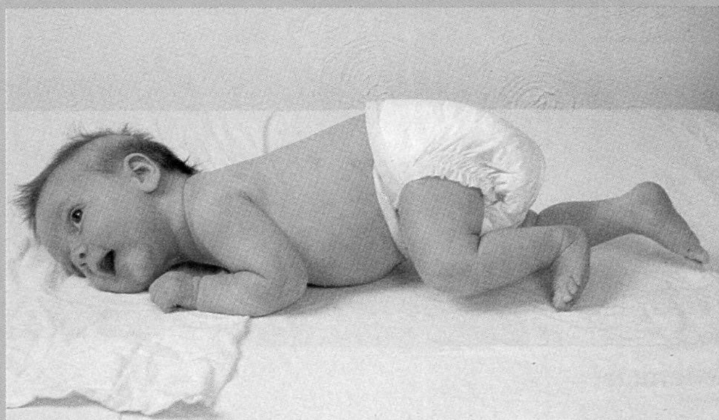

Figure 2.4.4 The chlorine-bleaching debate passed over the heads of many consumers of the nappies at the centre of the debate, but the change from chlorine to ozone bleaching may well result in a slightly less polluted world for them to grow up in.

Beginning in Scandinavia in the 1980s, media and consumer pressure built up against chlorine bleaching and its potentially damaging effects on the environment. As a result, disposable nappies and many other paper products now proudly claim to have been produced without the

use of chlorine bleaches, and babies seem none the worse for their slightly less than snow-white disposable nappies.

Water purification

The oxidising power of chlorine is also used in the treatment of drinking water piped to our homes. After the removal of solid matter like fine particles of dirt, water from reservoirs is treated with chlorine to kill any bacteria in it. Small amounts of chlorine remain in the treated water to prevent possible recontamination of the water by bacteria.

$$Cl_2(aq) + H_2O(l) \rightleftharpoons HCl(aq) + HClO(aq)$$

When chlorine dissolves in water, an equilibrium reaction is set up in which hydrochloric and chloric(I) acids are produced. It is the latter which is responsible for the oxidising power of aqueous chlorine.

Figure 2.4.5 In early Victorian times, many people drew their drinking water directly from rivers which were also used for sewage disposal. The cleaning up of water supplies, in which chlorine plays a vital part, has been one of the most significant advances in public health, eliminating the deaths of tens of thousands of people through diseases like cholera, spread through infected water. In less fortunate parts of the world, contaminated water still kills large numbers of people each year.

Reactions with metals

The halogens react strongly with all the more electropositive elements. When they react with metals, they remove some or all of the outer metal electrons and themselves become reduced to form negative halide ions. This is seen particularly clearly in the reactions with the alkali and alkaline–earth metals.

$$2Na(s) + F_2(g) \rightarrow 2NaF(s)$$

$$Mg(s) + Cl_2(g) \rightarrow MgCl_2(s)$$

The vigour of the reaction between a particular halogen and a metal is a function of the position of the metal in the activity series (in other words, how electropositive it is) and the particular halogen (fluorine always reacts more vigorously than bromine, for example). This is illustrated in figure 2.4.6 and table 2.4.2.

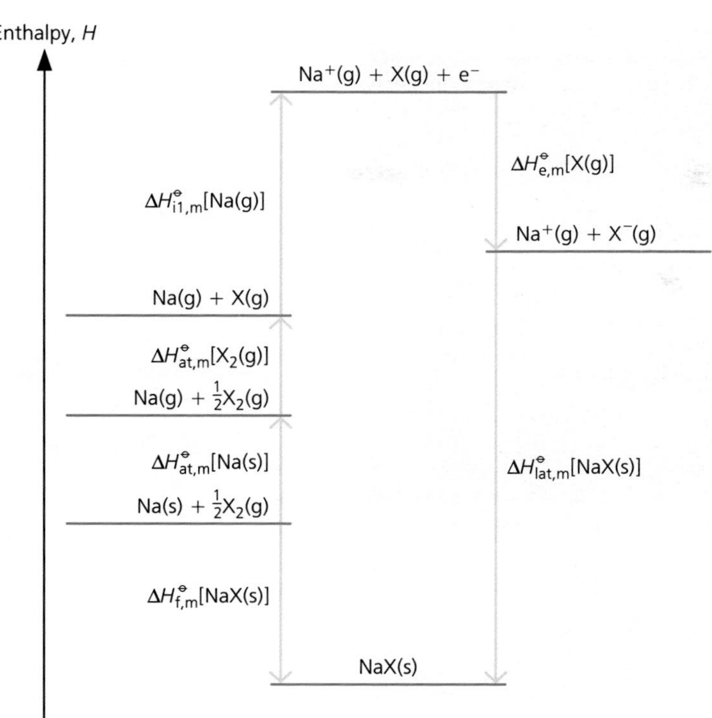

Figure 2.4.6 When the halogens react with electropositive metals, an oxidation reaction takes place, leaving positive metal ions and negative halide ions held together in a stable ionic lattice.

Element	$\Delta H^\ominus_{f,m}$ of sodium halide (NaX)/kJ mol^{-1}	$\Delta H^\ominus_{lat,m}$ of sodium halide (NaX)/kJ mol^{-1}
F	–573	–918
Cl	–411	–780
Br	–361	–742
I	–288	–705

Table 2.4.2 Standard molar enthalpy changes of formation and lattice enthalpies of sodium halides

As table 2.4.2 shows, the enthalpy changes of formation of the sodium halides fall from NaF to NaI. The Born–Haber cycle in figure 2.4.6 shows how the trends in atomisation energy, electron affinity (see table 2.4.1) and lattice enthalpy explain this decrease in $\Delta H^\ominus_{f,m}$ (NaX) down the group. The increasing size of the halide ion causes the lattice enthalpy of the solid to decrease, while in addition the particularly small enthalpy change of atomisation of fluorine contributes to the extremely favourable enthalpy change for the formation of NaF.

Fluorine reacts readily with every metal, even those such as gold and platinum which are usually regarded as being very unreactive. Iodine will also react with these elements, but only very slowly even at very high temperatures.

Reactions with non-metals

In reactions with the non-metals, the halogens usually achieve a noble gas configuration through covalent bonding. Fluorine forms only one covalent bond, but the other halogens can form three, five and seven bonds, as the box 'Halogen oxidation states' overleaf shows.

Halogen oxidation states

In covalent bonding, the halogens, with the exception of fluorine, can show oxidation states from +1 to +7 as well as –1. These higher oxidation states make use of the d orbitals of the outer shell in forming bonds, and are frequently seen in interhalogen compounds, where two different halogens join to form a compound. The range of oxidation numbers shown by the halogens is shown in table 2.4.3. Note that this is not an exhaustive list of the halogen compounds in each oxidation state. Four general points can be made:

- All the halogens form compounds in which their oxidation state is –1.
- Fluorine exists *only* in the –1 state in its compounds.
- Both chlorine and iodine form compounds in which their oxidation state is +7.
- Bromine's highest oxidation state is +5.

Oxidation number	Example
+7	Cl_2O_7, HIO_4
+6	ClO_3, I in IF_6
+5	Br in BrF_5, $HBrO_3$
+4	ClO_2, BrO_2
+3	Cl in ClF_3
+2	
+1	NaClO, Br in BrCl, I in ICl
0	F_2, Cl_2, Br_2, I_2
–1	NaF, KCl, NaBr, KI

Table 2.4.3 The oxidation states of the halogens

Reactions with hydrogen

Halogen reaction with hydrogen $[H_2(g) + X_2(g) \rightarrow 2HX(g)]$	Conditions
F	Explosive under all conditions
Cl	Explodes in direct sunlight, proceeds slowly in the dark
Br	300 °C and platinum catalyst
I	300 °C and platinum catalyst – reversible reaction, proceeds slowly and only partially

Table 2.4.4 Nowhere is the trend in the reactivity of the group VII elements seen more clearly than in their reactions with hydrogen.

Table 2.4.4 shows that under all normal conditions, a mixture of fluorine and hydrogen reacts explosively to form hydrogen fluoride, whilst iodine will only react slowly and incompletely with hydrogen even at 300 °C and with a platinum catalyst. Whilst the differences in the reactivities of the halogens with the non-metals are demonstrated particularly dramatically in their reactions with hydrogen, the pattern is similar for all the reactions with non-metals.

The hydrogen halides are acids, with properties summarised in table 2.4.5. The boiling points of the halides increase on passing down the group, with the exception of hydrogen fluoride. The very high boiling point of this compound is due to the relatively strong hydrogen bonding which results from the extremely electronegative character of the fluoride ion, as shown in figure 2.4.7.

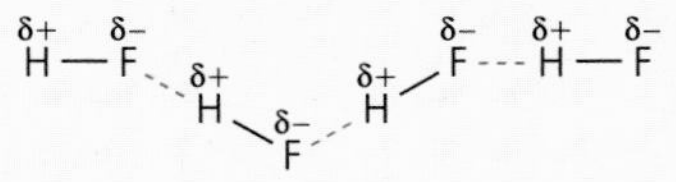

Figure 2.4.7 Hydrogen bonding is responsible for the high boiling point of hydrogen fluoride.

Hydrogen halide	Boiling point/°C	$\Delta H^\circ_{f,m}$/kJ mol^{-1}	$\Delta H^\circ_{d,m}$(H—X)/kJ mol^{-1}
HF	20	−271	568
HCl	−85	−92	432
HBr	−67	−36	366
HI	−35	+27	298

Table 2.4.5 The properties of the hydrogen halides

Explaining the trends in reactivity of the halogens

Fluorine is astoundingly reactive. It explodes in contact with hydrogen, even in the dark at −252 °C, when hydrogen is a liquid and fluorine is a solid; organic substances spontaneously burst into flame when placed in it; water burns in it, the equation for one of the combustion reactions being:

$$2F_2(g) + 2H_2O(g) \rightarrow 4HF(g) + O_2(g)$$

The reasons for fluorine's great reactivity can be understood if we look at some data from table 2.4.1, repeated in figure 2.4.8. When the halogens act as oxidising agents, the enthalpy changes occurring can be represented as in the first enthalpy cycle. The greater reactivity of fluorine in forming covalent bonds can be explained in a similar way, as in the second cycle.

Figure 2.4.8 The highly exothermic reactions of fluorine come in part from its low enthalpy change of atomisation, but also from the enormously strong bonds it forms with elements like hydrogen.

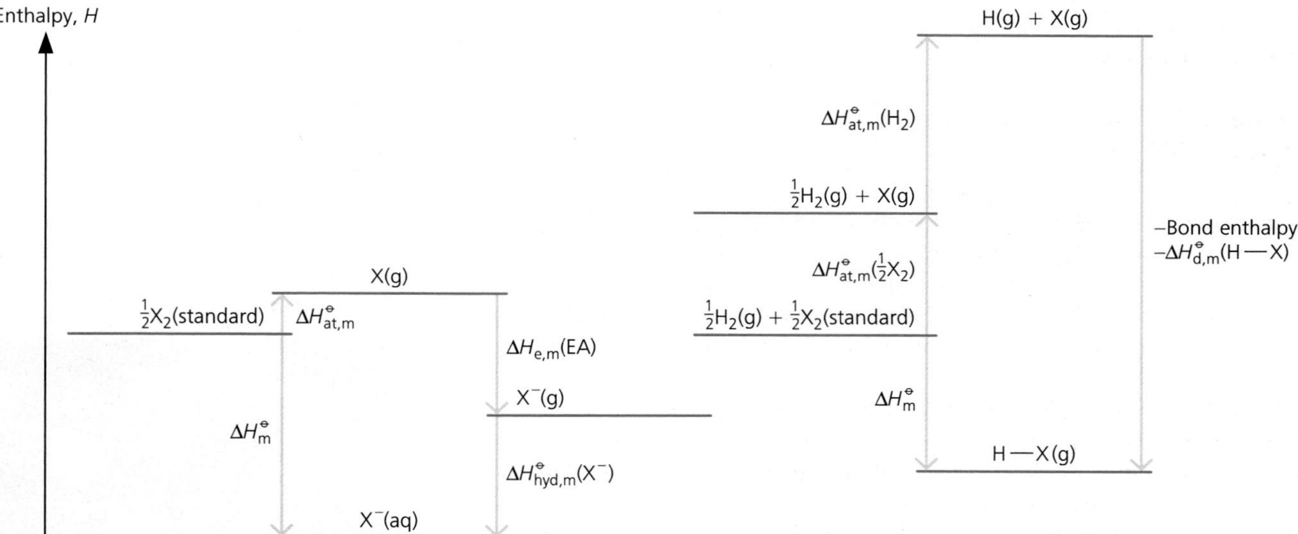

	F	Cl	Br	I
$\Delta H^\circ_{at,m}$/kJ mol^{-1}	79	122	112	107
$\Delta H_{e,m}$(electron affinity)/kJ mol^{-1}	−328	−349	−325	−295
$\Delta H^\circ_{hyd,m}$/kJ mol^{-1}	−506	−364	−335	−293
ΔH°_m for $\frac{1}{2}X_2$(standard) → X^-(aq) /kJ mol^{-1}	−755	−591	−548	−481

The great oxidising power of fluorine comes in part from its low enthalpy change of atomisation, but mainly from its small size, which produces a large enthalpy change of hydration (in the case of aqueous reactions) or large lattice enthalpies (in the case of the production of ionic compounds).

	F	Cl	Br	I
$\Delta H^\circ_{at,m}$/kJ mol^{-1}	79	122	112	107
$\Delta H^\circ_{at,m}$ for $\frac{1}{2}H_2$(s)/kJ mol^{-1}	218	218	218	218
$-\Delta H^\circ_{d,m}$(H—X)/kJ mol^{-1}	−568	−432	−366	−298
ΔH°_m for $\frac{1}{2}X_2$(standard) + $\frac{1}{2}H_2$(g) → H—X/kJ mol^{-1}	−271	−92	−36	+27

The covalent bond formed between fluorine and other atoms becomes weaker as the number of lone pairs in the outer shell of the other atom increases. This is due to repulsion between sets of lone pairs. The F—F bond is particularly weak, since each fluorine atom has three lone pairs. The strength of the bond between hydrogen and the halogens increases as shown here, with the H—F bond being exceptionally strong.

HALOGEN COMPOUNDS – FRIEND OR FOE?

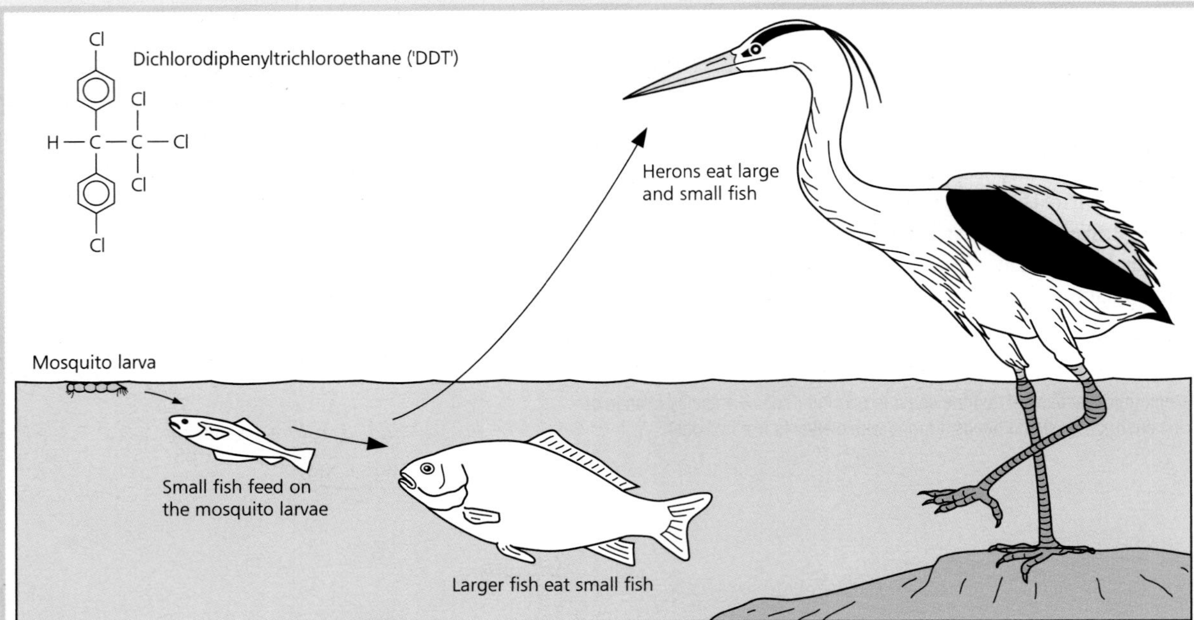

Dichlorodiphenyltrichloroethane ('DDT')
Herons eat large and small fish
Mosquito larva
Small fish feed on the mosquito larvae
Larger fish eat small fish
DDT used on mosquito larvae passed from prey to predator, reaching higher concentrations at each level. Although it killed some herons, its most important effect was on their reproduction. The birds became much less fertile and chicks were less likely to survive.

Figure 1 Halogenated compounds can be very effective in pest control, but their persistence in biological systems may result in unforeseen damage to other organisms within the food chain.

Halogen compounds are of great value both in industry and in the living world. However, some halogen compounds have also caused environmental problems whilst performing useful functions for the human race. One of the best known problems involving the halogens in recent years has been the damage caused to the ozone layer by compounds of chlorine, fluorine and carbon (chlorofluorocarbons or CFCs). The reason for the damaging effect of CFCs is discussed in section 2.6, pages 203–4.

Halogen compounds have also been used over the last few decades as weedkillers and pesticides. A variety of complex chlorinated organic molecules have been developed as effective pesticides. Some of these were targeted at the insects and other invertebrates that attack and destroy crops grown for food around the world. Others were aimed at the vectors of tropical diseases, such as the anopheles mosquito that carries malaria. One of the first pesticides to be used in this way was DDT, dichlorodiphenyltrichloroethane. DDT removed the threat of malaria for thousands of millions of people and saved millions of lives throughout the world, and also helped stop the spread of typhus and yellow fever.

Whilst DDT and other early halogen-containing compounds were very effective at their job, it was subsequently realised that they do not break down naturally but remain in the environment. The level of these toxic chemicals used originally was not high – it was sufficient to kill the pests but not a threat to larger organisms. However, because they are so stable, they have accumulated in the tissues of animals within food chains and have severely reduced the numbers of some carnivorous birds. This is illustrated in the diagram. These pesticides have also made their way into the human food chain. As many of the organisms they were used against have also developed a degree of resistance, the use of organohalogen pesticides is now very carefully regulated.

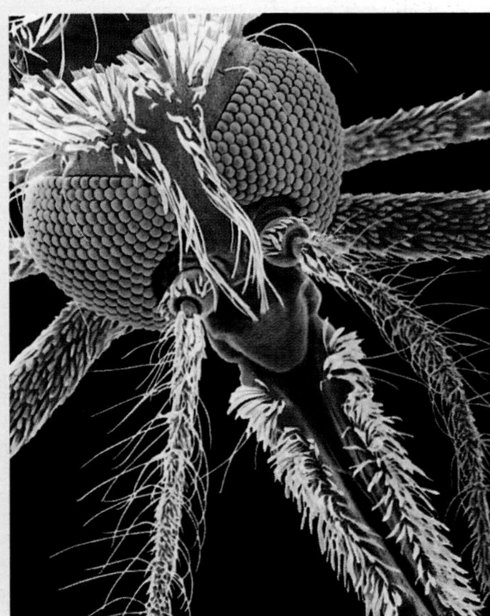

Figure 2 The female *Anopheles* mosquito spreads the malaria parasite when she bites. Halogen-containing compounds help control the mosquitoes and save millions from debilitating disease and death.

Reactions with alkalis

Fluorine reacts with alkalis in a different way to chlorine, bromine and iodine, always forming a mixture of fluoride ions and oxygen difluoride:

$$2F_2(g) + 2OH^-(aq) \rightarrow OF_2(g) + 2F^-(aq) + H_2O(l)$$

With chlorine, bromine and iodine, the products of the reaction depend on the temperature. In cold (15 °C) dilute alkali, a mixture of halide (X^-) and halate(I) (XO^-) ions is formed, for example:

$$Cl_2 + 2NaOH(aq) \rightarrow NaCl(aq) + NaClO(aq) + H_2O(l)$$

These halate(I) ions then **disproportionate** (see adjacent box) to give more halide ions and halate(V) ions (XO_3^-), the disproportionation occurring at different rates depending on both the halogen involved and the temperature:

$$3NaClO(aq) \rightarrow 2NaCl(aq) + NaClO_3(aq)$$

In the case of chlorine, the disproportionation occurs only very slowly at 15 °C, but at 70 °C it is very rapid. This provides a way of obtaining either of the two different products, simply by varying the temperature at which the reactions take place. The reaction of bromine with alkalis is the same, although in this case both reactions occur rapidly at 15 °C and the temperature needs to be as low as 0 °C to prevent disproportionation of the BrO^- ion. Both reactions occur rapidly even at 0 °C in the case of iodine, and so separating the different products is considerably less easy.

Disproportionation

The reactions of chlorine, bromine and iodine with alkali are examples of **disproportionation**, a redox reaction in which a molecule, atom or ion is simultaneously oxidised and reduced. When chlorine reacts with alkali, Cl^- ions and ClO^- ions are formed. This involves a change in oxidation number from 0 (in Cl_2) to –1 (in Cl^-) and to +1 (in ClO^-). Similarly, the disproportionation of ClO^- also results in both an increase and a decrease in the oxidation number of chlorine:

$$\underset{+1}{3ClO^-(aq)} \rightarrow \underset{-1}{2Cl^-(aq)} + \underset{+5}{ClO_3^-(aq)}$$

2

Reaction of halide ions with acids

When solid ionic halides are reacted with strong acids various products may be obtained, depending upon the oxidising power of the acid. With concentrated phosphoric(V) acid, the hydrogen halides are evolved as gases, while with concentrated sulphuric acid the products depend upon the halide (table 2.4.6).

Acid added	fluoride	chloride	bromide	iodide
Concentrated phosphoric(V)	HF(g) evolved	HCl(g) evolved	HBr(g) evolved	HI(g) evolved
Concentrated sulphuric	HF(g) evolved	HCl(g) evolved	HBr(g) evolved plus some Br_2(aq)	some HI(g) evolved plus I_2(aq) produced

Table 2.4.6 The reaction of halides with acids

If manganese(IV) oxide is added to the concentrated sulphuric acid, chloride ions are oxidised to chlorine, but fluoride is not oxidised, since fluorine is the strongest oxidising agent known.

Which halide?

Halides are extremely common, and therefore it is very useful to be able to identify their presence in an unknown solution. Analysis of the halide ions is based on the different solubilities of the silver halides. Dilute nitric acid is added to an unknown halide solution. Silver nitrate solution is then added. The dilute nitric acid prevents the precipitation of any other silver salts from the solution. The silver halides are formed, which have the characteristic colours shown.

$$Ag^+(aq) + X^-(aq) \rightarrow AgX(s)$$

Silver bromide and silver chloride react with ammonia solution:

$$AgX(s) + 2NH_3(aq) \rightarrow [Ag(NH_3)_2]^+(aq) + X^-(aq)$$

This reaction helps distinguish between the halide ions. Silver iodide does not dissolve in ammonia solution.

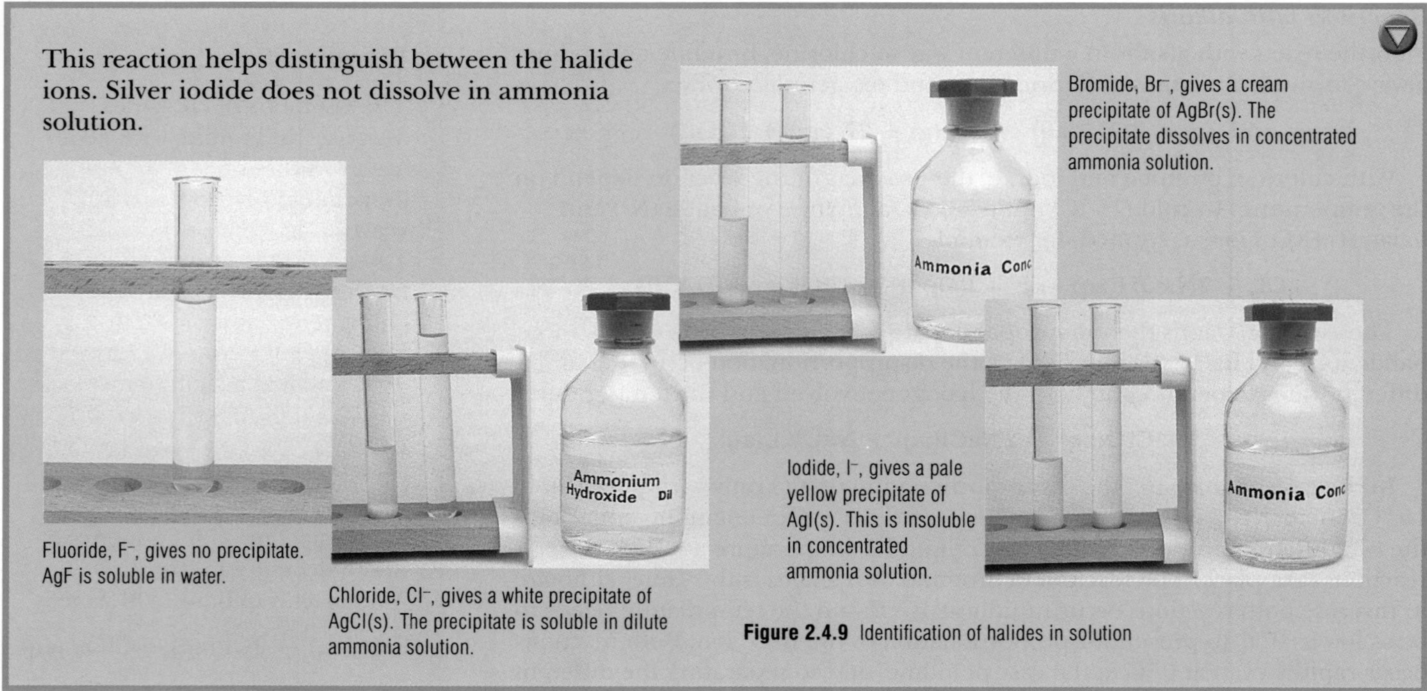

Fluoride, F^-, gives no precipitate. AgF is soluble in water.

Chloride, Cl^-, gives a white precipitate of AgCl(s). The precipitate is soluble in dilute ammonia solution.

Bromide, Br^-, gives a cream precipitate of AgBr(s). The precipitate dissolves in concentrated ammonia solution.

Iodide, I^-, gives a pale yellow precipitate of AgI(s). This is insoluble in concentrated ammonia solution.

Figure 2.4.9 Identification of halides in solution

SUMMARY

- The elements of group VII of the periodic table, fluorine, chlorine, bromine and iodine, are known as the **halogens**. Their chemical behaviour is determined by the seven electrons in their outer electron shells – two in the s subshell and five in the p subshell. The halogens tend to gain one electron and form X^- ions.
- The halogens are a family of non-metallic elements that exist as diatomic molecules showing strong similarities in their properties and reactions.
- As the atomic number increases down group VII, the elements become less reactive, less volatile and darker in colour.
- The halogens are so reactive that they are not found free in nature. They are obtained by the oxidation of halide ions.
- There is a decrease in volatility down the group due to increased van der Waals forces.
- The oxidising power of the halogens shows the trend fluorine>chlorine>bromine>iodine. Iodine is displaced from a solution of iodide ions by bromine, which is itself displaced by chlorine.
- The halogens react with all metals, including gold and platinum.
- Apart from fluorine, the halogens show oxidation states from +1 to +7 as well as –1.
- The reactions of chlorine, bromine and iodine with alkalis involve **disproportionation** – a redox reaction in which a molecule, atom or ion is simultaneously oxidised and reduced.
- Halide ions may be identified by reaction with acidified silver nitrate solution.

QUESTIONS

1 Draw up a table to illustrate the trends in the halogens of the following properties:

a the melting point of X_2
b the strength of X_2 as an oxidising agent
c the strength of X^- as an oxidising agent
d the strength of the X—X bond
e the electronegativity of X
f the polarity of the bond H—X
g the lattice enthalpy of NaX.

2 A student took four test tubes, each of which contained one of the hydrogen halides. She plunged a red-hot glass rod into each test tube in turn and made the observations shown in table 2.4.7.

Tube	Observation
Hydrogen chloride	No change
Hydrogen bromide	Slight brown coloration seen
Hydrogen iodide	Copious violet fumes seen

Table 2.4.7

a Suggest what substances are being formed in the tubes containing hydrogen bromide and hydrogen iodide.
b Write equations for these changes.
c Explain the pattern in these observations.
d How would you expect hydrogen fluoride to behave?

3 Chlorine forms four oxoacids, HOCl, $HClO_2$, $HClO_3$ and $HClO_4$.

a Find the oxidation number of chlorine in each of these compounds.
b These acids dissociate to give the anions ClO^-, ClO_2^-, ClO_3^- and ClO_4^-. From your answer to **a**, write down the systematic name of each of these chlorate ions.

$HClO_2$ cannot be isolated, as it reacts to form ClO_2 and HCl.

c Write an equation for this reaction.
d Show the oxidation number of chlorine in each substance on each side of the equation.
e What sort of reaction is this?

4 The analysis of a solution containing iron(III) ions can be carried out in the following way:

i Iron(III) is reduced to iron(II) by iodide ions:

$$2Fe^{3+} + 2I^- \rightarrow 2Fe^{2+} + I_2$$

ii The iodine formed is used to oxidise thiosulphate ions ($S_2O_3^{2-}$) to tetrathionate ions ($S_4O_6^{2-}$):

$$I_2 + 2S_2O_3^{2-} \rightarrow S_4O_6^{2-} + 2I^-$$

The endpoint of this reaction is seen by the addition of starch solution, which forms a blue complex with iodine molecules.

In the analysis of an unknown compound of iron(III), 10 cm^3 of a solution containing 0.748 g of the compound was taken and reacted with 15 cm^3 of potassium iodide solution (an excess). 5 cm^3 of this solution was then taken and titrated against 0.05 mol dm^{-3} sodium thiosulphate solution. It was found that 18.4 cm^3 of this solution was required to react exactly with the iodine.

a Why was excess potassium iodide solution used?
b Describe carefully how you would use starch as an indicator in this titration.
c Calculate the number of moles of sodium thiosulphate used.
d How many moles of iodine does this correspond to?
e How many moles of iron(III) were required to produce this amount of iodine?
f How many moles of iron(III) were present in the original solution?
g Calculate the percentage iron by mass in 0.748 g of the iron(III) compound.

Developing Key Skills

A new example of food chain contamination by halogen-containing pesticides has become public and there is a great deal of media interest in the topic. Using the focus material from this chapter and other resources such as the Internet and the library, prepare a report for the early evening television news:
either showing all the uses and advantages of halogen-containing compounds in society, including all the benefits which have come from the use of compounds such as DDT,
or condemning the use of such chemicals and showing some of the environmental damage the improper use of them has caused.

[Key Skills opportunities: C, IT]

FOCUS INDUSTRIAL DECISION MAKING

Figure 1 The majority of chemical plants function efficiently and relatively cleanly to provide jobs, generate wealth and produce the petrochemicals, plastics, detergents and pharmaceuticals that enhance our way of life.

Chemical industries are of vital importance to the economies of industrialised countries. Many of these industrial processes involve either extracting elements from their ores or producing compounds required by other industries. Before a company decides to carry out a particular industrial process there are many points which must be considered to determine whether the process is likely to be economically viable. If the raw materials require a large amount of preparation, or expensive catalysts and high temperatures and pressures are needed to maintain a reaction, then demand for the end product must be high enough to support the prices necessary to cover production costs. On the other hand, if little preparation is necessary and there are no extremes of temperature and pressure required, then balancing the books will be considerably easier.

Other factors will affect the cost of an industrial process, such as any safety measures needed to protect workers in the industry or the general public, and the amount of pollution or toxic waste produced as a by-product of the chemical process. The costs of conforming to set standards for pollution levels and of disposing of toxic waste all add to the final cost of the product.

The same kinds of questions have to be asked when considering recycling materials, for example aluminium, iron/steel and plastics. There may be clear environmental benefits of recycling (far less energy is need to produce a tonne of recycled aluminium compared to extracting a tonne of aluminium from its ore, for example) and this may also lead to economic advantages. However, the collection and sorting of plastics for recycling is an expensive business, and it may be difficult to produce a product with consistent properties where the make-up of the raw materials from recycling is not constant.

But there is a great deal more to setting up a new industrial plant than getting the economics right on paper. Most chemical technologies are capable of great benefit to people. They provide the chemicals to make many of the products we consider essential to a comfortable way of life. They provide many people with jobs. The financial success of chemical industries has a knock-on effect on both the local and the national economy. Because of all these factors, the development of new chemical plants in areas of high unemployment may be attractive to government and industry alike, not least because local objections to development in such areas are likely to be fewer. However, those same industries do have disadvantages which can particularly affect those people living close to the chemical plant.

Figure 2 In 1974 a chemical plant in Flixborough, Humberside, exploded. A faulty pipe was repaired incorrectly, allowing a blanket of the gas cyclohexane to leak out and fill the factory. When someone lit a cigarette, the whole site was destroyed in the subsequent blast. Although this was a one-in-a-million chance, the warnings of a disaster such as this one must be heeded and careful consideration given to the siting and design of all chemical plants.

Chemical processes may produce dirt and smoke. They may involve unpleasant or downright dangerous chemicals being transported to or from the chemical plant through the local community. They may work under operating conditions that could be potentially hazardous either to the individuals working in the complex, or to the community as a whole. They may be a source of pollution on a local, national or international scale.

When decisions are made to give planning permission for a new chemical industry to be set up, all these factors have to be weighed in the balance. Safeguards against potential hazards need to be put in place at the very beginning of a project. The problem for the chemical industry is that everyone wants the benefits, but nobody wants the factory in their own back garden.

2.5 The p-block elements: Aluminium

Group III

5 **Boron, B** $1s^22s^22p^1$
13 **Aluminium, Al** $1s^22s^22p^63s^23p^1$
31 **Gallium, Ga** (Ar) $3d^{10}4s^24p^1$
49 **Indium, In** (Kr) $4d^{10}5s^25p^1$
81 **Thallium, Tl** (Xe) $5d^{10}6s^26p^1$

The melting point of the metal gallium is only 29.8 °C – lower than that of chocolate! As human body temperature is 37 °C, gallium will melt in the palm of your hand. A major use is in the production of miniature gallium arsenide lasers used in fibre optic telecommunication systems.

Boron is used to produce borosilicate glass. This is heat resistant and so can be used to produce 'see-through' cookware and laboratory glassware.

Aluminium has a multitude of uses. One of the most spectacular is in the solid booster rockets used to lift the American space shuttle into orbit. The highly exothermic reaction as aluminium powder burns in oxygen to form aluminium oxide, Al_2O_3, provides the energy for the boosters' thrust.

Figure 2.5.1 Group III elements and some of their applications

Group III of the periodic table includes boron (a metalloid) and the metals gallium, indium and thallium. These elements are not very familiar to most people, but this low profile is more than balanced by the other member of the group – aluminium is a very well-known metal. It is the third most abundant element by weight in the Earth's crust after oxygen and silicon, which makes it the most abundant metal.

Occurrence and extraction of aluminium

Occurrence of aluminium

Aluminium is too reactive to occur free, but it is found in a number of compounds. Many rocks contain **aluminosilicates**, combinations of aluminium, oxygen and silicon, but as yet no economic way has been developed of extracting the aluminium from these compounds. Aluminium is also found in **cryolite**, Na_3AlF_6, but its chief ore is **bauxite** which contains aluminium in the form of a hydrated oxide, $Al_2O_3.xH_2O$.

As an anhydrous oxide, aluminium occurs in **corundum**. This oxide has been heated to over 1000 °C within the Earth, and has an immensely stable crystal structure as a result. When the corundum contains traces of impurities, gemstones are formed. Traces of iron give the golden colour of topaz, manganese results in the violet of amethyst, a mixture of iron and titanium produces the blue of sapphires and the red colour of rubies comes from a trace of chromium.

Aluminium is a metal of immense importance both as the element and in alloys. Its great value lies in its combination of lightness and strength, and it is used for a large number of applications. These range from the aluminium foil used for packaging food and other products, through aircraft construction, kettles and saucepans, to the electrical cables used in high-voltage overhead power lines.

Extraction of aluminium

Aluminium is considered today to be an extremely useful metal. However, until the end of the nineteenth century, it was very rare, affordable only by the rich. Charles Hall, a 21-year-old US student, began a series of experiments in an attempt to find a cheap way of extracting aluminium from its ore. Because aluminium is very reactive, it does not occur freely, and yet because it forms stable compounds it is difficult to produce in standard reactions. Electrolysis was the obvious method to choose, but this too raised difficulties as anhydrous aluminium salts were difficult to prepare, and aluminium oxide has such a high melting point (over 2000 °C) that no practicable method of melting it could be found. In 1886, Hall discovered that aluminium oxide would dissolve in molten cryolite, Na_3AlF_6, giving a conducting mixture with the relatively low melting point of 850 °C. From this mixture, pure aluminium could readily be produced by electrolysis. The **Hall process** is still the basis of modern aluminium production.

Aluminium is usually obtained from bauxite, its richest ore. Bauxite contains around 50% aluminium oxide, along with iron(III) oxide and various other impurities. The first step in the industrial production of aluminium is the separation of the aluminium oxide from the impurities in the bauxite. This depends on the amphoteric nature of aluminium oxide, which is explained on page 189. Concentrated sodium hydroxide dissolves the aluminium oxide, forming soluble sodium aluminate:

$$\underset{\text{aluminium oxide}}{Al_2O_3(s)} + 2NaOH(aq) + 3H_2O(l) \rightarrow \underset{\text{sodium aluminate}}{2NaAl(OH)_4(aq)}$$

Many of the other impurities do not dissolve and so can be removed by filtration, including iron(III) oxide which forms a 'red mud'. If the solution is then diluted with water or has carbon dioxide bubbled through it, aluminium hydroxide is precipitated:

$$NaAl(OH)_4(aq) \rightarrow Al(OH)_3(s) + NaOH(aq)$$

This is in turn removed by filtering and then heated to give pure anhydrous aluminium oxide:

$$2Al(OH)_3 \rightarrow Al_2O_3(s) + 3H_2O(g)$$

The oxide is then dissolved in molten cryolite and undergoes electrolysis to form molten aluminium at the carbon cathode. The aluminium is periodically removed.

$$Al^{3+}(l) + 3e^- \rightarrow 3Al(l)$$

Oxygen evolves at the carbon anodes. These slowly react to form carbon dioxide, so the anodes have to be replaced from time to time.

$$2O^{2-}(l) \rightarrow O_2(g) + 4e^- \qquad C(s) + O_2(g) \rightarrow CO_2(g)$$

The cell used in the production of aluminium is shown in figure 2.5.2.

Aluminium is one of our most plentiful resources, with large amounts present in the Earth's crust. However, the industrial production of the pure metal is an immensely costly procedure. Electrical energy is required not only

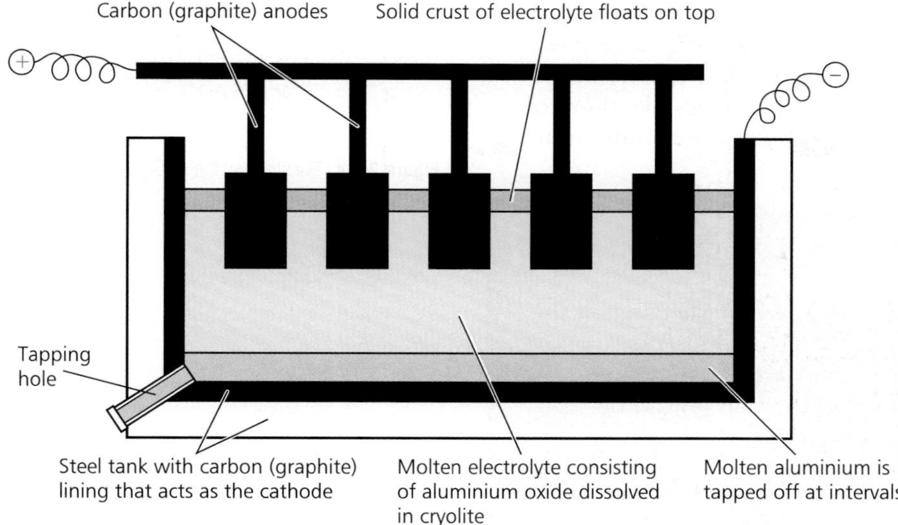

Figure 2.5.2 The industrial production of aluminium worldwide is still based on the electrolytic process devised by Charles Hall in the late nineteenth century.

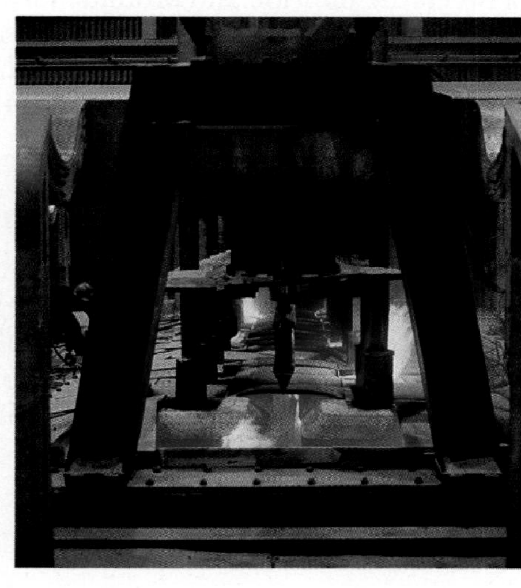

for the electrolysis itself, but also to maintain the electrolyte in a molten state at 850 °C. On the other hand, aluminium is readily recycled, as old aluminium artefacts such as cans can be crushed and melted for reuse. The cost of producing 'new' aluminium in this way is a fraction of the cost of extracting more aluminium from the Earth, in terms of both cash and energy resources.

Properties of aluminium

Physical properties

Aluminium is a shiny, reactive, light grey metal. It is a p-block element with two s electrons and one p electron in its outer shell. In common with the other group III elements, it usually shows an oxidation state of +3. In forming compounds, the three electrons are transferred from the outer shell of the aluminium atom to give a noble gas configuration, and so the ion has a charge of +3.

$$Al \rightarrow Al^{3+} + 3e^-$$

The first four ionisation energies in kJ mol^{-1} for aluminium are:

1st 578 2nd 1817 3rd 2745 4th 11 578

The fourth ionisation energy of aluminium is high compared with the first three, due to the fact that a noble gas configuration must be broken into in order to form the Al^{4+} ion. This means that aluminium forms ions with a charge of +3. These ions are small, because the electron shells contract as electrons are removed, due to increasing electrostatic attraction. The high charge and small size (ionic radius 0.053 nm) of the aluminium ion give it a very high charge density, a factor which dominates much of its chemistry.

The high charge density of the Al^{3+} ion tends to lead to distortion of the electron cloud around any neighbouring ion towards the positive aluminium ion. Thus a partial sharing of the electron cloud results, giving **partial covalent bonding**, and very stable compounds showing a degree of covalent character are the result. This tendency is demonstrated very clearly in the aluminium halides. The fluoride ion is very small and is not readily polarised. As a result aluminium fluoride, AlF_3, is highly ionic, consisting of symmetrical ionic crystals with a melting point of 1291 °C. Aluminium bromide, on the other hand, is largely covalent, as the large bromide ion is readily polarised.

Figure 2.5.3 The recycling of aluminium should perhaps become a matter of higher priority as we attempt to minimise our use of largely non-renewable energy sources.

The compound consists of simple dimers, Al_2Br_6, and has a melting point of only 98 °C. The chloride ion is intermediate in size and ease of polarisation between the fluoride and bromide ions, and thus aluminium chloride shows partially covalent character, subliming at 178 °C to form dimers like those of aluminium bromide, shown in figure 2.5.4. (The bonding in the aluminium chloride dimer was shown in figure 1.4.9, page 54.)

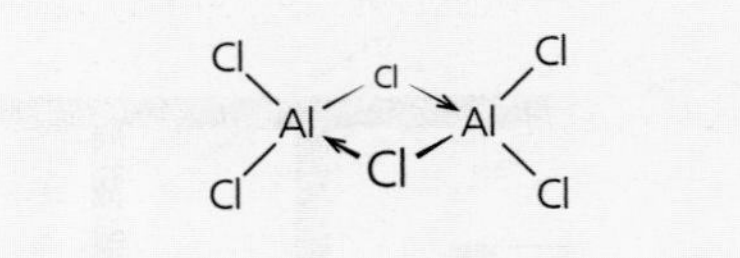

Figure 2.5.4 The shape of the Al_2Cl_6 molecule

The electronegativity of aluminium

The different properties of the aluminium halides are predictable on the basis of the differences in electronegativity of the elements. The electronegativities of aluminium and the halogens are given in table 2.5.1, from which the ionic/covalent character of the halides can be worked out as was described in figure 1.4.21, page 60.

Element	Electronegativity	Ionic/covalent character of aluminium halide
Al	1.5	—
F	4.0	Difference in electronegativity = (4.0 – 1.5) = 2.5 ⇒ 79% ionic
Cl	3.0	Difference in electronegativity = (3.0 – 1.5) = 1.5 ⇒ 43% ionic
Br	2.8	Difference in electronegativity = (2.8 – 1.5) = 1.3 ⇒ 34% ionic

Table 2.5.1

Chemical properties

Although aluminium is a very reactive metal, its uses suggest quite the opposite. It is frequently used to wrap substances and protect them from air and water. It is used for the bodies of aircraft which are exposed to the damp atmosphere during flight. The position of aluminium in the electrochemical series (see page 160) means that we would normally expect it to react with oxygen and the dilute acids in the atmosphere under these circumstances. However, on exposure to air aluminium rapidly forms a thin layer of extremely stable oxide which prevents any further attack by oxygen or water:

$$4Al(s) + 3O_2(g) \rightarrow 2Al_2O_3(s)$$

It is the formation of this oxide layer that makes aluminium so useful.

The thickness of the oxide layer can be increased by a process known as **anodising**. Sulphuric acid is electrolysed using the aluminium object as the anode. Oxygen is released at the anode during the electrolysis and reacts with the aluminium and thickens the oxide layer. Dyes can easily be absorbed into this anodised layer, colouring the article at the same time as protecting it.

The reactive nature of aluminium, a highly electropositive element, is often masked by the protective layer of oxide over the metal's surface.

Figure 2.5.5 The protective oxide layer can be increased on aluminium objects that need to be particularly resistant to corrosion by anodising.

With concentrated sulphuric acid and with dilute hydrochloric acid, aluminium reacts rapidly. The acid initially attacks the layer of oxide film on the metal surface, revealing the metal underneath. This then reacts with the acid, liberating hydrogen and forming simple salts:

$$2Al(s) + 3H_2SO_4(aq) \rightarrow Al_2(SO_4)_3(aq) + 3H_2(g)$$

$$2Al(s) + 6HCl(aq) \rightarrow 2AlCl_3(aq) + 3H_2(g)$$

No reaction is seen with either dilute or concentrated nitric acid, since the oxide film is resistant to attack by this acid. If the oxide film is removed (for example, by scratching it or by reacting it with mercury(II) chloride solution), aluminium reacts readily with all dilute mineral acids, and even with water.

Aluminium reacts with the halogens, forming aluminium halides. Aluminium chloride and bromide may be prepared by heating the metal in the gaseous halogen, giving a highly exothermic reaction. Aluminium and iodine react spontaneously at room temperature if a few drops of water are added to the mixture:

$$2Al(s) + 3I_2(s) \rightarrow 2AlI_3(s)$$

Anhydrous aluminium chloride produces fumes of hydrogen chloride in moist air, due to hydrolysis:

$$2AlCl_3(s) + 3H_2O(l) \rightarrow Al_2O_3(s) + 6HCl(g)$$

Even with its protective oxide layer, aluminium is attacked by alkalis such as sodium hydroxide, forming the aluminate and hydrogen gas. Notice that aluminium metal reacts with both acids and bases to yield a salt and hydrogen:

$2Al$

$6H^+$ ↙ ↘ **$2OH^- + 6H_2O$**

$2Al^{3+} + 3H_2$ **$2Al(OH)_4^- + 3H_2 \rightarrow 2AlO_2^- + 3H_2 + 4H_2O$**

(e.g. $Al(NO_3)_3$, aluminium nitrate)   **(e.g. $KAlO_2$, potassium aluminate)**

The aqueous chemistry of aluminium is discussed in more detail on page 190.

Aluminium oxide

As we have seen, aluminium oxides are the main source of available aluminium in the Earth's crust, and also make up a variety of precious and semiprecious gemstones. However, these are very different from the white powder which is the pure aluminium oxide made in the laboratory. The aluminium ion is able to distort the electron cloud around the oxide ion, and so aluminium oxide demonstrates partial covalent bonding.

Aluminium oxide and aluminium hydroxide are amphoteric – they react with both acids and bases. With acids, simple aluminium salts are formed, while bases produce the aluminates:

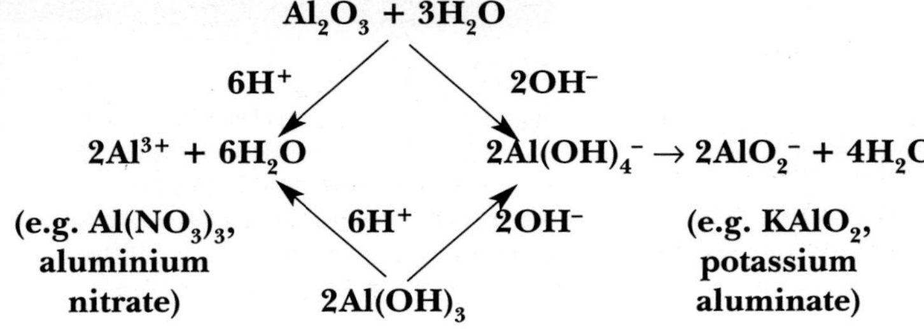

2

The Thermite reaction

Aluminium oxide is a very stable compound, with an enthalpy change of formation of -1676 kJ mol^{-1}. This great affinity of aluminium for oxygen is put to good use in a spectacular process known as the **Thermite reaction**.

In the Thermite reaction, a coarse mixture of aluminium powder and iron(III) oxide is ignited. The reaction that ensues is violently exothermic, proceeding with a shower of sparks and a brilliant white glare – the temperature of the reaction may reach 3000 °C. White-hot molten iron is formed as the aluminium reduces the iron(III) oxide, and the iron sinks to the bottom of the reaction vessel, with a layer of fused aluminium oxide floating on top of it. The equation for the reaction is:

$$\mathbf{Fe_2O_3(s) + 2Al(s) \rightarrow Al_2O_3(l) + 2Fe(l)}$$

Figure 2.5.6 The Thermite process is used to weld steel railway tracks. The molten iron produced runs down into a mould around the rail ends, joining them together.

Aluminium ions in solution

As we have already seen, aluminium Al^{3+} ions are small and highly charged. When aluminium salts are added to water, the Al^{3+} ions attract the δ- ends of the water molecules, so that six water molecules are held around the central aluminium ion. This is known as the **hexaaquaaluminium(III) ion** and has the formula $[Al(H_2O)_6]^{3+}$. Not surprisingly, this is more frequently written as $Al^{3+}(aq)$. The strong attractive forces of the aluminium ion draw the electron cloud in the O—H bonds of the water molecules towards the aluminium, and so allow the water molecules to readily become proton donors, losing H^+ ions. This means that solutions of aluminium salts are acidic, as they increase the hydrogen ion concentration. With water, only one proton is lost:

$$\mathbf{[Al(H_2O)_6]^{3+}(aq) \rightleftharpoons [Al(H_2O)_5(OH)]^{2+}(aq) + H^+(aq)}$$

However, more and more protons are removed as bases of increasing strength are added to solutions of aluminium salts. Ammonia, NH_3, sulphide, S^{2-} and carbonate, CO_3^{2-} are all strong enough bases to remove three protons from $[Al(H_2O)_6]^{3+}$, as is the stronger base aqueous sodium hydroxide, NaOH(aq). This removal of three protons results in the formation of insoluble aluminium hydroxide which appears as a gelatinous precipitate:

$$\mathbf{[Al(H_2O)_6]^{3+}(aq) + 3OH^-(aq) \rightarrow [Al(H_2O)_3(OH)_3](s) + 3H_2O(l)}$$

$$\mathbf{2[Al(H_2O)_6]^{3+}(aq) + 3CO_3^{2-}(aq) \rightarrow 2[Al(H_2O)_3(OH)_3](s) + 3CO_2(g) + 3H_2O(l)}$$

An excess of aqueous sodium hydroxide or other very strong bases will remove four protons, and this results in a soluble anion:

$$\mathbf{[Al(H_2O)_3(OH)_3](s) + OH^-(aq) \rightarrow [Al(H_2O)_2(OH)_4]^-(aq) + H_2O(l)}$$

A simpler way of representing these reactions of the aluminium ion is as follows:

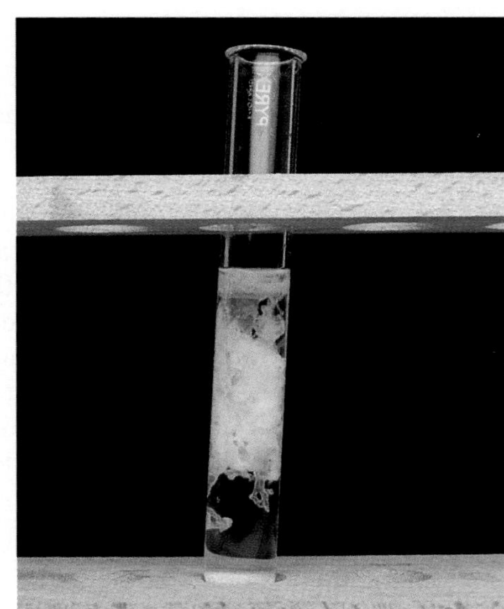

Figure 2.5.7 Insoluble aluminium hydroxide is frequently written in its simplest form as $Al(OH)_3$. This does not show the water molecules contained in its structure, which result in the gelatinous appearance of the precipitate.

$$Al^{3+}(aq) + 3OH^-(aq) \rightarrow Al(OH)_3(s)$$
white precipitate

$$Al(OH)_3(s) + OH^-(aq) \rightarrow Al(OH)_4^-(aq)$$
colourless solution

The white precipitate of aluminium hydroxide described here plays an important role as a mordant, a substance that fixes dyes, in the production of coloured textiles for clothing and furnishings.

Colour from aluminium hydroxide

Colouring fabrics has been part of human culture for thousands of years. However, the effective dyeing of fabrics is not always as simple as it might seem. If a piece of material is just immersed in certain dyes, the colour is poorly absorbed by the fabric, giving weak coloration and also a tendency for the dye to wash out. The process of dyeing is made far more efficient with the use of a **mordant** such as aluminium hydroxide. Cloth is soaked first in aluminium sulphate and then a dilute alkali is added. This precipitates the aluminium hydroxide between the interwoven fibres of the cloth. The gelatinous nature of the precipitate of aluminium hydroxide means that dyes penetrate readily and are held within the structure. The precipitate binds to the fabric and the result is that the dyestuff is taken up strongly and held firmly into the cloth, forming striking and long-lasting colours.

Figure 2.5.8 The mordant properties of aluminium hydroxide allow it to 'bite onto' (Latin *mordere*, to bite) both dye and fabrics.

SUMMARY

- Group III of the periodic table is the first group in the p block, containing the elements boron, aluminium, gallium, indium and thallium. The atoms of the group III elements have three electrons in their outer shell, two in the s subshell and one in the p subshell. The principal oxidation state of the group III elements is +3.
- Aluminium is found in ores including **bauxite**, hydrated aluminium oxide, $Al_2O_3.xH_2O$.
- Aluminium is extracted from its ore by the **Hall process**, which involves the electrolysis of the oxide dissolved in molten cryolite.
- Elemental aluminium is a shiny, reactive, light grey metal. Its reactivity is suppressed by a coating of the oxide which gives the metal a dull appearance. This oxide coating is valuable in protecting the metal, and can be increased by **anodising**.
- Aluminium ions have a high positive charge and a relatively small ionic radius. This tends to distort the electron cloud around a neighbouring anion, leading to partial covalent bonding.

- Aluminium is **amphoteric** – it reacts with both acids and alkalis to produce a salt and hydrogen. Aluminium oxide and hydroxide are also amphoteric.
- The aluminium ion is surrounded by six water molecules in aqueous solution, forming the hexaaquaaluminium(III) ion, $[Al(H_2O)_6]^{3+}$. Solutions of aluminium ions are acidic.

1 a Why is aluminium so unreactive?
b What is anodising?

2 a Why is the Al^{3+} ion acidic?
b Write down equations to show the amphoteric nature of aluminium hydroxide.

3 When finely powdered lithium hydride is added to a solution of aluminium chloride dissolved in ethoxyethane ('ether'), lithium tetrahydridoaluminate(III) is formed:

$$4LiH + AlCl_3 \rightarrow LiAlH_4 + 3LiCl$$

Lithium tetrahydridoaluminate(III) (lithium aluminium hydride) acts as a reducing agent, and is commonly used in organic chemistry to reduce carboxylic acids to alcohols (see page 494). The electronegativity of aluminium is 1.5, that of lithium is 1.0, and the electronegativity of hydrogen is 2.1.
a What is the oxidation number of hydrogen in lithium tetrahydridoaluminate(III)?
b Suggest why lithium tetrahydridoaluminate(III) can act as a reducing agent.
Hydrides of some elements can be made by reacting chlorides with lithium tetrahydridoaluminate(III). Silane, SiH_4, can be prepared by reacting silicon(IV) chloride in this way, when lithium chloride and aluminium chloride are also formed.
c Write an equation for this process.
d Show the oxidation numbers of each element on each side of the equation.
e Is lithium tetrahydridoaluminate(III) behaving as a reducing agent in this reaction?

4 Aluminium carbide, Al_4C_3, is formed when aluminium and carbon react together at temperatures above 1000 °C. Aluminium carbide reacts instantly with water to form methane, and X-ray studies show that it contains carbon atoms that are widely separated. It is usually regarded as an ionic compound.
a What ions does aluminium carbide contain?
b Write down an equation for the reaction of aluminium carbide with water.
c Is it reasonable to consider aluminium carbide as an ionic compound?

Developing Key Skills

Investigate the economic and environmental advantages of recycling aluminium. Use the information to plan a recycling programme for your school or local community. A major aspect of your plan must be a leaflet and flyers or posters to inform people about the scheme and why it is important to take part. Produce a leaflet and one small flyer for use either in school or in the whole local community.

[Key Skills opportunities: C, IT]

2.6 The p-block elements: Carbon and group IV

Carbon atoms are ubiquitous in living organisms – the proteins, carbohydrates and fats that are the structural and storage materials of cells are based on the unique properties of carbon. But the role of carbon is not confined to organic molecules – inorganic carbon compounds are important too, both in the laboratory and for industry.

Group IV

Group IV
6 **Carbon, C** $1s^2 2s^2 2p^2$
14 **Silicon, Si** $1s^2 2s^2 2p^6 3s^2 3p^2$
32 **Germanium, Ge** (Ar) $3d^{10} 4s^2 4p^2$
50 **Tin, Sn** (Kr) $4d^{10} 5s^2 5p^2$
82 **Lead, Pb** (Xe) $5d^{10} 6s^2 6p^2$

We often hear terms such as 'silicon chip' and 'silicon valley', illustrating the use of silicon as a semiconductor in the electronics industry. Germanium, another group IV element, is also used to make semiconductors, but has never become as well known as silicon.

Tin is used to coat steel in the manufacture of 'tin cans' and in a variety of alloys. It has played an interesting role in human civilisation. Bronze is an alloy of copper with 5–10% tin. The discovery of bronze enabled superior tools, weapons and utensils to be made, which took the human race forward from the Stone Age.

Lead has a variety of uses. An alloy containing 5–10% of lead in tin produces a metal which gives organ pipes particularly good tonal properties.

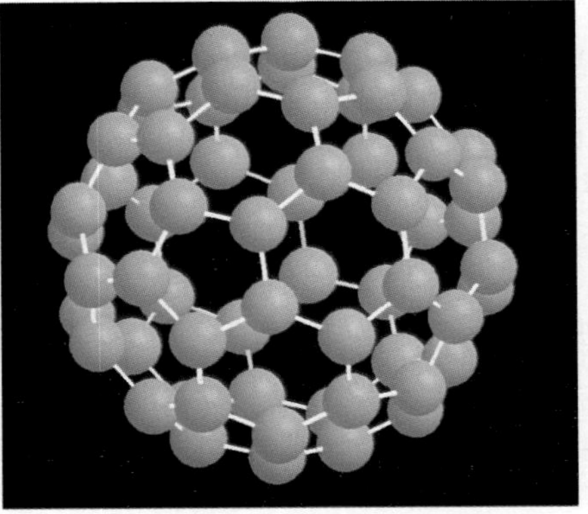

Carbon exists in a variety of allotropic forms. The most common are diamond and graphite along with the much rarer amorphous carbon which forms soot. In the 1980s a new family of carbon allotropes was discovered, more or less spherical molecules known as **fullerenes**. The example here was the first to be discovered and rejoices in the name of buckminsterfullerene.

Figure 2.6.1 Group IV elements and some of their applications

The elements of group IV

Group IV is very diverse – the elements do not form a cohesive family like the halogens. Instead, they show a great range and gradation in the properties of both the elements and their compounds from the non-metal carbon through the metalloids silicon and germanium to the metals tin and lead.

Carbon has the unusual feature of forming particularly strong bonds between its atoms. This means that carbon alone amongst the elements can form chain molecules thousands of atoms long, with the carbon atoms linked by single, double or triple bonds as well as in complex ring structures. Carbon thus forms an almost infinite number of compounds. This ability to form long molecules with many atoms of the same element bonded together is called **catenation**. The element carbon alone has a wide variety of uses depending on its allotropic (crystal) form, as we shall see later.

Silicon is the second most abundant element on the Earth (after oxygen), present in sand and sandstone as silicon dioxide, SiO_2. It is also found in rocks and clays as a variety of silicates, giant molecules made up of repeated SiO_4^{4-} ions coupled with various cations. Silicon undergoes catenation, to a lesser extent than carbon. In addition to its role in the microelectronics industry, silicon is a major component of glass and is used in the production of silicones (polymers used as waterproofing agents) and lubricants. Silicones are also used in plastic and cosmetic surgery, although the long-term health risks of injecting silicone to give fuller lips or implanting silicone sacs to enlarge breasts are under question. Germanium, tin and lead are substantially rarer elements, with tin and lead showing distinct metallic character.

Physical properties

The bonding between the atoms of the group IV elements changes from giant molecular lattices in carbon and silicon to giant metallic structures in tin and lead. The explanation for this is that as the atoms get larger and the atomic radius increases, the bonding between the atoms gets weaker and the attraction of neighbouring nuclei for electrons also gets weaker. Covalent bonds do not form between the larger atoms and the outer electrons are freed to form a 'sea'.

As we descend group IV, the elements have an increasing tendency to display an oxidation number of +2 in their compounds, until in lead compounds this is by far the more stable state. As an example of this behaviour, carbon forms the stable compound tetrachloromethane (in which the oxidation number of carbon is +4), while lead(IV) chloride decomposes readily to form lead(II) chloride and chlorine. The reason for this is sometimes given as the 'inert pair effect', which explains the trend as being due to the increasing tendency of the outer pair of s electrons to be inert as the group is descended, so that these electrons do not readily take place in bonding. The physical properties of the group IV elements are shown in table 2.6.1.

A fuller explanation

The 'inert pair effect' is really a name, not an explanation. A fuller explanation involves the decreasing strength of the M—X bond going down the group (for covalent compounds) or the decreasing lattice energies of compounds containing the M^{4+} ion (for ionic compounds). In this way the energy inputs needed to form compounds of the formula MX_4 are less likely to be balanced by the energy released when the four M—X bonds are formed, so the equilibrium

$$\mathbf{MX_2 + X_2 \rightleftharpoons MX_4}$$

favours the left-hand side.

Since there are few compounds in this group with the formula MX_2 or MX_4 which can be regarded as entirely ionic or covalent, this is a simplification, but it does at least describe the factors that should be considered.

Element	**C**	**Si**	**Ge**	**Sn**	**Pb**
Density/g cm^{-3}	2.25(graphite) 3.51(diamond)	2.33	5.35	7.28 (white)	11.34
Melting point/°C	3652(graphite) (sublimes)	1410	937	232	328
Boiling point/°C	4827	2355	2830	2270	1740
Atomic radius/nm	0.077	0.118	0.122	0.140	0.154
Crystal structure	Giant molecular	Giant molecular (similar to diamond)	Giant molecular (similar to diamond)	Giant metallic	Giant metallic
First ionisation energy/kJ mol^{-1}	1086	789	762	709	716
Electrical conductivity	Fair(graphite) None(diamond)	Semi-conductor	Semi-conductor	Good	Good

Table 2.6.1 A summary of some of the physical properties of the group IV elements. In spite of the change from non-metal to metal, many of the expected trends down a group are still in evidence.

Chemical properties

The transition from non-metal to metal down group IV has already been noted on the basis of the physical properties of the elements. The chlorides and oxides of the group IV elements in their +4 oxidation state illustrate further the change in character of the elements from non-metal to metal down the group.

		C	Si	Ge	Sn	Pb
Chlorides	**Formula**	CCl_4	$SiCl_4$	$GeCl_4$	$SnCl_4$	$PbCl_4$
	Structure	← Simple molecular →				
	Thermal stability	← Stable to high temperatures →			Decomposes on heating $\rightarrow SnCl_2 + Cl_2$	Decomposes at room temperature $\rightarrow PbCl_2 + Cl_2$
	Reaction with water	No reaction	← Hydrolysed readily to form hydroxides + HCl →			
Oxides (+4 oxidation state)	**Formula**	CO_2	SiO_2	GeO_2	SnO_2	PbO_2
	Structure	Simple molecular	Giant molecular	← Giant molecular/ionic →		
	Thermal stability	← Stable to high temperatures →				Decomposes on heating $\rightarrow PbO + \frac{1}{2}O_2$
	Reaction with bases	$CO_2 + 2OH^- \rightarrow CO_3^{2-} + H_2O$	$SiO_2 + 2OH^- \rightarrow SiO_3^{2-} + H_2O$	$GeO_2 + 2OH^- \rightarrow GeO_3^{2-} + H_2O$	$SnO_2 + 2OH^- \rightarrow SnO_3^{2-} + H_2O$	$PbO_2 + 2OH^- \rightarrow PbO_3^{2-} + H_2O$
		Dilute aqueous alkali	Concentrated aqueous alkali			Fused alkali
		Increasingly severe conditions →				
	Reaction with acids	—	—	$GeO_2 + 4HCl \rightarrow GeCl_4 + 2H_2O$	$SnO_2 + 4HCl \rightarrow SnCl_4 + 2H_2O$	$PbO_2 + 4HCl \rightarrow PbCl_4 + 2H_2O$

Table 2.6.2 The properties of the group IV chlorides and oxides

Table 2.6.2 shows some of the chemical properties of the chlorides and oxides of the group IV elements in their +4 oxidation state. There are several points to note about these compounds.

Chlorides

1 The chlorides are all simple molecular substances with tetrahedral molecules.

2 The stability of the chlorides decreases down the group and the +2 oxidation state becomes more stable than the +4 state. Only tin and lead form chlorides in which their oxidation state is +2, the other chlorides existing solely in the +4 state. Tin(II) chloride is a solid that is soluble in water, giving a solution which conducts electricity; it is also soluble in organic solvents. Its melting point is 246 °C. Lead(II) chloride is also a solid. It is sparingly soluble in water, giving a solution which conducts electricity, and melts at 501 °C. These observations suggest that tin(II) chloride has both covalent and ionic character, while lead(II) chloride is predominantly ionic.

3 All the chlorides with +4 oxidation state are readily hydrolysed by water, except tetrachloromethane (CCl_4). This hydrolysis is just as favourable thermodynamically as the others. The reason why it does not occur is explored in the box below.

Oxides

1 The oxides show a marked trend in structure from the molecules of carbon dioxide to giant structures intermediate between ionic and covalent lower down the group.

2 The +2 oxidation state is the more stable state in the case of lead oxide, and lead(IV) oxide decomposes on heating giving lead(II) oxide, a solid that melts at 886 °C. The structure of lead(II) oxide is predominantly ionic.

3 The oxides at the top of the group (CO_2 and SiO_2) have an acidic nature, the carbonate ion CO_3^{2-} being produced easily in dilute aqueous solutions. The ease of formation of oxoanions (SiO_3^{2-}, GeO_3^{2-}, etc.) decreases down the group as the acidic character decreases. The oxides of germanium, tin and lead are amphoteric, reacting to form simple salts with acids.

The resistance of tetrachloromethane to hydrolysis

e

Once again, the different behaviour of the elements in the second period from those in the periods below them is due to electronic structure. In the hydrolysis of the chlorides of the group IV elements, a lone pair of electrons from the oxygen atom of a water molecule attacks the group IV atom in the chloride, which has a partial positive charge due to the electron-withdrawing effect of the four chlorine atoms. (We shall learn more about this sort of reaction mechanism in section 4.5.) Figure 2.6.2 shows the hydrolysis of the $SiCl_4$ molecule.

$$SiCl_4 + 4H_2O \rightarrow Si(OH)_4 + 4HCl$$

In this reaction, the group IV atom briefly has five pairs of electrons round it. For this reaction to occur, there must be empty orbitals for these electrons to occupy. In the case of silicon and the group IV elements below it, the extra pair of electrons is accommodated using the empty d-orbitals available. No d-orbitals are available in the case of carbon, however, and so tetrachloromethane cannot be hydrolysed. This is an example of kinetic stability – even though the reaction is thermodynamically favourable, no reasonable mechanism exists for it to occur.

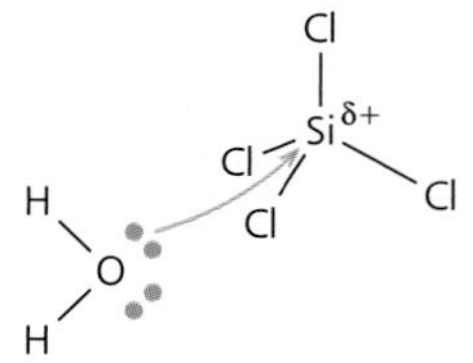

Figure 2.6.2 A lone pair on an oxygen atom attacks the silicon, which is partially positively charged.

Carbon – the element

The allotropes of carbon

Pure carbon exists as one of four **allotropes**, different crystal forms of the same element. These allotropes are diamond, graphite, amorphous carbon, and the recently discovered fullerenes. We shall look at each in turn.

Diamond

Pure diamond is composed entirely of carbon atoms in an interlocking tetrahedral crystal structure, each atom covalently bonded to its four nearest neighbours as shown in figure 2.6.3. A diamond is uniformly bonded throughout and may be thought of as a giant molecule. The exceptional strength of the carbon–carbon bond and the covalently interlocked crystal structure accounts for the physical properties of diamond. Diamond is the hardest natural substance known and most of its uses in drilling, cutting and grinding and as bearings are based on this exceptional hardness.

Although pure diamond is colourless and transparent, when contaminated with other minerals it may appear in various colours ranging from pastels to opaque black. The diamond crystal is chemically inert but may be induced to burn in air at high temperatures. Until 1955 the only sources of diamond were natural deposits of volcanic origin. Since then diamonds have been made artificially from graphite subjected to high pressures and temperatures. Before the 1970s, only diamonds of industrial quality could be made in this way – they had the hardness of natural diamond but did not have an attractive appearance and were not large enough to be used as gemstones. Diamonds of gem quality have now been produced artificially, but they are so expensive to create that they are not made for jewellery.

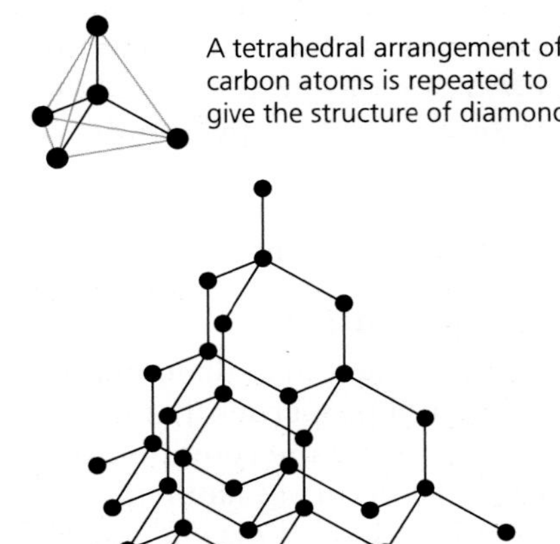

A tetrahedral arrangement of carbon atoms is repeated to give the structure of diamond.

Figure 2.6.3 The combined strength of the many carbon–carbon bonds within the structure of diamond give it both great hardness and a lack of chemical reactivity.

Graphite

Graphite is a black, lustrous substance that easily crumbles or flakes. It has a slippery feel because of the tendency of the crystal to cleave in thin layers. It is chemically inert, although somewhat less so than diamond, and is an excellent conductor of both heat and electricity. It occurs as a mineral in nature, usually in impure form, and can be produced artificially from amorphous carbon.

Graphite is composed entirely of planes of trigonal carbon atoms joined in a honeycomb pattern. Each carbon molecule is bonded to three others with a bond angle of 120°. These planes are arranged in sheets to form three-dimensional crystals as shown in figure 2.6.4. The layers are too far apart for covalent bonds to form between them. Instead, they are held in place only by weak van der Waals forces and so the layers are easily removed. Because each atom is covalently bonded to only three neighbouring atoms, the remaining valence electron (one in each atom) is free to circulate within each plane of atoms, contributing to graphite's ability to conduct electricity. Under normal conditions graphite is the more stable allotrope of carbon:

$$\textbf{C(diamond)} \rightarrow \textbf{C(graphite)} \qquad \Delta H^{\ominus}_{m} = -2\ \text{kJ mol}^{-1}$$

However, the rate of this reaction is so slow at normal temperatures as to be imperceptible.

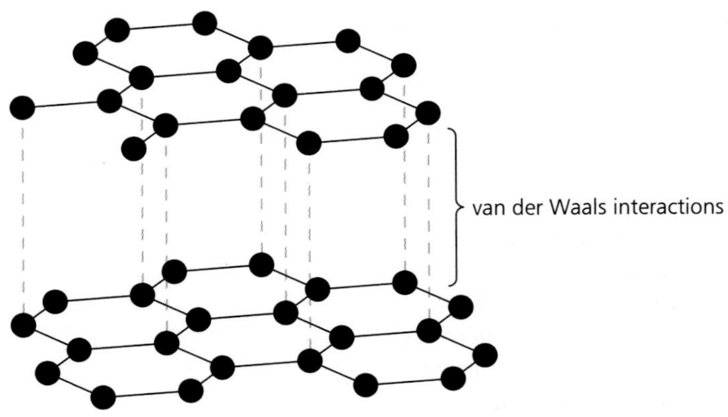

Figure 2.6.4 The structure of graphite. Whilst the bonds within the layers are strong, those between the layers are not and so they slide over each other easily.

One of the main uses for graphite – as a lubricant – results from the characteristic sliding of one layer over another within the crystal. The 'lead' in pencils is actually graphite, and the mark left on the page is the result of layers of graphite being scraped off the crystal. Graphite is also used as an electrical conductor and electrode material (in dry cells, for instance), and in nuclear reactors in order to absorb some of the energy of the neutrons, increasing their ability to cause fission.

Amorphous carbon

Less well defined than diamond or graphite, amorphous carbon has physical and chemical properties that may vary depending on its method of manufacture and the conditions to which it is later subjected. It is a deep black powder that occurs in nature as a component of coal, and is also seen frequently as soot. It may be obtained artificially from almost any organic substance by heating the substance to very high temperatures in the absence of air. In this way coke is produced from coal, and charcoal from wood. Burning organic vapours with insufficient oxygen produces such amorphous forms as carbon black and lampblack.

Amorphous carbon is the most reactive form of carbon. It burns relatively easily in air, thereby serving as a fuel, and is attacked by strong oxidising agents. Amorphous carbon is *not* finely divided graphite, but appears to have some of the structural features of graphite, such as local regions of sheets and layers. Its atomic structure, however, is much more irregular. The most important uses for carbon black are as a stabilising filler for rubber and plastics, and as a black pigment in inks and paints. Charcoal and coke are used as clean-burning fuels. Certain types of 'activated' charcoal are useful as absorbers of gases and of impurities from solutions.

Fullerenes

For some years, scientists predicted the existence of another allotrope of carbon. This was confirmed in the 1980s, when carbon atoms were found linked to form a more or less spherical molecule similar in structure to a football. A whole family of allotropes with differing numbers of atoms is now known to exist. The first to be identified and the most symmetrical of the family, with 60 atoms and 32 sides (20 hexagons and 12 pentagons), was nicknamed 'buckyball' and was then formally named **buckminsterfullerene**, because it resembles the geodesic domes developed by an American inventor called R. Buckminster Fuller (see figure 2.6.1, page 193). The group of spherical carbon molecules is called **fullerenes**.

Buckminsterfullerene is now being produced in large quantities for study by scientists. It may be quite widespread in carbon-containing materials on Earth, and some scientists suggested that it is a fairly common interstellar molecule. Its superconducting properties and its potential for opening new areas of chemistry have made study of the 'buckyball' one of the most rapidly expanding areas of chemical research.

Physical properties

It is almost impossible to summarise the physical properties of carbon as they vary depending on the allotrope. For example, diamond and amorphous carbon behave purely as non-metals, but graphite with its free electrons has many metallic characteristics. At present, until the full extent of the occurrence of the fullerenes is known, diamond and graphite are believed to be the most commonly occurring allotropes of carbon, and when physical properties for the group IV elements are compared it is usual to indicate which of these two allotropic forms is referred to.

FOCUS BUCKY BALLS AND BUCKY TUBES

For many years carbon was known to exist in three forms (**allotropes**) – diamond, graphite and amorphous carbon. Then, in 1985, a new allotrope of carbon was discovered by Richard Smalley and Robert Curl of Rice University, Texas working with Harry Kroto of Sussex University. This was buckminsterfullerene, soon nicknamed the 'buckyball'. Once discovered, it was not long before researchers were producing other types of fullerenes, because their superconducting properties and potential for opening up new fields of chemistry made the buckyball one of the most rapidly expanding areas of research chemistry.

Figure 1 Harry Kroto, Robert Curl and Richard Smalley, who were awarded the Nobel prize for chemistry in 1995 for discovering buckminsterfullerene and working out its structure.

Son of buckyballs

In 1991 a Japanese scientist, Sumio Iijima reported that in the soot produced as he was working to make fullerenes he had discovered elongated carbon cage-like structures. Other scientists rushed to duplicate his work, and the science of carbon nanotubes – or buckytubes – was born. These closed-cage carbon structures all contain 12 five-membered rings and almost any number of six-membered rings. They make highly complex shapes – onions, toroids (doughnut shapes), corkscrews and cones have been produced. The form which is currently causing the most interest is the nanotube.

Chemists have produced single-walled nanotubes (SWNTs) and multiwalled nanotubes MWNTs), and the possible uses for these complex carbon structures seem almost limitless. Reacting MWNTs with metal oxides results in the formation of carbide monorods, some of which are superconducting (they have zero resistance to electric currents). Studies of nanotubes show that they are stiffer than any other known material, and so could be embedded in polymer resins to produce new composite materials which would have good electrical conductivity as well as enormous strength and great lightness. The potential uses of these types of materials in the aircraft, space and car industries alone are huge.

MWNTs are also being used as tips for scanning tunnelling electron microscopes. Not only do these new tips give increased resolution in the pictures, they bend rather than breaking if they come into contact with the specimen, allowing scientists to probe ever deeper into the ultramicroscopic details of the world around us.

But it is the electronic properties of the carbon nanotubes which are causing most excitement. SWNTs can show both metallic and semiconductor properties. This and other work is leading many people to believe that buckytubes will be the cornerstone of twenty-first century molecular electronic devices. Perhaps in the future computers will be based not on silicon chips but on their carbon nanotube based equivalent – an organic entity in the making?

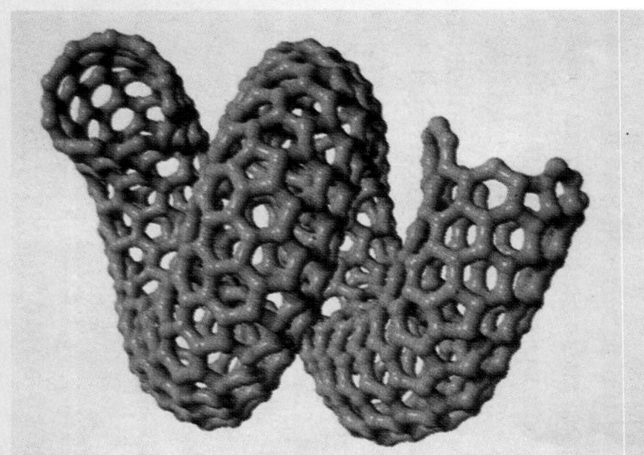

Figure 2 A single walled corkscrew shaped carbon nanotube – just one of the molecules which it is predicted will play an important role in the future of technology

Carbon compounds

Carbon compounds have been traditionally classified as either **inorganic** or **organic**. Inorganic carbon compounds are those in which carbon plays a role roughly analogous to those of its neighbouring elements in group IV. They include the binary compounds of carbon with other elements that lack carbon–carbon bonds, and also salts and complexes of metals that contain simple carbon species. Organic compounds are those reflecting carbon's tendency toward covalent bonding and catenation and they will not be considered here.

Carbides

The **metal carbides**, compounds of carbon with metals, have properties ranging from the reactive and salt-like, such as sodium, magnesium and aluminium carbides, to the unreactive and metallic, such as the carbides of the transition metals titanium and niobium.

Oxides

Carbon compounds containing non-metals are usually gases or low-boiling point liquids. Carbon forms two common oxides, carbon monoxide and carbon dioxide.

Figure 2.6.5 Carbide lanterns were once used by miners and as the headlights of bicycles and very early cars. The lamps contained a supply of calcium carbide and a reservoir of water. As water dripped onto the carbide, ethyne (or acetylene as it is sometimes called) was produced and burned to provide light:

$$\underset{\text{Calcium carbide}}{CaC_2(s)} + 2H_2O(l) \rightarrow \underset{\text{ethyne}}{C_2H_2(g)} + Ca(OH)_2(s)$$

Carbon monoxide

Carbon monoxide, CO, is a colourless, odourless flammable gas that forms during the incomplete combustion of carbon or any carbon-containing fuel. In a plentiful supply of oxygen, carbon monoxide will itself burn to form carbon dioxide.

$$2C(s) + O_2(g) \rightarrow 2CO(g) \qquad \Delta H^\ominus_m = -110.5\ \text{kJ mol}^{-1}$$

$$2CO(g) + O_2(g) \rightarrow 2CO_2(g) \qquad \Delta H^\ominus_m = -283.0\ \text{kJ mol}^{-1}$$

Carbon monoxide exists as simple covalent molecules which can be represented as shown in figure 2.6.6.

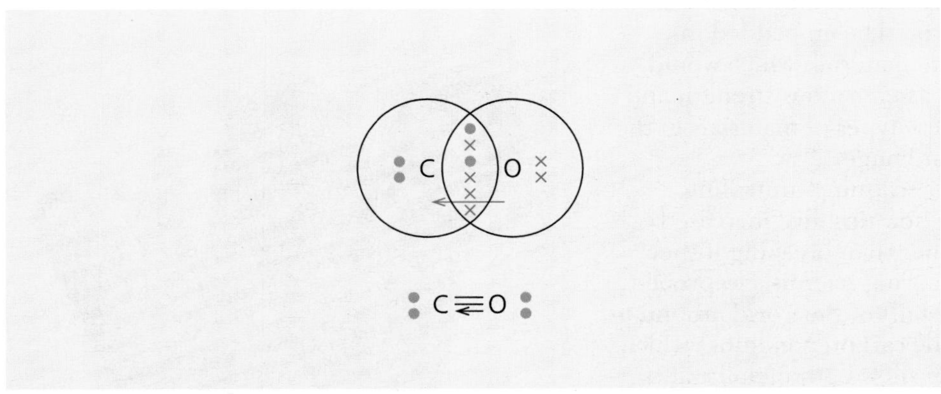

Figure 2.6.6 The structure of carbon monoxide – the triple bond is made up of two shared covalent bonds and one dative covalent bond.

Carbon monoxide is highly toxic to animals because of its ability to bind tightly to metal atoms. Carbon monoxide binds to the Fe^{2+} ions in haemoglobin about 200 times more strongly than oxygen does, forming a cherry-red complex. This inhibits the ability of haemoglobin to transport oxygen in the blood and so causes death. The fumes from inefficient gas and oil heaters as well as car exhaust fumes can be lethal because of the carbon monoxide they contain. Carbon monoxide is a good reducing agent at high temperatures, and is used industrially in the production of metals from their compounds (see page 224):

$$\mathbf{FeO(s) + CO(g) \xrightarrow{heat} Fe(s) + CO_2(g)}$$

$$\mathbf{CuO(s) + CO(g) \xrightarrow{heat} Cu(s) + CO_2(g)}$$

Carbon dioxide

Carbon dioxide, CO_2, is a colourless, nearly odourless gas that is formed by the combustion of carbon in a plentiful supply of oxygen. The gas is made up of discrete covalently bonded molecules, as shown in figure 2.6.8. It may be prepared in the laboratory by the action of dilute hydrochloric acid on marble chips:

$$\mathbf{2HCl(aq) + CaCO_3(aq) \rightarrow CO_2(g) + CaCl_2(aq) + H_2O(l)}$$

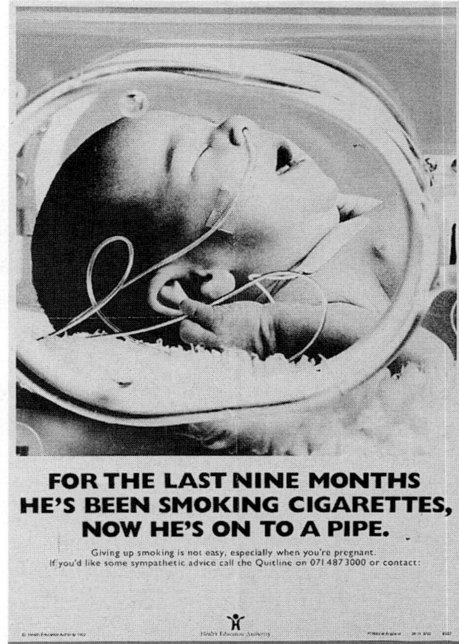

Figure 2.6.7 Tobacco smoke contains between 2% and 5% carbon monoxide, which binds to the haemoglobin in the blood of smokers, making it less efficient at carrying oxygen. In pregnant women, this lack of oxygen may affect the development of the fetus, which relies on the blood of its mother to supply it with all the oxygen and nutrients it needs.

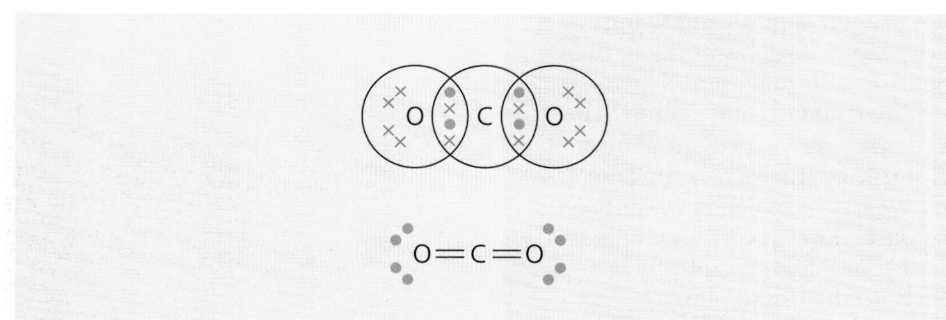

Figure 2.6.8 The structure of carbon dioxide

Carbon dioxide is a product of respiration in most living organisms, and is used by plants as a source of carbon for the process of photosynthesis. It is also produced naturally by forest fires. Its industrial uses range from fire extinguishers to the carbonation of soft drinks and beers. Around a quarter of the world production of carbon dioxide is used for this purpose. However, the largest use of the chemical – about 50% of the total production – is in the form of frozen carbon dioxide, known as dry ice, which is used as a refrigerant.

The greenhouse effect

Heat energy from the Sun reaches the Earth and is radiated back into space. To avoid a change in the temperature of the Earth, the heat needs to be radiated back at exactly the same average rate as it is received. An increased rate of radiation would lead to a cooling of the global climate, whilst a decrease in the radiation rate would lead to global warming.

Over the latter years of the twentieth century, human activities have produced a massive increase in the amounts of a number of gases in the upper atmosphere. These include the carbon-containing compounds carbon dioxide from the burning of fossil fuels (see figure 2.6.9), methane from rotting vegetation and chlorofluorocarbons which have been used as aerosol propellants and refrigerants. These gases are often called **greenhouse gases**. At the same time, people have removed great areas of the world's vegetation in the much-publicised destruction of the rainforests. Plants use carbon dioxide for photosynthesis, so this deforestation reduces the ability of the Earth as a whole to absorb carbon dioxide from the atmosphere. The trees are frequently replaced by cattle, which add to the greenhouse gases by producing methane.

Carbon dioxide makes the largest single contribution to what is frequently referred to as the **greenhouse effect**. The electromagnetic radiation from the Sun contains a wide range of wavelengths (see page 31). Radiation from the Sun reaching the Earth's surface is absorbed and then re-radiated, mainly as long-wavelength infra-red radiation. Greenhouse gases absorb this infra-red radiation and re-emit a large proportion of it back to Earth, so keeping the Earth rather warmer than it would otherwise be. The greenhouse gases in the atmosphere thus act as a kind of blanket, albeit one with holes in.

As the levels of the greenhouse gases steadily increase, the temperature of the surface of the Earth appears to be warming up slightly. The long-term effects of this global warming remain to be seen. It is predicted that the melting of the polar icecaps will cause a rise in sea level, resulting in the loss of low-lying regions of many countries. It is also suggested that relatively small changes in global temperature could be sufficient to bring about major shifts in global climate, thus destroying the present pattern of world food production as many currently fertile countries become barren deserts. Whether or not these predictions are accurate remains to be seen. At present, all we can do is endeavour to take action to reduce, or at least not add to, the current levels of greenhouse gases, for example, by replacing any trees cut down for use as fuels, so that there is no net production of carbon dioxide.

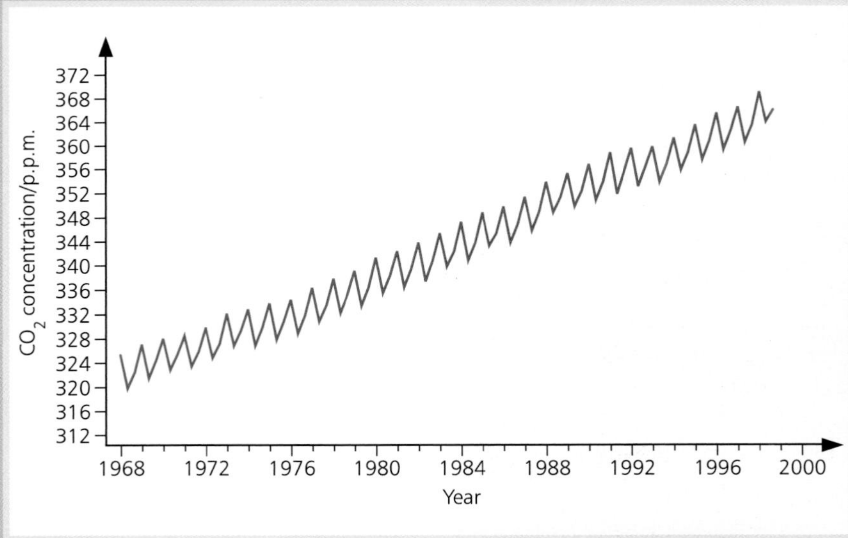

Figure 2.6.9 The burning of fossil fuels and deforestation are leading to a long-term increase in the level of carbon dioxide in the atmosphere. The readings for this graph were taken on a mountaintop in Hawaii. The annual fluctuations in the levels of carbon dioxide seem to be the result of seasonal differences in the fixation of carbon dioxide by plants, demonstrating the importance of plants as carbon dioxide sinks, and the need to stop the destruction of massive areas of forest.

Halides

The compounds carbon forms with the halogens will mainly be dealt with in sections 4 and 5, relating to organic chemistry. However, it is worth noting here the strength of the C—F bond (467 kJ mol^{-1}) compared with the C—H bond (413 kJ mol^{-1}) and also the C—Cl bond (346 kJ mol^{-1}). This high bond strength gives great stability to compounds containing C—F bonds. In addition, compounds containing many C—Cl bonds are found to be non-flammable. The combination of C—F and C—Cl bonds in a molecule therefore leads to the useful properties of great stability combined with non-flammability. An important group of compounds containing both C—Cl and C—F bonds are the **chlorofluorocarbons** or **CFCs**.

Within the CFCs, a set of molecules known by the trade name Freon are widely used. These are molecules based on methane, CH_4, and ethane, C_2H_6, in which some or all of the hydrogen atoms are replaced by chlorine or fluorine atoms. Halogenated carbon compounds such as these find an enormous range of uses, which include:

- refrigerants – the liquid in a refrigerator that is compressed and then expands, taking heat from the interior of the refrigerator as it does so. Dichlorodifluoromethane, CCl_2F_2, is the most common CFC in this application.
- aerosol propellants – the gases that propel the liquid out of a spray can, forming a fine mist of droplets (this mist is called an **aerosol**, which is where the term 'aerosol can' comes from). In this application, the fact that CFCs are inert and non-toxic means that they are ideal for this job.
- fire extinguishers – the inert nature and non-flammability of CFCs makes them ideal for use in fire extinguisher systems in buildings housing sensitive electronic equipment such as computers, where other fire-fighting agents cannot be used.
- solvents – tetrachloromethane, CCl_4, is probably the best known of these. It is a colourless, fairly inert but toxic liquid that serves as an excellent solvent for fats and oils.

The problem with CFCs

In the upper atmosphere of the Earth is a layer of ozone gas, O_3, that absorbs short-wavelength ultraviolet light very strongly. Such radiation has harmful effects on living organisms, and the ozone layer protects

Figure 2.6.10 The widespread use of halogen compounds in aerosol cans and refrigeration units has had unforeseen and potentially devastating effects on the Earth's atmosphere. We may soon need this kind of protection from ultraviolet radiation not only at high altitudes but also in lowland areas in our everyday lives.

life at the surface of the Earth from these effects. Because of their stability, chlorofluorocarbons that are released into the atmosphere do not break down, and so they find their way eventually into the upper atmosphere. When they reach the ozone layer, a series of complex reactions takes place, some examples of which are given here.

CFC molecules are broken down under the influence of ultraviolet light:

$$CCl_2F_2 \xrightarrow{UV} \cdot CClF_2 + Cl\cdot$$

This reaction produces a chlorine atom, which has seven electrons in its outer shell and is very reactive as a result. A species like this with an unpaired electron is called a **radical**. (Radical reactions are explored further on pages 381–2.) The unpaired electron is indicated by the dot. The chlorine atom may simply recombine with its parent molecule, or it may attack a molecule of ozone:

$$Cl\cdot + O_3 \rightarrow ClO + O_2$$

The ClO molecule may then react with an oxygen radical:

$$ClO + O\cdot \rightarrow Cl\cdot + O_2$$

Net reaction: $$O_3 + O\cdot \rightarrow 2O_2$$

Notice that chlorine atoms are not used up in these reactions, so one CFC molecule can destroy literally thousands of ozone molecules.

Decreases in ozone concentrations in the ozone layer were first detected by scientists in the late 1970s, and these have since been confirmed. The most massive decreases happen over Antarctica, where ozone concentrations temporarily fall each spring, resulting in an ozone 'hole'. The size of this hole appears to be increasing, and a similar hole now appears to be occurring in the Arctic as well. In 1987 an international treaty called the Montreal Protocol set out plans for a reduction in the use of CFCs. In 1989, 93 further nations agreed to phase out their production and use of CFCs. However, current levels of CFCs in the atmosphere will persist for some time, and it is likely to be well into the twenty-first century before the rate of ozone depletion starts falling.

Uses of carbon and its compounds

Carbon and its compounds play an enormous role in the global economy. Synthetic fibres and plastics, which are ultimately derived from petroleum-based chemicals (petrochemicals), have largely replaced natural substances such as cotton and wool. Other synthetic materials serve as the basis for numerous technological advances, including those in the fields of electronics and transport. The research and development of synthetic drugs has contributed enormously both to relieving suffering and to helping to understand the molecular machinery of life.

In addition, carbon-based fuels have provided most of the world's energy needs since 1900, although the twenty-first century will see these fuels becoming increasingly less readily available and more expensive. Replacing these fuels by processing coal and living matter together to produce other carbon-based synthetic fuels risks a further increase in already serious levels of air pollution due to the combustion of hydrocarbons, and also an increase in acid rain, a result of burning sulphur-containing fuels. However, the development of fuels from quick-growing plants is an approach that will help prevent levels of carbon dioxide rising, since the plants take in carbon dioxide and this counteracts the carbon dioxide produced when they are burned.

We shall return to the chemistry of this fundamental element in sections 4 and 5.

SUMMARY

- Group IV of the periodic table contains the elements carbon, silicon, germanium, tin and lead. The atoms of group IV have four electrons in their outer shell.
- The group IV elements and their compounds show a great range and gradation of properties from the non-metal carbon through the metalloids silicon and germanium to the metals tin and lead.
- Bonding between atoms of the group IV elements changes from the giant molecular lattices of carbon and silicon to the giant metallic structures of tin and lead.
- Carbon is a ubiquitous atom in living organisms and forms the basis of organic chemistry.
- Carbon exists as four allotropes – diamond, graphite, amorphous carbon and the fullerenes.
- Carbides (compounds of carbon with metals) may be reactive and salt-like, or unreactive and metallic.
- Carbon forms two common oxides, carbon monoxide and carbon dioxide.

2

QUESTIONS

1 The elements in group IV show a great range and diversity in the properties of both the elements and their compounds.
  a Compare the physical properties of the element carbon with those of the other elements in group IV, giving reasons for the similarities and differences.
  b Contrast the properties of the chlorides of carbon with those of the other elements in group IV. Explain the trends you observe.

2 Carbon exists in four main allotropic forms.
  a What is an allotrope?
  b Give a brief description of the four main allotropes of carbon.

3 Carbon monoxide is an extremely poisonous gas.
  a Under what conditions may carbon monoxide be formed?
  b Why is carbon monoxide so toxic?
  c Carbon monoxide is a good reducing agent. Give two examples of its use as a reducing agent, in each case describing the redox reaction in detail.

4 **a** Describe a laboratory preparation of carbon dioxide gas, giving chemical equations where possible.
  b By what means does carbon dioxide enter the atmosphere? Give chemical equations where possible.
  c In recent years, carbon dioxide has been implicated in global warming through the greenhouse effect. What is meant by the term 'greenhouse effect' and how is carbon dioxide involved?
  d What are the major difficulties in reducing the carbon dioxide emissions produced by the human population of the Earth?

2.7 The d-block elements: The transition metals

Most of the metals we meet on a day-to-day basis are not found in any of the groups of the periodic table we have considered so far – they are members of the transition metals.

The **transition elements** occupy the main body of the periodic table and are members of the d block of elements – their character is determined by the electrons in the filling d subshells. The **inner transition elements** are those shown below the main table itself, and are the f-block elements. They are divided into the **lanthanides** – the first row following the element lanthanum, and the **actinides** – following actinium. The lanthanides and actinides are very rare elements, and here we shall concentrate on the d-block metals, particularly those in the first series from titanium to copper.

Scandium, Sc
Titanium, Ti
Vanadium, V
Chromium, Cr
Manganese, Mn
Iron, Fe
Cobalt, Co
Nickel, Ni
Copper, Cu
Zinc, Zn

Iron is the second most abundant metal in the Earth's crust. It is one of the most widely used metals in the world of engineering, shown here in the form of stainless steel.

Now Dulux tempts you with two kinds of white.

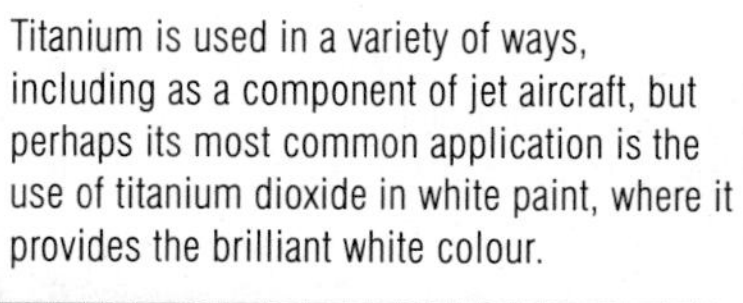

Titanium is used in a variety of ways, including as a component of jet aircraft, but perhaps its most common application is the use of titanium dioxide in white paint, where it provides the brilliant white colour.

The most familiar use of chromium is to prevent corrosion. Cars of years gone by were heavily adorned with chrome-plated trim, not only for aesthetic reasons but also to prevent rusting. Chrome compounds find many uses as pigments as a result of their strong colours.

Nickel is very resistant to corrosion. In combination with copper it forms the alloy monel, used to make propeller shafts resistant to attack by sea water.

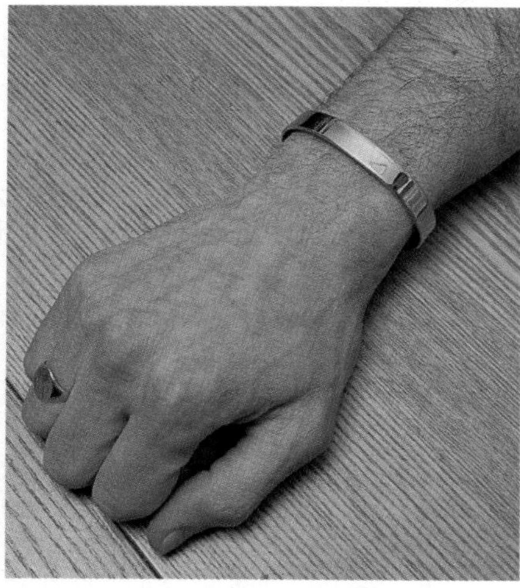

Copper has many uses based on its ability to conduct both heat and electricity. A more unusual use is its role in alternative medicine. Amongst other claims, wearing copper bracelets is supposed to relieve the pain of arthritis.

Figure 2.7.1 The transition metals of the first series have a wide range of uses both as elements and in combination as alloys or compounds.

FOCUS PHYTOMINING – METALS FOR THE FUTURE?

In the early 1970s, a tree was discovered in New Caledonia which produced a blue sap. This turned out to contain 26% nickel in its dry mass. Other plants can also accumulate large amounts of transition metals such as nickel, cobalt, cadmium, zinc and even gold (this process is sometimes called **hyperaccumulation**). However it was not until the 1980s that scientists began to make the serious suggestion that plants might be used to extract certain metals from the earth – and it was a decade later before it was first tried.

In California, the plant *Streptanthus polygaloides* is planted on nickel-rich soils. The plants take up nickel until it makes up as much as 1% of their dry mass. The plants are burned to ash which is then smelted to produce the metal, while the energy produced by burning the plants is used to generate electricity to power the extraction process, any excess electricity being sold to the local power company. The combination of the money raised from the sale of the metal extracted and the electricity generated means that a metal farmer can make considerably more money per hectare than the equivalent wheat farmer.

The South African nickel hyperaccumulator *Berkeya coddii* seems the best candidate at the moment for commercial **phytomining** (mining using plants) of nickel, as it has a high **biomass** (a lot of plant material is formed) and it grows easily from seed. It is also perennial, so in the right conditions it will grow year after year without more planting. Scientists are also researching the possibility of phytomining metals such as thallium, lead, cobalt and even gold. Sometimes special chemicals need to be sprayed on the soil to increase the solubility of these metals and make it easier for the plants to take them up.

There are a number of advantages to phytomining. It can be used to exploit ores and mineralised soils which are uneconomic for conventional mining. Bio-ores are almost sulphur free, which means they need less energy for smelting and cause much less acid rain pollution. The metal content of bio-ores is usually much greater than that of a mineral ore – for example, 22% zinc was found in the ash of the Zimbabwean hyperaccumulator *Peasonia metallifera*. And finally, phytomining is a very 'green' technology when viewed as an alternative to open cast mining of low grade ores.

As with many seemingly perfect schemes, there is a catch. Phytomining is only commercially viable if the price of the metal to be extracted remains high. But the price of a metal at the time of planting can be very different to the price by the time of harvest. Although the dried plants or the ash can be stored and saved until metal prices rise again if necessary, the whole process is something of a gamble – and so far, phytomining has not taken off in a big way, although in the future this may change.

However, there is another equally exciting use of the extraordinary ability of some plants to extract metals from soil. Many of the heavy metals such as thallium and lead are very toxic. If land becomes contaminated by these heavy metals it cannot be used for growing crops or grazing animals. But the use of hyperaccumulator plants mean that the toxic metals can be removed from the soil in a process called **phytoremediation**, making the soil safe to use again. The use of plants to extract metals from the soil will continue for the foreseeable future as a safety measure, if not as a way of making a fortune.

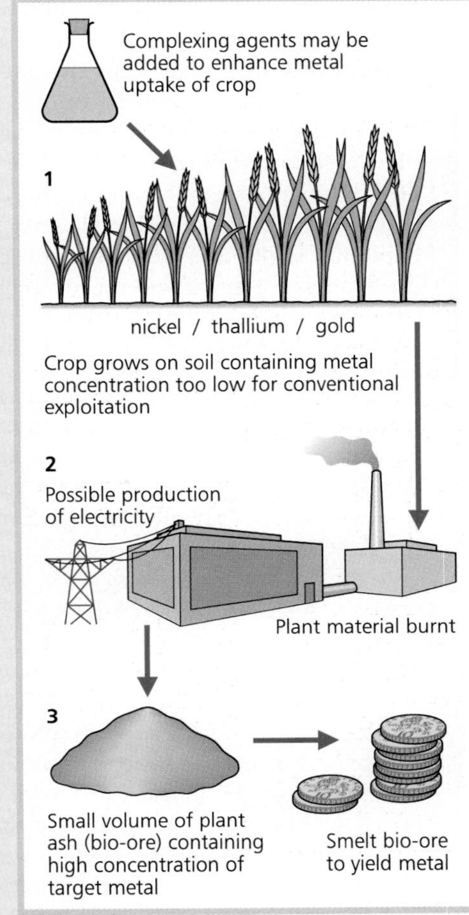

Figure 1 This is a model showing how the process of phytomining works.

Figure 2 A crop of nickel is to be found within this field of *Streptanthus polygaloides*.

Silver

Around 7000 years before the birth of Christ, people in the Middle East and Afghanistan were using gold and silver as ornamental jewellery, sometimes set with decorative stones. This is the first known use of metals by human beings. Silver was among the first metals to be used because, along with gold, it is unreactive and may be found in its native state. The value of silver as an ornamental metal continues to this day and it is this, combined with its relative rarity, that prevents the exploitation of some of its other properties. Silver has the highest electrical and thermal conductivity of any metal, yet demand for silver for decorative and coinage uses makes it too expensive for use in electric wires.

Silver may be plated onto other metals using a solution of potassium dicyanoargentate(I), $K[Ag(CN)_2]$. This ensures that the layer of silver deposited has a smooth finish, by maintaining a steady low concentration of Ag^+ ions in solution as the $Ag(CN)_2^-$ ion dissociates:

$$Ag(CN)_2^-(aq) \rightleftharpoons Ag^+(aq) + 2CN^-(aq)$$

Silver reacts with nitric acid to form silver(I) nitrate. With concentrated acid, nitrogen dioxide, NO_2, is produced in the reaction. With dilute acid nitrogen oxide, NO, is formed instead, although this rapidly reacts with oxygen in the air to form nitrogen dioxide. Silver does not form an oxide with the oxygen in the air, yet silver ornaments and cutlery tarnish. This is due to traces of the gas hydrogen sulphide, H_2S, in the air, and the reaction between the silver, hydrogen sulphide and oxygen causes the formation of a black deposit of silver(I) sulphide, Ag_2S. The same reaction occurs if a silver spoon is used to eat foods with a high sulphur content such as boiled eggs or a mustardy dressing.

$$4Ag(s) + 2H_2S(g) + O_2(g) \rightarrow 2Ag_2S(s) + 2H_2O(l)$$

The only important oxidation state of silver ions is +1, and most silver compounds are relatively unstable and break down rapidly.

One unusual feature of silver compounds is that in many cases their decomposition is speeded up by the absorption of light. The result of this light sensitivity is that almost all the photographic film used today depends on silver compounds to produce an image. Photographic film is made up of a light-sensitive emulsion spread on a supporting base, usually plastic. The light-sensitive emulsion consists of silver halides, usually silver bromide or a mixture of silver bromide and silver iodide. When the film is exposed to light, **photochemical decomposition** occurs. After development with appropriate chemicals, this leaves a deposit of silver on those areas where most light has fallen and less silver on the areas where less light has fallen. The decomposition reaction is summarised as:

$$2AgI(s) \rightarrow 2Ag(s) + I_2(s)$$

Unreacted silver halide must now be removed from the emulsion before the film can be exposed to light again. After it has been thoroughly washed, the film is immersed in a solution of sodium thiosulphate. This dissolves silver ions as thiosulphatoargentate(I) ions:

$$AgI(s) + S_2O_3^{2-}(aq) \rightarrow Ag(S_2O_3)^-(aq) + I^-(aq)$$

The film now contains silver deposits of varying densities, and is no longer photosensitive. It consists of a **negative image** – light areas appear dark and dark areas appear light, as shown in figure 2.7.2. When this negative is converted into a print, the process is essentially repeated in reverse, so that a positive image results reproducing the original scene accurately. Colour photographs are produced by a similar process, with dye particles incorporated into the photosensitive emulsion.

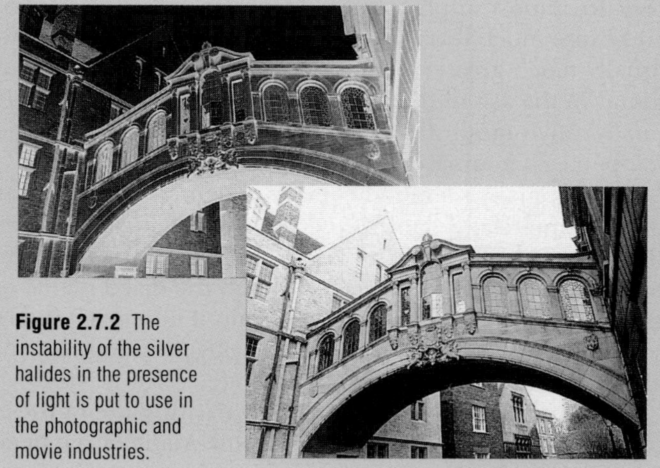

Figure 2.7.2 The instability of the silver halides in the presence of light is put to use in the photographic and movie industries.

The transition elements

Based on an examination of the periodic table, the simplest definition of a transition metal is that it is an element within the d block or the f block of the periodic table. However, this definition causes problems when considering structure, properties and reactions. A second, much more useful definition of a transition element is:

An element that forms at least one *ion* with a partially filled d subshell.

Based on this definition, the transition elements have a number of properties in common:

- They are all hard metals with high melting and boiling points.
- They show more than one oxidation state in their compounds.
- They tend to form coloured compounds and ions.
- Many show catalytic activity.
- They form **complex ions**, in which molecules called **ligands** form dative bonds with the metal ion.

The simplest definition of a transition element would include scandium and zinc in the first series of transition elements. Although these two elements are hard metals, they do not show multiple oxidation states, they form white rather than coloured compounds, and they do not act as catalysts. Using the second definition excludes scandium and zinc, for the following reasons:

- The scandium atom loses three electrons to give the ion Sc^{3+}, which has the same electronic configuration as argon.
- The zinc atom loses only the outer 4s electrons to give the Zn^{2+} ion, with an electronic configuration (Ar)$3d^{10}$.

In neither case is an ion formed with an incomplete d subshell and so scandium and zinc are not considered true transition elements. This reflects the fact that their chemical behaviour differs from that of the other elements in the d block. In accord with this definition, we shall look at the elements titanium to copper as the first transition series.

Properties of the transition elements

Electronic configuration and atomic radius

The points made here frequently apply to all the transition metals, but we shall refer specifically to the first transition series. The elements generally have a close-packed structure. They also have small and relatively similar atomic radii, due to the increasing number of electrons being used to fill d subshells rather than to add new shells. The atomic radius is very little affected until the point is reached where all the d orbitals contain one electron. When a second electron is put into a d orbital, the repulsive forces between the two electrons cause the radius of the atom to increase slightly, hence the trend seen in table 2.7.1.

Table 2.7.1 also shows the electronic configurations of the elements in the ground state. Note the configurations of chromium and copper, which have half-full 4s subshells, producing half-full and full 3d subshells. This 'electron borrowing' between subshells is associated with the relative stability of full and half-full subshells in comparison with incomplete subshells – see page 45.

Atoms of the d-block elements are smaller than those of the s-block elements of the same period, which results in the elements from scandium to zinc being more electronegative than the metals of groups I and II. The increase in nuclear charge from scandium to zinc produces the small increase in electronegativity shown in table 2.7.1. This is not as marked as the increase seen across a whole period, as the electrons added to the d orbitals are largely screened from the nuclear charge by inner s orbitals.

Influenced by the same factors as electronegativities, the ionisation energies of the d-block elements also increase by a small amount from scandium to zinc. Again, the increase is small compared with that seen across a complete period.

Element	Sc	Ti	V	Cr	Mn	Fe	Co	Ni	Cu	Zn
Electronic configuration of atom in ground state	(Ar) $3d^1 4s^2$	(Ar) $3d^2 4s^2$	(Ar) $3d^3 4s^2$	(Ar) $3d^5 4s^1$	(Ar) $3d^5 4s^2$	(Ar) $3d^6 4s^2$	(Ar) $3d^7 4s^2$	(Ar) $3d^8 4s^2$	(Ar) $3d^{10} 4s^1$	(Ar) $3d^{10} 4s^2$
Atomic (metallic) radius/nm	0.164	0.147	0.135	0.129	0.137	0.126	0.125	0.125	0.128	0.137
Electronegativity	1.3	1.5	1.6	1.6	1.5	1.8	1.8	1.8	1.9	1.6
First ionisation energy/kJ mol^{-1}	631	658	650	653	717	759	758	737	746	906
Second ionisation energy/kJ mol^{-1}	1235	1310	1414	1592	1509	1561	1646	1753	1958	1733
Third ionisation energy/kJ mol^{-1}	2389	2653	2828	2987	3249	2958	3232	3394	3554	3833
Density/g cm^{-3}	2.99	4.5	5.96	7.20	7.20	7.86	8.9	8.9	8.92	7.14

Table 2.7.1 The atomic radii, electronegativities and ionisation energies of the first transition series are a clear reflection of their electronic structures. Although we are not treating them as transition elements, scandium and zinc have been included in the table for completeness.

The first electrons to be lost by these elements are always those in the s subshell, with subsequent electrons being lost from the d subshell. The small fluctuations seen in the general trend of increasing ionisation energies can be explained by reference to the stability concerned with full and half-full subshells that we have noted before. Hence the second ionisation energy of chromium is higher than that of manganese because the process:

$$\mathbf{Cr^+ \rightarrow Cr^{2+} + e^-}$$

involves removing an electron from a half-full 3d subshell, whereas:

$$\mathbf{Mn^+ \rightarrow Mn^{2+} + e^-}$$

involves removing a single electron from the 4s subshell, which is of higher energy. Similar arguments explain why the second ionisation energy of copper is higher than that of zinc.

Bonding and physical properties

The combination of close packing and small atomic radius means that there are very strong metallic bonds between the atoms of a transition element. This is reflected in the properties of the elements. They all exhibit high melting and boiling points, as shown in figure 2.7.3, as well as higher densities and enthalpies of fusion and vaporisation than the equivalent elements in the s or p blocks.

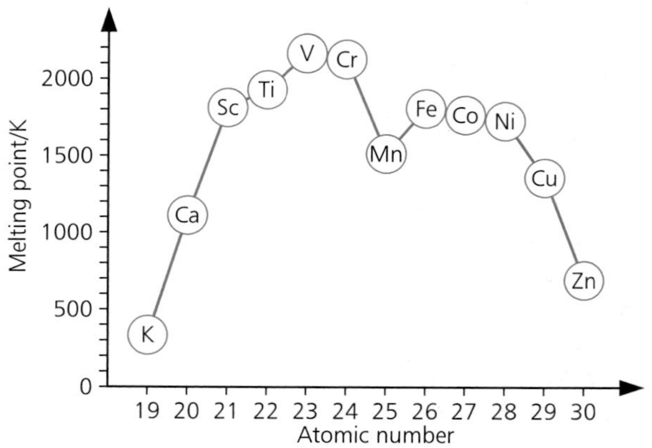

Figure 2.7.3 The effect of the strong metallic bonding between the small atoms of the transition metals is reflected in their melting points.

Many of the transition metals are of great value in engineering and technology, and this is due at least in part to the high tensile strength they show as a result of their strong metallic bonds.

Chemical properties

The transition elements are less reactive than the s-block metals, although they all (with the exception of copper) react with dilute solutions of acids to give hydrogen and a salt. However, very few react in this way under ordinary conditions, due to the formation of a protective oxide film over the metal surface, which acts to prevent further change. Chromium's half-full 3d and 4s subshells make it highly resistant to oxidation, so chromium is an extremely valuable non-corroding material which is used as a coating to protect other elements from the damaging effects of air and water.

Other transition metals are less well protected. Iron forms an oxide film (rust), but this is porous and allows air and water to penetrate, so that the rusting process continues beneath the surface. Iron is an extremely useful metal with valuable properties of tensile strength and malleability, but it needs to be either alloyed with other transition elements such as nickel, coated by an element such as chromium or zinc, or protected with a layer of paint or oil to reduce its tendency to corrode and thus extend its working life.

Transition metal compounds

Variable oxidation states

Another general characteristic of the transition metals is the occurrence of multiple oxidation states. The most common oxidation state of most of the elements is +2 resulting from the loss of the outer two 4s electrons, for example Mn^{2+}, Fe^{2+} and Co^{2+}. However, the underlying d subshells are very close in energy level to the 4s subshell, and so it is relatively easy to lose electrons from these as well. This means that several different ions of the same element are possible by losing different numbers of electrons – all of these different ions are approximately equally stable. The interconversions between one oxidation state and another are an important aspect of transition metal chemistry, and these interconversions are frequently reflected by colour changes in solutions of the complex ions. The oxidation states of some of the transition metal ions are shown in table 2.7.2.

Some general observations on the variable oxidation states of the transition metals titanium to copper are given below:

1 +1, +2 or +3 are among the most common oxidation states for each element. +3 is most common from titanium to chromium, then from manganese onwards +2 is more common.

Element	Most common oxidation states	Examples of compounds
Ti	+2, +3, +4	Ti_2O_3, TiO_2, $TiCl_4$
V	+1, +2, +3, +4, +5	V_2O_3, VCl_3, V_2O_5
Cr	+1, +2, +3, +4, +5, +6	Cr_2O_3, $CrCl_3$,CrO_3
Mn	+1, +2, +3, +4, +5, +6, +7	MnO, $MnCl_2$, MnO_2, Mn_2O_7
Fe	+1, +2, +3, +4, +6	FeO, $FeCl_2$, Fe_2O_3, $FeCl_3$
Co	+1, +2, +3, +4, +5	CoO, Co_2O_3, $CoCl_3$
Ni	+1, +2, +3, +4	NiO, $NiCl_2$
Cu	+1, +2, +3	Cu_2O, CuCl, CuO, $CuCl_2$

Table 2.7.2 Trends in the multiple oxidation states of the transition elements. The main stable oxidation states are shown in red.

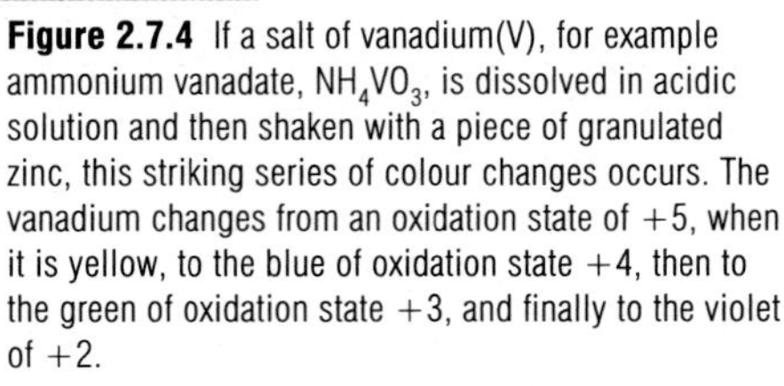
Figure 2.7.4 If a salt of vanadium(V), for example ammonium vanadate, NH_4VO_3, is dissolved in acidic solution and then shaken with a piece of granulated zinc, this striking series of colour changes occurs. The vanadium changes from an oxidation state of +5, when it is yellow, to the blue of oxidation state +4, then to the green of oxidation state +3, and finally to the violet of +2.

2 Transition metals usually show their highest oxidation states when combined with oxygen or fluorine, the most electronegative elements.

3 The highest oxidation states of all the elements up to and including manganese correspond to the loss in bonding of all the electrons outside the argon core. Thus the highest oxidation state of titanium is +4, whilst that of manganese is +7. Beyond manganese the d electrons are held more strongly as a result of the increasing nuclear charge, and so by and large the common oxidation states involve the 4s shell only.

4 When the elements exhibit very high oxidation states (+4 and above), they do not form simple ions. They are either involved in covalent bonding (for example, TiO_2, $TiCl_4$, CrO_3 and Mn_2O_7), or they form large ions such as chromate(VI), CrO_4^{2-}, or manganate(VII), MnO_4^-.

Transition metal complexes

Complex ions

Another characteristic of the transition metals is their tendency to form complex ions, also known as **coordination compounds**. Complex ions consist of a number of simpler species joined together. Usually, when metals form compounds they do so by forming ionic bonds. Complex ions are an exception to this, as the metals form **coordinate** or **dative covalent** bonds. These are bonds in which both electrons come from the same atom, as we saw in section 1.4.

As a result of their small size, the d-block ions have a strong electric field around them. This field attracts other species that are rich in electrons. Complex ions are formed when the central metal ion becomes surrounded either by anions or by molecules that act as electron pair donors. These electron pair donors are known as **ligands**, from the Latin *ligare* which means 'to bind'. The metal ion acts as an electron acceptor and the ligands as electron donors.

A simple example of complex formation involves the common laboratory compound copper(II) sulphate, $CuSO_4$. In aqueous solution, the copper is not present as simple Cu^{2+} ions – instead the copper ion becomes closely bonded to four water molecules arranged in a plane around it, with two more water molecules more loosely associated above and below the plane, as shown in figure 2.7.5. This arrangement produces the familiar blue colour of aqueous solutions containing Cu^{2+} ions.

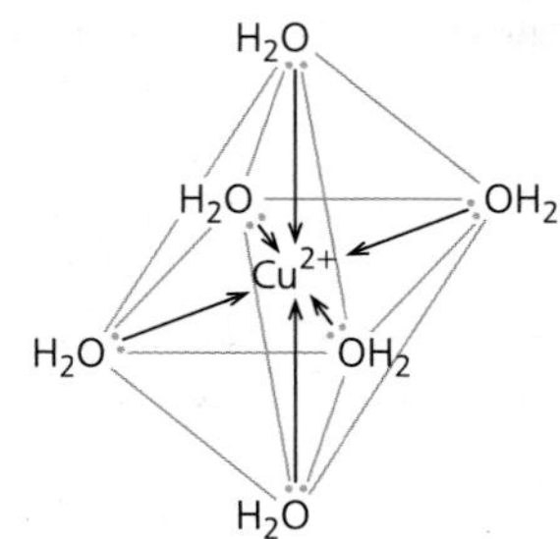

Figure 2.7.5 Cu^{2+} ions in solution are bonded to six water molecules, forming the hexaaquacopper(II) ion, $[Cu(H_2O)_6]^{2+}$.

Ligands

Clearly, ligands must be either anions or neutral molecules. Most importantly, they must have a lone pair of electrons available for donation. Some common anions that act as ligands include:

- the halides F^-, Cl^-, Br^- and I^-

- the sulphide ion S^{2-}
- the nitrite ion NO_2^-
- the cyanide ion CN^-
- the hydroxide ion OH^-
- the thiocyanate ion SCN^-
- the thiosulphate ion $S_2O_3^{2-}$.

By far the most common neutral compound to act as a ligand is water. Ammonia, NH_3, is another common neutral ligand.

In the formula of a complex ion, the metal ion is written first, followed by the ligands. The charge on the complex is the sum of the charge on the metal ion and the charge on the ligands. When the ligands are neutral molecules such as water and ammonia, the charge on the complex is obviously the same as the charge on the metal ion, but if the ligands are anions then the charge will be different from that on the metal ion.

Ligands may be displaced from a complex ion by other ligands. For example, if excess ammonia is added slowly to an aqueous solution of copper(II) sulphate, a deep blue solution results, as shown in figure 2.7.6. In copper(II) sulphate solution, Cu^{2+} ions exist as the complex ion $[Cu(H_2O)_6]^{2+}$, as we have seen. On the addition of excess ammonia solution, a precipitate of hydrated copper(II) oxide is initially formed:

$$[Cu(H_2O)_6]^{2+}(aq) + 2NH_3(aq) \rightarrow [Cu(H_2O)_4(OH)_2](s) + 2NH_4^+(aq)$$

This then dissolves in the ammonia solution to form the deep blue colour, with four ammonia molecules replacing the four water molecules lying in a plane around the copper ion:

$$[Cu(H_2O)_4(OH)_2](aq) + 4NH_3 \rightarrow [Cu(H_2O)_2(NH_3)_4]^{2+}(aq) + 2H_2O(l)$$

The shape of the $[Cu(H_2O)_2(NH_3)_4]^{2+}$ ion is shown in figure 2.7.6. Note that the two loosely bound water molecules in this ion are frequently ignored, so that it is often referred to as the tetraaminecopper(II) ion, $[Cu(NH_3)_4]^{2+}$. Similarly, the corresponding complex with water molecules is often written as $[Cu(H_2O)_4]^{2+}$, and called the tetraaquacopper(II) ion.

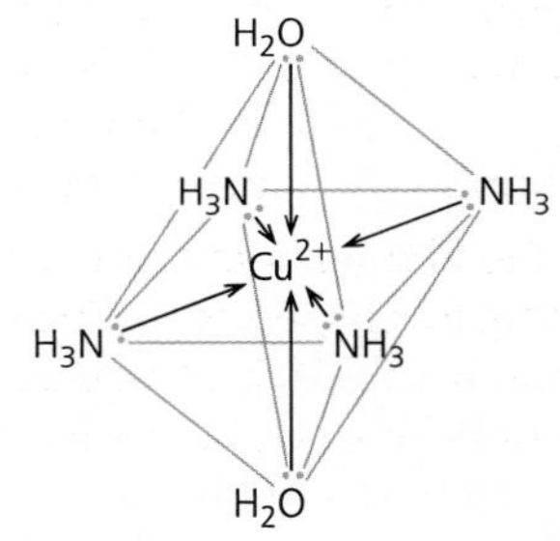

Figure 2.7.6 Ammonia displaces water as a ligand to the Cu^{2+} ion.

Copper

Copper-containing minerals such as malachite and turquoise were known and valued by civilisations up to 4000 years BC. It was then discovered that when these minerals were heated in a charcoal fire, they yielded the metal copper. The Egyptians exploited mines on the Sinai peninsula and produced thousands of tons of copper around 3200 BC. However, it was the development of bronze, a combination

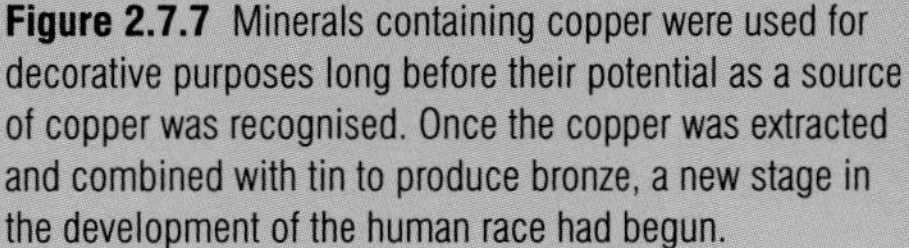

Figure 2.7.7 Minerals containing copper were used for decorative purposes long before their potential as a source of copper was recognised. Once the copper was extracted and combined with tin to produce bronze, a new stage in the development of the human race had begun.

of 90% copper and 10% tin to give a stronger, harder and more resistant alloy, that had a major effect on civilisations of the time and brought about the dawn of the Bronze Age. The new metal could be forged or cast to produce domestic artefacts, jewellery, tools and weapons, all superior to those that had gone before.

Figure 2.7.8 When exposed to oxygen in the presence of carbon dioxide (the conditions present in the air), copper reacts very slowly to form a green film of basic copper carbonate, $Cu_2(OH)_2CO_3$. The outer surface of the Statue of Liberty is made of copper and this reaction has resulted in the familiar green patina shown here.

Copper is an orange-red metal that is relatively unreactive and does not corrode readily, making it ideal for pipes carrying hot and cold water into homes and offices. It has very high thermal and electrical conductivity and it is also ductile, which is why it is widely used in electrical wiring.

Copper reacts with acids only under oxidising conditions. It reacts with moderately dilute or concentrated nitric acid to form copper(II) nitrate, and it is not affected by hydrochloric acid. In hot concentrated sulphuric acid, copper reacts to form the compound copper(II) sulphate, with its familiar bright blue hydrated ions, $CuSO_4.5H_2O$. Dry copper(II) sulphate is white – only the hydrated form shows the blue colour.

$$\mathbf{Cu(s) + 2H_2SO_4(conc) \rightarrow CuSO_4(aq) + SO_2(g) + 2H_2O(l)}$$

The most stable oxidation state of copper is +2, and most copper compounds contain the Cu^{2+} ion. This often occurs as a complex ion.

Coordination number

The number of ligands in a complex ion varies from ligand to ligand. For example, copper(II) combines with six water molecules, as we have seen, while it combines with four Cl^- ions to give the tetrachlorocopper(II) ion, $[CuCl_4]^{2-}$. The number of lone pairs bonded to the metal ion is known as the **coordination number**. The copper(II) ion shows a coordination number of 6 with water, and 4 with chloride ions. Different metal ions may show different coordination numbers with the same ligand.

The geometry of complex ions

Metal ions with a coordination number of 2 such as silver generally form complexes with a linear structure, for example $[Ag(NH_3)_2]^+$ and $[Ag(CN)_2]^-$:

$$\mathbf{[H_3N{:}\rightarrow Ag \leftarrow{:}NH_3]^+ \qquad [NC{:}\rightarrow Ag \leftarrow{:}CN]^-}$$

Complex metal ions with a coordination number of 4 may have one of two possible shapes – tetrahedral or planar. In general, any given ion with a particular type of ligand will always tend to have the same shape – for example, both $[CoCl_4]^{2-}$ and $[NiBr_4]^{2-}$ are always tetrahedral, while $[Ni(CN)_4]^{2-}$ is always

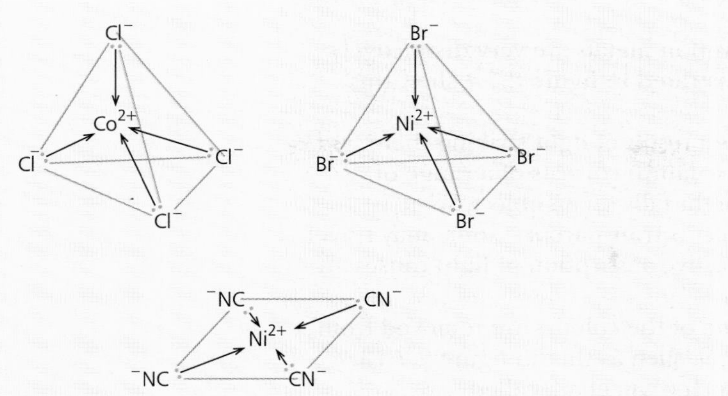

Figure 2.7.9 A coordination number of 4 may lead to a complex ion with one of two possible geometries, the tetrahedral arrangement shown for the tetrachlorocobalt(II) and tetrabromonickel(II) complexes, or the square planar structure shown for the tetracyanonickel(II) complex.

square planar, as shown in figure 2.7.9. Because of these two possible geometries, it is usually difficult to predict the shape of a complex ion which has a coordination number of 4.

The most common coordination number for the transition elements is 6, and these complexes are almost all octahedral, as shown by figure 2.7.10.

The geometric shapes of these complexes give rise to the phenomenon of **stereoisomerism** – isomerism in space.

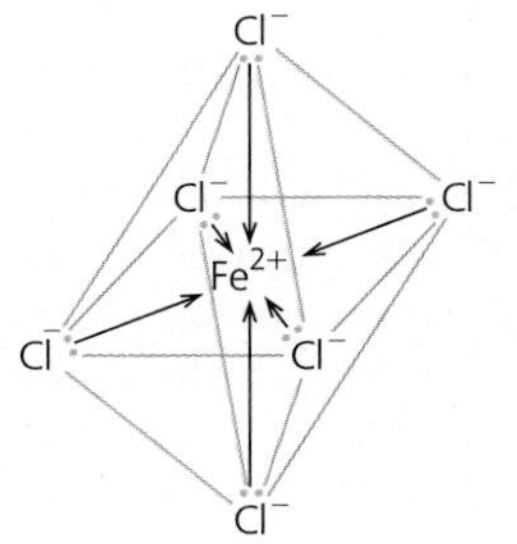

The hexachloroiron(II) complex

Figure 2.7.10 This octahedral geometry is common to almost all complexes with a coordination number of 6.

Stereoisomerism – the same but different

Stereoisomerism results when the bonds in a molecule or the ligands in a complex ion allow different possible orientations in space, resulting in two different molecular structures that cannot be superimposed one on the other. The different structures are called **stereoisomers**. The particular type of stereoisomerism shown on the right, *cis–trans* stereoisomerism, is known as **geometric isomerism**.

Biological systems are sensitive to the difference between geometric isomers, hence the specific sensitivity to cisplatin in figure 2.7.11. *Cis–trans* isomers also occur in octahedral complexes, as shown in figure 2.7.12.

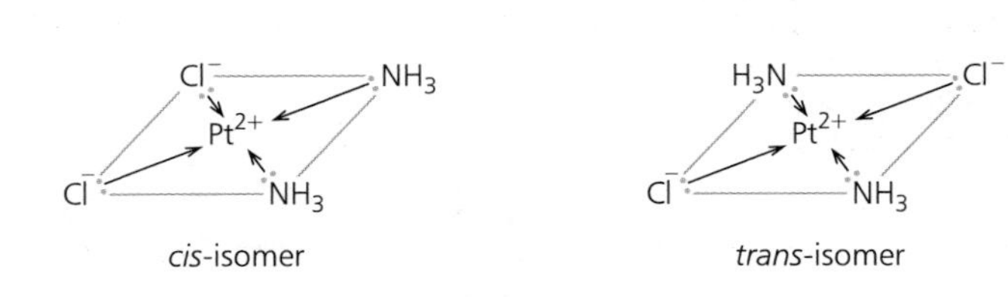

Figure 2.7.11 The isomer on the left, *cis*-$[Pt(NH_3)_2Cl_2]$, is known as cisplatin and it is an anticancer drug, active against tumours. The *trans*-isomer on the right is totally ineffective against cancer.

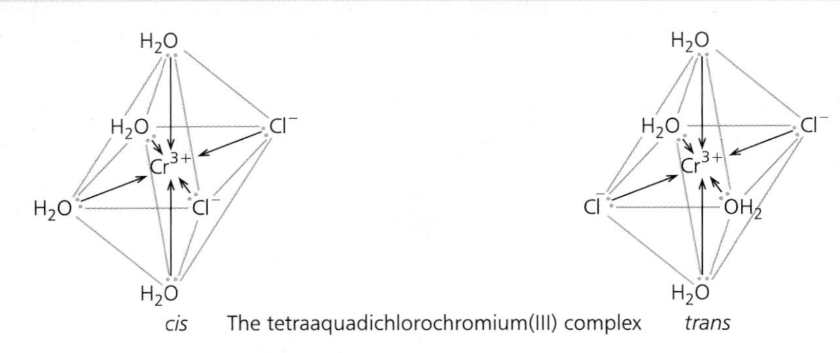

Figure 2.7.12 *Cis*- and *trans*-isomers of $[Cr(H_2O)_4Cl_2]^+$. *'Cis'* means 'on the same side' and *'trans'* means 'on opposite sides'.

The colours of complex ions

Solutions of the complex ions of many transition metals are very distinctively coloured – the compounds of vanadium illustrated in figure 2.7.4 show this very clearly. Why is this?

We see the colour of objects around us as a result of light striking them. White light (from the Sun or from an electric lamp) consists of a range of colours, from red through to violet. When light falls on an object, some is reflected, some is absorbed and – if the object is transparent – some may travel through it. Figure 2.7.13 shows how the selective absorption of light causes an object to look coloured.

The colour of light that results when some of the colours are removed from white light can be found using a colour wheel, such as that in figure 2.7.13. Colours that are opposite one another on such a wheel are called **complementary colours**. When light of a particular colour is absorbed from a beam of white light, the colour that results is the complement of the colour that is absorbed. Hence a solution or object that absorbs yellow light appears violet-blue, the complement of yellow.

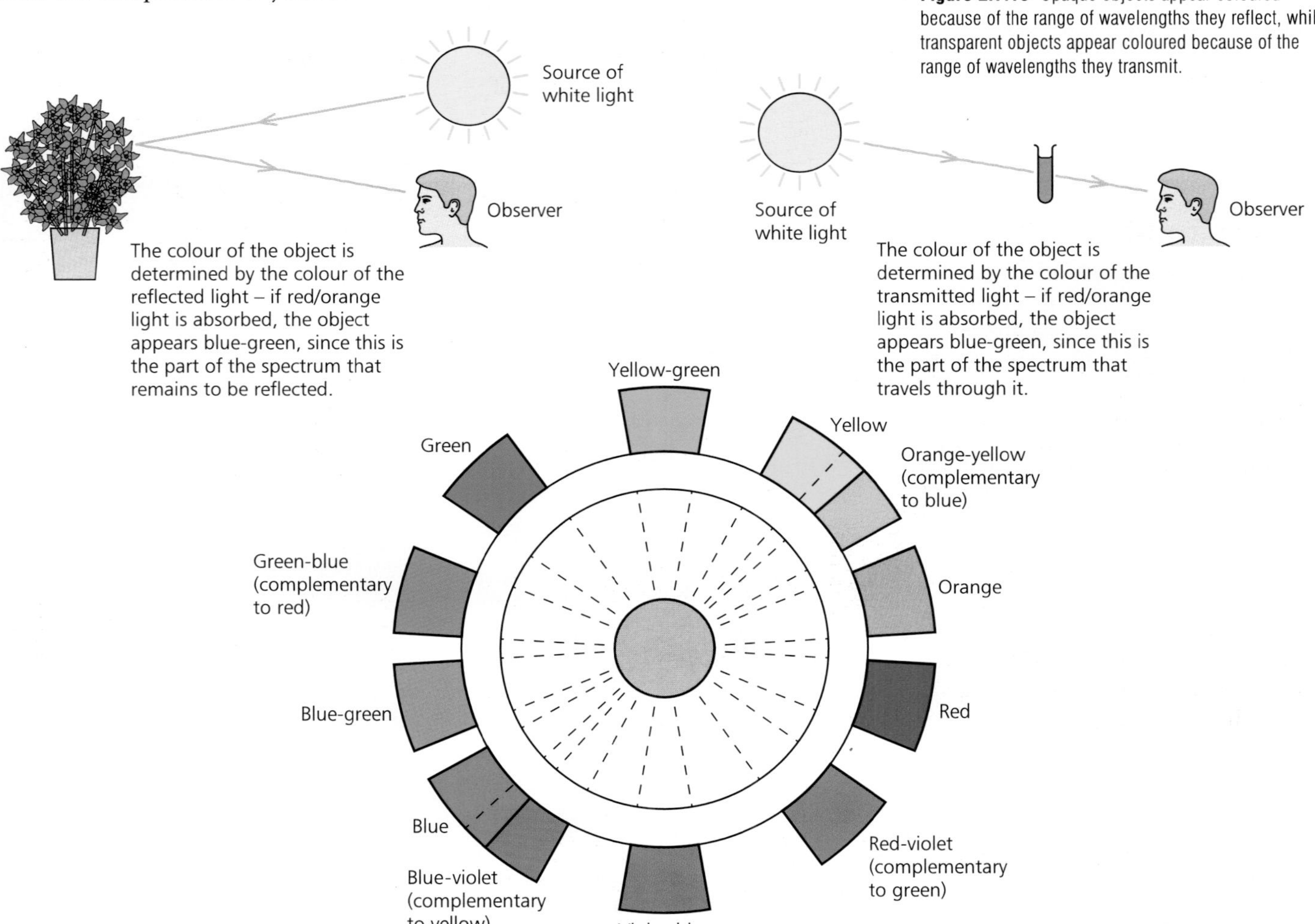

Source of white light
Observer
The colour of the object is determined by the colour of the reflected light – if red/orange light is absorbed, the object appears blue-green, since this is the part of the spectrum that remains to be reflected.
Source of white light
Observer
The colour of the object is determined by the colour of the transmitted light – if red/orange light is absorbed, the object appears blue-green, since this is the part of the spectrum that travels through it.
Yellow-green
Yellow
Orange-yellow (complementary to blue)
Orange
Red
Red-violet (complementary to green)
Violet-blue
Blue-violet (complementary to yellow)
Blue
Blue-green
Green-blue (complementary to red)
Green

Figure 2.7.13 Opaque objects appear coloured because of the range of wavelengths they reflect, while transparent objects appear coloured because of the range of wavelengths they transmit.

Section 1.3 discussed how electronic transitions are associated with the absorption of photons of different wavelengths. It is transitions of electrons between the partly filled d subshells in transition metal ions that cause their colours.

As an example, consider an aqueous solution containing Cr^{3+} ions, which is a violet colour due to absorption of light in the middle of the visible spectrum. The Cr^{3+} ion has the electronic configuration (Ar)$3d^3$. Simple calculations show that movement of an electron between the 3d subshell and other subshells (for example, the 4s subshell) cannot account for the absorption of photons of visible light, as the energy difference between the orbitals is too large. Instead, transitions *within* the 3d orbitals are responsible for the absorption.

An isolated Cr^{3+} ion will have five 3d orbitals, all with the same energy. However, when the ion is surrounded by six water molecules, this is no longer true. Those orbitals that are closer to the water molecules tend to have a higher energy than those which are further away, with the result that the five 3d orbitals are 'split' into three orbitals of lower energy and two orbitals of higher energy, as shown in figure 2.7.14. Absorption of visible light can now be explained in terms of a photon causing one of the three electrons to jump from an orbital with lower energy to one with higher energy.

2

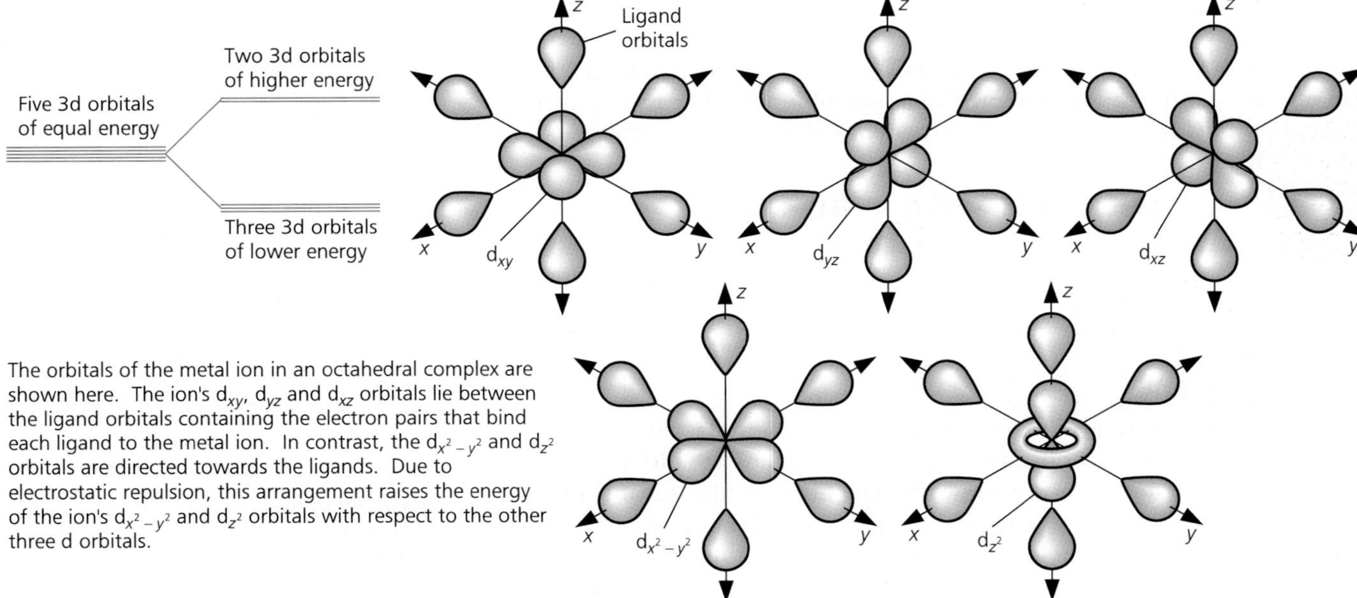
Five 3d orbitals of equal energy
Two 3d orbitals of higher energy
Three 3d orbitals of lower energy
Ligand orbitals
z
x
y
d_{xy}
d_{yz}
d_{xz}
$d_{x^2-y^2}$
d_{z^2}
The orbitals of the metal ion in an octahedral complex are shown here. The ion's d_{xy}, d_{yz} and d_{xz} orbitals lie between the ligand orbitals containing the electron pairs that bind each ligand to the metal ion. In contrast, the $d_{x^2-y^2}$ and d_{z^2} orbitals are directed towards the ligands. Due to electrostatic repulsion, this arrangement raises the energy of the ion's $d_{x^2-y^2}$ and d_{z^2} orbitals with respect to the other three d orbitals.

Figure 2.7.14 The octahedral arrangement of water molecules around the Cr^{3+} ion causes the 3d orbitals to split into two orbitals of higher energy and three of lower energy.

Although we have chosen to look at a simple example, the theory (known as **crystal field theory**) can also be applied to the ions of other transition elements. The presence of different numbers of electrons gives rise to different combinations of transitions, so that transition metal ions undergo changes of colour when changes of oxidation state occur. As we saw in the case of vanadium, these changes of colour may be quite spectacular.

A closer look at crystal field theory

From the example of Cr^{3+}, we have seen how an octahedral arrangement of ligands causes a splitting of the 3d orbitals. From the point of view of conservation of energy, it is necessary that this split happens in such a way that the *total* energy of the 3d orbitals remains unchanged. Figure

2.7.15 shows how this happens, and also shows how the orbitals split in the case of tetrahedral complexes.

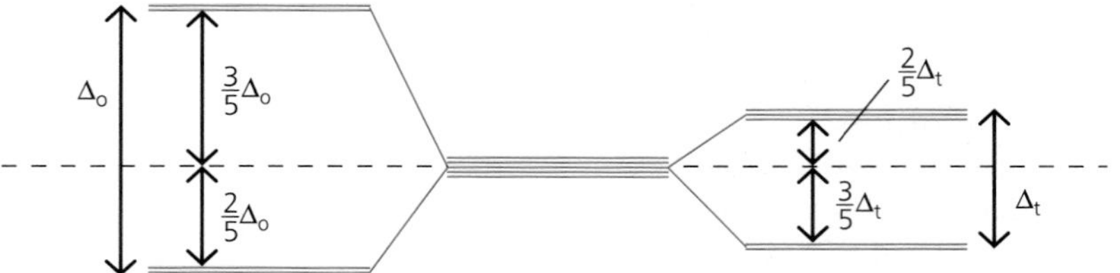

Figure 2.7.15 The splitting of the 3d orbitals in octahedral and tetrahedral complexes

Notice that the energy difference between the lower and higher orbitals in the octahedral complex is Δ_o, where 'o' simply stands for 'octahedral'. The comparable split for the tetrahedral complex is Δ_t, where 't' stands for 'tetrahedral'. It is found that $\Delta_t = \frac{4}{9}\Delta_o$. The sum of all the energy shifts of the orbitals in each case is zero.

The amount of splitting of the 3d orbitals depends on the metal ion, and also on the ligand. For a given ion, the ability of ligands to cause splitting of the d orbitals is given by the **spectrochemical series**. For the more common ligands, this is:

$$I^- < Br^- < Cl^- < H_2O < NH_3 < CN^-$$

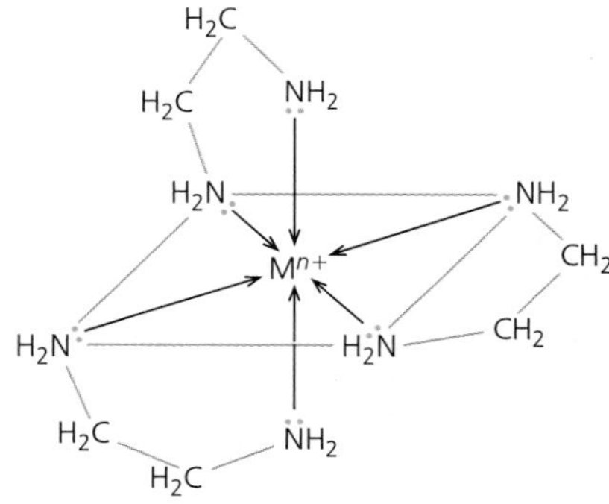

Figure 2.7.16 Bidentate ligands bind to the metal ion by means of two atoms.

Polydentate ligands

All of the ligands mentioned so far are **monodentate ligands**, which means that they join to the metal ion by one atom only. Some ligands are **bidentate**, binding to the metal ion by means of two atoms. Ligands like this include ethane-1,2-diamine and the ethanedioate ion (figure 2.7.16). Other ligands are **polydentate** – they may attach by two or more atoms leading to some complex ring structures. Chlorophyll, vitamin B_{12} and haemoglobin are all biological molecules involving transition metal ions within complexes. Figure 2.7.17 shows the structure of the porphyrin ring of haemoglobin, with its complexed iron(II) ion.

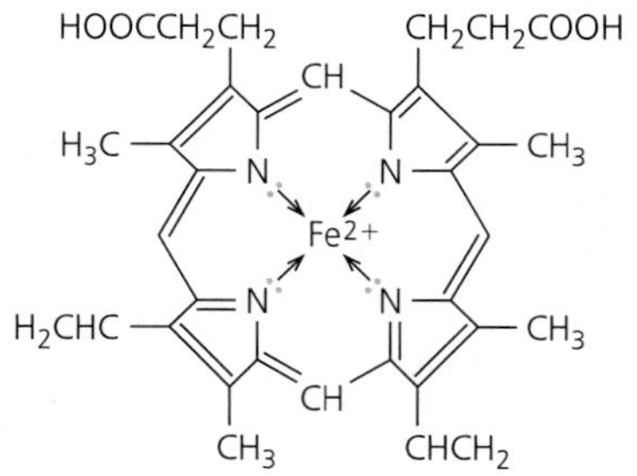

Figure 2.7.17 The 'business part' of the haemoglobin molecule. Haemoglobin is a complex protein with four subunits, each containing a porphyrin ring complexed with an Fe^{2+} ion like this – the arrangement is called a **haem group**. The protein backbone holds the haem group in such a way that an oxygen molecule can form another dative covalent bond with the iron ion, and so be carried round the bloodstream. Carbon monoxide (produced, for example, by faulty gas appliances) forms a stronger bond with the iron ion, effectively preventing the blood from carrying oxygen.

edta

One of the most common polydentate ligands is the compound ethylenediaminetetraacetic acid or **edta**. The anion $edta^{4-}$ has six available donor electron pairs, allowing it to wrap itself around metal ions and form very stable complexes. It is used in trace amounts in foods to prevent spoilage – it complexes with any traces of metal ions that might otherwise catalyse the reactions of oils in the food with oxygen from the air, making the product go rancid. Many shampoos also contain this anion to help soften the water, and it may be added in minute amounts to whole blood to mop up calcium ions and so prevent the blood from clotting.

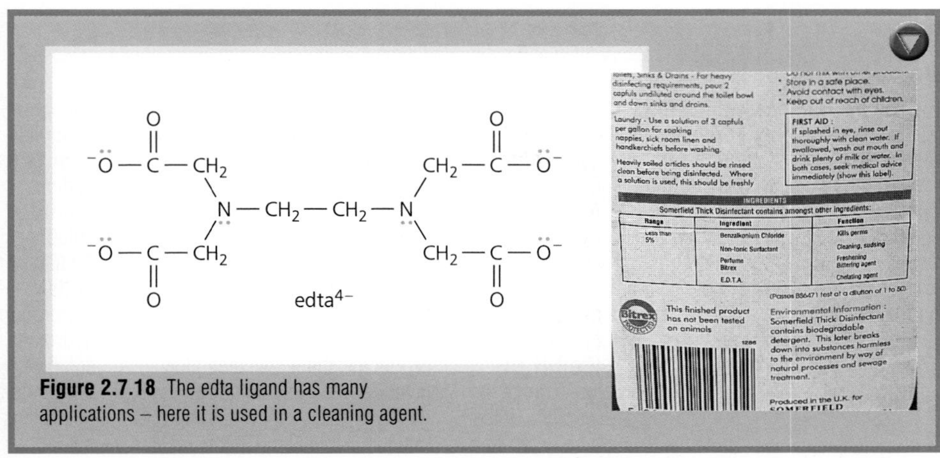

Figure 2.7.18 The edta ligand has many applications – here it is used in a cleaning agent.

Reactions of aqueous ions with alkali and ammonia

We have already seen that many transition metal ions display characteristic behaviour as the pH of their solutions is raised or lowered, and on the addition of other ligands that may displace water molecules. The addition of hydroxide ions, OH^-, in the form of sodium hydroxide and ammonium hydroxide solutions is commonly used as one means of simple identification of transition metal solutions. Table 2.7.3 describes the behaviour of some ions in the first row of the d-block elements under these conditions.

Reagent added	Cr^{3+}	Mn^{2+}	Fe^{2+}	Fe^{3+}	Co^{2+}	Ni^{2+}	Cu^{2+}	Zn^{2+}
Aqueous ion	The $[Cr(H_2O)_6]^{3+}$ ion is violet (see figure 2.7.20). Solutions containing this ion are slightly acidic.	The $[Mn(H_2O)_6]^{2+}$ ion is very pale pink.	The $[Fe(H_2O)_6]^{2+}$ ion is a pale blue-green.	The $[Fe(H_2O)_6]^{3+}$ ion is pale purple. However, aqueous solutions of iron(III) salts are usually yellow, due to the presence of the $[Fe(H_2O)_5OH]^{2+}$ ion.	The $[Co(H_2O)_6]^{2+}$ ion is pink.	The $[Ni(H_2O)_6]^{2+}$ ion is emerald green.	The $[Cu(H_2O)_6]^{2+}$ ion is blue. (See page 213.)	The $[Zn(H_2O)_6]^{2+}$ ion is colourless
Aqueous sodium hydroxide solution	Addition of a small amount of alkali forms a blue-violet precipitate of $Cr(H_2O)_3(OH)_3$. Addition of further alkali dissolves this to form the green $[Cr(H_2O)_2(OH)_4]^-$ ion.	A gelatinous white precipitate of manganese(II) hydroxide, $Mn(OH)_2$, forms. In the presence of oxygen, this rapidly darkens as it is oxidised to manganese(IV) oxide, MnO_2. This is insoluble when more alkali is added.	A gelatinous pale green precipitate of iron(II) hydroxide, $Fe(OH)_2$, forms. In the presence of oxygen, this rapidly darkens as it is oxidised to brown iron(III) hydroxide. This can be written $Fe(H_2O)_3(OH)_3$, but is better represented by the formula $Fe_2O_3{\cdot}xH_2O$.	Brown iron(III) hydroxide forms (see Fe^{2+} for details). This is insoluble when more alkali is added.	Addition of alkali precipitates a blue form of cobalt (II) hydroxide which gradually turns pink. The pink form is produced immediately if a solution containing Co^{2+} ions is added to alkali. Cobalt (II) hydroxide dissolves in highly concentrated alkali to produce a deep blue solution containing $[Co(OH)_4]^{2-}$ ions	An emerald green precipitate of nickel(II) hydroxide, $Ni(OH)_2$, forms, which is insoluble in excess alkali.	A gelatinous light blue precipitate of copper(II) hydroxide forms. The formula for this may be written as $Cu(OH)_2$ – see page 213. Further addition of alkali has no effect.	A white precipitate of zinc(II) hydroxide, $Zn(OH)_2$, forms. This dissolves on the addition of more alkali, to form the $[Zn(OH)_4]^{2-}$ ion.

Table 2.7.3 Reactions of some aqueous transition metal ions with aqueous sodium hydroxide and (overleaf) aqueous ammonia

Reagent added	Cr^{3+}	Mn^{2+}	Fe^{2+}	Fe^{3+}	Co^{2+}	Ni^{2+}	Cu^{2+}	Zn^{2+}
Aqueous ammonium hydroxide solution	A blue-violet precipitate of $Cr(H_2O)_3(OH)_3$ forms on addition of a small amount of ammonium hydroxide. Addition of further ammonium hydroxide dissolves this to form the yellow $[Cr(NH_3)_6]^{3+}$ ion.	A gelatinous white precipitate of manganese(II) hydroxide, $Mn(OH)_2$, forms In the presence of oxygen, this rapidly darkens as it is oxidised to manganese(IV) oxide, MnO_2. Addition of further ammonium hydroxide has no effect.	The same precipitates are seen as for sodium hydroxide solution. Addition of further ammonium hydroxide has no effect.	Brown iron(III) hydroxide forms (see Fe^{2+} for details). Addition of further ammonium hydroxide has no effect.	A blue precipitate of cobalt(II) hydroxide, $Co(OH)_2$, forms. Addition of excess ammonium hydroxide dissolves this to form the $[Co(NH_3)_6]^{2+}$ ion. This is readily oxidised by oxygen to a red solution containing a variety of complexes.	An emerald green precipitate of nickel(II) hydroxide, $Ni(OH)_2$, forms. This dissolves as more ammonium hydroxide is added, producing the lavender blue $[Ni(NH_3)_6]^{2+}$ ion.	The same precipitate is seen as for sodium hydroxide solution. Addition of further ammonium hydroxide solution produces a deep blue solution containing the $[Cu(NH_3)_4]^{2+}$ ion. (See page 213.)	The same precipitate is seen as for sodium hydroxide solution. Addition of further ammonium hydroxide solution produces a colourless solution containing the $[Zn(NH_3)_6]^{2+}$ ion.

Table 2.7.3 *(continued)*

Uses of the transition metals

Transition metals as catalysts

Many chemical reactions occur very slowly, or will only take place under extreme conditions of temperature, pressure or both. In the research laboratory, these extreme reacting conditions are inconvenient. More importantly, if reactions involved in an industrial process do not take place readily, there are many economic and safety implications. For example, a chemical plant that operates at continuous high pressures means not only extra capital costs in the engineering of the plant, but also an increased safety hazard. The transition metals can frequently solve these problems by their ability to act as **catalysts**.

A catalyst is a substance that speeds up a chemical reaction without affecting the proportion of the products which result. A catalyst is itself unchanged by the reaction and can be reused. Many of the transition elements act as effective catalysts as a result of their ability to exist in a variety of oxidation states. These variable oxidation states enable them to provide an alternative route for the reaction, lowering the activation energy (see section 3.6) and thus allowing the reaction to proceed more readily. As well as acting as catalysts in many industrial processes, the transition elements are also important as catalysts in **catalytic converters**, which reduce the pollution levels from car exhaust emissions.

There are many examples of transition elements acting as catalysts, and many of the reactions they catalyse are reversible. We shall study reaction rates and reversible reactions in some detail in section 3.6. Some reactions that transition metals may catalyse are given here:

- Titanium chloride, $TiCl_3$, is one of the substances used to catalyse the polymerisation of ethene to poly(ethene), also called polythene:

$$nC_2H_4 \rightarrow \left(\begin{array}{cc} H & H \\ | & | \\ -C- & C- \\ | & | \\ H & H \end{array} \right)_n$$

- Iron (Fe or Fe_2O_3) acts as a catalyst in the Haber process for producing ammonia:

$$N_2 + 3H_2 \rightleftharpoons 2NH_3$$

- Vanadium (V_2O_5 or vanadate VO_3^-) catalyses the Contact process in the manufacture of sulphuric acid:

$$2SO_2 + O_2 \rightleftharpoons 2SO_3$$

- Nickel is used as a catalyst in the addition of hydrogen to vegetable oils – this **hydrogenation** hardens the oils in the manufacture of vegetable margarine:

$$RCH{=}CH_2 + H_2 \rightarrow RCH_2CH_3$$

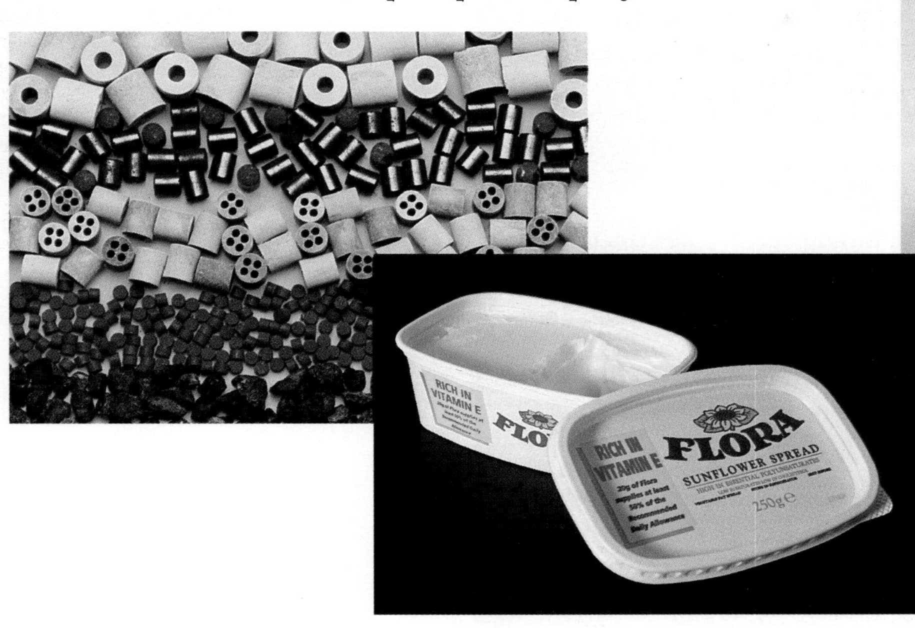

Figure 2.7.19 Transition metals are often pelleted or powdered to give a large surface area for catalytic reactions to take place. As a result of their ability to lower the activation energy of a variety of reactions, they enable many everyday substances to be produced at affordable prices.

Chromium

Chromium is a white, lustrous metal that is hard, rather brittle and very resistant to corrosion. It makes an excellent protective covering over other metals and is also widely used in the production of alloys. For example, stainless steel contains around 18% chromium along with 10% nickel and traces of other elements. Nichrome, as the name suggests, is an alloy of nickel and chromium, and is frequently used as the wire heating element in a variety of heating devices. Chromium's somewhat unexpected arrangement of electrons (see table 2.7.1, page 210) makes the element relatively unreactive, which is why it is useful for protecting other cheaper but more easily corroded metals.

Chromium exhibits a number of oxidation states in its compounds, the most common being +2, +3 and +6. One of the most distinctive features of the compounds of chromium is their colour, and colour changes frequently accompany changes in the oxidation state. The chromium(II) ion, Cr^{2+}, is pale blue in colour. It is easily oxidised to the most stable chromium oxidation state, +3, even by air. For this reason, preparation of chromium(II) salts must be carried out while air is excluded, the most common reducing agent used being zinc:

$$2Cr^{3+}(aq) + Zn(s) \rightarrow 2Cr^{2+}(aq) + Zn^{2+}(aq)$$

The Cr^{3+} ion forms a complex ion in water, $[Cr(H_2O)_6]^{3+}$, and the violet colour of this complex is typical of solutions of many chromium(III) salts. Chromium(III) ions also tend to exhibit amphoteric characteristics, in the same way as aluminium(III) ions.

The third oxidation state of chromium is +6. Two of the most common ions demonstrating this state are the chromate ion, CrO_4^{2-}, which is bright yellow in aqueous solution, and the dichromate ion, $Cr_2O_7^{2-}$, which is reddish orange in aqueous solution. These two species exist in an equilibrium that is shifted to the right in acidic solutions (when there are many hydrogen ions) and to the left in basic solutions (when there are fewer hydrogen ions available):

$$2CrO_4^{2-}(aq) + 2H^+(aq) \rightleftharpoons Cr_2O_7^{2-}(aq) + H_2O(l)$$

All the chromium(VI) species are good oxidising agents, particularly in acidic solutions.

Potassium chromate, K_2CrO_4, is used as an indicator for the titration of silver ions against a solution of chloride or bromide ions, in order to measure the concentration of halide ions in a solution. Initially the silver ions react with halide ions, forming a precipitate of silver halide:

$$Ag^+(aq) + Cl^-(aq) \rightarrow AgCl(s)$$

Once all the halide ions have been removed from solution, brick-red silver chromate is precipitated:

$$2Ag^+(aq) + CrO_4^{2-}(aq) \rightarrow Ag_2CrO_4(s)$$

Figure 2.7.20 Some of the brightly coloured solutions of chromium compounds. Left to right: when elemental chromium is dissolved in a dilute, non-oxidising acid that is oxygen free, the pale blue colour of the Cr^{2+} ion is seen. The violet colour of a solution of chrome alum, $KCr(SO_4)_2$, is typical of the $[Cr(H_2O)_6]^{3+}$ ion. The addition of an alkali to a solution of chrome alum gives a green solution of $[Cr(H_2O)_2(OH)_4]^-$ ions:

$$[Cr(H_2O)_6]^{3+}(aq) + 4OH^-(aq) \rightarrow [Cr(H_2O)_2(OH)_4]^-(aq) + 4H_2O$$

The dichromate(VI) ion that predominates in acidic solution is a reddish orange colour, whilst the chromate(VI) ion of more basic solutions is a bright yellow colour.

Elemental chromium is used in protective and decorative roles. Chromium compounds also have a variety of practical applications. Around 35% of the chromium compounds produced annually are used as green and yellow pigments in the building trade, as protective primers and by artists. A further 25% of chromium compound production, in particular chromium sulphate, is used in the process of tanning animal hides to produce leather for everyday products such as shoes and belts. Most of the remaining chromium compounds are used in various ways to prevent the corrosion of metal surfaces.

Iron

The importance of iron

Iron has been important to the human race for thousands of years, as it was one of the first metals people learned to extract from its ore and forge to make tools. It is thought that the large-scale use of iron in a society was first introduced by the Hittites of Anatolia in around 1400 BC, around 2000 years after bronze first appeared. This marked the ending of the Bronze Age and the beginning of the Iron Age. Before this, iron had been treated as a precious metal – Homer mentions it on a par with gold. Hardening iron by heating it and then thrusting it into cold water (quench-hardening) was a new technology that

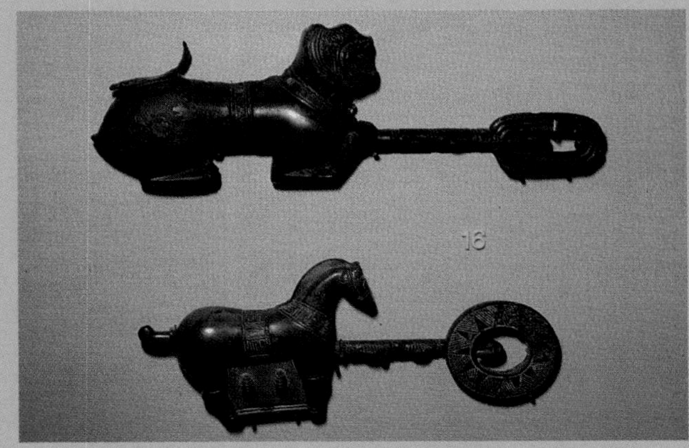

Figure 2.7.21 Iron garment hooks from Korea, 2–1 BC

Figure 2.7.22 The extraction of iron from its ore

Iron is extracted from iron ore in **blast furnaces**. The process depends on the reduction of iron oxide by carbon. Once commissioned, a blast furnace may work continuously for several years, and the largest furnaces can produce up to 10 000 tonnes of iron in a single day.

The material fed into the top of the blast furnace is known as the **charge**. It is a mixture of iron ore, limestone and coke. The iron oxides in the ore and the carbon in the coke react as shown in the diagram. The bottom part of the furnace is very hot, and here carbon reacts with oxygen to form carbon dioxide. This reacts with more carbon to give carbon monoxide in an endothermic reaction, causing a slight drop in temperature. The carbon monoxide reacts with the various iron oxides, producing molten iron as the final product. This trickles down through the furnace and is tapped off at the bottom.

In the intense heat of the furnace, limestone in the reaction mixture decomposes to give calcium oxide. This reacts with impurities in the ore, such as silica from sand, to form molten **slag**. Molten slag is less dense than molten iron. The slag also trickles down through the furnace and forms a layer of liquid slag on top of the layer of molten iron. The two liquids can then be drawn off separately. The iron formed in the blast furnace still contains impurities and is relatively brittle. It is known as **cast iron** or **pig iron**. The name 'pig iron' is derived from the old method of casting the molten metal in a line of sand moulds running off a central channel – this looked like piglets suckling from the sow.

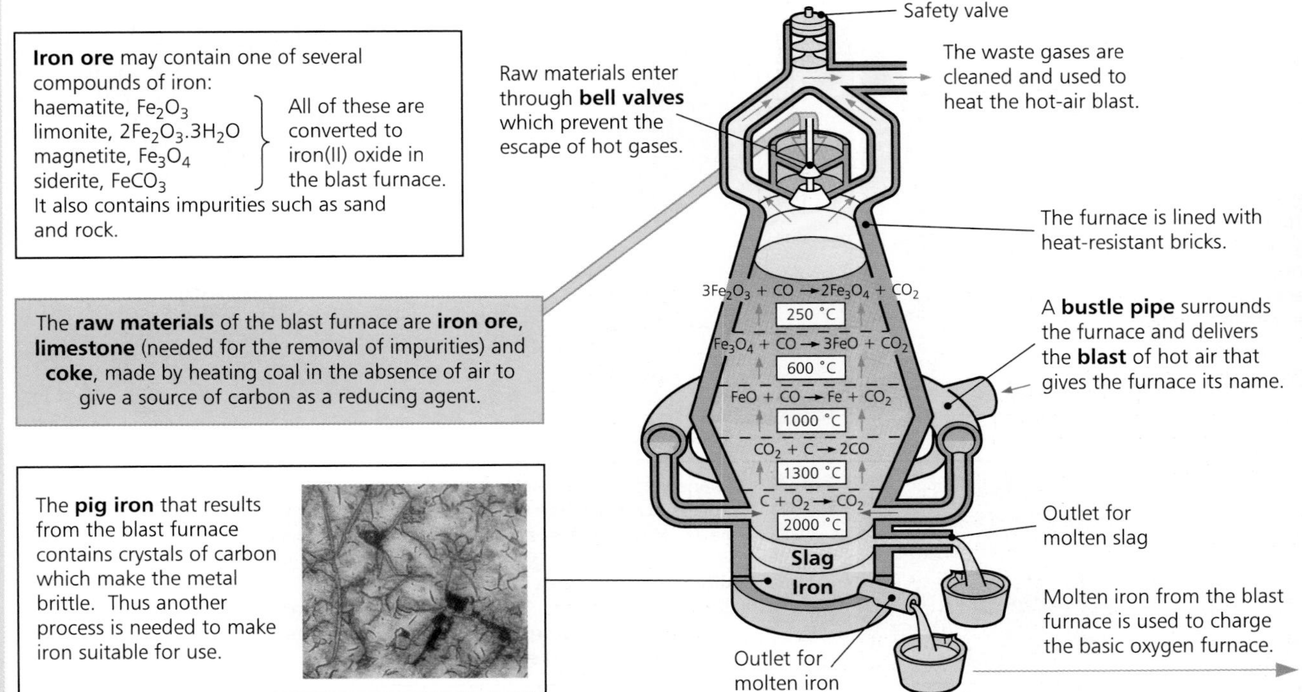
Iron ore may contain one of several compounds of iron:
haematite, Fe_2O_3
limonite, $2Fe_2O_3.3H_2O$
magnetite, Fe_3O_4
siderite, $FeCO_3$
All of these are converted to iron(II) oxide in the blast furnace.
It also contains impurities such as sand and rock.
The **raw materials** of the blast furnace are **iron ore**, **limestone** (needed for the removal of impurities) and **coke**, made by heating coal in the absence of air to give a source of carbon as a reducing agent.
The **pig iron** that results from the blast furnace contains crystals of carbon which make the metal brittle. Thus another process is needed to make iron suitable for use.
Safety valve
Raw materials enter through **bell valves** which prevent the escape of hot gases.
The waste gases are cleaned and used to heat the hot-air blast.
The furnace is lined with heat-resistant bricks.
$3Fe_2O_3 + CO \rightarrow 2Fe_3O_4 + CO_2$
250 °C
$Fe_3O_4 + CO \rightarrow 3FeO + CO_2$
600 °C
$FeO + CO \rightarrow Fe + CO_2$
1000 °C
$CO_2 + C \rightarrow 2CO$
1300 °C
$C + O_2 \rightarrow CO_2$
2000 °C
Slag
Iron
A **bustle pipe** surrounds the furnace and delivers the **blast** of hot air that gives the furnace its name.
Outlet for molten slag
Outlet for molten iron
Molten iron from the blast furnace is used to charge the basic oxygen furnace.

had a major impact on society, as hardened iron tools and weapons were far superior to those made of bronze. In Europe, iron artefacts were made by forging (hammering the red-hot metal) until the fourteenth century AD, as temperatures high enough to melt iron and mould (cast) it were not available until then. However, in China, cast iron was known from about the third century BC, as the Chinese recognised that the addition of carbon to the iron allowed it to melt at a lower temperature.

In modern times, iron is still of enormous importance, particularly when combined with other transition metals to form stainless steel. The second most abundant metal in the Earth's crust, its extraction and processing is the basis of major industries, as shown in figures 2.7.22 and 2.7.23.

Figure 2.7.23 Converting iron to steel – the basic oxygen furnace

Steel is the most useful form of iron. It contains considerably less carbon than pig iron, and also other elements in precisely controlled proportions. The conversion of pig iron to steel therefore involves removing the impurities and most of the carbon, and then adding other metals. This conversion used to be carried out in a **Bessemer converter**, but this gave steel of variable quality. The **open hearth furnace** replaced the Bessemer converter, but this was both slow and expensive.

Today, most steel is produced using a **basic oxygen furnace**, controlled by computer systems – a sophisticated modification of the old Bessemer process. The basic oxygen furnace is charged with about 30% scrap iron and steel along with 70% molten iron from the blast furnace, which melts the scrap metal. In addition, limestone is added to the furnace. The oxygen lance then blows pure oxygen through the molten metal mixture. This reacts with the carbon in the iron and also reacts with any impurities to form oxides. These in turn react with the calcium oxide formed from the limestone, and form a slag. Other metals may then be added to produce steel with the desired properties. Using the basic oxygen furnace, over 300 tonnes of steel can be made in less than an hour.

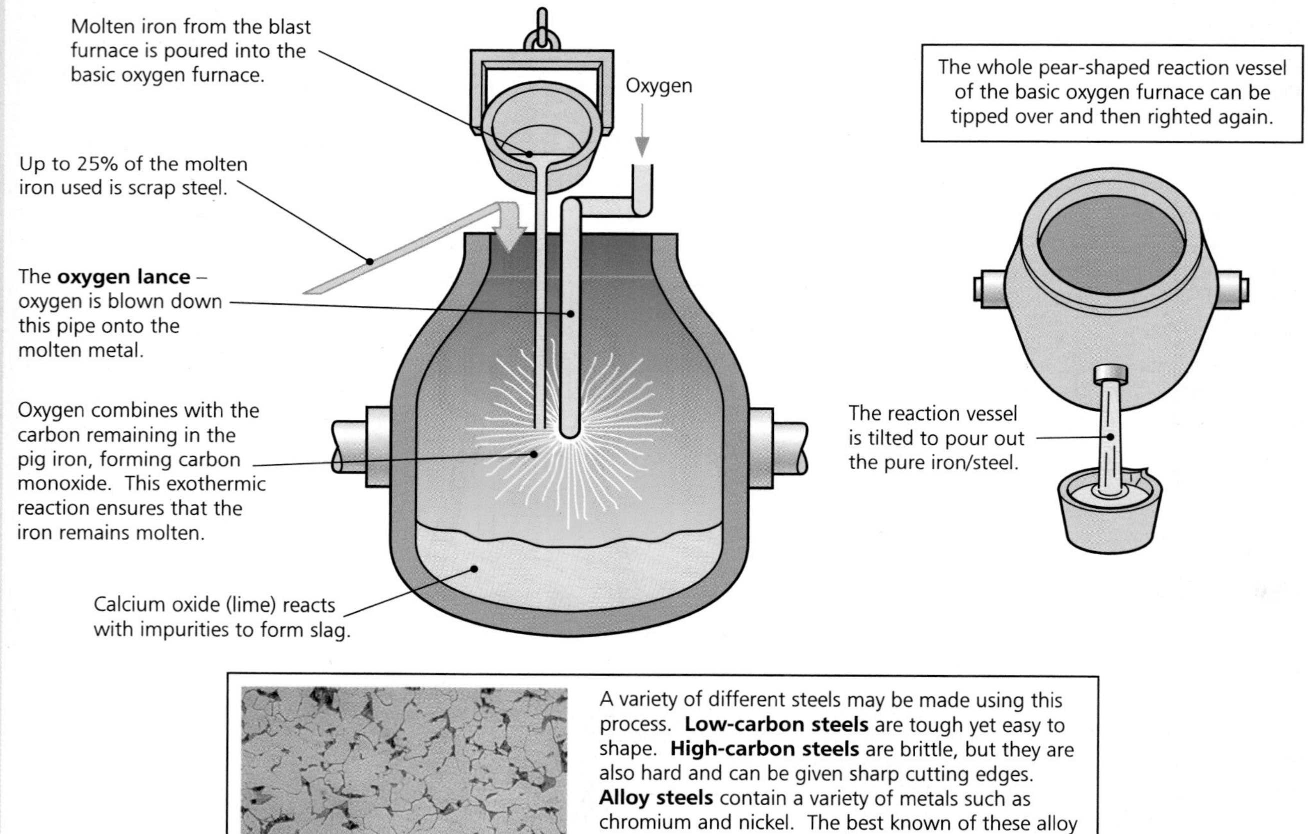

Molten iron from the blast furnace is poured into the basic oxygen furnace.
Oxygen
The whole pear-shaped reaction vessel of the basic oxygen furnace can be tipped over and then righted again.
Up to 25% of the molten iron used is scrap steel.
The **oxygen lance** – oxygen is blown down this pipe onto the molten metal.
Oxygen combines with the carbon remaining in the pig iron, forming carbon monoxide. This exothermic reaction ensures that the iron remains molten.
The reaction vessel is tilted to pour out the pure iron/steel.
Calcium oxide (lime) reacts with impurities to form slag.

A variety of different steels may be made using this process. **Low-carbon steels** are tough yet easy to shape. **High-carbon steels** are brittle, but they are also hard and can be given sharp cutting edges. **Alloy steels** contain a variety of metals such as chromium and nickel. The best known of these alloy steels is **stainless steel**. The properties of the steel are also affected by the way it cools down.

The reactions of iron

Iron forms compounds with most of the non-metals in both its common oxidation states, +2 and +3. In aqueous solution, the Fe^{2+} and Fe^{3+} ions usually exist as coloured complexes – the iron(II) ion forms the $[Fe(H_2O)_6]^{2+}$ complex that gives a pale blue-green colour both to the solutions and to solid Fe^{2+} salts, as the complex frequently persists in the crystals. Fe^{3+} gives a yellow colour due to the $[Fe(H_2O)_6]^{3+}$ complex.

Iron forms many complex ions with different substances – for example, when a solution containing Fe^{3+} is added to a solution containing the ferrocyanide ion, $Fe[(CN)_6]^{4-}$, a deep blue precipitate known as **Prussian blue** is formed. This coloration is the basis of the blueprint process by which engineers and architects used to produce copies of their plans.

The Fe^{2+} ion is more stable than the Fe^{3+} ion in acidic solutions. This situation is reversed in alkaline solutions, where Fe^{3+} is the more stable state. If sodium hydroxide solution is added to a solution containing Fe^{2+} ions, a precipitate of pale green iron(II) hydroxide forms:

$$Fe^{2+}(aq) + 2OH^-(aq) \rightarrow Fe(OH)_2(s)$$

In the presence of air, this precipitate rapidly darkens to a red-brown colour, as oxygen oxidises iron(II) to iron(III):

$$4Fe(OH)_2(s) + O_2(g) + 2H_2O(l) \rightarrow 4Fe(OH)_3(s)$$

In acidic solutions, stronger oxidising agents such as chlorine or dichromate ions, $Cr_2O_7^{2-}$, are required to oxidise Fe^{2+} to Fe^{3+}, while quite mild reducing agents like sulphur dioxide or iodide ions will reduce Fe^{3+} back to Fe^{2+}.

Iron(II) salts are formed when iron reacts with acids, liberating hydrogen, a typical example being the reaction with hydrochloric acid:

$$2Fe(s) + 2HCl(aq) \rightarrow FeCl_2(aq) + H_2(g)$$

The reactions of iron with nitric and sulphuric acids are more complex. Dilute sulphuric acid in the absence of air produces iron(II) sulphate, $FeSO_4$, while dilute nitric acid gives iron(II) nitrate, with the reduction of nitrate ions, NO_3^-, to ammonium ions, NH_4^+:

$$4Fe(s) + 10HNO_3(aq) \rightarrow 4Fe(NO_3)_2(aq) + NH_4NO_3(aq) + 3H_2O(l)$$

Hot concentrated sulphuric acid produces iron(III) sulphate, the acid being reduced to sulphur dioxide:

$$2Fe(s) + 6H_2SO_4(aq) \rightarrow Fe_2(SO_4)_3(aq) + 3SO_2(g) + 6H_2O(l)$$

Concentrated nitric acid reacts initially in the same way as the dilute acid, but after a short time the iron ceases to react and becomes passive. This is due to the formation of a thin layer of oxide, which is exceptionally inert and impervious, protecting the iron underneath from attack.

Iron is renowned for **rusting**, the common name given to the reaction between iron, oxygen and water. Rusting is in fact a complex chemical reaction. In pure, oxygen-free water, iron does not rust. In pure, dry oxygen, iron does not rust. In a mixture of oxygen and water, the metal corrodes. The phenomenon of this reaction, with all its implications for the use of iron the world over and the economics of engineering projects both large and small, will be explored further in section 3.5.

The extraction of titanium

The extraction of titanium presents problems, since in its molten state (above 1800 °C) titanium attacks the refractory linings of furnaces and absorbs gases. The formation of a **carbide** (in which carbon atoms are inserted into holes, or **interstices**, in the titanium lattice) is also possible if the ore is heated with carbon alone, which makes extraction by this means impossible. The metal has very high resistance to both corrosion and heat, and a density nearly half that of stainless steel – it was these properties that first stimulated interest in the metal in aircraft engineering, although titanium now finds many uses, including chemical and nuclear engineering. Titanium dioxide is also widely used as a brilliant white pigment in paint (page 206).

The **Kroll process** is used to extract titanium from its ores, rutile (TiO_2) and ilmenite ($FeO.TiO_2$). The first stage of this process involves the reaction of the ore with chlorine in the presence of carbon, at a temperature of between 700 and 800 °C:

$$TiO_2 + C + 2Cl_2 \rightarrow TiCl_4 + CO_2$$

$$TiO_2 + 2C + 2Cl_2 \rightarrow TiCl_4 + 2CO$$

The titanium(IV) chloride produced in this reaction is condensed as a colourless liquid. It is then reduced to titanium metal using either magnesium or sodium as the reducing agent, under an atmosphere of helium or argon at temperatures up to 800 °C:

$$TiCl_4 + 2Mg \rightarrow Ti + 2MgCl_2$$

$$TiCl_4 + 4Na \rightarrow Ti + 4NaCl$$

Titanium metal remains in the reaction vessel after the magnesium or sodium chloride has been removed. Electrolysis may then be used to regenerate the reducing metal and chlorine gas.

Figure 2.7.24 Titanium has many uses – in nuclear power plants, in chemical engineering, in bike parts – but its use on the Guggenheim roof in Bilbao must be one of the most spectacular!

SUMMARY

- The **transition metals** are elements that form at least one ion with a partially filled d subshell, and this determines their chemical character.
- The transition elements have the following common properties:

  They are all hard metals with high melting and boiling points.
  They show more than one oxidation state in their compounds.
  They tend to form coloured compounds and ions.
  Many show catalytic activity.
  They form **complex ions**.
- In the first transition series, the elements have relatively small atomic radii due to electrons filling the d subshell rather than adding new subshells.

- Transition metals are less reactive than the s-block metals, and most have protective oxide films.
- The most important oxidation states of the transition elements are usually +2 and/or +3. The ability to form compounds with a variety of oxidation states is due to the similarity in the energy levels of the 3d and 4s subshells.
- **Complex ions** are formed when a central cation is surrounded by a number of anions or electron pair donors known as **ligands**. The number of coordinate bonds formed gives the **coordination number**.
- Complex ions with a coordination number of 2 show linear geometry. Those with a coordination number of 4 may be planar or tetrahedral, while those with a coordination number of 6 are octahedral.
- The geometric shape of complex ions gives rise to **stereoisomerism**.
- The colours of complex ions are caused by transitions of electrons between partly filled d subshells.
- Transition metals play important roles in both industry and biological systems as **catalysts**.
- Chromium is extremely resistant to corrosion as it is unreactive, due to its electronic configuration with half-filled d and s subshells.
- Copper is relatively unreactive, and is a good conductor of heat and electricity. The Cu^{2+} complex ions give distinctive blue solutions.
- Iron is of great industrial importance in the production of steel. It is extracted from its ores by chemical reduction using carbon in the **blast furnace**.
- Titanium is extracted from its ore with difficulty using the Kroll process.

QUESTIONS

1 a 'Transition elements are d-block elements.' Do you agree with this statement? Give reasons for your answer.

b What are the main physical characteristics of the transition elements?

c Explain the trends in
i electronegativity and
ii first ionisation energy across the first row of transition elements.

d Summarise *briefly* the main chemical properties of the transition elements.

2 a Why do the transition elements tend to form complex ions?

b Why do many complex ions produce coloured solutions?

c What is meant by the 'coordination number' of a complex ion?

3 Some examples of the many complexes formed by the first series of transition elements are as follows:

$[TiF_6]^{2-}$ $[VO(H_2O)_5]^{2+}$ $[Cr(CO)_6]$ $[Mn(H_2O)_6]^{2+}$
$[Fe(H_2O)_5NO]^{2+}$ $[CoCl_4]^{2-}$ $[Ni(H_2O)_2(NH_3)_4]^{2+}$ $[Cu(CN)_4]^{3-}$

a Which of these complexes would you expect to be octahedral? Draw the structure of these complexes.

b The $[Cu(CN)_4]^{3-}$ ion is tetrahedral. What other shape may four-coordinate complex ions have?

c What is the oxidation number of chromium in $[Cr(CO)_6]$?

d Which complex(es) in the list may have more than one stereoisomer?

e Cobalt forms the ion $[Co(NO_3)_4]^{2-}$, in which the nitrate ions (NO_3^-) are bidentate. Suggest a structure for this ion.

QUESTIONS

4 a Describe the physical properties of chromium.
b Why is chromium such a chemically unreactive metal?
c What are the main industrial uses of chromium?
d 'Chromium reacts with the halogens on heating, forming chromium(III) chloride with chlorine. Reduction of a solution of chromium(III) chloride with zinc under a carbon dioxide atmosphere yields a bright blue solution. If this bright blue solution is filtered into a saturated solution of sodium ethanoate, a precipitate of chromium(II) ethanoate is formed.'
i Write equations for the reactions described here, leaving out any spectator ions.
ii What changes in oxidation number are happening here?
iii Why do you think the reduction of chromium(III) is carried out under an atmosphere of carbon dioxide?

5 Iron has been of immense importance to the human race for thousands of years.
a Describe how iron is extracted from iron ore, giving chemical equations where possible.
b Most of the iron produced by blast furnaces is converted to steel before it is used.
i What are the advantages of steel over iron?
ii How is pig iron modified to produce steel?

6 a How would you oxidise iron(II) to iron(III) under
i basic conditions, and
ii acidic conditions?
b Addition of colourless potassium iodide solution to an acidic solution of iron(III) chloride which is a pale yellow colour leads to a dark brown solution. Why?
c The amount of iron in a sample of ore may be determined by dissolving a sample of the ore in dilute sulphuric acid and reducing any iron(III) present with zinc metal. A sample of this solution is then titrated against potassium manganate(VII) solution:

$$5Fe^{2+}(aq) + MnO_4^-(aq) + 8H^+(aq) \rightarrow 5Fe^{3+}(aq) + Mn^{2+}(aq) + 4H_2O(l)$$

i How can you tell when all the iron(II) in the sample being titrated has been oxidised to iron(III)?
10 g of an ore containing iron was dissolved in sulphuric acid and converted to iron(II), when the volume of the solution was made up to 250 cm^3. A 25 cm^3 aliquot of this solution was titrated against 0.05 mol dm^{-3} potassium manganate(VII) solution. 14.3 cm^3 of this solution was required to oxidise all the iron(III) in the aliquot.
ii Calculate the number of moles of manganate(VII) ions in 14.3 cm^3 of 0.05 mol dm^{-3} potassium manganate(VII) solution.
iii How many moles of iron(II) ions does this correspond to?
iv How many moles of iron were present in the 10 g sample?
v What mass of iron is this?

Developing Key Skills

Iron is an element which has had a major impact on the development of human society for a very long time. Using electronic and library resources, find out as much as you can about iron and its long history of use by people. Produce a short book titled 'The story of iron' for use by pupils in key stage 3, helping to make cross-curricular links between science and history.

[Key Skills opportunities: C, IT]

2 Questions

1 The highest oxidation state of chromium exists as the yellow oxyanions CrO_4^{2-}.

a Deduce the oxidation number of chromium in CrO_4^{2-}. **(1 mark)**

b When a dilute acid is added to a solution of CrO_4^{2-}, the solution changes colour.

i State the new colour formed.

ii Write a balanced equation for the reaction that has taken place. **(2 marks)**

c When treated with sulphur dioxide in acidic solution, the oxyanion CrO_4^{2-} can be reduced to a lower oxidation state forming an ion **A**.

i What is the identity of ion **A** formed from CrO_4^{2-}?

ii State the new colour formed. **(2 marks)**

d Manganese forms an unstable green oxochloride, **B**, with the following composition by mass: Mn, 39.7%; O, 34.7%; Cl, 25.6%.

i Calculate the empirical formula of the oxochloride **B**.

ii Deduce the oxidation state of manganese in the oxyanion **B**.

(3 marks)

(Total 8 marks)

(OCR specimen)

2 1,2-Diaminoethane, $NH_2CH_2CH_2NH_2$, is a *bidentate ligand*.

a Explain the term *bidentate ligand*. **(2 marks)**

b There are three isomeric complexes with the formula $[Cr(NH_2CH_2CH_2NH_2)_2Cl_2]^+$, all having the same basic shape.

i State the shape of these complexes.

ii Draw structures of these three complexes to show the differences between them.

iii Which of the complexes you have drawn will have a dipole?

(5 marks)

(Total 7 marks)

(OCR specimen)

3 a Explain what is meant by ligand exchange. **(1 mark)**

B Describe the colour changes that take place when an aqueous solution of ammonia is gradually added to a solution of $Cu^{2+}(aq)$, until the ammonia is in excess. Write equations for these transformations. **(4 marks)**

c Blood gets its colour from oxygen-carrying molecules that consist of organic groups surrounding a transition metal ion. In humans this transition metal is iron, and the blood is red. In horseshoe crabs, the metal is copper and the blood is blue, and in sea squirts the metal is vanadium and the blood is green. The sketch below shows the major absorption peak for human blood. On a copy of this sketch show and label the corresponding absorption peaks for the blood of horseshoe crabs and sea squirts. **(2 marks)**

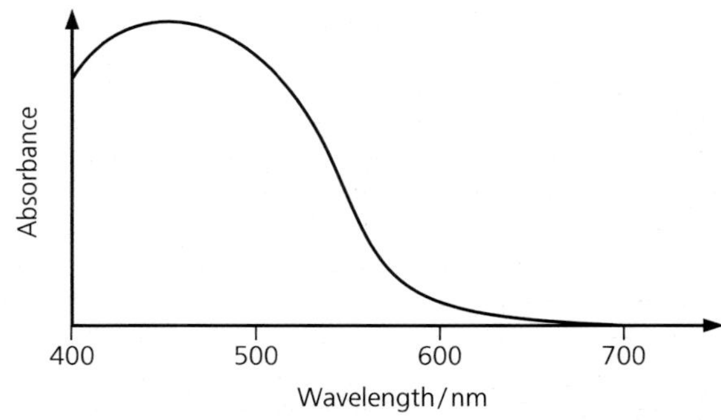

d A 0.0100 mol sample of an oxochloride of vanadium, $VOCl_x$ required 20.0 cm^3 of 0.100 mol dm^{-3} acidified potassium manganate(VII) for oxidation of the vanadium to its +5 oxidation state.

i Calculate how many moles of potassium manganate(VII) were reacted.

ii How many moles of electrons were removed by the MnO_4^- ions?

iii Determine the change in oxidation state of the vanadium.

iv Deduce the value of *x* in the formula $VOCl_x$. **(4 marks)**

(Total 11 marks)

(OCR specimen)

4 Cobalt(II) forms the following coloured complexes with water molecules and chloride ions: $[Co(H_2O)_6]^{2+}$ and $[CoCl_4]^{2-}$ Describe how the different ligands, H_2O and Cl^-, affect the stereochemistry and colour of these complexes. **(8 marks)**

(OCR specimen)

5 Copper is a transition metal. A typical property of a transition metal, such as copper, is the ability to form complex ions with ligands such as water molecules.

a State **two** other typical properties of copper or its compounds that are different from those of non-transition metals. **(2 marks)**

b Explain how a water molecule behaves as a ligand. **(2 marks)**

c Aqueous copper(II) sulphate contains the complex ion $[Cu(H_2O)_6]^{2+}$. When an excess of aqueous ammonia is added to aqueous copper(II) sulphate, ligand substitution takes place.

i What would you see?

ii Write an equation for this ligand substitution. **(3 marks)**

(Total 7 marks)

(OCR specimen)

6 A student prepared two chlorides of iron in the laboratory.

- In the first experiment, the student reacted iron with an excess of hydrogen chloride gas forming a chloride **A**, with the composition by mass, Fe: 44.0%; Cl: 56.0%.
- In the second experiment, the student formed 8.12 g of a chloride **B** by reacting 2.79 g of iron with an excess of chlorine. [A_r: Fe, 55.8; Cl, 35.5.]

a Identify compounds **A** and **B**. Include all your working and equations for the two reactions.

b What are the electronic configurations of Fe in its metallic form and in compounds **A** and **B**?

c The chloride **A** has a much higher melting point (672 °C) than that of the chloride **B** (220 °C). An aqueous solution of chloride **A** is neutral whereas that of **B** is acidic. Explain what this information suggests about the structure and bonding in **A** and in **B**.

(13 marks)

(OCR specimen)

7 Redox reactions are an important type of reaction in chemistry. Explain what is meant by a redox reaction. Illustrate your answer with **two** examples drawn from inorganic chemistry (one of which should involve a transition element) and **two** examples from organic chemistry. *(In this question, 2 marks are available for the quality of written communication.)* **(17 marks)**

(OCR specimen)

8 The atomic radii of some of the elements in groups 1–7 of the periodic table are shown in the table below. Some radii have been omitted.

		group						
		1	2	3	4	5	6	7
Period 2	Element Atomic radius/nm	Li 0.134	Be 0.125	B 0.090	C 0.077	N 0.075	O 0.073	F 0.071
Period 3	Element Atomic radius/nm	Na 0.154	Mg 0.145	Al 0.130	Si 0.118	P 0.110	S	Cl 0.099
Period 4	Element Atomic radius/nm	K 0.196	Ca 0.174	Ga	Ge 0.122	As 1.122	Se 0.117	Br 0.144

a **i** State the trend shown in atomic radius across a period.
ii Explain this trend. **(4 marks)**

b **i** State the trend shown in atomic radius down a group.
ii Explain this trend. **(4 marks)**

c Mendeleev studied periodic data to make predictions for the properties of elements which had yet to be discovered.
Use the data above to suggest values for the atomic radius of
i S **ii** Ga. **(2 marks)**

(Total 10 marks)

(OCR specimen)

9 A student carried out a series of two experiments with magnesium.

a In the first experiment, the student heated magnesium with oxygen, forming magnesium oxide.
i State what the chemist would see in this reaction.
ii Write an equation, including state symbols, for the reaction.
iii The chemist added water to the magnesium oxide. Some of the magnesium oxide reacted forming a solution. Suggest a value for the pH of this solution.
iv Magnesium oxide is a solid with a melting point of 2852 °C. Explain, in terms of structure and bonding, why its melting point is so high. **(8 marks)**

b In a second experiment, the student reacted 1.20 g of magnesium with 2.00 mol dm^{-3} hydrochloric acid. [A_r Mg, 24.0; Cl, 35.5].

$$Mg(s) + 2HCl(aq) \rightarrow MgCl_2(aq) + H_2(g)$$

i How many moles of Mg were used in the experiment?
ii Calculate the minimum volume of 2.00 mol dm^{-3} hydrochloric acid needed to react completely with this amount of magnesium.
iii Calculate the volume of H_2 gas that would be produced at room temperature and pressure (r.t.p.). [1 mole of gas molecules occupies 24 dm^3 at r.t.p.]
iv State the reagent(s) that you could use to show the presence of chloride ions in the aqueous magnesium chloride. State what you would expect to observe. **(6 marks)**

c The student repeated both experiments with calcium.
i What difference would you expect in reactivity?
ii Explain your answer to **i**. **(3 marks)**

(Total 17 marks)

(OCR specimen)

10 **a** Aluminium is extracted by the electrolysis of a 5% solution of aluminium oxide in molten cryolite, Na_3AlF_6. The temperature of the electrolyte is maintained at around 1200 K.
i Why is pure molten aluminium oxide not used as the electrolyte? **(1 mark)**
ii Write an equation for the reaction that occurs at the anode. Explain why if carbon anodes are used, they have to be replaced regularly. **(3 marks)**
iii Suggest **two** reasons why the extraction of aluminium requires a lot of energy. **(2 marks)**
iv Suggest **two** different properties of aluminium that have resulted in its economic importance. **(2 marks)**

b Aluminium/air electrochemical cells can be used to power golf trolleys and invalid carriages. One electrode is made of aluminium while the other is made by bubbling air through an inert porous material. The electrolyte is usually sodium hydroxide solution. The equations for the reactions taking place at the electrodes are:

I $Al(s) + 3OH^-(aq) \rightarrow Al(OH)_3(s) + 3e^-$
II $O_2(g) + 2H_2O(l) + 4e^- \rightarrow 4OH^-(aq)$

i Which electrode acts as the cathode of the cell? **(1 mark)**
ii Write an equation for the overall cell reaction. **(1 mark)**
iii Suggest **one** reason why the efficiency of the cell may be reduced over a period of time. **(1 mark)**

(Total 11 marks)

(Edexcel 1999)

11 **a** Complete the electronic configuration of a cobalt(II) ion.
Co^{2+} [27] **(1 mark)**

b When cobalt(II) chloride is treated, under certain conditions, with the bidentate ligand, $NH_2CH_2CH_2NH_2$, (which can be represented by the symbol "en"), the compound $[CoCl_2(en)_2]Cl$ is formed.
i What is the oxidation state of cobalt in the compound formed?
ii What is meant by the term *bidentate* as applied to a ligand?
iii What is the co-ordination number of cobalt in this compound?
iv When this compound is treated with aqueous silver nitrate, only one mole of silver chloride is produced per mole of compound. Explain this observation. **(5 marks)**

c When hydrazine, NH_2NH_2, reacts with cobalt(II) chloride in aqueous solution, the compound $CoCl_2(NH_2NH_2)_2$ is formed. This compound has a polymeric structure in which cobalt is six co-ordinate and the cobalt ions are linked by hydrazine molecules. Draw the structure of the repeating unit of the polymer. **(2 marks)**

(Total 8 marks)

(AQA 1999)

12 Study the passage below and answer the questions which follow. Crystalline iron(III) nitrate nonahydrate, $Fe(NO_3)_3.9H_2O$, has a very pale violet colour and contains the ion $[Fe(H_2O)_6]^{3+}$. When added to water, the crystals dissolve to form a brown solution. Treatment of this brown solution with concentrated nitric acid yields a very pale violet solution.

a Name the shape of the $[Fe(H_2O)_6]^{3+}$ ion. **(1 mark)**

b Write an equation to show the $[Fe(H_2O)_6]^{3+}$ ion behaving as an acid in aqueous solution. **(1 mark)**

c Deduce the formula of the species responsible for the brown colour of the solution described above. **(1 mark)**

d Explain why the addition of concentrated nitric acid causes the colour of the solution to change from brown to very pale violet. **(2 marks)**

e When concentrated hydrochloric acid is added to the brown solution of iron(III) nitrate, however, a yellow solution containing $[FeCl_4]^-$ ions is formed. Give **two** reasons for a colour change in this reaction. **(2 marks)**

f When an excess of magnesium metal is added to an aqueous solution of iron(III) nitrate, effervescence occurs, and a brown precipitate forms. Identify the gas evolved, give the formula of the brown precipitate and construct an equation, or equations, for the reaction occurring. **(3 marks)**

(Total 10 marks)

(AQA 1999)

13 a Explain why the boiling temperatures of the halogens increase down Group VII from fluorine to iodine. **(2 marks)**

b State how, and explain why, the reducing powers of the halide ions change down Group VII from fluoride to astatide. **(4 marks)**

c Use your knowledge of the reactions of solid sodium chloride, bromide and iodide to predict the gaseous products formed when concentrated sulphuric acid is warmed with solid sodium astatide. Identify the role of the astatide ion in the formation of each gaseous product and write equations for the reactions occurring. **(9 marks)**

d The composition of a mixture of two solid sodium halides was investigated in two separate experiments.

Experiment 1: When a large excess of chlorine gas was bubbled through a concentrated solution of the mixture, orange-brown fumes and a black precipitate were produced.

Experiment 2: 0.545 g of the solid mixture was dissolved in water and an excess of silver nitrate solution was added. The mass of the mixture of silver halide precipitates formed was 0.902 g. After washing the mixture of precipitates with an excess of concentrated aqueous ammonia, the mass of the final precipitate was 0.564 g.

Write equations for each of the reactions occurring in these experiments and explain how these results enable you to identify the halide ions present. Use the information given above to calculate the percentage by mass of each halide ion present in the solid mixture. **(15 marks)**

(Total 30 marks)

(AQA 1999)

14 a Why are the elements sodium to argon placed in Period 3 of the periodic table? Describe and explain the trends in electronegativity and atomic radius across Period 3 from sodium to sulphur. **(7 marks)**

b Describe the trend in pH of the solutions formed when the oxides of the Period 3 elements, sodium to sulphur, are added separately to water. Explain this trend by reference to the structure and bonding in the oxides and by writing equations for the reactions with water. **(19 marks)**

c Describe and explain any differences in the thermal stabilities of the carbonates of Group I metals. **(4 marks)**

(Total 30 marks)

(AQA 1999)

15 The elements of the third period are as follows.

Na Mg Al Si P S Cl Ar

All of your answers below should relate to these elements.

a Which elements can exist

i as diatomic molecules at room temperature,

ii as macromolecular structures? **(2 marks)**

b Which pairs of elements combine to produce compounds with formulae of the type *XY*? **(2 marks)**

c Two elements form chlorides with formula of the type XCl_3. Draw displayed formulae for these two chlorides, and suggest values for the bond angles. **(4 marks)**

d i One element combines with oxygen to form an oxide which reacts with water to give a strongly alkaline solution. Name the element, and write a balanced equation for the oxide reacting with water.

ii One element combines with oxygen to form an oxide of the type XO_2, which reacts with water to given an acidic solution. Name the element and write a balanced equation for the oxide reacting with water. **(4 marks)**

(Total 12 marks)

(OCR 1998)

16 Hydrogen iodide can be prepared by adding water to a mixture of red phosphorus and iodine, and then warming gently.

a Construct the following equations:

i phosphorus and iodine forming phosphorus tri-iodide.

ii phosphorus tri-iodide and water reacting to form hydrogen iodide and phosphoric(III) acid, H_3PO_3. **(2 marks)**

b What would you expect to see when hydrogen iodide reacts with

i aqueous silver nitrate, followed by aqueous ammonia,

ii warm concentrated sulphuric acid? **(3 marks)**

(Total 5 marks)

(OCR 1998)

2

3 HOW FAR? HOW FAST?

Introduction

When chemicals react we very often need to be able to predict what is going to happen and what will be the end result of the reaction. In this section we shall look at chemical reactions and ask whether they 'go' to completion – whether the reactants are fully converted to products – and what affects the rate at which the reaction occurs. This branch of chemistry is sometimes called **physical chemistry**.

Many chemical reactions are **equilibria**, in which the final result of a reaction is a mixture of the reactants and the products of the reaction. There are various types of equilibria, each important in its own right.

The reactions between acids and bases are of major importance both in the laboratory and in the world around us, so we need to understand acid–base equilibria and ways of measuring and calculating the strengths of acids and bases. The importance of acids and bases can also be seen in their roles as **buffers** and **indicators** – buffer solutions resist changes in pH, whilst indicators are substances (often weak acids) which show a dramatic colour change at a particular pH. Both buffers and indicators are examples of acid–base equilibria.

Another area of chemistry where an understanding of equilibria plays an important role is in the study of precipitation and complex ion formation. Precipitation involves the appearance of an insoluble solid from a solution during a chemical reaction. Complex ions, as their name suggests, are large, complicated ions, many of which have unusual properties and which may be beautifully coloured (as we saw in section 2.7). Redox reactions, which we met in section 1.4, are also examples of equilibria, and we shall look at reduction and oxidation in some detail.

However, in chemistry we don't simply need to know how far a reaction will go. We also want to know how fast it is going to happen – will it be explosively fast or can we come back next year? So in **reaction kinetics** we look at the factors which affect the rate of a reaction, and ways in which we can change the rate of a reaction when we want to.

Our final look at physical chemistry involves **thermodynamics**, in which we shall examine the interactions between energy and matter in chemical reactions. Thermodynamics has a lot to tell us about why chemical change takes place.

Figure 1 The formation of stalagmites and stalactites is an example of a very slow precipitation reaction.

FOCUS CLEANING UP THE ENGINE

The motor car is a fundamental part of modern life for most of us. Many households have one car, while some have two or more. People rely on cars for work and use them during leisure time. Yet more people are now questioning the use of the car than at any time since they were first given the free run of the roads at the turn of the century.

Figure 1 There are already around 600 million cars in the world, and as long as we continue to see cars as a way of improving our status with our peers, as well as the most convenient form of transport, then the numbers will continue to climb.

What's the problem?

Cars are powered by the burning of hydrocarbon fuel in an atmosphere containing oxygen and nitrogen. The complete combustion of an alkane in oxygen produces carbon dioxide and water, for example:

$$2C_8H_{18} + 25O_2 \rightarrow 16CO_2 + 18H_2O$$

However, in a car engine the combustion is not always complete, and so carbon monoxide is formed along with the carbon dioxide.

Cars also produce hydrocarbon pollution. Not all the hydrocarbons in the fuel undergo combustion, and some unburned hydrocarbon ends up in the exhaust gases. More hydrocarbon pollution – including some very toxic compounds such as benzene – results from vapours which escape from the petrol tank of the car itself as it is filled.

When air is drawn into the engine of a car, both oxygen and nitrogen are present. These gases can react to form nitrogen oxide:

$$N_2(g) + O_2(g) \rightleftharpoons 2NO(g)$$

At room temperature, the equilibrium constant for this reaction is 4.8×10^{-31}, which tells us that at equilibrium the reaction is far over to the left-hand side – in other words, nitrogen oxide is not formed at room temperature. However, the reaction is endothermic, and so at high temperatures this equilibrium will be shifted to the right, resulting in the production of nitrogen oxide. The gases inside a car engine and its exhaust system are at a temperature of several thousand degrees Celsius – easily high enough to shift the reaction equilibrium to the right so that some nitrogen oxide is formed. When the gas leaves the car exhaust system, it cools very rapidly. The kinetic stability of nitrogen oxide at ambient air temperatures means that the reverse reaction does not take place – nitrogen oxide does not decompose and so it is present in the exhaust gases leaving the car. Once in the atmosphere, it is rapidly oxidised to brown nitrogen dioxide, NO_2, which is a major contributor to air pollution and poor air quality.

Takes your breath away?

Pollution from cars does not stay in one place – it spreads throughout the atmosphere as a whole. We have now reached the point where the quantity of pollution produced is more than can be 'lost' into the atmosphere without an adverse effect on the environment. How is this effect felt?

Carbon dioxide is a **greenhouse gas**. It accumulates in the upper atmosphere and reduces heat loss by radiation from the surface of the Earth. This is thought to be causing global warming, with resultant changes in

FOCUS CLEANING UP THE ENGINE (continued)

weather patterns, desertification and rising sea levels. Any process that adds to the level of greenhouse gases is therefore a problem in need of attention.

The carbon monoxide also produced in the combustion of petrol not only adds to the greenhouse gases, but is highly toxic. It combines with the haemoglobin of the blood more effectively than oxygen, thus preventing haemoglobin from carrying oxygen around the body.

The hydrocarbon emissions contain compounds which are both toxic and carcinogenic. The concentrations in the atmosphere taken in by children are likely to be higher than those inhaled by adults due to the height difference, as children are nearer the exhausts of vehicles. What is more, the rapidly dividing and growing cells of small children are particularly susceptible to chemical damage.

The rise in oxides of nitrogen in the atmosphere has several damaging effects on the environment. Nitrogen dioxide smog is corrosive and irritating to the respiratory tract. In Britain, the numbers of people suffering from both asthma and hay fever have increased dramatically in recent years in spite of improvements in the medications available. The increase in air pollution, particularly oxides of nitrogen, is thought to be at least in part to blame and reports of air quality are now given on the national weather forecasts. In cities such as New York, Los Angeles, Tokyo and Athens, anyone suffering from respiratory complaints such as asthma is advised to remain indoors when the smog comes down. Oxides of nitrogen also contribute to the formation of acid rain. This rain, which contains nitric and nitrous acids, HNO_3 and HNO_2, is causing untold damage to the fertility of the soil, to life in streams, lakes and rivers and to trees. The resulting low pH (pH 2.4 has been measured in Scotland) is not conducive to life.

Figure 2 The effect of many thousands of cars in a relatively small area is demonstrated only too well in cities such as Montreal (top). When the weather conditions are good, the pollutant gases from car exhausts are carried away by the wind. But on still days the full extent of the pollution caused by emissions from car exhausts is only too clearly visible as smog – a mixture of smoke, fog and chemical fumes. In the UK the weather conditions rarely favour the formation of smog, though pollution from cars causes other problems.

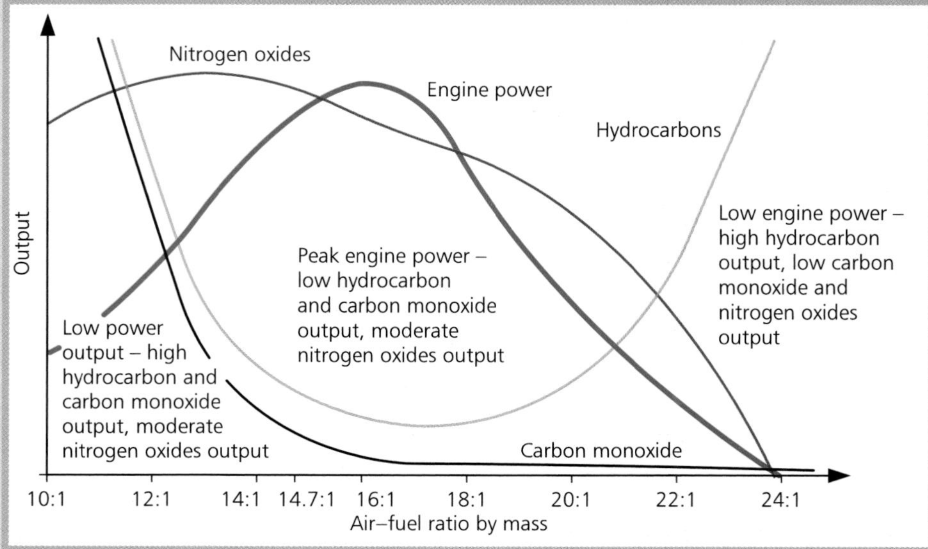

Figure 3 The effect of stoichiometry – engine emissions for a range of air–fuel mixtures

FOCUS CLEANING UP THE ENGINE (continued)

What can be done?

A whole range of measures is available which can and do reduce the pollution – but they need financial investment and public good-will to succeed. A very important approach is to tackle the source of the emission – the engine – and try to change conditions there to reduce or prevent the formation of the pollutants.

Cleaning up the engine

The biggest problem with reducing the pollution produced by the engines is that in many cases, altering the mechanics of the engine adversely affects the performance of the car. In spite of the fact that almost all cars on the road can now go far faster than is either legally possible or safe, and few drivers use the high accelerations boasted of in advertisements, the demand is for ever increasing rather than decreasing performance. For example, at the high compression ratios in the cylinders of modern cars, the air–fuel mixture is heated to very high temperatures before it explodes. The hot gases which result favour the production of nitrogen oxide, as we have seen. If the compression ratio of the engine is lowered, the combustion temperature is also lowered, shifting the equilibrium away from the production of nitrogen oxide and so reducing its emission in the exhaust gases. Unfortunately, the lower compression ratio also lowers the efficiency of the engine (increasing the fuel consumption) and reduces the power available, neither of which is a good selling point. Mixing water into the petrol–air mixture is another possible solution, the water absorbing some of the heat from the combustion and so keeping the temperature of the exhaust gases lower.

For many years, tetraethyl lead, $Pb(C_2H_5)_4$, has been in petrol to prevent 'knocking' and improve the performance of car engines. However, following wide-spread concern over the high levels of lead found in the bodies of people, particularly young children, in the late twentieth century, 'leaded' petrol was gradually phased out. It was replaced by 'unleaded' petrol which worked in slightly modified engines and removed at least one source of air pollution.

Research is currently proceeding on a new family of engines referred to as **lean-burn engines**. A lean-burn engine produces greatly reduced toxic emissions. The main technical problem is that there is a complex interaction between the air–fuel ratio, the engine temperature and the production of pollutants, as the graph shows. The theory is that if an engine can be designed to work with sustained power with a high air–fuel ratio, then any unburned hydrocarbon will be absorbed by the engine, which produces only low levels of nitrogen oxides, carbon monoxide and hydrocarbons.

Since the 1950s there has been a staggering increase in car ownership in countries as diverse as the USA and Greece, Australia and Japan, as well as Britain. Realistically this is a trend which will continue, as car ownership in the developing world also increases ever more rapidly. To tackle pollution, we must therefore work towards ever cleaner engines, even if extremes of performance are sacrificed along the way.

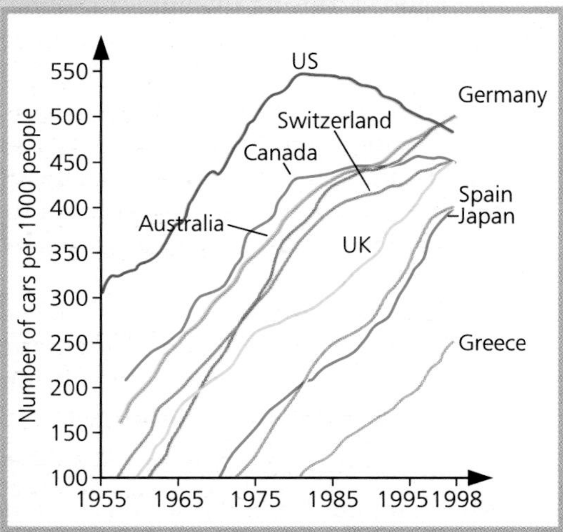

Figure 4 The trend in car ownership for each of these countries is very similar – ever onward and upward.

3.1 Chemical equilibria: How far?

There are many situations in the world around us in which two factors have contrasting effects on a system. A book resting on a table is pulled downwards by the gravitational attraction of the Earth acting on it, and this force is exactly balanced by the force of the table pushing upwards on the book. The number of individuals in a population of rabbits will stay reasonably constant as long as the amount of food available and the number of predators stay the same. Each of these examples can be described in terms of balancing two opposing tendencies. When the system is stable, we may say that it is in **equilibrium**. The idea of equilibrium is used by chemists to describe how chemical reactions moving in opposite directions may balance out.

In section 3 we shall be concerned with two main questions:

How far? – does a reaction go to completion and if not, why not? and

How fast? – at what rate do reactions happen and can we affect that rate?

We shall start by looking at what chemists mean by the term 'equilibrium' – answering the question 'How far?'

Figure 3.1.1 The right level of sugar in the blood is critical for the functioning of the brain. This level is affected by the amount of food eaten and the amount of insulin produced in the body. Blood sugar levels are kept steady in a healthy person because the amount of insulin produced varies in relation to the amount of sugar in the blood. The diabetic (who produces little or no insulin) must carry out this complicated biochemical balancing act by injecting insulin regularly.

Introducing equilibria

Equilibria in changes of state

As we saw in section 1.7, a change of state occurs when a substance is transformed from one state of matter to another – for example, from solid to liquid. Such changes provide a simple example of equilibrium, sometimes called a **phase equilibrium**.

Consider putting a liquid in an empty box and sealing the box. Some of the liquid molecules have more kinetic energy than the others and they escape from the liquid and collect in the empty space above it. These molecules move around and collide with each other, with the walls of the container and with the surface of the liquid. When a molecule collides with the liquid surface, it tends to rejoin the liquid, as during the impact its kinetic energy is transferred to the molecules in the liquid. This means that the molecule no longer possesses enough energy to escape from the liquid again. Figure 3.1.2 shows this situation.

At first, far more molecules leave the liquid than enter it, because there are very few molecules in the vapour phase. As the number of molecules in the vapour phase increases, the rate at which molecules rejoin the liquid increases, because there are more molecules in the vapour phase colliding with the liquid surface. Eventually a point is reached where molecules are leaving the liquid at the same rate as molecules are joining it.

When this happens, we have reached a point where the two opposing processes, evaporation and condensation, are occurring at equal rates, so that they cancel out. We have an equilibrium – but more importantly, this is an example of a **dynamic equilibrium**.

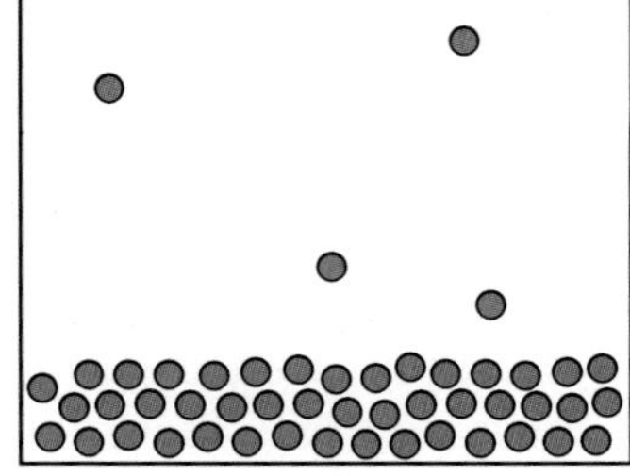

Figure 3.1.2 A liquid has just been poured into a box. Molecules may leave the liquid if they have sufficient energy, while molecules may leave the vapour and rejoin the liquid, giving up some of their energy to the other molecules in the liquid as they do so.

Dynamic equilibrium, changes in conditions and Le Chatelier

What happens when the sealed container in our example is heated? As the temperature of the liquid in the container rises, its molecules have more energy and so more of them leave the liquid. (How the rise in temperature causes the

Static and dynamic equilibrium

The ruler in figure 1.8.14 (page 113) is an example of a **static equilibrium**. It is at rest, in either an unstable or a stable state. In both cases, all physical processes that may disturb the ruler have ceased.

Our liquid–vapour equilibrium is very different. The system appears to have come to rest to an external observer, because the observer is large compared with the individual molecules, and sees simply a constant amount of vapour and liquid. However, the molecules in the system are still moving between the two phases, which is why this is described as a **dynamic equilibrium**.

One way of describing dynamic equilibrium is to say that it is characterised by constant **macroscopic** properties (those properties that an external observer can see and measure), while at the same time, **microscopic** processes (processes on a molecular scale) continue to occur, as shown in figure 3.1.3.

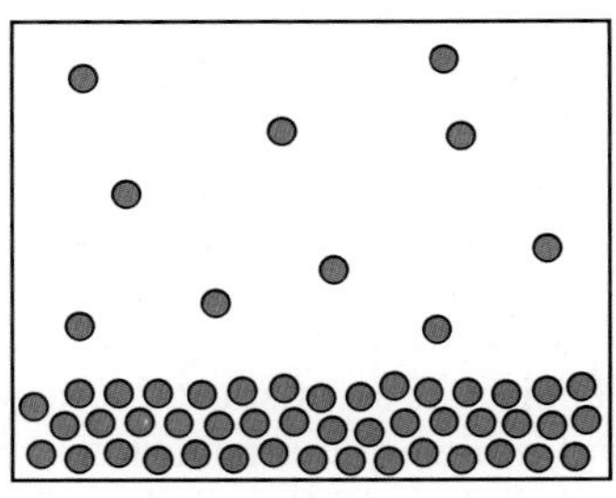

Figure 3.1.3 Although there appears to be no change occurring in a system in dynamic equilibrium, at a molecular level two opposing processes are going on, and these exactly cancel out.

increase in kinetic energy, and how the energy is distributed through the molecules, is described in section 3.6.) The system is no longer in equilibrium, since more molecules are now leaving the liquid than are entering it. Eventually, however, there will be a large enough number of molecules in the vapour for the rate of condensation to equal the rate of evaporation, and equilibrium will be established again.

This response of the liquid–vapour equilibrium to a change in conditions is an example of how equilibria in general respond to disturbances. When a system that is in dynamic equilibrium is upset, it responds in such a way as to return it to equilibrium again. The tendency of systems to behave in this way was noted by the French chemist, Henri Le Chatelier, who in 1888 proposed the principle which now bears his name:

Whenever a system which is in dynamic equilibrium is disturbed, it tends to respond in such a way as to oppose the disturbance and so restore equilibrium.

How does Le Chatelier's principle apply to the liquid–vapour equilibrium? To understand this, we need to note that in order to raise the temperature of the contents of the sealed vessel, we must transfer energy to them. We can represent this process as:

$$\textbf{Energy + liquid} \rightleftharpoons \textbf{vapour}$$

in which the double arrows $\rightleftharpoons$ represent the dynamic equilibrium of molecules moving from the liquid to the vapour and vice versa. Transferring energy to the system tends to make the equilibrium shift to the right, so that molecules leave the liquid and go into the vapour. This process of evaporation requires energy – so the system is behaving in such a way as to oppose the disturbance to it (the increase in energy). We shall see many examples of disturbances to dynamic equilibria during the course of section 3.

An equilibrium like this can only be established in a **closed system** – one in which matter is prevented from leaving or entering the system – since any change in the amount of matter in the system will disturb the equilibrium, as shown in figure 3.1.4.

Equilibria in chemical changes

Reversible reactions

A dynamic equilibrium exists when two opposing processes occur at equal rates. An equilibrium of this type can be established in chemical processes too, as the following example shows.

The gas responsible for the brown haze seen in heavily polluted air is the gas nitrogen dioxide, NO_2, formed by oxidation of the nitrogen oxide, NO, emitted by car engines. Nitrogen dioxide is in equilibrium with dinitrogen tetroxide, N_2O_4:

$$\mathbf{2NO_2(g) \rightleftharpoons N_2O_4(g)}$$

If we set up an investigation in which we seal 0.0800 mol of nitrogen dioxide in a container and keep it at a constant 25 °C, we can measure the amount of nitrogen dioxide and dinitrogen tetroxide in the container spectroscopically, since the gases absorb light at different frequencies (NO_2 is dark brown while N_2O_4 is pale yellow). Eventually we find that the amount of nitrogen dioxide drops to a steady 0.0132 mol, while at the same time, the amount of dinitrogen tetroxide rises to a steady 0.0334 mol. We can represent what has happened as shown in the left-hand side of figure 3.1.5.

Suppose that we now take a container of the same size and put 0.0400 mol of dinitrogen tetroxide in it. (This is exactly the amount of dinitrogen tetroxide that 0.0800 mol of nitrogen dioxide would produce if it completely reacted to form dinitrogen tetroxide.) If we keep this container at 25 °C, we find that once again we end up with 0.0132 mol of nitrogen dioxide and 0.0334 mol of dinitrogen tetroxide when a steady state has been reached. The right-hand side of figure 3.1.5 represents this situation.

Notice how the final situation is the same whether we start with pure nitrogen dioxide or pure dinitrogen tetroxide. In both cases, the amount of the starting substance (NO_2 or N_2O_4) falls and the amount of finishing substance (N_2O_4 or NO_2) rises until equilibrium is attained, when the rate at which nitrogen dioxide reacts to form dinitrogen tetroxide is exactly equalled by the rate at which dinitrogen tetroxide reacts to form nitrogen dioxide.

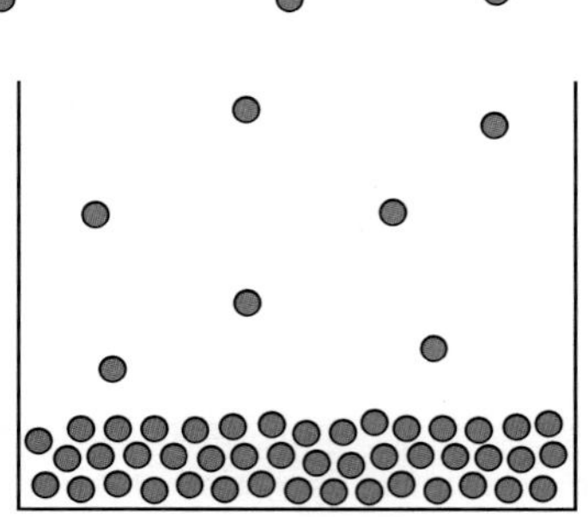

Figure 3.1.4 If the lid is removed from our box containing the liquid–vapour equilibrium, molecules of the vapour will escape. After a while, all the liquid will have evaporated, and no equilibrium will exist – equilibrium can only occur in a closed system.

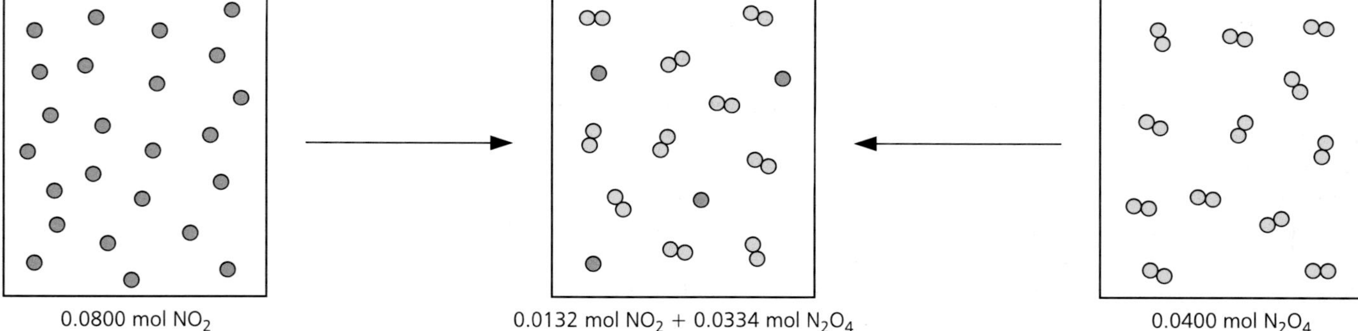

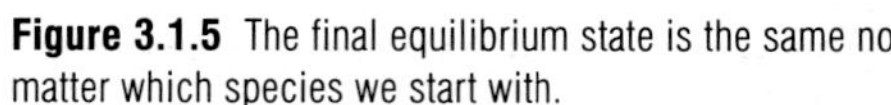

Figure 3.1.5 The final equilibrium state is the same no matter which species we start with.

This example shows how a dynamic equilibrium is established in a chemical process. As the reaction proceeds, the concentrations of the reactants and products can be plotted as in figure 3.1.6. These concentrations change fairly rapidly at first, and then reach steady values. At this point, the reactants are being converted into products at exactly the same rate as the products are being converted into reactants, and we have a **steady state**. Another way of looking at this is to say that at equilibrium, the rate of the forward reaction and the rate of the backward reaction are equal.

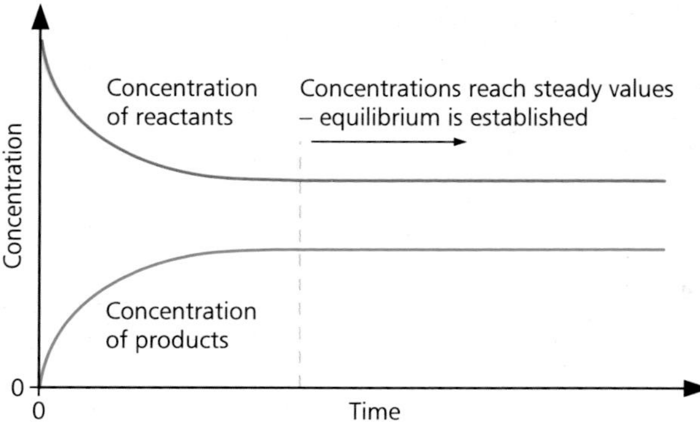

Figure 3.1.6 As equilibrium is reached, the concentrations of reactants and products in a chemical process reach steady values.

Almost all chemical processes eventually reach a state of dynamic equilibrium like this, although we cannot always see that this is the case. For example, in water vapour at room temperature the equilibrium:

$$2H_2O(g) \rightleftharpoons 2H_2(g) + O_2(g)$$

exists, although water molecules are extremely stable so we cannot easily detect any hydrogen or oxygen present – effectively, this reaction is one that 'doesn't go'. Similarly, the equilibrium:

$$HNO_3(aq) \rightleftharpoons H^+(aq) + NO_3^-(aq)$$

exists when nitric acid dissolves in water, although in practice the acid is completely ionised and no HNO_3 molecules can be detected – this reaction 'goes to completion'.

3

The equilibrium law

Concentrations of products and reactants

When a chemical system is in equilibrium, there is a simple relationship between the molar concentrations of the products and reactants. This relationship can be demonstrated if we look at the equilibrium in which bromine chloride is formed from bromine and chlorine in tetrachloromethane (CCl_4) solution:

$$Br_2(CCl_4) + Cl_2(CCl_4) \rightleftharpoons 2BrCl(CCl_4)$$

Several experiments are set up in which the initial (starting) amounts of bromine, chlorine and bromine chloride are varied, while the temperature in each case is kept constant (in this case, at 298 K). The results shown in table 3.1.1 are obtained.

The concentrations are different in each experiment, and at first sight there appears to be no relationship between them. However, we *can* make sense of these figures using a relationship that comes from the equation for the reaction. For each set of experimental results in table 3.1.1, if we take the

Experiment number	**Initial concentrations**/mol dm^{-3}			**Equilibrium concentrations**/mol dm^{-3}		
	[$Br_2(CCl_4)$]	**[$Cl_2(CCl_4)$]**	**[$BrCl(CCl_4)$]**	**[$Br_2(CCl_4)$]**	**[$Cl_2(CCl_4)$]**	**[$BrCl(CCl_4)$]**
1	0.0662	0.0548	0	0.0322	0.0208	0.0680
2	0.0662	0.0384	0	0.0389	0.0111	0.0546
3	0.0466	0.0384	0	0.0227	0.0145	0.0478
4	0	0	0.0526	0.0114	0.0114	0.0299

Table 3.1.1

equilibrium concentration of bromine chloride (written as $[BrCl(CCl_4)]_{eq}$ – the 'eq' subscript means 'at equilibrium') and square it, and then divide this by the equilibrium concentration of bromine, $[Br_2(CCl_4)]_{eq}$ and the equilibrium concentration of chlorine, $[Cl_2(CCl_4)]_{eq}$, we get a number. This is shown more clearly in table 3.1.2. The quantities in the second column of this table have no units, since the expression has units of $(mol\ dm^{-3})^2$ in the numerator (the top part) and $(mol\ dm^{-3})^2$ in the denominator (the bottom part), so the units simply cancel out.

Within experimental error, the numbers in the last column are constant. Experiments carried out on this equilibrium show that in any mixture of chlorine, bromine and bromine chloride dissolved in tetrachloromethane at a temperature of 298 K, careful measurement of the equilibrium concentrations results in a value for this expression of approximately 6.90.

Experiment number	$\frac{[BrCl(CCl_4)]^2_{eq}}{[Br_2(CCl_4)]_{eq}[Cl_2(CCl_4)]_{eq}}$
1	6.90
2	6.90
3	6.94
4	6.88

Table 3.1.2

Errors and uncertainties

Errors and mistakes are a fact of lofe – although that one was deliberate! All methods of measurement have limits to how precisely they can measure quantities. Whenever we make measurements using instruments there is some error or **uncertainty** in the result. Sometimes the uncertainty may be due to the experimenter (for example, reading a scale wrongly) and sometimes it may be due to the equipment (for example, a wrongly calibrated thermometer). The word 'uncertainty' is used because the problems that arise in making measurements are not always due to mistakes – they may often be due to the limits of the equipment being used, or of the experiment itself.

It is important in any experiment involving measuring physical quantities (mass, temperature, pH and so on) that some attempt is made to estimate the magnitude of these uncertainties, and to show how they are likely to affect the final result.

The equilibrium constant

The expression:

$$\frac{[BrCl(CCl_4)]^2_{eq}}{[Br_2(CCl_4)]_{eq}\ [Cl_2(CCl_4)]_{eq}} = 6.90$$

is called the **equilibrium law** for the reaction between bromine and chlorine to form bromine chloride. The law tells us that whenever we measure the concentrations of bromine, chlorine and bromine chloride in tetrachloromethane in a system at 298 K containing only these substances, the value of the expression must be 6.90 for the substances to be in equilibrium. If the value is not 6.90, then there is no equilibrium, and the substances will react further until an equilibrium is set up. The value 6.90 associated with this equilibrium is the **equilibrium constant**, K_c – the 'c' subscript stands for 'concentration'.

Predicting the equilibrium concentrations

The equilibrium concentrations for a particular reaction can always be predicted from its stoichiometric (balanced) equation. The general equilibrium for substances A, B, C and D in the equation:

$$a\mathbf{A} + b\mathbf{B} \rightleftharpoons c\mathbf{C} + d\mathbf{D}$$

has an equilibrium expression which may be written as:

$$\frac{[C]^c_{eq}\ [D]^d_{eq}}{[A]^a_{eq}\ [B]^b_{eq}} = K_c$$

Once again, K_c is the equilibrium constant for the reaction, which is constant at a given temperature. For convenience, the 'eq' subscript is often omitted, as it is assumed that concentrations in expressions like this are equilibrium concentrations.

Describing the equation

By convention, the concentrations of the *products* of an equilibrium always appear in the numerator of the equilibrium law expression, and the concentrations of the *reactants* always appear in the denominator. This means that we must be careful to specify exactly what reaction we are describing. For example, at 298 K the equilibrium constant K_c is 6.90 for the reaction:

$$\mathbf{Br_2(CCl_4) + Cl_2(CCl_4) \rightleftharpoons 2BrCl(CCl_4)}$$

Knowing this, we can write the equilibrium law for the reaction as:

$$\frac{[\mathrm{BrCl(CCl_4)}]^2}{[\mathrm{Br_2(CCl_4)}][\mathrm{Cl_2(CCl_4)}]} = K_c = 6.90$$

Units and the equilibrium constant

In the case of the bromine, chlorine and bromine chloride reaction, we saw that K_c had no units, as the units on the top and bottom of the expression simply cancelled out. This is not the case for the nitrogen dioxide/dinitrogen tetroxide equilibrium, however, for which the equilibrium expression is:

$$K_c = \frac{[\mathrm{N_2O_4(g)}]}{[\mathrm{NO(g)}]^2}$$

This expression contains the units:

$$\frac{\mathrm{mol\ dm^{-3}}}{(\mathrm{mol\ dm^{-3}})^2} = \frac{\mathrm{mol\ dm^{-3}}}{\mathrm{mol^2\ dm^{-6}}} = \frac{1}{\mathrm{mol\ dm^{-3}}} = \mathrm{dm^3\ mol^{-1}}$$

Using the same arguments, we can show that the units of K_c for the equilibrium:

$$\mathbf{N_2(g) + 3H_2(g) \rightleftharpoons 2NH_3(g)}$$

are $\mathrm{dm^6\ mol^{-2}}$.

Reversing the equation

What would happen if we were considering the dissociation of bromine chloride rather than its formation from bromine and chlorine? In this case, the reaction would be:

$$\mathbf{2BrCl(CCl_4) \rightleftharpoons Br_2(CCl_4) + Cl_2(CCl_4)}$$

and the equilibrium law would be written:

$$\frac{[\mathrm{Br_2(CCl_4)}][\mathrm{Cl_2(CCl_4)}]}{[\mathrm{BrCl(CCl_4)}]^2} = K'_c = \frac{1}{6.90} = 0.15$$

From this, it is clear that:

$$K'_c = \frac{1}{K_c}$$

3

Measuring equilibrium constants

The experimental determination of an equilibrium constant has three basic stages:

(1) Combine known amounts of reactants or products under carefully measured and controlled conditions and allow the reaction to come to equilibrium at a fixed temperature.

(2) Measure accurately the equilibrium concentration of one or more of the substances in the equilibrium mixture, and from this calculate the equilibrium concentrations of all substances in the mixture.

(3) Substitute these equilibrium concentrations in the equilibrium expression.

For accurate work, this procedure is then repeated for different starting concentrations in order to find an average value for the equilibrium constant.

This is how the method works for the reaction of ethene, C_2H_4, with steam to form ethanol, C_2H_5OH, a reaction we shall meet again in section 4.3:

$$C_2H_4(g) + H_2O(g) \rightleftharpoons C_2H_5OH(g)$$

To begin with, 0.1948 mol of ethene was mixed with 0.2136 mol of steam in a container with a volume of 1.00 dm^3. This mixture was allowed to come to equilibrium at constant temperature, when it was found that 0.0112 mol of ethanol had been formed.

The stoichiometric equation tells us that one mole of ethene reacts with one mole of steam to give one mole of ethanol, so we can work out the number of moles of ethene at equilibrium by using the relationship:

Moles of ethene at equilibrium = **moles of ethene at start** – **moles of ethanol formed at equilibrium**

The same process can be used to calculate the number of moles of steam at equilibrium. Table 3.1.3 below shows the concentration of each substance at equilibrium, calculated by dividing the number of moles by the volume of the container, in this case 1.00 dm^3.

These values can now be substituted into the equilibrium expression to calculate K_c:

$$K_c = \frac{[C_2H_5OH(g)]}{[C_2H_4(g)][H_2O(g)]}$$

$$= \frac{0.0112 \text{ mol dm}^{-3}}{0.1836 \text{ mol dm}^{-3} \times 0.2024 \text{ mol dm}^{-3}}$$

$$= 0.30 \text{ dm}^3 \text{ mol}^{-1}$$

K_c for the reaction between ethene and steam to form ethanol at this temperature is 0.30 dm^3 mol^{-1}.

	$C_2H_4(g)$	+	$H_2O(g)$	$\rightleftharpoons$	$C_2H_5OH(g)$
Initial concentration	$\frac{0.1948 \text{ mol}}{1.00 \text{ dm}^3} = 0.1948 \text{ mol dm}^{-3}$		$\frac{0.2136 \text{ mol}}{1.00 \text{ dm}^3} = 0.2136 \text{ mol dm}^{-3}$		0
Final concentration	$\frac{0.1948 \text{ mol} - 0.0112 \text{ mol}}{1.00 \text{ dm}^3}$ $= 0.1836 \text{ mol dm}^{-3}$		$\frac{0.2136 \text{ mol} - 0.0112 \text{ mol}}{1.00 \text{ dm}^3}$ $= 0.2024 \text{ mol dm}^{-3}$		$\frac{0.0112 \text{ mol}}{1.00 \text{ dm}^3} = 0.0112 \text{ mol dm}^{-3}$

Table 3.1.3

Equilibrium constants for reactions involving gases

So far we have used the **concentrations** of gaseous substances to calculate equilibrium constants. For gases, it is often more convenient to use **partial pressures** in the equilibrium expression, since the partial pressure of a gas in a mixture of gases is proportional to its mole fraction, which is a measure of its concentration (see section 1.7, page 99). We can write the equilibrium constant for a reaction involving gases:

$$a\text{A} + b\text{B} \rightleftharpoons c\text{C} + d\text{D}$$

as:

$$K_p = \frac{p_C{}^c\, p_D{}^d}{p_A{}^a\, p_B{}^b}$$

where p_A is the partial pressure of A in the mixture, and so on.

When calculating K_p it is important to express pressures in the SI unit of pressure, the pascal, *not* in non-standard units such as millimetres of mercury or atmospheres.

The relationship between K_c and K_p

The equation of state for an ideal gas states that:

$$pV = nRT$$

which can be rewritten as:

$$p = \frac{n}{V}RT$$

Now n/V is effectively the concentration of the gas, as it is the number of moles of the gas divided by the volume of the container it is in. For a gas A in a container of volume V, we can therefore write:

$$p_A = [A]RT$$

We can apply this to the equilibrium:

$$N_2O_4(g) \rightleftharpoons 2NO_2(g)$$

where:

$$K_c = \frac{[NO_2(g)]^2}{[N_2O_4(g)]}$$

and:

$$K_p = \frac{p_{NO_2}{}^2}{p_{N_2O_4}}$$

Since:

$$p_{NO_2} = [NO_2]RT \quad \text{and} \quad p_{N_2O_4} = [N_2O_4]RT$$

it follows that:

$$K_p = \frac{p_{NO_2}{}^2}{p_{N_2O_4}} = \frac{([NO_2]RT)^2}{[N_2O_4]RT} = \frac{[NO_2]^2}{[N_2O_4]}RT = K_cRT$$

We can generalise this expression to say that, if:

$$\Delta n = \begin{matrix}\textbf{number of moles}\\ \textbf{on right-hand side}\\ \textbf{of equation}\end{matrix} - \begin{matrix}\textbf{number of moles}\\ \textbf{on left-hand side}\\ \textbf{of equation}\end{matrix}$$

then the relationship between K_p and K_c is:

$$K_p = K_c(RT)^{\Delta n}$$

This shows that when $\Delta n = 0$, $K_p = K_c$. Note that for this equation, *all* of the quantities must be in SI units – that is, pressures in pascals, temperatures in kelvins and concentrations in moles per cubic metre, *not* moles per cubic decimetre.

What does the equilibrium constant tell us?

The equilibrium constant for a reaction, whether we choose to use K_c or K_p, is a measure of *how far* a reaction proceeds to completion. The reaction:

$$2H_2(g) + O_2(g) \rightleftharpoons 2H_2O(g)$$

has $K_c = 9.1 \times 10^{80}\ dm^3\ mol^{-1}$ at 298 K, so at equilibrium:

$$K_c = \frac{[H_2O(g)]^2}{[H_2(g)]^2[O_2(g)]} = \frac{9.1 \times 10^{80}\ dm^3\ mol^{-1}}{1}$$

The fact that the numerator in the expression is very much bigger than the denominator shows that the concentration of water at equilibrium is huge compared with the concentrations of hydrogen and oxygen. This means that at equilibrium, all of the hydrogen and oxygen in the system is to be found combined as water molecules – so the reaction 'goes to completion'.

For the equilibrium:

$$N_2(g) + O_2(g) \rightleftharpoons 2NO(g)$$

K_c at 298 K is 4.8×10^{-31}. In this case, at equilibrium we have:

$$K_c = \frac{[NO(g)]^2}{[N_2(g)][O_2(g)]} = \frac{4.8}{1 \times 10^{31}} \quad \left(\text{because } 10^{-31} = \frac{1}{10^{31}}\right)$$

The denominator is many, many times larger than the numerator, so the concentration of nitrogen oxide must be minute compared with the concentrations of nitrogen and oxygen. Fortunately for us, the reaction between nitrogen and oxygen 'doesn't go', at least not at 298 K! However, at the high temperatures inside the cylinders of a car engine, nitrogen oxide does form, since the equilibrium constant increases with temperature – we shall look at how temperature changes affect equilibria in more detail later in this section.

The relationship between the equilibrium constant and the extent to which a reaction proceeds to completion is summarised in table 3.1.4. The examples of equilibria given are at varying temperatures and standard pressure.

Value of equilibrium constant	Extent of reaction	Example
$\sim 10^{59}$ mol^{-1} dm^3 (at 570 K)	Reaction goes virtually to completion.	$2NO(g) + 2CO(g) \rightleftharpoons N_2(g) + 2CO_2(g)$
Around 1 (at 470 K)	Concentrations of reactants and products at equilibrium are nearly the same.	$PCl_3(g) + Cl_2(g) \rightleftharpoons PCl_5(g)$
$\sim 10^{-17}$ (at 300 K)	Reaction virtually does not go.	$2HCl(g) \rightleftharpoons H_2(g) + Cl_2(g)$

Table 3.1.4

The term **position of equilibrium** is often used to describe reactions. When a reaction has a large equilibrium constant, the position of the equilibrium is said to lie to the right-hand side of the equation, while for a reaction with a small equilibrium constant, the equilibrium lies well to the left.

Finally, note that the equilibrium constant tells us nothing about the *rate* of a reaction. Even if the equilibrium constant for a reaction is very large, the rate at which it happens may be extremely slow. The reaction of hydrogen and oxygen to form water is a good example of this. Even though K for the reaction is around 10^{80}, a mixture of hydrogen and oxygen at room temperature does not spontaneously react to form water – it is **kinetically stable**. We shall explore the reasons behind this in section 3.6.

Heterogeneous equilibria

Reactions in more than one phase

All the chemical equilibria that we have considered so far involve only one phase. This is the case for many reactions, where all the reactants and products are gases, or where all the substances concerned are in solution.

Equilibria involving more than one phase are called **heterogeneous equilibria**. Examples of this type of equilibrium include the liquid–vapour equilibrium that we met at the beginning of this section, and equilibria like:

$$CaO(s) + SO_2(g) \rightleftharpoons CaSO_3(s)$$

This reaction can reach equilibrium in just the same way as the other reactions we have studied, provided it is kept under constant conditions in a sealed container (to prevent any sulphur dioxide escaping). For the equilibrium constant we can write:

$$K = \frac{[CaSO_3(s)]}{[CaO(s)][SO_2(g)]}$$

However, this expression can be simplified. The expression contains two terms, $[CaSO_3(s)]$ and $[CaO(s)]$, each of which is the concentration of a solid. If we had a crystal containing one mole of calcium oxide, we know from the relative molecular mass and density of calcium oxide that the volume of the crystal would be 16.75 cm^3. Two moles of calcium oxide would have a volume of $(2 \times 16.75\ cm^3) = 33.5\ cm^3$, while four moles would occupy 67.0 cm^3. In each case, the ratio of moles to volume is the same, 59.7 mol dm^{-3}, so the concentration of calcium oxide is constant, no matter how much of it there is.

The same is true for calcium sulphite too, so the equilibrium expression contains two terms that are constants. To simplify the expression, these constants are combined with the other numerical constant in the expression, the equilibrium constant:

$$K_c = K\frac{[CaO(s)]}{[CaSO_3(s)]} = \frac{1}{[SO_2(g)]}$$

By using exactly the same arguments, we can show that the same is true for liquids. In general,

The equilibrium expression for a heterogeneous reaction is written in such a way that the concentrations of solids and liquids are included in the equilibrium constant.

Heterogeneous equilibria – worked example

The solubility of the silver halides in water decreases in the order AgCl > AgBr > AgI. Because of this difference in solubility, bromide ions will displace chloride ions from solid silver chloride. The equilibrium constant K_c for the reaction:

$$AgCl(s) + Br^-(aq) \rightleftharpoons AgBr(s) + Cl^-(aq)$$

is 360 at 298 K. If 0.100 mol dm^{-3} $Br^-(aq)$ is added to solid AgCl, what will be the equilibrium concentrations of $Br^-(aq)$ and $Cl^-(aq)$?

We know that when 1 mol of Br^- reacts, 1 mol of Cl^- is formed. This means that starting with 1 mol of Br^-, the amount of Br^- at equilibrium will be $(1 - x)$, where x is the amount of Cl^- formed. If we take a solution containing 0.100 mol dm^{-3} of Br^- ions and add excess solid silver chloride to it, then the equilibrium concentrations of ions can be calculated as shown in table 3.1.5.

	$AgCl(s) + Br^-(aq)$	$\rightleftharpoons$	$AgBr(s) + Cl^-(aq)$
Concentration of ions initially/mol dm^{-3}	0.100		0
Concentration of ions at equilibrium/mol dm^{-3}	$0.100 - x$		x

Table 3.1.5

Now:

$$K_c = \frac{[Cl^-(aq)]}{[Br^-(aq)]}$$

Substituting the value of K_c and the expressions for the ion concentrations gives:

$$360 = \frac{x}{(0.100 - x)}$$

Rearranging:

$$360\ (0.100 - x) = x$$
$$36 - 360x = x$$

so:

$$x = \frac{36}{361}$$
$$= 0.099\ 723\ \text{mol dm}^{-3}$$

The concentration of Cl^- ions in the equilibrium mixture is 0.099 723 mol dm^{-3}, while the concentration of Br^- ions is (0.100 000 – 0.099 723) mol dm^{-3} = 0.000 277 mol dm^{-3}.
Check: The ratio of chloride to bromide ions should be approximately 360:1:

$$\frac{[Cl^-(aq)]}{[Br^-(aq)]} = \frac{0.099\ 723\ \text{mol dm}^{-3}}{0.000\ 277\ \text{mol dm}^{-3}} = \frac{360}{1}$$

Changes in chemical equilibria

At the beginning of this section we saw that Le Chatelier's principle could be used to predict the effect of a change in conditions on a liquid–vapour system in equilibrium. The principle can be used equally well to make predictions about chemical systems, as we shall see now.

(1) Changes in reacting quantities

Adding or removing a substance from a chemical reaction in equilibrium changes the concentrations of the substances in the equilibrium. As a result, the value of the equilibrium expression is no longer equal to the equilibrium constant, and the concentrations must be allowed to change again in order for equilibrium to be restored. Le Chatelier's principle allows us to predict whether this change will occur through the reaction proceeding to the left or to the right. As an example, consider the equilibrium:

$$[Cu(H_2O)_6]^{2+}(aq) + 4Cl^-(aq) \rightleftharpoons CuCl_4^{2-}(aq) + 6H_2O(l)$$

The $[Cu(H_2O)_6]^{2+}$ ion is blue, and the $CuCl_4^{2-}$ ion is yellow, so a mixture of the two appears green. (These colours can be seen in figure 3.1.8, page 251.) Adding concentrated hydrochloric acid to the solution adds chloride ions, which means that the system is no longer in equilibrium. To restore the equilibrium, the concentration of Cl^- must fall, which can be achieved by the formation of more of the $CuCl_4^{2-}$ ion – in other words, the reaction moves to the right. In the same way, addition of water to the reaction mixture moves the equilibrium to the left.

The stability constant for the $CuCl_4^{2-}$ ion

When the $CuCl_4^{2-}(aq)$ ion is formed in aqueous solution, we can write the equilibrium constant for the reaction as:

$$K_{stab} = \frac{[CuCl_4^{2-}(aq)]}{[[Cu(H_2O)_6]^{2+}(aq)][Cl^-(aq)]^4}$$

This equilibrium constant is called the **stability constant** for the $CuCl_4^{2-}(aq)$ ion, and has the value $4.0 \times 10^5\ dm^{12}\ mol^{-4}$ at 298 K.

Think about a test tube containing $[Cu(H_2O)_6]^{2+}(aq)$ ions, to which concentrated hydrochloric acid is added. Once the equilibrium has settled down, the concentrations of the ions in the solution are measured. (This can be done by measuring the amount of light absorbed at the maximum of the absorption spectrum for each ion. The maxima of the absorption spectra of $[Cu(H_2O)_6]^{2+}(aq)$ and $CuCl_4^{2-}(aq)$ occur at different frequencies, which is what gives their solutions different colours.) The results obtained might be as follows:

$$[[Cu(H_2O)_6]^{2+}(aq)] = 1.00 \times 10^{-4}\ mol\ dm^{-3}$$

$$[Cl^-(aq)] = 0.22\ mol\ dm^{-3}$$

$$[CuCl_4^{2-}(aq)] = 0.094\ mol\ dm^{-3}$$

Substituting these values into the equilibrium expression shows that the system is in equilibrium:

$$\frac{[CuCl_4^{2-}(aq)]}{[[Cu(H_2O)_6]^{2+}(aq)][Cl^-(aq)]^4} = \frac{0.094\ mol\ dm^{-3}}{1.00 \times 10^{-4}\ mol\ dm^{-3} \times (0.22\ mol\ dm^{-3})^4}$$

$$= 4.0 \times 10^5\ dm^{12}\ mol^{-4}$$

$$= K_{stab}$$

Suppose some Ag^+ ions are added. These react with the Cl^- ions, producing a precipitate of AgCl. The concentration of chloride ions falls instantly to $0.10\ mol\ dm^{-3}$. This means that the system is no longer in equilibrium:

$$\frac{[CuCl_4^{2-}(aq)]}{[[Cu(H_2O)_6]^{2+}(aq)][Cl^-(aq)]^4} = \frac{0.094\ mol\ dm^{-3}}{1.00 \times 10^{-4}\ mol\ dm^{-3} \times (0.10\ mol\ dm^{-3})^4}$$

$$= 9.4 \times 10^6\ dm^{12}\ mol^{-4}$$

$$> K_{stab}$$

In order for the system to return to equilibrium, some $CuCl_4^{2-}(aq)$ must react to form $[Cu(H_2O)_6]^{2+}(aq)$. Noting that $[Cl^-(aq)]$ will remain at approximately $0.1\ mol\ dm^{-3}$ because of the presence of Ag^+ ions, we can write the new concentrations at equilibrium as:

$$[[Cu(H_2O)_6]^{2+}(aq)] = (1.00 \times 10^{-4} + x)\ mol\ dm^{-3}$$

$$[Cl^-(aq)] = 0.10\ mol\ dm^{-3}$$

$$[CuCl_4^{2-}(aq)] = (0.094 - x)\ mol\ dm^{-3}$$

Substituting these values into the equilibrium expression gives:

3

$$4.0 \times 10^5\ \text{dm}^{12}\ \text{mol}^{-4} = \frac{(0.094\ \text{mol dm}^{-3} - x)}{(1.00 \times 10^{-4}\ \text{mol dm}^{-3} + x) \times (0.10\ \text{mol dm}^{-3})^4}$$

$$= \frac{(0.094\ \text{mol dm}^{-3} - x)}{1.00 \times 10^{-8}\ \text{mol}^5\ \text{dm}^{-15} + 1.00 \times 10^{-4}\ \text{mol}^4\ \text{dm}^{-12} \times x}$$

Rearranging this equation and cancelling the units gives:

$$4.0 \times 10^5 \times (1.00 \times 10^{-8}\ \text{mol dm}^{-3} + (1.00 \times 10^{-4})x) = (0.094\ \text{mol dm}^{-3} - x)$$

or:

$$40x + 4.0 \times 10^{-3}\ \text{mol dm}^{-3} = 0.094\ \text{mol dm}^{-3} - x$$

which simplifies to:

$$x = \frac{0.090\ \text{mol dm}^{-3}}{41}$$

$$= 0.0022\ \text{mol dm}^{-3}$$

This may now be used to calculate the new equilibrium concentrations of complex copper ions:

$$[[Cu(H_2O)_6]^{2+}(aq)] = (1.00 \times 10^{-4} + x)\ \text{mol dm}^{-3}$$
$$= (0.0001 + 0.0022)\ \text{mol dm}^{-3}$$
$$= 0.0023\ \text{mol dm}^{-3}$$

$$[CuCl_4^{2-}(aq)] = (0.094 - x)\ \text{mol dm}^{-3} = (0.094 - 0.0022)\ \text{mol dm}^{-3}$$
$$= 0.0918\ \text{mol dm}^{-3}$$

Finally, a quick check confirms that the system is now in equilibrium again:

$$\frac{[CuCl_4^{2-}(aq)]}{[[Cu(H_2O)_6^{2+}](aq)][Cl^-(aq)]^4} = \frac{0.0918\ \text{mol dm}^{-3}}{0.0023\ \text{mol dm}^{-3} \times (0.10\ \text{mol dm}^{-3})^4}$$
$$= 4.0 \times 10^5\ \text{dm}^{12}\ \text{mol}^{-4}$$
$$= K_{stab}$$

For chemical equilibria in general:

- If a substance is added, the reaction tends to move in the direction that will consume the substance added.
- If a substance is removed, the reaction tends to move in the direction that will replace the substance removed.

(2) Changing the volume in reactions involving gases

Changing the volume of a system of gases in equilibrium changes the pressure. The effect of this can also be predicted using Le Chatelier's principle. The effect varies, according to whether the total number of moles and therefore the volume of the products differs from that of the reactants or not. (Remember that one mole of any gas at s.t.p. occupies the same volume, 22.4 dm^3.)

In the equilibrium:

$$N_2(g) + 3H_2(g) \rightleftharpoons 2NH_3(g)$$

there is a decrease in the number of moles of gas as we go from left to right. Decreasing the volume of the reaction mixture increases the pressure of the system. This causes the reaction to move to the right, because this will tend to oppose the change in pressure by reducing the volume of gas present (for every two moles of ammonia produced, four moles of gas on the left-hand side must react together).

In the case of the equilibrium:

$$N_2(g) + O_2(g) \rightleftharpoons 2NO(g)$$

changes of volume have no effect, however. This is because there are the same number of moles of gas (two) on each side of the equation, so the reaction cannot oppose the change, and so does not respond to changes in volume.

For chemical equilibria in general:

- For a decrease in volume or increase in pressure, the reaction tends to move in the direction that will reduce the number of moles of gas.
- For an increase in volume or decrease in pressure, the reaction tends to move in the direction that will increase the number of moles of gas.

Changes of volume and pressure and equilibria

If the total pressure P of a mixture of reacting gases in equilibrium is changed, the partial pressures of the gases change, and this may cause the mixture to be no longer in equilibrium. Consider the following two examples.

Example 1

$$CH_4(g) + H_2O(g) \rightleftharpoons CO(g) + 3H_2(g)$$

For this reaction:

$$K_p = \frac{p_{CO}\,p_{H_2}{}^3}{p_{CH_4}\,p_{H_2O}}$$

If we let $p_{CO} = w$, $p_{H_2} = x$, $p_{CH_4} = y$ and $p_{H_2O} = z$, then:

$$K_p = \frac{wx^3}{yz}$$

If the pressure of the reaction mixture now doubles, the partial pressures of all the gases will also double, so the equilibrium expression becomes:

$$\frac{2w(2x)^3}{2y\ 2z} = \frac{16wx^3}{4yz} = 4K_p$$

As this shows, the ratio of the partial pressures is now increased four-fold – the reaction must therefore move to the left (decreasing w and x, while increasing y and z) so restoring the ratio to its previous value. This reaction is used to produce hydrogen from natural gas as a feedstock in the production of ammonia – see section 3.6, page 330.

Example 2

$$N_2(g) + O_2(g) \rightleftharpoons 2NO(g)$$

For this reaction,

$$K_p = \frac{p_{NO}{}^2}{p_{N_2}\,p_{O_2}}$$

If we let $p_{NO} = x$, $p_{N_2} = y$ and $p_{O_2} = z$, then:

$$K_p = \frac{x^2}{yz}$$

Once again, doubling the overall pressure will double the individual partial pressures, so the equilibrium expression becomes:

$$\frac{(2x)^2}{(2y)(2z)} = \frac{x^2}{yz} = K_p$$

In this case, changing the pressure has no effect on the reaction.

(3) Catalysts

A catalyst affects the rate of a reaction. A catalyst has the same effect on both the forward and the reverse reactions in an equilibrium. This means that there is no overall effect on the composition of the equilibrium mixture, and the only result of the presence of a catalyst is that the system reaches equilibrium more quickly.

(4) Changing the temperature

The value of the equilibrium constant K_c or K_p for a particular reaction is unaffected by changes in the concentrations of reactants and products, changes in volume or pressure, or the presence of catalysts. However, temperature changes *do* affect the equilibrium constant, as figure 3.1.7 shows.

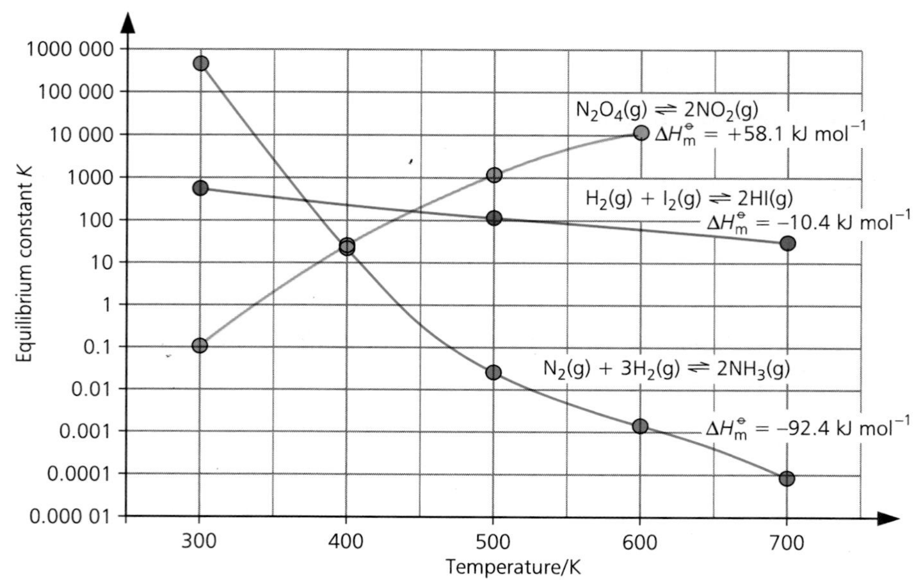

Figure 3.1.7 The effect of changing the temperature on the equilibrium constants for three different reactions

The effect of a temperature change on an equilibrium depends on the energy changes taking place in the chemical reaction concerned. Notice in figure 3.1.7 that for reactions that are exothermic in the forward direction, the equilibrium constants *decrease* with increasing temperature, while for the reaction that is endothermic in the forward direction, the equilibrium constant *increases* with increasing temperature.

The effect of temperature on a reaction is exactly as Le Chatelier's principle predicts. For example, the reaction:

$$\mathbf{CO(g) + 2H_2(g) \rightleftharpoons CH_3OH}$$

has a standard molar enthalpy change of −18 kJ mol^{-1}. If we remember that to increase the temperature of a system we must transfer energy to it, we can rewrite the equation for the reaction as:

$$\mathbf{CO(g) + 2H_2(g) \rightleftharpoons CH_3OH + energy}$$

and the effect of increasing the temperature becomes obvious. The reaction must move to the left, absorbing some of the added energy and so opposing the change. In the same way, the reaction:

$$\mathbf{N_2O(g) + NO_2(g) \rightleftharpoons 3NO(g)}$$

has a standard molar enthalpy change of +156 kJ mol^{-1}. Rewriting the equation as:

$$\mathbf{Energy + N_2O(g) + NO_2(g) \rightleftharpoons 3NO(g)}$$

shows that the reaction will respond to an increase in temperature by moving to the right, absorbing some of the added energy.

For chemical equilibria in general:

- If the temperature is raised, the equilibrium position of an exothermic reaction tends to move to the left (that is, *K decreases*). The equilibrium position of an endothermic reaction tends to move to the right (that is, *K increases*).

- If the temperature is lowered, the equilibrium position of an exothermic reaction tends to move to the right (that is, *K increases*). The equilibrium position of an endothermic reaction tends to move to the left (that is, *K decreases*).

The equilibrium we met on page 246,

$$[Cu(H_2O)_6]^{2+}(aq) + 4Cl^-(aq) \rightleftharpoons CuCl_4^{2-}(aq) + 6H_2O(l)$$

can be used to illustrate the effect of temperature changes on reactions, as figure 3.1.8 shows.

We shall see how these principles of equilibrium are applied in industrial chemistry in section 3.6.

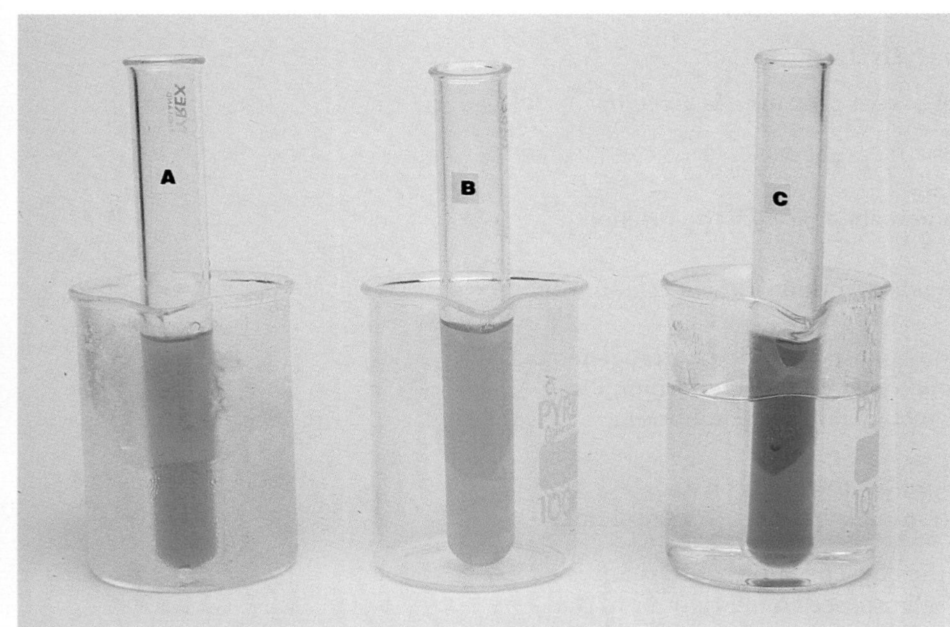

Figure 3.1.8 The tube in the middle is at room temperature, and contains a mixture of $[Cu(H_2O)_6]^{2+}$ and $CuCl_4^{2-}$ ions. Cooling the reaction mixture (left) shifts the equilibrium to the left, producing more of the blue $[Cu(H_2O)_6]^{2+}$ ions. Raising the temperature (right) shifts the equilibrium to the right, producing more of the $CuCl_4^{2-}$ ions. These changes in the equilibrium constant show that the reaction is endothermic in the forward direction.

SUMMARY

- A system in **dynamic equilibrium** shows constant **macroscopic properties** (properties an observer can see and measure).
- In a system in dynamic equilibrium, **microscopic processes** (processes happening at a molecular level) continue to occur. At equilibrium, microscopic processes occur at the same rate in the forward and backward directions, hence the system remains in a steady state.
- A dynamic equilibrium involving two states of matter in equilibrium with each other (for example, liquid water in equilibrium with ice) is called a **phase equilibrium**.
- Le Chatelier's principle states that whenever a system in dynamic equilibrium is disturbed, it tends to respond in such a way as to oppose the disturbance and so restore equilibrium.

- When a chemical system is in equilibrium, a simple relationship known as the **equilibrium law** exists between the molar concentrations of the products and reactants. For the reaction:

$$a\text{A} + b\text{B} \rightleftharpoons c\text{C} + d\text{D}$$

the equilibrium law may be expressed as:

$$K_c = \frac{[\text{C}]^c[\text{D}]^d}{[\text{A}]^a[\text{B}]^b}$$

K_c is the **equilibrium constant** for the reaction expressed in terms of concentrations. The value of K_c is constant at a given temperature.

- For reaction mixtures involving only gases, the **partial pressures** of the gases in the equilibrium mixture may be used in the equilibrium expression. For the equilibrium:

$$N_2O_4(g) \rightleftharpoons 2NO_2(g)$$

K_p, the equilibrium constant in terms of partial pressures, is given by:

$$K_p = \frac{{p_{NO_2}}^2}{p_{N_2O_4}}$$

- The equilibrium constant gives information about *how far* the reaction goes.
- The equilibrium constant for a reaction tells us nothing about *how fast* the reaction reaches equilibrium.
- Equilibria in which there is more than one phase are called **heterogeneous equilibria**. In the equilibrium law for a heterogeneous equilibrium, the concentrations of solids and liquids are included in the equilibrium constant.
- Le Chatelier's principle can be used to predict the effect on an equilibrium of changing the concentrations of substances or the volume:

  Changes in reacting quantities change the concentrations of substances in the equilibrium mixture. The reaction will proceed to the right or to the left in order to restore equilibrium.

  Changes in the volume of a reaction involving gases change the pressure of the equilibrium system. The reaction will proceed to the right or to the left in order to restore equilibrium.

- A catalyst affects the rate of the forward and backward reactions equally, so does not affect the position of equilibrium.
- Changes in temperature change the equilibrium constant for a given reaction. The change can be predicted using Le Chatelier's principle:

  If the forward reaction is exothermic, the equilibrium constant decreases with increasing temperature.

  If the forward reaction is endothermic, the equilibrium constant increases with increasing temperature.

QUESTIONS

1 Sketch a graph showing how the concentration of reactants and products varies with time during the course of a typical chemical reaction:

$$A \rightarrow B$$

a assuming no B is present at the start of the reaction

b assuming no A is present at the start of the reaction.

2 Write down the equilibrium law for each of the following reactions:

a $Ag^+(aq) + 2NH_3(aq) \rightleftharpoons [Ag(NH_3)_2]^+(aq)$

b $CH_3COOH(aq) + CH_3OH(aq) \rightleftharpoons CH_3COOCH_3(aq) + H_2O(l)$

c $Zn(s) + Cu^{2+}(aq) \rightleftharpoons Zn^{2+}(aq) + Cu(s)$

d $CH_4(g) + H_2O(g) \rightleftharpoons CH_3OH(g) + H_2(g)$

3 Methanol, a potential replacement for petrol as a motor fuel, can be made using the equilibrium:

$$CO(g) + 2H_2(g) \rightleftharpoons CH_3OH(g) \quad \Delta H^{\ominus}_m = -92 \text{ kJ mol}^{-1}$$

At 500 K, the value of the equilibrium constant K_p is 6.25×10^{-3}.

a In what units might K_p be measured?

How would the equilibrium be affected by the following changes:

b adding hydrogen

c removing carbon monoxide

d adding a catalyst

e increasing the temperature

f decreasing the pressure?

g Under what conditions might an industrial plant manufacturing methanol using this reaction operate?

4 The density of graphite is 2.25 g cm^{-3} while that of diamond is 3.51 g cm^{-3}. The standard molar enthalpy change for the reaction

$$C(\text{graphite}) \rightleftharpoons C(\text{diamond})$$

is +1.88 kJ mol^{-1}. What conditions of temperature and pressure favour the formation of diamond from graphite? Explain your answer.

5 A solution containing a mixture of ethanoic acid, ethanol, ethyl ethanoate and water was allowed to come to equilibrium in a stoppered bottle at 373 K, when the amount of each of the four substances was measured. These were found to be as in table 3.1.6.

Substance	**Amount**/mol
Ethanoic acid (CH_3COOH)	0.09
Ethanol (CH_3CH_2OH)	0.21
Ethyl ethanoate ($CH_3COOCH_2CH_3$)	0.18
Water (H_2O)	0.42

Table 3.1.6

The equation for the reaction is:

$$CH_3COOH(aq) + CH_3CH_2OH(aq) \rightleftharpoons CH_3COOCH_2CH_3(aq) + H_2O(l)$$

a Write down the equilibrium law for the reaction.

b Calculate the equilibrium constant for the reaction at this temperature.

c Calculate the equilibrium constant for the reaction:

$$CH_3COOCH_2CH_3(aq) + H_2O(l) \rightleftharpoons CH_3COOH(aq) + CH_3CH_2OH(aq)$$

d 0.20 mol of ethyl ethanoate is mixed with 0.60 mol of water in a stoppered bottle. The mixture is kept at 373 K until equilibrium is reached. How many moles of ethanoic acid are present in this mixture?

6 At a certain temperature, K_p = 20 kPa^{-1} for the reaction:

$$PCl_3(g) + Cl_2(g) \rightleftharpoons PCl_5(g)$$

A sealed vessel at this temperature initially contains these three gases with partial pressures as follows:

p_{PCl_3} = 0.82 kPa $\quad p_{Cl_2}$ = 0.29 kPa $\quad p_{PCl_5}$ = 3.80 kPa

a Write down the equilibrium law in terms of partial pressures for this reaction as written.

b Which way must the reaction go in order for equilibrium to be attained?

c Calculate the partial pressures of the three gases at equilibrium.

7 A special type of equilibrium constant known as the **partition coefficient** describes the way a solute is distributed between two immiscible solvents A and B (see figure 3.1.9).

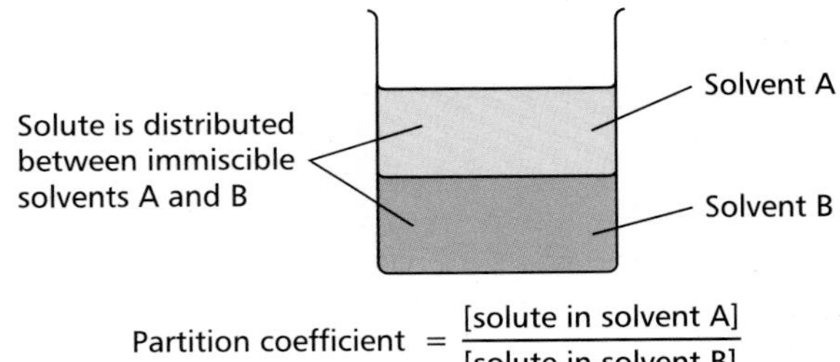

Figure 3.1.9

A solution contains 6.35 g of iodine and 83.0 g of potassium iodide dissolved in 1000 cm^3 of water. In this solution, the iodide ions and iodine molecules are in equilibrium with $I_3^-(aq)$ ions:

$$I^-(aq) + I_2(aq) \rightleftharpoons I_3^-(aq)$$

The value of K_c for this equilibrium is 730 at a certain temperature.

a What are the units of K_c for this equilibrium?

b If *x* represents the concentration of $I_3^-(aq)$ formed at equilibrium, write down an expression that relates *x* with K_c and the initial concentrations of $I^-(aq)$ and $I_2(aq)$.

The aqueous solution is now shaken with 1000 cm^3 of toluene, and some of the iodine molecules dissolve in the toluene layer. The partition coefficient for iodine distributed between a layer of toluene and a layer of water is 400 at this temperature (iodine is more soluble in toluene than in water).

QUESTIONS

c If the number of moles of iodine that dissolve in the toluene layer is *y*, write down an expression for the new value of $[I_2(aq)]_{eq}$ in terms of *x, y* and the initial concentration of $I_2(aq)$.

d Use your answers to parts **b** and **c** to write down an expression that relates *x* and *y* with K_c and the initial concentrations of $I^-(aq)$ and $I_2(aq)$.

e Now write down an expression that relates *y* with the partition coefficient and $[I_2(aq)]_{eq}$.

f Use your answers to **c** and **d** to calculate **i** the concentration of the iodine molecules dissolved in the toluene, and **ii** the concentration of $I_3^-(aq)$ ions dissolved in the aqueous layer.

Developing Key Skills

As part of the follow-up to the 1992 Rio summit on climate change and pollution control, there is to be a meeting of European Environment ministers to discuss possible solutions to the continuing increase in air pollution. Prepare a paper with as much data as possible (from the Focus material and other sources) proposing cleaner engines as a major factor in reducing air pollution in the future.

[Key Skills opportunities: C, A, IT]

ACID RAIN

Natural unpolluted rain has a pH of between about 5 and 5.6 as a result of carbon dioxide gas dissolved from the air. However, in many areas of the world rain is now falling with a pH of below 5, and in some areas of North America rain has been recorded with a pH of 2.3 (about 1000 times more acidic than natural rain). Rain with a pH below 5 is known as **acid rain**. It is the result of large quantities of nitrogen oxides (NO_x) and sulphur dioxide (SO_2) being released into the air by industrial processes and car exhausts, pollutants which then dissolve in the rain. Emissions of sulphur dioxide are responsible for the majority (60–70%) of acid rain.

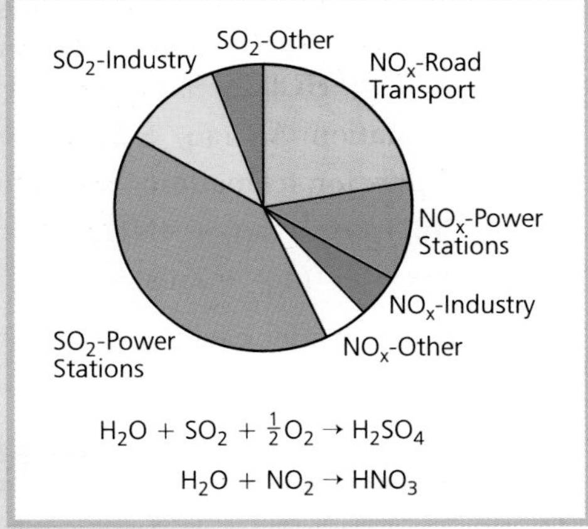

Figure 1 The sources of the two main gases which result in acid rain.

What does it do?

Acid rain causes lakes and rivers to become acidic, and the drop in pH can destroy the ecological balance. (In Minnesota all the fish in 140 lakes have been killed and the trout population of Norway's major rivers has fallen enormously.) It also causes the release of toxic metals held in rocks, especially aluminium, mercury and cadmium. Once the pH of a lake gets down to 4.5 almost everything will die. However, some lakes and streams are unaffected because they run on limestone beds, and this reacts with and neutralises the acid.

Acid rain can also damage and destroy trees. Many soils contain limestone and have natural buffering properties, but without this the acidified soil means plant nutrients are leached from the soil, aluminium is released damaging root systems, seeds stop germinating, helpful soil organisms are killed and the leaves of the trees and other plants are destroyed so that, unable to make food, they die.

Buildings, particularly those made with limestone, chalk or marble, are also very vulnerable to the ravages of acid rain, which reacts with the calcium carbonate of the stone, dissolving it away.

The effects of acid rain on our environment are given wide publicity, but acid rain also affects people directly. The toxic metals released into water sources by acid rain can cause poisonous levels of mercury to build up in the fish we eat. Acidic drinking water reacts with our water pipes – in Sweden at one stage levels of copper in the drinking water were so high that some people's hair started to turn green! Levels like this can also cause damage to the kidneys and liver, and diarrhoea in young children.

Figure 2 Large areas of forest across Europe, America and Canada are dead or dying as a result of acid rain.

What can be done?

Once a lake or area of land has been badly affected by acid rain, liming – the addition of chemicals with basic properties to neutralise the acid, such as hydrated lime or slaked lime – can help to restore the natural balance. But the best solution is to prevent the pollution at source. The sulphur emissions of industries and electrical generation plants must be limited; they can be cut by using low-sulphur fuels and **scrubbers** on the smoke stacks to absorb excess SO_2. The important thing is that *all* countries need to take these measures. Sulphur dioxide pollution travels in the atmosphere. Many European countries suffer considerable acid rain pollution not from their own relatively clean industries but from their dirtier neighbours. To do away with acid rain will need international effort – only time will tell if we are all willing to do our bit.

3.2 Acid–base equilibria

In sections 3.2–3.4 we shall be concerned with equilibria in aqueous solution involving ions. These equilibria fall into four main types:

1 Acid–base: $HCl(aq) + H_2O(l) \rightleftharpoons H_3O^+(aq) + Cl^-(aq)$

2 Precipitation: $Ag^+(aq) + Cl^-(aq) \rightleftharpoons AgCl(s)$

3 Complex ion formation:

$$[Cu(H_2O)_6]^{2+}(aq) + 4NH_3(aq) \rightleftharpoons [Cu(H_2O)_2(NH_3)_4]^{2+}(aq) + 4H_2O(l)$$

4 Redox: $Cu^{2+}(aq) + Zn(s) \rightleftharpoons Cu(s) + Zn^{2+}(aq)$

We shall consider each type of equilibrium in turn, starting with acid–base equilibria.

What is an acid? What is a base?

Acids and bases all around us

Acids and bases make up some of the most familiar chemicals in our everyday lives, as well as being some of the most important chemicals in laboratories and industries – see figure 3.2.1. For example, acids are used to remove the surface oxide film from metals before they are painted. Sodium bicarbonate, milk of magnesia and household ammonia are all bases. Bases are used in the manufacture of both paper and artificial fibres from plant material. The range and number of both acids and bases is enormous. What gives the chemicals in each of these groups their common properties?

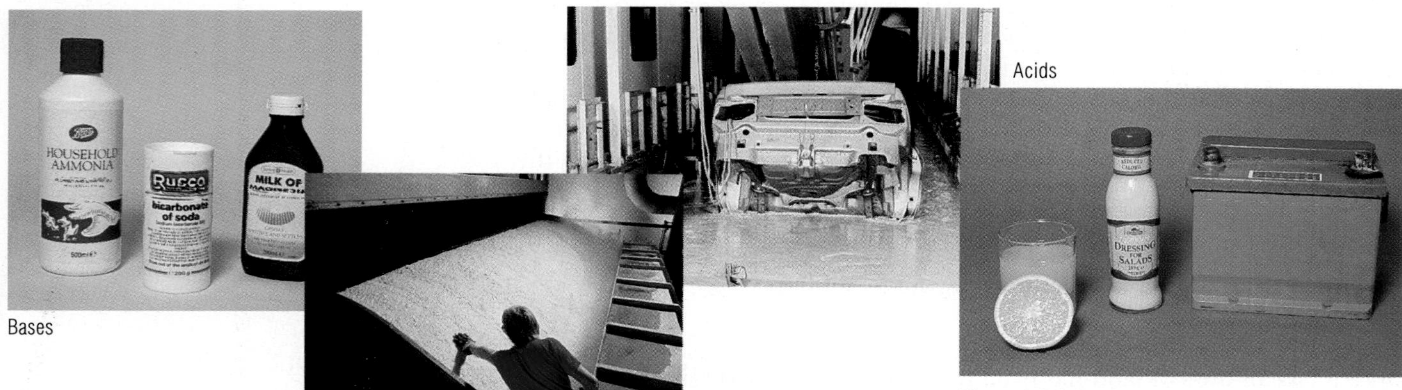

Figure 3.2.1 We use acids and bases in our everyday lives, both directly and indirectly.

You probably know that acids have a 'sharp' taste, while bases have a 'soapy' feel when rubbed between a damp forefinger and thumb. Such 'taste and feel' tests for acids and bases are not recommended, however, since many acids and bases can cause considerable harm – citric acid and sodium bicarbonate may be quite safe to treat in this way, but battery acid (sulphuric acid) and sodium hydroxide are very harmful.

A safer way to test acids and bases is to use **litmus**. A solution containing an acid will turn blue litmus red, while a solution containing a base will turn red litmus blue. We shall discuss other methods for testing acids and bases later in this section.

Theories of acids and bases

The first attempt to define the difference between acids and bases was in 1777, when Antoine Lavoisier suggested that acids were compounds that contained oxygen. Shortly after this it was discovered that hydrochloric acid, HCl, contained no oxygen. The English chemist Sir Humphry Davy then proposed in 1810 that hydrogen rather than oxygen was the important element in acids. In 1838 the German chemist Justus von Liebig offered the first useful definition of an acid, suggesting that acids were compounds containing hydrogen that can react with a metal to produce hydrogen gas.

Figure 3.2.2 Arrhenius' theory that aqueous solutions contain electrically charged particles meant that he very nearly failed his PhD, since the examiners were extremely critical of his ideas! This work eventually won him a Nobel prize in 1903.

Hydrogen ions

The first comprehensive theory of the behaviour of acids and bases in solution was put forward by the Swedish chemist Svante Arrhenius in his PhD thesis, published in 1884. Arrhenius proposed that when acids, bases and salts dissolve in water, they separate partially or completely into charged particles called ions. This process is called **dissociation**. Because solutions of ions are good electrical conductors, the substances that produce them are called **electrolytes**. Acids were considered to be electrolytes that produce the hydrogen ion, H^+, when they dissolve in water. According to Arrhenius' theory, acids have properties in common despite the obvious differences in their formulae, because they all produce H^+ ions when they dissolve in water:

$$HA(aq) \rightleftharpoons H^+(aq) + A^-(aq)$$

In the same way, Arrhenius' theory explained the common properties of bases as being due to the fact that they all produce the hydroxide ion in solution:

$$B(aq) + H_2O(aq) \rightleftharpoons BH^+(aq) + OH^-(aq)$$

Neutralisation

One of the most important properties of acids and bases is that they can cancel each other out when mixed together in the right proportions – a reaction called **neutralisation**. For example, when exactly equal volumes of hydrochloric acid and sodium hydroxide solution of exactly the same molar concentrations are mixed together, the solution formed has no effect on litmus. The reaction between these two chemicals is:

$$HCl(aq) + NaOH(aq) \rightarrow NaCl(aq) + H_2O(l)$$

The reaction between an acid and a base produces a salt (in this case sodium chloride, or 'common salt'). We say that the acid **neutralises** the base, or vice versa.

According to Arrhenius' theory, neutralisation occurs because the number of hydrogen ions from the acid is exactly equal to the number of hydroxide ions from the base, and so the two react completely to form water:

$$H^+(aq) + OH^-(aq) \rightarrow H_2O(aq)$$

Notice that these reactions have been written with a single arrow in the forward direction rather than with equilibrium arrows. As we shall see shortly, the equilibrium constants for such reactions are very large, so they effectively 'go to completion'.

The oxonium ion

Developments in chemists' understanding of the atom in the years following Arrhenius showed that it was almost inconceivable that the H^+ ion could exist in solution independently. The H^+ ion is simply a proton, with a diameter of about 70 000 times less than the diameter of a Li^+ ion. As a result, it was suggested that the H^+ ion exists in association with a water molecule as the H_3O^+ ion, called the **oxonium ion**:

$$HA(aq) + H_2O(l) \rightarrow H_3O^+(aq) + A^-(aq)$$

When talking about acidic solutions, the term 'hydrogen ion' is often used. Strictly speaking we should always remember that protons do not exist in solution, and should talk about the oxonium ion, writing H_3O^+ instead of H^+. In practice H^+ is commonly used, for simplicity's sake.

The importance of water in the behaviour of acids was recognised because substances like hydrogen chloride, HCl and ethanoic acid, CH_3COOH do not show acidic properties when dissolved in organic solvents like methylbenzene. They are non-electrolytes and do not affect dry litmus paper, since these solutions contain no H^+ (H_3O^+) ions.

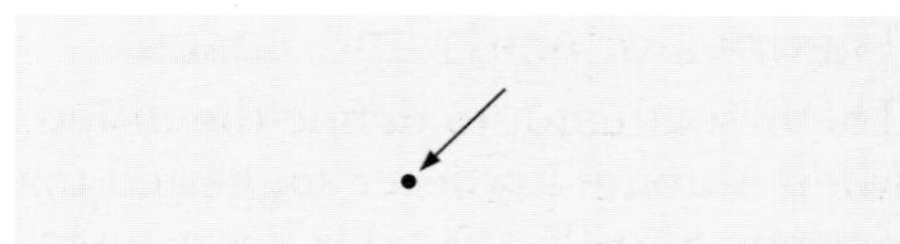

Figure 3.2.3 If a H^+ ion is represented by a dot 1 mm across, a Li^+ ion would need to be represented by a circle 70 metres in diameter!

The Brønsted–Lowry definition of acids and bases

Although useful, the Arrhenius definition of an acid was limited to situations in which water was present, since it defined acids and bases in terms of the ions produced in aqueous solutions. This definition was far too restrictive, since many reactions which appear to be acid–base reactions occur in solvents other than water, or even with no solvent at all. One such reaction occurs between hydrogen chloride and ammonia.

In aqueous solution, ammonia (a base) and hydrogen chloride (an acid) react to form a solution of ammonium chloride (a salt). According to the Arrhenius definition, this reaction occurs between the H^+ ions formed by the ionisation of the hydrogen chloride when it dissolves in the water, and the OH^- ions formed when the ammonia dissolves in water, as shown in figure 3.2.4.

$$NH_3 + H_2O \longrightarrow NH_4^+ + OH^-$$

$$OH^- + H_3O^+ \longrightarrow 2H_2O$$

$$HCl + H_2O \longrightarrow Cl^- + H_3O^+$$

Figure 3.2.4

In fact, the reaction between ammonia and hydrogen chloride does not need water or any other solvent to happen, as figure 3.2.5 shows. The white fumes are tiny crystals of ammonium chloride, formed as ammonia gas and hydrogen chloride gas react together:

$$\mathbf{NH_3(g) + HCl(g) \rightarrow NH_4Cl(s)}$$

This is clearly the same reaction as that occurring in aqueous solution, and it ought therefore to be defined as an acid–base reaction. However, the Arrhenius definition of an acid–base reaction does not allow us to do this, since there is no reaction between H_3O^+ and OH^- ions.

This problem was recognised in 1923, when the Danish chemist Johannes Nicolaus Brønsted and the English chemist Thomas Martin Lowry independently formulated a more general definition of acids and bases. They proposed that:

An acid is a proton donor;
a base is a proton acceptor.

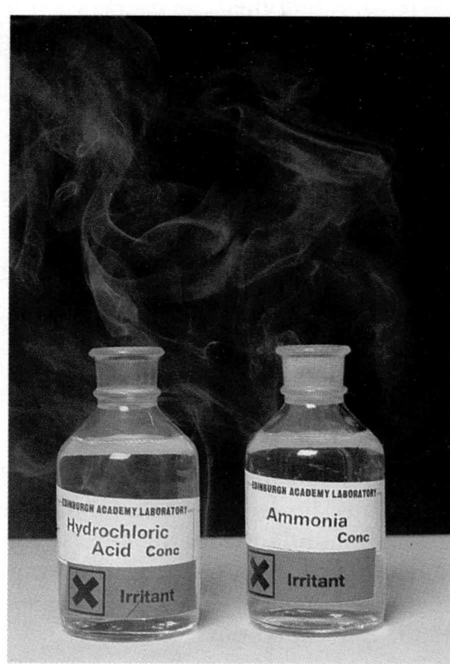

Figure 3.2.5

Using this definition, the reaction between ammonia and hydrogen chloride can be defined as an acid–base reaction, since the hydrogen chloride molecule can be seen to act as a proton donor, and the ammonia molecule as a proton acceptor:

Proton transferred

$$\mathbf{H_3N + HCl \rightarrow NH_4^+ + Cl^-}$$

Lewis acids

The Brønsted–Lowry definition of acids and bases views acid–base reactions as processes in which the transfer of a proton occurs. An alternative view of acid–base reactions was put forward by the American chemist G. N. Lewis in 1938, who suggested that:

- An acid is any species that can accept a pair of electrons in the formation of a coordinate covalent bond.
- A base is any species that can donate a pair of electrons in the formation of a coordinate covalent bond.
- Neutralisation is the formation of a coordinate covalent bond between a species that donates an electron pair (the base) and a species that accepts an electron pair (the acid).

Acids and bases described in this way are called **Lewis acids** and **Lewis bases**. The Lewis model obviously describes the situation where a proton becomes attached to a molecule of ammonia, in which the ammonia molecule (the Lewis base) donates a pair of electrons to the proton (the Lewis acid), forming a coordinate covalent bond (neutralisation). This is shown in figure 3.2.6(a).

However, the definition extends the concept of acid and base to cover reactions like the one between ammonia and boron trifluoride shown in figure 3.2.6(b), where ammonia is regarded as a Lewis base, reacting with the Lewis acid BF_3.

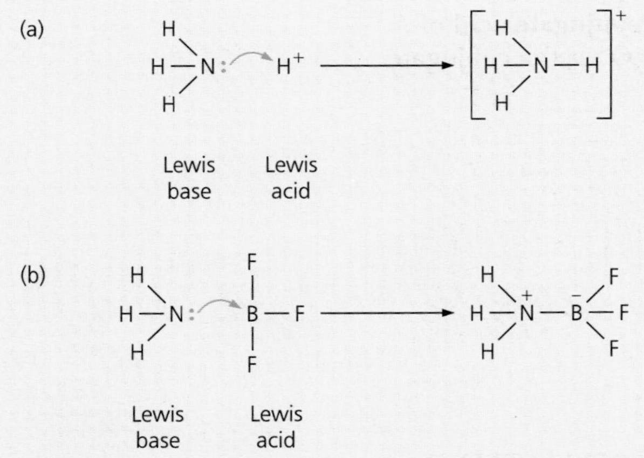

Figure 3.2.6 Ammonia as a Lewis base

The Lewis concept of an acid is thus much broader than the Arrhenius and Brønsted–Lowry definitions, which it includes as 'special cases'. The Lewis definition includes the formation of complexes of metal ions with ligands as acid–base reactions, and many reactions in organic chemistry can be understood in these terms too, as figure 3.2.7 illustrates.

Conjugate acids and bases

As a result of the Brønsted–Lowry definition, we can consider the dissociation of an acid as an acid–base reaction which is in equilibrium:

$$\mathbf{HA(aq) + H_2O(l) \rightleftharpoons H_3O^+(aq) + A^-(aq)}$$

In the forward reaction the HA acts as an acid, donating a proton to a water molecule. Water, accepting the proton, acts as a base. In the reverse reaction the H_3O^+ acts as an acid, donating a proton to A^-, which of course acts as a base.

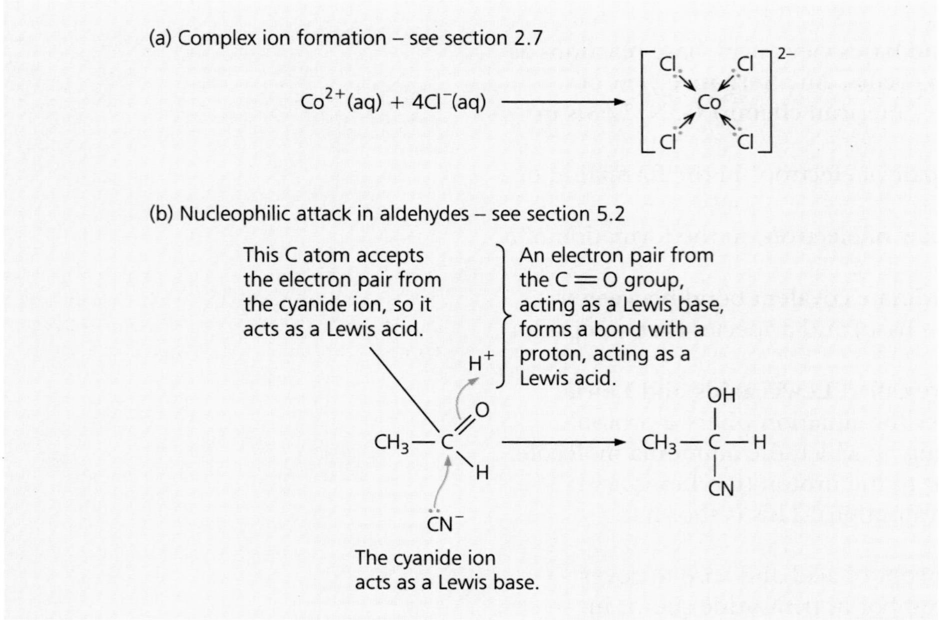

(a) Complex ion formation – see section 2.7

$$Co^{2+}(aq) + 4Cl^-(aq) \longrightarrow [CoCl_4]^{2-}$$

(b) Nucleophilic attack in aldehydes – see section 5.2

This C atom accepts the electron pair from the cyanide ion, so it acts as a Lewis acid.

An electron pair from the C$=$O group, acting as a Lewis base, forms a bond with a proton, acting as a Lewis acid.

$$CH_3-CHO + H^+ + CN^- \longrightarrow CH_3-CH(OH)-CN$$

The cyanide ion acts as a Lewis base.

Figure 3.2.7 The broad definition of Lewis acids and bases enables reactions like these to be understood in terms of acid–base behaviour too.

In acid–base reactions it is always possible to find *two* acids (in this case, HA and H_3O^+) and *two* bases (in this case, H_2O and A^-). The acid on one side of the equation is formed from the base on the other side of the equation. These are called **conjugate acid–base pairs**. In this example HA and A^- form one such pair, while H_2O and H_3O^+ form the other. HA is the **conjugate acid** of A^- (which means that A^- is the **conjugate base** of HA), and H_3O^+ is the conjugate acid of H_2O (so H_2O is the conjugate base of H_3O^+).

Conjugate pair

$$\underset{\textbf{Acid}}{HA} + \underset{\textbf{base}}{H_2O} \rightleftharpoons \underset{\textbf{base}}{A^-} + \underset{\textbf{acid}}{H_3O^+}$$

Conjugate pair

Acid – or base?

Some substances are able to act as both acids and bases. The most common example of this is water, which acts as an acid when it reacts with ammonia, donating a proton during the reaction:

$$H_2O(l) + NH_3(aq) \rightarrow OH^-(aq) + NH_4^+(aq)$$

However, it behaves as a base when it reacts with hydrogen chloride, accepting a proton in the reaction:

$$H_2O(l) + HCl(aq) \rightarrow H_3O^+(aq) + Cl^-(aq)$$

Substances that behave like this are called **amphoteric**. The amphoteric character of water is one of its many important properties met in section 2.2.

Strong acids and weak bases

We saw at the beginning of this section that different acids may have different strengths, making it possible to taste an acid like citric acid quite safely, whereas sulphuric acid will quickly damage your tongue!

Think about the equilibrium set up when a weak acid like ethanoic acid dissolves in water:

$$CH_3COOH(aq) + H_2O(l) \rightleftharpoons CH_3COO^-(aq) + H_3O^+(aq)$$

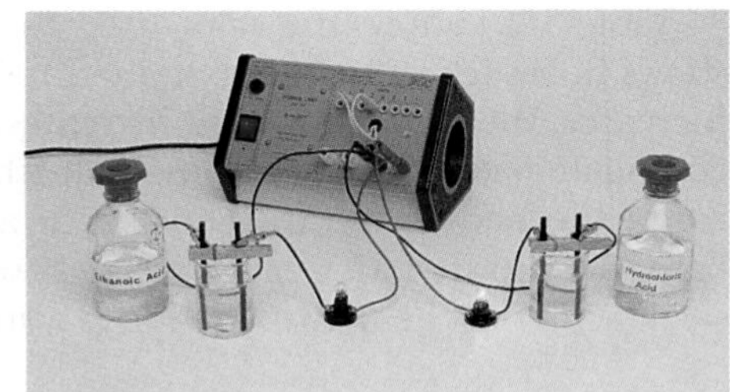

Figure 3.2.8 The low electrical conductivity of an aqueous solution of ethanoic acid (left) shows how few ions there are in the solution, and therefore that it is a weak acid. In contrast, hydrochloric acid (right) is almost completely ionised.

The two acids in this equilibrium are CH_3COOH and H_3O^+, and they are competing to donate a proton to a base. We can tell that this equilibrium lies well over to the left from the evidence in figure 3.2.8. This must mean that the oxonium ion is a better proton donor than ethanoic acid, since there is much more CH_3COOH than H_3O^+ in the solution. Therefore H_3O^+ is a stronger acid than CH_3COOH. Similarly, CH_3COO^- must be a stronger base than H_2O, since it is better at accepting protons.

Another point about conjugate acid–base pairs can be made if we look again at what happens when hydrogen chloride – an extremely powerful proton donor – dissolves in water:

$$H_2O(l) + HCl(aq) \rightarrow H_3O^+(aq) + Cl^-(aq)$$

The equilibrium constant for this reaction is so large that we can consider the reaction effectively to 'go to completion'. As well as telling us that hydrogen chloride is a very strong Brønsted acid, this also tells us that the chloride ion is a very weak Brønsted base, since it cannot attract protons from even a good proton donor like H_3O^+. In general:

Strong Brønsted acids have weak conjugate bases

and:

Weak Brønsted acids have strong conjugate bases.

These strong acid/weak base relationships can be summed up as in table 3.2.1. Note that any acid in the table will donate protons to any base which is lower down the list – the strength of the acids increases as we go up the list, while the strength of the bases increases as we go down the list.

Measuring acidity and basicity

How strong?

The strength of an acid or a base is a measure of how good a proton donor or acceptor it is. Consider the dissociation of the acid HA in aqueous solution:

$$HA(aq) \rightleftharpoons H^+(aq) + A^-(aq)$$

One way of measuring the strength of this acid is to measure the conductivity of a solution of the acid – as we saw in figure 3.2.8, a solution of a weak acid is only weakly ionised, so it conducts electricity very poorly.

The conductivity of a solution of an acid will obviously depend on the concentration of the acid. Because of this, a much better measure of its strength is the equilibrium constant for its dissociation into ions, which for the equilibrium above is written as:

$$K_a = \frac{[H^+(aq)]\,[A^-(aq)]}{[HA(aq)]}$$

The equilibrium constant for this reaction K_a is known as the **dissociation constant** of the acid. K_a provides a direct measure of the strength of the acid because it measures its dissociation – the larger the value of K_a, the more dissociated the acid, and the greater its strength as a proton donor. Because it is an equilibrium constant, the value of K_a is unaffected by concentrations, although it *is* affected by temperature.

Table 3.2.1 shows the value of K_a for some acids. (Question 4 on page 268 shows how such values are obtained.) Notice that as the strength of the acid increases, the value of K_a also increases. Notice too that the strength of the conjugate base of the acid–base pair (that is, A^-) increases as the strength of its conjugate acid decreases. Thus, nitric acid is a strong acid, but its conjugate base, the nitrate ion, is a very weak base indeed. In contrast, water is a very weak acid, but the hydroxide ion is a strong base.

Increasing strength of acid ↑

Acid	Equilibrium: Acid		Equilibrium: Base	K_a at 25 °C/mol dm^{-3}
Sulphuric acid	H_2SO_4	$\rightleftharpoons$	$H^+ + HSO_4^-$	Very large
Nitric acid	HNO_3	$\rightleftharpoons$	$H^+ + NO_3^-$	40
Trichloroethanoic acid	CCl_3CO_2H	$\rightleftharpoons$	$H^+ + CCl_3CO_2^-$	0.23
Sulphurous acid	H_2SO_3	$\rightleftharpoons$	$H^+ + HSO_3^-$	0.015
Hydrated iron(III) ion	$[Fe(H_2O)_6]^{3+}$	$\rightleftharpoons$	$H^+ + [Fe(H_2O)_5(OH)]^{2+}$	0.006
Hydrofluoric acid	HF	$\rightleftharpoons$	$H^+ + F^-$	0.000 56
Methanoic acid	HCO_2H	$\rightleftharpoons$	$H^+ + HCO_2^-$	0.000 16
Ethanoic acid	CH_3CO_2H	$\rightleftharpoons$	$H^+ + CH_3CO_2^-$	1.7×10^{-5}
Carbonic acid	H_2CO_3	$\rightleftharpoons$	$H^+ + HCO_3^-$	4.5×10^{-7}
Hydrogen sulphide	H_2S	$\rightleftharpoons$	$H^+ + HS^-$	8.9×10^{-8}
Ammonium ion	NH_4^+	$\rightleftharpoons$	$H^+ + NH_3$	5.6×10^{-10}
Phenol	C_6H_5OH	$\rightleftharpoons$	$H^+ + C_6H_5O^-$	1.3×10^{-10}
Hydrogen peroxide	H_2O_2	$\rightleftharpoons$	$H^+ + HO_2^-$	2.4×10^{-12}
Water	H_2O	$\rightleftharpoons$	$H^+ + OH^-$	1.0×10^{-14}

Increasing strength of conjugate base ↓

Table 3.2.1 The value of K_a for some acids. The K_a value for strong acids like sulphuric and nitric acids is rarely quoted – K_a is large, and we assume that the ionisation is complete.

We can also write an equilibrium constant for the dissociation of a base B that produces OH^- ions by reaction with water:

$$\mathbf{B(aq) + H_2O(l) \rightleftharpoons BH^+(aq) + OH^-(aq)}$$

The equilibrium constant for this reaction is the dissociation constant of the base, K_b:

$$K_b = \frac{[BH^+(aq)][OH^-(aq)]}{[B(aq)]}$$

Note that this expression assumes that we may take the concentration of water as constant, which is a reasonable assumption for dilute solutions.

K_a for bases

As an alternative to K_b, the strength of a base is sometimes indicated by giving a value of K_a for its conjugate acid, that is, for the equilibrium:

$$\mathbf{BH^+(aq) + H_2O(l) \rightleftharpoons B(aq) + H_3O^+(aq)}$$

the expression for K_a is:

$$K_a = \frac{[B(aq)][H_3O^+(aq)]}{[BH^+(aq)]}$$

(As before, this expression assumes that water is present in excess so that its concentration may be taken as constant.)

From what we have seen so far about conjugate acids and bases, it follows that K_a for the conjugate acid BH^+ of a weak base B will be large, as BH^+ is by definition a strong acid. Hence weak bases have larger values of K_a for their conjugate acids than do strong bases, as table 3.2.2 opposite shows.

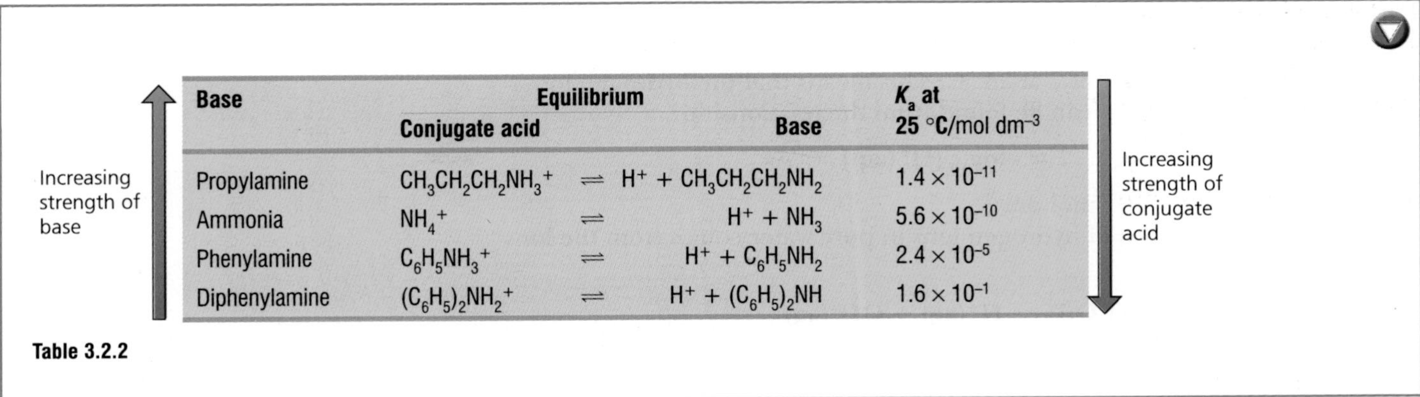

Base	Equilibrium: Conjugate acid		Equilibrium: Base	K_a at 25 °C/mol dm^{-3}
Propylamine	$CH_3CH_2CH_2NH_3^+$	⇌	$H^+ + CH_3CH_2CH_2NH_2$	1.4×10^{-11}
Ammonia	NH_4^+	⇌	$H^+ + NH_3$	5.6×10^{-10}
Phenylamine	$C_6H_5NH_3^+$	⇌	$H^+ + C_6H_5NH_2$	2.4×10^{-5}
Diphenylamine	$(C_6H_5)_2NH_2^+$	⇌	$H^+ + (C_6H_5)_2NH$	1.6×10^{-1}

Table 3.2.2

The pH scale

The degree of dissociation of an acid gives us information about its strength. However, we are often more interested in the concentration of protons (hydrogen ions) when thinking about the effect of an acid as a proton donor. The concentration of hydrogen ions in a solution is measured on the **pH scale**, in which:

$$\mathbf{pH = -log_{10} [H^+(aq)]}$$

The concentrations of hydrogen ions in most aqueous solutions lie between about 10^{-14} mol dm^{-3} and 1 mol dm^{-3}. The range of pH values for these solutions is therefore between 0 and 14, since $-\log_{10}(1) = 0$ and $-\log_{10}(10^{-14}) = 14$. The pH scale is shown in figure 3.2.9.

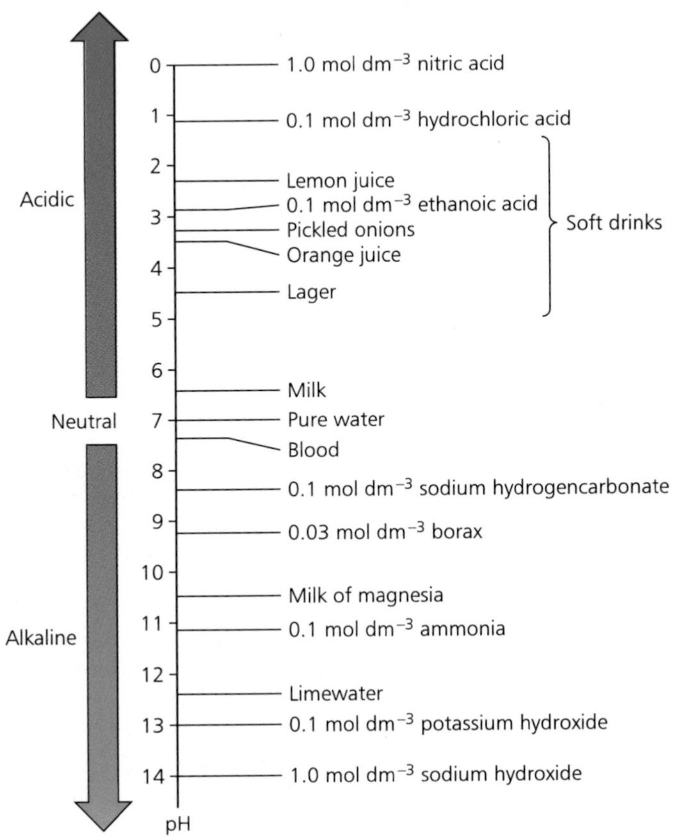

Figure 3.2.9 The pH scale

Measuring pH

The simplest way to measure pH is by using substances called **indicators**, which are discussed in section 3.3. More accurate work may require the use of a **pH meter**. How such a meter works is discussed in section 3.5, page 303.

The ionic product of water

A sample of pure water, obtained by the repeated distillation of water under an inert atmosphere, has a pH of 7 at 25 °C. This means that the hydrogen ion concentration in pure water can be found from the relationship:

$$7 = -\log_{10} [H^+(aq)]$$

from which $[H^+(aq)] = 10^{-7}$ mol dm^{-3}.

This tiny concentration of hydrogen ions in pure water comes from the ions formed by its dissociation:

$$H_2O(l) \rightleftharpoons H^+(aq) + OH^-(aq)$$

We can write an equilibrium constant for this dissociation:

$$K_c = \frac{[H^+(aq)][OH^-(aq)]}{[H_2O(l)]}$$

Since this equilibrium lies well to the left, the concentration of water can be regarded as constant, so we can write:

$$K_w = K_c[H_2O(l)] = [H^+(aq)][OH^-(aq)]$$

where K_w is called the **ionic product of water.**

From the equation for the dissociation of water, we see that:

$$[H^+(aq)] = [OH^-(aq)] = 10^{-7} \text{ mol dm}^{-3} \text{ at } 25\ °C$$

From this it follows that the value of K_w at 25 °C is given by:

$$\begin{aligned} K_w &= 10^{-7} \text{ mol dm}^{-3} \times 10^{-7} \text{ mol dm}^{-3} \\ &= 10^{-14} \text{ mol}^2 \text{ dm}^{-6} \end{aligned}$$

Like all equilibrium constants, K_w varies with temperature, as shown in figure 3.2.10.

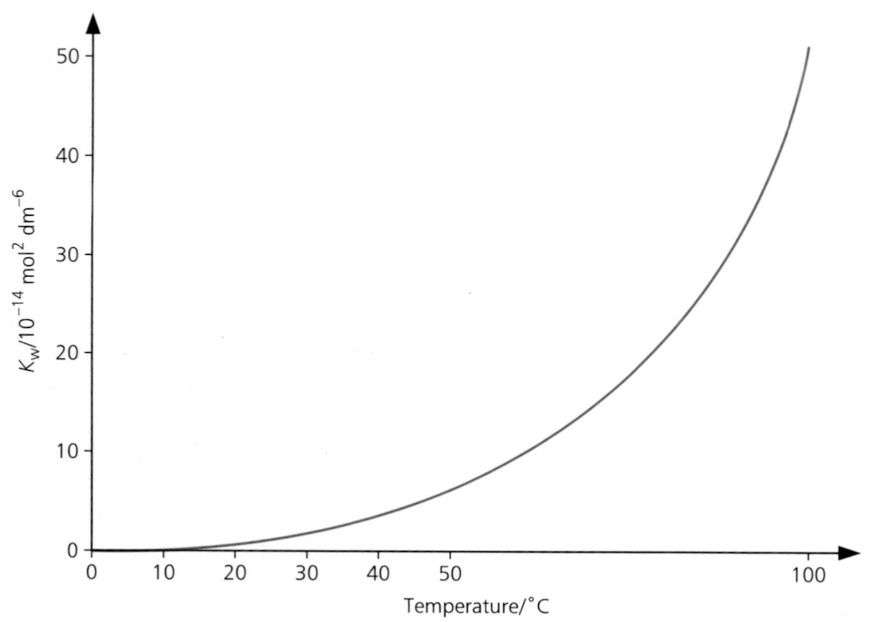

Figure 3.2.10 The variation of K_w with temperature. $\Delta H^{\ominus}_m$ for the reaction:

$$H_2O(l) \rightleftharpoons H^+(aq) + OH^-(aq)$$

is +58 kJ mol^{-1}. This explains why K_w increases with increasing temperature – Le Chatelier's principle leads us to expect that an endothermic reaction will move further to the right with an increase in temperature.

pK_a and pH

Instead of quoting the value of K_a for an acid, we may instead quote its pK_a, where:

$$pK_a = -\log_{10} K_a$$

For example, the K_a of ethanoic acid is 1.7×10^{-5} mol dm^{-3}, so its pK_a is given by:

$$pK_a = -\log_{10} (1.7 \times 10^{-5})$$
$$= 4.8$$

The stronger an acid, the *larger* its K_a and the *smaller* its pK_a.

An acid's pK_a is useful to us as follows. From the relationship between pK_a and K_a above, we can write:

$$pK_a = -\log_{10} \left(\frac{[H^+(aq)][A^-(aq)]}{[HA(aq)]}\right)$$

$$= -\left(\log_{10} [H^+(aq)] + \log_{10}\left(\frac{[A^-(aq)]}{[HA(aq)]}\right)\right)$$

so:

$$pK_a = pH - \log_{10} \left(\frac{[A^-(aq)]}{[HA(aq)]}\right)$$

This is significant, because it means that pK_a = pH when:

$$\log_{10} \left(\frac{[A^-(aq)]}{[HA(aq)]}\right) = 0$$

that is, when:

$$\frac{[A^-(aq)]}{[HA(aq)]} = 1 \text{ (because } \log_{10} 1 = 0)$$

In other words, at a pH equal to its pK_a, an acid is exactly 50% dissociated. This fits exactly with the Brønsted–Lowry theory, since by definition weaker acids have stronger conjugate bases, which will tend to be protonated at higher pH values. Stronger acids have weaker conjugate bases, which will tend not to be protonated except at lower pH values.

In the same way as pK_a can be quoted for acids, we can quote values of pK_b for bases – a weak base will have a *small* K_b, and consequently a *large* pK_b.

3

Calculating pH from K_a and K_b

It is quite straightforward to calculate the pH of a solution containing an acid from the acid's dissociation constant, as the following examples show.

(1) Strong acid

In a solution of a strong acid like hydrochloric acid, we assume that the acid is 100% dissociated, so that $[H^+(aq)]$ is equal to the concentration of the acid in the solution. For example, in a solution containing 1 mol dm^{-3} of HCl, we assume that $[H^+(aq)]$ = 1 mol dm^{-3}. The pH can then be found:

$$pH = -\log_{10} [H^+(aq)]$$
$$= -\log_{10} 1 = 0$$

In dilute solutions, the assumption that strong acids are completely ionised is a reasonable one. This may not be the case in very concentrated solutions of acids, and this may need to be taken into account when calculating $[H^+(aq)]$.

(2) Weak acid

For a weak acid, we use the value of K_a to calculate the pH of the solution. The K_a of methanoic acid is 1.6×10^{-4} mol dm^{-3}. To calculate the pH of a solution containing 0.5 mol dm^{-3} of methanoic acid, assume that $[H^+(aq)] = [A^-(aq)]$. It follows that we can then write down the concentrations of undissociated methanoic acid and methanoate ions as in table 3.2.3.

	$HCOOH(aq)$	$\rightleftharpoons H^+(aq)$	$+ HCOO^-(aq)$
Initial concentration/ mol dm^{-3}	0.5	0	0
Equilibrium concentration/ mol dm^{-3}	0.5 – $[H^+(aq)]$	$[H^+(aq)]$	$[H^+(aq)]$

Table 3.2.3

We can now write down the relationship between these concentrations and K_a:

$$K_a = \frac{[H^+(aq)][A^-(aq)]}{[HA(aq)]}$$

$$= \frac{[H^+(aq)]^2}{0.5 \text{ mol dm}^{-3} - [H^+(aq)]}$$

Since the value of K_a is small for this acid, it is only weakly ionised, so we can assume that

$[H^+(aq)] << 0.5\ mol\ dm^{-3}$, and we can write:

$$K_a \approx \frac{[H^+(aq)]^2}{(0.5\ mol\ dm^{-3})}$$

Substituting for K_a:

$$1.6 \times 10^{-4}\ mol\ dm^{-3} = \frac{[H^+(aq)]^2}{(0.5\ mol\ dm^{-3})}$$

Rearranging this gives:

$$1.6 \times 10^{-4}\ mol\ dm^{-3} \times 0.5\ mol\ dm^{-3} = [H^+(aq)]^2$$

from which:

$$[H^+(aq)] = \sqrt{(1.6 \times 10^{-4}\ mol\ dm^{-3} \times 0.5\ mol\ dm^{-3})}$$
$$= 8.9 \times 10^{-3}\ mol\ dm^{-3}$$

The pH of this solution is therefore found from:

$$pH = -\log_{10} [H^+(aq)]$$
$$= -\log_{10} (8.9 \times 10^{-3})$$
$$= -(-2.1) = 2.1$$

(3) Strong base

To find the pH of a solution containing a base, we must first find the concentration of OH^- ions, and then use the ionic product of water to find the concentration of hydrogen ions.

We may assume that a strong base is completely ionised, in the same way as a strong acid. The calculation is similar whether the base produces OH^- ions directly as it dissolves in water (as in the case of sodium hydroxide, for which $NaOH(s) \rightarrow Na^+(aq) + OH^-(aq)$), or whether they are produced as a result of the reaction of the base with water (as in the case of potassium oxide, for which $K_2O(s) + H_2O(l) \rightarrow 2K^+(aq) + 2OH^-(aq)$).

For example, in a solution containing $0.2\ mol\ dm^{-3}$ of sodium hydroxide, $[OH^-(aq)]$ can be taken to be $0.2\ mol\ dm^{-3}$. Writing the relationship for K_w:

$$K_w = [H^+(aq)][OH^-(aq)]$$

we can substitute for $[OH^-(aq)]$ and K_w to find $[H^+(aq)]$:

$$1.0 \times 10^{-14}\ mol^2\ dm^{-6} = [H^+(aq)] \times 0.2\ mol\ dm^{-3}$$

so:

$$[H^+(aq)] = \frac{1.0 \times 10^{-14}\ mol^2\ dm^{-6}}{0.2\ mol\ dm^{-3}}$$
$$= 5.0 \times 10^{-14}\ mol\ dm^{-3}$$

Once again, the pH of this solution is found from:

$$pH = -\log_{10} [H^+(aq)]$$
$$= -\log_{10} (5.0 \times 10^{-14})$$
$$= -(-13.3)$$
$$= 13.3$$

(4) Weak base

The pH of a solution containing a weak base is calculated using K_b, in a way similar to that using K_a to find the pH of a solution of a weak acid.

For example, K_b for ammonia is $1.8 \times 10^{-5}\ mol\ dm^{-3}$. To calculate the pH of a solution containing $0.1\ mol\ dm^{-3}$ of ammonia, assume that $[OH^-(aq)] = [NH_4^+(aq)]$. It follows that we can then write down the concentrations of undissociated ammonia and ammonium ions as in table 3.2.4, ignoring the concentration of water which can be regarded as being constant.

	$NH_3(aq) + H_2O(l) \rightleftharpoons$	$NH_4^+(aq)$ +	$OH^-(aq)$
Initial concentration/ $mol\ dm^{-3}$	0.1	0	0
Equilibrium concentration/ $mol\ dm^{-3}$	$0.1 - [OH^-(aq)]$	$[OH^-(aq)]$	$[OH^-(aq)]$

Table 3.2.4

We can now write down the relationship between these concentrations and K_b:

$$K_b = \frac{[NH_4^+(aq)][OH^-(aq)]}{[NH_3(aq)]}$$
$$= \frac{[OH^-(aq)]}{(0.1\ mol\ dm^{-3} - [OH^-(aq)])}$$

Since the value of K_b is small for this base, it is only weakly ionised, so we can assume that $[OH^-(aq)] << 0.1\ mol\ dm^{-3}$, and we can write:

$$K_b \approx \frac{[OH^-(aq)]^2}{(0.1\ mol\ dm^{-3})}$$

Substituting for K_b:

$$1.8 \times 10^{-5}\ mol\ dm^{-3} = \frac{[OH^-(aq)]^2}{(0.1\ mol\ dm^{-3})}$$

Rearranging this gives:

$$1.8 \times 10^{-5}\ \mathbf{mol\ dm^{-3}} \times 0.1\ \mathbf{mol\ dm^{-3}} = [\mathbf{OH^-(aq)}]^2$$

from which:

$$[\mathbf{OH^-(aq)}] = \sqrt{(1.8 \times 10^{-5}\ \mathbf{mol\ dm^{-3}} \times 0.1\ \mathbf{mol\ dm^{-3}})}$$
$$= 1.34 \times 10^{-3}\ \mathbf{mol\ dm^{-3}}$$

We may now use this to calculate the pH of the solution, using the ionic product of water:

$$1.0 \times 10^{-14}\ \mathbf{mol^2\ dm^{-6}} = [\mathbf{H^+(aq)}] \times 1.34 \times 10^{-3}\ \mathbf{mol\ dm^{-3}}$$

so:

$$[\mathbf{H^+(aq)}] = \frac{1.0 \times 10^{-14}\ \mathbf{mol^2\ dm^{-6}}}{1.34 \times 10^{-3}\ \mathbf{mol\ dm^{-3}}}$$
$$= 7.46 \times 10^{-12}\ \mathbf{mol\ dm^{-3}}$$

Once again, the pH of this solution is found from:

$$\mathbf{pH} = -\log_{10} [\mathbf{H^+(aq)}]$$
$$= -\log_{10} (7.46 \times 10^{-12})$$
$$= -(-11.1)$$
$$= 11.1$$

We shall see in section 3.3 how this knowledge of acids and bases can be put to practical use.

SUMMARY

- According to the **Brønsted–Lowry theory** of acids and bases, an acid is a **proton donor**, and a base is a **proton acceptor**.
- Acids and bases react together in a **neutralisation** reaction to produce salts.
- Protons exist in aqueous solutions as **oxonium ions**, H_3O^+.
- When an acid HA dissolves in water and **dissociates** or ionises, an acid–base equilibrium is set up:

$$\mathbf{HA(aq) + H_2O(l) \rightleftharpoons H_3O^+(aq) + A^-(aq)}$$

  A^- is the **conjugate base** of the acid HA. H_3O^+ is the **conjugate acid** of the base H_2O.

- An **amphoteric** substance may behave both as an acid and as a base. Water is an amphoteric substance when it dissociates:

$$\mathbf{\underbrace{\underset{Acid}{H_2O(l)} + \overbrace{\underset{}{\overset{base}{H_2O(l)}} \rightleftharpoons \overset{acid}{H_3O^+(aq)}}^{Conjugate\ acid\text{–}base\ pair} + \underset{base}{OH^-(aq)}}_{Conjugate\ acid\text{–}base\ pair}}$$

- The strength of an acid or base is a measure of how good it is at donating or accepting protons.
- The equilibrium constant K_a for the dissociation of an acid into ions is known as its **dissociation constant** and is a direct measure of its strength as an acid. For the acid HA:

$$K_a = \frac{[\mathbf{H^+(aq)}][\mathbf{A^-(aq)}]}{[\mathbf{HA(aq)}]}$$

- The dissociation constant K_b for the dissociation of a base B:

$$B(aq) + H_2O(l) \rightleftharpoons BH^+(aq) + OH^-(aq)$$

is given by:

$$K_b = \frac{[BH^+(aq)][OH^-(aq)]}{[B(aq)]}$$

- The **pH scale** is used to measure the concentration of hydrogen ions in a solution. $pH = -\log_{10}[H^+(aq)]$.
- At 25 °C, $K_w = [H^+(aq)][OH^-(aq)] = 10^{-14}\ mol^2\ dm^{-6}$ for all aqueous solutions. For a neutral solution at this temperature, $[H^+(aq)] = [OH^-(aq)] = 10^{-7}\ mol\ dm^{-3}$.
- The pH of solutions containing known concentrations of acids and bases may be calculated using their dissociation constants.

QUESTIONS

1 a Describe how the **i** Arrhenius **ii** Brønsted–Lowry and **iii** Lewis theories define acids and bases.

b Use each of these three theories to classify the behaviour of the substances in the following reactions:
- **i** sodium oxide, Na_2O, dissolving in water
- **ii** phosphine, PH_3, reacting with hydrogen iodide to form phosphonium oxide, PH_4I
- **iii** copper(II) ions forming a complex with edta^{4-} ions.

2 A sample of anhydrous sodium carbonate, Na_2CO_3, was contaminated with sodium chloride. In order to determine the amount of contamination, a chemist dissolved 0.546 g of the contaminated sodium carbonate in water. The resulting solution required 23.8 cm^3 of 0.42 mol dm^{-3} hydrochloric acid to neutralise it.

a How many moles of HCl were required to neutralise the sodium carbonate present?

b To how many moles of Na_2CO_3 is this equivalent?

c What mass of sodium carbonate was present in the initial 0.546 g of contaminated product?

d What is the percentage contamination of the sodium carbonate by mass?

3 The three substances HX, HY and HZ are all weak acids.

a Copy and complete table 3.2.5.

	HX	HY	HZ
Value of K_a	10^{-10}	10^{-7}	10^{-4}
$[H^+(aq)]$ in 1.0 mol dm^{-3} solution			
pH of 1.0 mol dm^{-3} solution			
pK_a			

Table 3.2.5

b When an aqueous solution containing HZ is diluted, it is noticed that the pH of the solution rises until eventually a constant value is reached. Explain this.

c An aqueous solution contains 0.1 mol dm^{-3} HX. What happens to [HX(aq)] when
- **i** concentrated hydrochloric acid is added
- **ii** a salt NaX is added
- **iii** water is added?

Explain your answers.

4 The dissociation constant of a weak acid may be found by measuring the pH of a solution of the acid whose concentration is known. Carry out the following calculations to find the K_a for *para*-aminobenzoic acid (PABA, systematic name 4-aminobenzoic acid), the active ingredient in many sunscreens.

a Representing the parent acid as HPABA and its conjugate base as PABA$^-$, write down:
- **i** the equilibrium involved when HPABA dissociates
- **ii** an expression for K_a for *para*-aminobenzoic acid.

A solution of HPABA in water was made, and brought to a steady temperature of 25 °C. The pH of this solution was found to be 3.22 when measured using a pH electrode and meter.

b Calculate $[H^+(aq)]$ for this solution.

c Write down the value of $[PABA^-(aq)]$.

A 24.00 cm^3 portion of the HPABA solution was titrated against 0.05 mol dm^{-3} sodium hydroxide solution. It was found that 14.4 cm^3 of the sodium hydroxide solution exactly neutralised the HPABA solution (1 mol of sodium hydroxide reacts exactly with 1 mol of HPABA).

d Calculate the number of moles of *para*-aminobenzoic acid present in 24.00 cm^3 of the solution.

e Hence calculate the concentration of HPABA in the solution, stating any assumptions you make.

QUESTIONS

  f From your answers to **b**, **c** and **e**, calculate K_a for *para*-aminobenzoic acid.

5 Using the figures from table 3.2.6 below, plot a graph of the pH of a neutral solution for temperatures between 0 °C and 100 °C.

6 Heavy water, D_2O, ionises in exactly the same way as water. The ionic product of heavy water at 20 °C is 8.9×10^{-16} mol^2 dm^{-6}. What is: **a** $[D^+]$ **b** $[OD^-]$ **c** pD at this temperature?

7 Deuteroammonia, ND_3, is a weak base, with $pK_b = 4.96$ at 25 °C. Calculate:

a K_b for deuteroammonia

b **i** $[H^+(aq)]$

ii the pH

of a solution containing 0.1 mol dm^{-3} deuteroammonia in water.

Temperature/°C	0	10	20	30	40	50	100
K_w/10^{-14} mol^2 dm^{-6}	0.114	0.293	0.681	1.471	2.916	5.476	51.3

Table 3.2.6

Developing Key Skills

A new scheme of work is being developed for pupils in Key Stage 3. The plan is to teach these 11–14 year olds about acids and bases through work on acid rain. Plan a booklet for the students to use with this module along with two practical investigations which they could carry out to further their understanding of acids, bases and acid rain.

[Key Skills opportunities: C, IT]

3

3.3 Buffers and indicators

We saw in section 3.2 how we can define a number of useful measurements in acid–base equilibria. We shall now look in some detail at two of the uses to which acids and bases are put – **buffers** and **indicators.**

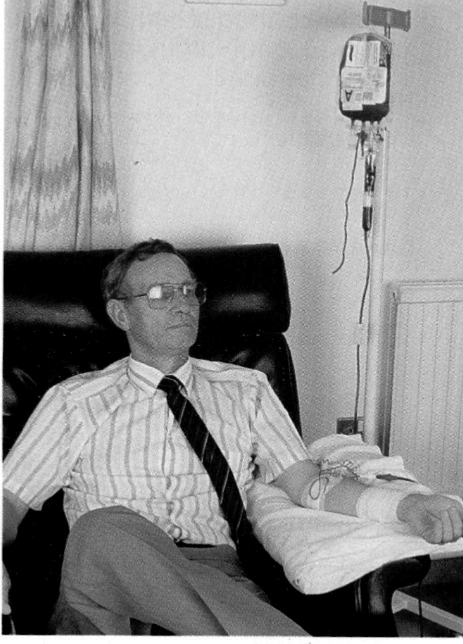

Figure 3.3.1 Blood contains hydrogencarbonate ions, HCO_3^-, which play an important part in controlling its pH.

Buffer solutions

What is a buffer solution?

A small change in the pH of a system can often have dramatic, sometimes undesired, results. For example, a small amount of lemon juice added to milk or cream causes considerable changes in the structure of the proteins, resulting in 'curdling'. Much more seriously, if the pH of your blood were to change from 7.35 (its normal value) to 7.00 or to 8.00, you would die. (You can find out more about this in the Focus page 'Body buffers' on page 274.

Natural systems have mechanisms to prevent large changes in pH happening, and chemists have developed similar ways of stabilising the pH of solutions in which reactions are occurring. Natural and artificial solutions containing a mix of solutes which together resist changes in pH are known as **buffer solutions** – these solutions are said to be **buffered** against changes in pH. The principles of buffer solutions can be understood using the ideas met in section 3.2.

A buffer solution usually consists of two solutes. One of these is a weak Brønsted acid, and the other is its conjugate base – for example, ethanoic acid (the weak acid) and sodium ethanoate (supplying its weak conjugate base, the ethanoate ion). If we represent the weak acid as HA, the equilibrium concerned can be written as:

$$\mathbf{HA(aq) \rightleftharpoons H^+(aq) + A^-(aq)}$$

The salt, represented as MA, is fully dissociated, so:

$$\mathbf{MA(s) \rightarrow M^+(aq) + A^-(aq)}$$

From these two equations, we can see that the buffer solution contains a large amount of A^-. This pushes the acid–base equilibrium to the left, so the solution also contains a large amount of HA. The solution thus has a reservoir of the weak acid HA, and a reservoir of its conjugate base A^-. Figure 3.3.2 overleaf shows how this helps to prevent changes in pH.

As figure 3.3.2 shows, the stability of the pH of a buffer solution is due to two factors:

- a reservoir of HA, which supplies H^+ ions if they are removed from the solution, e.g. by the addition of OH^- ions
- a reservoir of A^-, which reacts with any H^+ ions added, removing them from the solution.

The pH of a buffer solution

Consider a buffer solution made by dissolving a salt MA in a solution of the weak acid HA. We know that the dissociation constant of the acid K_a is given by:

$$K_a = \frac{\mathbf{[H^+(aq)][A^-(aq)]}}{\mathbf{[HA(aq)]}}$$

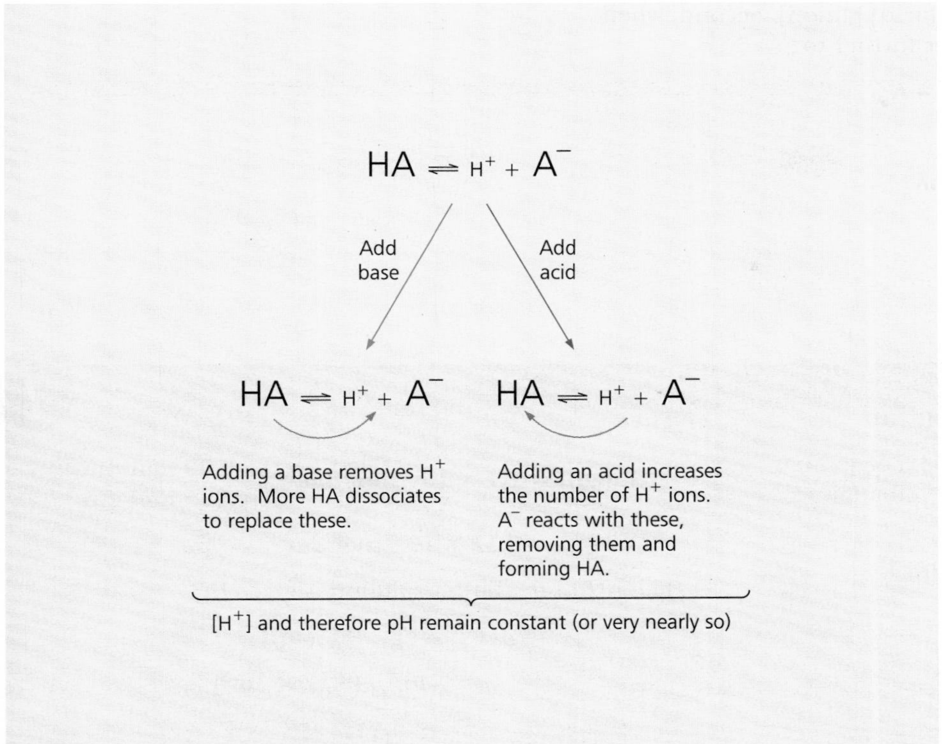

Figure 3.3.2 How a buffer solution works

3

This relationship can be rearranged so that $[H^+(aq)]$ is on the left-hand side of the equation:

$$[H^+(aq)] = K_a \times \frac{[HA(aq)]}{[A^-(aq)]}$$

Strictly speaking, the concentrations appearing in this expression are *equilibrium* concentrations. However, remember that we are dealing with a solution of a weak acid, containing a large reservoir of its conjugate base A^-. Because of this, the equilibrium

$$HA(aq) \rightleftharpoons H^+(aq) + A^-(aq)$$

will lie well to the left-hand side, and the acid will be hardly dissociated. The concentration of the acid at equilibrium will therefore effectively be equal to the initial concentration of the acid. In the same way, the concentration of the ion A^- will effectively be the concentration supplied by the salt. We can represent these simplifying assumptions as:

$$[HA(aq)]_{eq} \approx [HA(aq)]_{initial} = [acid]$$

$$[A^-(aq)]_{eq} \approx [A^-(aq)]_{initial} = [anion]$$

Substituting these simplifications into the equation for $[H^+(aq)]$ gives:

$$[H^+(aq)] = K_a \times \frac{[acid]}{[anion]}$$

Notice two things about this relationship. First, because the concentration of hydrogen ions in the solution (and thus its pH) depends on the *ratio* of [acid]:[anion], the pH of such a solution will be very little affected by dilution, since diluting it will not affect the ratio of [acid]:[anion]. Second, when [acid] = [anion], the relationship simplifies further to:

$$[H^+(aq)] = K_a$$

and:

$$pH = pK_a$$

Calculating the pH of a buffer solution

A buffer solution is made by adding 2.05 g of sodium ethanoate to 1000 cm^3 of 0.09 mol dm^{-3} ethanoic acid. What is the pH of the solution produced?

(K_a for ethanoic acid is 1.7×10^{-5} mol dm^{-3}.)

We know that:

$$[H^+(aq)] = K_a \times \frac{[acid]}{[anion]}$$

In this case, [acid] = 0.09 mol dm^{-3}, and [anion] is found as follows:

$$[CH_3COO^-(aq)] = \frac{2.05\ g}{82\ g\ mol^{-1}} \times \frac{1}{1.0\ dm^3}$$

$$= 0.025\ mol\ dm^{-3}$$

Therefore:

$$[H^+(aq)] = 1.7 \times 10^{-5}\ mol\ dm^{-3} \times \frac{0.09\ mol\ dm^{-3}}{0.025\ mol\ dm^{-3}}$$

$$= 6.12 \times 10^{-5}\ mol\ dm^{-3}$$

The pH of the buffer solution is given by:

$$pH = -\log_{10} [H^+(aq)]$$

$$= -\log_{10} (6.12 \times 10^{-5})$$

$$= -(-4.21)$$

$$= 4.21$$

The effectiveness of a buffer solution

How effective is a buffer at stabilising the pH of a solution? Consider the solution in the box above. What happens if 1 cm^3 of 1.0 mol dm^{-3} sodium hydroxide solution is added to the solution containing the buffer?

The initial pH of the buffer solution is 4.21. When 1 cm^3 of 1.0 mol dm^{-3} sodium hydroxide solution is added, the concentration of acid falls and the concentration of anion rises, due to the reaction:

$$OH^-(aq) + CH_3COOH(aq) \rightarrow CH_3COO^-(aq) + H_2O(l)$$

Now the amount of OH^- in 1.0 cm^3 of 1.0 mol dm^{-3} sodium hydroxide is given by:

$$\text{Number of moles of } OH^- = \frac{1\ cm^3}{1000\ cm^3\ dm^{-3}} \times 1.0\ mol\ dm^{-3}$$

$$= 0.001\ mol$$

The solution initially contained 0.09 mol of acid (1.0 dm^3 × 0.09 mol dm^{-3}). Since 1 mol of OH^- removes 1 mol of CH_3COOH, the new concentration of acid is given by:

$$[CH_3COOH(aq)] = (0.090 - 0.001)\ mol\ dm^{-3}$$

$$= 0.089\ mol\ dm^{-3}$$

Similarly, the concentration of CH_3COO^- rises from its initial value of 0.025 mol dm^{-3}:

$$[CH_3COO^-(aq)] = (0.025 + 0.001)\ mol\ dm^{-3}$$

$$= 0.026\ mol\ dm^{-3}$$

(Notice that in both these calculations we have neglected the slight increase in volume of the solution due to adding the sodium hydroxide solution.)

The new concentration of hydrogen ions in the solution is therefore given by:

$$[H^+(aq)] = 1.7 \times 10^{-5}\ \text{mol dm}^{-3} \times \frac{0.089\ \text{mol dm}^{-3}}{0.026\ \text{mol dm}^{-3}}$$

$$= 5.8 \times 10^{-5}\ \text{mol dm}^{-3}$$

The pH of this solution is thus:

$$\text{pH} = -\log_{10} [H^+(aq)]$$

$$= -\log_{10} (5.8 \times 10^{-5})$$

$$= -(-4.24)$$

$$= 4.24$$

This is an increase in pH of 0.03. What would the change in pH have been if the same amount of sodium hydroxide solution had been added to a solution of hydrochloric acid with a pH of 4.21?

From the buffer solution calculation we know that $[H^+(aq)]$ in a solution with a pH of 4.21 is 6.12×10^{-5} mol dm^{-3}. When 0.001 mol of sodium hydroxide are added to 1 dm^3 of hydrochloric acid containing 6.12×10^{-5} mol dm^{-3} of H^+ ions, the final concentration of OH^- ions will be given by:

$$[OH^-(aq)] = (1.00 \times 10^{-3} - 6.12 \times 10^{-5})\ \text{mol dm}^{-3}$$

$$= 9.39 \times 10^{-4}\ \text{mol dm}^{-3}$$

The new value of $[H^+(aq)]$ can be found using the ionic product of water:

$$K_w = [H^+(aq)][OH^-(aq)]$$

so:

$$1.00 \times 10^{-14}\ \text{mol}^2\ \text{dm}^{-6} = [H^+(aq)] \times 9.39 \times 10^{-4}\ \text{mol dm}^{-3}$$

from which:

$$[H^+(aq)] = \frac{1.00 \times 10^{-14}\ \text{mol}^2\ \text{dm}^{-6}}{9.39 \times 10^{-4}\ \text{mol dm}^{-3}}$$

$$= 1.065 \times 10^{-11}\ \text{mol dm}^{-3}$$

The pH of the solution is given by:

$$\text{pH} = -\log_{10} [H^+(aq)]$$

$$= -\log_{10} (1.065 \times 10^{-11})$$

$$= -(-11.0)$$

$$= 11.0$$

This is a pH change of nearly 7 units!

3

The stoichiometry of acid–base reactions

Titrations

Titration is an important laboratory technique used in chemical analysis. A solution containing an unknown quantity of a known substance is placed in a conical flask, and a solution of accurately known strength (called a **standard solution**) is then added from a burette. The addition of this solution continues until the **end point** of the titration is reached. This is signified by a colour change in an **indicator** or some other visible effect, and at this point the titration is stopped. The **equivalence point** of the titration occurs when the two solutions have reacted exactly. If we have chosen our indicator well, the end point of the titration should exactly coincide with its equivalence point.

An indicator is a substance that changes colour when the reaction is complete. In an acid–base titration, the indicator is one colour at one pH and a different colour at another pH. We shall study how indicators work very shortly.

FOCUS BODY BUFFERS

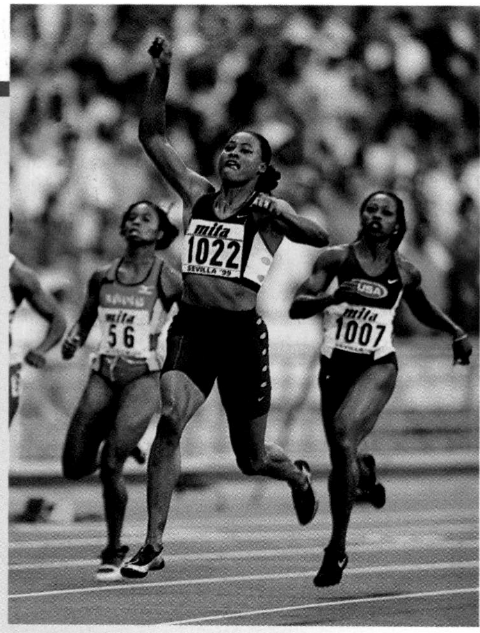

Figure 1 The more exercise we do, the more carbon dioxide we produce and the greater the risk of our blood pH rising.

The human body operates effectively within a very narrow pH range. The normal pH level of the plasma of the blood is 7.4. If this changes significantly, particularly if it becomes more acidic, it affects the functioning of almost every part of the body and can rapidly bring about cell death. Not surprisingly, systems have evolved to prevent such pH changes and in many cases these involve buffers.

All the cells of the body need oxygen to function. This is used in respiration to provide energy for the cells, with carbon dioxide being produced as a waste product. But carbon dioxide in solution forms carbonic acid, which causes a drop in pH.

Buffers and the blood

Oxygen is carried in the blood bound to haemoglobin, a protein containing iron. The key equilibrium for this process is:

$$\mathbf{HHb + O_2 \rightleftharpoons H^+ + HbO_2^-}$$

The presence of H^+ on the right-hand side of this equilibrium means that the transport of oxygen is very sensitive to pH. If the pH falls ($[H^+]$ increases), oxygen will tend to be displaced from the haemoglobin, while if the pH rises ($[H^+]$ decreases), oxygen will tend to bind to haemoglobin more tightly. Both these situations can be potentially life-threatening if the changes in pH are uncontrolled.

Carbon dioxide produced in the body tissues represents the greatest threat to the stability of the blood pH under normal circumstances. Figure 2 shows how the ability of haemoglobin to act as a buffer helps to control the pH of the blood, absorbing hydrogen ions produced by carbon dioxide.

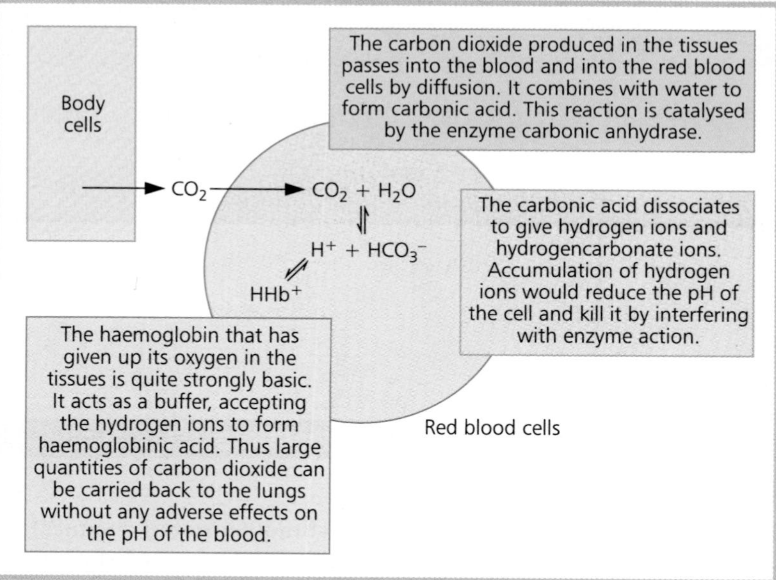

Body cells
The carbon dioxide produced in the tissues passes into the blood and into the red blood cells by diffusion. It combines with water to form carbonic acid. This reaction is catalysed by the enzyme carbonic anhydrase.
CO_2
$CO_2 + H_2O$
$H^+ + HCO_3^-$
HHb^+
The carbonic acid dissociates to give hydrogen ions and hydrogencarbonate ions. Accumulation of hydrogen ions would reduce the pH of the cell and kill it by interfering with enzyme action.
The haemoglobin that has given up its oxygen in the tissues is quite strongly basic. It acts as a buffer, accepting the hydrogen ions to form haemoglobinic acid. Thus large quantities of carbon dioxide can be carried back to the lungs without any adverse effects on the pH of the blood.
Red blood cells

Figure 2 The pH of the blood is controlled in part through the buffering effect of haemoglobin.

Acid and the kidney

The maintenance of a steady blood pH is too important to rely on only one control system. If there are excessive changes in the blood chemistry the kidneys kick in to prevent the pH from shooting up. They excrete hydrogen ions and retain the hydrogencarbonate ions if the pH falls, and retain hydrogen ions if the pH rises. This means that the pH of the urine is very variable, with a normal range from 4.5 to 8.5. A fall in blood pH also stimulates the kidneys to produce ammonia, which combines with hydrogen ions and is excreted in the urine as ammonium salts.

By a combination of the blood buffers and the kidneys the pH of the blood can be maintained within a narrow range, meaning that your body biochemistry stays in balance regardless of whether you are running a marathon or simply watching it on TV!

Titration calculations

A student carried out a titration of an unknown solution of sodium hydroxide. 25.0 cm^3 of the solution were exactly neutralised by the addition of 15.8 cm^3 of 0.025 mol dm^{-3} hydrochloric acid. What was the strength of the sodium hydroxide solution?

The equation for this reaction is:

$$HCl(aq) + NaOH(aq) \rightarrow NaCl(aq) + H_2O(l)$$

The amount of HCl added is given by:

$$\text{Number of moles of HCl} = \frac{15.8 \text{ cm}^3}{1000 \text{ cm}^3 \text{ dm}^{-3}} \times 0.025 \text{ mol dm}^{-3}$$

$$= 3.95 \times 10^{-4} \text{ mol}$$

From the stoichiometry of this reaction we know that this must be the amount of sodium hydroxide present in 25.0 cm^3 of the sodium hydroxide solution. The amount of sodium hydroxide in 1 dm^3 of the solution is then given by:

$$\text{Number of moles of NaOH} = \frac{1000 \text{ cm}^3}{25.0 \text{ cm}^3} \times 3.95 \times 10^{-4} \text{ mol}$$

$$= 1.58 \times 10^{-2} \text{ mol}$$

The concentration of the sodium hydroxide solution is therefore 0.0158 mol dm^{-3}.

Figure 3.3.3 Phenolphthalein is being used as an indicator in this titration of a base against an acid. The pink colour is formed as the pH of the solution rises and the end point is approached.

3

Acid–base indicators

Acid–base indicators are generally weak acids, with a dissociation which can be represented as:

$$HIn(aq) \rightleftharpoons H^+(aq) + In^-(aq)$$

The weak acid HIn and/or its conjugate base In^- is coloured. A change in pH causes a shift in the equilibrium above, which causes a change in colour. For example, the indicator bromothymol blue is yellow in its protonated form (HIn) and blue in its unprotonated form (In^-). In acidic solution below pH 6, bromothymol blue is yellow. Adding OH^- ions removes protons from the right-hand side of the equilibrium, pulling it to the right. This causes more of the unprotonated form of the indicator to be produced, and causes the colour to become blue.

As figure 3.3.4 shows, any given indicator changes colour over a *range* of pH, rather than sharply at one particular pH. Due to the limits of sensitivity of the human eye, this range is normally about 2 pH units. The box 'The pH range of indicators' opposite gives more details about this.

Indicators and end points

When titrating one solution against another, the end point of the reaction occurs when the two solutions have exactly reacted. The ideal indicator for a given acid–base titration is one that is in the middle of its colour change at the pH of the equivalence point of the titration. This pH depends on the acid and base being titrated.

What makes a good indicator?

A good indicator should show a dramatic colour change in order to make it easy to detect the end point of a titration. One of the best indicators from this point of view is phenolphthalein, with a colour change from colourless to pink. In contrast, the colour change of methyl orange (red to yellow) may be difficult to see in dilute solutions.

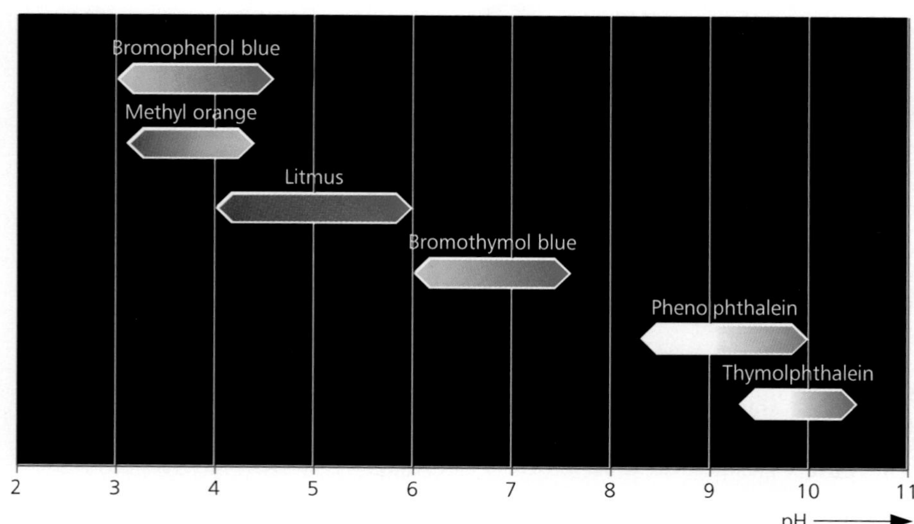

Figure 3.3.4 The colour range and pH range of some indicators

The reason for this can be understood if we think carefully about the equilibrium constant for the dissociation of the indicator K_{in}, which is given by:

$$K_{in} = \frac{[H^+(aq)][In^-(aq)]}{[HIn(aq)]}$$

In the middle of the indicator's colour change, the two forms of the indicator HIn and In^- are present in equal amounts, so $[HIn(aq)] = [In^-(aq)]$. The expression above then simplifies at this point to:

$$K_{in} = [H^+(aq)]$$

If we define pK_{in} in the same way as we defined pH (and pK_a) in section 3.2, then:

$$pK_{in} = pH$$

In other words:

The pK_{in} of an indicator should be equal to (or as close as possible to) the pH at the equivalence point of the titration.

The pH range of indicators

For the two forms of an indicator to be distinguishable, the ratio of HIn to In^- is around 10:1. If, for example, HIn is yellow and In^- is blue, then $[In^-]/[HIn]$ must be at least as small as 1/10 for the solution to appear yellow, and at least as large as 10/1 for it to appear blue.

We know that:

$$K_{in} = \frac{[H^+(aq)][In^-(aq)]}{[HIn(aq)]}$$

When the eye perceives the indicator start to change from yellow to blue,

$$K_{in} = [H^+(aq)] \times \frac{1}{10}$$

Taking logarithms gives:

$$pK_{in} = pH - \log_{10} 0.1$$

or:

$$pH = pK_{in} - 1$$

In the same way, at the other end of the indicator's colour change when the eye sees the indicator as blue,

$$K_{in} = [H^+(aq)] \times \frac{10}{1}$$

Taking logarithms gives:

$$pK_{in} = pH - \log_{10} 10$$

or:

$$pH = pK_{in} + 1$$

Hence the range over which the eye perceives the indicator to change colour is:

$$pH = pK_{in} \pm 1$$

that is, 2 pH units.

3

Choosing an indicator for a titration

At the equivalence point of a titration, the pH must change sharply by several units if the end point of the titration is to coincide accurately with the equivalence point. The pH change during a titration is greatly dependent on the acid and base being titrated, and indicators must be selected accordingly. Figure 3.3.5 gives details of this.

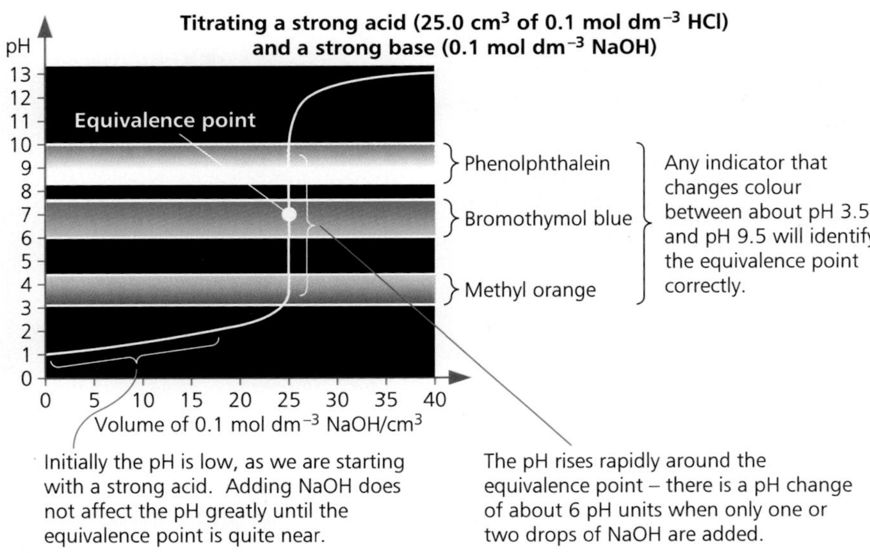

Figure 3.3.5 *(continues)*

Titrating a weak acid (25.0 cm^3 of 0.1 mol dm^{-3} ethanoic acid) and a strong base (0.1 mol dm^{-3} NaOH)

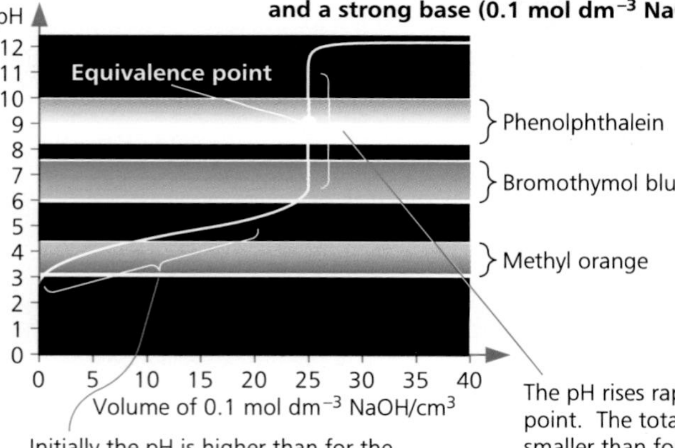

Any indicator that changes colour between about pH 6.5 and pH 11 will identify the equivalence point correctly. Both phenolphthalein and bromothymol blue may be used, but methyl orange will have changed colour long before the equivalence point is reached.

The pH rises rapidly around the equivalence point. The total change is about 4.5 pH units, smaller than for the strong acid/strong base titration. Notice that the pH at the equivalence point is greater than 7 – this is because we have a solution containing a strong base and the salt of a weak acid.

Initially the pH is higher than for the titration of a strong acid. The pH also rises more quickly when the alkali is added.

Titrating a strong acid (25.0 cm^3 of 0.1 mol dm^{-3} HCl) and a weak base (0.1 mol dm^{-3} NH_3)

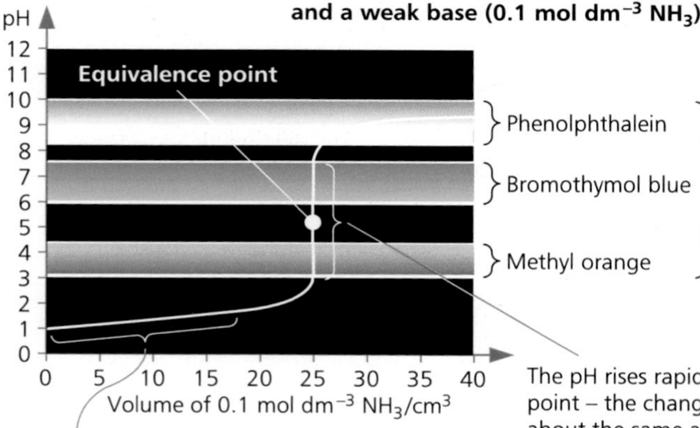

Any indicator that changes colour between about pH 3 and pH 7.5 will identify the equivalence point correctly. Both methyl orange and bromothymol blue may be used, but phenolphthalein will not change colour until long after the equivalence point has been passed.

The pH rises rapidly around the equivalence point – the change is about 4.5 pH units, about the same as for the weak acid/strong base titration. Notice that the pH at the equivalence point is less than 7 – this is because we have a solution containing a weak base and the salt of a strong acid.

Initially the pH is low, as we are starting with a strong acid. Adding NH_3 does not affect the pH greatly until the equivalence point is quite near.

Titrating a weak acid (25.0 cm^3 of 0.1 mol dm^{-3} ethanoic acid) and a weak base (0.1 mol dm^{-3} NH_3)

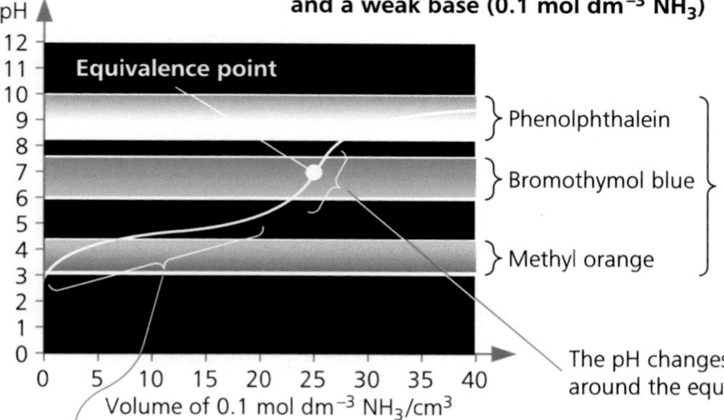

No indicator will change colour dramatically when a small volume of alkali is added at the equivalence point – weak acid/weak base titrations cannot have their equivalence points shown satisfactorily by indicators. Other methods (e.g. a pH meter) must be used.

The pH changes quite slowly around the equivalence point.

Initially the pH is the same as for the other titration involving a weak acid. Once again, the pH rises quite rapidly as alkali is added.

Figure 3.3.5 *(continued)* Selecting indicators for different titrations

SUMMARY

- A **buffer solution** contains a mixture of solutes which resists changes in the pH of the solution.
- Buffer solutions with a pH less than 7 can be made from a weak Brønsted acid and its conjugate base.
- **Titrations** provide an important way to determine practically the stoichiometry of reactions.
- The end point of a titration is shown using an **indicator** – a substance that changes colour when a reaction is complete.
- In acid–base titrations, the indicator changes colour with pH. pK_{in}, the point at which $[HIn(aq)] = [In^-(aq)]$, should be as close as possible to the pH of the solution at the equivalence point of the titration.

QUESTIONS

1 Aspirin is a weak monobasic acid (that is, a substance in which each molecule may donate no more than one proton). Representing aspirin as HAsp and its conjugate base as Asp^-, show how a solution containing aspirin and its sodium salt can function as a buffer solution.

2 The acid dissociation constant K_a for aspirin is 3.27×10^{-4} mol dm^{-3} at 25 °C. A chemist wishes to make a buffer solution with a pH of 4.00 using aspirin and its sodium salt. The chemist starts with a solution containing 0.05 mol dm^{-3} aspirin.

a Calculate: **i** $[H^+(aq)]$ **ii** the pH of this solution.
b Calculate $[H^+(aq)]$ for a solution with pH = 4.00.
c Calculate the concentration of the sodium salt of aspirin required to produce a solution with a pH of 4.00.
d By how much will the pH of the buffer solution in **c** change if 1.00 cm^3 of 0.5 mol dm^{-3} NaOH is added to it?

3 Copy and complete table 3.3.1 for the titration of 0.10 mol dm^{-3} sodium hydroxide solution against 0.10 mol dm^{-3} hydrochloric acid, and then plot a graph like those in figure 3.3.5.

4 Repeat the exercise in question **3**, but using 0.10 mol dm^{-3} ethanoic acid instead of hydrochloric acid. Take the dissociation constant of ethanoic acid as 1.74×10^{-5} mol dm^{-3}.

Volume of 0.10 mol dm^{-3} NaOH added/cm^3	Total volume of solution/cm^3	$[H^+(aq)]$/mol dm^{-3}	pH
	Initial volume of 0.10 mol dm^{-3} hydrochloric acid = 25.00 cm^3		
0	25.00	0.10	1.00
10.00			
15.00			
20.00			
24.90			
24.99			
25.00			
25.01			
25.10			
30.00			
35.00			
40.00			
50.00			

Table 3.3.1

3

FOCUS THE CHEMISTRY OF CORAL REEFS

Coral reefs are some of the most spectacular and ancient structures in the living world – modern coral reefs can be up to 2.5 million years old. Coral is made by tiny organisms known as polyps, which are rather like tiny sea urchins. They build huge colonies, and to protect themselves from attack each polyp produces a hard limestone skeleton around it. It is these limestone skeletons which make up what we know as a coral reef.

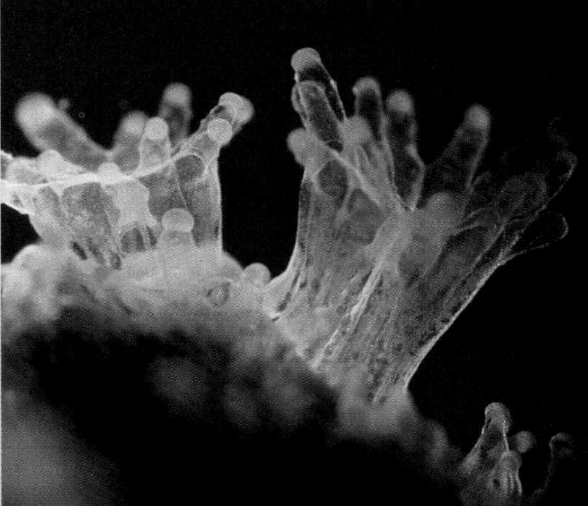

Figure 1 These microscopic creatures carry out the chemistry needed to build coral reefs miles long.

How is a reef formed?

Coral polyps make calcium carbonate which they precipitate out of solution to form their limestone skeleton. Polyps make use of a series of reactions, but the most crucial stage is their ability to capture calcium ions from sea water. Amazingly, although corals cover a tiny fraction (less than 0.2%) of the bottom of the ocean, they capture about half of all the calcium which flows into the oceans each year.

Carbon dioxide from the air dissolves in sea water, forming hydrogen ions and hydrogencarbonate ions:

$$CO_2 + H_2O \rightleftharpoons H^+ + HCO_3^-$$

Coral polyps combine calcium ions with hydrogencarbonate ions from the water around them to form calcium hydrogencarbonate, which they break down to form calcium carbonate, water and carbon dioxide, thus forming their limestone skeleton. The overall equation for this reaction is:

$$Ca^{2+}(aq) + 2HCO_3^-(aq) \rightleftharpoons CaCO_3(s) + H_2O(l) + CO_2(g)$$

Coral is often covered in a growth of algae and plants. These organisms photosynthesise, using up carbon dioxide from the water surrounding the coral. This helps to push the equilibrium producing the limestone skeleton to the right, resulting in a greater production of calcium carbonate.

The growth of coral reefs is very slow: although corals put on new material at different rates, the overall growth of a reef is calculated at about 2.5 cm per year. The growth rate is affected by many things – for example, the same brain coral grows six times faster in the West Indies than it does in Bermuda, which is more northerly and cooler.

Coral reefs act as carbonate factories and to some extent as carbon dioxide sinks, using up some of the CO_2 from the atmosphere. But they are vulnerable to both physical damage and chemical pollution: changes in the purity of the water, particularly changes in pH, damage to the polyps from soil or sewage sediments, commercial fishing tearing up great chunks of coral and global warming. Coral reefs in 93 of the 109 countries which are graced by these ecosystems have been damaged directly or indirectly by human activity. Coral reefs contain a quarter of all known marine species of living organisms. With their complex chemistry and their rich variety of life they are the rain forests of the sea – we destroy them at our peril.

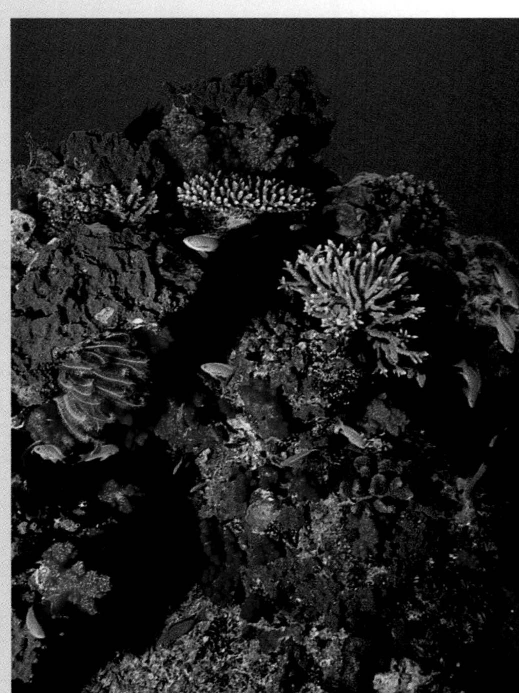

Figure 2 It seems astonishing that coral reefs are produced by polyps using simple precipitation – the Great Barrier Reef in Australia is one of the few creations of living organisms which is visible from space.

3.4 Precipitation and complex ion formation

In this section we shall examine equilibria of ions in solution, in particular, the processes of precipitation and the formation of complex ions. These areas of chemistry, like many others, are applied in a wide variety of situations, and are important in the living world too.

Sparingly soluble salts

Solubility product

How soluble is a substance? This question is rather similar to the question 'How strong is an acid?'. To answer it, we need to look at the situation when the ions from a sparingly soluble salt are in equilibrium with the solid:

$$\mathbf{MX(s) \rightleftharpoons M^+(aq) + X^-(aq)}$$

As an example, consider the equilibrium involving silver chloride:

$$\mathbf{AgCl(s) \rightleftharpoons Ag^+(aq) + Cl^-(aq)}$$

We can write an equilibrium expression for this process in the usual way:

$$K = \frac{[Ag^+(aq)][Cl^-(aq)]}{[AgCl(s)]}$$

Now [AgCl(s)] represents the concentration of a pure solid, which we know is constant (see page 245), so we can rewrite this expression to combine this constant with the equilibrium constant:

$$K_s = K[AgCl(s)] = [Ag^+(aq)][Cl^-(aq)]$$

This combined equilibrium constant K_s is called the **solubility product**. The solubility product of a substance varies with temperature, just like other equilibrium constants. Some typical values of K_s are given in table 3.4.1.

Salt	$M_xN_y(s)$	Equilibrium $\rightleftharpoons$	$xM^{m+}(aq) + yN^{n-}(aq)$	K_s at 25 °C/ $(mol\ dm^{-3})^{(x+y)}$
Halides				
Silver chloride	$AgCl(s)$	$\rightleftharpoons$	$Ag^+(aq) + Cl^-(aq)$	2.0×10^{-10}
Silver bromide	$AgBr(s)$	$\rightleftharpoons$	$Ag^+(aq) + Br^-(aq)$	5.0×10^{-13}
Silver iodide	$AgI(s)$	$\rightleftharpoons$	$Ag^+(aq) + I^-(aq)$	8.0×10^{-17}
Carbonates				
Magnesium carbonate	$MgCO_3(s)$	$\rightleftharpoons$	$Mg^{2+}(aq) + CO_3^{2-}(aq)$	1.0×10^{-5}
Calcium carbonate	$CaCO_3(s)$	$\rightleftharpoons$	$Ca^{2+}(aq) + CO_3^{2-}(aq)$	5.0×10^{-9}
Barium carbonate	$BaCO_3(s)$	$\rightleftharpoons$	$Ba^{2+}(aq) + CO_3^{2-}(aq)$	5.5×10^{-10}
Hydroxides				
Aluminium hydroxide	$Al(OH)_3(s)$	$\rightleftharpoons$	$Al^{3+}(aq) + 3OH^-(aq)$	1.0×10^{-32}
Gold hydroxide	$Au(OH)_3(s)$	$\rightleftharpoons$	$Au^{3+}(aq) + 3OH^-(aq)$	5.5×10^{-46}

Table 3.4.1 Some solubility products

3

In general, for the equilibrium:

$$M_xN_y(s) \rightleftharpoons xM^{m+}(aq) + yN^{n-}(aq)$$

the solubility product expression is:

$$K_s = [M^{m+}(aq)]^x[N^{n-}(aq)]^y$$

Notice that the values of K_s in table 3.4.1 are all very small. The concept of solubility product is only valid for solutions that are quite dilute, so it is not possible to write solubility product expressions for salts like sodium chloride and copper sulphate, which are very soluble.

Calculations involving solubility products

(1) Calculating K_s from solubility data

Given the solubility of a substance (that is, the amount of that substance needed to make a saturated solution), we can calculate K_s. For example, the solubility of silver chromate, Ag_2CrO_4, is 2.22×10^{-2} g dm^{-3} at 25 °C. To calculate K_s at this temperature, we must first find the concentration of silver chromate in solution:

$$\text{Concentration of silver chromate} = \frac{2.22 \times 10^{-2}\ \text{g dm}^{-3}}{332\ \text{g mol}^{-1}}$$
$$= 6.69 \times 10^{-5}\ \text{mol dm}^{-3}$$

The equation for this equilibrium is:

$$Ag_2CrO_4(s) \rightleftharpoons 2Ag^+(aq) + CrO_4^{2-}(aq)$$

so:

$$[Ag^+(aq)] = 2 \times 6.69 \times 10^{-5}\ \text{mol dm}^{-3}$$
$$= 1.338 \times 10^{-4}\ \text{mol dm}^{-3}$$

and:

$$[CrO_4^{2-}(aq)] = 1 \times 6.69 \times 10^{-5}\ \text{mol dm}^{-3}$$
$$= 6.69 \times 10^{-5}\ \text{mol dm}^{-3}$$

The solubility product of silver chromate is given by:

$$K_s = [Ag^+(aq)]^2[CrO_4^-(aq)]$$
$$= (1.338 \times 10^{-4}\ \text{mol dm}^{-3})^2 \times (6.69 \times 10^{-5}\ \text{mol dm}^{-3})$$
$$= 1.20 \times 10^{-12}\ \text{mol}^3\ \text{dm}^{-9}$$

K_s for silver chromate is 1.20×10^{-12} mol^3 dm^{-9} at 25 °C.

(2) Calculating solubility from K_s

We can calculate solubility data from K_s in a similar way to that used for (1) above. K_s for magnesium hydroxide, $Mg(OH)_2$, is 2.0×10^{-11} mol^3 dm^{-9} at 25 °C. To calculate the solubility of magnesium hydroxide at this temperature we must first write down the equation for the equilibrium:

$$Mg(OH)_2(s) \rightleftharpoons Mg^{2+}(aq) + 2OH^-(aq)$$

From this, we can see that:

$$K_s = [Mg^{2+}(aq)][OH^-(aq)]^2$$

If the solubility of magnesium hydroxide is x, then the chemical equation for the equilibrium shows that $[Mg^{2+}(aq)]$ at equilibrium is equal to x, and $[OH^-(aq)]$ is equal to $2x$. This means that we can write:

$$K_s = (x)(2x)^2$$
$$= 4x^3$$

Substituting the value of K_s, we get:

$$2.0 \times 10^{-11}\ \text{mol}^3\ \text{dm}^{-9} = 4x^3$$

from which:

$$x^3 = 5.0 \times 10^{-12}\ \text{mol}^3\ \text{dm}^{-9}$$

giving:

$$x = \sqrt[3]{(5.0 \times 10^{-12}\ \text{mol}^3\ \text{dm}^{-9})}$$
$$= 1.71 \times 10^{-4}\ \text{mol dm}^{-3}$$

This is the **molar solubility** of magnesium hydroxide at 25 °C. The mass of 1 mol of magnesium hydroxide is 58 g, so its solubility can also be expressed as:

$$1.71 \times 10^{-4}\ \text{mol dm}^{-3} \times 58\ \text{g mol}^{-1}$$
$$= 9.92 \times 10^{-3}\ \text{g dm}^{-3}$$

The common ion effect

If calcium carbonate is shaken or stirred with water for long enough, the following equilibrium may be established:

$$CaCO_3(s) \rightleftharpoons Ca^{2+}(aq) + CO_3^{2-}(aq)$$

If we now add a soluble salt of calcium like calcium chloride, the additional Ca^{2+} ions will cause the equilibrium above to shift to the left. Eventually a new equilibrium will be established, but with a lower concentration of carbonate ions than before.

After the addition of calcium chloride, the calcium ion is common to both the original solution and the added solution. Ca^{2+} is therefore known as the **common ion**. The lowering of the solubility of an ionic compound by the addition of a common ion to the solution is known as the **common ion effect**.

Using solubility products and the common ion effect

What is the solubility of magnesium hydroxide in 0.1 mol dm^{-3} sodium hydroxide solution?

We know that the equilibrium involved here is:

$$Mg(OH)_2(s) \rightleftharpoons Mg^{2+}(aq) + 2OH^-(aq)$$

and from the box on the previous page that:

$$K_s = [Mg^{2+}(aq)][OH^-(aq)]^2$$
$$= 2.0 \times 10^{-11}\ mol^3\ dm^{-9} \text{ at } 25\ °C$$

Let the solubility of magnesium hydroxide in 0.1 mol dm^{-3} sodium hydroxide solution be z. The concentration of Mg^{2+} ions in such a solution will be z, while the concentration of hydroxide ions will be $(2z + 0.1\ mol\ dm^{-3})$, that is, $2z$ from the magnesium hydroxide and 0.1 mol dm^{-3} from the sodium hydroxide. Since we know that z will be very small (magnesium hydroxide is very insoluble), we can make the simplification that:

$$[OH^-(aq)] \approx 0.1\ mol\ dm^{-3}$$

The solubility product expression can now be used to find $[Mg^{2+}(aq)]$, and hence the solubility of magnesium hydroxide under these circumstances:

$$2.0 \times 10^{-11}\ mol^3\ dm^{-9} = z \times (0.1\ mol\ dm^{-3})^2$$

so:

$$z = \frac{2.0 \times 10^{-11}\ mol^3\ dm^{-9}}{(0.1\ mol\ dm^{-3})^2}$$
$$= 2.0 \times 10^{-9}\ mol\ dm^{-3}$$

The solubility of magnesium hydroxide in 0.1 mol dm^{-3} sodium hydroxide solution is 2.0×10^{-9} mol dm^{-3} at 25 °C. This compares with a solubility in water of 1.7×10^{-4} mol dm^{-3} at the same temperature which we calculated on page 281, over 100 000 times less.

Figure 3.4.2 The common ion effect can be used to soften water, using the carbonate ion as the common ion. Hard water often contains calcium ions which form insoluble salts with soap. These salts float on the surface of the water, and are usually called 'scum'. Adding washing soda ($Na_2CO_3.10H_2O$) precipitates the calcium ions out as calcium carbonate, pushing the equilibrium:

$$CaCO_3(s) \rightleftharpoons Ca^{2+}(aq) + CO_3^{2-}(aq)$$

to the left and dramatically reducing the concentration of calcium ions remaining dissolved. Bath salts contain sodium sesquicarbonate ($Na_2CO_3.NaHCO_3.2H_2O$), which works in the same way but is not so basic.

Predicting precipitation

Solubility products may be used to predict whether a precipitate will form when two or more solutions are mixed. For example, what will happen if a solution of 0.030 mol dm^{-3} sodium chloride is mixed with an equal volume of 0.30 mol dm^{-3} lead(II) nitrate?

The first thing to note in a situation like this is that after mixing equal volumes of two solutions, the concentrations of the solutes in them will be *halved*. Hence we have a solution in which initially:

$$[NaCl(aq)] = 0.015\ mol\ dm^{-3} \quad \text{and} \quad [Pb(NO_3)_2(aq)] = 0.15\ mol\ dm^{-3}$$

This solution is liable to precipitate lead chloride, the least soluble of the salts that could be formed. The solubility product of lead chloride at 25 °C is $1.7 \times 10^{-5}\ mol^3\ dm^{-9}$.

The equation for the equilibrium between lead chloride and its aqueous ions is:

$$PbCl_2(s) \rightleftharpoons Pb^{2+}(aq) + 2Cl^-(aq)$$

for which:

$$K_s = [Pb^{2+}(aq)][Cl^-(aq)]^2$$
$$= 1.7 \times 10^{-5}\ mol^3\ dm^{-9}$$

In the solution,

$$[Cl^-(aq)] = 0.015\ mol\ dm^{-3} \quad \text{and} \quad [Pb^{2+}(aq)] = 0.15\ mol\ dm^{-3}$$

so:

$$[Pb^{2+}(aq)][Cl^-(aq)]^2 = (0.15\ mol\ dm^{-3}) \times (0.015\ mol\ dm^{-3})^2$$
$$= 3.4 \times 10^{-5}\ mol^3\ dm^{-9}$$
$$> K_s$$

This is larger than the solubility product, so a precipitate *will* form.

Selective precipitation

Oxides and water

The oxide ion does not exist in aqueous solution.

Most metal oxides that are insoluble in water will dissolve in acidic solutions. For example, the oxide of iron, Fe_2O_3, is readily soluble in hydrochloric acid:

$$Fe_2O_3(s) + 6H^+(aq) \rightarrow 2Fe^{3+}(aq) + 3H_2O(l)$$

This is the basis of many rust treatments – the acid reacts with the oxide ion and 'unlocks' the Fe^{3+} ion from the rust, allowing it to be washed away.

Other metal oxides react with water as the acid:

$$K_2O(s) + H_2O(l) \rightarrow 2K^+(aq) + 2OH^-(aq)$$

The reason for this reaction is that the O^{2-} ion is simply too basic to exist in aqueous solution (its K_b is estimated as about $10^{22}\ mol\ dm^{-3}$!), so it reacts with water, forming hydroxide ions.

The reverse of this process is occasionally seen – mercury forms an insoluble *oxide* when Hg^{2+} ions react with hydroxide ions:

$$Hg^{2+}(aq) + 2OH^-(aq) \rightarrow HgO(s) + H_2O(l)$$

Effectively the Hg^{2+} ion is able to react with two hydroxide ions, removing O^{2-} and leaving water. The same behaviour is seen with silver ions, Ag^+.

Sulphides and water

Sulphur is in the same group of the periodic table as oxygen, and there are many similarities to the oxides in the behaviour of the sulphides. Like the oxide ion, the sulphide ion does not exist in aqueous solution, even in 8 mol dm^{-3} sodium hydroxide solution. Sodium sulphide dissolves in water by reacting with it, in the same way as sodium oxide:

$$Na_2S(s) + H_2O(l) \rightarrow 2Na^+(aq) + HS^-(aq) + OH^-(aq)$$

Limewater and carbon dioxide – the disappearing precipitate

Limewater is a colourless solution produced when calcium hydroxide (slaked lime) is mixed with water and the resulting milky suspension is filtered. Limewater is useful as a test solution for carbon dioxide. As carbon dioxide bubbles through the colourless solution it turns milky due to the precipitation of calcium carbonate. However, in an excess of carbon dioxide the solution turns clear again, as the calcium carbonate reacts to form soluble calcium hydrogencarbonate and redissolves.

$$Ca^{2+}(aq) + 2OH^-(aq) + CO_2(g) \rightarrow CaCO_3(s) + H_2O(l)$$

$$CaCO_3(s) + H_2O(l) + CO_2(g) \rightarrow Ca^{2+}(aq) + 2HCO_3^-(aq)$$

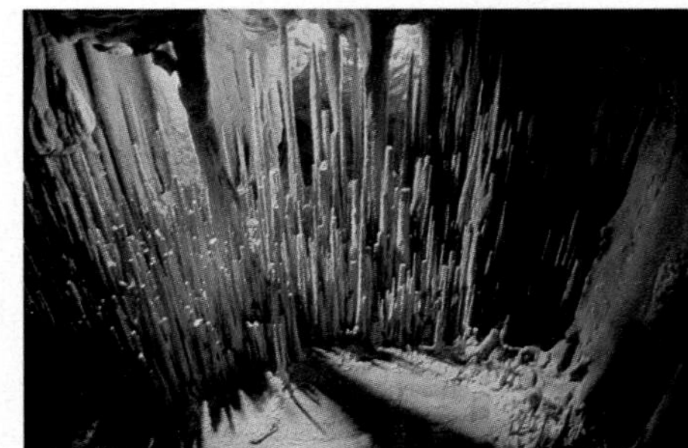

Figure 3.4.3 Drop by drop, over millions of years, dissolved calcium ions and carbonate ions have precipitated and formed these spectacular formations in California.

A few metal oxides may be formed by the reverse reaction, of the metal ion with hydroxide ions, as we have just seen. The HS^- ion behaves in a similar way, forming a precipitate of the sulphide with many metal ions.

A small amount of the HS^- ion is formed when hydrogen sulphide is bubbled through water:

$$\mathbf{H_2S(aq) + H_2O(l) \rightleftharpoons HS^-(aq) + H_3O^+(aq) \quad (K_a = 8.9 \times 10^{-8}\ mol\ dm^{-3})}$$

The presence of H_3O^+ on the right-hand side of this equilibrium means that the precipitation of metal sulphides is sensitive to pH. Under acidic conditions, the reverse of this reaction happens – the equilibrium is pushed to the left, the concentration of HS^- is lowered, and only the more insoluble sulphides are precipitated. These are the **acid insoluble** sulphides in table 3.4.2. As the pH is raised, the more soluble sulphides are precipitated as the concentration of HS^- ions rises – these are the **acid soluble** sulphides in table 3.4.2.

Before more sophisticated spectroscopic techniques of analysis were developed, this behaviour provided a way of selectively precipitating and identifying metal ions from a solution containing a mixture of metal ions, by carefully controlling the pH at which the precipitation occurs. Identification of the metal ion was then possible from the colour of the precipitate formed.

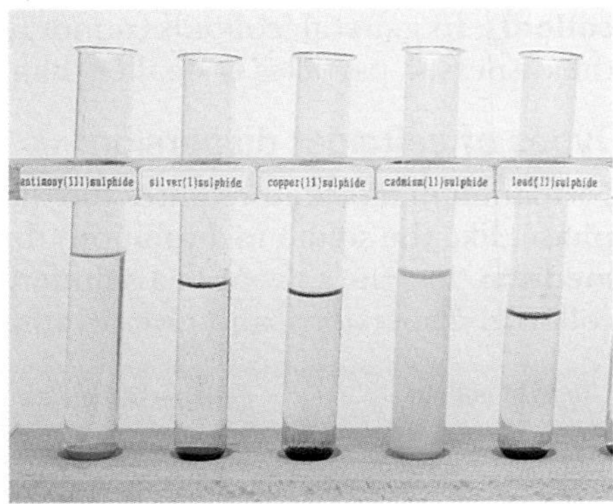

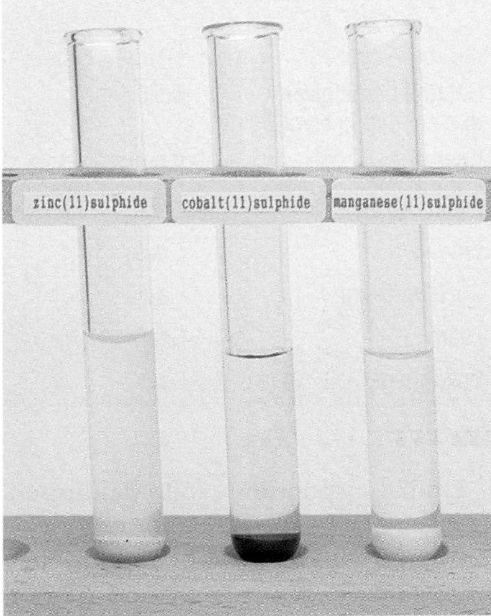

Table 3.4.2 The solubility products and colours of some metal sulphides

Metal ion	K_s for M_xS_y/ (mol dm^{-3})$^{(x+y)}$	Colour of precipitate
Acid insoluble sulphides		
Sb^{3+}	1.7×10^{-93}	Orange-red
Ag^+	6.3×10^{-51}	Black
Cu^{2+}	6.3×10^{-36}	Black
Cd^{2+}	1.6×10^{-28}	Yellow
Pb^{2+}	1.3×10^{-28}	Black
Acid soluble sulphides		
Zn^{2+}	1.6×10^{-24}	White
Co^{2+}	4.0×10^{-21}	Black
Mn^{2+}	2.5×10^{-10}	Pink

Solutions, colloids and suspensions

The importance of particle size

So far in this section we have been concerned with ions and molecules in solution. Experience tells us that no matter how long a solution is left, it remains **homogeneous** – that is to say, the ions or molecules in it remain completely mixed with the water. Compare this with how very fine sand behaves when mixed with water. The sand and water mixture remains more or less homogeneous as long as it is shaken, but begins to settle once the shaking stops.

The shaken mixture of sand and water is an example of a **suspension**. It behaves differently from a solution because it contains particles that are much larger than the ions or molecules of solute contained in a solution, which have dimensions of the order of 0.1 nm (10^{-10} m). The particles making up a suspension are larger in at least one dimension than about 1000 nm (10^{-6} m). Because of this, suspensions can usually be separated by filtration, since the particles are larger than the pores in a piece of filter paper.

If the size of the suspended particles lies between about 1 nm (10^{-9} m) and 1000 nm, the mixture is a **colloidal dispersion** (often referred to simply as a

3

colloid). In general, colloids cannot be separated by filtration, since the size of the dispersed particles is smaller than that of the pores in filter paper.

Types of colloidal dispersion

Formation of a colloid requires the dispersion of one phase (the **dispersed phase**, like the solute in a solution) through another phase (the **dispersing medium**, like the solvent in a solution). Table 3.4.3 shows the different types of colloidal dispersions, and gives examples of each.

Type of colloid	Dispersed phase	Dispersing medium	Examples
Smoke	Solid	Gas	Dust, smoke from fires
Sol	Solid	Liquid	'Emulsion' paint, starch in water, dessert jelly (when not set)
Gel (the name given to a sol that is semi-rigid)	Solid	Liquid	Dessert jelly (when set)
Solid sol	Solid	Solid	Metal alloys, pearls
Liquid aerosol	Liquid	Gas	Clouds
Emulsion	Liquid	Liquid	Milk, hand cream
Solid emulsion	Liquid	Solid	Cheese
Foam	Gas	Liquid	Whipped cream
Solid foam	Gas	Solid	Pumice, marshmallow

Table 3.4.3

Unlike suspensions, colloidal dispersions in the fluid state do not tend to separate out, as the influence of gravity on the dispersed particles is small compared with the jostling motion they experience from the fluid surrounding them. (This jostling was discussed briefly in the box 'Evidence for the existence of atoms' on page 4.) Even so, some colloidal dispersions do tend to separate out with time, as anyone who has made salad dressing from oil and vinegar knows. The initial shaking of oil and vinegar produces a mixture containing oil droplets that are of the right size to form a colloidal dispersion. Once shaking has ceased, however, these tiny droplets gradually coalesce and merge to form larger drops that are sufficiently massive for gravity to pull them to the bottom of the container.

To produce a stable colloid we must not only produce dispersed particles of the right size – we must also prevent these particles from joining back together. One way of doing this is to ensure that all the particles have the same kind of electrical charge. This is the case in many sols, where solid particles attract ions with a particular charge to their surface, as figure 3.4.4(a) shows. Some large protein particles form stable colloidal dispersions due to the presence of charged groups on the outside of the protein, shown in figure 3.4.4(b).

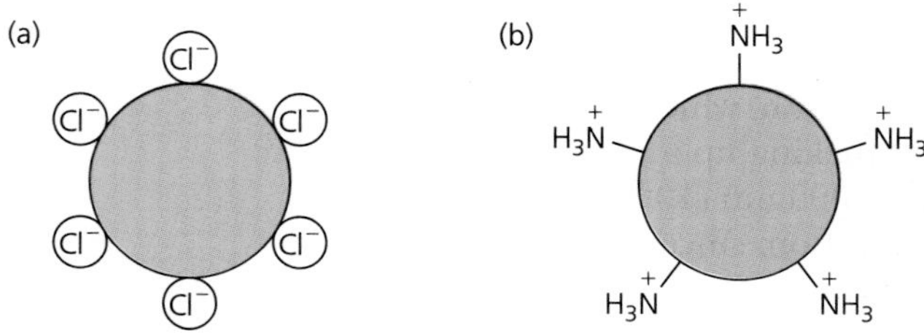

Figure 3.4.4 The presence of similar electrical charges around colloidal particles stabilises the dispersion. Diagram (a) shows a colloidal particle that has attracted chloride ions to itself, while (b) shows the charged groups around the outside of a large protein molecule. In neither case can the particles merge with other particles to form larger particles which would eventually separate out from the dispersing medium.

Emulsions are often stabilised by an **emulsifying agent**, which serves to prevent the droplets in the dispersed phase from coalescing. Milk is an emulsion in which droplets of fat are coated with the protein casein. The protein forms a layer around each droplet, and acts as an emulsifying agent, keeping the droplets of oil dispersed throughout the dispersing medium, water. Margarine is an example of a water-in-oil emulsion, in which the organic molecules lecithin and glyceryl monostearate act as emulsifying agents.

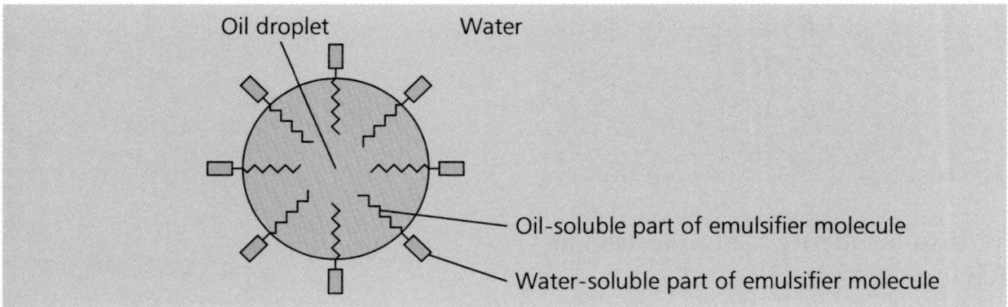

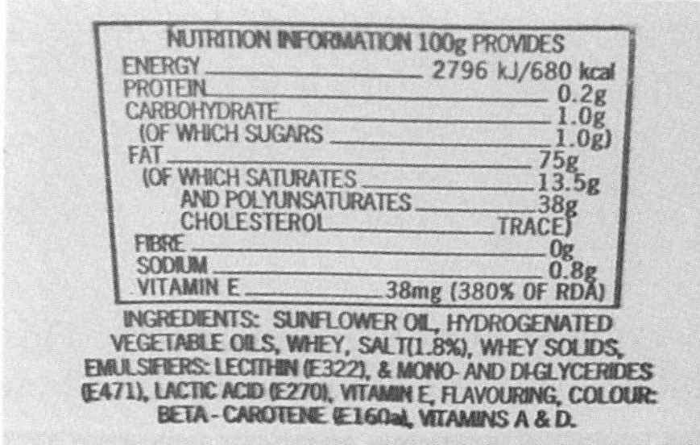

Figure 3.4.5 Emulsifier molecules have one end that interacts strongly with water, while the other end interacts strongly with oils.

Destroying colloids

If we wish to destroy a colloidal suspension we need to bring the dispersed particles together in order to separate them out from the dispersing medium, a process known as **flocculation**. One common method of doing this involves adding ionic substances such as magnesium chloride or aluminium chloride to the colloid. The ions interact strongly with the dispersed particles in the colloid, forming particles large enough either to settle out under the influence of gravity, or to be filtered out. This is the principle by which water is treated to remove suspended solids, where aluminium sulphate is added to the water prior to filtering.

Complex ions

Stability constants

We saw in section 2.7 how the transition metals form complex ions when species called ligands attach themselves to metal ions in solution. An example of this is the $CuCl_4^{2-}$ ion – we studied the equilibrium expression for the formation of this ion in section 3.1. In fact, the formation of such ions is a stepwise process involving many species, so the equilibrium expression is potentially quite complicated. Fortunately, the situation is quite simple if the ligand is present in excess, which is the usual situation. For the formation of $CuCl_4^{2-}$, the equilibrium is:

$$[Cu(H_2O)_6]^{2+}(aq) + 4Cl^-(aq) \rightleftharpoons CuCl_4^{2-}(aq) + 6H_2O(l)$$

The concentration of water can be taken as constant and the stability constant is written:

$$K_{stab} = \frac{[CuCl_4^{2-}(aq)]}{[[Cu(H_2O)_6]^{2+}(aq)][Cl^-(aq)]^4}$$

The larger the value of this constant, the more stable the complex ion. Table 3.4.4 overleaf shows the stability constants for some complex ions.

A ligand may be displaced from a metal ion in the presence of another ligand that can form a complex with a larger stability constant. For example, water may be displaced as the ligand from the $[Cu(H_2O)_6]^{2+}$ ion by ammonia, forming the $[Cu(H_2O)_2(NH_3)_4]^{2+}$ ion. This in turn may be displaced by the polydentate ligand $edta^{4-}$ to form $[Cu(edta)]^{2-}$. Finally, $edta^{4-}$ may be displaced

Figure 3.4.6 Colloidal dispersions scatter light. Dust and other small solid and liquid particles suspended in the Earth's atmosphere scatter light by amounts that vary according to its wavelength. Such scattering leads to spectacular sunsets, like that shown here.

Ligand	Equilibrium $M^{m+} + xL^{l-}$		$ML_x^{(m-xl)+}$	Stability constant at 25 °C/$(mol\ dm^{-3})^x$
NH_3	$Ag^+ + 2NH_3$	⇌	$[Ag(NH_3)_2]^+$	1.7×10^7
	$Cu^{2+} + 4NH_3$	⇌	$[Cu(NH_3)_4]^{2+}$	1.4×10^{13}
	$Co^{3+} + 6NH_3$	⇌	$[Co(NH_3)_6]^{3+}$	4.5×10^{33}
Cl^-	$Cu^{2+} + 4Cl^-$	⇌	$CuCl_4^{2-}$	4.0×10^5
	$Hg^{2+} + 4Cl^-$	⇌	$HgCl_4^{2-}$	1.7×10^{16}
CN^-	$Cd^{2+} + 4CN^-$	⇌	$[Cd(CN)_4]^{2-}$	7.1×10^{16}
	$Fe^{2+} + 6CN^-$	⇌	$[Fe(CN)_6]^{4-}$	1.0×10^{24}
	$Hg^{2+} + 4CN^-$	⇌	$[Hg(CN)_4]^{2-}$	2.5×10^{41}

Table 3.4.4 The stability constants for some complex ions. The large stability constants for complexes with cyanide ions (CN^-) as ligands show why cyanide is such a powerful poison – it binds almost irreversibly to Fe^{2+} ions in the body. Fe^{2+} ions are essential for the transport of oxygen in the blood. Cyanide binds more strongly than oxygen to the Fe^{2+} ion in haemoglobin, and this stops the carriage of oxygen in the blood.

by cyanide, forming $[Cu(CN)_4]^{2-}$. This series of displacements is summarised as follows, with the stability constant beneath each ion:

$$[Cu(H_2O)_6]^{2+} \xrightarrow{\text{excess } NH_3} \underset{1.4 \times 10^{13}\ mol^4\ dm^{-12}}{[Cu(H_2O)_2(NH_3)_4]^{2+}} \xrightarrow{\text{excess edta}^{4-}} \underset{6.3 \times 10^{18}\ mol\ dm^{-3}}{[Cu(edta)]^{2-}} \xrightarrow{\text{excess } CN^-} \underset{2.0 \times 10^{27}\ mol^4\ dm^{-12}}{[Cu(CN)_4]^{2-}}$$

Complex ions and the solubility of salts

The formation of complex ions in solution may have a dramatic effect on the solubility of a sparingly soluble salt, as the following example shows.

Silver bromide, AgBr, is only sparingly soluble in water, the equilibrium being:

$$AgBr(s) \rightleftharpoons Ag^+(aq) + Br^-(aq) \qquad \textbf{solubility equilibrium}$$

K_s for this equilibrium is 5.0×10^{-13} $mol^2\ dm^{-6}$, which tells us that the equilibrium lies well over to the left.

Addition of ammonia to the solution produces an additional equilibrium reaction:

$$Ag^+(aq) + 2NH_3(aq) \rightleftharpoons [Ag(NH_3)_2]^+(aq) \qquad \textbf{complex equilibrium}$$

K for this reaction is 1.7×10^7 $mol^2\ dm^{-6}$ – the equilibrium lies well to the right.

Adding ammonia to a solution of silver bromide in equilibrium with its aqueous ions will result in the formation of the complex ion $[Ag(NH_3)_2]^+(aq)$ from the trace amount of aqueous silver ions present. The complex equilibrium lies well to the right, so the large majority of the silver ions that are in solution will form complexes. This complexing of $Ag^+(aq)$ ions in the solution will cause the solubility equilibrium to move to the right, dissolving more silver bromide. This reaction is used to distinguish the halides, as we saw in section 2.4.

In general:

The solubility of a sparingly soluble salt is increased when one of its ions forms a complex ion in solution.

Acid–base behaviour of complex ions with water as a ligand

We already know from section 2.5 that the aluminium ion Al^{3+} may behave as an acid in water:

$$[Al(H_2O)_6]^{3+}(aq) \rightleftharpoons [Al(H_2O)_5(OH)]^{2+}(aq) + H^+(aq) \qquad K_a = 10^{-5}\ mol\ dm^{-3}$$

Table 3.2.1 on page 262 showed that the Fe^{3+} ion behaves like this too:

$$[Fe(H_2O)_6]^{3+}(aq) \rightleftharpoons [Fe(H_2O)_5(OH)]^{2+}(aq) + H^+(aq) \quad K_a = 6 \times 10^{-3}\ mol\ dm^{-3}$$

This behaviour is general to cations that are surrounded by water molecules. The positive charge on the ion increases the polarity of the already polar O—H bond, favouring the forward reaction in the equilibrium:

$$[M(H_2O)_n]^{m+} + B(aq) \rightleftharpoons [M(H_2O)_{n-1}(OH)]^{(m-1)+} + BH^+(aq)$$

in which $M(H_2O)_n{}^{m+}$ acts as the Brønsted acid and B as the Brønsted base.

As we would expect, ions that are small and highly charged tend to increase the polarisation of the O—H bond most. This description obviously applies to Fe^{3+} and Al^{3+}, and also to other ions like Mg^{2+} and Cu^{2+} to some extent.

The position – and hence the value of K_a – for the equilibrium above will also depend on the strength of the base B. When M is Fe^{3+} or Al^{3+}, even water is a strong enough base to remove a proton from one of the ligand water molecules. In contrast, ions like Zn^{2+} will only show acidic behaviour in the presence of stronger bases like ammonia.

The charge on the complex ion becomes decreasingly positive as more protons are removed. In principle such ions may have up to six values of K_a for the loss of successive protons, although in practice it is likely that only the first two or three such dissociations will happen under normal conditions. The example of aluminium in section 2.5, page 190, illustrates this.

SUMMARY

- The solubility of a sparingly soluble substance may be expressed as a combined equilibrium constant known as the **solubility product**, K_s. For the sparingly soluble solute AB_2 dissolving in water:

$$AB_2(s) \rightleftharpoons A^{2+}(aq) + 2B^-(aq)$$

K_s is given by:

$$K_s = [A^{2+}(aq)][B^-(aq)]^2$$

$[AB_2(s)]$ is a constant, which is incorporated in K_s.

- The solubility product of a substance varies with temperature.
- The addition of a **common ion** to a solution, causing the lowering of the solubility of another ionic compound, is known as the **common ion effect**.
- Solubility products may be used to predict whether a precipitate will form when two or more solutions are mixed.
- Complex ions result when ligands form coordinate bonds to other ions in solution. For the equilibrium:

$$M^{m+}(aq) + 6L^-(aq) \rightleftharpoons ML_6(aq)^{(6-m)-}$$

the **stability constant** for the complex ion formed can be written as:

$$K_{stab} = \frac{[ML_6(aq)]^{(6-m)-}}{[M^{m+}(aq)][L^-(aq)]^6}$$

as long as L^- is present in excess.

- The solubility of a sparingly soluble salt is increased when one of its ions forms a complex ion in solution.

QUESTIONS

1 At 25 °C the solubility of calcium sulphate in water is 4.50×10^{-3} mol dm^{-3}. What is its solubility product at this temperature?

2 Barium ions are very effective absorbers of X-rays, and are used in 'barium meals', despite the fact that they are extremely poisonous. A barium meal is safe because the barium is in the form of barium sulphate, $BaSO_4$, which has a solubility product of 1.0×10^{-10} mol^2 dm^{-6}. What is the solubility of barium sulphate in g dm^{-3}?

3 The solubility product of magnesium hydroxide, $Mg(OH)_2$, is 2.0×10^{-11} mol^3 dm^{-9}. What is its solubility in 0.1 mol dm^{-3} sodium hydroxide solution?

4 The solubility products for the silver halides are given in table 3.4.5.

AgCl	AgBr	AgI
2.0×10^{-10} mol^2 dm^{-6}	5.0×10^{-13} mol^2 dm^{-6}	8.0×10^{-17} mol^2 dm^{-6}

Table 3.4.5

The stability constant for the formation of the $[Ag(NH_3)_2]^+$ ion is 1.7×10^7 mol^{-2} dm^6. Different halide ions in solution can be distinguished by:
(1) acidifying the solution with dilute nitric acid
(2) adding silver nitrate solution
(3) testing the precipitate by adding concentrated ammonia solution, when a silver chloride precipitate dissolves readily, a silver bromide precipitate dissolves sparingly, and a silver iodide precipitate is insoluble.
Explain the principles of steps (1), (2) and (3).

5 The stability constant for the formation of $[Hg(CN)_4]^{2-}$ in aqueous solution is 2.5×10^{41} mol^{-4} dm^{12}, for which the equilibrium may be written:

$$Hg^{2+}(aq) + 4CN^-(aq) \rightleftharpoons [Hg(CN)_4]^{2-}(aq)$$

If the concentration of free cyanide ions in a solution of $[Hg(CN)_4]^{2-}$ is 0.01 mol dm^{-3}, what is the ratio of the number of Hg^{2+}(aq) ions to $[Hg(CN)_4]^{2-}$(aq) ions in solution?

Developing Key Skills

Coral reefs are dying as a result of both global warming and pollution. The structure of the reefs is being damaged and in some cases the coral is dissolving away. BBC's Newsround is planning a feature on the formation of coral reefs and the reasons why global warming and pollution are having such devastating effects. Produce a suitable script for a $2\frac{1}{2}$ to 3 minute feature. It is important that the chemistry of the reef is explained carefully so that the viewers understand what is really going on in the oceans of the world.

[Key Skills opportunities: C]

3.5 Electrochemistry

Many chemical systems involve reactions in which reduction and oxidation occurs – the **redox** reactions already met in sections 1.4 and 2.3. The chemistry of redox reactions helps us to understand such diverse processes as rusting, the manufacture of chlorine, and the production of electricity from batteries, as well as enabling us to monitor the progress of chemical reactions using electrical measurements. Redox reactions form a vital part of the complex web of life too – in nerve impulses, in the reactions that produce energy from our food, and ultimately in the process on which all life on Earth depends – photosynthesis.

In this section we shall look more closely at **electrochemistry** – the chemistry that all these **electrochemical processes** share.

Figure 3.5.1 Photosynthesis is a redox reaction which is energetically unfavourable – the transfer of electrons from water to carbon dioxide. The presence of the green pigment chlorophyll in the leaves of plants makes the reaction possible, through the capture of energy from sunlight.

Electrolysis

The electrolysis of molten salts

When electricity flows through a molten (or **fused**) salt or through a solution of an electrolyte, the salt or electrolyte is split up in a chemical process called **electrolysis**. (The substance undergoing electrolysis must be molten or in solution so that its ions can move, since they are the particles that carry the current. In a metallic solid, the charge-carrying particles are electrons.) The electrolysis of fused sodium chloride is shown in figure 3.5.2.

We know from section 2.3 that oxidation involves the loss of electrons, while reduction involves a gain of electrons. At the anode of an electrolytic cell, negatively charged ions (called **anions** because they are attracted to the positive anode) are oxidised – electrons are removed from them. The direct current source connected to the cell pumps these electrons round the circuit to the cathode, where positively charged ions (**cations**, attracted to the negative cathode) are reduced – electrons are added to them. For the electrolysis of sodium chloride, the chemical changes happening at the two electrodes can be summarised by writing two **half-equations**, in the same way as we did for the reaction between sodium and oxygen on page 159:

$$2Cl^-(l) \rightarrow Cl_2(g) + 2e^- \qquad \textbf{anode (oxidation)}$$

$$Na^+(l) + e^- \rightarrow Na(l) \qquad \textbf{cathode (reduction)}$$

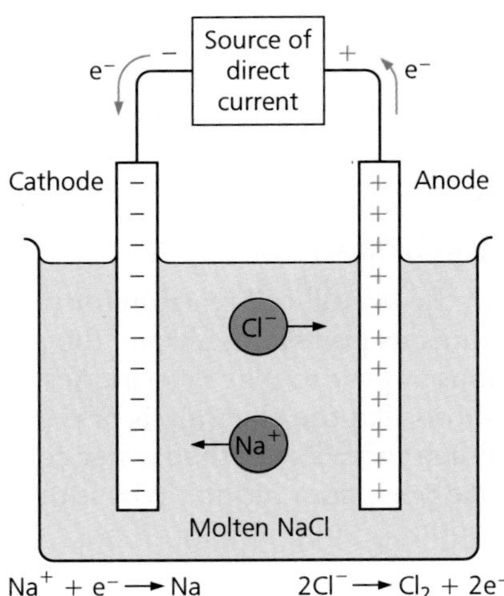

Figure 3.5.2 An electrolytic cell in which molten sodium chloride is being electrolysed to form metallic sodium and gaseous chlorine

The electrolysis of aqueous solutions

The electrolysis of aqueous solutions is not as straightforward as the electrolysis of molten salts, since water may be split in the process. This makes it difficult to predict exactly what products are likely to be formed. For example, in the electrolysis of aqueous potassium nitrate, KNO_3, the products are hydrogen at the cathode and oxygen at the anode, formed by the electrolysis of water:

$$2H_2O(l) \rightarrow O_2(g) + 4H^+(aq) + 4e^- \qquad \textbf{anode (oxidation)}$$

$$4H_2O(l) + 4e^- \rightarrow 2H_2(g) + 4OH^-(aq) \qquad \textbf{cathode (reduction)}$$

(Remember that oxidation involves a gain of electrons and reduction a loss of electrons.) This problem of predicting exactly what will happen during electrolysis of aqueous solutions is due to the very complicated processes that occur at the surface of an electrode, and the best guide to making such predictions is likely to be experience.

The cell reaction

The overall reaction taking place in an electrolytic cell is called the **cell reaction**. This can be found by adding together the two half-equations for the cell, ensuring that the number of electrons in each is the same. In the case of the electrolysis of fused sodium chloride, the half-equations are combined like this:

$$\mathbf{2Cl^-(l) \rightarrow Cl_2(g) + 2e^-}$$

$$\mathbf{2Na^+(l) + 2e^- \rightarrow 2Na(l)}$$

Overall reaction:

$$\mathbf{2Na^+(l) + 2Cl^-(l) \xrightarrow{electrolysis} 2Na(l) + Cl_2(g)}$$

We already know that this change does not happen spontaneously – quite the reverse, since sodium and chlorine react to form sodium chloride in a highly exothermic process. The fact that the splitting of sodium chloride requires energy is indicated by the word ‘electrolysis’ above the arrow in the equation.

The overall cell reaction for the electrolysis of potassium nitrate solution can be found in the same way:

$$\mathbf{2H_2O(l) \rightarrow O_2(g) + 4H^+(aq) + 4e^-}$$

$$\mathbf{4H_2O(l) + 4e^- \rightarrow 2H_2(g) + 4OH^-(aq)}$$

Overall reaction:

$$\mathbf{6H_2O(l) \rightarrow 2H_2(g) + O_2(g) + \underbrace{4H^+(aq) + 4OH^-(aq)}_{4H_2O(l)}}$$

which simplifies to:

$$\mathbf{2H_2O(l) \xrightarrow{electrolysis} 2H_2(g) + O_2(g)}$$

The net cell reaction for the electrolysis of aqueous potassium nitrate therefore involves the splitting of two water molecules to form two molecules of hydrogen and one molecule of oxygen.

The missing potassium nitrate

In the overall cell reaction for the electrolysis of aqueous potassium nitrate, the potassium and nitrate ions appear to play no part. Yet a simple experiment shows that the electrolysis of pure water proceeds much more slowly than water containing an electrolyte like potassium nitrate. Why is this?

The presence of ions in the solution ensures that there is always electrical neutrality around the electrodes. Without nitrate ions present, the production of H^+ ions at the anode rapidly leads to an excess of positive charge in this region. In the presence of potassium nitrate, nitrate ions diffuse towards the anode as H^+ ions are formed, producing overall electrical neutrality. The reverse happens at the cathode, where potassium ions ensure electrical neutrality as OH^- ions are produced.

FOCUS THE BIRTH OF ELECTROCHEMISTRY

In 1791 James Faraday (a rather hard-up blacksmith) and his wife Margaret quietly celebrated the birth of their third child and second son. No-one could have imagined the impact that their new baby, Michael, was going to have on the world.

Michael Faraday received little more than a primary education – by his own admission he was far happier out playing on the streets of London with his friends! At the age of 14 he was apprenticed to a bookbinder, and it was by reading the pages of the texts he was supposed to be binding that Faraday became fascinated by science. After hearing a lecture by the famous chemist Sir Humphrey Davy, Faraday made a bound copy of the notes he had taken and sent them to the great man himself. At the age of 21 Faraday was appointed assistant to Davy in the laboratory of the Royal Institution in London. He went on to make many important discoveries, including the first electric motor, the dynamo which still underpins all electricity generation, the chemistry of benzene – and what we now call electrochemistry.

Figure 1 Faraday's understanding of electrolysis shed light on a subject which was puzzling many of his contemporaries.

Electrolysis

Around 1820 a major controversy broke out in the scientific community. It had been noticed that if an electric current flowed through water two gases, hydrogen and oxygen, were produced. This observation caused mayhem. Theories abounded among the great scientists of the day, but none stood up to testing and none of them was right!

In 1832 Michael Faraday entered the arena and began to apply his mind to the observations that had been made. He worked by passing electricity through acidified water and the conclusions he had reached by 1833 completely revolutionised the way that **electrolysis** was understood, and built the foundations for all our modern understanding of the science. Not only did Michael Faraday explain what was happening during electrolysis, he also introduced the terminology – electrolysis, electrolyte, electrode – that we still use today.

Faraday's laws

As part of his careful work on electrolysis, Faraday measured the amounts of different substances given off during the process. As a result of all his observations and measurements he formulated two laws which can be used to predict the outcome of any electrolysis. They are known as **Faraday's laws** and they apply as much today as they did over 150 years ago.

Faraday's first law states that 'the quantity of a substance deposited, evolved or dissolved at an electrode during electrolysis is directly proportional to the quantity of electricity passed through the electrolyte'.

Faraday's second law states that 'the quantities of different substances deposited, evolved or dissolved at electrodes by the passage of the *same* quantity of electricity is directly proportional to the combining weights of the substances'.

Electrolysis is widely used in modern industry. Aluminium, one of the most widely used metals, is extracted from its ore by electrolysis, as are fluorine and sodium. Copper is refined by electrolysis, brine is split to form sodium hydroxide, chlorine and hydrogen and processes such as silver, gold and copper plating all use the same method. The calculations for these techniques still rely on Faraday's own laws.

Figure 2 Electrolysis provides us both with useful objects and with things simply to enjoy.

The stoichiometry of electrochemical processes – more electronic book-keeping

In a series of now famous experiments on electrochemistry, Michael Faraday discovered that the extent of the chemical change during electrolysis is directly proportional to the amount of electrical charge passed through the electrolytic cell. For example, to deposit one mole (108 g) of silver atoms from the reduction of a solution containing silver ions requires one mole of electrons:

$$Ag^{+}(aq) + e^{-} \rightarrow Ag(s)$$

The production of one mole (63.5 g) of copper atoms requires two moles of electrons:

$$Cu^{2+}(aq) + 2e^{-} \rightarrow Cu(s)$$

The SI unit of electrical charge is the **coulomb** (C). One coulomb is the charge that flows past a given point in an electrical circuit when a current of one ampere flows for one second. By making chemical and electrical measurements, Faraday was able to calculate the charge on 1 mole of electrons as 96 500 C.

Calculations using currently accepted values for the Avogadro constant N_A and the charge on the electron e give the value of the charge on 1 mole of electrons as:

$$N_A \times e = 6.022\,137 \times 10^{23}\ \text{mol}^{-1} \times 1.602\,177 \times 10^{-19}\ \text{C}$$
$$= 96\,485\ \text{C mol}^{-1}$$

Electrolysis and refining copper

Copper is about 99% pure when first obtained from its ore, the main impurities being silver, platinum, iron, gold and zinc. This level of impurities is sufficient to reduce the electrical conductivity of the copper, and so it must be purified before it can be used in, for example, electrical wires. This is carried out using electrolysis, as figure 3.5.3 shows.

At the anode:

$$\underset{\textbf{Impure copper}}{Cu(s)} \rightarrow Cu^{2+}(aq) + 2e^{-}$$

At the cathode:

$$Cu^{2+}(aq) + 2e^{-} \rightarrow \underset{\textbf{Pure copper}}{Cu(s)}$$

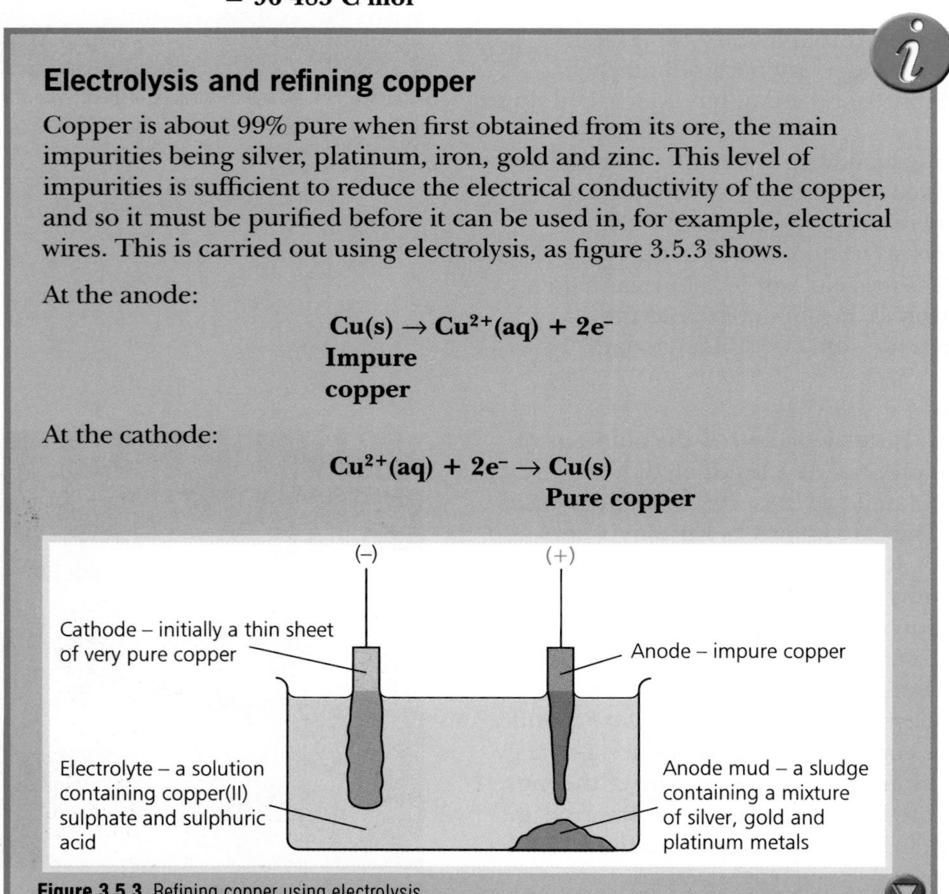

Figure 3.5.3 Refining copper using electrolysis

The voltage across the cell is controlled very carefully. At the appropriate operating voltage, only copper and those impurities more easily oxidised than copper (iron and zinc) are oxidised. Other impurities fall to the base of the cell and form **anode mud**. Over a period of time the anode dissolves and copper is deposited on the cathode, producing copper of about 99.96% purity. The precious metals reclaimed from the anode mud are sold, making a substantial financial contribution to the refining operation.

Electrochemical cells

Electricity from chemical change

In 1791 the Italian Luigi Galvani announced the discovery of 'animal electricity', having observed that the muscles in severed frogs' legs twitch when touched by two different metals. Three years later, Alessandro Volta showed that the effect described by Galvani was unconnected with living things, and that electricity can be produced whenever two metals are immersed in a conducting solution.

If a piece of zinc metal is placed in a solution of copper(II) sulphate, the blue colour of the copper sulphate slowly fades, the zinc dissolves and pink copper metal takes its place. The ionic equation for the overall reaction taking place is:

$$Zn(s) + Cu^{2+}(aq) \rightarrow Zn^{2+}(aq) + Cu(s)$$

We can show that this is a redox reaction if we separate it into two half-equations:

$$Zn(s) \rightarrow Zn^{2+}(aq) + 2e^- \qquad \textbf{oxidation}$$

$$Cu^{2+}(aq) + 2e^- \rightarrow Cu(s) \qquad \textbf{reduction}$$

This reaction produces energy, which is lost as heat if we simply carry out the process in a single reaction vessel. However, if we separate the reactions in two **half-cells**, we can harness the energy through the flow of electrons taking place between the cells, as figure 3.5.5 shows. This combination of half-cells is an **electrochemical cell** – the chemical reactions produce a flow of electrons, in contrast to an electrolytic cell, where the opposite happens. Electrochemical cells may also be known as **galvanic cells**, or sometimes **voltaic cells**.

Figure 3.5.4 Luigi Galvani (top) carried out careful experiments on the effects of static electricity on nerves and muscles. Using prepared frogs, Galvani observed muscle contractions when the spinal cords were connected by brass hooks to an iron railing. Galvani took this as confirmation of the theory that animal nerve and muscle tissue contained an electric fluid. Alessandro Volta (bottom) later demonstrated that the electricity did not come from the animal tissue but rather from the brass and iron coming into contact with each other via an electrolyte.

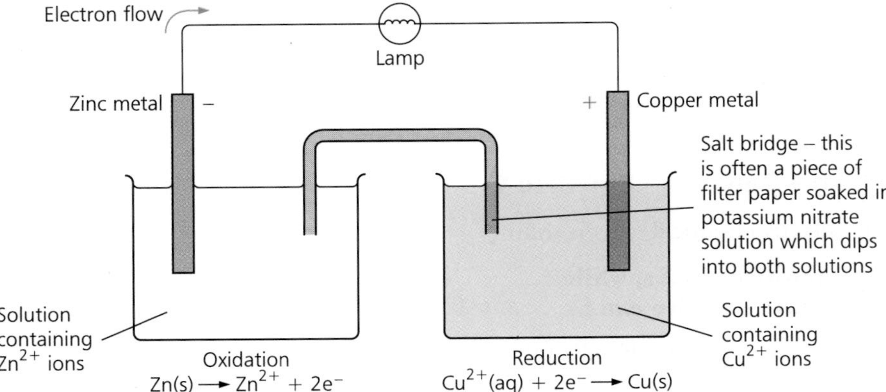

Figure 3.5.5 An electrochemical cell

In the electrochemical cell shown in figure 3.5.5, the copper electrode slowly increases in mass as copper ions leave the solution around it, while the zinc electrode slowly decreases in mass as the zinc dissolves. As this happens, electrons flow along the wire joining the two electrodes. This current can be detected using a small lamp. In the overall reaction, copper is reduced and zinc is oxidised, but in this redox process we can harness the electron transfer that occurs.

The lamp in figure 3.5.5 may be replaced by a high-resistance voltmeter. In this way the maximum potential difference across the cell – its **electromotive force** or e.m.f. – can be measured. The e.m.f. of a cell can be thought of as the 'push' it is able to provide to a current flowing through it, and e.m.f. is measured in volts. As we shall see in section 3.7, the e.m.f. of a cell is related to the maximum amount of useful work we can get out of it.

What does the salt bridge do?

The role of the potassium nitrate in the salt bridge is very similar to that of the potassium nitrate in the electrolysis of water on page 306. Without the salt bridge, the half-cell containing the zinc would slowly become positively charged as electrons left it, while the copper half-cell would become negatively charged.

With the salt bridge present, ions are able to move in and out of the solutions, keeping both half-cells electrically neutral.

The salt used in a salt bridge is chosen so that it does not react with the ions in either half-cell.

Half-cell notation

Chemists use a system of 'shorthand' notation to represent the half-cells that make up an electrochemical cell. Using this notation, the zinc–copper cell would be represented as in figure 3.5.6.

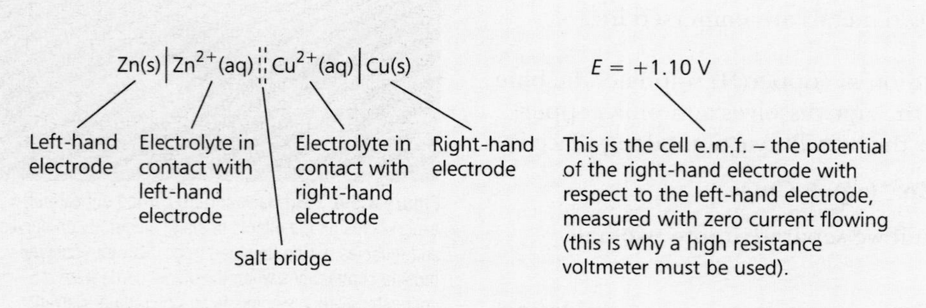
$Zn(s)|Zn^{2+}(aq) \vdots\vdots Cu^{2+}(aq)|Cu(s)$
$E = +1.10$ V
Left-hand electrode
Electrolyte in contact with left-hand electrode
Salt bridge
Electrolyte in contact with right-hand electrode
Right-hand electrode
This is the cell e.m.f. – the potential of the right-hand electrode with respect to the left-hand electrode, measured with zero current flowing (this is why a high resistance voltmeter must be used).

Figure 3.5.6

Notice that when we measure the e.m.f. for this cell, the right-hand electrode is *positive* with respect to the left-hand electrode, since the electrons flow from the zinc to the copper. By convention, the cell notation always refers to the reaction taking place from left to right, which in this case is:

$$Zn(s) \rightarrow Zn^{2+}(aq) + 2e^-$$

$$Cu^{2+}(aq) + 2e^- \rightarrow Cu(s)$$

Overall: $$Zn(s) + Cu^{2+}(aq) \rightarrow Zn^{2+}(aq) + Cu(s) \qquad E = +1.10\text{ V}$$

If the cell is reversed, so that zinc is now the right-hand electrode and copper is the left-hand electrode, electrons still flow from zinc to copper, so now the cell e.m.f. is *negative*. The cell notation is now $Cu(s)|Cu^{2+}(aq) \vdots\vdots Zn^{2+}(aq)|Zn(s)$ and refers to the process:

$$Cu(s) \rightarrow Cu^{2+}(aq) + 2e^-$$

$$Zn^{2+}(aq) + 2e^- \rightarrow Zn(s)$$

Overall: $$Cu(s) + Zn^{2+}(aq) \rightarrow Cu^{2+}(aq) + Zn(s) \qquad E = -1.10\text{ V}$$

The result of the convention for representing electrochemical cells is that:

Changes that occur spontaneously have positive e.m.f.s, while changes that do not occur spontaneously have negative e.m.f.s.

This is easily demonstrated if we replace the $Zn^{2+}(aq)|Zn(s)$ half-cell with other half-cells, as shown in table 3.5.1.

Left-hand half-cell	Right-hand half-cell	Cell e.m.f./V
$Cu(s)\,|\,Cu^{2+}(aq)$	$Ag^{+}(aq)\,|\,Ag(s)$	+0.46
$Cu(s)\,|\,Cu^{2+}(aq)$	$Cu^{2+}(aq)\,|\,Cu(s)$	0.00
$Cu(s)\,|\,Cu^{2+}(aq)$	$Ni^{2+}(aq)\,|\,Ni(s)$	−0.59
$Cu(s)\,|\,Cu^{2+}(aq)$	$Zn^{2+}(aq)\,|\,Zn(s)$	−1.10
$Cu(s)\,|\,Cu^{2+}(aq)$	$Mn^{2+}(aq)\,|\,Mn(s)$	−1.53

Table 3.5.1 The e.m.f.s of different cells with the $Cu(s)\,|\,Cu^{2+}(aq)$ cell as the left-hand half-cell

The cells in table 3.5.1 represent the following reactions:

$$\mathbf{Cu(s) + 2Ag^{+}(aq) \rightarrow Cu^{2+}(aq) + 2Ag(s)} \qquad E = +0.46\ \mathrm{V}$$

$$\mathbf{Cu(s) + Cu^{2+}(aq) \rightarrow Cu^{2+}(aq) + Cu(s)} \qquad E = 0.00\ \mathrm{V}$$

$$\mathbf{Cu(s) + Ni^{2+}(aq) \rightarrow Cu^{2+}(aq) + Ni(s)} \qquad E = -0.59\ \mathrm{V}$$

$$\mathbf{Cu(s) + Zn^{2+}(aq) \rightarrow Cu^{2+}(aq) + Zn(s)} \qquad E = -1.10\ \mathrm{V}$$

$$\mathbf{Cu(s) + Mn^{2+}(aq) \rightarrow Cu^{2+}(aq) + Mn(s)} \qquad E = -1.53\ \mathrm{V}$$

$Ag^{+}(aq)$ ions are a stronger oxidising agent than $Cu^{2+}(aq)$ ions, oxidising copper metal to copper(II). The first reaction therefore happens spontaneously in the direction written, with a decrease in the energy of the system. The last three reactions are different. $Mn^{2+}(aq)$, $Zn^{2+}(aq)$ and $Ni^{2+}(aq)$ ions are weaker oxidising agents than $Cu^{2+}(aq)$ ions, so the reactions written here do not happen spontaneously. Instead, the reverse reactions are the energetically favourable ones, in which $Cu^{2+}(aq)$ ions oxidise manganese metal to $Mn^{2+}(aq)$ ions, nickel metal to $Ni^{2+}(aq)$ ions and zinc metal to $Zn^{2+}(aq)$ ions.

The strength of oxidising and reducing agents

An alternative way of explaining the reactions in table 3.5.1 is in terms of reducing power. Copper metal is a stronger reducing agent than silver metal, but a weaker reducing agent than manganese, zinc and nickel metals. Hence manganese, zinc and nickel metals will all reduce $Cu^{2+}(aq)$ ions to copper metal, while copper metal will reduce $Ag^{+}(aq)$ ions to silver metal.

These two explanations are entirely equivalent. By definition, for the equilibrium:

$$M^{n+}(aq) + ne^{-} \rightleftharpoons M(s)$$

if M(s) is a strong reducing agent, $M^{n+}(aq)$ is a weak oxidising agent, while if $M^{n+}(aq)$ is a strong oxidising agent, M(s) is a weak reducing agent. Compare this with the strength of an acid and its conjugate base in section 3.2.

Cell measurements

Electrode potentials

When we measure the e.m.f. of a cell, we can imagine that the electron flow arises from competition for electrons between the two half-cells. Each half-cell reaction will have its own tendency to attract electrons, a tendency measured by the **electrode potential** of the half-cell. This is the e.m.f. measured when that half-cell forms an electrochemical cell with a reference half-cell. The more positive a half-cell's electrode potential, the greater its tendency to attract electrons. This is more clearly seen if we arrange the half-cells in table 3.5.1 in a diagram to show their potentials relative to one another. In figure 3.5.7 the half-cells that tend to attract electrons most strongly are at the top of the diagram – $Ag^{+}(aq)$ is the strongest oxidising agent here as it is the best competitor for electrons. The manganese half-cell is at the bottom of the diagram – Mn(s) is the strongest reducing agent, as it is the best species at giving away electrons.

We have said that the electrode potential of a half-cell is the e.m.f. when that half-cell is connected to a reference half-cell – in figure 3.5.7, this is the $Cu(s)\,|\,Cu^{2+}(aq)$ half-cell. In measuring the e.m.f. of a cell, we are comparing the electrode potentials of the two half-cells that make up the cell. We can therefore use the electrode potentials of two different half-cells to work out what would happen if they were connected together. Using this idea and the data in figure 3.5.7, we can calculate the e.m.f. of the cell:

$$\mathbf{Ni(s)\,|\,Ni^{2+}(aq)\ \vdots\vdots\ Ag^{+}(aq)\,|\,Ag(s)}$$

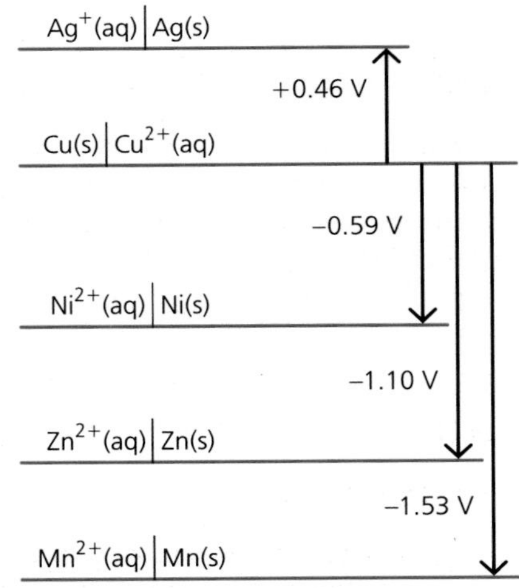

Figure 3.5.7 The relationships between electrode potentials. The electrode potential of each half-cell is compared with that of the $Cu(s)\,|\,Cu^{2+}(aq)$ half-cell.

The e.m.f. of this cell is the difference between the electrode potentials of the two half-cells, that is:

$$+0.46\ \text{V} + 0.59\ \text{V} = +1.05\ \text{V}$$

Because $Ag^+(aq)$ ions are reduced in this cell, the cell e.m.f. is positive, $E = +1.05$ V. The positive value of the e.m.f. confirms that the reaction:

$$Ni(s) + 2Ag^+(aq) \rightleftharpoons Ni^{2+}(aq) + 2Ag(s)$$

proceeds spontaneously in the forward direction. This agrees with our previous deduction that metals lower down the table (like nickel) are reducing agents, capable of being oxidised by ions at the top of the table (like Ag^+).

Standard electrode potentials

So far we have chosen to use the $Cu(s)|Cu^{2+}(aq)$ half-cell as our reference point against which to judge other electrode potentials. In practice, a different reference half-cell is used to measure all other electrode potentials. It is itself assigned an electrode potential of 0.00 V under standard conditions of 10^5 Pa, 25 °C and unit concentrations. The measurement of electrode potentials against a standard half-cell is similar to the way in which heights above sea level are measured in relation to one agreed point, figure 3.5.8.

The reference half-cell used in practice is the **standard hydrogen half-cell** (sometimes called the **standard hydrogen electrode**), shown in figure 3.5.9. This consists of hydrogen gas at 10^5 Pa and 25 °C bubbling around a platinum electrode in 1.00 mol dm^{-3} $H^+(aq)$ ions. The reaction that occurs in this half-cell is written:

$$2H^+(aq,\ 1.00\ \text{mol dm}^{-3}) + 2e^- \rightleftharpoons H_2(g,\ 10^5\ \text{Pa}) \qquad E^\ominus = 0.00\ \text{V}$$

The double arrows in this equation show that the reaction is reversible, not that it is an equilibrium. The direction in which the reaction goes will depend on the other half-cell to which the hydrogen half-cell is connected.

Figure 3.5.8 The reference point against which all other heights in the UK are judged is mean sea level at Newlyn, Cornwall, where there is an Ordnance Survey Tidal Observatory (this is marked on the map). Heights above this level are taken as positive, those below it are negative.

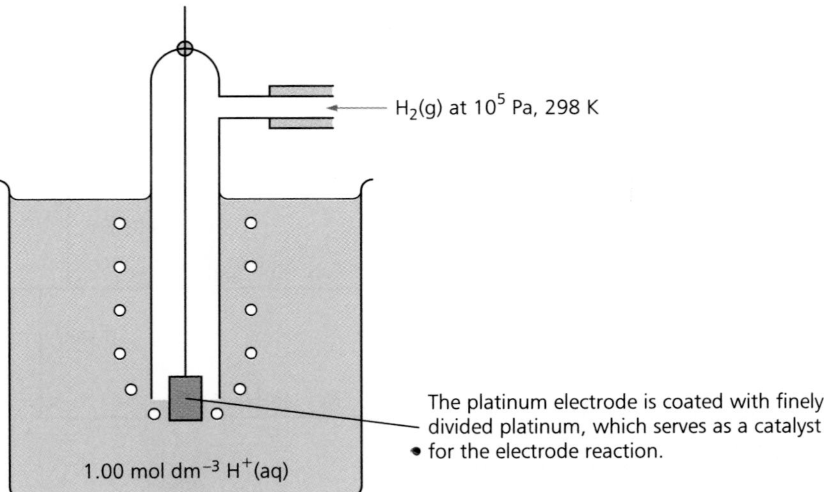

Figure 3.5.9 The standard hydrogen half-cell

The electrode potential of a half-cell measured using the standard hydrogen electrode under standard conditions of temperature, pressure and concentration is the **standard electrode potential** of that half-cell, given the symbol $E^\ominus$. Figure 3.5.10 shows how the standard electrode potential of the $Cu^{2+}(aq)|Cu(s)$ half-cell is measured.

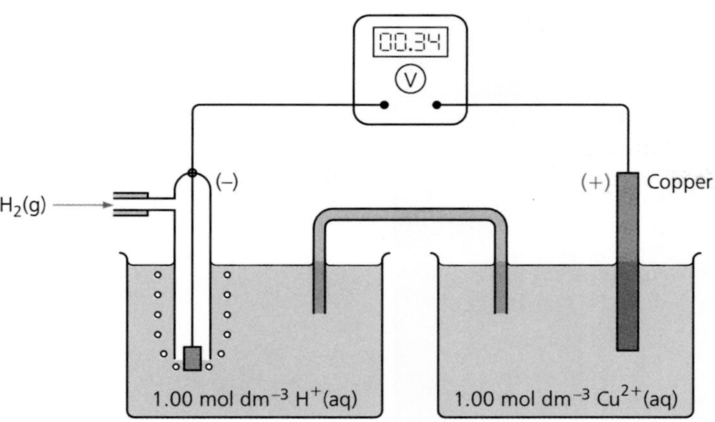

Figure 3.5.10 The electrochemical cell required to measure the standard electrode potential of the $Cu^{2+}(aq)\,|\,Cu(s)$ half-cell. By definition, $E^{\ominus}$ for the hydrogen half-cell is zero, whichever way the half-cell is written. If the cell is arranged in this way, the standard electrode potential of the right-hand half-cell is equal in magnitude to the standard e.m.f. of the cell, and has the same sign.

Measurement of this cell e.m.f. gives $E^{\ominus} = +0.34$ V, the copper half-cell being positive with respect to the hydrogen half-cell. Since the standard electrode potential of the hydrogen half-cell is zero, the standard electrode potential of the system $Cu^{2+}(aq)\,|\,Cu(s)$ must be $+0.34$ V, that is,

$$\mathbf{Cu^{2+}(aq) + 2e^- \rightleftharpoons Cu(s) \qquad E^{\ominus} = +0.34\ V}$$

In the same way as for the $Cu^{2+}(aq)\,|\,Cu(s)$ system, we may measure the standard electrode potentials of the other systems we considered on page 297:

$$\mathbf{Ag^{+}(aq) + e^- \rightleftharpoons Ag(s) \qquad E^{\ominus} = +0.80\ V}$$

$$\mathbf{Ni^{2+}(aq) + 2e^- \rightleftharpoons Ni(s) \qquad E^{\ominus} = -0.25\ V}$$

$$\mathbf{Zn^{2+}(aq) + 2e^- \rightleftharpoons Zn(s) \qquad E^{\ominus} = -0.76\ V}$$

$$\mathbf{Mn^{2+}(aq) + 2e^- \rightleftharpoons Mn(s) \qquad E^{\ominus} = -1.19\ V}$$

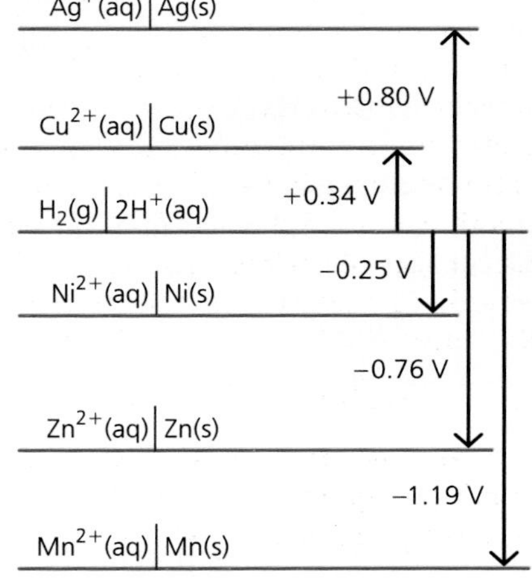

Figure 3.5.11

Notice that, although we are no longer measuring the electrode potentials against the copper half-cell, the *difference* between electrode potentials remains the same, as shown by figure 3.5.11. Therefore the e.m.f. of the cell:

$$\mathbf{Ni(s)\,|\,Ni^{2+}(aq) \;\vdots\vdots\; Ag^{+}(aq)\,|\,Ag(s)}$$

is given by:

$$\mathbf{E = +0.80\ V + 0.25\ V}$$
$$\mathbf{= +1.05\ V}$$

which is exactly the same as the e.m.f. from our previous calculation on page 298.

Conditions and conventions in electrochemistry

The conditions chosen to measure standard electrode potentials are the same as the conditions chosen for other standard measurements, for example the standard enthalpy changes we met in section 1.8. These conditions are as follows:

(1) All solutions have unit activity (effectively 1.00 mol dm^{-3} for our purposes).

(2) All measurements are made at 10^5 Pa pressure.
(3) All measurements are made at 25 °C.

In addition, we must also specify a way of making measurements for half-cells in which there are reactions like:

$$Fe^{3+}(aq) + e^- \rightleftharpoons Fe^{2+}(aq)$$

No metal exists as an electrode in this half-cell. Electrical connection is made to it via a platinum electrode dipping into the solution. The half-cell is then represented as:

$$Fe^{3+}(aq), Fe^{2+}(aq)|Pt(s)$$

For any cell such as this containing an inert electrode, the convention is to show the least oxidised of the species present in the half-cell next to the electrode. The hydrogen half-cell is thus represented as:

$$2H^+(aq), H_2(g)|Pt(s)$$

Another convention applies when writing redox reactions for half-cells. These are written in such a way that the more oxidised species appears on the left-hand side of the half-equation, as in the half-equation involving Fe^{3+} and Fe^{2+} above. The result of this convention is that the standard e.m.f. of any cell can be found from the relationship:

$$E^\ominus(\text{cell}) = E^\ominus(\text{right-hand electrode}) - E^\ominus(\text{left-hand electrode})$$

Competition for electrons

The standard electrode potential of a half-cell is a measure of the oxidising or reducing power of the species in it – in other words, their ability to compete for electrons. In general the stronger an oxidising agent, the more positive its electrode potential. Ozone gas is a more powerful oxidant than chlorine gas:

$$O_3(g) + 2H^+(aq) + 2e^- \rightleftharpoons O_2(g) + H_2O(l) \qquad E^\ominus = +2.08\ V$$

$$Cl_2(g) + 2e^- \rightleftharpoons 2Cl^-(aq) \qquad E^\ominus = +1.36\ V$$

A strong reducing agent has a large negative electrode potential. Calcium metal is a more powerful reducing agent than lead metal:

$$Pb^{2+}(aq) + 2e^- \rightleftharpoons Pb(s) \qquad E^\ominus = -0.13\ V$$

$$Ca^{2+}(aq) + 2e^- \rightleftharpoons Ca(s) \qquad E^\ominus = -2.87\ V$$

From the equations for these reversible reactions it follows that, for example,

$$2Cl^-(aq) \rightleftharpoons 2e^- + Cl_2(g) \qquad E^\ominus = -1.36\ V$$

$$Pb(s) \rightleftharpoons 2e^- + Pb^{2+}(aq) \qquad E^\ominus = +0.13\ V$$

since reversing the direction in which we write the reaction (electrons on the right rather than on the left) means that we must reverse the sign of the standard electrode potential too.

Table 3.5.2 shows standard electrode potentials for a number of half-reactions. The substances to the left of the double arrows in the table are oxidising agents, becoming reduced when they react in the direction left to right. The best oxidising agents are located at the top left of the table –

fluorine, for example. Substances to the right of the double arrows in the table are reducing agents, becoming oxidised when they react in the direction right to left. The best reducing agents are located at the bottom right of the table – lithium, for example.

Half-reaction	$E^\ominus$/V
$F_2(g) + 2e^- \rightleftharpoons 2F^-(aq)$	+2.87
$O_3(g) + 2H^+(aq) + 2e^- \rightleftharpoons O_2(g) + H_2O(l)$	+2.08
$H_2O_2(aq) + 2H^+(aq) + 2e^- \rightleftharpoons 2H_2O(l)$	+1.77
$MnO_4^-(aq) + 8H^+(aq) + 5e^- \rightleftharpoons Mn^{2+}(aq) + 4H_2O(l)$	+1.51
$Cl_2(g) + 2e^- \rightleftharpoons 2Cl^-(aq)$	+1.36
$Br_2(aq) + 2e^- \rightleftharpoons 2Br^-(aq)$	+1.09
$Ag^+(aq) + e^- \rightleftharpoons Ag(s)$	+0.80
$Fe^{3+}(aq) + e^- \rightleftharpoons Fe^{2+}(aq)$	+0.77
$I_2(aq) + 2e^- \rightleftharpoons 2I^-(aq)$	+0.54
$Cu^+(aq) + e^- \rightleftharpoons Cu(s)$	+0.52
$Cu^{2+}(aq) + 2e^- \rightleftharpoons Cu(s)$	+0.34
$Cu^{2+}(aq) + e^- \rightleftharpoons Cu^+(aq)$	+0.15
$2H^+(aq) + 2e^- \rightleftharpoons H_2(g)$	0.00
$Pb^{2+}(aq) + 2e^- \rightleftharpoons Pb(s)$	–0.13
$Ni^{2+}(aq) + 2e^- \rightleftharpoons Ni(s)$	–0.25
$Fe^{2+}(aq) + 2e^- \rightleftharpoons Fe(s)$	–0.44
$Zn^{2+}(aq) + 2e^- \rightleftharpoons Zn(s)$	–0.76
$Al^{3+}(aq) + 3e^- \rightleftharpoons Al(s)$	–1.66
$Na^+(aq) + e^- \rightleftharpoons Na(s)$	–2.37
$Li^+(aq) + e^- \rightleftharpoons Li(s)$	–3.03

Increasing strength of oxidising agent (upwards arrow, left). Increasing strength of reducing agent (downwards arrow, right).

Table 3.5.2 The standard electrode potentials of some half-cells

The reducing power of halide ions

The increase in reducing power of the halide ions as we go down the group from fluorine to iodine explains their behaviour with acids which we met on page 181. Iodide ions reduce concentrated sulphuric acid, while neither chloride ions nor fluoride ions do – so hydrogen chloride is the product formed when sodium chloride reacts with concentrated sulphuric acid, while sodium iodide produces mainly iodine. Bromide ions are slightly stronger reducing agents than chloride ions, and react with concentrated sulphuric acid to produce mainly hydrogen bromide.

3

Measuring quantities of oxidising and reducing agents

Reducing agents

Amounts of oxidising agents can be measured by titration, in the same way as titrations are used to measure amounts of acids and bases.

Reducing agents such as iron(II) ions (Fe^{2+}) can be measured using an oxidising agent like manganate(VII) ions (MnO_4^-):

$$5Fe^{2+}(aq) + MnO_4^-(aq) + 8H^+(aq) \rightarrow 5Fe^{3+}(aq) + Mn^{2+}(aq) + 4H_2O(l)$$

This equation shows that, for every mole of manganate(VII) ions reacting completely, there are five moles of iron(II). (You should be able to show why this is by assigning oxidation numbers to the manganese and the iron, as we saw in section 1.5.) It also shows that the reaction must take place under acidic conditions.

As with acid–base titrations, we need to be able to tell when the reaction has finished – this will be when we have added enough manganate(VII) ions to react with all of the iron(II) ions. Since acidic solutions of iron(II) are a very pale blue-green and manganate(VII) solutions are a deep magenta colour, the end point of the titration is taken as the point at which a faint pink colour persists in the solution when a small quantity of manganate(VII) solution is added.

As an alternative to manganate(VII) ions, chromate(VI) ions may be used:

$$6Fe^{2+}(aq) + Cr_2O_7^{2-}(aq) + 14H^+(aq) \rightarrow 6Fe^{3+}(aq) + 2Cr^{3+}(aq) + 7H_2O(l)$$

Here the colour change at the end of the reaction is hard to judge, as the orange colour of the chromate(VI) ions is replaced by the green of the chromium(III) ions produced in the reaction. A **redox indicator** is therefore used in this reaction – just like an acid–base indicator, this changes colour when the end point is reached. One such indicator is *diphenylamine*. When chromate(VI) ions are reacted with iron(II) ions in the presence of diphenylamine, the indicator remains colourless until all the iron(II) has reacted. At the endpoint, chromate(VI) then oxidises the diphenylamine, which changes from colourless to deep blue.

Oxidising agents

Rather than titrating them directly against reducing agents, oxidising agents such as the chlorate(I) ion in bleach are usually measured indirectly. The first stage in measurement is to use the oxidising agent to liberate iodine from a solution containing excess iodide ions:

$$ClO^-(aq) + 2I^-(aq) + 2H^+(aq) \rightarrow Cl^-(aq) + I_2(aq) + H_2O(aq)$$

The iodine formed in this reaction is then measured by titration against thiosulphate ions:

$$I_2(aq) + 2S_2O_3^{2-}(aq) \rightarrow 2I^-(aq) + S_4O_6^{2-}(aq)$$

As thiosulphate solution is run into the solution containing iodine, the brown colour of the iodine gradually turns through yellow to colourless. Just before the colourless stage is reached, a few drops of fresh starch solution are added, producing an intense blue colour. Further thiosulphate solution is added until this colour disappears.

Worked example

A 2.00 cm^3 sample of bleach containing sodium chlorate(I) is added to excess potassium iodide solution. The iodine liberated is titrated against 0.05M sodium thiosulphate solution. 32.4 cm^3 of this solution are required to discharge the colour of this iodine, using starch as an indicator. What is the strength of the bleach solution, expressed as mols of sodium chlorate(I) per dm^3?

The equation for the reaction of sodium chlorate(I) with iodide ions is :

$$ClO^-(aq) + 2I^-(aq) + 2H^+(aq) \rightarrow Cl^-(aq) + I_2(aq) + H_2O(aq)$$

Since excess iodide ions are present, we can assume that all of the chlorate(I) ions react with iodide ions to form iodine. The iodine is then titrated against sodium thiosulphate:

$$I_2(aq) + 2S_2O_3^{2-}(aq) \rightarrow 2I^-(aq) + S_4O_6^{2-}(aq)$$

Thus, for every mole of thiosulphate ions reacting with iodine, one mole of chlorate(I) ions was originally present in the bleach.

$$\textbf{Moles of thiosulphate ions added} = \frac{32.4\,cm^3}{1000\,cm^3} \times 0.05\ mol$$

$$= 0.001\ 62\ mol$$

This is the number of moles of chlorate(I) ions in 2.00 cm^3 of bleach, so in 1000 cm^3 of bleach there are:

$$\frac{1000\ \mathbf{cm^3}}{2.00\ \mathbf{cm^3}} \times \mathbf{0.001\,62\ mol}$$

$$= \mathbf{0.81\ mol}$$

So there are 0.81 moles of sodium chlorate(I) in 1000 cm^3 of bleach.

Using standard electrode potentials

Measuring pH

pH is measured by a **glass electrode**, which is used with a pH meter to convert an e.m.f. into a pH reading. The electrode has a hollow glass bulb made of very thin-walled glass. This bulb is filled with dilute hydrochloric acid, into which dips a platinum wire coated with silver and silver chloride. The potential of this electrode depends on the difference in the concentration of hydrogen ions between the two sides of the glass membrane. The potential therefore depends on the pH of the solution surrounding the bulb.

The glass electrode is used in conjunction with another (reference) electrode to produce an electrochemical cell. The potential of this cell depends on the pH of the solution in which the glass electrode is immersed. For compactness the two half-cells are normally enclosed in a single container, as the photograph in figure 3.5.12 shows.

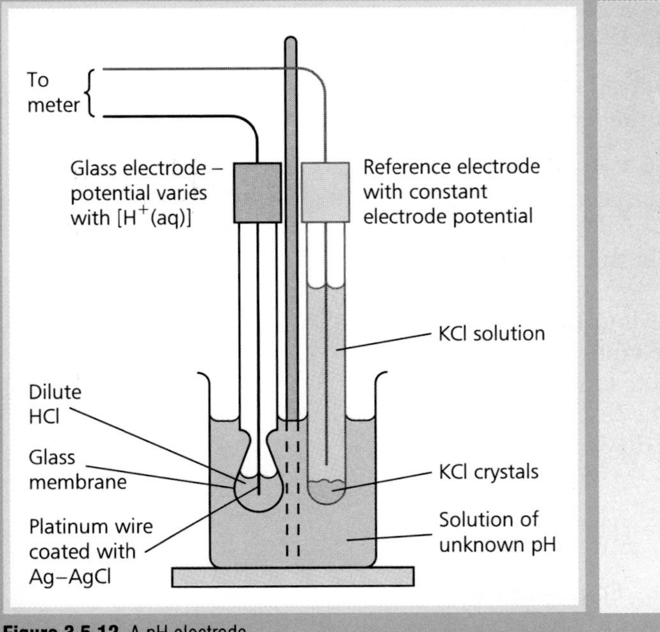

Figure 3.5.12 A pH electrode

3

Predicting spontaneous reactions

The standard electrode potentials in table 3.5.2 enable us to predict whether a redox reaction is spontaneous or not. To make a prediction about a reaction, we must first write down the two half-equations for it. The following example using the halogens shows how such predictions are made.

Is aqueous iodine able to oxidise bromide ions to aqueous bromine?

1 Write down the reaction:

$$I_2(aq) + 2Br^-(aq) \rightarrow 2I^-(aq) + Br_2(aq)$$

2 Split the reaction into two half-equations (taking care with the direction of each reaction) and show the electrode potentials:

$$I_2(aq) + 2e^- \rightarrow 2I^-(aq) \qquad E^\ominus = +0.54\ V$$

$$2Br^-(aq) \rightarrow Br_2(aq) + 2e^- \qquad E^\ominus = -(+1.09\ V)$$

(Notice that we have reversed the sign of the $Br_2(aq)|2Br^-(aq)$ half-cell since we have written it with the electrons on the right instead of on the left.)

3 Add the half-reactions and the electrode potentials together:

$$I_2(aq) + 2Br^-(aq) \rightarrow 2I^-(aq) + Br_2(aq) \qquad E^\ominus = +0.54\ V + (-1.09\ V)$$

$$= -0.55\ V$$

The negative value of the cell e.m.f. indicates that the reaction will *not* occur spontaneously – in fact, it will occur in the reverse direction, and bromine will oxidise iodide ions to iodine:

$$Br_2(aq) + 2I^-(aq) \rightarrow 2Br^-(aq) + I_2(aq) \qquad E^\ominus = -(-0.55\ V)$$

$$= +0.55\ V$$

The oxidation of bromide ions by aqueous chlorine does occur, however:

1

$$Cl_2(aq) + 2Br^-(aq) \rightarrow 2Cl^-(aq) + Br_2(aq)$$

2

$$Cl_2(aq) + 2e^- \rightarrow 2Cl^-(aq) \qquad E^\ominus = +1.36\ V$$

$$2Br^-(aq) \rightarrow Br_2(aq) + 2e^- \qquad E^\ominus = -(+1.09\ V)$$

(Once again the sign of the $Br_2(aq)|2Br^-(aq)$ half-cell is reversed since we have written it with the electrons on the right instead of on the left.)

3

$$Cl_2(aq) + 2Br^-(aq) \rightarrow 2Cl^-(aq) + Br_2(aq) \qquad E^\ominus = +1.36\ V + (-1.09\ V)$$

$$= +0.27\ V$$

The positive cell e.m.f. shows that this reaction *is* spontaneous in the forward direction.

When electrode potentials are shown as in table 3.5.2, the behaviour of a particular redox system is easy to spot. Reactions higher up the table have a tendency to go to the right, while those at the bottom of the table tend to go to the left. Because of this, if we pick a pair of reactions from the table, the one higher up the table tends to go to the right, and the one lower down the table tends to go to the left:

$$Fe^{3+}(aq) + e^- \rightleftharpoons Fe^{2+}(aq) \qquad E^\ominus = +0.77\ V$$

$$Ni^{2+}(aq) + 2e^- \rightleftharpoons Ni(s) \qquad E^\ominus = -0.25\ V$$

For obvious reasons, this is sometimes known as the 'clockwise rule'. However, it is better not to rely too heavily on rules like this, since electrode potentials may not always be written in the way you might expect!

Oxidation states of vanadium

In aqueous solution, vanadium has four common oxidation states. The electrode potentials for the interconversion of these are as follows:

$VO_2^+(aq) + 2H^+(aq) + e^- \rightleftharpoons VO^{2+}(aq) + H_2O(l)$ $E^\ominus = +1.00\,V$

$VO^{2+}(aq) + 2H^+(aq) + e^- \rightleftharpoons V^{3+}(aq) + H_2O(l)$ $E^\ominus = +0.34\,V$

$V^{3+}(aq) + e^- \rightleftharpoons V^{2+}(aq)$ $E^\ominus = -0.26\,V$

$V^{2+}(aq) + 2e^- \rightleftharpoons V(s)$ $E^\ominus = -1.18\,V$

We can use these electrode potentials to predict what reagents might be used to change the oxidation state of aqueous vanadium ions.

- The electrode potential for the reaction $Zn^{2+}(aq) + 2e^- \rightleftharpoons Zn(s)$ is $-0.76\,V$, so zinc metal will reduce V^{3+} to V^{2+}, VO^{2+} to V^{3+} and VO_2^+ to VO^{2+}.
- The electrode potential for the reaction $Fe^{3+}(aq) + e^- \rightleftharpoons Fe^{2+}(aq)$ is $+0.77\,V$, so Fe^{2+} ions will reduce VO_2^+ to VO^{2+}, while Fe^{3+} will oxidise V^{3+} to VO^{2+}, V^{2+} to V^{3+} and solid vanadium to V^{2+}.

The behaviour of copper in aqueous solutions

The positive standard electrode potentials for both $Cu^{2+}(aq)|Cu(s)$ ($E^\ominus = +0.34$ V) and $Cu^+(aq)|Cu(s)$ ($E^\ominus = +0.52$ V) show that copper metal will not displace hydrogen from aqueous solutions of acids under normal conditions. Copper *does* react with concentrated nitric acid, however, as the standard electrode potential of the following half-reaction shows:

$2NO_3^-(aq) + 4H^+(aq) + 2e^- \rightleftharpoons N_2O_4(g) + 2H_2O(l)$ $E^\ominus = +0.80$ V

Oxidation of copper metal to Cu^{2+} ions may also occur with oxygen in the presence of water, although this reaction is less favourable:

$O_2(g) + 2H_2O(aq) + 4e^- \rightleftharpoons 4OH^-(aq)$ $E^\ominus = +0.40$ V

Examination of the electrode potentials for $Cu^{2+}(aq)|Cu(s)$ and $Cu^+(aq)|Cu(s)$ provides some interesting information about the behaviour of copper in aqueous solutions:

$Cu^+(aq) + e^- \rightleftharpoons Cu(s)$ $E^\ominus = +0.52$ V

$Cu^{2+}(aq) + e^- \rightleftharpoons Cu^+(aq)$ $E^\ominus = +0.15$ V

Using the clockwise rule, these electrode potentials show that Cu^+ ions in aqueous solution disproportionate to form solid copper and Cu^{2+} ions according to the equation:

$2Cu^+(aq) \longrightarrow Cu(s) + Cu^{2+}(aq)$ $E^\ominus = +0.52\text{ V} - (+0.15\text{ V})$

$= +0.37$ V

Electrochemistry and the rusting of iron

Iron and the steel derived from it have been used for hundreds of years, because the metal is plentiful and relatively easily extracted and worked. However, iron has one major drawback – in the presence of oxygen and water it will corrode, reacting to form the orange-red oxide known as rust. The corrosion is an electrochemical process, as shown by figure 3.5.13. At one site on the surface, iron becomes oxidised in the presence of water and enters solution as $Fe^{2+}(aq)$. The iron itself acts as an anode:

$$Fe(s) \rightarrow Fe^{2+}(aq) + 2e^-$$

The electrons released when the iron is oxidised travel through the metal to a site where the metal surface is exposed to oxygen. A cathodic reaction then takes place on the metal surface, with oxygen reduced to give hydroxide ions.

$$\tfrac{1}{2}O_2(aq) + H_2O(l) + 2e^- \rightarrow 2OH^-(aq)$$

The Fe^{2+} ions formed react with the OH^- ions to form a precipitate of insoluble iron(II) hydroxide, $Fe(OH)_2$. In air this is readily oxidised to give iron(III) hydroxide, $Fe(OH)_3$, which in turn loses water. Partial dehydration results in rust, $Fe_2O_3.xH_2O$. This has a lower density than iron, which is why rust causes 'bubbles' to form under paintwork.

There are two different ways to protect iron from rusting. One involves coating the iron with a layer that prevents oxygen and water coming into contact with it. This can be done by painting the iron, or by covering it with a layer of oil or grease, or by plating it with a layer of a less reactive metal. 'Tin cans' are really made of steel, coated with a thin layer of tin.

The second way to prevent iron from rusting is known as **cathodic protection**. It involves placing the iron in contact with a metal that is more easily oxidised – zinc is often used. In this situation, iron acts as a cathode rather than an anode and so the other metal corrodes instead. The metal used to protect the iron is known as the **sacrificial anode** because it is sacrificed to protect the iron. The sacrificial anode has to be replaced at regular intervals to maintain the protection. Iron objects may also be covered in zinc or **galvanised**. The great advantage of this is that if the surface layer is scratched, the iron does not corrode because it is in contact with a metal which is more easily oxidised.

Figure 3.5.14 The blocks of zinc prevent the steel structure of this oil rig from corroding by cathodic protection.

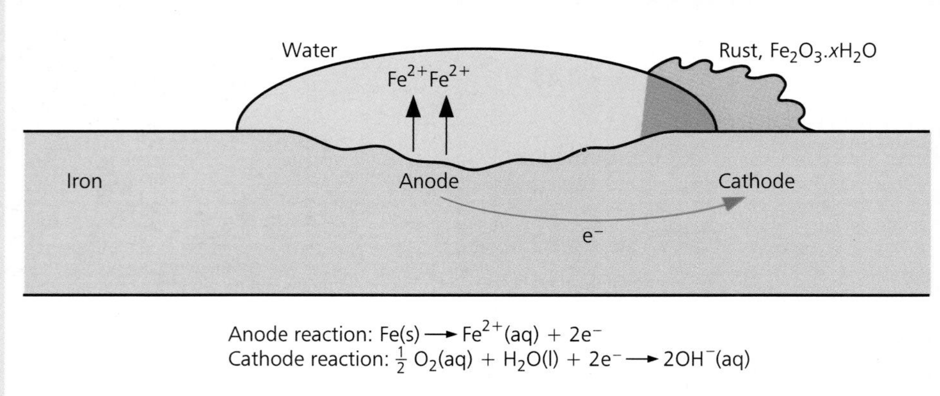

Figure 3.5.13 The diffusion of Fe^{2+} ions away from the site of corrosion explains why the damage caused by a scratch in the paintwork of a car may extend for some way underneath the paint.

Redox reactions and life

As food is broken down in the cells of your body, electrons are removed from glucose and from other food molecules. In the mitochondria of the cells, these electrons are transported along a series of molecules which make up the **electron transport chain**, in a process called **respiration**.

The various elements of the chain have different electrode potentials, with the first member of the chain having the most negative electrode potential, making it a relatively weak oxidising agent. The electrons pass along the chain via a series of molecules having increasingly positive electrode potentials, so their oxidising power increases along the chain. At the end of the chain, oxygen oxidises the final member of the chain, forming water in the process. The transport of electrons along the chain releases energy, which is used to drive the synthesis of a molecule called ATP (adenosine triphosphate), the universal 'energy carrier molecule' in living things.

The series of redox reactions known as **respiration** occurs in almost all living things. Respiration involves the oxidation of food and the transport of electrons through the electron transport chain, with the subsequent reduction of oxygen to water. **Photosynthesis** involves another series of redox reactions, in which electrons are *removed* from water and used to reduce carbon dioxide to produce sugars. This process occurs mainly in green plants. It requires energy to drive it, and this comes from light, captured by the green pigment chlorophyll.

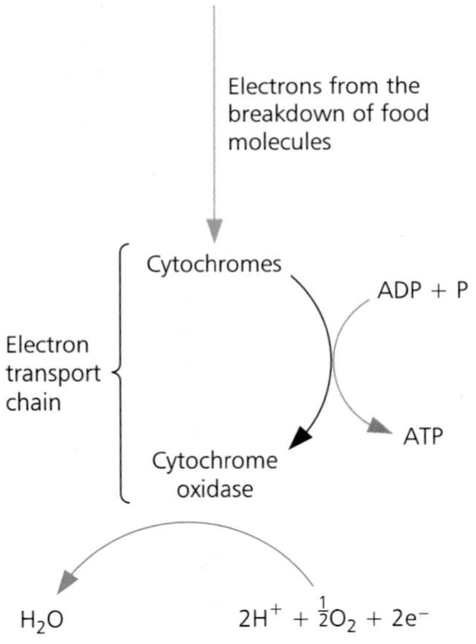

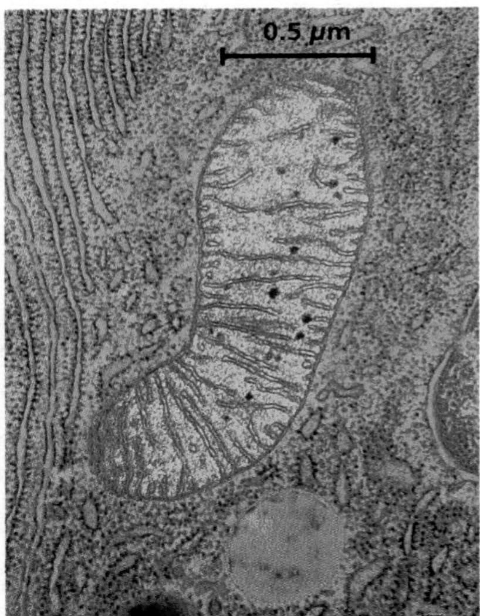

Figure 3.5.15 The electron transport chain (left). The redox reactions in this chain take place in the mitochondria (right), often called 'the powerhouse of the cell'.

3

The limitations of standard electrode potentials

Just like enthalpy changes, standard electrode potentials must be used with care when predicting the likelihood of a reaction. Both $\Delta H^\ominus$ and $E^\ominus$ are concerned with the *energetic* stability of a substance, not its *kinetic* stability – so they can tell us whether a reaction is *possible*, but not whether it happens at an appreciable rate. The oxidation of iron(II) salts in acidic solutions is an example of this:

$$\mathbf{Fe^{2+}(aq) \rightleftharpoons Fe^{3+}(aq) + e^-} \qquad E^\ominus = -0.77\ \text{V}$$

$$\mathbf{O_2(g) + 4H^+(aq) + 4e^- \rightleftharpoons 2H_2O(l)} \qquad E^\ominus = +1.23\ \text{V}$$

Overall:

$$4Fe^{2+}(aq) + O_2(g) + 4H^+(aq) \rightleftharpoons 4Fe^{3+}(aq) + 2H_2O(l) \quad E^\ominus = -0.77\ V + 1.23\ V$$
$$= +0.46\ V$$

The positive value here suggests that the reaction should proceed spontaneously. However, the oxidation of Fe^{2+} under these conditions is extremely slow because kinetic factors increase the stability of Fe^{2+} in acidic solutions.

Another important point about standard electrode potentials is that the conditions in the reaction under consideration may be very different from standard conditions. Temperature and concentrations are two factors that may be particularly important in influencing a reaction. For example, photographic film uses a light-sensitive emulsion containing silver bromide. The silver in the emulsion comes from silver ingots, which are dissolved in nitric acid:

$$2Ag(s) + 2NO_3^-(aq) + 4H^+(aq) \rightleftharpoons 2Ag^+(aq) + N_2O_4(g) + 2H_2O(l)$$

The half-equations and standard electrode potentials for this reaction are:

$$Ag^+(aq) + e^- \rightleftharpoons Ag(s) \quad E^\ominus = +0.80\ V$$

$$2NO_3^-(aq) + 4H^+(aq) + 2e^- \rightleftharpoons N_2O_4(g) + 2H_2O(l) \quad E^\ominus = +0.80\ V$$

This indicates that there should be no reaction between these two substances. However, if the temperature is raised to 60 °C, the electrode potentials become +0.765 V for the $Ag^+(aq)|Ag(s)$ half-cell reaction and +0.804 V for the $[2NO_3^-(aq) + 4H^+(aq)], [N_2O_4(g) + 2H_2O(l)]|Pt$ half-cell, so silver is oxidised under these conditions as it has the more positive electrode potential. Besides raising the temperature, the reaction is also made more favourable by using concentrated acid. This increases the hydrogen ion concentration, pushing the bottom reaction to the right.

Cells and batteries

Cells for portable power

Batteries store energy in the form of chemicals, the energy being released when a conductor is connected between the terminals of the battery. Batteries are composed of electrochemical cells, connected together in series in order to give a larger e.m.f. In this way a car battery consists of six cells, each of which has an e.m.f. of 2 V, so the total e.m.f. of the battery is 6 × 2 V = 12 V. The **lead–acid battery** used in a car is an example of a **secondary cell**. When the chemicals inside it have reacted and no more electricity can be drawn from it (the battery is 'flat'), the battery can be recharged by passing a current through it in the reverse direction. A **primary cell** cannot be recharged in this way.

The half-equations for a lead–acid cell as it discharges are:

$$PbO_2(s) + 4H^+(aq) + SO_4^{2-}(aq) + 2e^- \rightarrow PbSO_4(s) + 2H_2O(l) \quad \textbf{(cathode)}$$

$$Pb(s) + SO_4^{2-}(aq) \rightarrow PbSO_4(s) + 2e^- \quad \textbf{(anode)}$$

Overall:

$$PbO_2(s) + Pb(s) + 4H^+(aq) + 2SO_4^{2-}(aq) \rightarrow 2PbSO_4(s) + 2H_2O(l)$$

Notice how this reaction consumes sulphuric acid – when a lead–acid cell discharges, the concentration of sulphuric acid in it decreases. This provides a way of measuring the state of charge in the cell, by measuring the density of

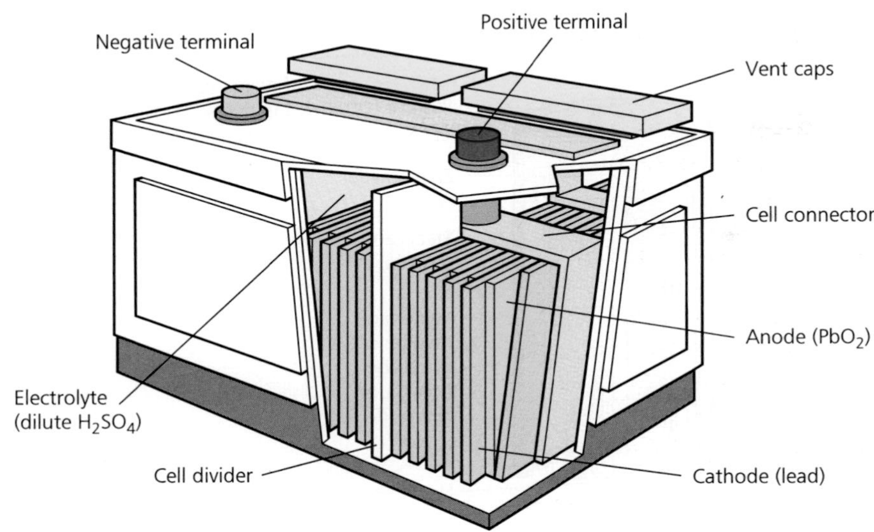

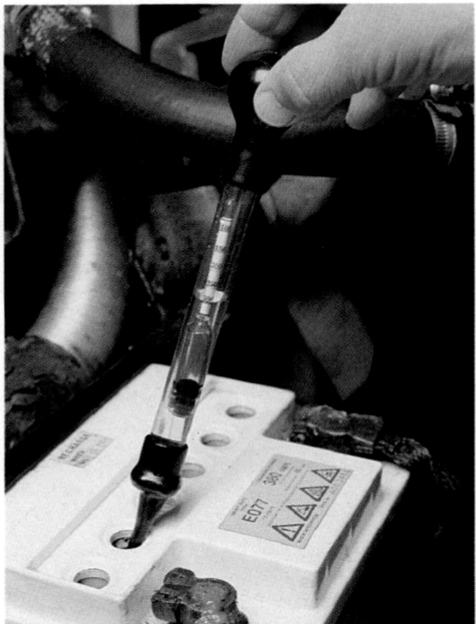

Figure 3.5.16 A 12 V lead–acid battery has six cells in series, each producing an e.m.f. of 2 V. The hydrometer can be used to check the state of each of the cells by measuring the density of the sulphuric acid in each cell – the higher the density, the greater the state of charge of the battery.

the electrolyte in it using a hydrometer. This gives an indication of the concentration of the acid. The change can be reversed by passing a current through the cell:

$$2PbSO_4(s) + 2H_2O(l) \xrightarrow{\text{electrolysis}} PbO_2(s) + Pb(s) + 4H^+(aq) + 2SO_4^{2-}(aq)$$

The range of cells available is very large, and is summarised in table 3.5.3.

Fuel cells

A fuel cell is very similar to a battery, in that it makes use of the energy stored in chemical substances. However, unlike the battery, in the fuel cell neither the high-energy reactants nor the low-energy products are stored inside the cell. The reactants are supplied from outside the cell and the products once formed are removed from it.

Probably the best known type of fuel cell is that used in spacecraft. These cells 'burn' hydrogen and oxygen, forming water as they do so. The energy transferred from the hydrogen and oxygen in this process is collected as electrical energy rather than as heat:

Anode: $2H_2(g) + 4OH^-(aq) \rightarrow 4H_2O(l) + 4e^-$ $E^\ominus = +0.83\ V$

Cathode: $O_2(g) + 2H_2O(l) + 4e^- \rightarrow 4OH^-(aq)$ $E^\ominus = +0.40\ V$

Overall: $2H_2(g) + O_2(g) \rightarrow 2H_2O(l)$ $E^\ominus = +0.83\ V + 0.40\ V$

$= +1.23\ V$

Fuel cells are compact and clean – the water produced by the fuel cells on the US space laboratory Skylab was used for drinking and washing.

Figure 3.5.17 Although fuel cells are still at an early stage in their development, they have been used with great success in space missions, where their small size and weight have obvious advantages.

Batteries under development

The search for more efficient batteries that carry a large amount of energy yet are very light (so they have a large 'energy density') is driven at least in part by the need to find an alternative to the internal combustion engine which powers our cars. The design of an electric car presents a problem to the engineer in terms of packing as much energy as possible into a small space – and the battery must recharge quickly too.

Cell	Reactions as cell discharges		Uses
Alkaline dry cell (primary cell – not rechargeable) e.m.f. ≈ 1.54 V	$Zn(s) + 2OH^-(aq) \rightarrow ZnO(s) + H_2O(l) + 2e^-$ (anode) $2MnO_2(s) + H_2O(l) + 2e^- \rightarrow Mn_2O_3(s) + 2OH^-(aq)$ (cathode)	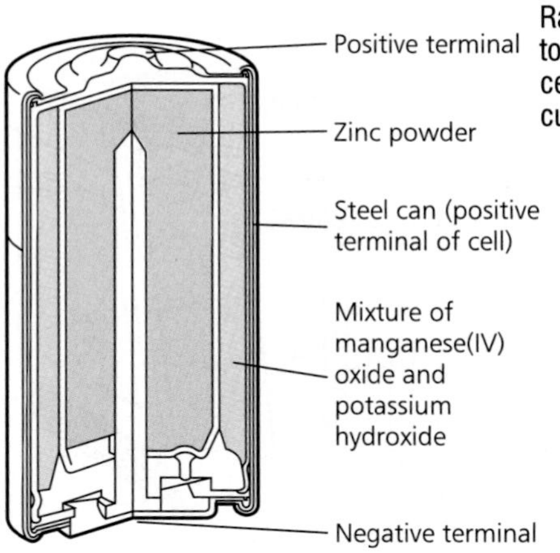 Positive terminal Zinc powder Steel can (positive terminal of cell) Mixture of manganese(IV) oxide and potassium hydroxide Negative terminal	Radios, cassette players, torches – this type of cell can supply quite large currents for long periods.
Nickel–cadmium (secondary cell – rechargeable) e.m.f. ≈ 1.4 V	$Cd(s) + 2OH^-(aq) \rightarrow Cd(OH)_2(s) + 2e^-$ (anode) $NiO_2(s) + 2H_2O(l) + 2e^- \rightarrow Ni(OH)_2(s) + 2OH^-(aq)$ (cathode)	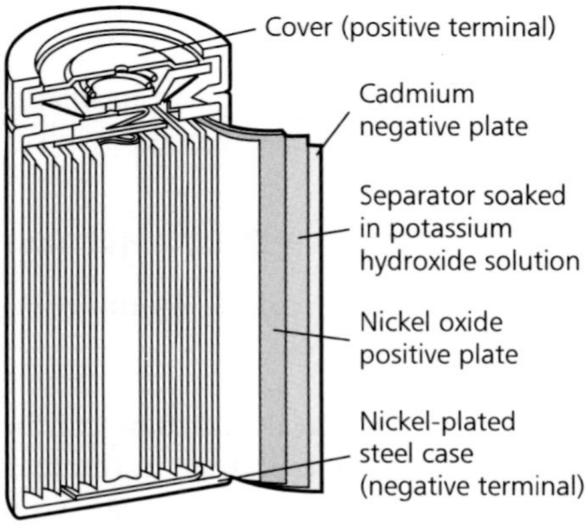 Cover (positive terminal) Cadmium negative plate Separator soaked in potassium hydroxide solution Nickel oxide positive plate Nickel-plated steel case (negative terminal)	Rechargeable batteries where a lead–acid cell would be too large – radios, torches, etc.
Mercury (primary cell – not rechargeable) e.m.f. ≈ 1.35 V	$Zn(s) + 2OH^-(aq) \rightarrow ZnO(s) + H_2O(l) + 2e^-$ (anode) $HgO(s) + H_2O(l) + 2e^- \rightarrow Hg(l) + 2OH^-(aq)$ (cathode)	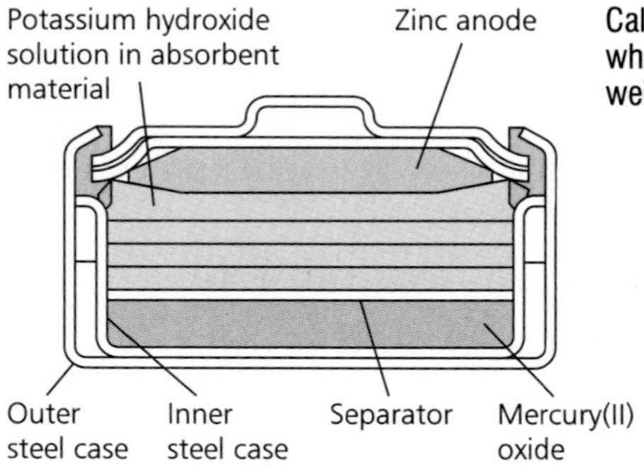 Potassium hydroxide solution in absorbent material Zinc anode Outer steel case Inner steel case Separator Mercury(II) oxide	Calculators and cameras – where small size and light weight are important.
Silver oxide (primary cell – not rechargeable) e.m.f. ≈ 1.5 V	$Zn(s) + 2OH^-(aq) \rightarrow ZnO(s) + H_2O(l) + 2e^-$ (anode) $Ag_2O(s) + H_2O(l) + 2e^- \rightarrow 2Ag(s) + 2OH^-(aq)$ (cathode)	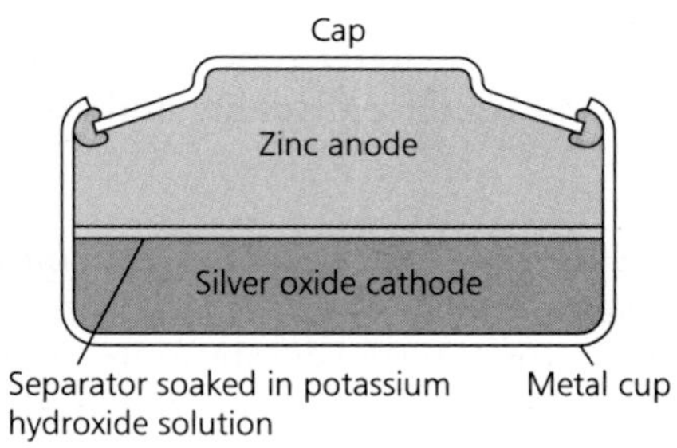 Cap Zinc anode Silver oxide cathode Separator soaked in potassium hydroxide solution Metal cup	Watches, miniature cameras and calculators – very small size.

Table 3.5.3 Some of the many types of cell available

Chemists have played a vital part in finding electrochemical reactions to offer possibilities for such batteries. One of these is the sodium–sulphur battery, in which the source of e.m.f. is the reaction:

$$2Na(l) + 5S(l) \rightarrow Na_2S_5(l)$$

The sodium–sulphur battery has a large energy density, but it must be operated at between 300 and 350 °C, presenting some practical problems for its use. Despite this, it seems to be one possibility in reducing the pollution caused by petrol- and diesel-powered vehicles.

SUMMARY

- **Redox reactions** involve competition for electrons. When a species is **reduced** it **gains** electrons; when a species is **oxidised** it **loses** electrons.
- When electricity flows through a molten salt, or through a solution of an electrolyte, the substance is split up in a chemical process called **electrolysis**. At the **anode**, **anions** (negatively charged ions) are oxidised. At the **cathode**, **cations** (positively charged ions) are reduced.
- The chemical changes at each electrode are summarised using **half-equations**.
- The overall **cell reaction** can be found by adding together the two half-equations.
- Electricity may be produced whenever two metals are immersed in conducting solution, forming an **electrochemical cell.**
- The **electromotive force** (**e.m.f.**) of a given electrochemical cell is the maximum potential difference that can exist across the terminals of that cell.
- Each half-cell reaction has its own tendency to attract electrons, which is measured by the **electrode potential** E of the half-cell.
- The half-cell against which all other electrode potentials are measured is the **standard hydrogen half-cell** or **standard hydrogen electrode**.
- The **standard electrode potential** $E^\ominus$ of a half-cell is the electrode potential of that half-cell relative to the standard hydrogen electrode under standard conditions. $E^\ominus$ is a measure of the oxidising or reducing power of the species in it.
- The standard conditions for electrochemical measurements are similar to those for standard enthalpy changes:
  All solutions have unit activity (effectively 1.00 mol dm^{-3}).
  All measurements are made at 25 °C and 10^5 Pa pressure.
- The reaction occurring in any half-cell is written with the more oxidised species on the left-hand side of the half-equation:

$$Cu^{2+}(aq) + e^- \rightleftharpoons Cu^+(aq)$$

- Standard electrode potentials can be used to predict the likelihood of a reaction proceeding spontaneously. However, care should be taken in doing this, since:
  Standard electrode potentials give no information about the *rate* of a reaction.
  Standard electrode potentials apply to standard conditions only.
- Batteries store chemical energy which can be transferred into electrical energy. They are made up of cells connected in series to increase the e.m.f.

QUESTIONS

1 a Describe what chemists mean by the terms **oxidation** and **reduction**.

b In the following reactions, state which species have been oxidised and which have been reduced:

i $Zn^{2+}(aq) + Cu(s) \rightarrow Zn(s) + Cu^{2+}(aq)$

ii $Cl_2(g) + 2Br^-(aq) \rightarrow 2Cl^-(aq) + Br_2(l)$

iii $Cr_2O_7^{2-}(aq) + 6Fe^{2+}(aq) + 14H^+(aq) \rightarrow 2Cr^{3+}(aq) + 6Fe^{3+}(aq) + 7H_2O(l)$

iv $2MnO_4^-(aq) + 6H^+(aq) + 5HCOOH(l) \rightarrow 2Mn^{2+}(aq) + 8H_2O(l) + 5CO_2(g)$

2 Write half-equations to show what happens when the following substances react:

a Na and Cl_2

b Fe^{3+} and I^-

c I_2 and $S_2O_3^{2-}$

d Cu and Ag^+.

3 When magnesium metal is dropped into a solution containing copper(II) ions, copper metal is formed and the magnesium dissolves.

a Write half-equations for this process.

b How could you use this reaction to make an electrochemical cell?

c Which metal would be positive in such a cell?

d What is the function of the salt bridge in an electrochemical cell?

4 Using the data in table 3.5.2 (page 301), write down the cell notation for cells made from the following pairs of half-cells and calculate the standard electrode potential for each cell:

a $Cu^{2+}(aq) + 2e^- \rightleftharpoons Cu(s)$
$Al^{3+}(aq) + 3e^- \rightleftharpoons Al(s)$

b $Cl_2(g) + 2e^- \rightleftharpoons 2Cl^-(aq)$
$Fe^{3+}(aq) + e^- \rightleftharpoons Fe^{2+}(aq)$

c $I_2(aq) + 2e^- \rightleftharpoons 2I^-(aq)$
$Ni^{2+}(aq) + 2e^- \rightleftharpoons Ni(s)$

d $Pb^{2+}(aq) + 2e^- \rightleftharpoons Pb(s)$
$H_2O_2(aq) + 2H^+(aq) + 2e^- \rightleftharpoons 2H_2O(l)$

5 Consider the following standard electrode potentials:

$Ce^{3+}(aq) + 3e^- \rightleftharpoons Ce(s)$ $E^\ominus = -2.33$ V
$V^{2+}(aq) + 2e^- \rightleftharpoons V(s)$ $E^\ominus = -1.18$ V
$Hg_2^{2+}(aq) + 2e^- \rightleftharpoons 2Hg(l)$ $E^\ominus = +0.79$ V
$Mn^{3+}(aq) + e^- \rightleftharpoons Mn^{2+}(aq)$ $E^\ominus = +1.49$ V

a Against which half-cell are all these electrode potentials measured?

b Which of these substances is:

i the strongest oxidising agent

ii the strongest reducing agent?

Use table 3.5.2 to answer the following:

c Which of these substances could oxidise chloride ions to chlorine?

d Which of these substances could reduce zinc ions to zinc metal?

e Which of these substances would be reduced by iodide ions?

6 Use the data in table 3.5.2 to decide which of these reactions will proceed spontaneously:

a $Ni^{2+}(aq) + Fe(s) \rightarrow Ni(s) + Fe^{2+}(aq)$

b $I_2(aq) + 2Fe^{2+}(aq) \rightarrow 2I^-(aq) + 2Fe^{3+}(aq)$

c $Pb^{2+}(aq) + 2Ag(s) \rightarrow Pb(s) + 2Ag^+(aq)$

d $5H_2O_2(aq) + 2Mn^{2+}(aq) \rightarrow 2MnO_4^-(aq) + 2H_2O(l) + 6H^+(aq)$

7 Calculate the standard electrode potentials for the following cells from the data in table 3.5.2. For each cell, write down the reaction that occurs when current is drawn from the cell.

a $Ni(s)|Ni^{2+}(aq) \,\vdots\vdots\, Pb^{2+}(aq)|Pb(s)$

b $Pt|Fe^{2+}(aq), Fe^{3+}(aq) \,\vdots\vdots\, Ag^+(aq)|Ag(s)$

c $Fe(s)|Fe^{3+}(aq) \,\vdots\vdots\, Cl_2(g), 2Cl^-(aq)|Pt$

Developing Key Skills

Using the information in the Focus material as a starting point, research and produce an obituary for Michael Faraday to be placed in *The Times*. As in all good obituaries, make sure that you refer to his work, his importance to society and his personal life.

[Key Skills opportunities: C, IT]

FOCUS CONTROL BY CATS!

Cars pollute the atmosphere, guzzle petrol and are the single largest cause of premature death in the developed world. Yet cars are also perceived as vital to our way of life and many people around the world find it difficult if not impossible to imagine life without them.

But the fact is that cars are one of the major polluting agents of the planet Earth (see page 233). The waste materials that issue from the car exhaust damage the atmosphere in a number of ways. If we cannot substantially reduce the numbers of cars in the world – and at the moment the number is still increasing on a massive scale – then we must either clean up the engine, or leave the engine unchanged but remove the pollutants before the waste gases pass out of the exhaust.

Back-end emission control

The development of the lean-burn engine has not been rapid enough so far to play a role in reducing the pollution produced by cars. The main thrust of the battle against pollution has been on back-end emission control. All new cars in

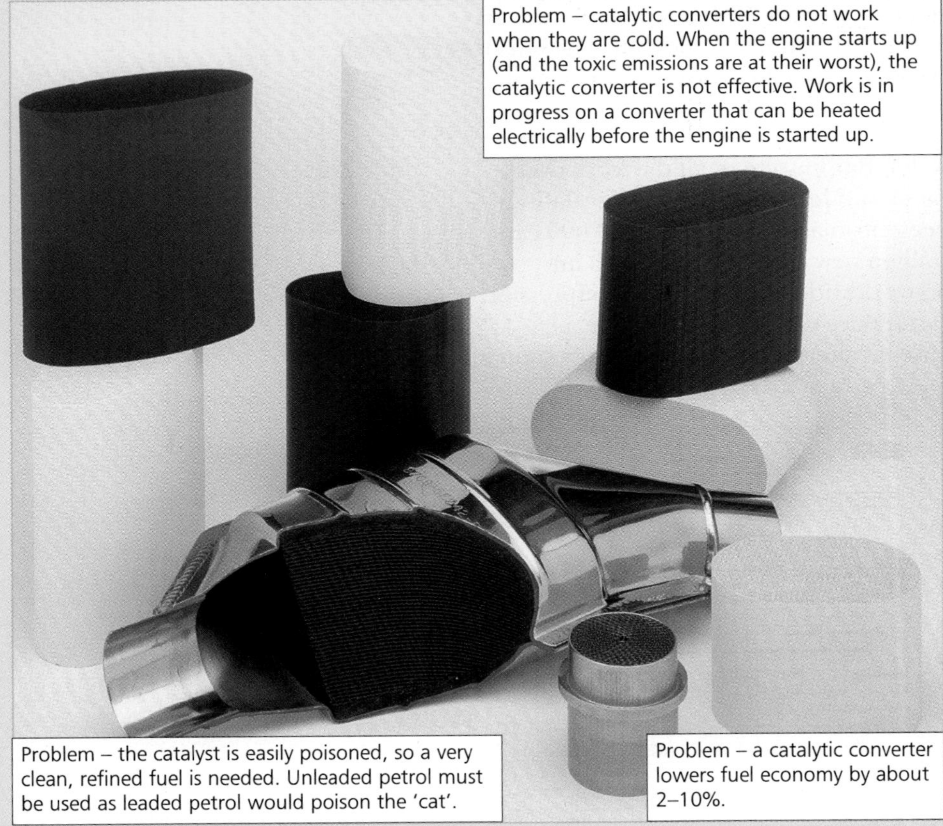

Problem – catalytic converters do not work when they are cold. When the engine starts up (and the toxic emissions are at their worst), the catalytic converter is not effective. Work is in progress on a converter that can be heated electrically before the engine is started up.
Problem – the catalyst is easily poisoned, so a very clean, refined fuel is needed. Unleaded petrol must be used as leaded petrol would poison the 'cat'.
Problem – a catalytic converter lowers fuel economy by about 2–10%.

Figure 1 Catalytic converters go a long way towards cleaning up back-end emissions at relatively low cost. However, there are still a few problems which may be solved in the next few years. This photograph shows a variety of catalytic converters. The catalysts used are platinum and/or palladium and rhodium.

3

FOCUS CONTROL BY CATS! (continued)

Europe must now be fitted with a 'cat'. A three-way **catalytic converter** is fitted onto a car exhaust, in front of the silencer. This removes most of the pollutant gases before they reach the atmosphere. A mixture of air and exhaust gases is passed though the converter, which contains a porous ceramic material that supports the catalyst. The porous nature of this medium means that the surface area of the catalyst is greater than that of two football pitches. In the converter, nitrogen oxide dissociates into nitrogen and oxygen atoms, and the oxygen molecules from the air also dissociate into oxygen atoms. Nitrogen molecules form and the oxygen is used to oxidise carbon monoxide to carbon dioxide. Any unburned hydrocarbons are also oxidised to carbon dioxide and water. As a result, the exhaust gases from a car with a catalytic converter contain simply a mixture of nitrogen, carbon dioxide and water. This is far less polluting than before, but still does not solve the problem of the greenhouse gas carbon dioxide.

To prevent the catalyst being poisoned, no lead may be added to the petrol. In unleaded petrol, another way of preventing knocking has to be found, which is usually done by increasing the amount of aromatic hydrocarbons in unleaded petrol. Cars fitted with catalytic converters are less efficient than those which are not. In addition, the reformulation necessary to produce 'unleaded' petrol means that effectively less petrol is manufactured for each tonne of crude oil. The effect of the trade-off between cleaner engines versus more efficient use of fuel is still under debate.

Modern cars are certainly cleaner than their predecessors. Six modern cars produce less pollution than one 1970s model. But in spite of all the work over the last few years to make car engines cleaner and less polluting, the escalating number of vehicles outweighs the improvements made. More than 100 000 cars are produced per day. That is nearly 40 million new cars each year, with an estimated 600 million vehicles on the road worldwide. To make a real impression on the pollution these cars produce we *must* reduce car usage, and make a concerted effort to produce vehicles that use an alternative cleaner energy source – but that's another story.

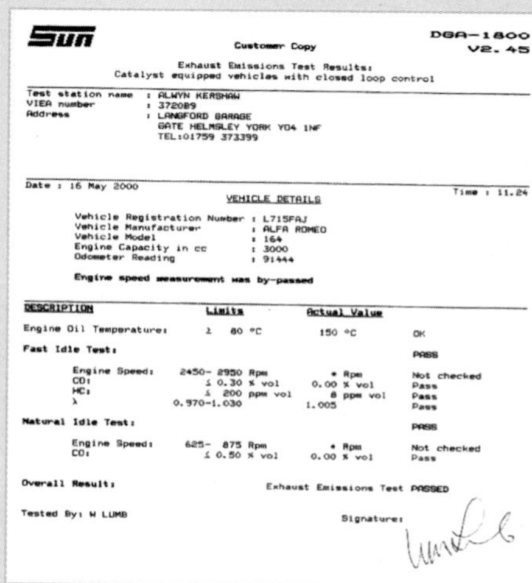

Sun — Customer Copy — DGA-1800 V2.45

Exhaust Emissions Test Results:
Catalyst equipped vehicles with closed loop control

Test station name : ALWYN KERSHAW
VIEA number : 372089
Address : LANGFORD GARAGE
GATE HELMSLEY YORK YO4 1NF
TEL:01759 373399

Date : 16 May 2000 — Time : 11.24

VEHICLE DETAILS

Vehicle Registration Number : L715FAJ
Vehicle Manufacturer : ALFA ROMEO
Vehicle Model : 164
Engine Capacity in cc : 3000
Odometer Reading : 91444

Engine speed measurement was by-passed

DESCRIPTION	Limits	Actual Value	
Engine Oil Temperature:	≥ 80 °C	150 °C	OK
Fast Idle Test:			PASS
Engine Speed:	2450- 2950 Rpm	* Rpm	Not checked
CO:	≤ 0.30 % vol	0.00 % vol	Pass
HC:	≤ 200 ppm vol	8 ppm vol	Pass
λ	0.970-1.030	1.005	Pass
Natural Idle Test:			PASS
Engine Speed:	625- 875 Rpm	* Rpm	Not checked
CO:	≤ 0.50 % vol	0.00 % vol	Pass

Overall Result: Exhaust Emissions Test PASSED

Tested By: W LUMB — Signature:

Figure 2 In the UK, all cars over 3 years old have their exhaust emissions checked during the yearly MOT test.

3.6 Reaction kinetics: How fast?

We have looked at equilibria in chemical reactions, and the factors that affect those equilibria, answering the question 'How far?'. This tells us whether a particular reaction is possible. However, for a reaction to 'go' it is not enough just for it to be possible – it must occur rapidly enough for us to observe it and to collect the products. In this section we shall look at how rapidly chemical reactions occur, and how they may be speeded up or slowed down – answering the question 'How fast?'.

Figure 3.6.1 Raising the temperature increases the rate of most reactions, and conversely, lowering the temperature slows them down. This man died of exposure on the slopes of the Austrian–Italian Alps around 5000 years ago. His body temperature was very low just before he died, and after death he became covered in snow and ice. This slowed down the chemical reactions that normally cause a body to decompose, to the extent that he remains almost perfectly preserved.

Reaction rates

Reaction rates are of fundamental importance to the research chemist, in chemical industry and in the living world. The **rate of reaction** describes the speed at which reactants are converted to products for a particular reaction. Knowledge about reaction rates allows industrial operating conditions to be chosen such that a process takes place with maximum efficiency and economy. Studying the rate of a reaction also gives us information about how the reaction is taking place.

Expressing the rate of chemical change

In order to quantify the progress of a chemical reaction, we measure the **reaction rate**. This is the change in concentration of either the reactants or the products with time. The concentration of a particular substance is measured at the beginning and end of a measured time period:

Change in concentration = final concentration – initial concentration

The symbol Δ (Greek delta) is sometimes used to indicate a change, so that 'change in concentration' may be written as Δ(concentration).

The rate of a chemical reaction can therefore be expressed as:

$$\textbf{Rate of reaction} = \frac{\textbf{change in concentration}}{\textbf{change in time}}$$

or:

$$\textbf{Rate of reaction} = \frac{\Delta\textbf{(concentration)}}{\Delta\textbf{(time)}}$$

Thus if the concentration of a product P of a reaction increases by 0.25 mol dm^{-3} each second, then the rate of reaction with respect to P is 0.25 mol dm^{-3} s^{-1}. The rate for a product, a substance whose concentration is *increasing*, always has a positive number. The rate for a reactant, whose concentration is *decreasing*, has a negative number, because Δ(concentration) is obviously negative. For example, if 0.5 mol dm^{-3} of a reactant R is used up every second, then the rate of reaction with respect to R is –0.5 mol dm^{-3} s^{-1}. The overall rate of a reaction is always positive, so if it is measured using the change in concentration of a reactant, a minus sign is added to produce a positive number for the rate.

3

Expressing rates of change

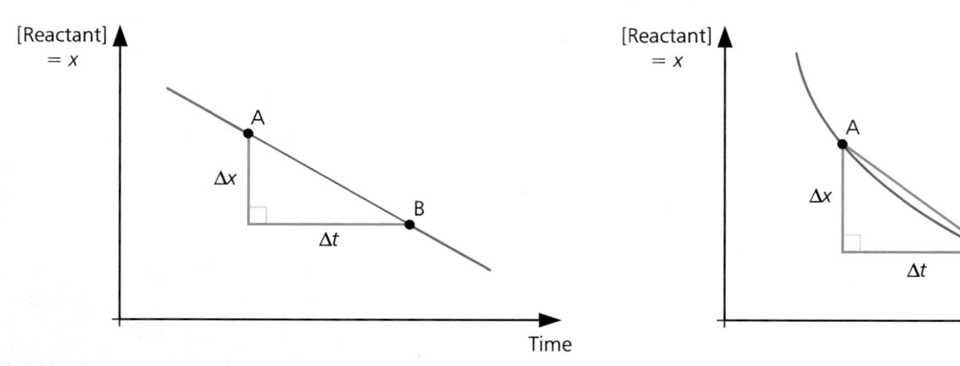

Figure 3.6.2

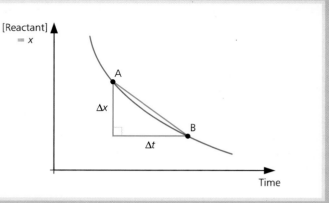

Figure 3.6.2 shows two plots of concentration of a reactant against time. The line AB is a measure of the gradient of each graph. In the first graph, the slope of AB is exactly equal to the slope of the graph, since the graph is a straight line between A and B. This graph represents a constant reaction rate – at any point between A and B, the **instantaneous rate** and the **average rate** between A and B are the same, $\Delta = x/\Delta t$. But this is not the case with the second graph. Here the rate of reaction is decreasing, and there will only be one point between A and B where the instantaneous rate and the average rate are the same. In this case, to find the instantaneous rate at a point (say point A), we need to find Δx over the smallest time interval possible, so Δt needs to become *very* small.

Figure 3.6.3 shows that the closer we move point B towards point A, the nearer the slope of line AB will get to the slope of the graph. If we keep moving point B towards point A, eventually the two points will be one on top of the other. At this point, the slope of line AB will be exactly equal to the slope of the graph, and AB will be a tangent to the graph at this point. Using mathematical language:

$$\textbf{Slope of line} = \frac{dx}{dt} = \lim_{\Delta t \to 0} \frac{\Delta x}{\Delta t}$$

In other words, as we move B closer and closer to A, Δt gets smaller and smaller ('the limit as Δt tends to zero'). Eventually Δt is so small that the slope of line AB is the same as the slope of the graph, which we write as dx/dt.

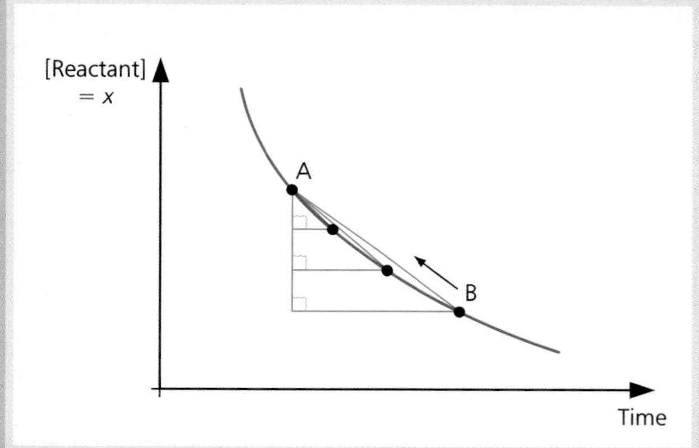

Figure 3.6.3

Factors affecting the rate of a reaction

In sections 3.1–3.4 we saw that a variety of factors can affect the equilibrium position of a reaction, driving it towards the right or left and so decreasing or increasing the likelihood that it will go to completion. In a similar way, a variety of factors can influence the rates of chemical reactions, speeding them up or slowing them down. There are five main factors which may affect reaction rate, and we shall begin by looking at each in turn.

(1) The chemical nature of the reactants

As we have seen in our studies of the periodic table in section 2, some elements are considerably more reactive than others, as illustrated in figure 3.6.4. This fundamental reactivity, based on the ease with which an element will gain or lose an electron, must affect the rate of any reaction involving the elements.

Figure 3.6.4 Freshly cut sodium tarnishes almost immediately in air, as an oxide film forms rapidly. Iron also reacts with the oxygen in the air, as the owners of many rusty cars will testify, but the rate of the reaction is considerably slower.

(2) The state of the reactants and their ability to meet

Most reactions involve a minimum of two reactants. It is obvious that no reaction will take place unless the reactants are brought into contact with each other. There are two different factors here which may speed up or slow down the rate of a reaction.

If the reactants are liquids or gases, their reacting particles can intermingle freely, giving them the maximum opportunity to react. This is why the majority of chemical reactions are carried out in aqueous or organic solution, or in the gaseous state. However, if one of the reactants is a solid, then the reacting particles can only meet at the surface of the solid. If the surface area of the solid is increased, then the rate at which the reaction can take place will increase too. The effect of particle size on surface area is demonstrated in figure 3.6.5.

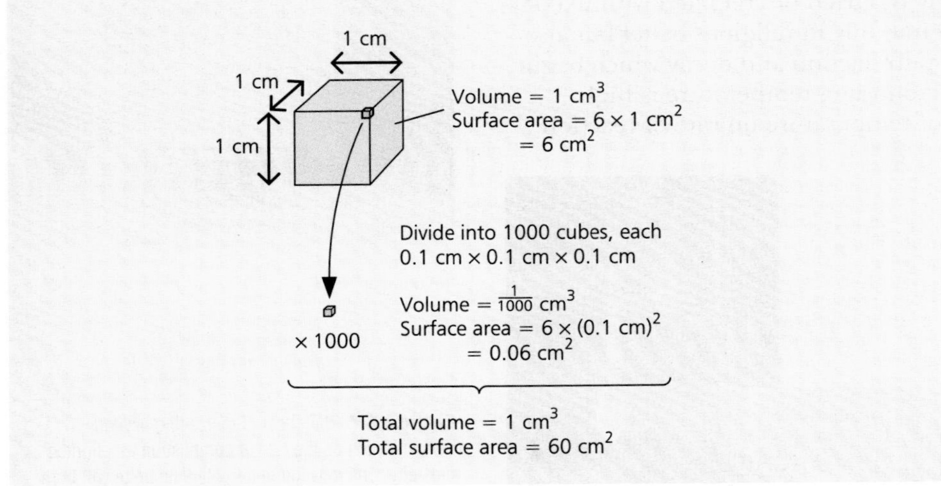

Figure 3.6.5 If you divide a 1 cm^3 cube into 0.1 cm^3 cubes, the surface area increases from 6 cm^2 to 60 cm^2.

Surface area can also be a factor in reactions between liquids and gases. When the reactants are in the same phase – when two gases or two liquids are reacting together – then the reaction is known as a **homogeneous reaction** and the reacting particles intermingle freely. However, when the reactants are in different phases, for example, one is a gas and the other is a liquid or a solid, the reactants can only meet at the interface between the two phases. The surface area of this interface then becomes important in determining the rate of the reaction. Such reactions between two substances in different phases are known as **heterogeneous reactions**.

Figure 3.6.6 gives some examples of heterogeneous reactions. However, most of the reactions we shall be considering in this section will be homogeneous.

(3) The concentrations of the reactants

The rate of most reactions is affected by the concentrations of the reactants (or the pressures if the reactants are gases). Increase or decrease their

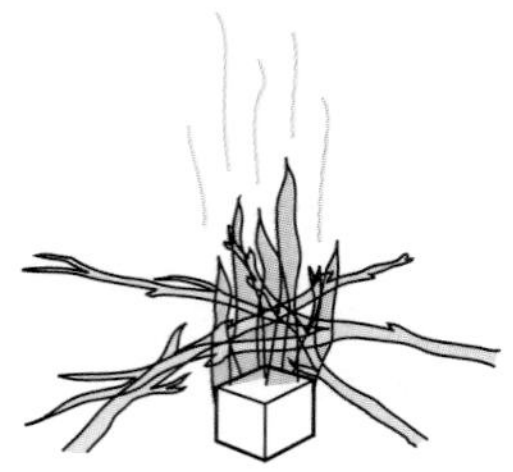

Figure 3.6.6 In the heterogeneous reaction between wood in the solid state and oxygen in the gaseous state, the surface area of the wood exposed to the air makes all the difference to the rate of combustion. The end result of an explosive heterogeneous reaction is shown in the photograph – the tiny size of the particles of sawdust allowed the combustion reaction between the solid wood and the gaseous oxygen to progress so rapidly that the woodchip storage container was completely destroyed.

concentration and the reaction rate will change, as illustrated in figure 3.6.7 below. An increase in the concentration of the reactants usually leads to an increase in the rate of the reaction, although this is not always the case, especially in biological reactions.

(4) The temperature of the system

Most chemical reactions occur more rapidly when the temperature is raised. This increase in the rate of chemical reactions with an increase in temperature has been recognised for centuries. Milk turns sour on the doorstep far more rapidly on a hot summer's day than in the cold winter. In very hot countries, funeral rites have evolved in which the body is buried or cremated within a day or so of death. These traditions are rooted not only in religious belief but also in sound common sense – the reactions of putrefaction and decay which begin after death occur very rapidly when the surrounding temperature is high. Figure 3.6.8 opposite illustrates the effect of temperature on rate of reaction.

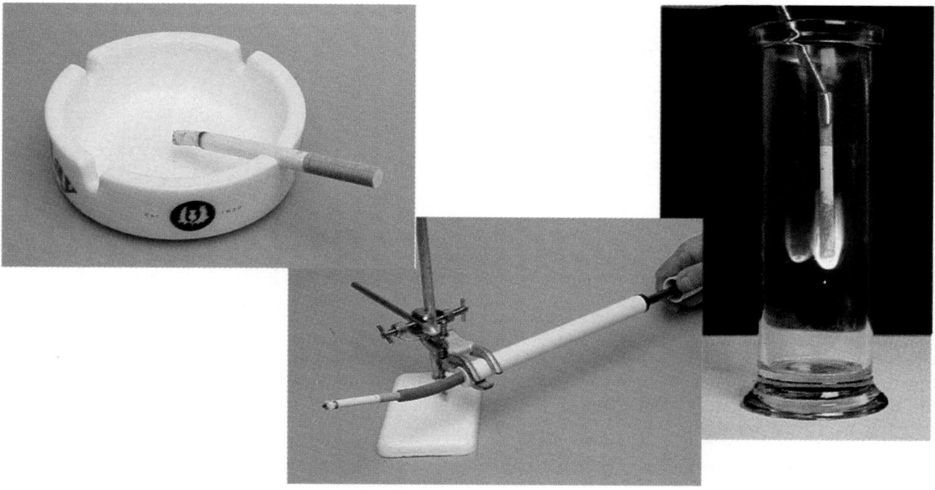

Figure 3.6.7 The effect of increasing oxygen concentration on the rate of combustion in a lighted cigarette. The tobacco in a lighted cigarette will burn slowly in the air. If air is drawn through the cigarette by a pump, more oxygen is available to the tobacco and the rate of burning increases. Placed in a gas jar full of pure oxygen, the lighted cigarette burns rapidly in a single violent flare.

(5) Catalysts

A **catalyst** is a substance that will alter the rate of a chemical reaction (usually speeding it up) without itself being used up or undergoing any permanent chemical change. Catalysts can enable reactions to occur that would be too slow to observe without them, as shown in figure 3.6.9 opposite. Almost every major industrial chemical process depends on catalysts to enable a profitable reaction rate to be maintained without resorting to excessive and expensive conditions of temperature and pressure. The enzymes that control the biochemistry of every living cell are also specialised catalysts made of protein.

Measuring reaction rates

Following changes in concentration

As we have seen, to arrive at an expression for the rate of a reaction we need to know the change in concentration of a reactant or product over a measured time period. Sometimes we can take an observed change in the reaction mixture as an approximation to a change in concentration. For example, if the reaction results in the production of bubbles of gas, such as:

$$\mathbf{Mg(s) + 2HCl(aq) \rightarrow MgCl_2(aq) + H_2(g)}$$

we can count the bubbles given off in a measured time period. This does not give us an accurate expression for the rate of reaction, but it allows rates to be compared under different conditions, such as at different temperatures. In any such investigation, it is of course essential that only one factor is varied at a time – if temperature is being investigated then concentration, particle size, etc. must be kept the same. More accurate measurements can be obtained by measuring the volume of gas given off during periods of time throughout the reaction, and then plotting a graph of the volume of gas produced against time.

Figure 3.6.8 The chemical reactions in the bodies of insects are greatly affected by the temperature of their surroundings. This enormous goliath beetle can only produce sufficient energy to fly when its temperature is over 30°C.

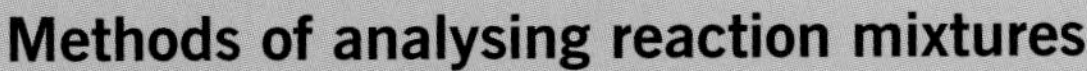

Methods of analysing reaction mixtures

As mentioned above, if one of the products of a reaction is a gas, then this can be collected in a gas syringe, giving a measure of the rate of reaction. Several commonly used methods of determining reaction rate in other types of reaction are summarised below.

Titrimetric analysis

Small portions or **aliquots** of a reacting mixture are removed at regular intervals. These aliquots are usually added to another reagent which immediately stops or **quenches** the reaction so that there are no further changes to the concentrations before the mixture is analysed. The quenched aliquots are then titrated to find the concentrations of known compounds in them.

Colorimetric analysis

This is particularly valuable where one of the reactants or products of a reaction is coloured. The colour changes throughout the reaction can be detected using a photoelectric colorimeter, and these colour changes subsequently used to give changes in concentration, by calibration of the colorimeter with solutions of known concentration.

Conductimetric analysis

The conductivity of a reacting mixture can be measured at regular time intervals. Changes in conductivity reflect changes in the ions present in the solution, and so can be used to measure the changes in concentration of the various components of the mixture.

Initial rates method

Analysis of a reaction mixture during the course of a reaction is not always possible. To compare rates of reaction under different conditions, a number of experiments may be set up in which the initial concentrations of reactants are known. The initial rate of reaction is then measured for each experiment – question 5 on page 339 shows how this is done.

Figure 3.6.9 Hydrogen peroxide is thermodynamically unstable with respect to water and oxygen, as we saw in section 1.8 (page 112). However, the rate at which hydrogen peroxide decomposes at room temperature is normally very slow in the absence of a catalyst. In the presence of a catalyst such as manganese(IV) oxide, the reaction is very rapid.

Finding the rate of reaction

Once the reaction mixture has been analysed, the reaction rate can be obtained by plotting a graph of the concentration of a reactant or product against time. A tangent to this curve drawn at any point on the graph gives the reaction rate at that point. The rate of reaction is not usually constant throughout a reaction, because the rate usually depends on the concentration of the reactants, and these change during the course of the reaction. Table 3.6.1 and figure 3.6.10 give data for a reaction with one reactant, sulphur dichloride oxide, SO_2Cl_2.

$$\mathbf{SO_2Cl_2(g) \rightarrow SO_2(g) + Cl_2(g)}$$

The rate of this reaction can be followed by monitoring the pressure of the gases in the reaction vessel, since the number of moles of gas doubles on going from left to right. The concentration can be readily calculated from the pressure.

Time/s	$[SO_2Cl_2(g)]$/mol dm^{-3}
0	0.50
500	0.43
1000	0.37
2000	0.27
3000	0.20
4000	0.15

Table 3.6.1

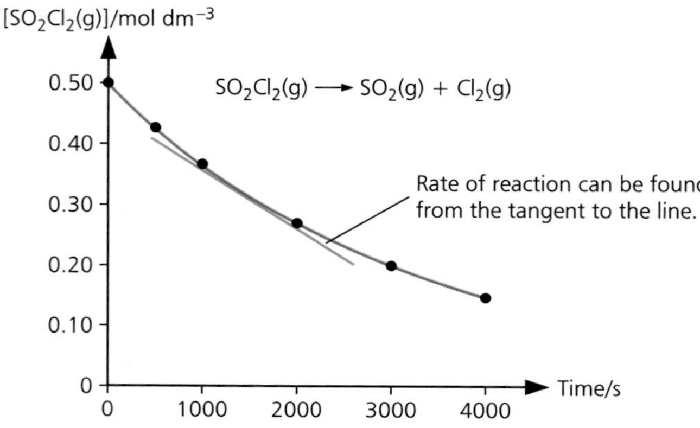

Figure 3.6.10 The rate of dissociation of SO_2Cl_2 into SO_2 and Cl_2 decreases as the concentration of SO_2Cl_2 falls. The rate of reaction at any given time can be found by drawing the tangent to the graph.

The graph in figure 3.6.10 shows how $[SO_2Cl_2(g)]$ falls during the reaction. We can write an expression for the rate of this reaction with respect to SO_2Cl_2:

$$\textbf{Rate} = \frac{-\Delta[SO_2Cl_2(g)]}{\Delta t} = \frac{-\,d[SO_2Cl_2(g)]}{dt}$$

$[SO_2Cl_2(g)]$ *decreases* with time, so $d[SO_2Cl_2(g)]/dt$ is *negative*. The negative sign ensures that the rate for the overall reaction is positive.

Now look at table 3.6.2 and figure 3.6.11 opposite. The numbers in the table have been obtained by taking tangents at various points along a graph like that shown in figure 3.6.10.

Straight-line graphs

Graphs are extremely useful in chemistry, for finding and confirming relationships between different variables (for example, the variation of the volume of a gas with its temperature). The simplest type of relationship is **linear**, where a graph of one variable against another is a straight line.

The graph in figure 3.6.11 is an example of a linear graph. The reaction rate is plotted on the vertical axis (referred to as the '*y*-axis' or the 'ordinate'), and the concentration of SO_2Cl_2 is plotted on the horizontal axis (referred to as the '*x*-axis' or the 'abscissa').

The general form of the equation for a straight line is:

$$y = mx + c$$

where m is the slope or gradient of the line and c is the intercept on the y-axis (where the line crosses the y-axis). In this case $c = 0$, so the equation has the form:

$$y = mx$$

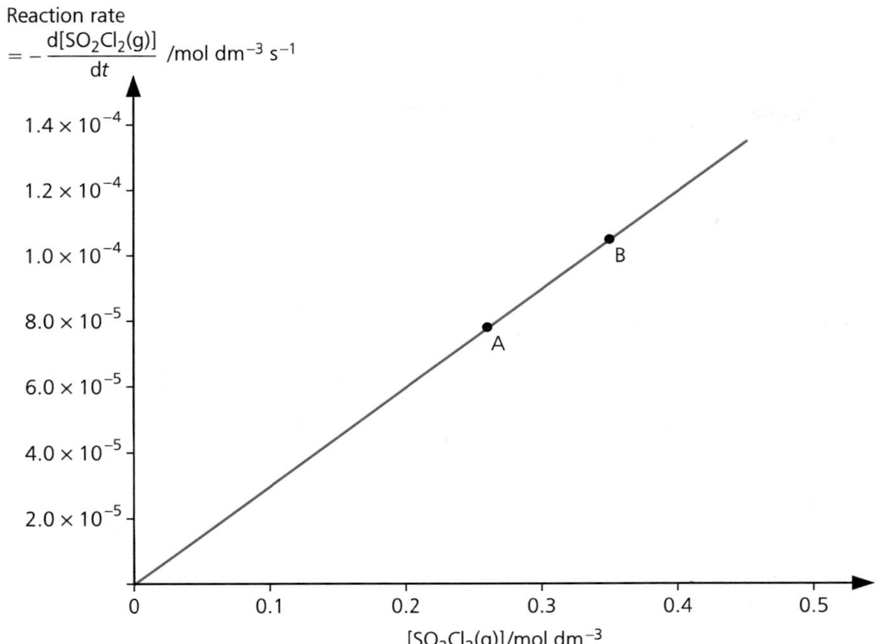

Figure 3.6.11 The variation of reaction rate with concentration of SO_2Cl_2 for the dissociation of SO_2Cl_2. Points A and B will be used for calculations later.

$[SO_2Cl_2]$/ mol dm^{-3}	$-d[SO_2Cl_2(g)]/dt$/ mol dm^{-3} s^{-1}
0.45	1.35×10^{-4}
0.39	1.17×10^{-4}
0.34	1.02×10^{-4}
0.28	8.40×10^{-5}
0.23	6.9×10^{-5}
0.18	5.4×10^{-5}

Table 3.6.2

The rate constant

The graph of $-d[SO_2Cl_2(g)]/dt$ against $[SO_2Cl_2(g)]$ in figure 3.6.11 is a straight line. Since reaction rate is plotted on the y-axis of the graph and $[SO_2Cl_2(g)]$ is plotted on the x-axis, it follows that:

$$\textbf{Reaction rate} = \frac{-\,d[SO_2Cl_2(g)]}{dt} = k[SO_2Cl_2(g)]$$

where k is a constant, known as the **rate constant** or the **velocity constant** of the reaction. The rate constant has a particular value for a given reaction at a given temperature.

Rate expressions

The rate of dissociation of SO_2Cl_2 follows a law which can be written as:

$$\textbf{Rate} = k[SO_2Cl_2(g)]$$

This relationship is the **rate expression** or **rate law** for the reaction.

We can calculate the rate constant k for the reaction from the gradient of the graph in figure 3.6.11. The coordinates of points A and B are (0.26 mol dm^{-3}, 7.8×10^{-5} mol dm^{-3} s^{-1}) and (0.35 mol dm^{-3}, 1.05×10^{-4} mol dm^{-3} s^{-1}) respectively, so:

$$k = \frac{\Delta y}{\Delta x}$$

$$= \frac{1.05 \times 10^{-4}\ \text{mol dm}^{-3}\ \text{s}^{-1} - 7.8 \times 10^{-5}\ \text{mol dm}^{-3}\ \text{s}^{-1}}{0.35\ \text{mol dm}^{-3} - 0.26\ \text{mol dm}^{-3}}$$

$$= 3.0 \times 10^{-4}\ \text{s}^{-1}$$

3

The rate constant for a reaction varies with temperature, so the value of *k* calculated from a particular investigation applies *only* at the temperature at which the investigation was carried out.

Order of reaction

The rate expression just given could be written as:

$$\textbf{Rate} = k[SO_2Cl_2(g)]^1$$

The reaction is said to be **first order** with respect to SO_2Cl_2, because the concentration of SO_2Cl_2 appears in the rate expression raised to the power of 1.

Unlike equilibrium expressions, rate expressions *cannot* be predicted from the stoichiometric equation, but must be determined by experiment in the way we have just seen. For example, the oxidation of nitrogen oxide to nitrogen dioxide in automobile emissions (important in the formation of smog) may involve carbon monoxide, according to the stoichiometric equation:

$$NO(g) + CO(g) + O_2(g) \rightarrow NO_2(g) + CO_2(g)$$

This reaction may be followed using colorimetry, since one of the products of the reaction is coloured while the other products and the reactants are not. Investigation shows that the rate of this reaction is proportional to $[NO(g)]^2$, but is independent of $[CO(g)]$ and $[O_2(g)]$, so the rate expression for the reaction is:

$$\textbf{Rate} = k[NO(g)]^2[CO(g)]^0[O_2(g)]^0$$

From this rate expression we can say that this reaction is **second order** with respect to nitrogen oxide and **zero order** with respect to both carbon monoxide and oxygen. Since any number raised to the power zero is equal to one, the rate expression can be simplified to:

$$\textbf{Rate} = k[NO(g)]^2$$

In general, for a reaction in which:

$$\textbf{A + B} \rightarrow \textbf{products}$$

the rate expression has the form:

$$\textbf{Rate} = k[A]^x[B]^y$$

where:

x is the order of the reaction with respect to A
y is the order of the reaction with respect to B

and:

$(x + y)$ is the overall order of the reaction

For some reactions the rate expression can be quite complex. For example, the reaction:

$$2HCrO_4^-(aq) + 3HSO_3^-(aq) + 5H^+(aq) \rightarrow 2Cr^{3+}(aq) + 3SO_4^{2-}(aq) + 5H_2O(l)$$

has the rate expression:

$$\textbf{Rate} = k[HCrO_4^-(aq)][HSO_3^-(aq)]^2[H^+(aq)]$$

Notice that this reaction is **fourth order** overall – **first order** with respect to $HCrO_4^-$ and to H^+, and **second order** with respect to HSO_3^-.

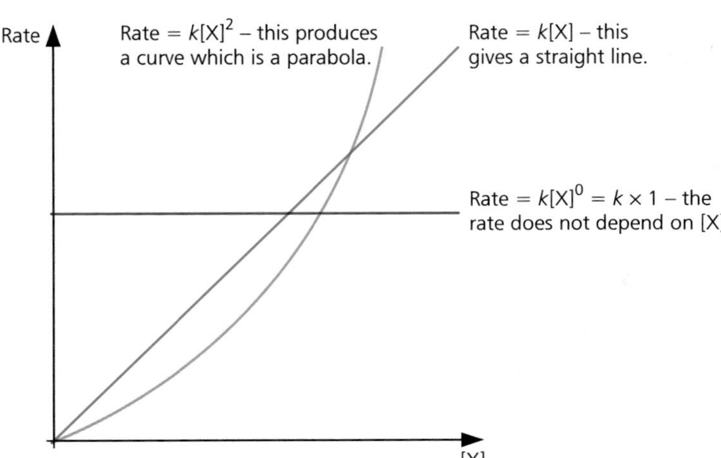
Rate
Rate = k[X]² – this produces a curve which is a parabola.
Rate = k[X] – this gives a straight line.
Rate = k[X]⁰ = k × 1 – the rate does not depend on [X].
[X]

Figure 3.6.12 The variation of reaction rate with [X] for reactions which are zero, first and second order with respect to X.

A plot of reaction rate against concentration of reactant X gives rise to a very characteristic graph, which varies according to the order of the reaction with respect to X. Figure 3.6.12 shows such plots for reactions which are zero, first and second order with respect to X.

Predicting the rate expression

As a general rule, it is not possible to obtain the rate expression for a reaction from its stoichiometric equation. This is because a great many reactions happen in several steps – for example, the decomposition of ozone, O_3 to oxygen, O_2 is known to occur in two steps, the first of which is rapid, the second much slower:

$$O_3 \rightleftharpoons O_2 + O \qquad \textbf{Equilibrium constant} = K$$

$$O + O_3 \rightarrow 2O_2 \qquad \textbf{Rate constant} = k$$

The second step is slow, and therefore determines the rate of the reaction – it is called the **rate determining step**. We *can* write a rate expression for a reaction that occurs in a single step – in this reaction the rate of formation of O_2 will obviously depend on [O] and $[O_3]$, so that the rate expression is:

$$\textbf{Rate} = k[O][O_3]$$

Because the second step is slow, the first step causes an equilibrium concentration of O to build up.

We can obtain an expression for [O] from the equilibrium expression for this step:

$$K = \frac{[O_2][O]}{[O_3]}$$

from which:

$$[O] = K\frac{[O_3]}{[O_2]}$$

Hence the rate expression for this reaction is:

$$\textbf{Rate} = k'\frac{[O][O_3]}{[O_2]} \qquad \textbf{where } k' = kK$$

The decomposition of ozone shows how kinetic studies of reactions can be useful in determining their mechanisms – if a rate expression for a reaction is known, it is possible to use it to see which of several proposed mechanisms is most likely for the reaction. We shall see an example of this in section 4.5.

The half-life of first order reactions

Figure 3.6.13 is an extended version of figure 3.6.10, showing the time taken for the concentration of SO_2Cl_2 to halve. Notice that this time is constant – in other words, it takes the same time for the concentration to fall from 0.50 mol dm^{-3} to 0.25 mol dm^{-3} as it takes for it to fall from 0.25 mol dm^{-3} to 0.125 mol dm^{-3}. This time is known as the **half-life** $t_{\frac{1}{2}}$ of the reaction.

All first order reactions have constant half-lives.

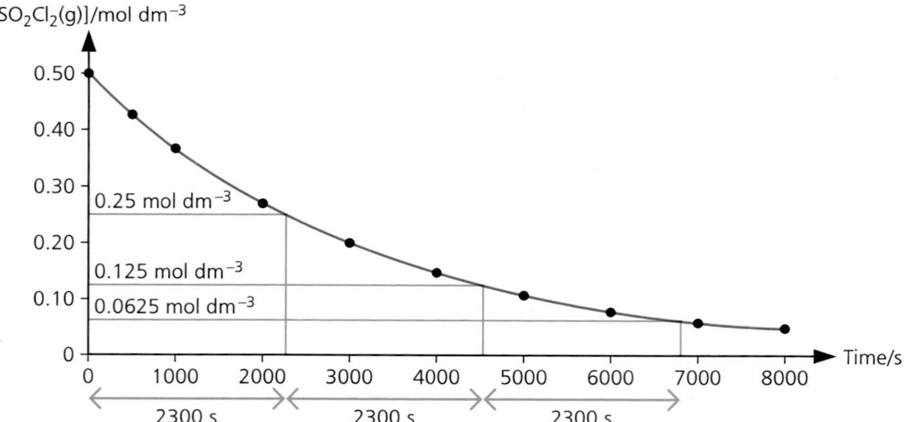

Figure 3.6.13 This graph of $[SO_2Cl_2]$ shows that the reaction has a constant half-life – that is, the time taken for the concentration of SO_2Cl_2 to halve is always constant (at a given temperature). All first order reactions behave like this.

Radioactive decay is an example of a first order reaction, in which the half-life may vary from a tiny fraction of a second (the decay of polonium-212 has $t_{\frac{1}{2}} = 3 \times 10^{-7}$ s) to millions of years (the decay of uranium-238 has $t_{\frac{1}{2}} = 4.5 \times 10^9$ years).

Theories about reaction rates

So far we have looked at the rates of different types of reactions and the factors that affect those rates. In particular, we have seen that the rate of almost all reactions is speeded up by an increase in concentration, and also by an increase in temperature. A 10 °C rise in temperature usually gives an approximate doubling of the rate of reaction. To understand how concentration and temperature affect the rate of a reaction, we must have an effective model of what happens to the molecules in a reacting system.

Collisions between reacting particles

Collision theory is one of the simplest models for the molecular events of a reaction. This says that for a reaction between two particles to occur, an **effective collision** must take place – that is, a collision that results in the formation of product molecules. The reaction rate is a measure of how frequently effective collisions occur. Any factor that increases the rate of effective collisions will also increase the rate of the reaction.

This theory explains why an increase in the concentration of reactants will tend to lead to an increase in the rate of reaction – there are more particles present in the same volume, and so they are more likely to collide. However, not all collisions between particles result in reactions – if they did, all reactions would be virtually instantaneous – but the greater the number of collisions, the higher the chance that some of them will be effective. Thus to develop a closer understanding of how reactions come about and how factors such as temperature and concentration affect the rate, we need to consider what makes a collision an effective collision.

Molecular orientation

When two reactants collide, their atoms must be correctly orientated for a reaction to occur. If the orientation is wrong then the two reactant particles simply bounce off each other. This is shown in the reaction between nitrogen oxide and ozone, as illustrated in figure 3.6.14.

$$NO(g) + O_3(g) \rightarrow NO_2(g) + O_2(g)$$

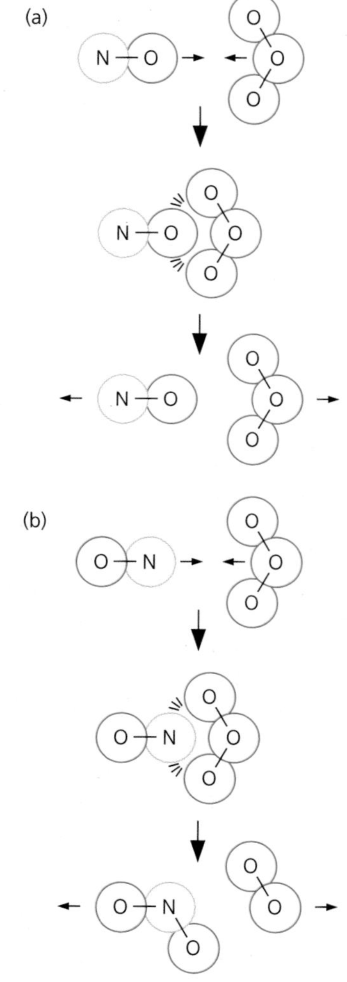

Figure 3.6.14 The importance of molecular orientation during a collision. The collision shown in (a) will not produce a reaction, but that shown in (b) results in the formation of NO_2 and O_2.

Molecular kinetic energy

Even when the orientation of the molecules is correct, few collisions between molecules actually lead to a chemical change. This is because in order to react, the molecules must collide with a certain minimum kinetic energy known as the **activation energy** E_A. If two slow-moving molecules collide, even if they are in the right orientation, they will simply bounce apart as a result of the repulsion of their negative electron clouds. To meet with sufficient energy for bonds to be broken and reformed, the molecules must be moving quickly. An increase in temperature will increase the kinetic energy of the molecules and so increase the rate of reaction, as a greater proportion of the molecules will have sufficient energy to react, as shown in figure 3.6.15.

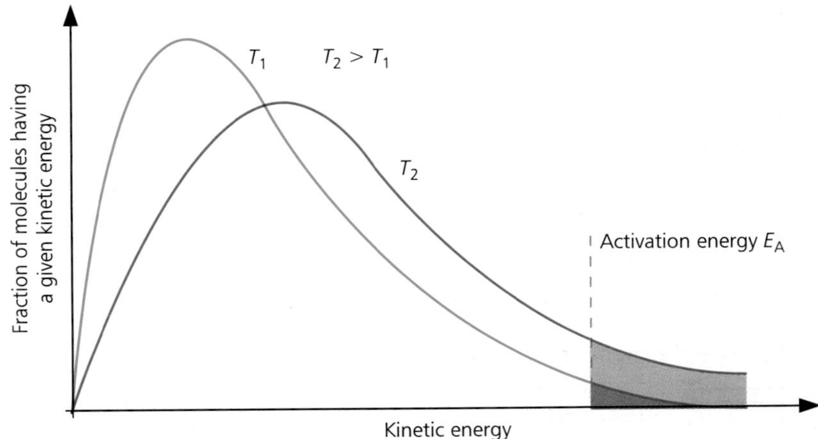

Figure 3.6.15 These curves show the kinetic energy distributions for a reaction mixture at two different temperatures. The shaded areas are proportional to the total fraction of molecules having the minimum activation energy.

Kinetic stability and activation energy

In section 1.8 we saw that the enthalpy change for a reaction can be used as some sort of guide as to whether or not a reaction is likely to happen, while in section 3.1 we met the idea of the equilibrium constant for a reaction. In both sections we saw that, although enthalpy changes and equilibrium constants may provide information about reactions, even reactions which are favourable according to their enthalpy changes and equilibrium constants may not happen at an observable rate. In both cases, we said that 'kinetic stability' was involved – and kinetic stability is linked with activation energy.

The activation energy of a reaction can be thought of as a sort of 'energy barrier'. The reactant molecules must have sufficient energy to get over this barrier if a reaction is to occur, or we shall be left simply with a mixture of unreacted chemicals, no matter how favourable the reaction is in terms of its overall enthalpy change or its equilibrium constant. Kinetic stability happens when the activation energy of a reaction is so large that there are virtually no molecules in the reaction mixture with sufficient energy to overcome the energy barrier, and so the reaction effectively does not happen.

What happens during a reaction?

When molecules collide with the correct orientation they slow down, stop and then fly apart again. This occurs regardless of whether they have the required activation energy to react. In an unsuccessful collision the molecules separate unchanged, whereas in an effective collision the activation energy barrier is crossed and the particles that separate are chemically different from those

The distribution of molecular kinetic energies

In the mid-nineteenth century, Ludwig Boltzmann and James Clerk Maxwell were working quite independently of each other in Vienna, Austria and Cambridge, England. The two scientists developed a statistical treatment of the distribution of energy amongst a collection of particles, resulting in distributions like those shown in figure 3.6.15. This led to a greater understanding of how the macroscopic behaviour of matter may be related to the microscopic particles of which it is composed, and to the development of the kinetic theory through a branch of the sciences now called statistical mechanics.

The work of Maxwell and Boltzmann on the distribution of the speeds of particles in a gas underpins much of the understanding of the rate of chemical reactions. Yet Boltzmann's theories of the behaviour of matter were far from accepted at the time. Coming soon after the work of Darwin, many scientists saw Boltzmann's work as threatening the purposeful, God-given workings of the Universe, for once it could be shown that the behaviour of matter on a grand scale could be understood by studying its behaviour on a much smaller scale, what scope was left for the Creator? Stung by the scorn of his fellow scientists, Boltzmann committed suicide.

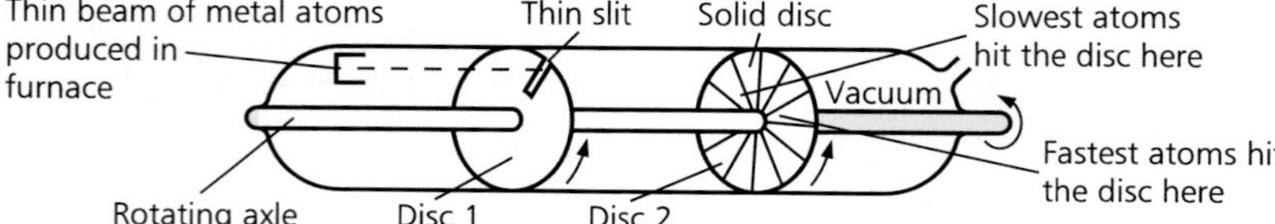

Metal atoms pass through the slit in disc 1 and travel along until they strike disc 2, to which they stick. The time taken for an atom to travel between disc 1 and disc 2 depends on its speed. The relative distribution of velocities can be found by analysing the number of metal atoms which have struck the disc at different points, which can be done by cutting the disc into sectors (like slices of a cake) and weighing them.

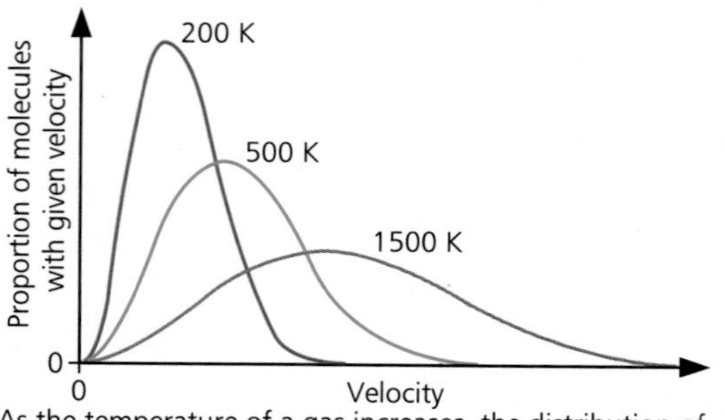

As the temperature of a gas increases, the distribution of molecular speeds within it changes. At higher temperatures the average speed of the molecules increases (the internal energy of the gas is greater), although the proportion of molecules with the most probable speed decreases.

Ludwig Boltzmann (1844–1906)

Figure 3.6.16 Seventy years after Boltzmann published his work, Zartmann and Ko showed that the distribution of velocities in the molecules of a gas agreed with Boltzmann's predictions.

which came together. As the molecules slow down, they possess less and less kinetic energy and their potential energy increases. Figure 3.6.17 shows two examples of a **potential energy diagram**, which follows the path of a reaction as reactant molecules come together, collide and form product molecules which then separate. Such a diagram expresses graphically the relationship between the activation energy and the total potential energy of the colliding reactants.

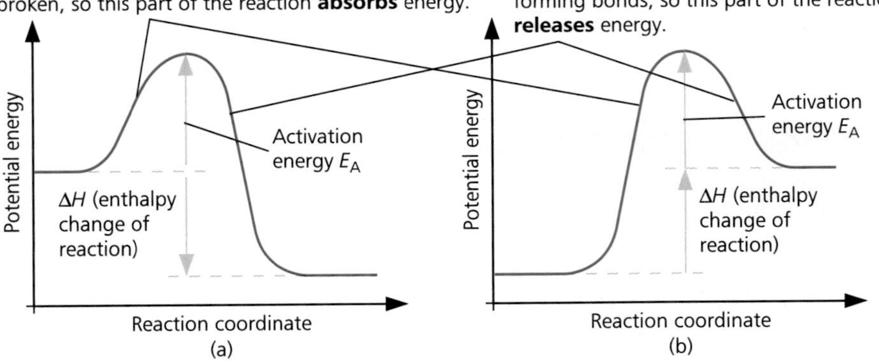
In this part of the reaction, the reactant molecules are coming together and breaking apart. Separating atoms in the reactant molecules requires bonds to be broken, so this part of the reaction absorbs energy.
In this part of the reaction the product molecules are forming and moving apart. Producing product molecules involves forming bonds, so this part of the reaction releases energy.
Potential energy
Activation energy E_A
ΔH (enthalpy change of reaction)
Reaction coordinate
(a)
Potential energy
Activation energy E_A
ΔH (enthalpy change of reaction)
Reaction coordinate
(b)

Figure 3.6.17 Potential energy diagrams such as these provide us with a great deal of information about a reaction. Graph (a) is for an exothermic reaction and graph (b) shows the path of an endothermic reaction. The progress of the reaction is indicated along the horizontal axis, which is called the **reaction coordinate**.

The activated complex

Chemists use a model of chemical reactions in which bonds are broken and then reformed in collisions between reacting molecules. In this model, colliding molecules with sufficient energy and the correct orientation to react form an **activated complex** as they reach the peak of the potential energy curve for a reaction, like that shown in figure 3.6.17.

The activated complex is not a chemical substance which can be isolated, but consists of an association of the reacting molecules, in which bonds are in the process of being broken and formed. Figure 3.6.18 shows how the activated complex in the decomposition of hydrogen iodide to hydrogen and iodine might look.

Notice that H—H bonds and I—I bonds are in the process of forming at the same time as the H—I bonds are breaking in this complex. This is a simple single-step reaction with the rate expression:

$$\textbf{Rate} = k[\textbf{HI}]^2$$

The idea of the activated complex can be used to help explain how a catalyst is able to reduce the activation energy of a reaction, as we shall see shortly.

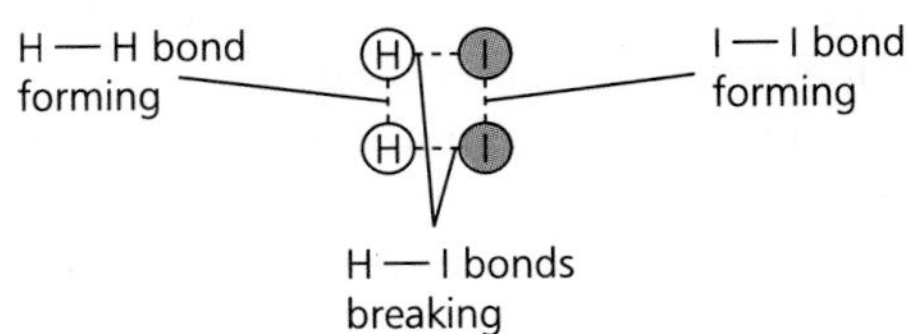
H — H bond forming
I — I bond forming
H — I bonds breaking

Figure 3.6.18 The activated complex in the decomposition of hydrogen iodide

The activation energy barrier

As two reacting molecules collide, they slow down and their kinetic energy is converted to potential energy – they begin to climb the activation energy hill. If their combined initial kinetic energies are less than E_A, then their potential energy will never reach E_A. They cannot get to the top of the hill so they fall back as reactants, gaining their original kinetic energy again. On the other hand, if the combined kinetic energies of the reactants are equal to or greater than E_A and the molecules are correctly orientated, then they can overcome the activation energy barrier and form product molecules.

The difference between the potential energy of the reactants and the potential energy of the products indicates the enthalpy change of reaction. When the products have a lower potential energy than the reactants, the reaction is exothermic, as the kinetic energy of the system and thus its temperature has increased. When the products have a higher potential energy than the reactants, the reaction is endothermic, as a net input of energy is needed to form in the products. The average kinetic energy of the system drops, and so does the temperature.

Figure 3.6.19 shows the implications of this for a reversible reaction.

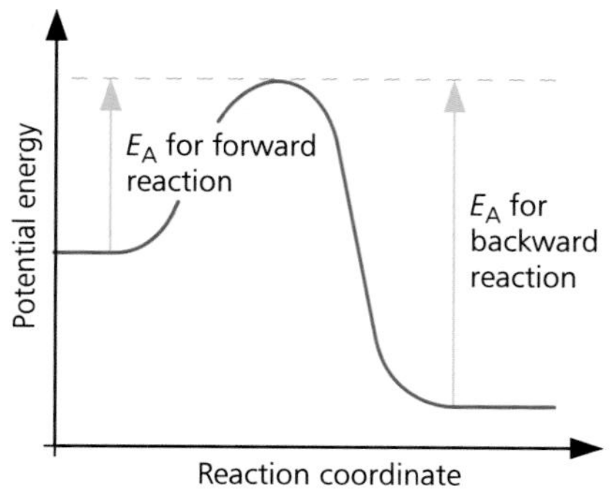

Figure 3.6.19 This potential energy diagram demonstrates the situation in a reversible reaction – if it is exothermic in one direction, it will be endothermic in the reverse direction. The activation energy for one of the reactions will be lower than for the reaction in the reverse direction.

Catalysts

As we have seen, the rate of a reaction can be increased both by increasing the concentrations of the reactants and by increasing the temperature of a reacting mixture. Both these measures increase the likelihood that molecules will collide with the correct orientation and with sufficient kinetic energy to overcome the activation energy barrier of the reaction. In many reactions, however, the activation energy is very high. This means that in order to convert reactants to products at a reasonable rate, the reaction mixture would need to be maintained at an impractically high temperature. However, this difficulty can be overcome in both the laboratory and chemical industry by the use of **catalysts**.

Catalysts and activation energies

A catalyst is a substance that increases the rate of a chemical reaction without itself being used up or permanently changed. In theory at least, we should be able to retrieve at the end of a reaction the exact amount of catalyst added at the beginning of the reaction. **Homogeneous catalysts** exist in the same phase as the reactants, for example, the H^+ ions or OH^- ions used to catalyse the esterification of carboxylic acids (see section 5.1). **Heterogeneous catalysts**, which are more common, exist in a separate phase to that of the reactants. The iron used as a catalyst in the Haber–Bosch process (see box on pages 330–2) is an example of a heterogeneous catalyst.

Catalysts increase the rate of a reaction by providing an alternative mechanism with a lower activation energy than that of the uncatalysed reaction. Because the activation energy of the catalysed mechanism is lower, a greater proportion of the colliding molecules will achieve the minimum energy needed to react, and so the rate of product formation will be increased, as shown in figure 3.6.20 opposite.

Catalysts are of immense importance to the chemical industries. The catalytic cracking of oil provides the raw materials for an enormous range of products, and processes such as the production of ammonia in the Haber–Bosch process and the synthesis of sulphuric acid by the Contact process would not be possible without catalysts – the operating temperatures and pressures needed would require feats of engineering initiative, and the cost of maintaining the conditions would be prohibitive.

Heterogeneous catalysts are frequently transition metals, for example, iron, platinum and nickel. Inorganic catalysts such as these are used to catalyse a wide range of different reactions. The reacting mixtures are usually adsorbed

Activation energy E_A
Activation energy E_A
Potential energy
Reactants
Uncatalysed path
Catalysed path
Products

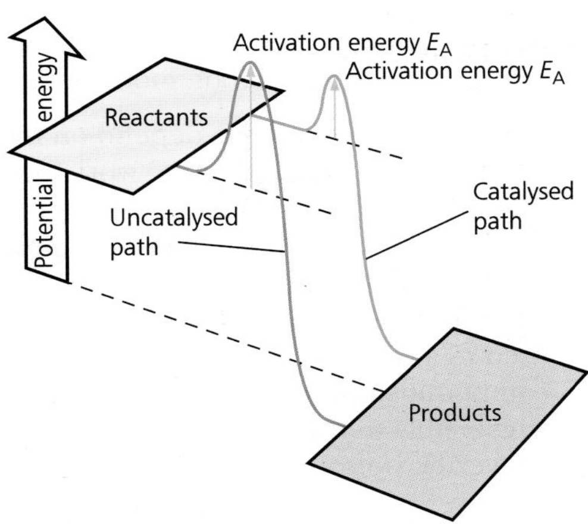

Figure 3.6.20 A catalyst provides an alternative low-energy route for a reaction and thus increases the rate of production of products from reactants.

onto the surface of the catalyst. For this reason, the catalyst is usually given a very large surface area to maximise its effect on the reaction rate – a large ingot of iron would have little effect compared with the same mass in finely divided form.

3

Heterogeneous catalysis and the activated complex

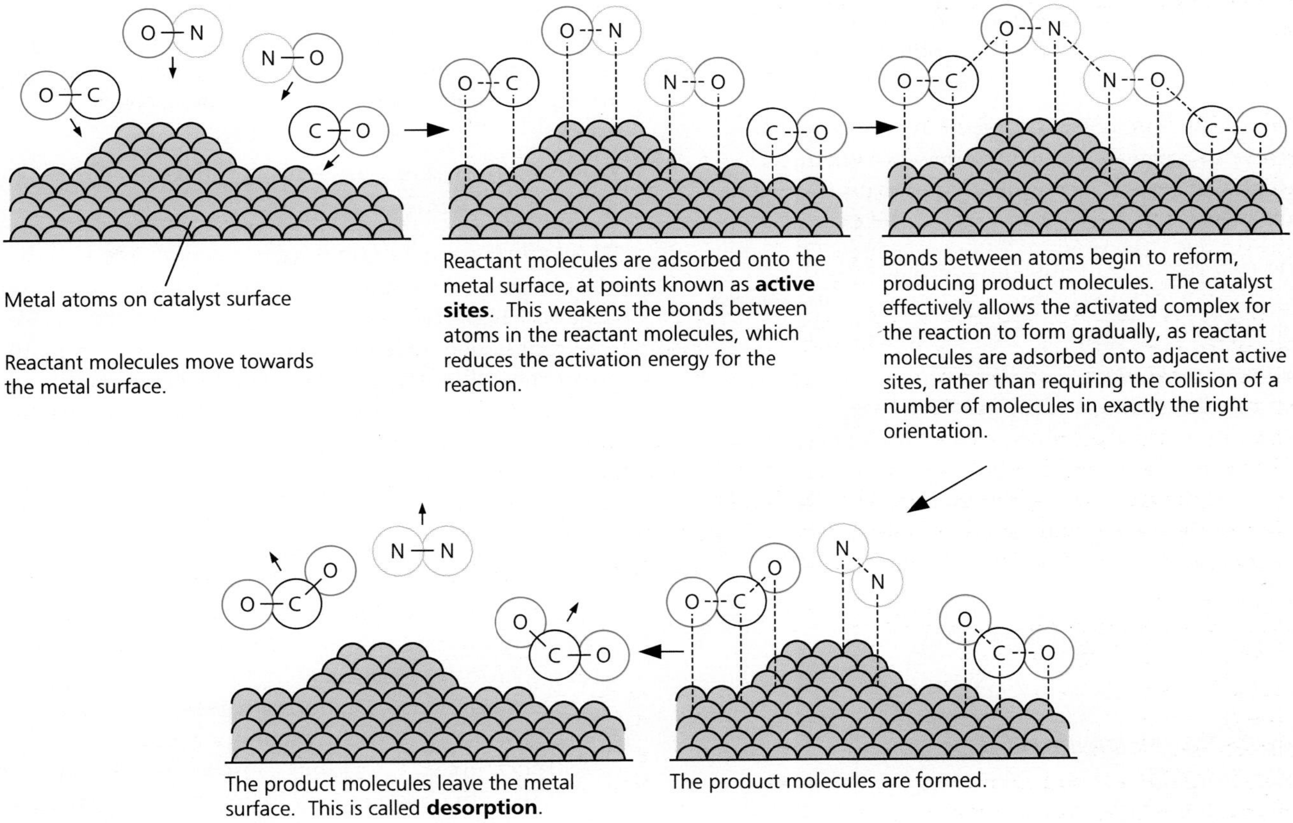

O—N
N—O
O—C
C—O
Metal atoms on catalyst surface
Reactant molecules move towards the metal surface.
Reactant molecules are adsorbed onto the metal surface, at points known as **active sites**. This weakens the bonds between atoms in the reactant molecules, which reduces the activation energy for the reaction.
Bonds between atoms begin to reform, producing product molecules. The catalyst effectively allows the activated complex for the reaction to form gradually, as reactant molecules are adsorbed onto adjacent active sites, rather than requiring the collision of a number of molecules in exactly the right orientation.
N—N
The product molecules leave the metal surface. This is called **desorption**.
The product molecules are formed.

Figure 3.6.21 The mechanism of action of a heterogeneous catalyst

It is thought that a heterogeneous catalyst increases the rate of a reaction by loosely binding the reactant molecules to its surface, in a process called **adsorption**. Adsorption is believed to increase the reaction rate in two ways. First, as reactant molecules are adsorbed onto the surface of the catalyst, the bonds between the atoms in them are weakened, reducing the energy required to break the reactant molecules apart. Second, the adsorbed molecules are held closely together on the metal surface, so the likelihood of a reaction occurring is much greater. Figure 3.6.21 shows how the catalyst in the catalytic converter of a car's exhaust system is able to promote the formation of carbon dioxide and nitrogen from nitrogen oxide and carbon monoxide.

Heterogeneous catalysts depend on the availability of active sites to enable them to work. Some molecules bind irreversibly to the active sites of heterogeneous catalysts. As a result, the catalyst is **poisoned** and no longer able to catalyse the reaction, since it cannot adsorb the reactant molecules. When leaded petrol is used in a car with a catalytic converter, the lead compounds in the combustion products bind irreversibly to the active sites of the catalyst in the converter, poisoning it and destroying its catalytic activity.

Balancing equilibrium position and reaction rate – the Haber–Bosch process

The economic production of any chemical on an industrial scale depends on being able to produce reasonable quantities of the chemical in a reasonable time. This means that any company that uses a chemical reaction to produce a product needs to know something about:

(1) the equilibrium constant for the reaction (which gives information about the *amount* of chemical which will be produced when equilibrium is reached)

(2) the rate expression and rate constant for the reaction (which gives information about *how quickly* equilibrium will be reached).

In order to make decisions about the best conditions under which to carry out the reaction, the information about equilibrium and rate constants will need to be available at different temperatures. The synthesis of ammonia provides a good illustration of how the conditions are chosen for an industrial chemical reaction.

Ammonia is a very important chemical, playing a major role in the manufacture of fertilisers and explosives (see page 333).

Optimising the reaction conditions

The formation of ammonia involves the following equilibrium:

$$N_2(g) + 3H_2(g) \rightleftharpoons 2NH_3(g) \quad \Delta H^\circ_m = -92.4 \text{ kJ mol}^{-1}$$

The position of equilibrium lies to the left, and for many years it was impossible to synthesise ammonia in sufficient quantities to be of use industrially. In 1908 Fritz Haber, a young German physical chemist, found a set of conditions which gave greatly increased yields of ammonia. By 1913 Carl Bosch, a brilliant chemical engineer, had designed a plant capable of producing ammonia industrially using a process that has since become known as the **Haber–Bosch process**.

There are three key factors in the physical chemistry of the manufacture of ammonia.

(1) A look at the equation for the reaction shows us that 4 volumes of reactants produce 2 volumes of product. Higher pressures push the equilibrium to the right, favouring the production of ammonia. In practice, the pressure of the system is limited by both engineering and financial constraints, and an operating pressure of around 250 atm is used.

(2) The reaction is exothermic, so heating the reaction mixture is not at first sight a good idea. Cooling the mixture increases the equilibrium constant so increases the yield of ammonia. However, many bonds need to be broken in the reactant molecules to allow the reaction to occur, and so the activation energy for the reaction is high. At low temperatures the reaction occurs so slowly that it is not economic, so the normal operating temperature in the Haber–Bosch process is around 450 °C. This increases the reaction rate, although it also makes the equilibrium constant less favourable.

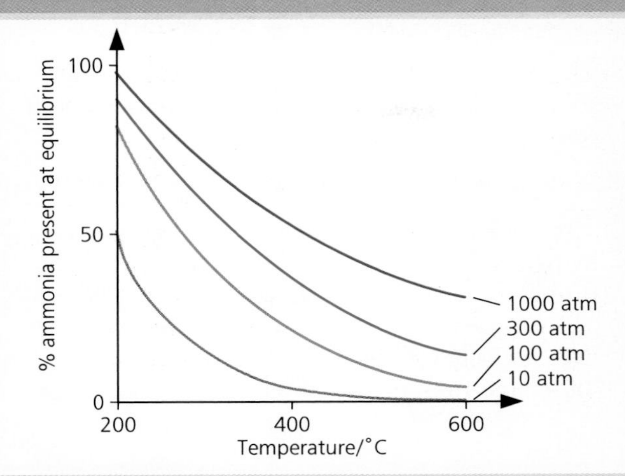

% ammonia present at equilibrium
100
50
0
200
400
600
Temperature/°C
1000 atm
300 atm
100 atm
10 atm

Figure 3.6.22 Haber carried out hundreds of experiments to determine the amount of ammonia present in equilibrium mixtures of nitrogen, hydrogen and ammonia. This graph shows the percentage of ammonia in equilibrium mixtures formed from equal amounts of nitrogen and hydrogen under different conditions of temperature and pressure.

(3) Even at the temperature and pressure described above, the yield of ammonia in this process would be minimal without a catalyst to speed the reaction up. The catalyst used for the production of ammonia is finely divided iron with traces of aluminium oxide, potassium hydroxide and other salts. Bosch carried out 6500 experiments to find the best catalyst for his chemical plant!

The industrial process

The raw materials of the industrial process are not the simple mixture of nitrogen and hydrogen that the equation for the reaction suggests. Water, methane and air are used in a sophisticated process that generates a substantial part of the heat needed to make ammonia. The steps of the process are outlined below.

(1) Methane and steam are passed over a nickel catalyst at high pressure (30 atm) and temperature (750 °C), converting around 91% of the methane to hydrogen, along with carbon monoxide and carbon dioxide:

$$CH_4(g) + H_2O(g) \rightleftharpoons CO(g) + 3H_2(g)$$

$$CH_4(g) + 2H_2O(g) \rightleftharpoons CO_2(g) + 4H_2(g)$$

(2) Air is injected into the flowing gas stream. Some hydrogen in the stream reacts with oxygen, forming water and nitrogen:

$$2H_2(g) + \underbrace{O_2(g) + 4N_2(g)}_{\text{from the air}} \rightleftharpoons 2H_2O(g) + 4N_2(g)$$

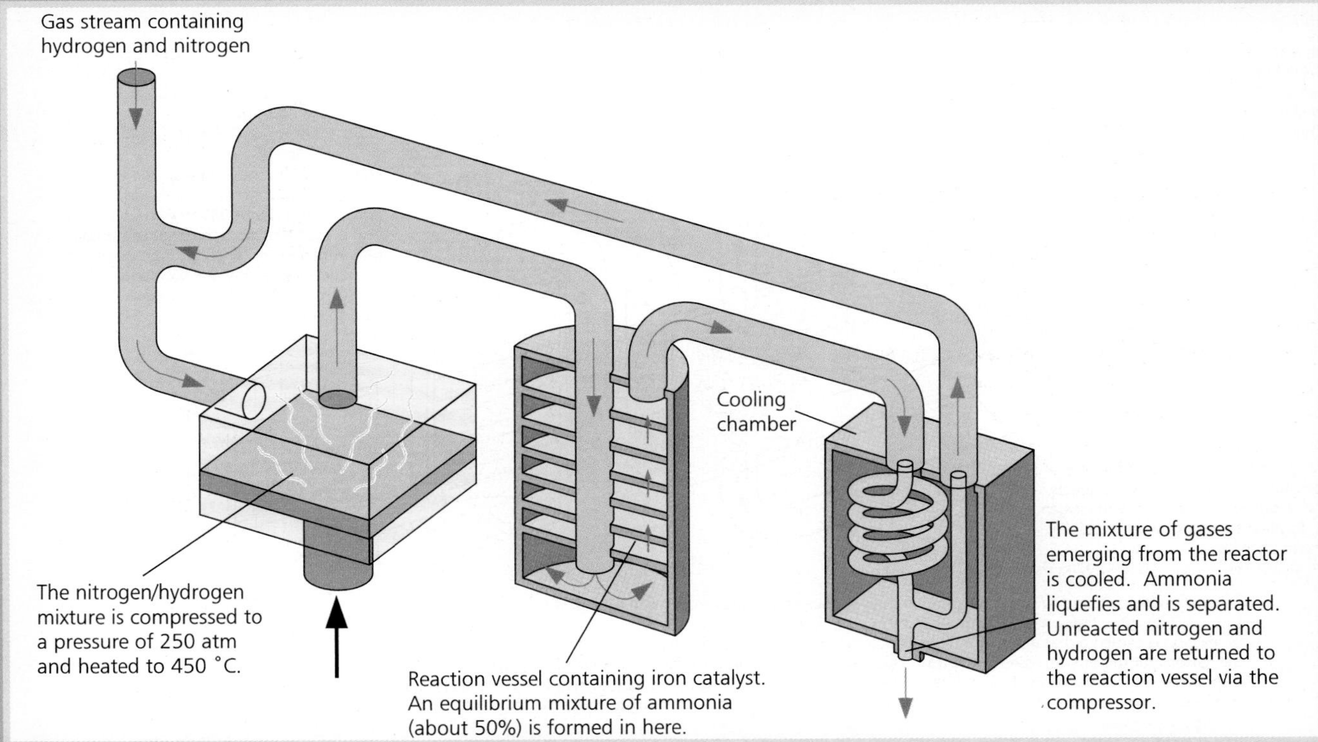

Gas stream containing hydrogen and nitrogen
Cooling chamber
The nitrogen/hydrogen mixture is compressed to a pressure of 250 atm and heated to 450 °C.
Reaction vessel containing iron catalyst. An equilibrium mixture of ammonia (about 50%) is formed in here.
The mixture of gases emerging from the reactor is cooled. Ammonia liquefies and is separated. Unreacted nitrogen and hydrogen are returned to the reaction vessel via the compressor.

Figure 3.6.23 Producing ammonia – nitrogen and hydrogen are recycled through the reaction vessel.

3

(3) Step 2 is strongly exothermic and raises the internal temperature to about 1100 °C. This heat is used in three ways:
(a) it causes the remaining methane to react with water, giving more hydrogen
(b) it provides energy to operate the gas compressors
(c) it is used to produce the high-temperature steam introduced at the beginning of step 1.
As a result of these heat transfers, the gas mixture cools.

(4) If left in the mixture, carbon monoxide would poison the final iron catalyst. It is removed by reacting it with steam over a series of catalysts:

$$CO(g) + H_2O(g) \xrightarrow{\text{catalysts}} CO_2(g) + H_2(g)$$

(5) The gas stream passes through a concentrated solution of potassium carbonate to remove both water and carbon dioxide:

$$CO_2(g) + H_2O(g) + K_2CO_3(aq) \rightarrow 2KHCO_3(aq)$$

(6) The gas stream now consists of nitrogen and hydrogen in approximately the correct proportions. These are reacted to form ammonia, which is cooled to –50 °C so that it liquefies. It is then collected and stored. This final step is illustrated in figure 3.6.23.

Balancing equilibrium position and reaction rate – the Contact process

About 150 million tonnes of sulphuric acid are produced worldwide each year. Sulphuric acid is so widely used by industry that its per capita use can be regarded as quite a good measure of a country's industrial activity and of the prosperity of its inhabitants.

The parent compound of sulphuric acid is sulphur trioxide, SO_3, in which sulphur has the oxidation state +6. Sulphur trioxide exists in equilibrium with molecules of the trimer S_3O_9 – this mixture is colourless, melts at 16.9 °C and boils at 44.6 °C. Sulphur trioxide, and thus sulphuric acid, is manufactured by the **Contact process** from sulphur dioxide, which in turn is formed from the direct

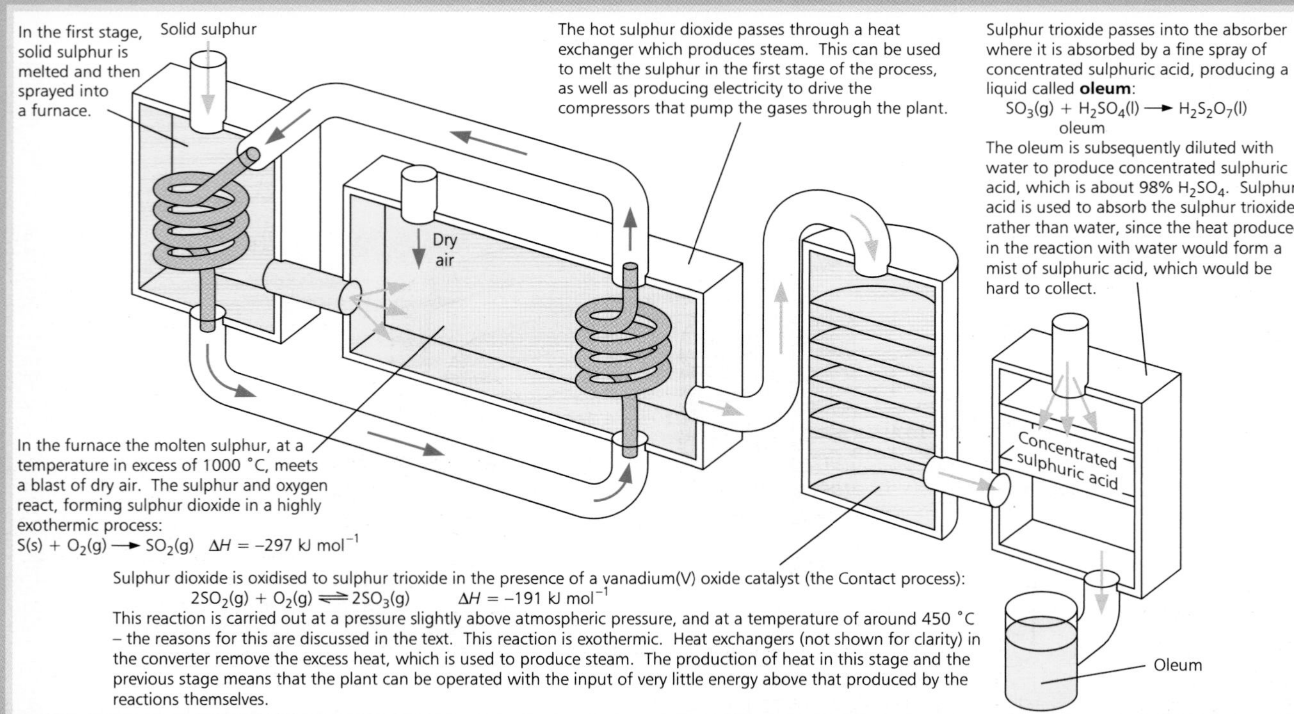

In the first stage, solid sulphur is melted and then sprayed into a furnace.
Solid sulphur
The hot sulphur dioxide passes through a heat exchanger which produces steam. This can be used to melt the sulphur in the first stage of the process, as well as producing electricity to drive the compressors that pump the gases through the plant.
Sulphur trioxide passes into the absorber where it is absorbed by a fine spray of concentrated sulphuric acid, producing a liquid called **oleum**:
$SO_3(g) + H_2SO_4(l) \rightarrow H_2S_2O_7(l)$
oleum
The oleum is subsequently diluted with water to produce concentrated sulphuric acid, which is about 98% H_2SO_4. Sulphuric acid is used to absorb the sulphur trioxide rather than water, since the heat produced in the reaction with water would form a mist of sulphuric acid, which would be hard to collect.
Dry air
Concentrated sulphuric acid
In the furnace the molten sulphur, at a temperature in excess of 1000 °C, meets a blast of dry air. The sulphur and oxygen react, forming sulphur dioxide in a highly exothermic process:
$S(s) + O_2(g) \rightarrow SO_2(g) \quad \Delta H = -297 \text{ kJ mol}^{-1}$
Sulphur dioxide is oxidised to sulphur trioxide in the presence of a vanadium(V) oxide catalyst (the Contact process):
$2SO_2(g) + O_2(g) \rightleftharpoons 2SO_3(g) \quad \Delta H = -191 \text{ kJ mol}^{-1}$
This reaction is carried out at a pressure slightly above atmospheric pressure, and at a temperature of around 450 °C – the reasons for this are discussed in the text. This reaction is exothermic. Heat exchangers (not shown for clarity) in the converter remove the excess heat, which is used to produce steam. The production of heat in this stage and the previous stage means that the plant can be operated with the input of very little energy above that produced by the reactions themselves.
Oleum

Figure 3.6.24 The manufacture of sulphuric acid

reaction of sulphur with oxygen. The steps in the production of sulphuric acid are shown in figure 3.6.24.

The conditions under which the Contact process is carried out are determined by the equilibrium constant for the reaction. Since the reaction is exothermic, the value of K decreases as temperature increases, favouring the production of SO_3 at low temperatures – see figure 3.6.25.

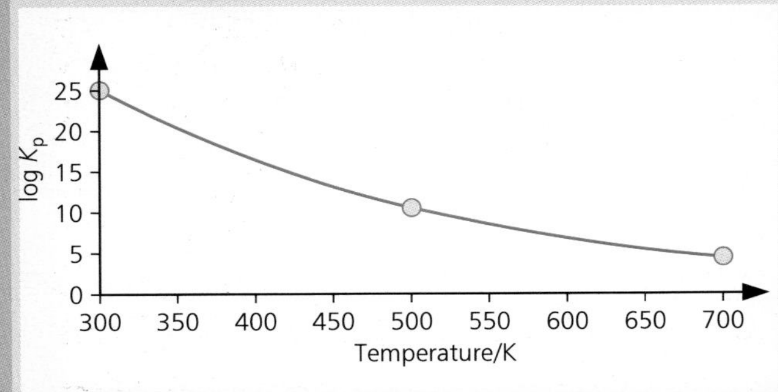

Figure 3.6.25 The variation of K with temperature for the equilibrium $2SO_2(g) + O_2(g) \rightleftharpoons 2SO_3(g)$

At 300 K, the equilibrium constant is about 10^{25} atm^{-1}, overwhelmingly favouring the production of sulphur trioxide. Unfortunately, the reaction proceeds only very slowly at this temperature, so a higher temperature must be chosen. At 450 °C the equilibrium constant is still large (more than 10^{10} atm^{-1}) and the reaction proceeds rapidly in the presence of a vanadium(V) oxide catalyst. In addition, the problems of corrosion of the plant by the chemicals involved in the process are manageable at this temperature, whereas this would become an increasing concern at higher temperatures. The reaction is carried out at a pressure just above atmospheric pressure. Increasing the pressure would increase the yield of the reaction as it would push the equilibrium to the right. However, the greater cost of building and operating a plant at higher pressures would not be justified by the increase in yield involved, since the proportion of sulphur trioxide at equilibrium at 450 °C and 1 atm is around 97% anyway.

Chemicals, fertilisers and food

The Haber process and the Contact process have made possible the production of large amounts of relatively cheap ammonia and sulphuric acid – but what do we need them for? The major use of both chemicals is in the production of inorganic fertilisers. All over the world farmers apply these chemicals to the soil to supply the minerals needed by growing crop plants. Growing plants need a number of minerals from the soil, and two of the most important of these are nitrates and phosphates.

The nitrogen cycle

Nitrogen is probably the limiting factor in the production of new living material in most circumstances. It is a vital constituent of amino acids and therefore of proteins, which are made up of amino acids. Although almost 80% of the Earth's atmosphere is nitrogen, relatively little is available in a form that can be used by living organisms because nitrogen is so inert. It does not naturally combine with other substances in the soil and the air, but has to be actively **fixed** by microorganisms. This fixing converts nitrogen from the atmosphere into nitrate, that can be utilised by plants and animals to produce proteins. The nitrogen is eventually released from the tissues of living things and returned to the atmosphere by the action of more microorganisms, the decomposers. The process of fixing nitrogen, its passage through the bodies of plants and animals and its eventual return to the atmosphere make up the **nitrogen cycle**, shown in figure 3.6.26 overleaf.

In modern agriculture, crop plants are grown intensively and removed from the soil before they die and decompose. This interferes with the natural nitrogen cycle which maintains soil fertility. As a result, enormous sums of money are spent in the industrial production of nitrate fertilisers designed to restore and even enhance the natural fertility of the soil.

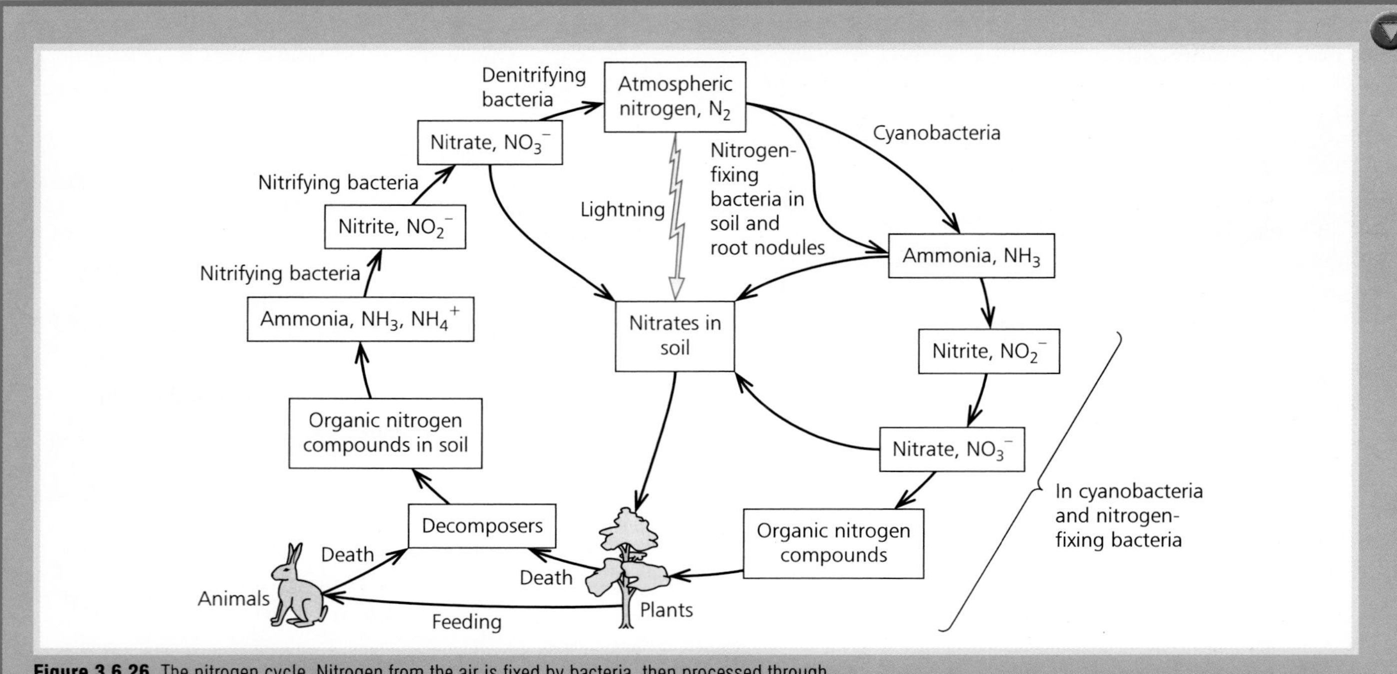

Figure 3.6.26 The nitrogen cycle. Nitrogen from the air is fixed by bacteria, then processed through plants, animals and more bacteria before it is returned to the atmosphere whence it came.

Almost all the inorganic nitrogen-containing fertilisers applied to the soil throughout the world are based on the ammonia produced by the Haber process. A quarter of all the ammonia produced is used directly on the soil. A large proportion of the remainder is fed into another industrial process, the Ostwald process for producing nitric acid. This involves another series of catalysts and temperatures designed to make the acid as quickly, safely and cheaply as possible. The Ostwald process involves three stages, which may be summarised as follows:

- Dry ammonia and air are passed over a platinum gauze catalyst at 7 atmospheres pressure and 900 °C. The ammonia is oxidised to nitrogen oxide, NO.

$$\mathbf{4NH_3(g) + 5O_2(g) \rightleftharpoons 4NO(g) + 6H_2O(g) \qquad \Delta H^\ominus_m = -950\ kJ\ mol^{-1}}$$

  This is a strongly exothermic reaction, and before the next stage the gases are cooled down to 25 °C.

- The cooled gases from the first stage are mixed with more air and are oxidised to the reddish brown gas nitrogen dioxide, NO_2.

$$\mathbf{2NO(g) + O_2(g) \rightleftharpoons 2NO_2(g) \qquad \Delta H^\ominus_m = -114\ kJ\ mol^{-1}}$$

- Finally, the nitrogen dioxide reacts with water in large absorption towers designed to ensure that the reacting gases are thoroughly mixed. The final mixture contains about 60% nitric acid, which can then be concentrated by distillation with concentrated sulphuric acid:

$$\mathbf{3NO_2(g) + H_2O(l) \rightarrow 2HNO_3(aq) + NO(g) \qquad \Delta H^\ominus_m = -117\ kJ\ mol^{-1}}$$

75% of the nitric acid made in this way is used in the production of fertilisers, particularly ammonium nitrate, NH_4NO_3. A further 15% of the nitric acid is used in the production of explosives.

Another vital mineral for the growth of plants is phosphate, but although phosphate containing rock is not rare, it usually contains $Ca_5(PO_4)_3F$. This is very insoluble in water and therefore of little use as a source of phosphate ions for plants. This is why 70% of the sulphuric acid from the Contact process is used in the production of phosphate fertilisers. Reacting phosphate rocks with sulphuric acid gives a mixture known as **superphosphate**, which is much more soluble in water and is used internationally as a fertiliser.

As a result of the Haber, Ostwald and Contact processes, cheap inorganic fertilisers have been produced for many years now. They have resulted in both the enormous quantity and variety of relatively cheap food available in the developed world, and in an increase in food production in the developing world. Although artificial fertilisers can cause problems such as nitrate pollution with overuse, we do well to remember the enormous benefits they have brought so far and which, with careful use, they will continue to bring for years to come.

Enzymes

Biological catalysts

In biological systems, 100 or more chemical reactions may be going on at any one time in the minute volume of a single cell. These reactions need to be controlled so that they do not interfere with each other, and speeded up so that they occur quickly enough for life to continue. This control and catalysis is brought about by a group of biological catalysts known as **enzymes**. Enzymes are complex globular protein molecules (see section 5.5) which function in a very similar way to inorganic catalysts by providing an alternative route of lower activation energy. Enzymes are very effective catalysts, giving greater increases in reaction rate than inorganic catalysts.

3

Summary of enzyme characteristics

The following characteristics are shared by all enzymes.

(1) Enzymes are globular proteins with an **active site** contained in their three-dimensional structure.
(2) Like all catalysts, enzymes increase the rate of a reaction but do not affect the reaction in any other way.
(3) Enzymes are required in very small amounts – they are very effective catalysts.
(4) Unlike inorganic catalysts, enzymes are very specific to a particular reaction or type of reaction.
(5) Enzyme activity is affected by **substrate concentration** (reactant concentration) until it reaches a maximum, when all the active sites are **saturated**.
(6) Enzyme activity is affected by temperature – it increases until the protein denatures.
(7) Enzyme activity is affected by pH – different enzymes have differing optimum pH levels.

Enzymes catalyse reactions by lowering the activation energy needed for the reaction to take place. To bring this about, the enzyme forms a **complex** with the substrate or substrates (reactants) of the reaction. Thus a simple picture of enzyme action is:

Substrate + enzyme → enzyme/substrate complex → enzyme + products

Once the products of the reaction are formed, they are released and the enzyme is free to form a new complex with more substrate.

How does this relate to the three-dimensional structure of the enzyme, which is vital to the successful functioning of the enzyme? The basic picture is summarised in a model called the **lock and key mechanism**, shown in figure 3.6.27.

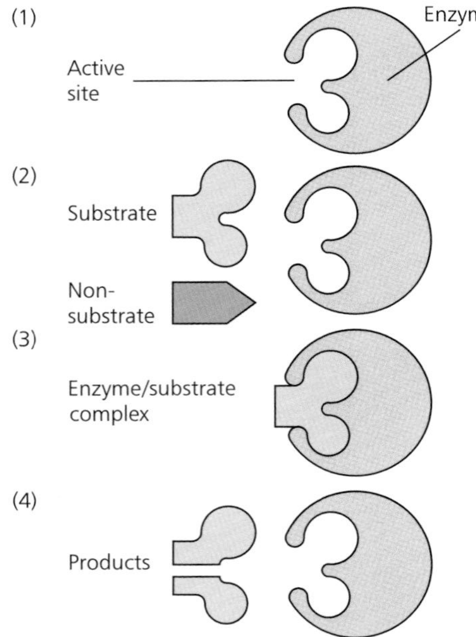

(1) Within the structure of each enzyme is an area known as the **active site**. It may involve only a small number of amino acids. It has a very specific shape, which gives each enzyme its specificity, as only one substrate or type of substrate will fit into the gap.

(2) Here we can see the difference between the shape of the enzyme substrate and another biological molecule. Only a molecule of the right shape can be a substrate for the enzyme.

(3) The enzyme and substrate slot together to form a complex, as a key fits into a lock. In this complex the substrate is enabled to react at a lower activation energy. This may be due to bonds within it being deformed and stressed in the complex, so making them more likely to react.

(4) Once the reaction has been catalysed, the products are no longer the right shape to stay in the active site and the complex breaks up, releasing the products and freeing the enzyme for further catalytic action.

Figure 3.6.27 The lock and key mechanism – the basis of our understanding of enzyme action

The lock and key mechanism fits most of our evidence about enzyme characteristics. We can see how enzymes can become saturated when the concentration of substrate molecules rises above a certain level – all the active sites become bound in enzyme/substrate complexes. It also explains how any change in the protein structure, such as those brought about by changes in pH or temperature, can affect enzyme action by altering the shape of the active site.

However, it is now thought that the lock and key mechanism is a slight over-simplification. Evidence from X-ray crystallography, chemical analysis of active sites and other techniques suggests that the active site of enzymes is not the rigid shape that was once supposed. In the **induced fit theory**, which is generally accepted as the best current model, the active site is still thought of as having a very distinctive shape and arrangement, but a rather more flexible one. Thus, once the substrate enters the active site, the shape of that site is modified around it to form the active complex. Once the products have left the complex, the enzyme reverts to its inactive, relaxed form until another substrate molecule binds, as we can see in figure 3.6.28.

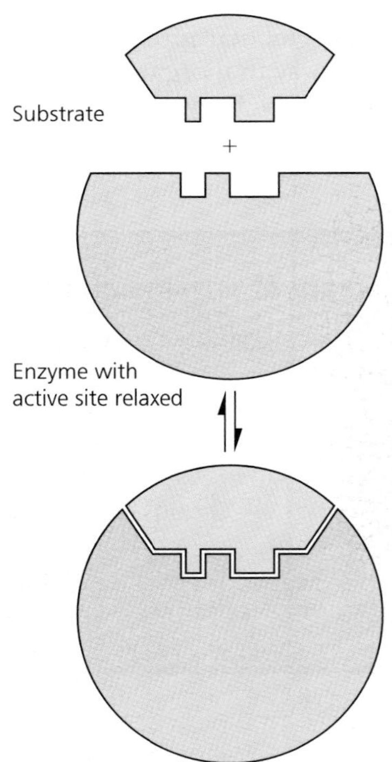

Enzyme/substrate complex showing the induced form of the active site, fitting snugly round the substrate

Figure 3.6.28 The induced fit theory of enzyme action proposes that the catalytic groups of the active site are not brought into their most effective positions until a substrate molecule is bound onto the site, **inducing** a change in conformation.

Enzyme technology

Enzymes are obviously of vital importance in biological systems. What is perhaps more surprising is their increasingly important role in industry. Enzymes have much to offer industry. Unlike most inorganic catalysts, they work at low temperatures, normal pressures and at easily achieved pH levels. This means that an industrial process which can use an enzyme catalyst will be relatively cheap to run. With advances in genetic engineering, we can increasingly use microorganisms to produce specific enzymes, and this widens the scope of enzyme use in industry. Already much enzyme-based technology exists, and enzymes are used for a wide variety of processes.

Concentration, temperature and enzyme-catalysed reactions

These graphs show the effect of increased reactant concentration and increased temperature on typical enzyme-catalysed reactions. The protein structure of an enzyme limits its capabilities as a catalyst.

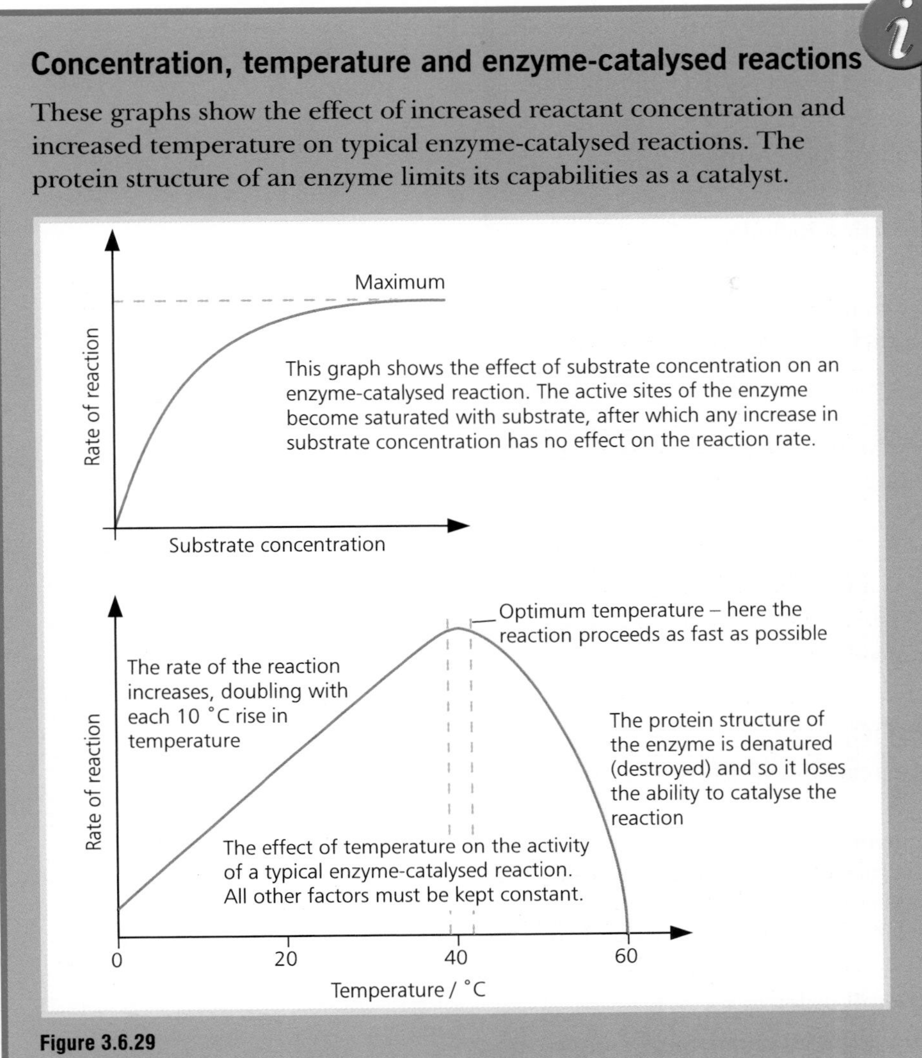
Maximum
Rate of reaction
This graph shows the effect of substrate concentration on an enzyme-catalysed reaction. The active sites of the enzyme become saturated with substrate, after which any increase in substrate concentration has no effect on the reaction rate.
Substrate concentration
Optimum temperature – here the reaction proceeds as fast as possible
The rate of the reaction increases, doubling with each 10 °C rise in temperature
Rate of reaction
The protein structure of the enzyme is denatured (destroyed) and so it loses the ability to catalyse the reaction
The effect of temperature on the activity of a typical enzyme-catalysed reaction. All other factors must be kept constant.
0
20
40
60
Temperature / °C

Figure 3.6.29

Not surprisingly, the food industry is one of the main users of enzyme technology. Rennin is used to clot milk for cheese making, and enzymes from yeast are used widely in both the brewing and the baking industries, as they have been for many years past. Cellulases and pectinases are used to clear hazes in fruit juice production. Other uses of enzymes have more recent histories. Trypsin is used to predigest baby foods, proteases are used in biscuit manufacture to lower the protein content of the flour, and a variety of enzymes are used to make sweet syrups from starch.

Enzymes are used in detergents to digest particular types of dirt, particularly the protein elements of food and sweat. Enzymes are also being developed which will nip off 'pilling' – the little bobbles that form on cotton and woollen clothing when they are washed – and leave the fabric smooth. Enzymes are used in the rubber industry to produce the oxygen needed to convert latex to foam rubber, in the paper industry and in the photographic industry. They also have many uses in medical fields – for example, glucose oxidase is used for detecting excess sugar in the urine, a symptom called glycosuria. With improvements in technology and advances in genetic engineering making 'designer enzymes' possible, the use of enzymes in industry can only increase in the future.

Figure 3.6.30 The enzymes that are effective at attacking protein dirt also attack people's skin. Until these tiny capsules were developed to contain the enzymes, many workers in detergent factories suffered from allergic reactions.

SUMMARY

- The rate of a chemical reaction is the rate at which reactants are used up and products are formed:

$$\textbf{Rate of reaction} = \frac{\textbf{change in concentration}}{\textbf{change in time}}$$

- Factors affecting the rate of a chemical reaction include:

  the chemical nature of the reactants

  the state of the reactants and their ability to meet – particles intermingle freely in **homogeneous** reactions but the surface area of the interface is important in **heterogeneous** reactions

  the concentrations of the reactants

  the temperature of the system – for most chemical reactions a 10 °C rise in temperature gives an approximate doubling of reaction rate at room temperature

  the presence of a catalyst.

- The rate of a reaction may be expressed as a **rate expression** or **rate law**. For the theoretical reaction:

$$a\text{A} + b\text{B} + c\text{C} \rightarrow \text{products}$$

  the rate law may be written as:

$$\text{Rate} = k[\text{A}]^x[\text{B}]^y[\text{C}]^z$$

  The constant k is the **rate constant** for the reaction, which is constant for a given reaction at a given temperature.

- The **order** of a reaction with respect to a given reactant is the power to which the concentration of that reactant is raised in the rate expression. For the reaction above, the order with respect to reactant A is x, the order with respect to reactant B is y, and the order with respect to reactant C is z.
- The overall order of a reaction is the sum of the powers in the rate expression. For the reaction above, the overall order $= x + y + z$.
- The rate expression gives information about the **mechanism** of a chemical reaction. The rate expression *cannot* be predicted from the stoichiometric equation for a reaction.
- A first order reaction has a constant half-life. This is similar to the radioactive decay of an unstable isotope.
- For a chemical reaction to occur successfully, collision theory states that the reacting particles must collide with their atoms correctly orientated and with a minimum kinetic energy known as the **activation energy** E_A.
- A catalyst increases the rate of a chemical reaction without itself being used up or permanently changed. It provides an alternative reaction mechanism with a lower activation energy than that of the uncatalysed reaction.
- In biological systems globular protein molecules known as **enzymes** act as catalysts.
- Industrial processes may require a trade-off between reaction rate and equilibrium constant if a reaction is to be economic.

QUESTIONS

1 Give an example of an everyday reaction that happens:

a extremely slowly

b extremely rapidly

c rapidly at a high temperature but very slowly, if at all, at a low temperature.

Give reasons why you think each of these reactions behaves like this.

2 For the reaction:

$$X + Y \rightarrow \text{products}$$

the rate expression is:

$$\frac{d[X]}{dt} = -k\,[X]\,[Y]^2$$

where d[X]/d*t* means 'the rate of change of [X] with time'.

a Is d[X]/d*t* positive or negative?

b What is the sign of the rate constant, *k*? Explain your answer.

What would be the effect of:

c doubling the concentration of X while keeping that of Y constant

d halving the concentration of Y while keeping that of X constant

e increasing the temperature at which the reaction happens?

3 Radon-220 is a naturally occurring radioactive gas with a half-life of 56 s. What fraction of a sample of this gas remains after:

a 56 s

b 112 s

c 560 s?

4 All naturally occurring water contains tritium, a radioactive isotope of hydrogen, formed by bombardment of hydrogen atoms in water vapour high up in the atmosphere. Tritium decays with a half-life of 12.3 years. The age of a bottle of wine may be estimated by comparing the concentration of tritium in the wine with that in naturally occurring water.

a How does this method work?

b The amount of tritium in the wine in a particular bottle is found to be 25% of that in naturally occurring water. How long ago was the wine bottled?

5 At a certain temperature, the kinetics of the reaction

$$2ICl(g) + H_2(g) \rightarrow I_2(g) + 2HCl(g)$$

were studied, and the results in table 3.6.3 were obtained:

Initial concentration/mol dm^{-3}		Initial rate of formation of I_2(g)/mol dm^{-3} s^{-1}
ICl	H_2	
0.20	0.10	0.042
0.20	0.20	0.168
0.40	0.10	0.084

Table 3.6.3

a What is the order of the reaction with respect to ICl? Explain your answer.

b What is the order of the reaction with respect to H_2? Explain your answer.

c Determine the rate constant for this reaction, with the correct units.

6 In the gas phase, molecule A decomposes to molecules B and C at high temperatures. A chemist suspects that this reaction is first order with respect to A. In an experiment to explore the kinetics of this reaction, the data in table 3.6.4 were obtained for the decomposition of A at 800 K:

Time/s	Partial pressure of A/kPa
0	1300
20	1051
40	849
60	685
80	554
100	448
120	361
140	292
160	236
180	191
200	154

Table 3.6.4

a If the reaction is first order with respect to A, write down the rate expression.

b By using a graph, or otherwise, find out if the reaction is first order with respect to A.

c Calculate the rate constant for the reaction. What are its units?

d What does the rate expression for this reaction tell you about its mechanism?

Developing Key Skills

A new low-cost run-about car is about to be launched on the market. As one of its selling features it boasts a catalytic converter to control back-end emissions. Write part of the manual to accompany the car, explaining how the catalytic converter works and why it is so important to use the right grade of fuel when driving a car with a cat.

[Key Skills opportunities: C, IT]

3

FOCUS THE HEAT DEATH OF THE UNIVERSE?

Thermodynamics is the branch of science which deals with the laws of heat and the conversion of heat into other types of energy. The first law of thermodynamics tells us that energy can never be created or destroyed. The second states that entropy always increases in the transformation of energy. In everyday understanding, entropy can be thought of as a tendency for everything to move from a state of order to chaos as objects (whether they are atoms or the things in our sitting room) left to themselves, mix and randomise themselves as much as they can.

Figure 1 The tidy bedroom becomes a jungle of clothes and other debris, dead bodies decay, iron goes rusty – in life, as in science, chaos rules.

Scientists use ideas from thermodynamics to try and explain not just what happened when our universe came into being (see page 344) but also what will happen when the universe ends. In the middle of the seventeenth century Archbishop James Ussher of Ireland made the startling revelation that God created Heaven and Earth on October 22, 4004 BC, at 8 o'clock in the evening! This was later modified by the English biblical scholar Dr John Lightfoot, who gave the date for the creation of Adam as October 23, 4004 BC, at 9 o'clock in the morning! Our understanding of the physical world has increased immeasurably since then, thanks mainly to advances made in the physical sciences. However, scientists are now engaged in sometimes heated debate about the way in which the universe will end.

One of the strongest predictions based on the second law of thermodynamics is that the whole universe must inevitably move towards a state of maximum entropy. Seen as an isolated system, the entire universe must end up at the same temperature. Everything will be in a state of equilibrium and the temperatures involved will mean that all life (as we know it or envisage it) in the universe will cease to exist. Our universe will become a vast expanse of identical nothingness. This scenario is known as the 'heat-death of the universe' – it was described by Sir James Jeans in 1928, and by many other scientists since.

However, some scientists now argue that because the second law of thermodynamics takes no account of forces such as electromagnetism and gravity it gives a somewhat restricted picture of physical processes. It describes what happens in an ideal and isolated system but, they argue, the universe is neither of these things. Led by the Belgian Nobel Prize winner Ilya Prigogine, scientists have developed new interpretations of the second law. These recognise that everything affects everything else. Atoms and molecules are not 'left to themselves'. They are usually exposed to a flow of energy and material which can at least partially reverse the descent into chaos which seems to be inevitable as a result of the second law of thermodynamics. Prigogine suggests that self-organising structures, reversing entropy, are common in nature and so in his analysis the heat death of the universe simply will not happen. Different groups of scientists are working on a number of different scenarios for alternative endings! The implications of these various ideas for the end of the universe are still well in the realms of theoretical, if not imaginary, science. But one thing is fairly certain – whether the world ends with a bang or a whimper, we are most unlikely to be there to see it!

Figure 2 'The second law of thermodynamics compels materials in the universe to move ever in the same direction along the same road which ends only in death and annihilation' (The end of the universe as predicted by Sir James Jeans)

3.7 Thermodynamics

The interaction of energy and matter governs all our lives. We take for granted our ability to use the chemicals in batteries to produce sound from a radio, or to get in a car and use petrol to quickly travel distances that would have taken our ancestors hours if not days. Energy and matter interact on other scales too – from the remnants of the explosion of a star, hundreds of light years in diameter, to the nucleus of the atom, one billionth of a millimetre in diameter, the way energy and matter behave affects the entire Universe.

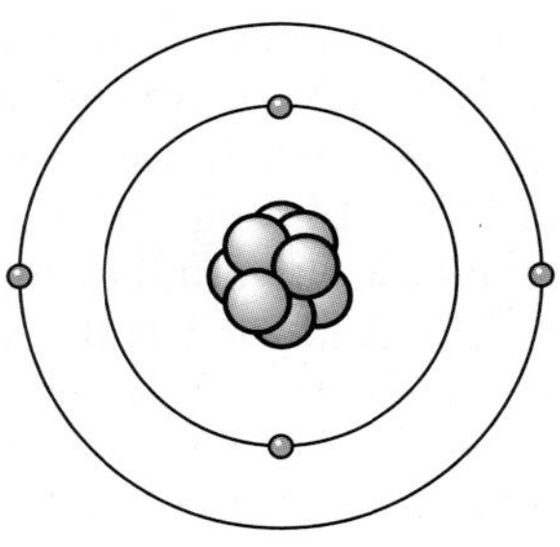

Figure 3.7.1 The behaviour of energy and matter is probably the most fundamental interaction in the Universe. We take it for granted, yet everything we do depends on it.

3

The importance of this interaction between energy and matter has not always been recognised. The understanding of this area of science grew out of the Industrial Revolution and the rise of the steam engine as a means of power. The nineteenth century saw the growth and development of industry in Great Britain, and a similar growth in the understanding of the world around us. Between them, the steam engine and the atom gave rise to a branch of science that is a powerful tool for interpreting what we see around us, from the scale of atoms to a scale which encompasses the entire Universe, its birth and its death. This branch of science is called **thermodynamics**, and it can tell us a great deal about why chemical changes take place.

The behaviour of energy

Energy dissipation

Imagine making a short film, lasting three or four seconds, of the two processes in figure 3.7.2. Now imagine the film running backwards. Which would look strange?

The film of the swinging pendulum would look the same whether run forwards or backwards, but the film of the chemical reaction would seem very unreal indeed! Both of these processes involve energy changes, in which the *total* amount of energy remains constant. The gravitational potential energy of the pendulum is transferred to kinetic energy and back to gravitational potential energy again – a 'two-way' process. Yet the energy transferred from the chemicals as they react, producing heat energy as they do so, never gets

Aluminium powder reacting with iodine – an exothermic reaction in which clouds of iodine vapour are given off

Figure 3.7.2

transferred back to the chemicals again – the energy is dispersed away from the reaction, and the energy flow here seems to be a 'one-way' process. And even apparently 'two-way' processes like a swinging pendulum eventually come to a stop. Why does energy behave in this way?

Before we answer this question, we need to look at two apparently unrelated questions:

1 Why does diffusion happen?
2 How is energy like the particles of a gas?

Figure 3.7.3

(1) Why does diffusion happen?

Imagine what happens when the cover slide between the gas jars shown in figure 3.7.3 is removed. Our experience tells us that, given long enough, we shall find the bromine vapour spread evenly throughout the two gas jars – we should not expect to find all of it in one jar. In order to convince ourselves of why this is so, it is best to look at a much simpler situation, where many fewer particles are involved. We can imagine starting off with only five bromine particles in the left-hand jar, rather than the 10^{22} or so there would actually be in the jar in the photograph, and figure 3.7.4 shows the result.

If we increase the number of particles in the gas jars to 50, W becomes 2^{50}, or about 10^{15}. And for a real situation with something like 10^{22} particles, W has an enormous value – about $2^{10^{22}}$. Only two of these arrangements have all the

Figure 3.7.4 Each particle has two possible ways of being arranged – in the left-hand jar or in the right-hand jar. If we represent the number of possible arrangements for particle 1 as W_1, for particle 2 as W_2 and so on, then the total number of ways W that the five particles can be arranged between the two jars is given by:
$W = W_1 \times W_2 \times W_3 \times W_4 \times W_5 =$
$2 \times 2 \times 2 \times 2 \times 2 = 2^5 = 32$
Since only one of these 32 ways results in all the particles being in the left-hand gas jar, it would be surprising if this arrangement were to happen very often.

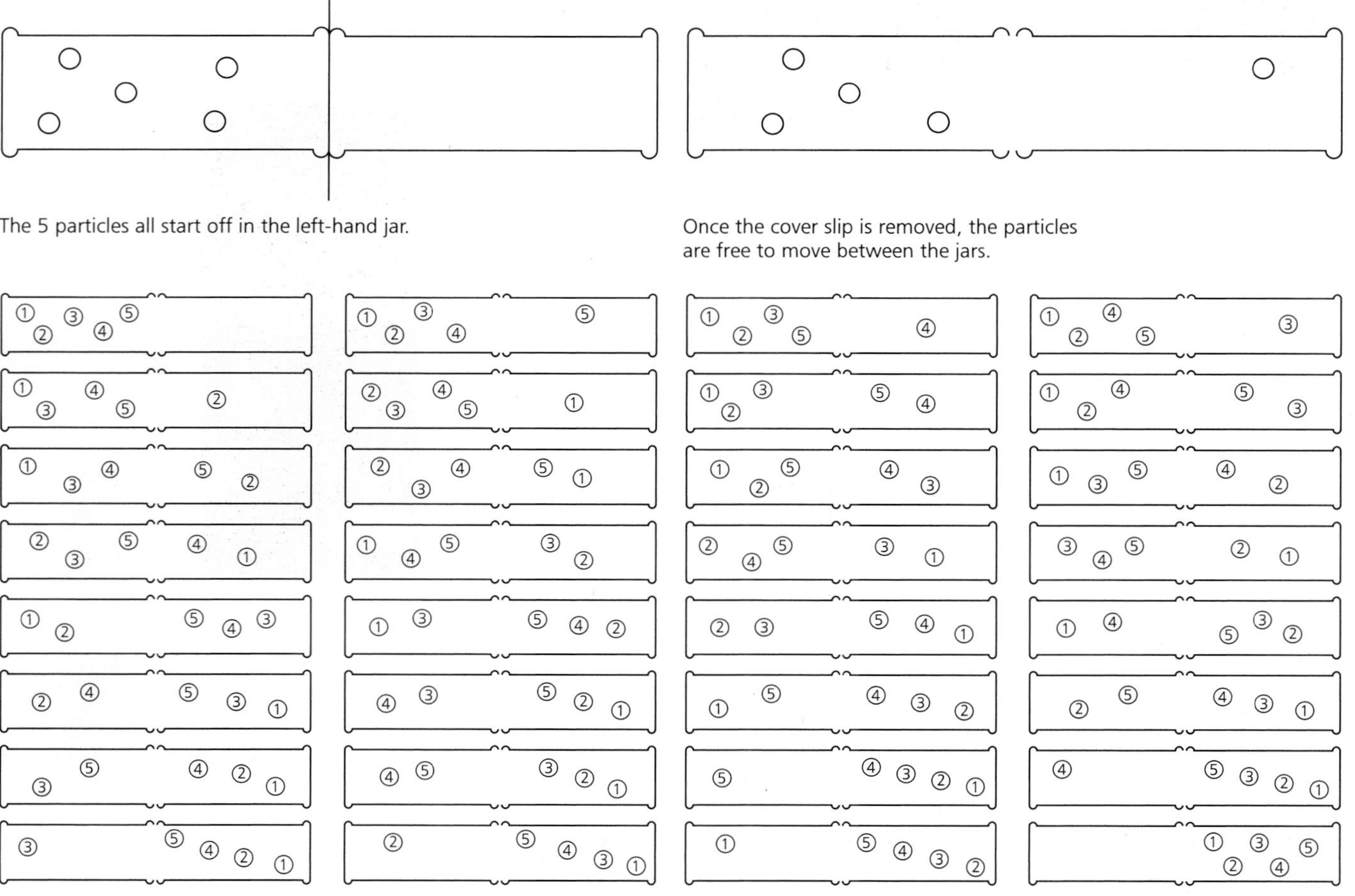

The 5 particles all start off in the left-hand jar.
Once the cover slip is removed, the particles are free to move between the jars.

particles in one jar, whereas the majority of them have the particles spread more or less evenly between the jars.

To have some idea of what this means, imagine that it were possible to make a note of the position of all the particles in the two jars once every second. Only *once* in every $2^{10^{22}}$ seconds would we be likely to see all the particles in the left-hand jar. The age of the Earth is thought to be something like 10^{17} seconds – so we should have a very long wait indeed, and we are quite justified in saying that gases *always* spread out.

Chaos and entropy

In the model of the bromine particles, W represents the number of possible arrangements of particles that exist. Values of W are combined by multiplying them together, which results in extremely large numbers when there are many particles involved. For practical purposes it is useful to have a property of a system which measures the number of possible arrangements, and expresses this as a number which is not too difficult to handle. The property that is defined in order to do this is **entropy**. The entropy S of a system is defined by the relationship:

$$S = k \log_e W$$

The entropy of a system is a measure of the disorder or chaos in it – the larger the entropy, the more disorder. So the changes that occur when a gas is allowed to diffuse into a larger volume represent an increase in entropy, since the number of ways of arranging the particles (that is, the disorder) increases.

The constant of proportionality k in the relationship is called the Boltzmann constant. Entropy and the Boltzmann constant both have units of J K^{-1} in the SI system.

3

Why the even distribution?

The argument we have just seen explains why it is that we never see all the bromine particles in one of the gas jars – not that it *cannot* happen, but that it is *extremely unlikely* to happen. What it does not explain is why there is always (once a steady state has been reached) an even distribution of particles between the two gas jars.

A simple way of answering this is to say that as there are only two ways of making all the N particles appear in one gas jar or the other out of 2^N ways of arranging them, we do not see either of these arrangements – that is, there are few ways of making this arrangement happen, so it happens only infrequently. Conversely, there are many ways of distributing the particles more or less evenly between the two gas jars, so this arrangement is found frequently.

The graphs in figure 3.7.5 show that as the number of particles N increases, the likelihood of finding a given departure from $N/2$ particles in each gas jar decreases too. By the time N has reached the number of particles in the gas jars, the likelihood of a departure from a distribution of $N/2$ in each gas jar by even a tiny amount is so small as to be negligible.

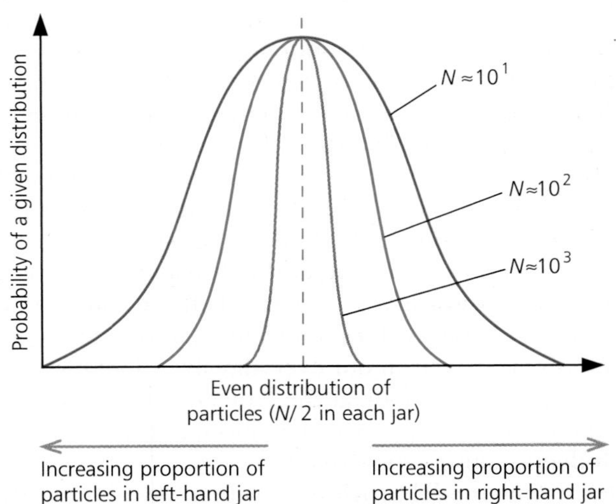

Figure 3.7.5

(2) How is energy like the particles of a gas?

Figure 3.7.6 Unlikely as it may seem, the same idea is needed to explain why oil from a sinking tanker causes pollution many miles away, how photocopiers work and why energy behaves as it does!

Investigations about the way atoms behave when they interact with energy sources led scientists to believe that atoms do not deal in energy in 'any old amounts', but only in set quantities or **quanta**, as we saw in section 1.3 (page 32).

Putting together the idea of particles distributing themselves between gas jars with energy coming in quanta gives us a way of explaining the behaviour of energy in the examples of the pendulum and the chemical reaction.

A simple model for the behaviour of energy

The world around us is a complicated place, and the Universe even more so, but it is possible to model what is happening in them by virtue of the fact that there are some simple processes going on in both of them. A very simple model universe is shown in figure 3.7.7. In it, the universe consists of 1600 atoms, each of which may have either one quantum of energy or none. An atom which has one quantum we shall call *on*, and one with no quanta we shall call *off*. The universe we are modelling has two regions, one with 100 atoms in the bottom left-hand corner of the screen, the other with 1500 atoms. The universe has 40 quanta, all of which are to be found in the smaller region when we start our observations.

Figure 3.7.7 The situation at the 'birth' of our universe. The region in the bottom left-hand corner of the screen has many atoms which are on, while the rest of the universe has none. This corresponds to a great deal of energy stored in the small region, and none in the rest of the universe.

What will happen if we allow this universe to evolve? By the mechanism of thermal conduction, as the atoms in the universe oscillate they transfer energy from one to another. This results in an atom which is *on* passing its quantum of energy on to its neighbour which then becomes *on*, the first atom becoming *off* as this happens. This process continues throughout the universe so that energy flows from the bottom left-hand corner of the screen into the rest of the universe. As this happens, the total amount of energy in the first region decreases, and that in the rest of the universe increases. Figure 3.7.8 shows this process.

Is there an 'end result' to this evolution? On the scale that we have been observing, that is, to an observer looking at individual atoms, the answer is no – the quanta are constantly shuffling between atoms. But to a more distant observer who is only looking at the distribution of energy throughout the universe the answer is yes – eventually there comes a point where the distribution of energy does not change, and a steady state is reached. The second observer is us in our everyday lives, looking at our simple universe in the way we normally look at the Universe in which we live. This observer sees only the pattern of energy distribution, not the constant shuffling of quanta which deeper observation perceives.

Our simple model provides us with a way of understanding how energy behaves. Energy localised in one area spreads out and finds its way throughout the Universe, just as the particles of bromine spread out through the two gas jars. Just as the laws of probability govern the behaviour of particles of matter,

(a)

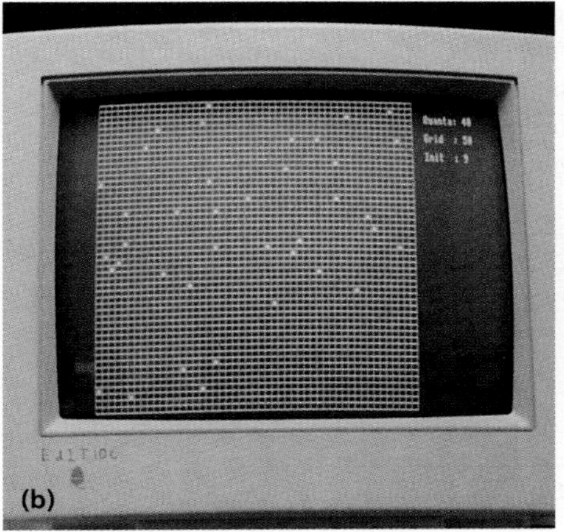

(b)

(c)

Figure 3.7.8 Our universe (a) shortly after its birth, (b) and (c) once a steady state has been reached. In (b) and (c) note that although the *arrangement* of quanta is different, there is a similar *distribution* of quanta throughout the universe in each case.

so they govern the behaviour of quanta of energy, making the concentration of energy from a 'spread out' distribution into a 'localised' distribution unlikely in a system as small and simple as our model, and effectively impossible in a universe as large and complex as the real one.

Entropy and energy

In the same way as a spreading out of particles represented an increase in entropy (see box 'Chaos and entropy' on page 343), the spreading out of energy represents an overall increase of entropy too. Although quanta leaving the smaller region of the universe reduce the entropy of that region, because there are now fewer ways of arranging the smaller number of quanta within it, the quanta entering the rest of the universe more than compensate by raising its entropy, since there are many more ways of arranging the quanta in the large and small regions together than there were in the small region alone. In this way, the overall entropy of the universe increases.

The concept of temperature

Notice that our simple model of the universe provides a concept of temperature. Once a steady state is reached and there is no net flow of energy between the two regions, a state of thermal equilibrium exists between the two regions. Our understanding of temperature says that for this to happen, the two regions must both be at the same temperature. The implication of this is that temperature is a measure of the 'concentration' of energy in an object – although the large region has more quanta than the small region at thermal equilibrium, they both have the same concentration, and it is the concentration of quanta which governs their movement, just like the particles of bromine.

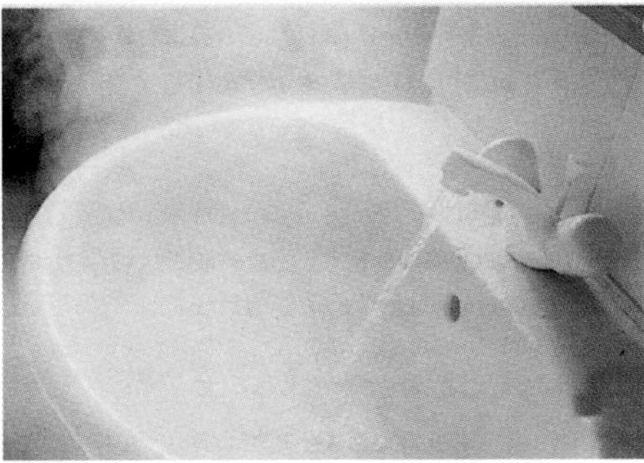

Figure 3.7.9 Although the basin of water contains more quanta, the concentration of quanta in it is much lower than the concentration in the red-hot pin. Putting your hand into the basin won't hurt you, because the rate at which quanta move between the water and your skin is quite slow. The red-hot pin is a different matter, however!

The second law of thermodynamics

Some simple rules for the behaviour of matter and energy as described by our model universe were first set out by the physicists William Thomson (later Lord Kelvin) and Rudolf Clausius in the middle of the nineteenth century. Between them, Kelvin and Clausius formulated a law known as the **second law of thermodynamics**. Perhaps the easiest way to state the law is to say that:

No process is possible in which there is an overall decrease in the entropy of the Universe

which can be simplified into statements such as 'you can't unscramble scrambled eggs' and 'heat energy always spreads out'. The important thing to realise is that, even when we see a process in which entropy decreases, it is always linked to an increase in entropy somewhere else in the Universe which more than offsets the entropy decrease, or at best matches it.

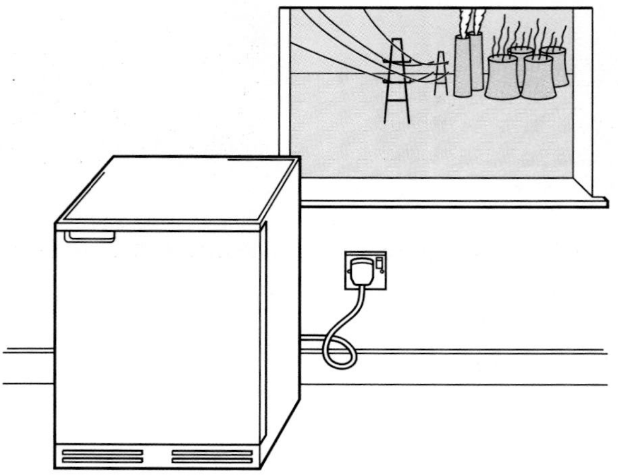

Figure 3.7.10 A refrigerator appears to contravene the second law by moving energy from a cold place to a warmer one, causing a decrease in entropy. However, it does this only because elsewhere in the Universe a lump of coal or a stream of water is involved in a process which increases the entropy of the Universe by a greater amount.

The second law and chemical change

We have seen how diffusion and the spreading out of energy result in an increase in entropy. These ideas can be applied to chemical systems too.

First of all, consider the entropy of the three states of matter. In general, we find that gases have higher entropies than liquids, which in turn have higher entropies than solids. The standard molar entropies $S^{\ominus}_{m}$ of water in its three states in table 3.7.1 illustrate this well. The **standard molar entropy** of a substance is the entropy of one mole of the substance in its standard state at a specified temperature.

	$H_2O(s)$	$H_2O(l)$	$H_2O(g)$
$S^{\ominus}_{m}$/J mol^{-1} K^{-1}	48.0	69.9	188.7

Table 3.7.1

Just as we looked at enthalpy changes in chemical reactions in section 1.8, we can also consider entropy changes in chemical reactions. The reaction:

$$2O_3(g) \rightarrow 3O_2(g)$$

involves an *increase* in entropy for the reaction in a forward direction, since the right-hand side contains more moles of gas than the left-hand side. On the other hand, the reaction:

$$2Na(s) + Cl_2(g) \rightarrow 2NaCl(s)$$

involves a *decrease* in entropy, as the particles in a gas which are free to move throughout the reaction vessel become confined in an ionic solid. (Changes in entropy have the same sign convention as changes in enthalpy – for increases in entropy, $\Delta S > 0$, while for decreases in entropy, $\Delta S < 0$.)

Experiments concerning changes in entropy show that in general:

1 Gases usually have greater entropies than solids, so for most reactions in which the number of moles of gases increases, $\Delta S > 0$.

Entropy and temperature

The entropy of a substance depends on temperature (the units of entropy are J K^{-1}), since as the temperature of a substance increases the motion of its particles increases, and so its disorder increases. Standard molar entropies, like standard electrode potentials, are quoted for 298 K and 10^5 Pa. The standard molar entropy of a substance is calculated from knowledge of its heat capacity, together with the fact that the entropy of any element or its compounds is equal to zero at 0 K.

2 Solids dissolving usually have $\Delta S > 0$.

3 When molecules split up, $\Delta S > 0$.

Entropy changes, the system and the surroundings

We have just seen that the reaction of sodium and chlorine to form sodium chloride involves a decrease in entropy:

$$2Na(s) + Cl_2(g) \rightarrow 2NaCl(s)$$

We know that this reaction definitely *does* occur, and that it releases a considerable amount of energy too, provided that there is a flame to supply the activation energy for the reaction. How can a change in which there is a decrease in entropy be spontaneous?

The answer to this question is that we have only considered the entropy change of the **system**, ΔS_{system}. The system for this reaction might be the gas jar containing the reactants. We must also think about the change in entropy of the **surroundings**, $\Delta S_{surroundings}$. Energy leaves the system during an exothermic reaction (ΔH is negative), as illustrated in figure 3.7.11.

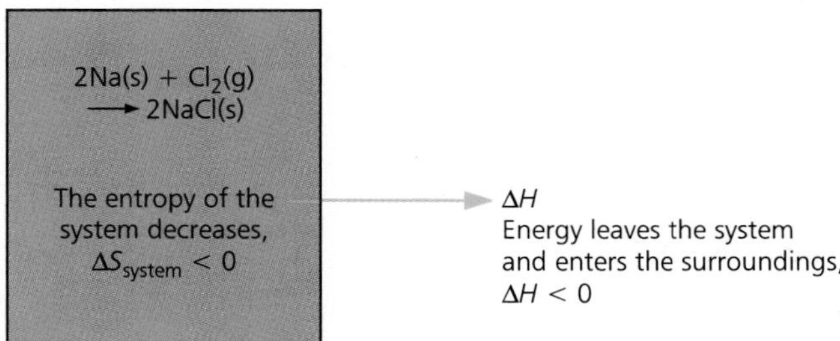

Figure 3.7.11 In this reaction, energy leaves the system. This spreading out of energy brings about an increase in entropy that more than offsets the decrease in the entropy of the system.

Despite the fact that there is a decrease in entropy of the system in this reaction, the energy leaving the system produces a substantial increase in entropy of the surroundings, since there are more ways of arranging the energy quanta in the rest of the Universe than there are of arranging them in the system alone. It is the *total* entropy change that determines whether the reaction is spontaneous or not:

$$\Delta S_{total} = \Delta S_{system} + \Delta S_{surroundings}$$

In the case of the reaction of sodium and chlorine, the release of large amounts of heat into the Universe makes $\Delta S_{surroundings}$ large and positive, which is more than enough to offset the negative value of ΔS_{system}.

Calculating changes in entropy

Finding $\Delta S^{\ominus}_{system}$ for a chemical change is generally quite easy – we can look up the standard molar entropy for the substances on each side of the equation and calculate the difference, using data like those given in table 3.7.2.

Substance	$S^{\ominus}_m$/J mol^{-1} K^{-1}	Substance	$S^{\ominus}_m$/J mol^{-1} K^{-1}
Ag(s)	42.6	Fe(s)	27.3
C(diamond)	2.4	Hg(l)	77.4
C(graphite)	5.7	Hg(g)	175
$CH_3CH_2OH(l)$	161	Na(s)	51.2
$Cl_2(g)$	223	NaCl(s)	72.4

Table 3.7.2 The standard molar entropies of some elements and compounds

For the formation of one mole of sodium chloride, the standard molar entropy change of the *system* is given by:

$$\mathbf{Na(s) + \tfrac{1}{2}Cl_2(g) \rightarrow NaCl(s)}$$

$$\Delta S^{\ominus}_{system} = 72.4\ \mathrm{J\ mol^{-1}\ K^{-1}} - (51.2\ \mathrm{J\ mol^{-1}\ K^{-1}} + \tfrac{1}{2} \times 223\ \mathrm{J\ mol^{-1}\ K^{-1}})$$

$$= -90.3\ \mathrm{J\ mol^{-1}\ K^{-1}}$$

To calculate the standard entropy change of the *surroundings*, we need to know the energy transferred to them, which is given by $\Delta H^{\ominus}$. The entropy change is calculated from the relationship:

$$\Delta S_{surroundings} = -\frac{\Delta H}{T}$$

or, under standard conditions,

$$\Delta S^{\ominus}_{surroundings} = -\frac{\Delta H^{\ominus}}{T}$$

where T is the absolute temperature, measured in kelvins.

$\Delta H^{\ominus}_{f,m}$, the standard molar enthalpy change for the formation of sodium chloride, is $\Delta H^{\ominus}$ for this reaction. $\Delta H^{\ominus}_{f,m}$(NaCl) is –411 kJ mol^{-1}. This gives:

$$\Delta S^{\ominus}_{surroundings} = -\frac{-411\ 000\ \mathrm{J\ mol^{-1}}}{298\ \mathrm{K}}$$

$$= +1379\ \mathrm{J\ mol^{-1}\ K^{-1}}$$

(Notice that $\Delta H^{\ominus}$ must be in J mol^{-1}, because $\Delta S^{\ominus}$ is in J mol^{-1} K^{-1}.)

Finally, we can now calculate $\Delta S^{\ominus}_{total}$:

$$\Delta S^{\ominus}_{total} = \Delta S^{\ominus}_{system} + \Delta S^{\ominus}_{surroundings}$$

$$= -90.3\ \mathrm{J\ mol^{-1}\ K^{-1}} + 1379\ \mathrm{J\ mol^{-1}\ K^{-1}}$$

$$= 1289\ \mathrm{J\ mol^{-1}\ K^{-1}}$$

Overall, then, the change in entropy is positive – this is a **favourable** process under standard conditions, and the reaction happens spontaneously.

The entropy change of the surroundings

Why does $\Delta S_{surroundings} = -\Delta H/T$? The entropy change of the surroundings is temperature sensitive. The transfer of a given quantity of energy to surroundings that are at a low temperature produces a greater increase in entropy than the transfer of the same amount of energy to surroundings at a higher temperature.

The reason for the difference can be understood if we think about the effect of transferring quanta between the regions of the model universe in figure 3.7.7. Transferring 10 quanta to a region already containing 5 quanta causes a large change in the number of ways of arranging quanta, while transferring 10 quanta to a region already containing 20 quanta has a much smaller effect.

Another way of looking at this is to think about the sound of someone shouting. This is barely noticeable in the roar of a football crowd, but in the quiet of a library . . .

Entropy, free energy and spontaneous change

The relationships:

$$\Delta S_{total} = \Delta S_{system} + \Delta S_{surroundings}$$

and:

$$\Delta S_{surroundings} = -\frac{\Delta H}{T}$$

can be combined to produce the relationship:

$$\Delta S_{total} = \Delta S_{system} - \frac{\Delta H}{T}$$

Multiplying this relationship by T gives:

$$T\Delta S_{total} = T\Delta S_{system} - \Delta H$$

or:

$$-T\Delta S_{total} = \Delta H - T\Delta S_{system}$$

The quantity ($-T\Delta S_{total}$) is called the **Gibbs function** or **Gibbs free energy change**, ΔG, named after the American scientist Josiah Willard Gibbs. Since ΔS_{total} must be positive for a change to occur spontaneously, it follows that spontaneous changes have $\Delta G < 0$.

How can we use the relationship:

$$\Delta G = \Delta H - T\Delta S_{system}$$

to predict whether changes will be spontaneous? There are four cases to consider.

1 Exothermic changes that are accompanied by an increase in the system's entropy will *always* happen spontaneously:

ΔH is negative

ΔS_{system} is positive, so $-T\Delta S_{system}$ is negative (remember T is always positive)

so:

ΔG is always negative

Example:

$$CH_3CH_2OH(l) + 3O_2(g) \rightarrow 2CO_2(g) + 3H_2O(g)$$

2 Endothermic changes that are accompanied by a decrease in the system's entropy will *never* happen spontaneously:

ΔH is positive

ΔS_{system} is negative, so $-T\Delta S_{system}$ is positive

so:

ΔG is always positive

Example: $CO_2(g) \rightarrow C(s) + O_2(g)$

3 Endothermic changes that are accompanied by an increase in the system's entropy will be spontaneous if the temperature is sufficiently high:

ΔH is positive

ΔS_{system} is positive, so $-T\Delta S_{system}$ is negative

so:

ΔG is negative if the magnitude of $T\Delta S_{system}$ is greater than the magnitude of ΔH

Example: $H_2O(l) \rightarrow H_2O(g)$

Figure 3.7.12 Josiah Willard Gibbs, the American scientist after whom the Gibbs function was named

3

4 Exothermic changes that are accompanied by a decrease in the system's entropy will be spontaneous if the temperature is sufficiently low:

ΔH is negative

ΔS_{system} is negative, so $-T\Delta S_{system}$ is positive

so:

ΔG is negative if the magnitude of ΔH is greater than the magnitude of $T\Delta S_{system}$

Example: $H_2O(g) \rightarrow H_2O(l)$

This is summarised in table 3.7.3, and illustrated in figure 3.7.13.

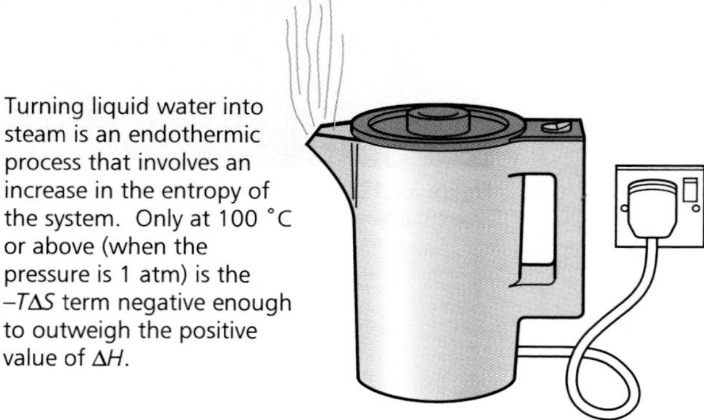

Turning liquid water into steam is an endothermic process that involves an increase in the entropy of the system. Only at 100 °C or above (when the pressure is 1 atm) is the $-T\Delta S$ term negative enough to outweigh the positive value of ΔH.
Turning steam into liquid water is an exothermic process which involves a decrease in the entropy of the system. Only below 100 °C (when the pressure is 1 atm) is the $-T\Delta S$ term small enough to ensure that it is outweighed by the negative value of ΔH.

Figure 3.7.13

		ΔH	
		+	–
ΔS_{system}	+	Spontaneous only at high temperatures	Always spontaneous
	–	Never spontaneous	Spontaneous only at low temperatures

Table 3.7.3 The effect of ΔH and ΔS_{system} on the spontaneity of changes

The activation energy barrier

The value of ΔG must be used carefully when deciding whether a change is spontaneous, as when using enthalpy changes, standard electrode potentials and equilibrium constants. Each of these quantities gives us information about the thermodynamic feasibility of a change, but tells us nothing about the activation energy for the change – that is, whether it occurs at a reasonable rate or not. Even if we succeed in finding a temperature at which a chemical process has a negative value of ΔG, this does not guarantee success in turning reactants into products.

Making diamonds

The manufacture of synthetic diamonds requires the change:

C(graphite) $\rightarrow$ C(diamond)

Figure 3.7.14 shows how the value of ΔG for this reaction varies with

pressure and temperature. The lines in the shaded area on this graph show the temperatures and pressures at which the conversion of graphite to diamonds is *theoretically* possible. In practice, much higher temperatures and pressures are needed to ensure the conversion takes place at a reasonable rate – reaction conditions involving pressures of up to 10^5 atm and 3000 K may be used.

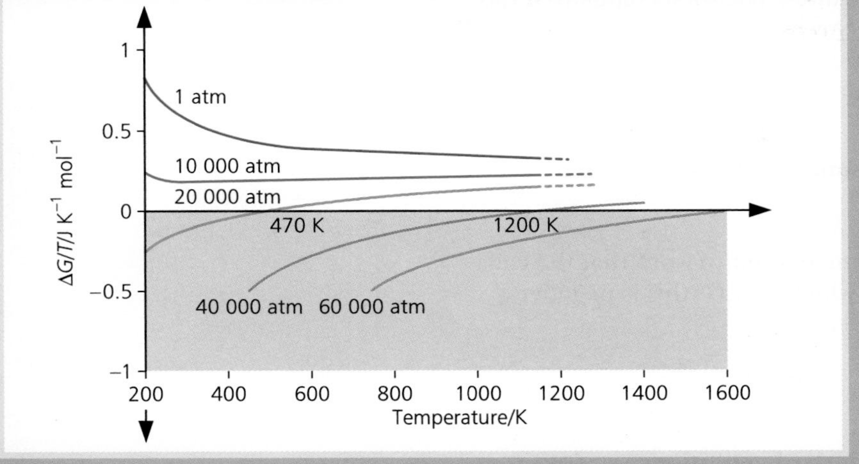

Figure 3.7.14 The variation with pressure and temperature of ΔG for the conversion of graphite into diamond

Free energy and maximum work

Chemical reactions are frequently used to do work – the combustion of a mixture of petrol and air to power cars, the reaction of chemicals in batteries to power pocket calculators and the chemical reactions in our bodies which keep our hearts, brains and other organs functioning are all examples of this.

The science of thermodynamics tells us that the maximum amount of useful work we can get from a chemical process is equal to the value of ΔG – this is the energy that is *free* to do work, and is why ΔG is referred to as **free energy**.

The maximum amount of work that an electrochemical cell can do is related to its e.m.f., and so E and ΔG are related, as we shall now see.

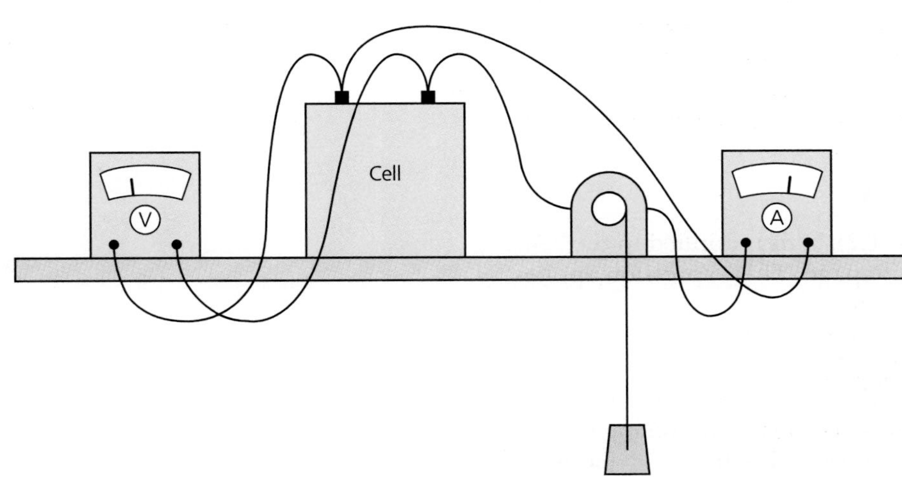

Figure 3.7.15 Using chemical change to do useful work

3

If we wish to raise the weight in figure 3.7.15 quickly, we shall need to draw a large current from the cell. Because of this large current, we shall lose a large proportion of the energy from the cell, which will be dissipated as heat in the resistance of the cell and the circuit. We must therefore carry out this change very, very slowly (thermodynamicists use the term **quasistatically**) if we are to get the maximum amount of work from the cell.

Suppose that in raising the weight, reactants in the cell are converted into products, driving z moles of electrons round the circuit. The charge on 1 mole of electrons is 96 500 C mol^{-1}, which we shall call F. The total charge that has moved round the circuit will therefore be zF. Now when a charge q moves through a potential difference E, the energy transferred (the work done) is qE. From this it follows that the maximum amount of work W_{max} that the cell can do is given by:

$$\begin{aligned} W_{max} &= \textbf{charge moved round circuit} \times \textbf{cell e.m.f.} \\ &= zFE \end{aligned}$$

Now we have already said that the maximum amount of work that the cell can supply is the free energy for this change, which is $-\Delta G$ (this is negative because the cell loses energy in this change). So:

$$\Delta G = -zFE$$

or:

$$\Delta G^{\ominus} = -zFE^{\ominus} \quad \textbf{under standard conditions}$$

ΔG⊖ for the nickel–cadmium cell

The cell reaction in the nickel–cadmium cell is:

$$Cd(s) + NiO_2(s) + 2H_2O(l) \rightarrow Cd(OH)_2(s) + Ni(OH)_2(s)$$

for which the cell e.m.f. is about 1.4 V under standard conditions. The number of electrons transferred in this reaction is 2, so the e.m.f. corresponds to a value of ΔG given by:

$$\begin{aligned} \Delta G^{\ominus} &= -zFE^{\ominus} \\ &= -2 \times 96\,500 \text{ C mol}^{-1} \times 1.4 \text{ V} \\ &= -270 \text{ kJ mol}^{-1} \end{aligned}$$

Free energy and the equilibrium constant

The equilibrium constant K tells us to what extent a reaction goes, and is related to ΔG by the simple expression:

$$\Delta G = -RT \log_e K$$

where R is the molar gas constant (see section 1.7). This is applied to the nickel–cadmium cell in the box on the opposite page. This expression applies to all reactions, not just to those involving galvanic cells.

In conclusion – spontaneous change, growth and decay

The laws of thermodynamics provide a powerful tool for understanding how the Universe behaves. The crumbling of an ancient house, the cooling of a cup of tea and the finite nature of our resources of fossil fuels – thermodynamics

K for the nickel–cadmium cell

Using the relationship $\Delta G = -RT \log_e K$ for the nickel–cadmium cell gives:

$$-270\,000 \text{ J mol}^{-1} = -8.31 \text{ J K}^{-1} \text{ mol}^{-1} \times 298 \text{ K} \times \log_e K$$

Rearranging:

$$\log_e K = \frac{270\,000 \text{ J mol}^{-1}}{8.31 \text{ J K}^{-1} \text{ mol}^{-1} \times 298 \text{ K}}$$

$$= 109$$

so:

$$K \approx 10^{47}$$

This enormous value of *K* shows that the cell reaction goes overwhelmingly to completion.

explains all these as the result of increasing chaos in the Universe as a whole. Thermodynamics suggests that the order we see in the world around us has emerged out of chaos, and that it will one day merge into chaos again.

Yet for all its power, we should remember that thermodynamics is only one of many ways of understanding the Universe and its mysteries. Instead of talking of chaos and entropy when explaining why a chemical change occurs, we may equally well speak of enthalpy changes and equilibrium constants. The two statements:

1. N_2O_4 dissociates into NO_2 at high temperatures because the increase in entropy of the surroundings more than compensates for the decrease in entropy of the system
2. The equilibrium:

$$N_2O_4(g) \rightleftharpoons 2NO_2(g)$$

lies well to the right at high temperatures because *K* increases with temperature

are both correct – they simply represent different ways of describing the same phenomenon.

SUMMARY

- All spontaneous changes result in an increase in the number of ways *W* in which quanta of energy and particles of matter can be arranged.
- The entropy *S* of a system is a measure of the disorder or chaos in it – the larger the entropy, the more disorder. $S = k \log_e W$.
- **Standard entropies** are quoted at 298 K and 10^5 Pa pressure.
- In general, gases have higher entropies than liquids, which in turn have higher entropies than solids.
- The total entropy change ΔS_{total} for a process is the sum of the entropy changes for the system and its surroundings:

$$\Delta S_{total} = \Delta S_{system} + \Delta S_{surroundings}$$

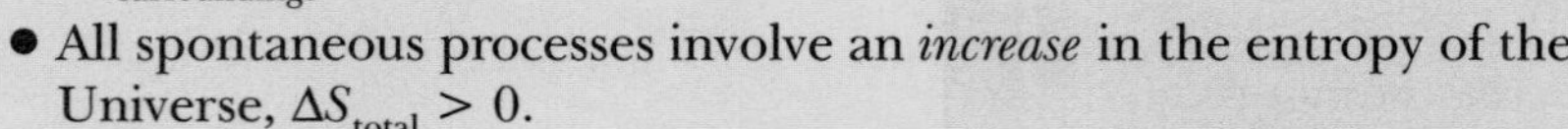

- $\Delta S_{surroundings} = -\Delta H/T$.
- All spontaneous processes involve an *increase* in the entropy of the Universe, $\Delta S_{total} > 0$.
- The **Gibbs free energy change** ΔG for a reaction is given by:

$$\Delta G = \Delta H - T\Delta S_{system}$$

  ΔG must be *negative* if the reaction is to happen spontaneously.
- Reactions for which ΔG is positive may sometimes be made to go by changing the reaction conditions, particularly the temperature.
- ΔG provides no information about kinetic factors affecting a reaction.
- $\Delta G = -zFE$.
- $\Delta G = -RT \log_e K$.

QUESTIONS

1 Is it reasonable to expect a cold cup of tea to warm up spontaneously? Explain your answer.

2 Tidying an untidy room reduces its entropy. Does this process contravene the second law of thermodynamics?

3 Entropy is a measure of the disorder in a system. The concept of **negentropy** has been developed, which is a measure of the order in a system. Do you agree with the statement 'In our lives we do not so much consume energy, but consume negentropy'? Give your reasons.

4 The enthalpy change of vaporisation, $\Delta H^{\ominus}_{vap}$, of trichloromethane ($CHCl_3$, chloroform) is +29.7 kJ mol^{-1}. The change:

$$CHCl_3(l) \rightarrow CHCl_3(g)$$

has $\Delta S^{\ominus}_{system}$ = +88.7 J mol^{-1} K^{-1}. At what temperature will trichloromethane boil at atmospheric pressure? (*Hint:* This is the temperature at which the liquid and vapour are in equilibrium – what will be the value of ΔG for the change of state at this temperature?)

5 For each of the following changes, suggest whether $\Delta S^{\ominus}_{system}$ is likely to be:

a positive
b negative
c approximately zero:

i $KI(s) \rightarrow K^+(aq) + I^-(aq)$
ii $4Al(s) + 3O_2(g) \rightarrow 2Al_2O_3(s)$
iii $Zn(s) + CuSO_4(aq) \rightarrow ZnSO_4(aq) + Cu(s)$
iv $N_2(g) + 3H_2(g) \rightarrow 2NH_3(g)$
v protein $\rightarrow$ amino acids
vi $6CO_2(g) + 6H_2O(g) \rightarrow C_6H_{12}O_6(aq) + 6O_2(g)$

6 The standard e.m.f. of the cell:

$$Ni(s)|\,Ni^{2+}(aq) \,\vdots\vdots\, Fe^{3+}(aq),Fe^{2+}(aq)|\,Pt$$

is +1.02 V.

a What chemical reaction occurs when current is drawn from the cell?
b What does the cell's e.m.f. tell you about the equilibrium constant for this reaction?
c What is the maximum amount of energy that can be obtained from such a cell?
d A battery manufacturer proposes to use this reaction in a new type of cell and wants to know if there is any way in which the cell's e.m.f. may be increased. What advice would you give?

3 Questions

1 Nitrogen oxides such as nitrogen monoxide, NO, and nitrogen dioxide, NO_2, are formed unintentionally by man and cause considerable harm to the environment.

a The oxidation of nitrogen monoxide in car exhausts may involve the following reaction:

$$NO(g) + CO(g) + O_2(g) \rightarrow NO_2(g) + CO_2(g)$$

This reaction was investigated in a series of experiments. The results are shown below in the table below.

Experiment	[NO(g)] /mol dm^{-3}	[CO(g)] /mol dm^{-3}	[O_2(g)] /mol dm^{-3}	Initial rate /mol dm^{-3}s^{-1}
1	1.00×10^{-3}	1.00×10^{-3}	1.00×10^{-1}	0.44×10^{-3}
2	2.00×10^{-3}	1.00×10^{-3}	1.00×10^{-1}	1.76×10^{-3}
3	2.00×10^{-3}	2.00×10^{-3}	1.00×10^{-1}	1.76×10^{-3}
4	2.00×10^{-3}	2.00×10^{-3}	4.00×10^{-1}	7.04×10^{-3}

i For each reactant, deduce the order of reaction. Show your reasoning.

ii Deduce the rate equation and calculate the rate constant for this reaction.

iii Suggest, with a reason, what would happen to the value of the rate constant, k, as the car's exhaust gets hotter. **(11 marks)**

b State **two** environmental consequences of nitrogen oxides and outline their catalytic removal from car exhaust gases. **(5 marks)**

c Not all nitrogen compounds are harmful: some, such as nitrogen fertilisers, are beneficial to man.
A nitrogen fertiliser, **D**, was analysed in the laboratory and was shown to have the composition by mass: Na, 27.1%; N, 16.5%; O, 56.4%. On heating, 3.40 g of **D** was broken down into sodium nitrite, $NaNO_2$, and oxygen gas.
Showing your working, suggest an identity for the fertiliser, **D**, and calculate the volume of oxygen that was formed.
[Under the experimental conditions, 1 mole of gas molecules occupy 24 dm^3.] **(4 marks)**

(Total 20 marks)

(OCR specimen)

2 A student set up the following electrochemical cell.

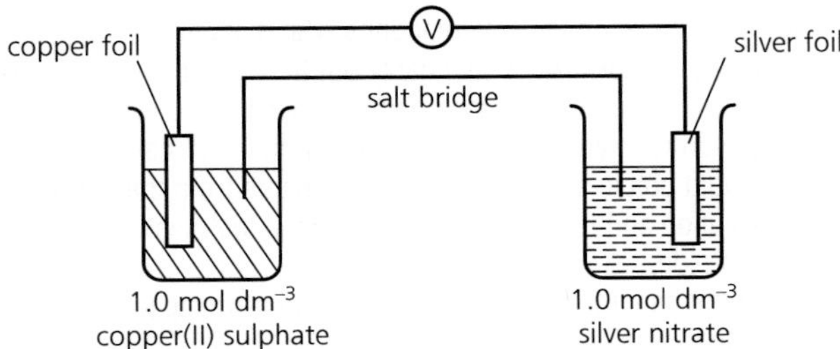

a How could the student have made the salt bridge? **(1 mark)**

b Write half-equations showing the reactions that occurred in

i the Cu/Cu^{2+} half cell,

ii the Ag/Ag^+ half cell. **(2 marks)**

c Write an equation for the overall cell reaction. **(1 mark)**

d i Using data from table 3.5.2 on page 301, calculate the standard cell potential for this cell.

ii Identify the electrode at which reduction occurs. Explain your answer. **(4 marks)**

e The student found that the e.m.f. obtained for this cell was less than the calculated value. Suggest a reason for this. **(1 mark)**

(Total 9 marks)

(OCR specimen)

3 Ethanol, C_2H_5OH, is an important industrial chemical with about 200 000 tonnes manufactured in the UK each year. The usual method of manufacture is by the hydration of ethene with steam in the presence of a phosphoric acid catalyst at 550 K and a pressure of about 7000 kPa.

$$C_2H_4(g) + H_2O(g) \rightleftharpoons C_2H_5OH(g) \quad \Delta H = -46\ \text{kJ mol}^{-1}$$

a i Predict, with justification, the optimum conditions for this reaction.

ii Explain why the actual conditions used may be different from the optimum conditions.

iii The boiling points of the three chemicals involved in this equilibrium are shown in the table below.

Compound	C_2H_4	H_2O	CH_3CH_2OH
Boiling point/°C	–104	100	78

Suggest how the ethanol could be separated from the equilibrium mixture. **(8 marks)**

b Write an expression for K_p of this reaction and explain, with a reason in each case, whether you would expect the value of K_p to alter if any of the external variables were changed. **(5 marks)**

c Alcohols such as ethanol can be used as alternative fuels to petrol. The combustion of ethanol tends to be more complete than the combustion of the alkanes present in petrol, partly because less oxygen is required for combustion.
Use equations to compare the amount of oxygen required per gram of fuel combusted. Suggest why there is this difference. **(5 marks)**

(Total 18 marks)

(OCR specimen)

4 Using knowledge, principles and concepts from different areas of chemistry, explain and interpret, as fully as you can, the data given in the table below. In order to gain full credit, you will need to consider each type of information separately and also to link this information together.

Compound	Boiling point/K	Properties of a 0.1 mol dm^{-3} solution Electrical conductivity	$[H^+]$/mol dm^{-3}
NaCl	1686	good	1.0×10^{-7}
CH_3COOH	391	slight	1.3×10^{-3}
CH_3CH_2OH	352	poor	1.0×10^{-7}
$AlCl_3$	451	good	3.0×10^{-1}

(19 marks)
(Quality of Written Communication: 3 marks)
(Total 22 marks)
(OCR specimen)

5 Sulphur trioxide, SO_3 is made industrially by the Contact process. This is an example of dynamic equilibrium:

$$2SO_2(g) + O_2(g) \rightleftharpoons 2SO_3(g) \qquad \Delta H = -197\ \text{kJ mol}^{-1}$$

a State **two** features of a reaction with a *dynamic equilibrium*. **(2 marks)**

b Use le Chatelier's principle to explain what happens to the **equilibrium** position of this reaction as
i the temperature is raised;
ii the pressure is increased. **(4 marks)**

c Use your answer to **b** to deduce the theoretical conditions for this equilibrium to provide a high yield. **(1 mark)**

d Explain what happens to the **rate** of this reaction as
i the temperature is raised;
ii the pressure is increased. **(4 marks)**

e The conditions used often in the Contact process are 400°C and normal atmospheric pressure.
Using your anwers to **b**, **c** and **d**, comment on this choice of
i temperature,
ii pressure. **(2 marks)**

(Total 13 marks)
(OCR specimen)

6 Ammonia, NH_3 is made industrially by the Haber process:

$$N_2(g) + 3H_2(g) \rightleftharpoons 2NH_3(g)$$

In the conditions often used in the Haber process, there is only a 15% yield of ammonia. This ammonia is removed by rapidly cooling the equilibrium mixture to –40°C. The boiling points of N_2, H_2 and NH_3 are shown below.

Compound	Boiling point/°C
N_2	–196
H_2	–253
NH_3	–33

a How does cooling to –40°C allow the ammonia to be removed?
b Suggest what happens to any unreacted nitrogen and hydrogen. **(4 marks)**
c Much of the ammonia produced is used to make fertilisers such as ammonium nitrate, NH_4NO_3. This is prepared by an acid–base reaction between nitric acid, HNO_3, and ammonia..
i How does nitric acid behave as an acid?
ii Construct an equation for the acid–base reaction of ammonia with nitric acid.
iii Farmers use ammonium nitrate for its nitrogen content. Calculate the percentage of nitrogen in NH_4NO_3.
[Ar: H, 1.0; N, 14.0; O, 16.0] **(4 marks)**

(Total 8 marks)
(OCR specimen)

7 a Describe, using **one** example in each case, the different modes of action of homogeneous and heterogeneous catalysis. **(7 marks)**
b Many catalysts are very expensive but their use does allow the chemical industry to operate more profitably. Explain why the use of catalysts provides economic benefits to this industry. **(3 marks)**

(Total 10 marks)
(Quality of written communication: 2 marks)
(OCR specimen)

8 The diagram below shows the energy distribution of reactant molecules at a temperature T_1.

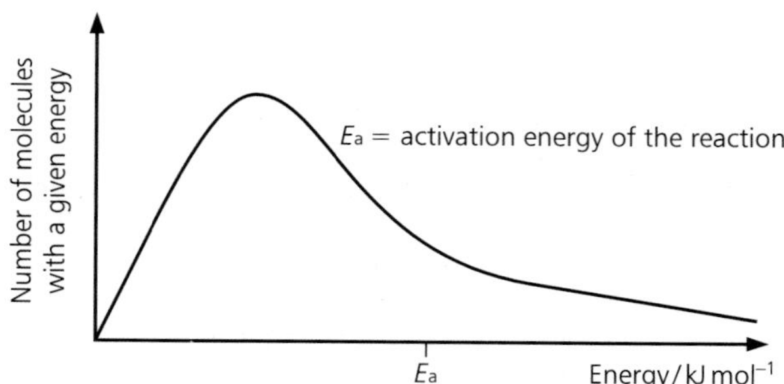

a Explain what you understand by the term *activation energy*. **(1 mark)**
b Copy the diagram and mark on it the activation energy, E_c, in the presence of a catalyst. **(1 mark)**
c Explain, in terms of the distribution curve in the diagram, how a catalyst speeds up the rate of a reaction. **(2 marks)**
d Raising the temperature can also increase the rate of this reaction.
i Sketch on your diagram a second curve to represent the energy distribution at a higher temperature. Label your curve T_2.
ii Explain, in terms of the diagram, how an increase in temperature can cause an increase in the rate of a reaction. **(4 marks)**

(Total 8 marks)
(OCR specimen)

9 You are required to plan an experiment to determine the percentage by mass of bromine in a bromoalkane. The bromoalkane, which boils at 75° C, can be hydrolysed completely by heating with an appropriate amount of boiling, aqueous sodium hydroxide for about 40 minutes. The bromide ion released can be estimated by converting it into a silver bromide precipitate which is subsequently weighed.
a Write equations for the reactions which occur.
b Describe how you would carry out the hydrolysis, giving details of the apparatus and the conditions which you would use.
c Describe, giving details of the apparatus and reagents, how you would obtain a silver bromide precipitate from the hydrolysis solution and how you would determine the mass of the silver bromide.
d Show how the percentage by mass of bromide ion, in the original haloalkane, can be calculated. **(15 marks)**
(AQA specimen)

10 Hydrogen peroxide decomposes in the presence of metal oxide catalysts to form oxygen and water according to the following equation.

$$2H_2O_2 \rightarrow 2H_2O + O_2$$

You are required to plan an experiment to compare which of two metal oxides is the more effective catalyst by measuring the amount of oxygen formed during a 2 minute period immediately after a solution of hydrogen peroxide is added to a catalyst.

a Calculate the volume of oxygen formed (measured at 20 °C and 100 kPa) when 100 cm^3 of 1.00 M hydrogen peroxide decomposes. **(5marks)**

b Describe, giving details of the apparatus and procedure, how you could carry out experiments to compare the two metal oxides. Use your answer to part **a** to help you choose a suitable scale of apparatus and amount of hydrogen peroxide solution to use. **(10 marks)**

(Total 15 marks)

(AQA specimen)

11 a i Define the term *Bronsted–Lowry acid*.

ii What is meant by the term *strong* when describing an acid?

iii Give the value of the ionic product of water, K_w, measured at 298 K, and state its units. **(4 marks)**

b At 298 K, 25.0 cm^3 of a solution of a strong acid contained 1.50×10^{-3} mol of hydrogen ions.

i Calculate the hydrogen ion concentration in this solution and hence its pH.

ii Calculate the pH of the solution formed after the addition of 50.0 cm^3 of 0.150 M NaOH to the original 25.0 cm^3 of acid. **(8 marks)**

c A solution of a strong acid was found to have a pH of 0.5

i Calculate the hydrogen ion concentration in this solution.

ii Calculate the volume of water which must be added to 25.0 cm^3 of this solution to increase its pH from 0.5 to 0.7. **(5 marks)**

(Total 17 marks)

(AQA 1999)

12 a Write an equation for the reaction which occurs when the weak acid HA is added to water. **(1 mark)**

b Write an expression for the dissociation constant, K_a, for the weak acid HA. **(1 mark)**

c The dissociation of the acid HA is an endothermic process. Deduce the effect, if any, of

i an increase in temperature on the value of the dissociation constant, K_a

ii an increase in temperature on the pH of an aqueous solution of the acid

iii an increase in the concentration of the acid on the value of K_a **(3 marks)**

d Identify a compound which could be added to aqueous ethanoic acid so that the pH of the resulting solution would not change significantly if a small volume of dilute hydrochloric acid were added. State the name given to solutions which behave in this way. **(2 marks)**

(Total 7 marks)

(AQA 1999)

13 When ammonia gas is heated, a homogeneous, dynamic equilibrium is established between ammonia and its constituent elements. This decomposition is endothermic.

a Explain the terms *homogeneous*, *dynamic* and *equilibrium*. Write an equation for this decomposition and derive an expression for the equilibrium constant, K_c. **(5 marks)**

b State and explain the conditions under which a high equilibrium concentration of hydrogen would be obtained. **(4 marks)**

c The decomposition of ammonia might in the future be used as an industrial method for the manufacture of hydrogen.
Explain why an industrial chemist might decide to use conditions different from those you have given in part **b** if large quantities of hydrogen were to be produced by this decomposition. Discuss the effect that using a catalyst would have on the equilibrium yield and on the amount of hydrogen which could be produced in a given time. **(6 marks)**

(Total 15 marks)

(AQA 1999)

14 a The ammonium ion content of fertilisers can be found by heating the fertiliser with sodium hydroxide and passing the ammonia produced into an excess of hydrochloric acid.

i Write an equation for the reaction between ammonium sulphate and sodium hydroxide. **(1 mark)**

ii 3.00 g of a fertiliser mixture containing ammonium sulphate was made to 250 cm^3 of solution. 25.0 cm^3 portions of this were added to an excess of sodium hydroxide solution, and the ammonia produced passed into 50.0 cm^3 of 0.100 mol dm^{-3} hydrochloric acid solution.
The residual acid was then titrated with 0.100 mol dm^{-3} sodium hydroxide solution, 25.4 cm^3 being required.
Find the percentage by mass of ammonium sulphate in the fertiliser mixture. **(4 marks)**

b Some of the nitrogen content of a fertiliser could be present as nitrate ions, which are not detected by sodium hydroxide solution. Nitrate ions can be reduced to ammonia by the use of aluminium in sodium hydroxide solution.

i Complete the half equation below for the reduction of nitrate ions:

$NO_3^-(aq) + 6H_2O(l) +$ ....$e^- \rightarrow NH_3(g) +$ .... **(2 marks)**

ii The reaction occurs in strongly alkaline solution; suggest a formula for the aluminium-containing species present at the end of the reaction. **(1 mark)**

c In cold aqueous sodium hydroxide ammonium salts produce the following equilibrium:

$$NH_4^+(aq) + OH^-(aq) \rightleftharpoons NH_3(aq) + H_2O(l)$$

i Identify the two acid–base conjugate pairs in this equilibrium. **(2 marks)**

ii Explain the effect of raising the temperature on the equilibrium. **(2 marks)**

d Pure **liquid** ammonia ionises as follows:

$$NH_3(l) + NH_3(l) \rightleftharpoons NH_4^+(am) + NH_2^-(am)$$

where (am) represents solutions in liquid ammonia. The Bronsted–Lowry theory of acid–base behaviour applies to solutions in liquid ammonia.

i Suggest why ammonium salts behave as acids in liquid ammonia. **(1 mark)**

ii The salt sodium amide $Na^+NH_2^-$ reacts with ammonium chloride in liquid ammonia in an acid–base reaction. Write an equation to represent the reaction. **(2 marks)**

(Total 15 marks)

(Edexcel 1999)

3

4 ORGANIC CHEMISTRY 1

Introduction

The original organic chemists believed that the chemicals they were studying could only be created by organic living things. In contrast, modern organic chemists frequently synthesise useful molecules first found in animals or plants! Organic chemistry is now recognised as the study of chemicals based around carbon atoms with their unique ability to form double and triple bonds, long chains, branched molecules and complex rings.

The number of organic compounds has been calculated at around 7 million – an astonishing number to deal with. Fortunately they fall into a number of clearly defined and recognisable families with characteristic bonding and atomic arrangements. By looking at these families and the way they react we can build up a set of mental tools which will enable us to predict the behaviour of almost any organic chemical we may come across. A wide variety of technologies exist for clarifying the structure of organic chemicals, and we shall look in some detail at mass, infra-red and nuclear magnetic resonance spectroscopy. In this first section of organic chemistry we shall be concentrating on some of the simpler organic families.

The **saturated hydrocarbons** contain only hydrogen and carbon joined by simple single bonds. They include the **alkanes** and the **cycloalkanes**. The patterns within these groups are quite regular and clear to see.

The **unsaturated hydrocarbons** also contain only carbon and hydrogen, but in these groups there are double and even triple bonds between the carbon atoms. The **alkenes** have a double bond between two of their carbon atoms, whilst the **alkynes** have a triple bond in place. These unsaturated bonds have a major effect on the chemistry of the compounds, making them much more reactive.

One of the most important properties of the alkenes is their ability to form polymers, long chain molecules which make new and useful materials such as polythene and PVC.

Another important group of unsaturated hydrocarbons is the arenes, ring molecules all related to benzene, founder member of the group. Phenol, the first recognised antiseptic, is an important member of the arenes. Finally in this section we shall look at the organohalogen compounds. These are chemicals which are almost all synthesised by chemists. They have a hydrocarbon skeleton with a halogen functional group and we make use of them in an immense variety of ways.

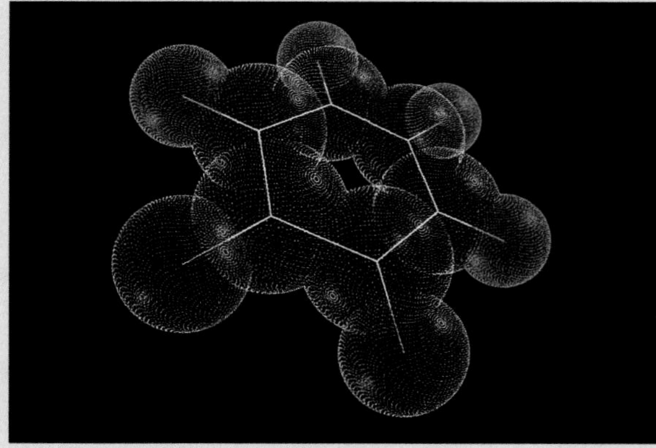

Figure 1 Organic molecules can be very small, containing just one carbon atom and four hydrogens. On the other hand, they can be enormous – but developing an understanding of the way the smaller organic molecules like benzene work enables us to model much large ones.

FOCUS WHAT IS ORGANIC CHEMISTRY?

From the very beginnings of science people have heated substances to see what happened to them. In 1807 Jöns Jacob Berzelius decided that chemicals could all be divided into one of two groups based on their behaviour when heated. Any substances which burnt or charred on heating – and these were mostly from living things – he called **organic chemicals**. Any which melted or vaporised when they were heated but then returned to their original state, Berzelius said were **inorganic**. Although we now recognise this classification as somewhat shaky, it gave chemists of the day something to focus their work on.

At the same time a variety of different inorganic chemicals was being synthesised. However, there was a widespread belief that it was impossible to synthesise organic compounds, that they were formed by animals and plants under the influence of a vital force within the living body. Friedrich Wöhler put an end to this erroneous belief in 1828, when he heated ammonium cyanate (NH_4OCN) and produced urea (NH_2CONH_2). Wöhler showed that the synthesised urea was exactly the same chemically as urea extracted from dog's urine – this was the first synthesis of an organic chemical from an inorganic source.

Figure 1 A classification based on whether a substance melts or burns when it is heated is fine if you test it on materials such as wood and ice, but not surprisingly it has since been replaced by a much more rigorous definition of organic and inorganic chemicals.

Organic synthesis and exploding aprons

Once Wöhler had shown that it was possible to synthesise organic molecules from inorganic ingredients, others followed. The first recorded synthesis of a new organic compound, as opposed to a naturally occurring one, is credited to Christian Schönbein in 1846. Working on his experiments, he spilled a mixture of acids. He grabbed the nearest thing to mop them up, which happened to be his wife's apron. The apron duly exploded and vanished in a puff of smoke! The acids had reacted with the cellulose fibres in the cotton of the apron to produce **nitrocellulose** (known at the time as guncotton), a powerful and unstable explosive chemical which caused the deaths of several other chemists who attempted its synthesis. Eventually a modified and safer form known as **cordite** was produced, and nitrocellulose also formed the basis of the first plastics. The comments of Schönbein's wife have unfortunately not been recorded.

Although the model of the role of the vital force was wrong, the idea that organic chemicals were in some way 'different' was certainly correct and valuable. Organic chemicals are all based around carbon atoms and their particular ability to form double and triple bonds, long chains and rings. The study of carbon chemistry is of major importance both in the science of chemistry and in the wider world today. Organic chemists are involved in the production of new materials, in the development of drugs to help in the battle against disease, in the development of safe flavours, colours and preservatives for use in our foods and in the production of pesticides, weedkillers and fungicides to help ensure that healthy crops are grown and that they are not damaged after harvesting.

Organic chemists produce dyes to colour our clothes and furniture and work on new antiseptics and anaesthetics to make surgery even safer. They also play a major role in analysing the way some organic compounds have polluted our Earth and its atmosphere, and are at the forefront of looking for ways to minimise the damage and prevent it from happening again.

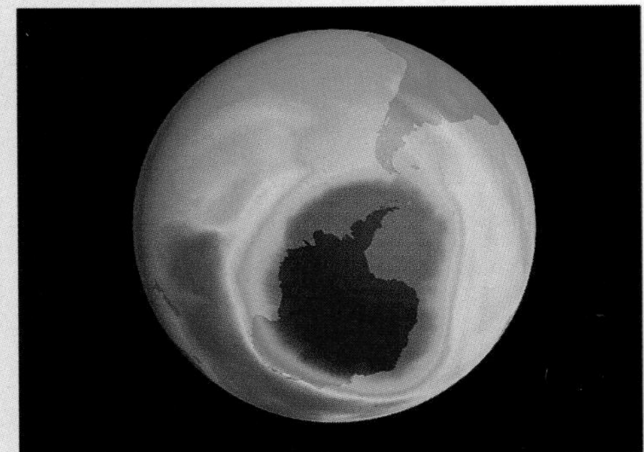

Figure 2 For better or worse: organic chemistry has brought great benefits to the human race, but has also been responsible for much damage. When chlorofluorocarbons (CFCs) were first synthesised and used as propellants in aerosols and as refrigerants, it meant that chemicals could be used more conveniently in spray form and the problems of drug storage and food wastage through decay could be substantially reduced. No-one had any idea that years later the reactions of these same chemicals would cause a hole to develop in the protective ozone layer surrounding the Earth...

4

4.1 Organising the range of organic molecules

Inorganic chemistry looks at the compounds of the 91 naturally occurring elements and their reactions with each other, while organic chemistry simply considers the compounds of carbon. The number of these organic compounds is quite awesome – in the region of seven million – far outstripping the number of compounds of all of the other elements known. Almost all plastics, synthetic and natural fibres such as polyester, wool and cotton, dyes, drugs, pesticides, flavourings and foodstuffs consist largely of organic compounds. The complex structural molecules that make up living cells and the enzymes that control the reactions within them are also organic chemicals, as is crude oil and all the oil-based products we use. It might appear that the enormous scope of carbon chemistry would make it almost impossible to study in any meaningful way – how can we hope to get to grips with these millions of different chemicals? Fortunately the very properties of carbon that make possible such diversity of chemicals also ensure that the compounds that result fall into distinct types or families. We can make sense of the diversity by considering family traits.

Figure 4.1.1 From the alcohol in the glasses to the charcoal burning in the barbecues, from the synthetic and natural fibres in clothing to the cells of our bodies, almost everything in this picture consists of organic chemicals, chains of carbon atoms combined in various ways with hydrogen and other elements.

The singular carbon atom

We have already considered the inorganic chemistry of carbon in section 2.6, and so here we shall concentrate solely on the organic chemistry of carbon. Carbon atoms are unique in their ability to form covalent bonds both with other carbon atoms and with other non-metals. Carbon forms very strong bonds with other carbon atoms, and may form single, double or triple bonds as shown in figure 4.1.2. This ability leads to the formation of chains of carbon atoms which may be thousands of carbon atoms long, and also to complex ring structures. This chain formation is known as **catenation**. Carbon also forms very strong bonds with hydrogen, and almost all organic molecules contain at least one carbon–hydrogen bond.

Bond	Bond dissociation enthalpy/kJ mol^{-1}
C—C	347
C═C	612
C≡C	838
C—H	413
C—O	358
Si—Si	226
Si—H	318

Table 4.1.1 The average bond enthalpies of carbon atoms with other carbon atoms, and with hydrogen and oxygen. These bond enthalpies account for the wide variety of organic compounds. The bond enthalpies of silicon with silicon and silicon with hydrogen are given for contrast – silicon is the most similar element to carbon in the periodic table and yet the bond enthalpies are very different.

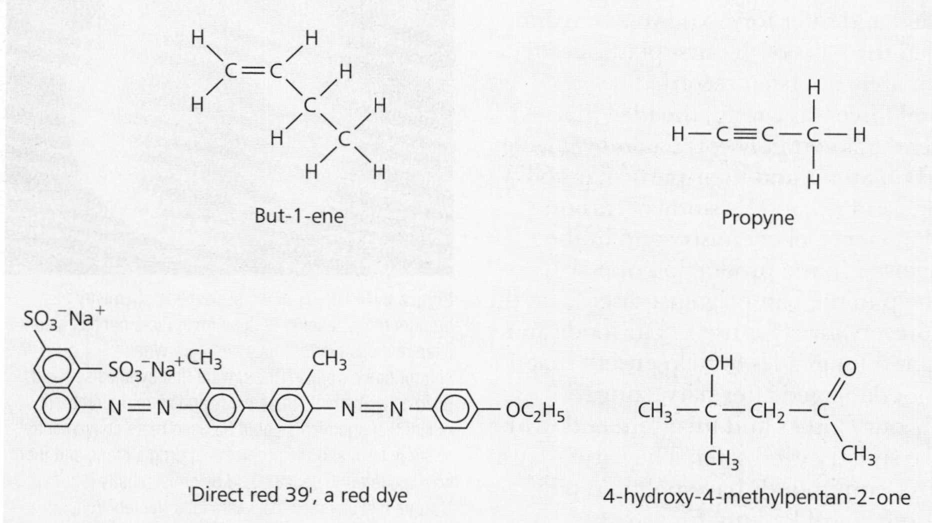

Figure 4.1.2 As these few examples show, organic molecules come in a variety of shapes and sizes.

The electronic structure of carbon in organic compounds

Carbon tends to form four covalent bonds by sharing electrons with other atoms, and the arrangement of the four bonds around a carbon atom is tetrahedral, or very nearly so. The electronic structure of carbon in the ground state is $1s^2 2s^2 2p^2$. The formation of four covalent bonds is at first sight surprising, as only the electrons in the 2p orbital are unpaired in this arrangement, so we might expect carbon to form two bonds rather than four. Why does carbon form four bonds?

The answer to this question lies in a rearrangement of the orbitals of the carbon atom in a process known as **hybridisation**, shown in figure 4.1.3. In one type of hybridisation, the s and p orbitals mix to form four identical new orbitals known as sp^3 **hybrid orbitals**. In carbon, each of these orbitals contains a single electron, which can then form a single bond. Although hybridisation requires energy, this is more than compensated for by the formation of four bonds rather than two, since the outer electrons then form an octet.

Due to electrostatic repulsion between the electron pairs, the sp^3 orbitals are oriented tetrahedrally. Overlap between each of the four sp^3 orbitals with the 1s orbital of a hydrogen atom produces the methane molecule shown in figure 4.1.4(a).

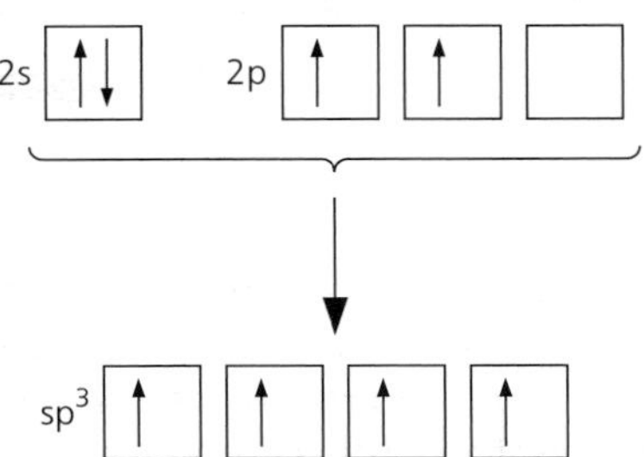

Figure 4.1.3 By combining the s orbital with the three p orbitals, four sp^3 orbitals are produced, each containing a single electron.

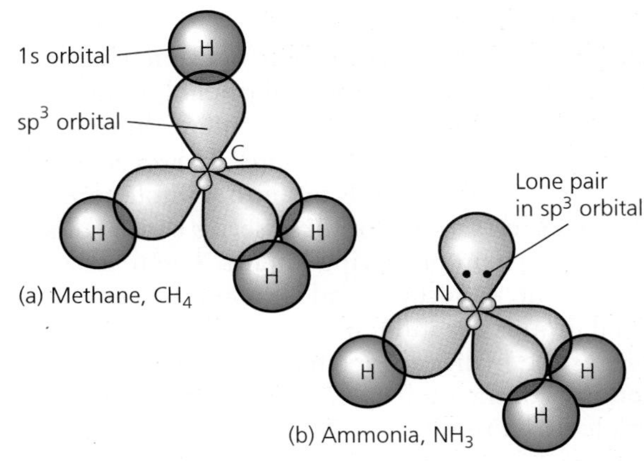

Figure 4.1.4 Hybridisation explains the tetrahedral structure of methane, and also explains the shape of the ammonia molecule and its ability to act as an electron pair donor.

Carbon forms four covalent bonds. This ability, along with the strength of the carbon–carbon bond, allows carbon chains to be built up with many different groups attached, and so produces a great diversity of organic compounds.

Organic families

Classifying organic compounds

Just as the members of the living world may be classified into large divisions such as the plants and animals, and then into smaller divisions of particular species – horses, daffodils, human beings – so organic compounds can be classified in a hierarchy of groups. The mass of organic compounds may first be divided into three main groups on the basis of the arrangement of the carbon chain. **Aliphatic** molecules contain straight or branched chain carbon skeletons, as shown in figure 4.1.5(a) overleaf. **Alicyclic** molecules consist of closed rings of carbon atoms. The rings may contain single or multiple bonds. An example is shown in figure 4.1.5(b). **Arenes** are aromatic hydrocarbons often derived from the organic molecule **benzene**, and many arenes contain a

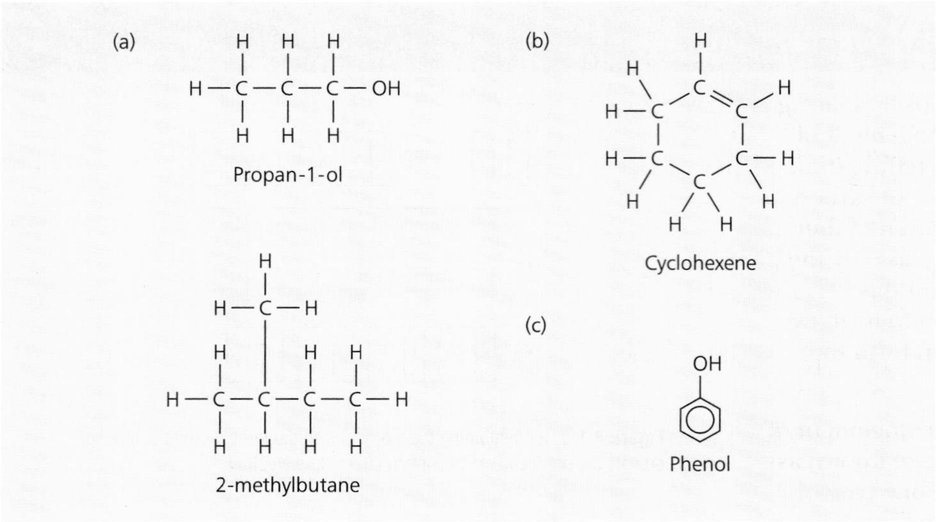

(a) Propan-1-ol

2-methylbutane

(b) Cyclohexene

(c) Phenol

Figure 4.1.5 Aliphatic compounds may have straight or branched chains. Alicyclic compounds contain rings, and aromatic compounds contain benzene rings.

benzene ring in their structure. The benzene ring with its six carbon atoms can be seen in the structural formula of phenol, in figure 4.1.5(c).

These different molecular arrangements have a great effect on how organic compounds react, as we shall see later. However, as well as these three major divisions, organic compounds fall into a whole range of families or **homologous series**. Some of these series consist of a reactive group attached to a hydrocarbon chain of varying length. The reactive group determines the chemistry of the molecules, and is called the **functional group**. However, the functional groups are influenced by their environment to some extent – for example, as the hydrocarbon chain gets bigger it has an increasing effect on the chemistry of the molecule, and the influence of the functional group becomes less as a result.

Some homologous series

The simplest organic molecule is a chain of carbon atoms joined by single bonds, with only hydrogen atoms attached to the chain. The family with this structure is the **alkanes**, and an example of an alkane is propane, shown in figure 4.1.6(a).

Each homologous series has a **general formula** describing the number of carbon atoms and their relationships with the other atoms in the molecule. The general formula of the alkanes is C_nH_{2n+2}.

Another family of organic compounds is similar to the alkanes, except that each member has an -OH group. This family is known as the **alcohols**, and an example is propan-1-ol, figure 4.1.6(b). (The way chemists name organic molecules such as this is discussed on pages 370–3.)

If the $-CH_2OH$ of the alcohols is replaced by the -COOH functional group, we have another organic family – the **carboxylic acids**. An example of this group is propanoic acid, figure 4.1.6(c).

As we have seen, the properties of organic families are determined first and foremost by their functional group, with the shape and size of the carbon chain also affecting the way the compound reacts. The final aspect of the molecular structure of an organic family that has a major impact on the characteristics of the series is the number of double and triple bonds between the carbon atoms in the carbon chain. If there are single bonds between all the carbon atoms in the chain, then the compounds will be relatively stable. If there are double or triple bonds in the carbon chain, then there is greater scope for reactions with other substances

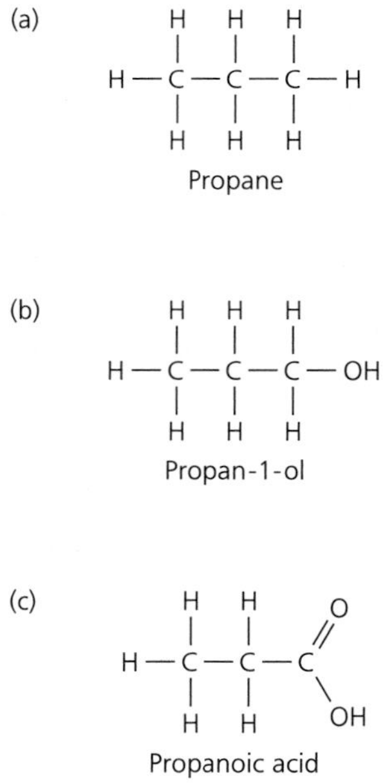

(a) Propane

(b) Propan-1-ol

(c) Propanoic acid

Figure 4.1.6

and the compound will be more reactive. An example is the difference between the **alkanes**, the **alkenes** and the **alkynes**. All the carbon–carbon bonds in the alkanes are single, whilst the alkenes have at least one double carbon–carbon bond and the alkynes have a triple carbon–carbon bond, as shown by figure 4.1.7. The full implications of these differences in bonding will become clear as we consider the chemistry of these groups in greater detail later.

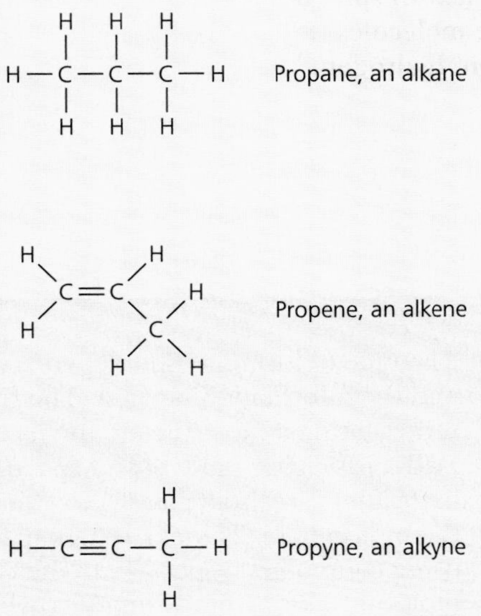

Propane, an alkane

Propene, an alkene

Propyne, an alkyne

Figure 4.1.7

Table 4.1.2 summarises some of the main groups of organic families.

Functional or distinguishing group	Organic family	Example
—C=C—	Alkenes	CH_2CH_2 (ethene)
—C≡C—	Alkynes	CHCH (ethyne)
C—OH	Alcohols	CH_3CH_2OH (ethanol)
$C—NH_2$	Amines	CH_3NH_2 (methylamine)
C—halogen	Halogeno compounds	$CHCl_3$ (trichloromethane)
C—O—C	Ethers	$CH_3CH_2OCH_2CH_3$ (ethoxyethane)
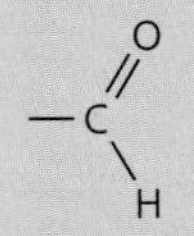 —C(=O)—H	Aldehydes	CH_3CHO (ethanal)
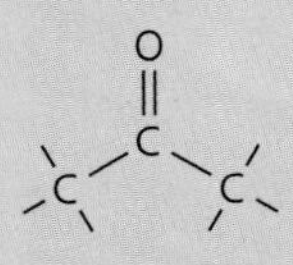 C—C(=O)—C	Ketones	CH_3COCH_3 (propanone)
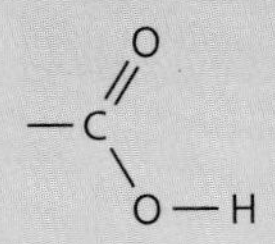 —C(=O)—O—H	Carboxylic acids	CH_3CH_2COOH (propanoic acid)

Table 4.1.2 Some common functional groups in organic families

Describing organic compounds

Unravelling the formulae of organic compounds

In order to understand the chemistry of a compound, we need to know its chemical make-up – the numbers of atoms of different elements present in a molecule, and how they are arranged. This information is available to us if we know the structural formula of a compound (see box below). Organic molecules are not, on the whole, simple – they tend to be large and complex. In spite of this we can obtain their structural formulae quite readily as the molecules usually contain only a small number of different atoms – carbon, hydrogen, oxygen and perhaps one or two others.

Representing organic compounds

In section 1.1 we met the terms **empirical formula** and **molecular formula**, which describe how many atoms of each type are in a molecule. The **structural formula** of an organic molecule is often more useful, as it shows not only the numbers of atoms present but also their arrangement relative to each other, and so shows how they are most likely to react.

- The structural formula can simply show the groups of atoms without much information about their arrangement, for example for propan-1-ol:

$$\mathbf{CH_3CH_2CH_2OH \text{ or } C_3H_7OH}$$

- The **displayed formula** of propan-1-ol shows both the relative placing of the atoms and the number of bonds between them, as in figure 4.1.9(a).
- Finally, the **stereochemical formula** shows the shape of the propan-1-ol molecule in three dimensions, indicating the orientation of the bonds, as in figure 4.1.8(b).

```
(a)      H   H   H
         |   |   |
     H — C — C — C — O — H
         |   |   |
         H   H   H
```

```
(b)        H   H
       H    \ /      O — H
        \    C      /
         C  / \    C
     H ◄ |        ◄ \
         H       H   H
```

▲ indicates a bond sticking out of the plane of the paper

⦙ indicates a bond sticking into the plane of the paper

Figure 4.1.8

The type of formula used depends on the context, and the information needed. The first three ways of representing a molecule are the most commonly used. **Ball-and-spring models** and **space-filling models** of molecules can also be made, to gain some idea of how the molecule really looks. These may be produced using computers (especially in the case of complex molecules), or by modelling kits, which use wooden or plastic spheres that connect together.

A ball-and-spring model represents the molecule in a similar way to the stereochemical formula.

Figure 4.1.9 Space-filling models like this, whether produced by a computer or simply by fitting coloured spheres together, accurately represent the size of the electron cloud in a molecule.

Finding the empirical formula

For many years, determining the structural formula of an organic compound involved burning the substance in oxygen and analysing the products in order to determine the empirical formula. In recent years, new technology has meant that the structural and empirical formulae of compounds are now much more readily found, using a combination of **infra-red spectroscopy**, **nuclear magnetic resonance spectroscopy** and **mass spectrometry**. We shall look at these methods shortly, but first we should examine how the empirical formula of a compound can be found using much simpler means.

Combustion analysis

In combustion analysis, the empirical formula of an unknown organic compound is found by completely burning a known mass of the compound in oxygen. The products of the reaction – which consist of carbon dioxide, water and perhaps other products – are collected and measured. These measurements are then used to calculate the empirical formula, as the example in the box below shows.

An example of combustion analysis

A sample of an organic compound with a mass of 0.816 g was completely burned in oxygen and found to produce 1.559 g of carbon dioxide and 0.955 g of water only. What is its empirical formula?

From these results we can say that the compound certainly contains carbon and hydrogen, and *may* contain oxygen too. We cannot tell whether it contains oxygen without further calculations.

The first step is to calculate the number of moles of carbon dioxide and water formed. From this we can then find the number of moles of carbon and hydrogen, and then the mass of carbon and hydrogen in the compound. (For this calculation we shall take the relative molecular mass of CO_2 as 44.0 g mol^{-1} and the relative molecular mass of H_2O as 18.0 g mol^{-1} – more accurate work would need greater precision in these figures.)

$$\frac{\mathbf{1.559\ g\ CO_2}}{\mathbf{44.0\ g\ mol^{-1}}} = \mathbf{0.0354\ mol\ CO_2}$$

$$\frac{\mathbf{0.955\ g\ H_2O}}{\mathbf{18.0\ g\ mol^{-1}}} = \mathbf{0.0531\ mol\ H_2O}$$

For each mole of carbon dioxide formed, the organic compound must have contained one mole of carbon, while for each mole of water formed, it must have contained two moles of hydrogen. The figures we have just calculated show that the sample contained:

0.0354 mol C

and:

(2 × 0.0531) mol H = 0.1062 mol H

We can now find the masses of carbon and hydrogen in the sample:

$$\mathbf{Mass\ C = 0.0354\ mol\ C \times 12.0\ g\ mol^{-1} = 0.4248\ g\ C}$$

$$\mathbf{Mass\ H = 0.1062\ mol\ H \times 1.0\ g\ mol^{-1} = 0.1062\ g\ H}$$

The total mass of carbon and hydrogen in the sample is:

0.4248 g + 0.1062 g = 0.531 g

The remainder must be oxygen:

0.816 g – 0.531 g = 0.285 g

The number of moles of oxygen in the sample is then:

$$\frac{\mathbf{0.285\ g\ O}}{\mathbf{16\ g\ mol^{-1}}} = \mathbf{0.0178\ mol\ O}$$

The number of moles of carbon, hydrogen and oxygen in the sample looks like this:

0.0354 mol C: 0.1062 mol H: 0.0178 mol O

The first formula we can suggest for this compound is:

$$C_{0.0354}H_{0.1062}O_{0.0178}$$

To get this into whole numbers, we must divide by the smallest number, 0.0178. This gives:

$$C_{\frac{0.0354}{0.0178}}H_{\frac{0.1062}{0.0178}}O_{\frac{0.0178}{0.0178}} = C_{1.989}H_{5.966}O_{1.000}$$

which suggests that the empirical formula is:

$$C_2H_6O$$

4

Finding the molecular formula

Once the empirical formula has been found, the next step is to discover whether this is the molecular formula of the compound or not. For example, the compound in the example may have a molecular formula of C_2H_6O, $C_4H_{12}O_2$, $C_6H_{18}O_3$ or even $C_{60}H_{180}O_{30}$! To decide on the correct molecular formula, we must find the relative molecular mass of the compound. A relative molecular mass of 46 tells us that C_2H_6O is correct ($2 \times 12 + 6 \times 1 + 1 \times 16 = 46$), while if the relative molecular mass is 92, then the molecular formula is $C_4H_{12}O_2$.

How do we find the relative molecular mass? There are a variety of methods. The use of the gas equations to find the relative molecular mass of a gas was explained in section 1.7. Far more common nowadays is the use of the mass spectrometer, details of which are given in section 1.1. The relative mass of the heaviest particle shown on a mass spectrum is usually taken to be that of the intact molecule with a single positive charge – the **molecular ion**. The mass of the molecular ion can be considered as the relative molecular mass, as shown by figure 4.1.10.

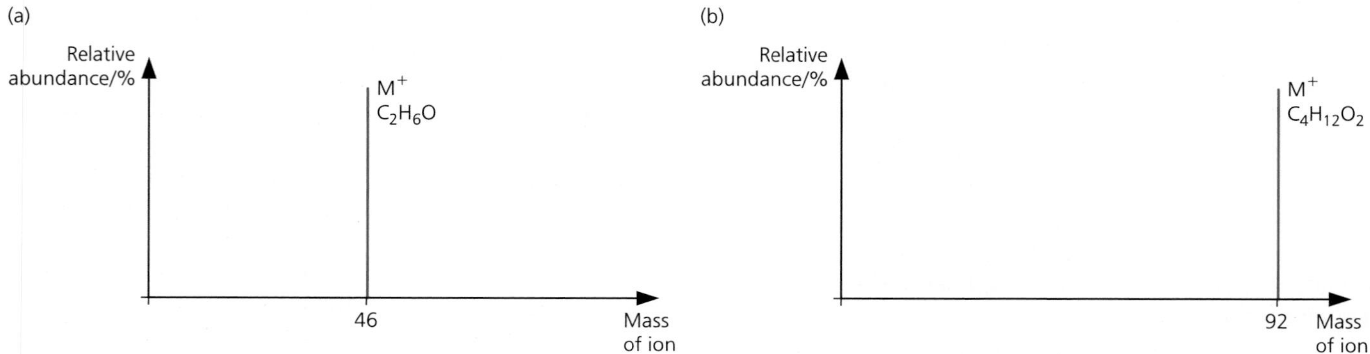

Figure 4.1.10 If we take our compound and run it through a mass spectrometer, the mass spectrum will tell us the relative molecular mass and hence the molecular formula. If we get a spectrum containing the peak shown in (a), then the compound is C_2H_6O. If the spectrum contains the peak shown in (b), then the compound is $C_4H_{12}O_2$.

Reaching the structural formula

The molecular formula of a compound gives us a considerable amount of information about the atoms contained in the molecule, but it does not necessarily tell us how they are arranged and therefore how it is likely to react. For some simple compounds such as methane, CH_4, the molecular formula tells us all we need to know. But for other more complex compounds, the structure is very important, and until we know the structural formula we cannot decide to which organic family the molecule belongs. There are two main ways in which the structural formula of a substance may be determined.

One is to observe and measure physical and chemical properties. The physical properties such as the state of the compound under given conditions, its melting point or boiling point, or a characteristic odour, can all be used to help to place a molecule in a particular homologous series and so determine its structural formula. Similarly, the chemical properties of the different organic families vary quite markedly. By observing the reactions of our unknown substance with, for example, the halogens, a fair idea of the family to which it belongs can be obtained and therefore the appropriate structural formula worked out.

Such chemical methods for finding the structure of complex molecules are extremely time-consuming, and may not always give the structural formula unambiguously. Fortunately, technology comes to the rescue of the modern chemist, with a variety of techniques. Three of the most important of these are described in the box 'Instruments and chemical structures'.

Instruments and chemical structures

The use of instruments in determining the structural formulae of compounds is universal. This branch of chemistry is developing rapidly, and only the very simplest account of the methods can be given here.

Mass spectra

The mass spectrum of a compound, which as we have seen reveals the relative molecular mass, also shows other peaks. These correspond to various fragments of the molecule, which undergoes a series of rearrangements as it passes through the machine after being ionised. By putting together the pieces of the 'jigsaw', the structure of the molecule may be deduced. As an example, figure 4.1.11 shows the main peaks in the mass spectrum of a compound with a molecular formula of C_4H_8O.

The molecular formula suggests that the structural formula may be:

$$\underset{\textbf{butanal}}{\mathbf{CH_3CH_2CH_2CHO}} \quad \textbf{or} \quad \underset{\textbf{butanone}}{\mathbf{CH_3COCH_2CH_3}}$$

The spectrum shows the molecular ion M^+ at 72, with two strong peaks at 43 and 57, coming from the fragments $(CH_3CO)^+$ and $(CH_3CH_2CO)^+$ respectively. Fragments at 15 and 29 correspond to $(CH_3)^+$ and $(C_2H_5)^+$. This pattern suggests that the compound is butanone, since butanal would give a peak at 71, corresponding to $(CH_3CH_2CH_2CO)^+$ – this peak is absent.

Infra-red spectroscopy

The frequency of infra-red radiation is lower than that of visible light (see figure 1.3.2 on page 31), and infra-red radiation may be absorbed by a molecule, causing the bonds in it to vibrate. The frequency of vibration of a particular bond is characteristic of the two atoms it is joining, and is also influenced by the other bonds around it in the molecule. The infra-red spectrum of a compound therefore gives information about the functional groups in a molecule, together with an indication of how these are positioned relative to one another. Figure 4.1.12 shows the infra-red spectrum of phenylethanone, and shows how the various regions of the spectrum correspond to the vibration of different bonds in the molecule.

The position of an absorption band in an infra-red spectrum is identified by its **wavenumber**. This is simply 1/(wavelength of radiation absorbed), and has units cm^{-1}.

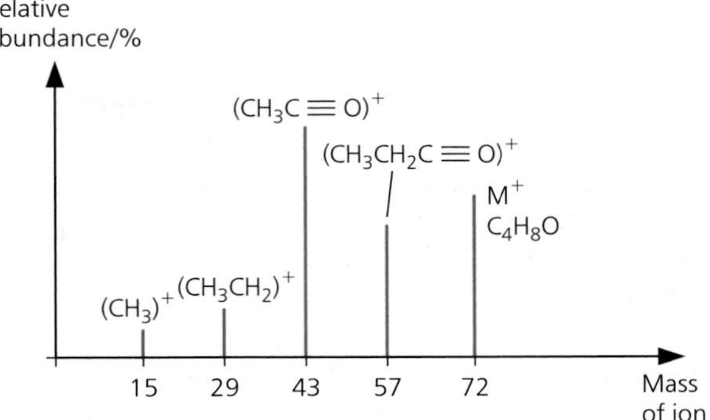

Figure 4.1.11 The mass spectrum of C_4H_8O, showing the main peaks

Nuclear magnetic resonance (NMR) spectroscopy

Like electrons, protons also possess a property called **spin** (see section 1.3, page 41). Because of this, the nuclei of certain atoms, including hydrogen, behave like tiny magnets. When these atoms are placed in a magnetic field, the nuclei align themselves with the magnetic field, just as a bar magnet aligns itself with the Earth's magnetic field. Energy is needed to change this alignment, in the same way as energy is needed to change the alignment of a compass needle. In the case of the compass needle, this energy can be supplied by a push from your finger. The nuclei of atoms can be 'pushed' from one alignment to another by energy supplied by radio waves: see figure 4.1.13, page 369.

The energy difference ΔE between the aligned and non-aligned states of the nucleus depends to some extent on its environment. Organic chemists use NMR spectroscopy particularly to find out about the hydrogen atoms in a molecule. When a hydrogen atom in a molecule absorbs radio waves as it 'flips' between states, the frequency of radiation it absorbs depends to some extent on its environment. This means that the spectrum of frequencies absorbed by a molecule provides a great deal of information about the position of the hydrogen atoms in it.

When a magnetic field is applied to a molecule, its hydrogen atoms do not all experience the same magnetic field. This is because the proton within a hydrogen atom's nucleus is shielded by the electrons surrounding it, modifying the magnetic field it 'feels'. The degree of this modification depends on the electron density around the proton, which is influenced by the position of the hydrogen atom within the molecule and on the other atoms

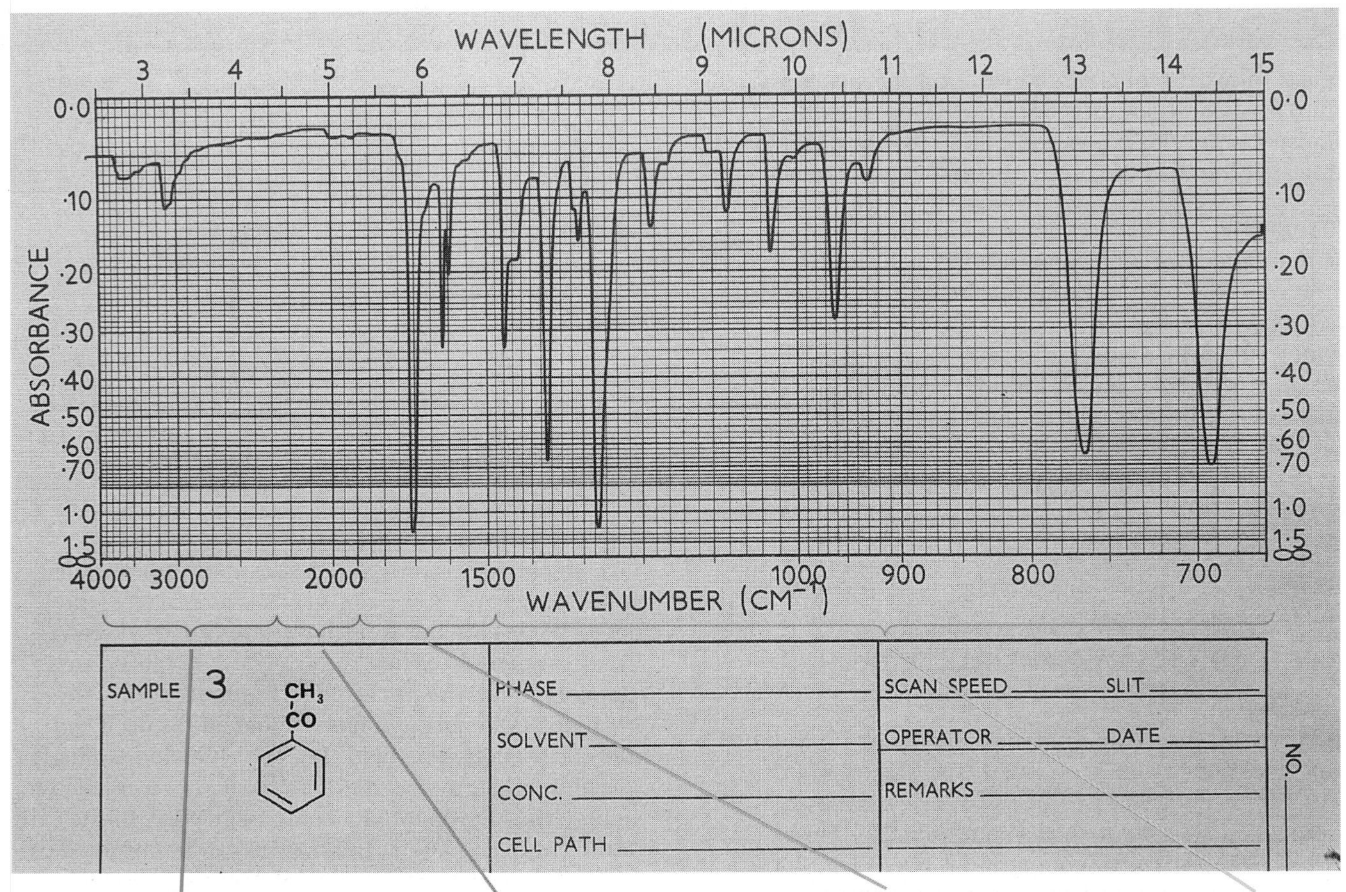

Bands in this region are due to bonds between heavy atoms (e.g. C, N, O) and hydrogen. The energy absorbed is due to the bonds stretching.

Hydrogen bonded O—H
3600–3200 cm^{-1} (usually broad)

Free O—H 3650–3590 cm^{-1}
(usually sharp)

Primary amines (RNH_2)
3500–3300 cm^{-1} (two bands)

Amides ~ 3500 cm^{-1} (medium intensity)

R—C≡C—H ~3300 cm^{-1} (sharp)

R^1R^2C=C—H 3095–3075 cm^{-1}
(medium intensity)

$R^1R^2CH_2$ and RCH_3
2960–2850 cm^{-1} (usually sharp)

Bands in this region are due to triple bonds stretching.

R^1—C≡C—R^2
2260–2150 cm^{-1}
(variable intensity, seen only if $R^1 \neq R^2$)

R—C≡N 2260–2200 cm^{-1}
(variable intensity)

Bands in this region are due to C=C, C=O, C=N and N=O stretching, and N—H bending. This region is particularly useful for identifying carbonyl compounds, which have C=O stretching frequencies that are very characteristic of the environment of the C=O bond. Typical wavenumbers for C=O stretching in carbonyl compounds are given below. Carboxylic acids generally absorb more strongly than esters, which in turn absorb more strongly than aldehydes and ketones. Amide absorption is very variable.

Acid anhydrides
(R^1—CO—O—CO—R^2)
two bands, 1850–1800 cm^{-1}

Acid chlorides (saturated)
1815–1790 cm^{-1}

Saturated esters
1750–1735 cm^{-1}

Saturated aldehydes
1740–1720 cm^{-1}

Saturated ketones
1725–1705 cm^{-1}

Saturated carboxylic acids
1725–1700 cm^{-1}

Amides ~ 1690 cm^{-1}

Aromatic rings – three bands at 1600, 1580 and 1500 cm^{-1}

The presence of C=C bonds next to the carbonyl group lowers these values by about 20 cm^{-1}

Below about 1500 cm^{-1} lies a region which consists of a variety of bands due to stretching and bending. Individual compounds show highly characteristic absorption in this region, which may be used to positively identify a sample. For obvious reasons this is known as the 'fingerprint region'.

Figure 4.1.12 The infra-red spectrum of phenylethanone

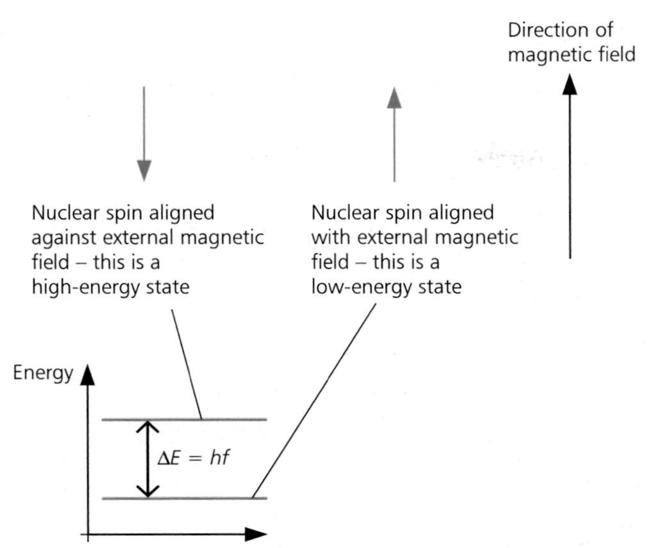

Figure 4.1.13 When a nuclear 'magnet' aligns against the direction of the applied field, it is less stable than when the two are aligned in the same direction. The energy difference ΔE between these two states corresponds to a particular radio frequency f, according to the relationship $\Delta E = hf$, where h is a constant.

surrounding it. The frequency at which magnetic resonance occurs is therefore an indication of the chemical environment of a proton within a molecule.

NMR spectra are usually obtained using samples dissolved in solvents such as tetrachloromethane, CCl_4, or $CDCl_3$, which is trichloromethane with a deuterium (^2H) atom rather than a hydrogen atom. These solvents do not show nuclear resonance – their atoms do not behave like tiny magnets.

The magnetic environment of a proton is measured as its **chemical shift** on a scale that is sometimes referred to as the δ scale. For reasons concerned with how this shift is measured, the units of δ are p.p.m. – parts per million. A small amount of tetramethylsilane (TMS, $Si(CH_3)_4$) is added to the solution to act as a reference substance for this scale. All the protons in this molecule are shielded to exactly the same extent, so it gives a single strong peak. These twelve identical protons of TMS are arbitrarily assigned a value of δ equal to zero. TMS is chosen because:

(1) It is a volatile liquid which may be added in trace amounts to the solution and then evaporated off if the pure sample is to be recovered.

(2) Molecules of TMS do not readily interact with the sample molecules. Such interactions would be undesirable, since they would modify the magnetic environment of the TMS and sample protons and so affect their chemical shifts.

(3) Protons in the majority of organic compounds show nuclear resonance at values of δ that are positive with respect to the protons of TMS, making the comparison of chemical shifts easy.

Figure 4.1.14 shows an NMR spectrum of ethanol.

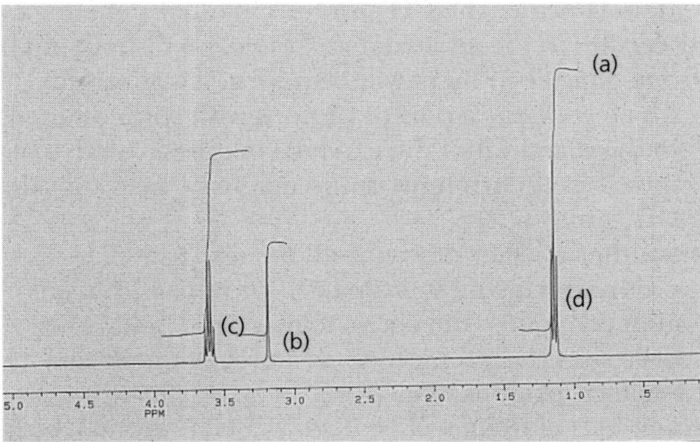

Figure 4.1.14 An NMR spectrum of ethanol, CH_3CH_2OH. Each set of peaks corresponds to one group of hydrogen atoms in the molecule.

The chemical shifts of the ethanol protons are shown relative to TMS, which is assigned a chemical shift of zero (δ = 0 p.p.m.). The spectrum of ethanol shows a number of features:

(a) The solid line above each set of peaks is produced by the spectrometer, which has calculated the area below the trace (this is sometimes called the *integrated* spectrum). This corresponds to the number of hydrogen atoms represented by each set of peaks. These lines show which set of peaks corresponds to which hydrogen atoms.

(b) The single peak at δ = 3.2 p.p.m. is due to the resonance of the proton of the single hydrogen atom of the ethanol OH group.

(c) The set of peaks centred at δ = 3.6 is due to the resonance of the two protons in the hydrogen atoms of the ethanol CH_2 group.

(d) The set of peaks centred at δ = 1.2 is due to the resonance of the three protons of the ethanol CH_3 group.

(e) Careful examination of the set of peaks at (c) shows that it consists of four peaks, with intensities in the ratio 1:3:3:1, while the set at (d) consists of three peaks, with intensities in the ratio 1:2:1. These patterns come about as a result of **spin–spin coupling**, in which the protons on neighbouring carbon atoms interact.

4

↑↑	↑↓	↓↓
	↓↑	

Table 4.1.3 The ways in which the spins of two protons can align. An arrow ↑ indicates a spin aligned in the direction of the external field, while an arrow ↓ indicates a spin opposing the external field.

If we consider the CH_3 protons in all of the ethanol molecules in the spectrometer, table 4.1.3 shows that:

- one quarter of them will be next to a CH_2 group which contains a pair of protons with spins aligned in the direction of the external magnetic field. This increases the strength of the magnetic field that the CH_3 protons 'see'.
- another quarter of them will be next to a CH_2 group which contains a pair of protons with spins aligned against the external magnetic field. This decreases the strength of the magnetic field that the CH_3 protons 'see'.
- one half of them will be next to a CH_2 group which contains a pair of protons with spins in opposite direction. This does not affect the strength of the magnetic field that the CH_3 protons 'see'.

In this way, the signal due to the CH_3 protons is split into three peaks (a 'triplet') with intensities in the ratio 1:2:1.

The explanation of why the signal for the CH_2 protons is split into four peaks with the ratio 1:3:3:1 is just the same, based on the possible combinations of spins between the three CH_3 protons (table 4.1.4).

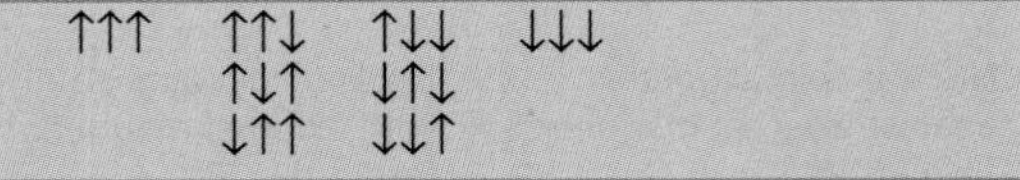

↑↑↑	↑↑↓	↑↓↓	↓↓↓
	↑↓↑	↓↑↓	
	↓↑↑	↓↓↑	

Table 4.1.4 The ways in which the spins of three protons can align. An arrow ↑ indicates a spin aligned in the direction of the external field, while an arrow ↓ indicates a spin opposing the external field.

Use of these and other methods of structural determination enables chemists to quickly determine chemical structures which would otherwise take a very long time indeed. For example, figure 4.1.15 shows the structure of γ-linolenic acid, an essential fatty acid present in oil of evening primrose. The oil has been used for hundreds of years for various conditions including eczema, and more recent research has revealed its active ingredient. Oil of evening primrose is now also being used to treat premenstrual tension.

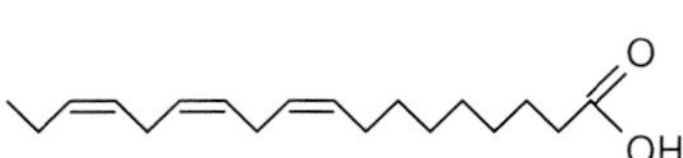

Figure 4.1.15 γ-linolenic acid, the molecule responsible for the action of oil of evening primrose (right). Without modern instruments, finding the structure of a molecule like this would be much more difficult.

Naming organic molecules

Naming inorganic molecules is not usually a problem, as the numbers of atoms in the molecules are small, and there are no equivalents of the homologous series found in organic chemistry. However, a considerable amount of information needs to be given in the name of an organic compound. This includes the organic family to which it belongs, the number of carbon atoms in the molecule, whether the chain is straight, branched or in a ring, along with information about any additional types of atom present in the molecule. To make sense of all the organic compounds and to include as much information as possible, a rigorous system of naming is applied, according to rules drawn up by IUPAC, in exactly the same way as for inorganic compounds. Some of the more complex organic compounds can end up with names which are longer and more complex than the formula of the compound! Often these very large molecules, such as carbohydrates, amino acids and fats, have standardised common names which are easier to remember. It is well worth taking time to understand how systematic names are built up, because the name of an organic compound will give you a considerable amount of information about the structure of the molecule and therefore how it is likely to react.

The prefix – the number of carbons in the chain

The first part of the name of an aliphatic compound refers to the number of carbon atoms in the carbon chain or, if the molecule is branched, in the longest carbon chain. This is shown in table 4.1.5.

Prefix	Number of carbon atoms in the main carbon chain
Meth-	1
Eth-	2
Prop-	3
But-	4
Pent-	5
Hex-	6
Hept-	7
Oct-	8
Dec-	10
Dodec-	12
Eicos-	20

Table 4.1.5 Naming carbon skeletons

The suffix – the functional group

The second part of the name refers to the functional group or the organic family to which the molecule belongs. Thus prop*ane* is an **alkane** with three carbon atoms in the chain, and prop*yne* is an **alkyne** which also has three carbon atoms. Propan*ol* is a three-carbon **alcohol**, and propan*al* is a three-carbon **aldehyde**. Equally eth*ene* has two carbon atoms, whilst pent*ene* has five carbons and dodec*ene* has twelve carbons, all members of the **alkene** family.

Numbering the chain

As we have seen, the alkanes are a group of saturated hydrocarbons (they contain only carbon and hydrogen, and only single bonds). Other organic families contain either double or triple bonds, or functional groups. Numbers are used to indicate whereabouts in the carbon chain the double or triple bond is situated, or where the functional group goes. For example, the parent alkene chain is counted as the chain in which the double bond occurs, and it is numbered from whichever end gives the lower number to the first carbon of the double bond. In other words, it is conventional to use the name involving the lowest numbers. These principles are followed for other unsaturated molecules as shown in figure 4.1.16.

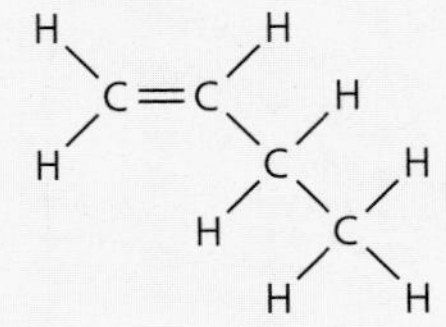

But-1-ene (*not* but-3-ene)

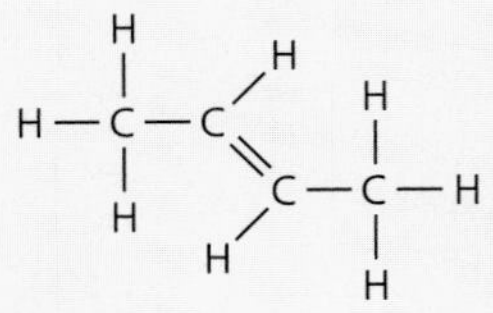

But-2-ene

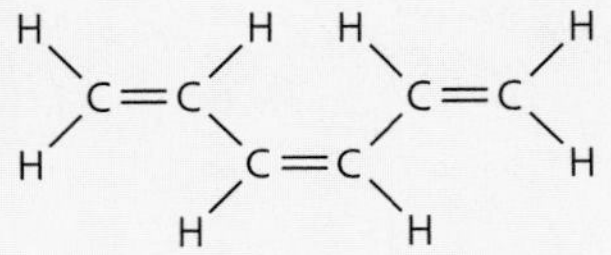

Hexa-1,3,5-triene

Figure 4.1.16

Side chains

Many organic molecules are not straight chains but have a variety of branches. These branches are often **alkyl groups**. An alkyl group is an alkane molecule that has lost a hydrogen to enable it to join to another carbon chain, so it has the general formula C_nH_{2n+1}. Typical alkyl groups are:

Methyl	CH_3—
Ethyl	CH_3CH_2—
Propyl	$CH_3CH_2CH_2$—
etc.	

The position of a side chain is indicated by a number to show to which carbon atom in the parent chain it is attached. Again, this numbering is taken from the end which gives the lower of two numbers, as shown in figure 4.1.17.

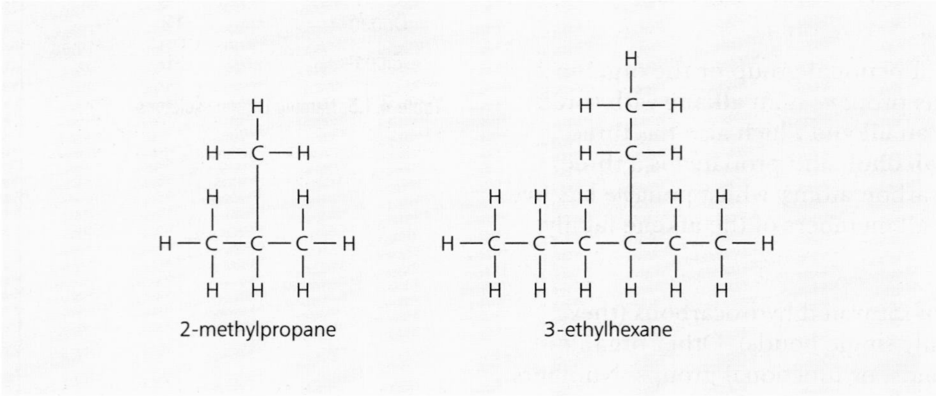

2-methylpropane

3-ethylhexane

Figure 4.1.17

If more than one side chain is attached, then the name of the compound includes the side groups in alphabetical order, regardless of the number of the carbon atom on which they are found. Thus the order for referring to the smaller alkyl groups is:

Butyl, ethyl, hexyl, methyl, pentyl, propyl

If two side chains are attached to the same carbon atom then the number of the carbon atom is repeated in the name, as shown in figure 4.1.18.

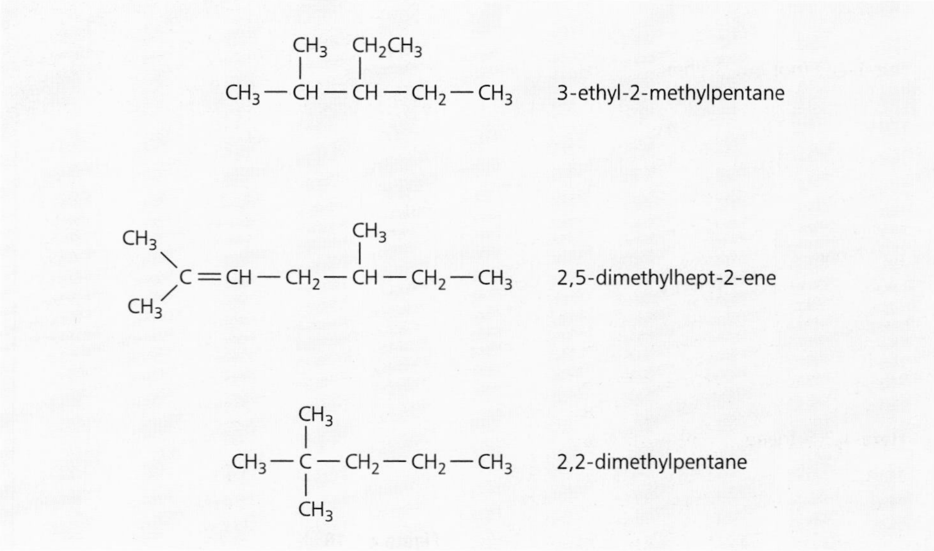

$CH_3-CH(CH_3)-CH(CH_2CH_3)-CH_2-CH_3$ 3-ethyl-2-methylpentane

$(CH_3)_2C=CH-CH_2-CH(CH_3)-CH_2-CH_3$ 2,5-dimethylhept-2-ene

$CH_3-C(CH_3)_2-CH_2-CH_2-CH_3$ 2,2-dimethylpentane

Figure 4.1.18

This systematic naming of organic compounds allows us not only to identify the carbon chain and functional group accurately, but also to know what side chains are present and where they are found within the molecule. In most cases throughout sections 4 and 5, IUPAC names will be used, although there will be times, particularly when large molecules are considered, when common names will be used instead.

The shapes of molecules

Isomerism in organic molecules

The vast majority of organic compounds have two or more isomers. As we saw in section 2.7, isomerism occurs when two or more compounds have the same molecular formula. Our dealings with organic compounds are complicated by the fact that all organic molecules containing four or more carbon atoms show isomerism, and many smaller molecules do too. There are several types of isomerism, and organic chemicals exhibit all of the different types.

Formula	Number of isomers
C_8H_{18}	18
$C_{10}H_{22}$	75
$C_{20}H_{42}$	366 319
$C_{40}H_{82}$	6.25×10^{13} (estimated!)

Table 4.1.6 The number of possible isomers of some of the larger organic compounds is vast, and goes some way to explaining the enormous number of organic chemicals in existence.

Structural isomerism

Structural isomerism is probably the simplest form of isomerism to understand. For example, the molecular formula C_4H_{10} gives two isomers, butane and 2-methylpropane. These isomers remain part of the same organic family, although their boiling points differ considerably, as can be seen in figure 4.1.19.

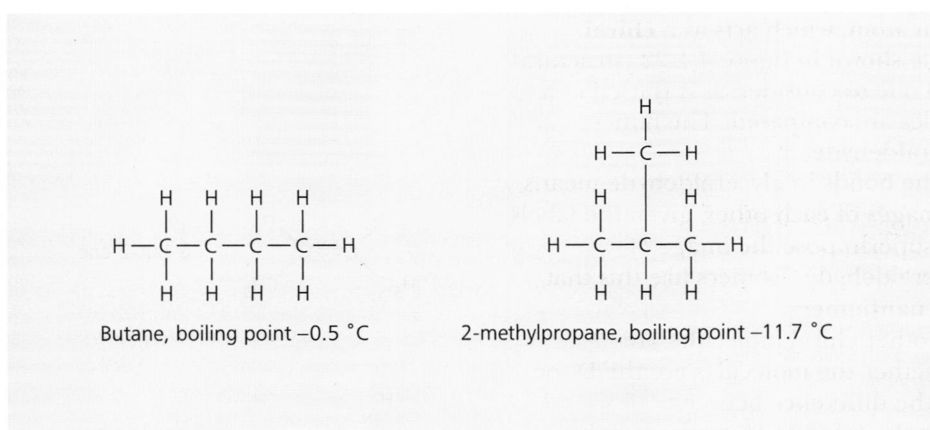

Figure 4.1.19 The two isomers of C_4H_{10} are members of the same organic family, but have different physical properties. The difference in boiling point is particularly striking. The different shapes of the molecules mean that they fit together in different ways, which affects the strengths of the intermolecular forces.

Another example of structural isomerism is seen in figure 4.1.20. In this case, the two structural isomers of C_2H_6O are members of different organic families – ethanol is an alcohol and methoxymethane is an ether.

Stereoisomerism

Another form of isomerism is **stereoisomerism** (isomerism in space), mentioned in section 2.7. It arises when the three-dimensional arrangement of the bonds in a molecule (or the ligands in a complex ion) allow different possible orientations in space. This results in two different forms, and however hard you try you cannot superimpose the image of one onto the other. Two such forms of a compound are called **stereoisomers**.

Geometric isomerism is one type of stereoisomerism. It arises due to the lack of free rotation around a bond, frequently a double bond. As we saw in section 2.7, geometric isomers occur when components of the molecule are arranged on different sides of the molecule. The difference between *cis*- and

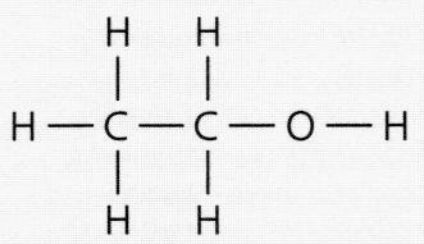

$$\begin{array}{c} H\ \ H \\ |\ \ \ | \\ H-C-C-O-H \\ |\ \ \ | \\ H\ \ H \end{array}$$ Ethanol, boiling point 78.5 °C

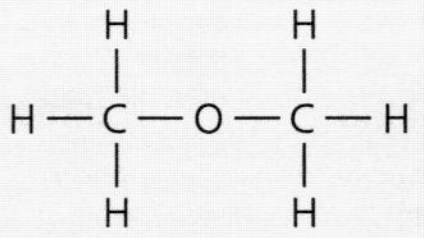

$$\begin{array}{c} H\ \ \ \ \ \ \ \ H \\ |\ \ \ \ \ \ \ \ \ | \\ H-C-O-C-H \\ |\ \ \ \ \ \ \ \ \ | \\ H\ \ \ \ \ \ \ \ H \end{array}$$ Methoxymethane, boiling point −23 °C

Ethanol

Methoxymethane

Figure 4.1.20 The two isomers of C_2H_6O are members of different organic families, and have very different physical and chemical properties.

trans-isomers was shown on page 215 with the example of cisplatin. Another example is but-2-ene – the two possible structures are shown in figure 4.1.21 overleaf. As the figure shows, in *trans*-but-2-ene the two methyl groups are on opposite sides of the double bond. In the *cis*-isomer both methyl groups are on one side of the bond, and both hydrogen atoms are on the other. (*Cis* means 'on the same side' and *trans* means 'on opposite sides'.) Geometric isomers frequently exhibit different physical properties, and this type of isomerism can influence their chemical reactions too.

Optical isomerism is another type of stereoisomerism. If two objects are the mirror image of each other and they are not superimposable, then they are said to be **chiral**. Chirality is an exclusive property of asymmetrical objects, and organic molecules are no exception. The simplest type of chiral molecule has four different groups attached to one carbon atom, which acts as a **chiral centre**. The 2,3-dihydroxypropanal molecule shown in figure 4.1.22 (structural formula $CH_2(OH)CH(OH)CHO$) is a useful one to consider, as it is used as a standard with which all other chiral molecules are compared. The non-systematic name for this compound is glyceraldehyde.

The three-dimensional arrangement of the bonds in glyceraldehyde means that there are two different forms, mirror images of each other, given the labels D and L. However hard you try, you cannot superimpose the image of D-glyceraldehyde onto the image of L-glyceraldehyde. Isomers like this that are mirror images of each other are called **enantiomers**.

The positions of the hydroxyl groups on other chiral molecules are compared with glyceraldehyde to decide whether the molecules are the D- or L- form. Biological systems are sensitive to the difference between stereoisomers. For example, it is largely only the D-forms of sugars and the L-forms of amino acids that are found in living things.

Enantiomers are remarkably similar in both their physical and their chemical properties. Unlike structural or geometric isomers, they cannot be told apart by discrepancies in their physical properties or chemical reactivities. However, they do have one difference that allows us to distinguish them – they have an effect on plane-polarised light. Compounds such as these are said to

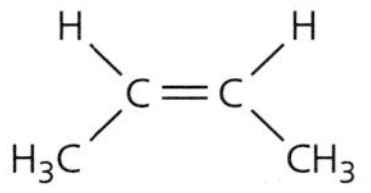

H, H, C=C, H_3C, CH_3 — *Cis*-but-2-ene

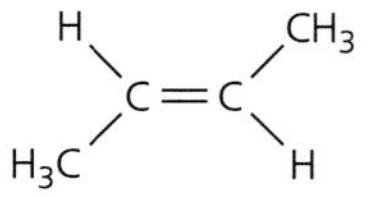

H, CH_3, C=C, H_3C, H — *Trans*-but-2-ene

Figure 4.1.21 *Cis*- and *trans*-isomers of the same molecule often have different properties – for example, the boiling point of *cis*-but-2-ene is 3.72 °C, while the boiling point of *trans*-but-2-ene is 0.88 °C.

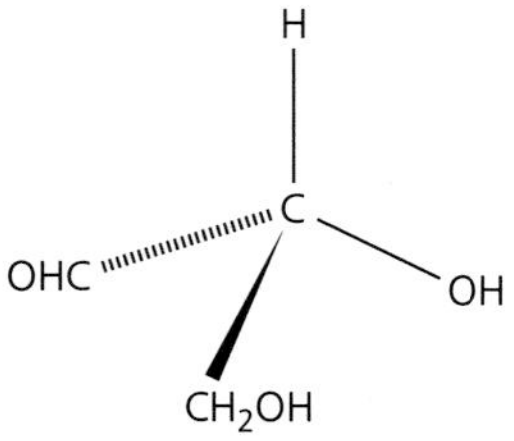

H, C, OHC, OH, CH_2OH — D-glyceraldehyde

H, C, HO, CHO, $HOCH_2$ — L-glyceraldehyde

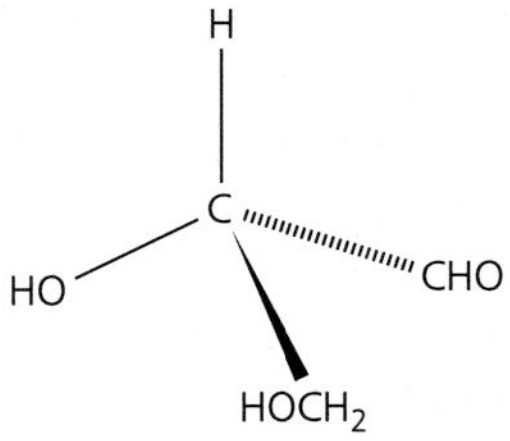

Figure 4.1.22 The stereoisomers of glyceraldehyde (2,3-dihydroxypropanal). The two forms cannot be superimposed – they are mirror images of each other.

be **optically active**, and their isomers are known as **optical isomers**. One enantiomer will rotate the polarised light to the right, indicated by (+), and the other will rotate it to the left (–). However, knowing whether the molecule is the D- or L-form does not tell you whether it is (+) or (–).

The complex nature of organic compounds – their tendency to form very large molecules and their ability to occur as a wide variety of different types of isomers – leads to the enormous variety of organic chemicals currently known to exist. In the rest of the book we shall look in more detail at just some of the wide range of organic families.

What is plane-polarised light?

The wave model of light envisages light as a form of electromagnetic radiation travelling in a **transverse wave**. This means that the electric and magnetic fields oscillate at right angles to the direction in which the wave travels.

The oscillations can occur in any plane perpendicular to the direction of travel. Figure 4.1.23 shows some of these planes. A wave like this in which the oscillations take place in a number of planes is called **unpolarised**, and represents the situation in 'normal' light. A wave in which the oscillations occur in one plane only is said to be **plane polarised** in that direction. Unpolarised light may be polarised by passing it through a material that only allows through oscillations in one plane.

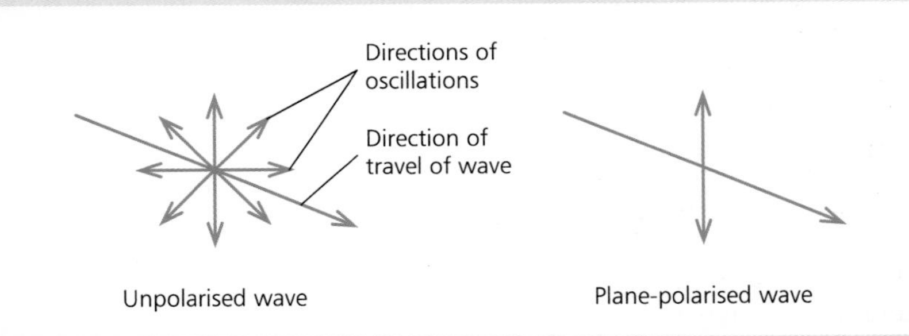

Figure 4.1.23 In an unpolarised wave, oscillations may occur in any plane, while in a plane-polarised wave they occur in only one plane.

The optical activity of an enantiomer is measured by seeing how plane-polarised light is rotated as it passes through a solution containing the enantiomer.

Chirality

Chirality is a very subtle difference between molecules which can be described in a less technical way as 'handedness'. A simple analogy is your own hands. Whilst similar in appearance, your left hand and your right hand are not the same. A left-handed glove simply does not fit the right hand. However, if you hold your left hand up to the mirror it looks exactly like your right hand. We describe them as being **mirror images**

4

of each other. The vital difference between your two hands is that the image of one cannot be superimposed on the other. It is only when two objects pass this test of superimposability that they can truly be described as identical.

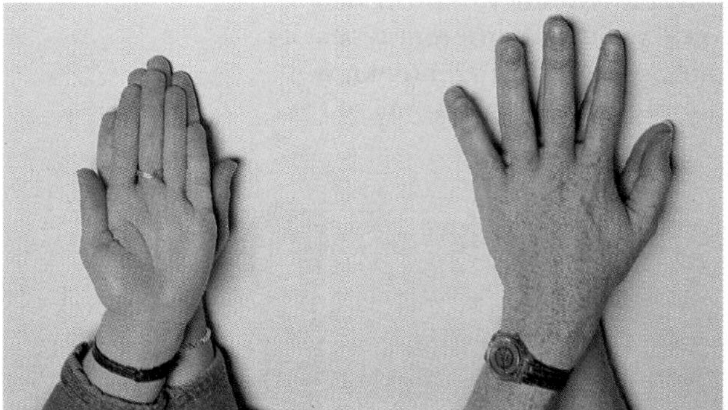

Figure 4.1.24 It is impossible to superimpose the image of one hand onto the other so that they match exactly. With both palms facing you, the mismatch is obvious. With palms together you can see that if the images were merged, the digits would match, but the palm of one hand would be where the back of the other is and vice versa.

SUMMARY

- Organic chemistry is largely the chemistry of carbon compounds. The vast range of organic molecules is the result of the ability of carbon atoms to form single, double and triple bonds as well as long chains and rings.
- The ability of carbon atoms to form long chains held together by strong carbon–carbon bonds is known as **catenation**.
- Carbon has the electronic structure $1s^2 2s^2 2p^2$ and forms four covalent bonds to obtain a stable electronic configuration.
- Organic compounds are made up of carbon skeletons, often with functional groups attached. Compounds with the same functional group are arranged in organic families known as **homologous series**.
- The structure of a compound may be determined by the use of **infra-red spectroscopy**, **nuclear magnetic resonance spectroscopy** and **mass spectrometry**.
- The **structural formula** of an organic compound shows the groups of atoms present, without much information about their arrangement.
- The **displayed formula** shows both the relative placing of the atoms and the number of bonds between them.
- The **stereochemical formula** shows the shape of a molecule in three dimensions.
- Organic compounds are named in a systematic way based on the number of carbon atoms in the carbon chain, and the functional groups and their positions in the molecule.
- **Structural isomers** are compounds with the same molecular formula but with different physical properties, as the atoms are arranged differently.

- **Stereoisomerism** is isomerism that results from the three-dimensional arrangement of the bonds in space. **Geometric isomerism** arises due to lack of rotation round a double bond and gives rise to *cis* and *trans* forms. In **optical isomerism**, molecules exist which are mirror images of each other. They are **optically active** – they have opposing effects on plane-polarised light.

QUESTIONS

1 a With reference to the electronic structure of the carbon atom, explain why such a wide range of organic compounds exists.

b Give the empirical, molecular, structural and displayed formulae of:

i butan-2-ol

ii butanoic acid.

2 Strychnine (a deadly poison) contains only carbon, hydrogen, nitrogen and oxygen. By mass, strychnine contains 75.42% carbon, 6.63% hydrogen and 9.57% oxygen.

a What is the empirical formula of strychnine?

b What is the molecular formula of strychnine, given that its relative molecular mass is 334?

3 a What is a **functional group**?

b Give the structural formulae of the following compounds:

i but-1-ene

ii ethyne

iii 2,4-dimethylhexane

iv 3-methylbutanone

v ethylbenzene.

4 a What is an isomer?

b Why are isomeric forms so common in organic chemistry?

c Decide whether the members of the pairs in figure 4.1.25 are identical compounds, are isomers or are chemically unrelated.

5 a Give all the possible structural formulae for C_7H_{16}.

b Combustion analysis has shown that a compound containing only carbon, hydrogen and oxygen has an empirical formula of CH_2O. How would a chemist set about determining the molecular and structural formulae of the compound?

Developing Key Skills

Make a poster aimed at GCSE Science/Chemistry students to show what organic chemistry is all about and how it differs from other areas of chemistry.

[Key Skills opportunities: C]

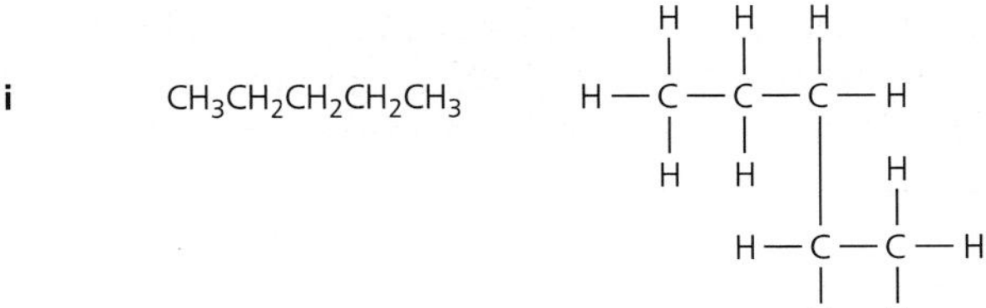

Figure 4.1.25

4.2 The saturated hydrocarbons

The hydrocarbons

The **hydrocarbons** are molecules that contain only the elements carbon and hydrogen. This group of organic compounds contains several families. They may be divided into the **aliphatic** and **alicyclic hydrocarbons**, and the **aromatic hydrocarbons** (molecules containing benzene rings), and in this section we shall look at the aliphatic and alicyclic hydrocarbons in detail. However, all the hydrocarbons have certain properties in common.

All hydrocarbons are insoluble in water. They all burn, and if there is sufficient oxygen available all hydrocarbons give carbon dioxide and water as the sole products of their combustion. Virtually all of our usable supplies of hydrocarbons come from fossil fuels – coal, petroleum and natural gas – which yield pure hydrocarbons after varying degrees of processing. The aliphatic hydrocarbons that result can be classified as belonging to one of three families – the **alkanes**, the **alkenes** and the **alkynes**. They have a very wide variety of uses, from fuels to the raw materials for an enormous number of industrial processes.

The alkanes

General properties

The alkanes are a family of **saturated** hydrocarbons. This means that they have no double or triple bonds between the carbon atoms – they contain the maximum amount of hydrogen possible. The general formula for the family is C_nH_{2n+2}, and the alkanes can occur as both straight- and branched-chain molecules. Straight-chain alkanes form a classic homologous series. As table 4.2.1 shows, their boiling points generally increase as the number of carbon atoms increases, largely due to increasing van der Waals forces as the molecular size increases.

The **viscosity** of the alkanes also increases as the molecules get larger. This means that the liquids become 'thicker' and flow less easily – they become more resistant to pouring. This is due to the increase in the length of the carbon chains – long chains are much more likely to get tangled up than short chains of 3, 4 or 5 carbon atoms. Alkanes are less dense than water, and therefore liquid alkanes float. This leads to difficulties when dealing with fuel fires – most common liquid fuels are alkanes. If water is used to douse the flames, it will simply spread the fire as the burning hydrocarbon floats and is carried along with the spreading water.

This very regular pattern in the physical properties of the alkanes makes it easy to predict the behaviour of unknown members of the series. The situation is not always so straightforward, however. The physical properties of the branched chain alkanes such as 3-methylhexane, 4-ethyl-2-methylheptane or 2,2-dimethylpentane are extremely difficult to predict. Variations in both the parent chain and the side chains affect the physical properties, and no real trend or order can be discerned. In very general terms, branching of the molecule increases the volatility (lowers the boiling point), and also lowers the density of the compound, due to side chains affecting the interactions between molecules.

Name	Molecular formula	Displayed/structural formula	Melting point/°C	Boiling point/°C	Density/ g cm^{-3}
Methane	CH_4	H–C(H)(H)–H	–182	–164	0.424
Ethane	C_2H_6	H–C(H)(H)–C(H)(H)–H	–183	–88	0.546
Propane	C_3H_8	H–C(H)(H)–C(H)(H)–C(H)(H)–H	–190	–42	0.501
Butane	C_4H_{10}	H–C(H)(H)–C(H)(H)–C(H)(H)–C(H)(H)–H	–138	–0.5	0.579
Pentane	C_5H_{12}	$CH_3CH_2CH_2CH_2CH_3$	–130	36	0.626
Hexane	C_6H_{14}	$CH_3CH_2CH_2CH_2CH_2CH_3$	–95	69	0.660
Heptane	C_7H_{16}	$CH_3CH_2CH_2CH_2CH_2CH_2CH_3$	–90	98	0.684
Octane	C_8H_{18}	$CH_3CH_2CH_2CH_2CH_2CH_2CH_2CH_3$	–57	126	0.703
Nonane	C_9H_{20}	$CH_3CH_2CH_2CH_2CH_2CH_2CH_2CH_2CH_3$	–51	151	0.718
Decane	$C_{10}H_{22}$	$CH_3CH_2CH_2CH_2CH_2CH_2CH_2CH_2CH_2CH_3$	–30	174	0.730

Table 4.2.1 The naming of the alkanes conforms to the IUPAC system described in section 4.1. Clear trends can be seen in the physical properties of the members of this homologous series, which become even more apparent when displayed on a graph as in figure 4.2.1.

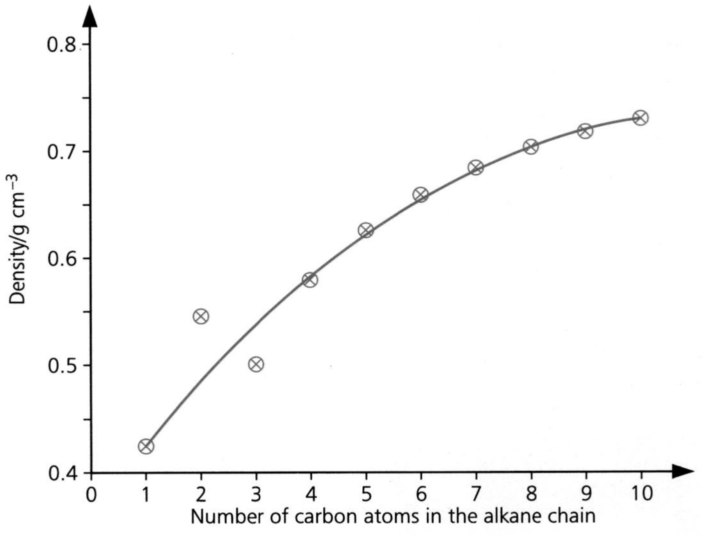

Figure 4.2.1 The density of the straight-chain alkanes plotted against the number of carbon atoms in the molecule

4

The formation of a fossil fuel

Millions of years ago, a far greater proportion of the surface of the Earth was covered in water than is the case today. Untold numbers of minute animal and plant life forms lived in these seas. As they died and sank to the bottom, they formed deep layers of rich, decomposing material, which became encapsulated in rock. With the passage of time, this trapped decaying matter has formed what we know today as **crude oil** or **petroleum** ('rock oil'). It is frequently found with another product of the same process, **natural gas**, which is largely methane.

Just as the fossils of marine organisms are the basis of oil and gas, so the fossilised remains of land plants form **coal**. Massive forests of plants, similar to but much larger than the ferns and horsetails of today, covered large parts of the land surface. In a mysterious extinction, not unlike that which brought an end to the age of the dinosaurs, these giant 'fern forests' were largely destroyed. The remains of these plants, fossilised after millions of years buried underground, form coal. Occasionally a piece of coal is seen with the clear imprint of one of these ancient fern leaves.

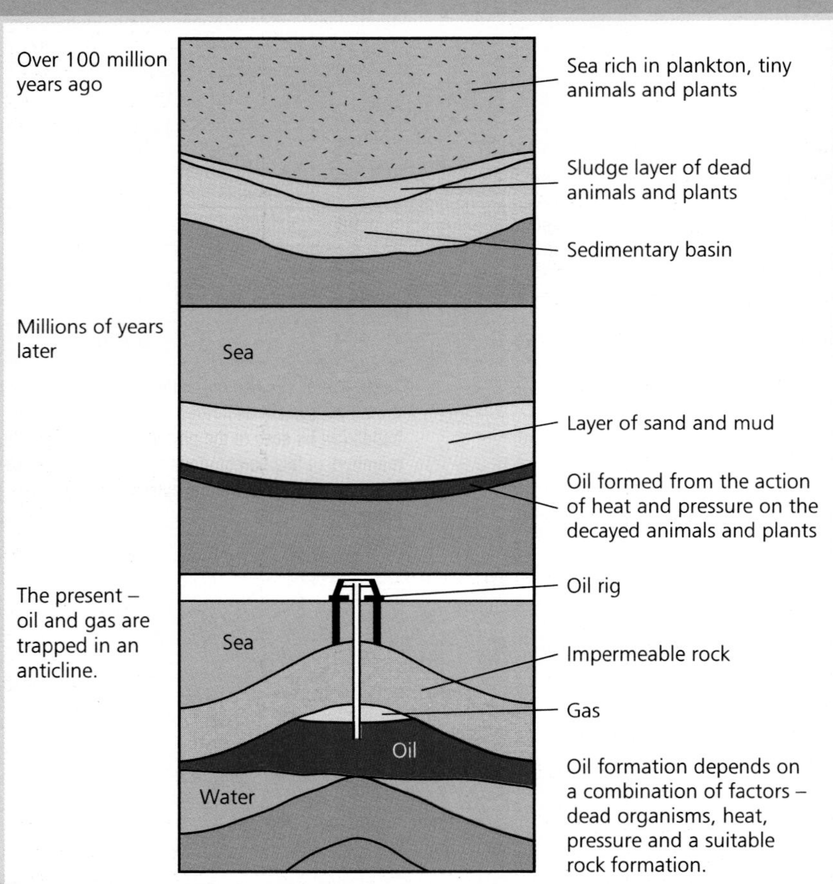

Figure 4.2.2 In a process taking millions of years to complete, the dead bodies of countless living organisms have been converted into the fossil fuels. Today we use these fuels not only as a source of warmth in our homes, but also to generate electricity and as the raw material of the plastics, pharmaceuticals and diverse other chemicals that are part of our everyday lives.

FOCUS ENERGY FOR THE FUTURE

Society in the developed world relies heavily on energy such as electricity for lighting, heating and computers and fuel for our cars, lorries, trains and planes. In the less developed areas of the world energy demands are lower but people are rapidly beginning to demand a more 'western' way of life, so energy use is increasing right across the globe. The major source of all this energy is from fossil fuels – coal, oil and natural gas (see opposite).

Natural gas, oil and coal are used to heat buildings, to cook food and to produce electricity. Crude oil is split into many different fractions, some of which we use as fuel whilst other fractions are used in the petrochemical industry as the basis for plastics and other synthetic materials and as the starting point for the manufacture of a wide range of other chemicals.

However, there are two major problems with the use of fossil fuels. One is that the burning of fossil fuels pollutes the atmosphere, generating the 'greenhouse gases' (see pages 201–2) which are bringing about global warming. The other is that the fuels are a non-renewable resource. New fossil fuels are not being made, so that once we have used up all the oil, the coal or the natural gas which are currently present in the Earth's crust there will be no more. The levels of these resources are falling ever lower as world demand increases. Society needs to invest money in developing alternative, renewable energy sources – yet while fossil fuels continue to be available the political will for such investment seems to be sadly lacking. One alternative energy source is hydroelectric power, but this is limited by the need for vast reserves of water. Other alternative fuel sources include solar power, wind power and tidal power, all of which have great potential but as yet have technological limitations on their use on a large scale. Yet another potential renewable energy source is **biofuel**. Plants use the energy from the Sun to make food and grow. If we can then use that plant material as a fuel, we can grow some more to replace it. When animals such as cattle and sheep eat plants they produce large quantities of dung. This dung can be collected and placed in a digester, where bacteria feed on the dung, breaking it down to release methane gas, a clean and efficient alkane fuel.

Another way of producing fuel from plants is by the fermentation of sugar-rich crops to produce ethanol. This alcohol can then be used as a fuel for cars instead of petrol. Not only is the ethanol a renewable resource, it also produces far less pollution when it is burned. There has been a big investment in this type of process in Brazil which has cut oil imports by more than 20%. Yet another example of a renewable biofuel is biodiesel. Conventional diesel fuel is a by-product of the crude oil distillation process and the USA alone uses well over 50 billion gallons of diesel each year. Biodiesel is produced from vegetable oil from sources such as soybeans and oil seed rape. Biodiesel performs almost identically in engines to conventional diesel, engines need little or no modification to use it, it is relatively clean when it burns and biodegradable when it spills. Most importantly of all, it is another renewable resource. The answers to the fossil fuel crisis are out there. We just need the will and determination to recognise and cultivate these sources of energy for the future.

Figure 1 Made from the remains of animals and plants which lived millions of years ago, fossil fuels like this coal are a wonderful resource.

Figure 2 So far biogas generators have been used mainly on a small scale for individual families or villages in the developing world. However, more and more countries are experimenting with such digesters on a larger scale to deal with the problems of sewage and rubbish tips, and to reduce their dependence on fossil fuels.

Obtaining the alkanes

The single most important source of the alkanes is, as we have already mentioned, the fossil fuels. However, crude oil is a mixture of many different hydrocarbons, many of them alkanes but also some alkenes and alkynes. How can we extract from this complex mixture the specific alkane we may need for a particular purpose?

The basic process by which crude oil is turned from a dark, thick, smelly liquid into a whole variety of useful chemicals is known as **primary distillation**, an industrial version of the fractional distillation we can carry out in the laboratory. Petroleum is boiled, and the vapours are cooled and liquefied at particular temperatures. The liquid collected over each range of temperatures is known as a **fraction**. This process, shown in figure 4.2.3, provides five major fractions which are each very useful, and which can also be separated further in more precise fractional distillation processes to give a pure yield of a particular alkane.

Figure 4.2.3 The oil industry locates deposits of petroleum, extracts it from under the ground and processes it to provide us with fuels and many other organic source materials.

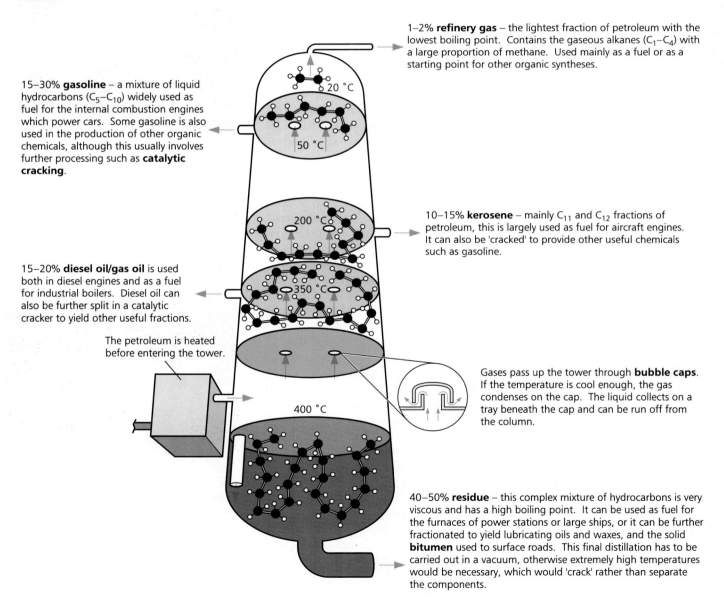

1–2% **refinery gas** – the lightest fraction of petroleum with the lowest boiling point. Contains the gaseous alkanes (C_1–C_4) with a large proportion of methane. Used mainly as a fuel or as a starting point for other organic syntheses.

15–30% **gasoline** – a mixture of liquid hydrocarbons (C_5–C_{10}) widely used as fuel for the internal combustion engines which power cars. Some gasoline is also used in the production of other organic chemicals, although this usually involves further processing such as **catalytic cracking**.

20 °C

50 °C

200 °C

10–15% **kerosene** – mainly C_{11} and C_{12} fractions of petroleum, this is largely used as fuel for aircraft engines. It can also be 'cracked' to provide other useful chemicals such as gasoline.

15–20% **diesel oil/gas oil** is used both in diesel engines and as a fuel for industrial boilers. Diesel oil can also be further split in a catalytic cracker to yield other useful fractions.

350 °C

The petroleum is heated before entering the tower.

Gases pass up the tower through **bubble caps**. If the temperature is cool enough, the gas condenses on the cap. The liquid collects on a tray beneath the cap and can be run off from the column.

400 °C

40–50% **residue** – this complex mixture of hydrocarbons is very viscous and has a high boiling point. It can be used as fuel for the furnaces of power stations or large ships, or it can be further fractionated to yield lubricating oils and waxes, and the solid **bitumen** used to surface roads. This final distillation has to be carried out in a vacuum, otherwise extremely high temperatures would be necessary, which would 'crack' rather than separate the components.

Thermal cracking and the 'cat cracker'

Some of the heavier fractions produced during the fractional distillation of petroleum are less useful than the lighter fractions. If these long-chain alkanes are heated to high temperatures, the molecules split and form more useful shorter-chain molecules, a process known as **thermal cracking**. Some of the smaller molecules are alkanes which can then be used in gasoline. Others are alkenes, particularly ethene, which is extremely useful. The alkanes are very unreactive and so are of very limited use as a starting point for the petrochemical industry. Ethene, on the other hand, contains a double bond which makes it much more reactive – an ideal starting molecule for synthetic reactions.

Producing the high temperatures needed to crack the heavy petroleum fractions would be very expensive. By using catalysts, the process can be carried out at a much lower temperature. This is known as **catalytic cracking (cat cracking** for short), and is carried out in a **cat cracker**.

Reactions of the alkanes

Unreactive molecules

The old, non-systematic name for the alkanes was the **paraffins**. This came from the Latin *parum affinis* which means 'little affinity', and it described the alkanes as a family very well. They are a very unreactive family – they have little or no affinity for other elements or compounds. At room temperature the alkanes are unaffected by concentrated acids such as sulphuric acid or concentrated alkalis such as sodium hydroxide solution. They are not affected by oxidising agents such as potassium manganate(VII) and they do not react even with the most reactive metals. The reason for this lack of reactivity is that both C—C and C—H bonds involve a very even sharing of electrons, since the electronegativities of carbon and hydrogen are very close (2.5 for carbon, 2.1 for hydrogen). Therefore the bonds in the alkane molecules are not polar, and so there are no charges to attract other polar or ionic species.

Almost all the reactions of the alkanes that do occur are due to the formation of **radicals**. This is explained in the box below. As a result the reactions have high activation energies, but once this barrier is overcome, the reactions proceed very rapidly in the gas phase.

The only common reactions of the alkanes occur when they are heated, or in the presence of halogens. We shall go on to consider both these types of reaction in turn.

Breaking bonds – homolytic and heterolytic fission

Chemical reactions involve the breaking and making of bonds between atoms, also known as **bond fission**. Organic reactions almost always involve the breaking and making of covalent bonds, with varying degrees of polarity or ionic character. Within a covalent bond, two electrons are shared between two atoms. When that bond is broken during a chemical reaction, there are two ways in which the electrons may be shared out.

Homolytic fission involves the equal sharing of the electrons in the bond, so that each atom receives one electron when the bond splits:

$$\mathbf{H{-}Cl \rightarrow H\cdot + Cl\cdot}$$

The presence of the unpaired electron gained by each atom is indicated

by a dot. An atom or molecule with an unpaired electron is called a **radical** (or sometimes a **free radical**).

Although neither atom has an overall charge, they are *extremely* reactive. This is because the unpaired electron has a very strong tendency to pair up with an electron from another molecule. The reaction of one radical with another substance usually results in the formation of another radical:

$$\mathbf{H\cdot + Cl{-}Cl \rightarrow H{-}Cl + Cl\cdot}$$

(The effect of reactions like this involving CFCs in the ozone layer was discussed in section 2.6, on pages 203–4.) Homolytic fission usually occurs where there is little or no ionic character in the covalent bond.

The other type of bond fission is **heterolytic fission**, which involves an unequal sharing of the electrons of the covalent bond, so that both electrons go to one atom. This results in two charged particles, the atom receiving the electrons gaining a negative charge and the other atom gaining a positive charge. Heterolytic fission is usually seen when the covalent bond already has a degree of polarity.

$$\mathbf{H{-}Cl \rightarrow H^+ + Cl^-}$$

Heating the alkanes

Cracking

When alkanes are heated to high temperatures in the absence of air, they 'crack' or split into smaller molecules. The cracking of methane yields finely powdered carbon, which is used in car tyres and for the formation of artificial diamond coatings, and hydrogen, used as a raw material for the chemical industry.

$$\mathbf{CH_4(g) \rightarrow C(s) + 2H_2(g)}$$

The cracking of ethane gives ethene, one of the most important raw materials of the chemical industry, as well as hydrogen.

$$\mathbf{C_2H_6(g) \rightarrow CH_2{=}CH_2(g) + H_2(g)}$$

Combustion

When alkanes are heated in a plentiful supply of air, combustion occurs. Alkanes are energetically unstable with respect to their oxidation products of water and carbon dioxide, so when the necessary activation energy is supplied they will burn completely. Alkanes only burn in the gaseous state, so solid and liquid alkanes must be vaporised before they will burn. This is seen clearly in the burning of a candle flame – the solid wax (which consists mainly of a carboxylic acid with a very long hydrocarbon tail) melts and then vaporises from the wick, where it burns.

The combustion of the alkanes is of great importance in our way of life – it is used to produce electricity, to fuel fires in the home, to provide central heating, for cooking and as the source of power for our vehicles. In all these examples, the important feature is the transfer of energy from the exothermic reaction into heat and light energy, as well as into the bonds in the combustion products. As with all exothermic changes, the release of energy results from the fact that the energy required to break the bonds in the chemical reaction of combustion is less than the energy returned when the new bonds are made in the products of combustion, illustrated in figure 4.2.5 opposite.

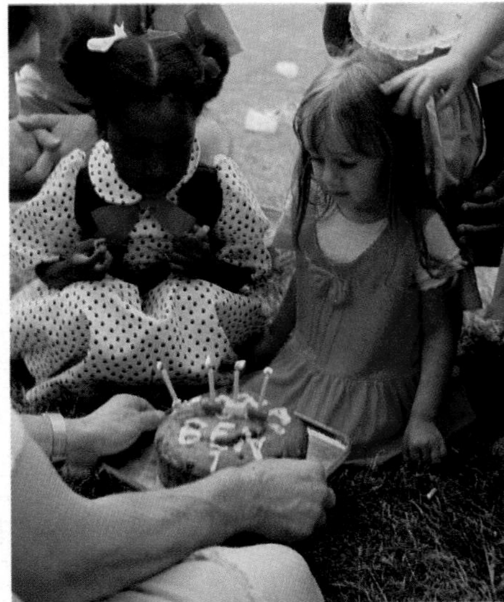

Figure 4.2.4 At the top of a burning candle, the heat from the flame melts the wax. The liquid wax soaks up through tiny channels in the wick. It becomes a gas, mixes with the surrounding air and then burns. This releases more heat, which keeps these processes going, and carries the products of combustion away from the flame, allowing more air to mix with the gaseous hydrocarbon.

Enthalpy, H

$3C(g) + 8H(g) + 10O(g)$

$5 \times \Delta H^{\ominus}_{d,m}(O{=}O)$
$= 5 \times 498\ \text{kJ mol}^{-1}$
$= 2490\ \text{kJ mol}^{-1}$

$3C(g) + 8H(g) + 5O_2(g)$

$2 \times \Delta H^{\ominus}_{d,m}(C{-}C) + 8 \times \Delta H^{\ominus}_{d,m}(C{-}H)$
$= 2 \times 347\ \text{kJ mol}^{-1} + 8 \times 413\ \text{kJ mol}^{-1}$
$= 3998\ \text{kJ mol}^{-1}$

$6 \times \Delta H^{\ominus}_{d,m}(C{=}O)$
$+ 8 \times \Delta H^{\ominus}_{d,m}(O{-}H)$
$= 6 \times 743\ \text{kJ mol}^{-1}$
$+ 8 \times 463\ \text{kJ mol}^{-1}$
$= 8162\ \text{kJ mol}^{-1}$

$C_3H_8(g) + 5O_2(g)$

$\Delta H_m = 3998\ \text{kJ mol}^{-1}$
$+2490\ \text{kJ mol}^{-1}$
$-8162\ \text{kJ mol}^{-1}$
$= -1674\ \text{kJ mol}^{-1}$

$3CO_2(g) + 4H_2O(g)$

(Note that this is not the *standard* enthalpy of combustion, since the water formed is in the gaseous state, not a liquid.)

Figure 4.2.5 When propane burns in air, the energy required to break apart the propane and oxygen molecules is more than compensated for by the energy released when the bonds in carbon dioxide and water are formed.

Figure 4.2.6 This gas fire is well fitted, and perfectly safe. However, if the supply of oxygen to a heater is insufficient, then incomplete combustion of alkanes occurs, and the products may include carbon or the poisonous gas carbon monoxide. Every year there are tragic accidents when ill-fitted domestic appliances with neither an adequate air supply nor satisfactory arrangements for venting the waste gases cause the deaths of the unsuspecting people using them.

Liquefied petroleum gas, LPG

Many of us take the natural gas piped into our homes for granted. However, there are areas of Britain and many other countries in the world where piped gas is not available. However, alkanes – generally propane rather than methane – can still be used for cooking and heating. The advantage of propane over methane is that it can be liquefied readily under pressure, so that a large quantity of gas can be stored in a very small space. The liquid will not burn until it is returned to the gaseous state, and so it can be stored and transported relatively easily and safely. This **liquefied petroleum gas** or **LPG** is transported to countries such as Korea in enormous refrigerated ships, and stored under pressure in vast tanks underground. It is then transferred to pressurised cylinders for purchase by people who use it for cooking and heating in the home. In Britain, the main use of LPG is for leisure purposes such as camping.

Figure 4.2.7 Cylinders of liquid butane are a common sight on caravan and tenting holidays as well as on boats. In cold climates, propane (boiling point −42 °C) is used in preference to butane. The relatively high boiling point of butane (−0.5 °C) means that on a cold morning, the liquid would not vaporise so would not burn.

'Knocking' and octane numbers

1 During the **air-intake stroke** the descending piston draws in a fuel–air mixture through an inlet valve.

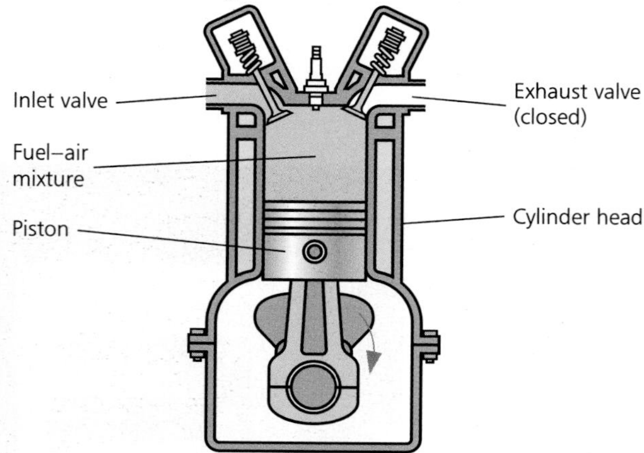

2 In the **compression stroke**, both valves are closed, and the piston rises.

3 The **power stroke** begins just before the piston reaches the top. The spark plug ignites the fuel–air mixture. The combustion products force the piston down, turning the crankshaft.

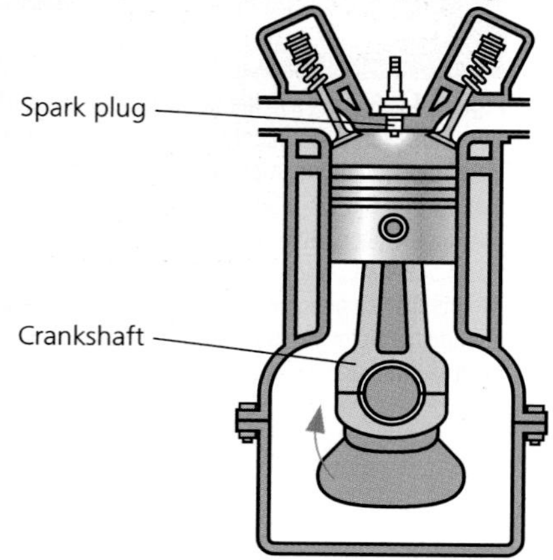

4 In the **exhaust stroke**, the exhaust valve opens, and the combustion gases escape, clearing the cylinder.

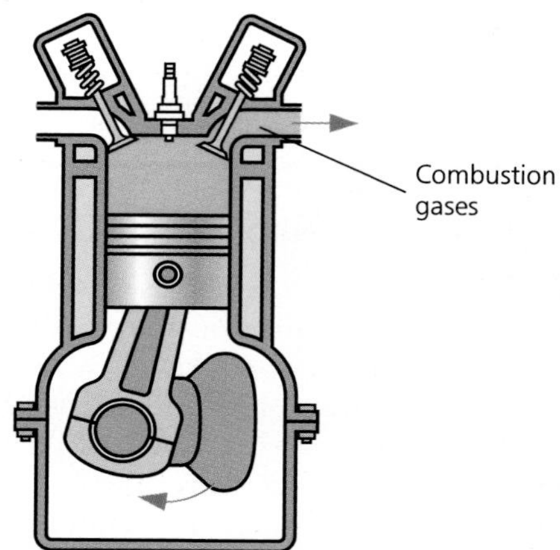

Figure 4.2.8 In the four-stroke petrol engine, a mixture of hydrocarbons and air is drawn into the cylinder, compressed and then ignited by a spark. The gases produced by the combustion drive the piston down, turning the crankshaft and starting the cycle over again.

The combustion of alkanes is applied in the use of petrol in car engines. The alkanes in gasoline, the petroleum fraction used as motor car fuel, have carbon chains that are usually 5–10 carbon atoms long. The power to turn the wheels is produced in the cylinders by the explosive combustion of gasoline and air, as shown in figure 4.2.8.

The smooth running of the engine depends on the explosion occurring in the cylinder at exactly the right moment, so that the minimum energy is wasted, and the maximum energy is transferred to the pistons. If the explosion occurs too early, the pistons are jarred and the engine makes disconcerting 'knocking' noises and also loses power. 'Knocking' is most likely to occur in high-performance cars, which have a high compression ratio in the cylinders that can cause premature ignition of the fuel–air mixture. It is also likely to occur with fuels that have a large proportion of straight-chain alkanes. The straight-chain alkanes such as heptane, octane and nonane tend to ignite very easily, and so can explode prematurely. Branched-chain alkanes such as 2,2,4-trimethylpentane are ignited less readily – their combustion is much more controllable. Gasoline mixtures that are high in these branched-chain alkanes make much more efficient fuels than mixtures high in straight-chain molecules.

To give an indication of the proportion of branched- to straight-chain molecules, and so the likelihood of a fuel to cause 'knocking', different mixtures are given an **octane rating**. The octane rating of 2,2,4-trimethylpentane (old name isooctane) is set at 100, whilst that of heptane (a straight-chain molecule that ignites very readily) is set at 0. A fuel can be tested by comparing it with known mixtures of heptane and 2,2,4-trimethylpentane in a test engine. The ratio of heptane to 2,2,4-trimethylpentane is varied until the mixture has the same ignition properties as the fuel sample. The percentage of 2,2,4-trimethylpentane in the mixture is then taken as the octane number of the fuel. Other methods for measuring octane ratings involve detailed chemical analysis of fuels. Octane ratings are indicated on petrol pumps by stars.

Most modern cars have relatively high-performance engines and require high-octane fuels. To meet this demand there are two methods. One is to prevent 'knocking' by adding a compound to the fuel which will retard its ignition. For many years tetraethyl-lead(IV) has been added to gasoline to perform this function, but awareness of the potential damage caused by lead pollution from car exhausts has led to a move in many countries to eliminate the use of 'leaded' petrol. The other way of preventing 'knocking' is to produce gasoline mixtures which are artificially high in branched-chain alkanes. This can be done by the **reforming process**. Very similar to catalytic cracking, this involves breaking up straight-chain molecules in the heavier oil fractions, and reforming them into molecules with branched chains. Cycloalkanes and arenes can also be added to the petrol to promote efficient combustion. Using other organic chemicals in this way provides a high-octane fuel without the addition of lead compounds.

Figure 4.2.9 High-performance cars like this have always needed high-octane fuels to prevent premature ignition in their high-compression cylinders. But as cars in every part of the market have become more sophisticated, and the number of cars with catalytic converters increases, most engines require plenty of branched-chain alkanes in their fuel to prevent loss of power.

Reactions with halogens

The alkanes react with the halogens, though only with an input of energy in the form of light. For example, hexane reacts with bromine in the vapour phase, decolorising it. A **substitution** reaction takes place, with bromine atoms substituting for the hydrogen atoms in the hexane molecule. This decolorisation does not occur in the dark – the reaction only takes place in the light.

$$\mathbf{CH_3CH_2CH_2CH_2CH_2CH_3(g) + Br_2(g) \xrightarrow{light} CH_3CH_2CH_2CH_2CH_2CH_2Br(g) + HBr(g)}$$

Similarly, methane and chlorine do not react in the dark, but in sunlight they react explosively to produce chloromethane and hydrogen chloride. This, combined with the very rapid reaction that follows, suggests that the reaction takes place by a radical mechanism, and this is indeed the case. This is another example of a substitution reaction, with chlorine atoms substituting for the hydrogen atoms in the methane molecule.

How does light cause chlorine to react with methane? As the following data show, the Cl—Cl bond is easier to break than the C—H bond. Light provides the energy needed to split the chlorine molecules into atoms – in other words, to **initiate** the reaction.

$$\mathbf{Cl_2 \rightarrow Cl\cdot + Cl\cdot \qquad \Delta H^\ominus_m = +243\ kJ\ mol^{-1}}$$

$$\mathbf{CH_4 \rightarrow CH_3\cdot + H\cdot \qquad \Delta H^\ominus_m = +435\ kJ\ mol^{-1}}$$

4

The chlorine atoms formed by this homolytic fission are radicals, and so are extremely reactive. The chlorine radicals react with methane molecules, combining with one of the hydrogen atoms to form HCl and another radical:

$$\mathbf{CH_4 + Cl\cdot \rightarrow CH_3\cdot + HCl \qquad \Delta H^{\ominus}_{m} = +4\ kJ\ mol^{-1}}$$

The methyl radical then reacts with another chlorine molecule to form chloromethane and a chlorine radical. This is a **propagation step**, as it produces another radical. The process is then repeated hundreds of times, which makes the reaction explosive:

$$\mathbf{CH_3\cdot + Cl_2 \rightarrow \underset{chloromethane}{CH_3Cl} + Cl\cdot \qquad \Delta H^{\ominus}_{m} = -97\ kJ\ mol^{-1}}$$

The propagating steps of this reaction continue until there is a **termination** step. This is the name given to the reaction between two radicals – a very exothermic process. (Compare the enthalpy changes for these reactions with the bond enthalpies in table 1.8.1 on page 115 – why are they similar?)

$$\mathbf{Cl\cdot + Cl\cdot \rightarrow Cl_2 \qquad \Delta H^{\ominus}_{m} = -243\ kJ\ mol^{-1}}$$

$$\mathbf{CH_3\cdot + Cl\cdot \rightarrow CH_3Cl \qquad \Delta H^{\ominus}_{m} = -338\ kJ\ mol^{-1}}$$

$$\mathbf{CH_3\cdot + CH_3\cdot \rightarrow \underset{ethane}{C_2H_6} \qquad \Delta H^{\ominus}_{m} = -347\ kJ\ mol^{-1}}$$

If the supply of chlorine is limited, the net result of this reaction is lots of chloromethane and hydrogen chloride, and little ethane. However, if there is a plentiful supply of chlorine, then further substitutions of the methane will take place to give di-, tri- and tetrachloromethane as follows:

$$\mathbf{CH_4 + Cl_2 \rightarrow CH_3Cl + HCl}$$

$$\mathbf{CH_3Cl + Cl_2 \rightarrow \underset{dichloromethane}{CH_2Cl_2} + HCl}$$

$$\mathbf{CH_2Cl_2 + Cl_2 \rightarrow \underset{trichloromethane}{CHCl_3} + HCl}$$

$$\mathbf{CHCl_3 + Cl_2 \rightarrow \underset{tetrachloromethane}{CCl_4} + HCl}$$

Trichloromethane or chloroform was one of the first anaesthetics to be used for surgical operations and to ease the pain of childbirth, whilst tetrachloromethane was widely used as a solvent until its carcinogenic properties were recognised. We shall look at the halogenoalkanes in section 4.5.

The reactions described here are typical of the reactions of the alkanes with the halogens. With chlorine and bromine, an input of energy is needed to overcome the activation energy and so initiate the reactions. With fluorine a vigorous reaction occurs even without an input of energy – a mixture of fluorine and a hydrocarbon will explode at room temperature even in the dark.

Radicals also play an important part in the reaction of the alkanes with oxygen, and in both cracking and reforming hydrocarbon molecules.

Burnt toast, radicals and the cancer connection

Radicals are an important element of our model for many organic reactions, and are obviously of great importance for that reason alone. But radicals also have a role in several common human problems.

Cancer is one of the major causes of death in the developed world. Cancer occurs when the normal growth-controlling mechanisms of a cell break down, so that rapid proliferation of small unspecialised cells takes place. This results in the formation of a tumour or growth, which may itself cause serious illness or death by filling up and destroying a vital organ. More commonly, small pieces will break off the tumour, travel around the body in the bloodstream and lodge in other places to grow again. Many different chemicals are thought to be responsible for the cellular changes that bring about the loss of control over cell growth – they are known as **carcinogenic substances**. Radicals in the body are now considered to be one of the culprits. The number of radicals in the body can be reduced by cutting down the intake of food high in radicals – burnt toast and charred barbecued food, amongst others. Equally, certain vitamins play a role in enabling the body to 'mop up' any radicals that may form. A plentiful supply of fresh fruit and vegetables containing vitamins A, C and E help an enzyme called superoxide dismutase to inactivate radicals, and so prevent them causing damage.

Figure 4.2.10 The same extreme reactivity of radicals that enables so many organic reactions to occur is the reason why they are potentially damaging to cells when they occur in excess.

The cycloalkanes

Whilst many of the alkanes present in crude oil are aliphatic, having straight- or branched-chain molecules, some of them form rings – they are alicyclic. These **cycloalkanes** have the general formula C_nH_{2n} instead of C_nH_{2n+2} for the chain molecules. Cycloalkanes behave very similarly to the other alkanes, but they tend to have higher melting and boiling points. They have the same name as the corresponding straight-chain molecule, but with the prefix cyclo-. These ring molecules are not aromatic compounds – the aromatic ring structure is based on the benzene ring, as we shall see in section 4.4.

The smallest cycloalkane is cyclopropane, which forms a triangular molecule which is much more reactive than straight-chain propane. Cyclobutane forms a square molecule which is less reactive than cyclopropane, but is more reactive than butane. Cyclopentane and higher cycloalkanes have a similar reactivity to their straight-chain equivalents. The structures of these cycloalkanes are shown in figure 4.2.11.

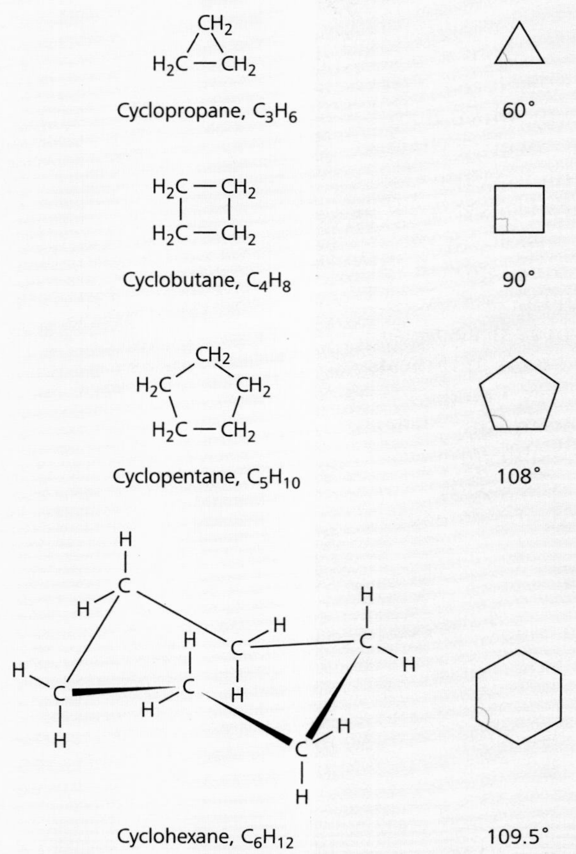

Cyclopropane, C_3H_6 — 60°

Cyclobutane, C_4H_8 — 90°

Cyclopentane, C_5H_{10} — 108°

Cyclohexane, C_6H_{12} — 109.5°

Figure 4.2.11 Skeletal formulae such as these are often used for cyclic molecules.

Reactivity of the alkanes and cycloalkanes

In methane, carbon bonds to hydrogen by the formation of sp^3 hybridised orbitals, as we saw on page 361. The formation of carbon–carbon bonds in the alkanes results from the overlap of an sp^3 orbital from each carbon atom, forming a bond known as a **σ bond (sigma bond)** – figure 4.2.12 shows such a bond in a molecule of ethane. Notice that the four orbitals around each carbon atom are arranged tetrahedrally.

The trends in reactivity of the cycloalkanes can be explained on the basis of the bond angles within them, shown in figure 4.2.11. Just as in the straight-chain alkanes, the four orbitals around each carbon atom in the cycloalkanes are arranged tetrahedrally. However, the angle between the carbon atoms in cyclopropane is 60°, which is considerably less than the tetrahedral angle of 109.5°. This leads to much weaker C—C bonds in cyclopropane than in straight-chain propane, since there is much less overlap possible between the orbitals of the carbon atoms in cyclopropane than in propane. Because they are weaker, the C—C bonds in cyclopropane are sometimes said to be **strained** compared with those in straight-chain

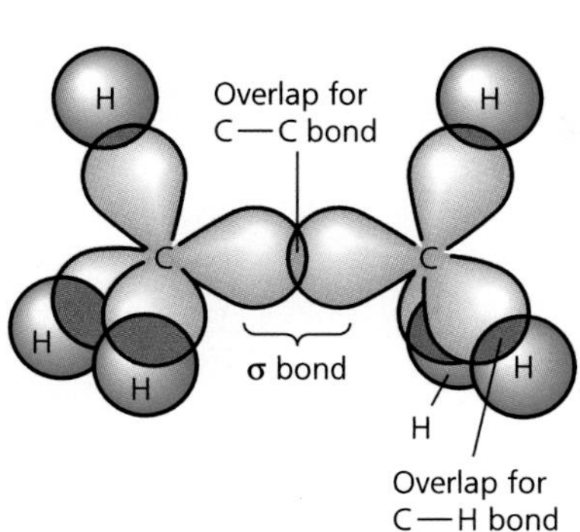

Figure 4.2.12 The overlap of two sp^3 orbitals producing the carbon–carbon σ bond in ethane

propane. This effect is less pronounced in cyclobutane, which has a bond angle of 90°, while the bonds in cyclopentane are almost entirely free from strain. Cycloalkanes with six or more carbon atoms are able to relieve the strain by puckering the ring, as figure 4.2.11 shows.

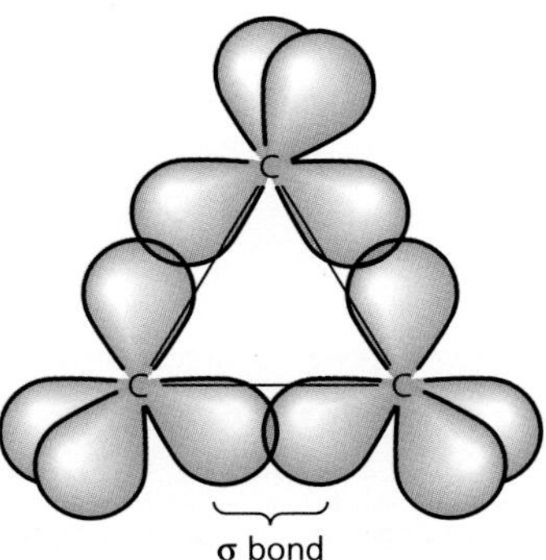

Figure 4.2.13 Overlap of sp^3 orbitals to form C—C σ bonds in cyclopropane. Compare this with the overlap of sp^3 orbitals forming the C—C σ bond in ethane.

Consideration of the alkanes gives us a good introduction to the chemistry of the hydrocarbons, and to some of the reaction mechanisms that are important in all the families of organic chemicals. In section 4.3 we shall look at the chemistry of the unsaturated hydrocarbons and see how this differs from that of the saturated alkanes.

SUMMARY

- **Hydrocarbons** are organic compounds containing only carbon and hydrogen. **Aliphatic** hydrocarbons have open chain molecules and **alicyclic** hydrocarbons consist of hydrocarbon rings, whilst **aromatic** hydrocarbons generally involve benzene rings.
- Aliphatic hydrocarbons belong to one of three families – **alkanes**, **alkenes** and **alkynes**.
- Hydrocarbons are largely derived from fossil fuels such as coal, crude oil and natural gas.

4

- The **alkanes** are **saturated** hydrocarbons – they contain no double or triple bonds. Their general formula is C_nH_{2n+2}.
- Alkanes are obtained from crude oil by a process of **primary distillation** (a form of fractional distillation).
- The straight-chain alkanes show very clear patterns in their physical properties as the size of the molecule increases. This makes it relatively easy to predict the properties of an unknown alkane. The reactions of alkanes with branched chains and cycloalkanes are less easy to predict.
- **Homolytic fission** is the breaking of a covalent bond such that the electrons that formed the bond are shared equally between the two atoms. **Heterolytic fission** involves an unequal sharing of the electrons of the covalent bond, such that one atom gains a negative charge and the other a positive charge.
- The alkanes are very stable compounds that undergo few chemical reactions.
- Alkanes burn in air to form carbon dioxide and water.
- In the absence of air, larger alkane molecules split when heated to give smaller alkanes. This is the basis of the **cracking** process used in the oil industry.
- Alkanes react with reagents such as oxygen or halogens, undergoing **substitution** reactions. The reactions involve homolytic fission, forming **radicals** – reactive species with an unpaired electron.

QUESTIONS

1 Look at the five alkanes shown in figure 4.2.14.

i CH_4

ii CH_3CH_3

iii $CH_3CH_2CH_2CH_3$

iv

```
      H     H
       \   /
  H     C       H
   \   / \    /
     C ——— C
   /         \
  H           H
```

v

```
  CH3
   |
CH3CHCH3
```

Figure 4.2.14

a What is an alkane?
b By what process would these alkanes have been extracted from crude oil?
c In which fraction of crude oil would you expect to find alkanes **i** and **ii**?
d Which alkane does not conform to the general formula of the alkanes? Why not?
e Which two of the alkanes shown are isomers?

2

IUPAC name	Number of carbon atoms *n*	Boiling point/°C
Methane	1	–163.4
Ethane	2	–88.5
Propane	3	–42.0
Butane	4	–0.4
Pentane	5	36.2
2-methylpropane	4	–11.6
2-methylbutane	5	28
2,2-dimethylpropane	5	9.6
Cyclopropane	3	–32.6
Cyclobutane	4	12.1
Cyclopentane	5	49.3

a Plot a graph of boiling point against *n* for the first five members of the straight-chain alkanes.
b What is the effect on boiling point of increasing the length of the carbon chain?

QUESTIONS

c Plot on the graph the boiling points of the branched-chain alkanes given. Is any obvious trend observable?
d Plot on the graph the boiling points of the cycloalkanes given. Do they appear to follow a trend?
e Compare the boiling point of pentane with those of 2-methylbutane and 2,2-dimethylpropane. Explain the differences in terms of intermolecular forces.

3 a The alkanes are a very unreactive group of chemical compounds. Explain why they are so unreactive in terms of C—C and C—H bond enthalpies.
b What happens to alkanes when they are heated to high temperatures (give chemical equations for the reactions):
i in the absence of air
ii in the presence of oxygen?
c Under what conditions do the alkanes react with halogens? Explain, using chemical formulae where appropriate, the mechanism of the reaction between hexane and bromine, showing why the reaction conditions are important.

4 Combustion in car engines is an important use of the alkanes.
a What fractions of crude oil make up gasoline?
b Giving chemical equations, describe the reaction that powers the car.
c What is meant by the term 'knocking'?
d What is the 'octane rating' of a fuel?
e There are two main ways of overcoming the problem of knocking. Outline each method, explaining how it works and any advantages or disadvantages.

Developing Key Skills

The local authority is planning to introduce a new scheme whereby all the sewage produced by the community will be fed into a giant biogas digester. This will be used to produce methane gas which will then be piped into the local authority buildings, hospitals and schools to provide heating. Any excess production will be fed into the town supply to reduce the cost of domestic heating.

Produce a leaflet to be given to homes and schools throughout the community. The leaflet should aim to explain the process and the value of the gas produced. It should outline all the advantages of this process and aim to answer any of the concerns which are likely to be raised.

[Key Skills opportunities: C, IT]

FOCUS THE PLASTICS PREDICAMENT

The boom in the development of synthetic polymers in the last fifty years has led to a great many useful materials. Plastics and other polymers have improved the quality of life in many ways, making things lighter, cheaper and often more durable. Plastics are an integral part of our everyday life, from the wrapping on the sandwiches of a packed lunch to the artificial valve working in someone's heart, from the storage containers in the kitchen to the dashboard of the car, and they have brought many benefits.

However, there are some disadvantages to these amazingly useful compounds. Firstly, some of the compounds used in the production of plastics are poisonous and explosive, so the potential for accidents in their production is high. Secondly, many of them are very difficult to dispose of in a way that does not damage the environment. Plastic articles can be disposed of on rubbish tips along with all our other household waste, but the polymers themselves, being synthetic, are frequently not broken down by living organisms in the way that natural compounds are – they are not **biodegradable**. This means that plastic dumped today will still be there in the landfill sites in centuries to come.

All plastics will burn, but when they do so their combustion often produces a variety of toxic gases. These include HCl (from halogen-containing polymers like PVC) and HCN which are damaging to both human health and other living organisms so burning is no solution to the problem.

The quantity of non-biodegradable polymers around the world is untold, and growing all the time. But as people have become increasingly aware of the problem of plastic disposal, so some possible solutions are appearing.

Recycling is one way in which plastics can safely be disposed of – and it has the added advantage that it reduces the need for the production of more plastics from ever dwindling supplies of fossil fuels. So far only thermoplastics (plastics which soften on heating) can be recycled. In many countries there are now collection points for the plastic bottles used for fizzy drinks, fruit juices, ketchup, etc. which are made of a plastic called PET (polyethyleneterephthalate). When this plastic is recycled it can be used for fibres in the manufacture of carpets and textiles, and new uses as a waterproofing material are being developed. In the USA almost a quarter of all plastic bottles are now recycled – that means almost a quarter of a million tonnes of plastic reused each year! In the UK and across Europe similar recycling is taking place. And it isn't just plastic bottles that can be recycled – many car manufacturers now label plastic components in their vehicles to make reclamation and recycling easier when the car is eventually scrapped.

Another growing solution to the plastics dilemma is the development of **biodegradable plastics**. Increasingly manufacturers are producing polymers which can be broken down by bacteria. This may involve modifying natural polymers such as starch or cellulose, or incorporating new subunits in existing polymers, but the end result is a plastic which can be attacked and broken down by microbes. The degradation may take twenty to thirty years, but eventually the product will be destroyed. Plastics and other polymers are here to stay, but hopefully in future we will be able to get rid of them when we need to!

Figure 1 Plastics dumped in landfill sites will cause problems for future generations.

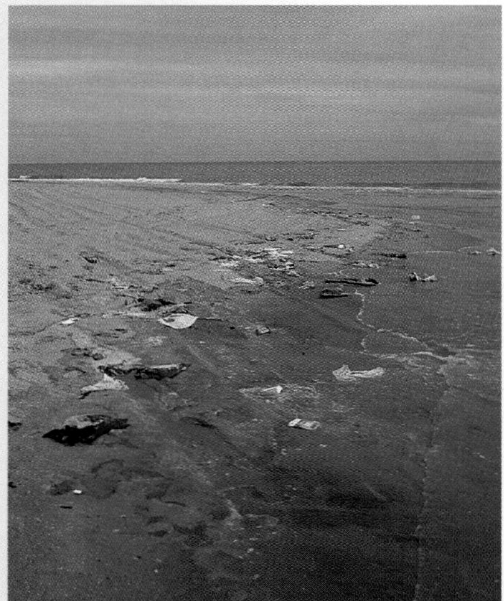

Figure 2 It is said that on shores all over the world, however remote, plastic debris is washed up by the sea. The development of plastics has revolutionised our lifestyles in many ways, but has produced a 'throw-away' society in its wake.

4.3 The unsaturated hydrocarbons: The alkenes and alkynes

The alkenes

The **alkenes** are another family of aliphatic hydrocarbons, but unlike the alkanes, they are **unsaturated** – they do not contain the maximum amount of hydrogen possible. The monoalkenes contain one double bond between two of the carbon atoms, and the family has the general formula C_nH_{2n}. As we saw in section 4.1, they are named on the basis of the number of carbon atoms in the chain, and the position of the carbon–carbon double bond, for example, pent-2-ene and 2-methylbut-2-ene, shown in figure 4.3.1. Like the alkanes, the unbranched alkenes form a family with regular trends in reactivity. The compounds are physically very similar to the equivalent alkanes, although melting and boiling points are a little lower.

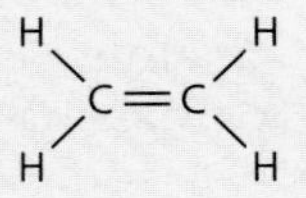

Ethene, C_2H_4

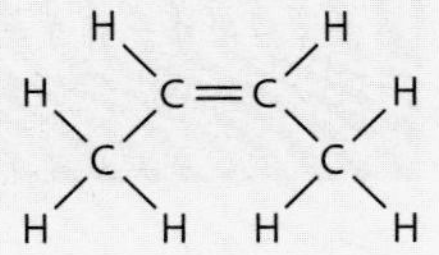

But-2-ene, C_4H_8

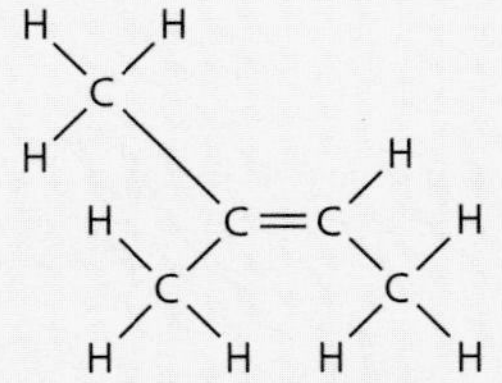

2-methylbut-2-ene, C_5H_{10}

Figure 4.3.1 Three members of the alkene family

The alkenes, and ethene in particular, are of immense importance to the chemical industry. They occur naturally only in very small quantities, but are derived from crude oil by the process of cracking as described in section 4.2. Ethene is the starting molecule for a great many synthetic processes, a large proportion of its annual production supplying the plastics industry, amongst others.

Figure 4.3.2 Ethene provides the starting point for the manufacture of many everyday products.

Reactions of the alkenes

Reactions of the double bond are the main feature of the chemistry of the alkenes, and these make their chemistry rather different from that of the alkanes. The presence of the double bond has two main effects on the chemistry of the alkenes:

- It means that alkenes exhibit *cis-trans* isomerism. There is no rotation around the double bond – rotation would place great strain on one of the bonds and break it. As a result, geometric isomers frequently occur amongst the alkenes, as we saw in section 4.1.
- It causes the alkenes to undergo **addition** reactions, rather than the substitution reactions of the alkanes. In an addition reaction, two substances react together to form a single product.

Alkenes are more reactive than alkanes, because the energy required to break the double bond is *less* than twice the energy required to split one single bond (347 kJ mol^{-1} for C—C, 612 kJ mol^{-1} for C═C). The energy input needed to break one component of the double bond is therefore less than that needed to break a single bond, resulting in increased reactivity.

The reactions of the alkenes do not involve radicals. Instead, heterolytic fission of the double bond occurs, and it is this that determines how the molecules behave. The high electron density associated with the double bond means that the alkenes are attacked by both **electrophiles** (species that 'love' negative charge) and oxidising agents.

Electrophiles and nucleophiles

The attack on the C═C double bond in an alkene occurs when a species containing an atom with a whole or partial positive charge approaches the electron-rich double bond. Such a species is said to be **electrophilic**, since it is attracted to areas of high electron density. An example of an electrophile is the nitronium cation, NO_2^+, which we shall meet in section 4.4.

Other reactions occur in organic chemistry when a **nucleophilic** species attacks an atom that has a whole or partial positive charge. Nucleophiles have an electron pair which can be donated to form a coordinate bond. An example of such a species is ammonia, NH_3.

The electronic structure of carbon in the alkanes and the alkenes

We saw in section 4.2, pages 390–1, that carbon forms σ bonds with other carbon atoms by the overlap of two sp^3 hybridised orbitals. Because the σ bond lies along a line joining the two carbon atoms, it allows rotation to occur about this axis, and the two ends of the ethane molecule are free to rotate.

An alternative way of hybridising the s and p orbitals explains the properties of the alkenes. If only two of carbon's p orbitals are hybridised with its s orbital, three identical sp^2 hybrid orbitals are produced, as figure 4.3.3 shows. These three orbitals lie in a plane at an angle of 120° to each other, with the unhybridised p orbital perpendicular to the plane, as shown in figure 4.3.4. The C═C bond in ethene is formed as a result of overlap between one sp^2 orbital from each of the two carbon atoms, giving a σ bond, plus overlap of the two 2p orbitals forming another bond. This bond has its electron density concentrated in two regions on either side of the axis of the σ bond, and is called a **π bond**. Figure 4.3.5 shows the formation of the C═C bond in ethene in stages.

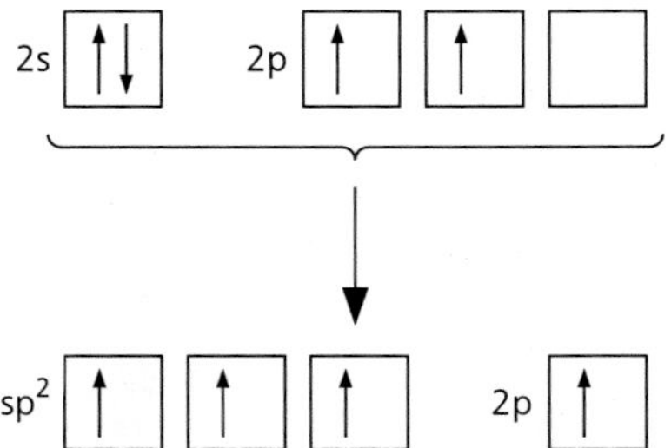

Figure 4.3.3 By combining the s orbital with two p orbitals, three sp^2 orbitals plus a single p orbital are produced, each containing a single electron.

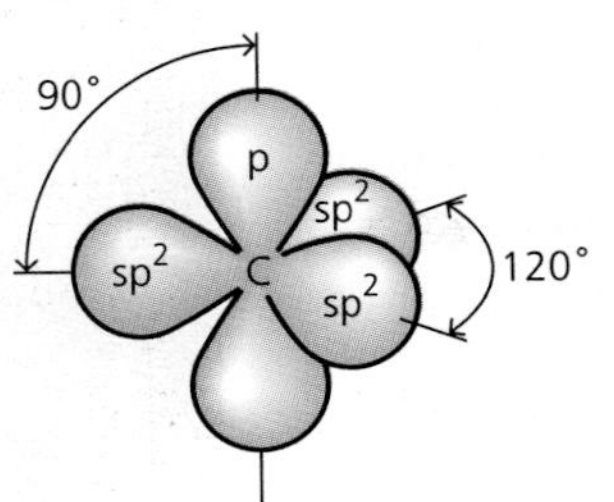

Figure 4.3.4 The orientation of the orbitals in an sp² hybridised carbon atom

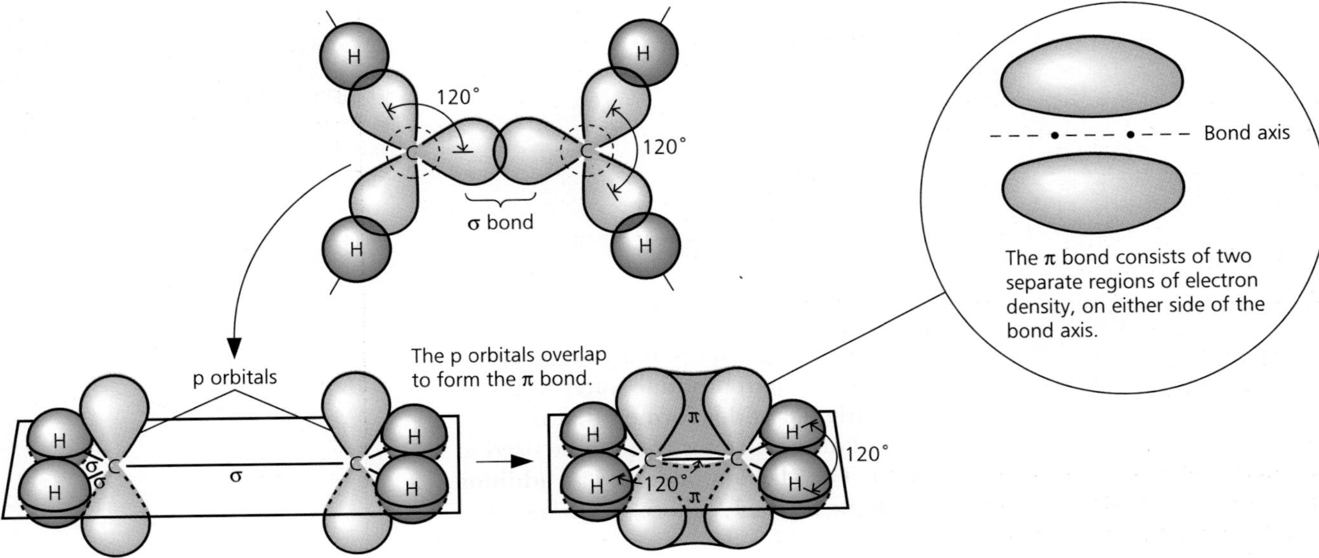

Figure 4.3.5 The formation of the double bond in ethene

The electron density in the π bond is on average further from the nuclei of the two carbon atoms than the electron density in a σ bond. This is why the double bond is less than twice as strong as a σ bond. This arrangement of electron density also explains two other properties of molecules containing C=C bonds:

- their reactivity, due to the attraction of electrophiles to the electron-rich double bond
- the *cis-trans* isomerism shown by many such compounds, since rotation about the bond axis requires the breaking of the π bond, which requires a large amount of energy.

Reactions with oxygen

The alkenes, just like all the other hydrocarbon families, burn in air to produce carbon dioxide and water. Ethene will react explosively with oxygen in a highly exothermic reaction. However, the alkenes are not used as fuels for two main reasons – they produce a lot of carbon when they burn, and they are too valuable as a feedstock for the chemical industry.

$$\mathbf{C_2H_4(g) + 3O_2(g) \rightarrow 2CO_2(g) + 2H_2O(l)} \quad \Delta H^{\ominus}_{m} = \mathbf{-1411\ kJ\ mol^{-1}}$$

There are a few instances where the reaction of an alkene with oxygen is useful, however. Industrially, oxygen is added to ethene in the presence of a finely divided silver catalyst and at 180 °C to produce a ring compound called

epoxyethane, shown in figure 4.3.6. This is the starting point for several other industrial syntheses, because the three-membered ring contains bonds that are under a lot of strain, and so epoxyethane is very reactive. These industrial processes include the formation of ethane-1,2-diol, an important component of the antifreeze mixtures that prevent the water in cars freezing in winter, and also a precursor in the manufacture of polyester for the textile market. Reactions between epoxyethane and alcohols are used in the production of solvents, plasticisers and detergents. (The non-systematic name for ethane-1,2-diol is ethylene glycol, a term still used for antifreeze in the motor trade.)

Figure 4.3.6 Car engines need water as a coolant to prevent them overheating. However, if the water freezes in the engine during cold weather, its expansion bursts joints and can also be very damaging to the engine. The addition of antifreeze mixtures containing ethane-1,2-diol lowers the freezing point of the water substantially, depending on the amount of antifreeze added. In Britain the freezing point of the water in car radiators is usually lowered from around 0 °C to –20 °C, preventing freezing on the coldest winter nights.

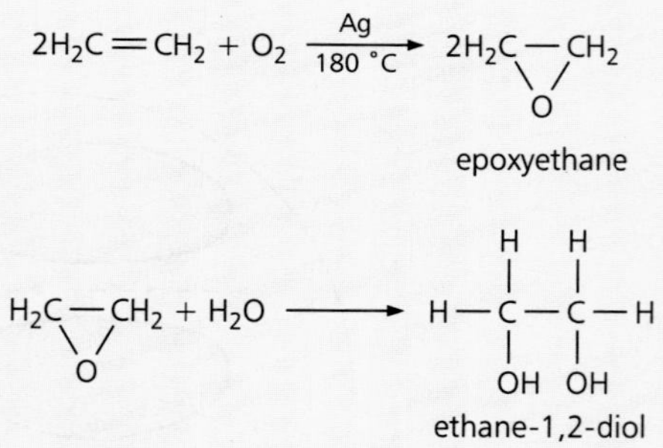

$$2H_2C{=}CH_2 + O_2 \xrightarrow[180\,°C]{Ag} 2H_2C{-}CH_2 \text{ (with O bridging, ring)}$$

epoxyethane

$$H_2C{-}CH_2 \text{ (O ring)} + H_2O \longrightarrow H{-}\underset{OH}{\overset{H}{C}}{-}\underset{OH}{\overset{H}{C}}{-}H$$

ethane-1,2-diol

Reactions with halogens

The alkenes react with the halogens very differently from the reaction of the alkanes. For example, when ethene is bubbled through bromine water in the dark, the bromine is decolorised and a colourless liquid is formed that is immiscible with water. This reaction is typical of the alkenes and can be used to demonstrate the presence of a double bond. The alkene undergoes an addition reaction with the bromine; for example, ethene forms 1,2-dibromoethane as shown in figure 4.3.7.

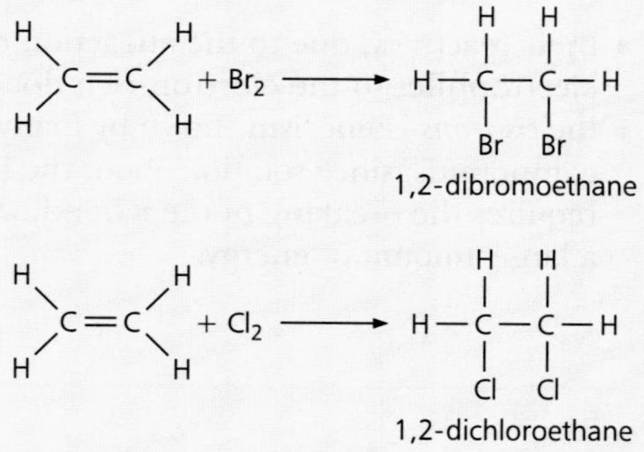

$$H_2C{=}CH_2 + Br_2 \longrightarrow H{-}\underset{Br}{\overset{H}{C}}{-}\underset{Br}{\overset{H}{C}}{-}H$$

1,2-dibromoethane

$$H_2C{=}CH_2 + Cl_2 \longrightarrow H{-}\underset{Cl}{\overset{H}{C}}{-}\underset{Cl}{\overset{H}{C}}{-}H$$

1,2-dichloroethane

Figure 4.3.7 The alkenes react with the halogens to form addition products such as those shown here.

All alkenes react vigorously with fluorine. The vigour of the reactions with the halogens decreases down the group, and reactions with iodine are relatively slow. Ethene and fluorine react explosively to form two molecules of tetrafluoromethane, while with iodine a very slow reaction takes place to form 1,2-diiodoethane. The reaction of ethene with chlorine shown in figure 4.3.7 is particularly important in the manufacture of chloroethene (non-systematic name vinyl chloride) which is used to make polyvinylchloride or PVC, a widely used plastic. (The systematic name of PVC is poly(chloroethene).)

The reactions of the alkenes with the halogens all proceed quite rapidly at room temperature, which suggests that the addition reaction does not involve radicals. This conclusion is borne out by further studies, as the box 'Addition across a double bond – introducing reaction mechanisms' shows.

Reactions with hydrogen halides

The double bond in the alkenes reacts readily with the hydrogen halides, producing the corresponding halogenoalkane. For example, ethene reacts as shown in figure 4.3.8(a). This reaction proceeds rapidly at room temperature, forming bromoethane.

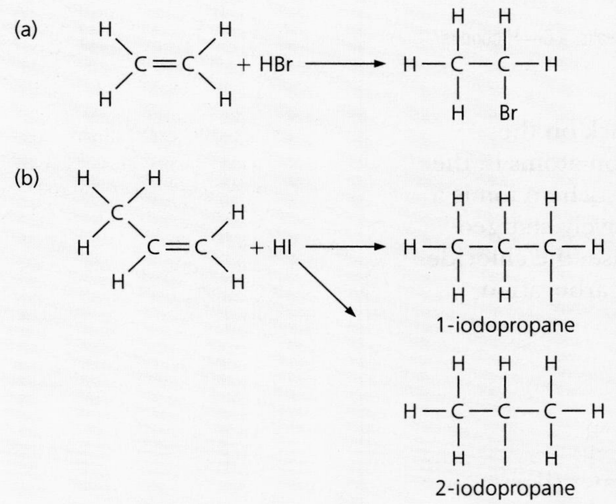

(a) $H_2C{=}CH_2 + HBr \longrightarrow H{-}CH_2{-}CH_2Br$

(b) $CH_3{-}CH{=}CH_2 + HI \longrightarrow CH_3{-}CH_2{-}CH_2I$ (1-iodopropane) and $CH_3{-}CHI{-}CH_3$ (2-iodopropane)

1-iodopropane

2-iodopropane

Figure 4.3.8

The addition of hydrogen halides to asymmetrical alkenes like propene can lead to two possible products, as the reaction of hydrogen iodide with propene in figure 4.3.8(b) shows. In this case, the major product formed is 2-iodopropane, with a smaller amount of the alternative 1-iodopropane. The likely products of the addition of hydrogen halides to asymmetrical alkenes can be predicted using **Markovnikov's rule**:

> **When HX adds across an asymmetrical double bond, the major product formed is the molecule in which hydrogen adds to the carbon atom in the double bond with the greater number of hydrogen atoms already attached to it.**

Like all simple rules in science, this one has its exceptions – however, Markovnikov's rule does provide a useful way of deciding the most likely product of this type of reaction.

Addition across a double bond – introducing reaction mechanisms

Chemists use **reaction mechanisms** to understand how reactions between molecules occur. Chemical reactions are concerned with breaking and making bonds, and this process is shown using curly arrows, which represent the movement of a *pair* of electrons. Figure 4.3.9 overleaf shows how curly arrows are used to indicate the breaking and making of bonds in the first stage of the reaction of a hydrogen halide with ethene.

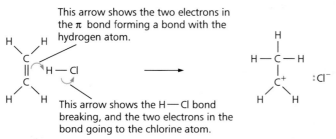

This arrow shows the two electrons in the π bond forming a bond with the hydrogen atom.

This arrow shows the H — Cl bond breaking, and the two electrons in the bond going to the chlorine atom.

Figure 4.3.9 In the first stage of the reaction of HCl with an alkene, the π bond breaks, a C—H bond is formed, and the H—Cl bond breaks, liberating a Cl^- ion.

The first stage of this reaction involves electrophilic attack on the electron-rich π bond. As a result of this, one of the carbon atoms in the ethene molecule gains a positive charge. The name for such an ion is a **carbocation**, or sometimes a **carbonium ion**. This positively charged carbon is then open to attack by nucleophiles. In this case, the chloride ion formed in the first stage of the reaction attacks the carbocation, forming a new C—Cl bond, as shown in figure 4.3.10.

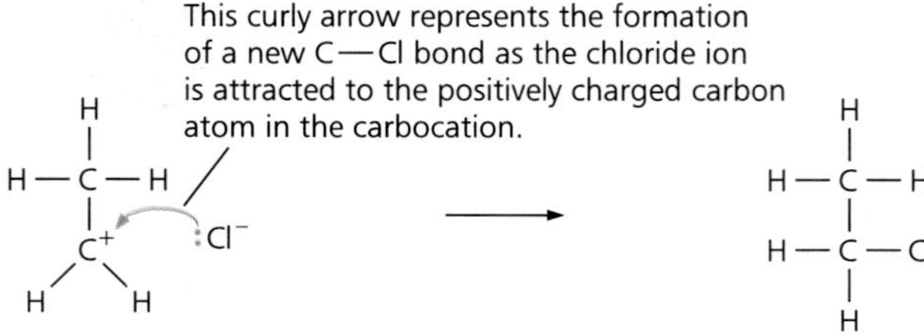

This curly arrow represents the formation of a new C — Cl bond as the chloride ion is attracted to the positively charged carbon atom in the carbocation.

Figure 4.3.10

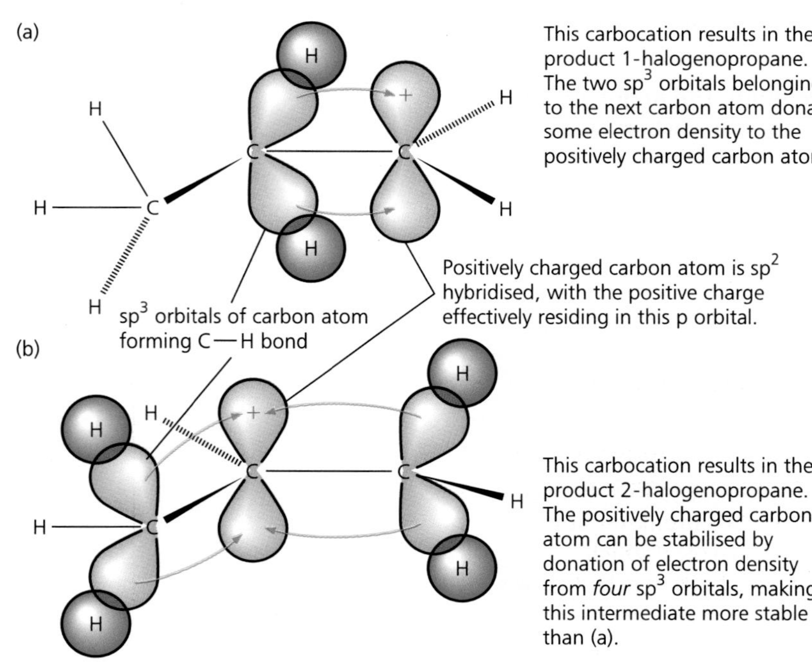

Figure 4.3.11

The existence of the intermediate carbocation helps to explain why Markovnikov's rule applies to the addition of HX to asymmetrical alkenes, as figure 4.3.11 shows.

Both intermediate carbocations are highly unstable. However, that in (b) is more stable than its alternative because of donation of electron density from four sp^3 orbitals and it tends to be formed in preference. The major product of this reaction is therefore the **secondary** compound, CH_3CHXCH_3, with less of the **primary** compound $CH_3CH_2CH_2X$. (A primary halogenoalkane has two hydrogens and one carbon attached to the halogen-carrying carbon. A secondary halogenoalkane has two carbons attached to the halogen-carrying carbon. There is more about this in section 4.5.)

This mechanism also applies to the addition of halogens across a double bond, as shown in figure 4.3.12. Evidence for the existence of the intermediate carbocation in both of these reactions comes from the observation that if competing nucleophiles are present in the reaction mixture (for example, if Cl^- ions are present when ethene reacts with bromine), a mixture of products is formed, as figure 4.3.13 shows.

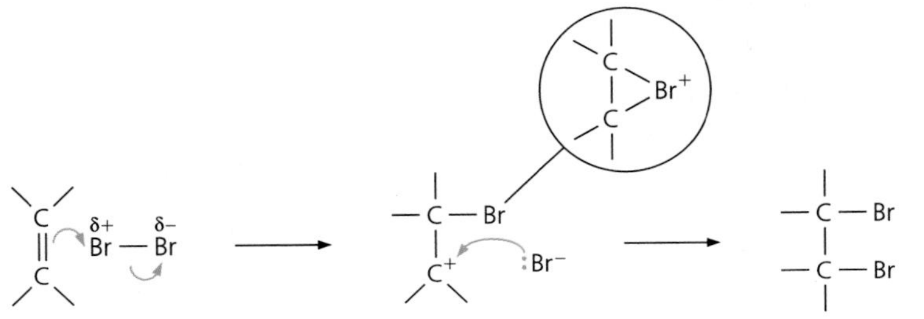

Figure 4.3.12 The mechanism of attack of bromine on a double bond involves electrophilic attack by a Br_2 molecule with an instantaneous dipole. There is some evidence that the intermediate carbocation has the cyclic structure shown in the inset.

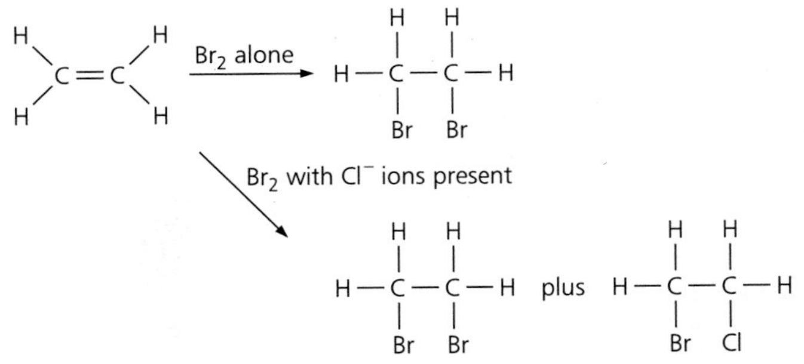

Figure 4.3.13 When ethene and bromine react, 1,2-dibromoethane is formed. In the presence of chloride ions, 1-chloro-2-bromoethane is formed too, as the Cl^- ions compete with the Br^- ions to attack the carbocation.

Some reaction mechanisms involve radicals, with the movement of a *single* electron rather than a pair of electrons – an example of this kind of reaction is the reaction of a chlorine atom with a molecule of methane, which we met on page 387. In reaction mechanisms with radicals, the movement of a single electron is represented by an arrow with a *single* head (figure 4.3.14).

$$H_3C{-}H \; {\cdot}Cl \rightarrow H_3C{\cdot} + H{-}Cl$$

Figure 4.3.14 A chlorine radical forms a bond with a hydrogen atom in a molecule of methane. This leads to a further electron movement, producing a methyl radical.

Reactions with hydrogen

The alkenes do not react with hydrogen under normal conditions of temperature and pressure. However, in the presence of a finely divided nickel catalyst and at a moderately high temperature (around 200 °C), alkenes undergo an addition reaction with hydrogen to form the equivalent alkane. The reaction of ethene with hydrogen shown in figure 4.3.15 is a typical example.

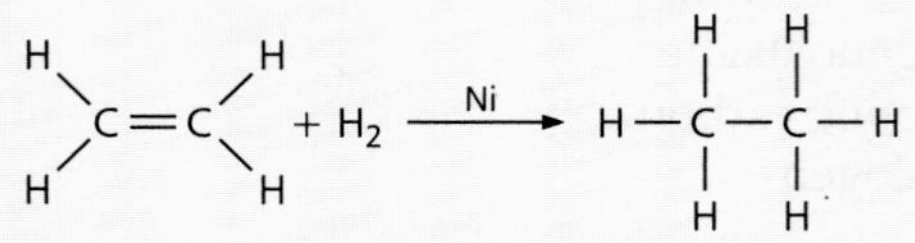

$$CH_2{=}CH_2 + H_2 \xrightarrow{Ni} CH_3{-}CH_3$$

Figure 4.3.15 The addition of hydrogen across a double bond is called **hydrogenation**, a process widely used in the manufacture of margarine.

Reactions with sulphuric acid

The alkenes will react with concentrated sulphuric acid at room temperature, although the reaction does not proceed very rapidly. When ethene reacts with cold concentrated sulphuric acid, the compound ethyl hydrogensulphate is the result. This can be further reacted with water to produce ethanol. This procedure used to be carried out industrially to manufacture ethanol, but nowadays the direct catalytic hydration of ethene (addition of water across the double bond) is more commonly used. Similarly, propan-2-ol can be produced from the reaction between propene and concentrated sulphuric acid, which is

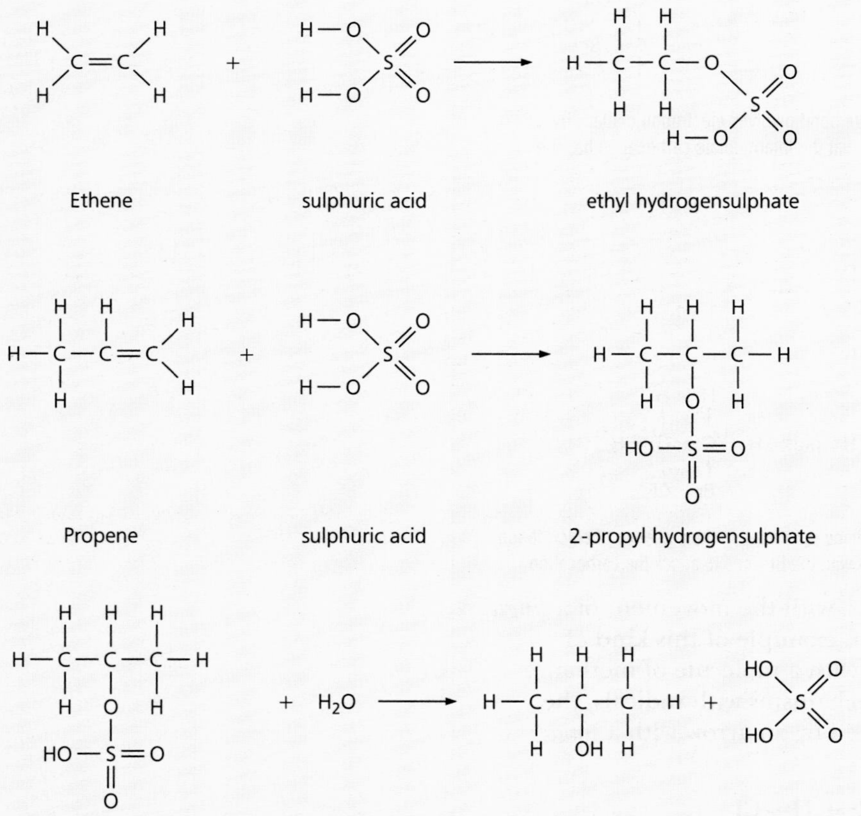

Ethene + sulphuric acid → ethyl hydrogensulphate

$$CH_2{=}CH_2 + (HO)_2SO_2 \longrightarrow CH_3CH_2{-}O{-}SO_2{-}OH$$

Propene + sulphuric acid → 2-propyl hydrogensulphate

$$CH_3{-}CH{=}CH_2 + (HO)_2SO_2 \longrightarrow CH_3{-}CH(OSO_2OH){-}CH_3$$

2-Propyl hydrogensulphate + water → propan-2-ol + sulphuric acid

$$CH_3{-}CH(OSO_2OH){-}CH_3 + H_2O \longrightarrow CH_3{-}CH(OH){-}CH_3 + (HO)_2SO_2$$

Figure 4.3.16 The addition of water across a double bond produces an alcohol. Sulphuric acid acts as a catalyst for the reaction.

still an industrial process of some importance. The reactions are shown in figure 4.3.16. Again, Markovnikov's rule says that propan-2-ol rather than propan-1-ol is the preferred product.

Reactions with acidified potassium manganate(VII)

The reaction of the alkenes with acidified potassium manganate(VII) involves both addition across the double bond and oxidation. The products of the reaction are alkanediols, and the purple manganate(VII) solution is decolorised in the process. For example, when ethene is bubbled through an acidified solution of potassium manganate(VII), the solution decolorises and ethane-1,2-diol is formed.

$$\underset{\text{Ethene}}{5CH_2{=}CH_2} + 2H_2O + 2MnO_4^- + 6H^+ \rightarrow \underset{\text{ethane-1,2-diol}}{5HOCH_2CH_2OH} + 2Mn^{2+}$$

Ethane-1,2-diol, an extremely useful chemical in a variety of industrial processes, is not made in this way industrially because potassium manganate(VII) is a relatively expensive chemical and it is cheaper and more efficient to produce ethane-1,2-diol from epoxyethane, produced in the catalysed reaction of ethene with oxygen described earlier.

The reaction of ethene with potassium manganate(VII) can be used to identify the alkenes from the alkanes and alkynes, as the alkanes do not react and the alkynes react to form organic acids.

Polymerisation – the most important reactions of the alkenes

What is a polymer?

A **polymer** is made up of very large molecules, which are long chains of repeated smaller units joined together. These smaller units are known as **monomers**. Molecules are generally regarded as polymers rather than just very large molecules when there are around 50 or more monomer units in the chain.

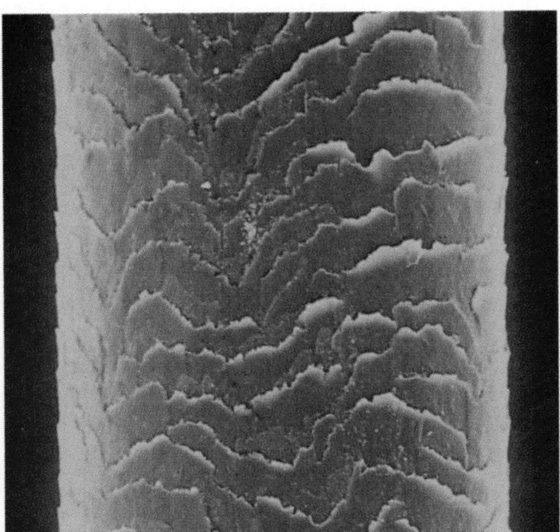

Human hair magnified 400 times

```
  H   H   H   H   H   H
  |   |   |   |   |   |
—C—C—C—C—C—C—
  |   |   |   |   |   |
  H   H   H   H   H   H
```

Part of a poly(ethene) chain

Figure 4.3.17 The living world is full of polymers – for example proteins, an important part of cell structure and hair, are polymers made up of long chains of monomers called amino acids. For the last 50 years or so, there has been a rapid increase in our ability to produce synthetic polymers such as the poly(ethene) used for carrier bags. Alkenes and alkene derivatives, with their reactive double bonds, play a large part in polymer manufacture.

4

Polymers are formed in one of two main ways.

- **Addition/elimination** or **condensation reactions**, in which a small molecule, often water, is eliminated as a bond is formed between the monomer units. Most natural polymers are formed in this way and so are some synthetics.

Monomer + monomer + monomer + ... → polymer + water

- **Addition reactions**, in which two (usually identical) monomer units react together and an addition reaction takes place across a double or triple bond. The polymer is the only product.

Monomer + monomer + monomer + ... → polymer

Designing new materials

Synthetic polymers are an excellent example of how chemists can take chemicals from naturally occurring materials such as oil and coal, and create novel materials of great usefulness. The double bonds of the alkenes and their tendency to undergo addition reactions means that they and their derivatives are frequently used in polymerisation reactions. Hundreds of alkene monomers react with each other to form the long-chain polymers which have become a fundamental part of our everyday life.

Ethene, the simplest of the alkenes, undergoes polymerisation to form poly(ethene), more commonly referred to as polythene. The double bonds 'break open' and the carbon atoms link together to form a long chain, as shown in figure 4.3.19.

Figure 4.3.18 The polymers of alkenes and their derivatives have found a multitude of uses in modern life. Polymer chemists can now develop 'designer polymers' with exactly the properties wanted for a particular task.

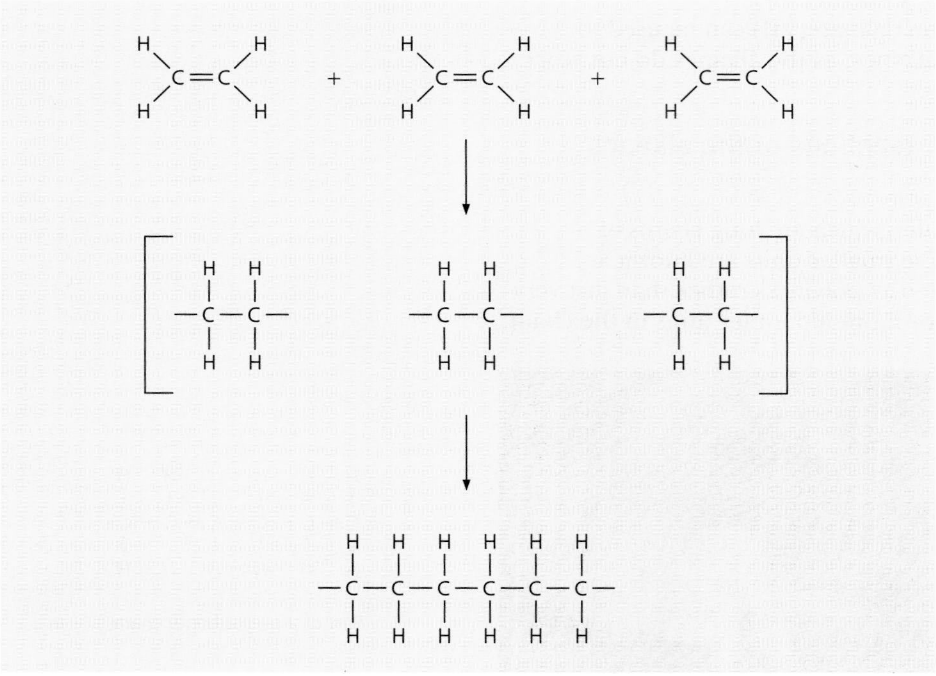

$$H_2C{=}CH_2 + H_2C{=}CH_2 + H_2C{=}CH_2$$

$$\downarrow$$

$$\left[-CH_2-CH_2- \quad -CH_2-CH_2- \quad -CH_2-CH_2- \right]$$

$$\downarrow$$

$$-CH_2-CH_2-CH_2-CH_2-CH_2-CH_2-$$

Figure 4.3.19 The formation of poly(ethene)

Poly(ethene) was first discovered in 1933 by the British chemical company ICI. The first poly(ethene), made by the polymerisation of ethene at 200 °C and 1200 atm with traces of oxygen, had highly branched chains and so was relatively soft and malleable with a fairly low melting point. This material is useful for packaging and for forming utensils, and it is still the main form of poly(ethene) made today. Because of its branching molecules it is known as **low-density poly(ethene)**. Karl Ziegler, a German chemist who won a Nobel prize in 1963 for his work on polymers and plastics, developed a different way of producing poly(ethene) that uses catalysts and as a result needs only low temperatures (around 60 °C) and atmospheric pressure. The poly(ethene) produced by the Ziegler process is called **high-density poly(ethene)**, as it has few branched chains and is thus more rigid, denser and melts at a higher temperature than low-density poly(ethene) (135 °C as opposed to 105 ° C).

Figure 4.3.20

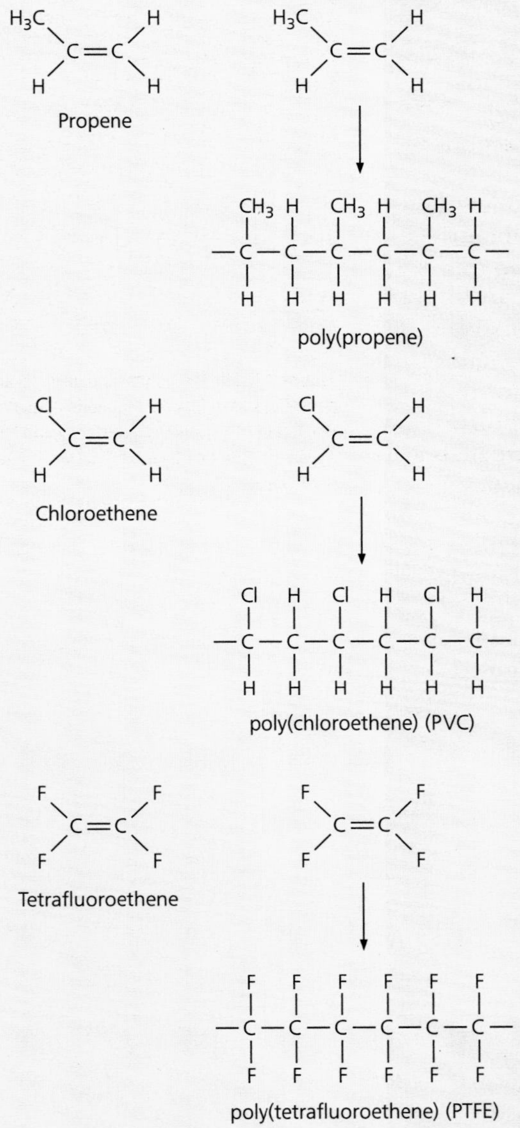

Propene

poly(propene)

Chloroethene

poly(chloroethene) (PVC)

Tetrafluoroethene

poly(tetrafluoroethene) (PTFE)

Other plastics made from the alkenes include poly(propene), which is also made using the Ziegler process. Poly(propene) has closely and regularly packed chains. It can be moulded, used in films or turned into fibres to make, for example, ropes that do not rot. Chloroethene polymerises to form poly(chloroethene), better known as PVC from its non-systematic name of polyvinylchloride. These reactions are shown in figure 4.3.20.

As a result of the polar nature of the C—Cl bond, there are many intermolecular forces in PVC, and so the polymer is rather rigid. The addition of other compounds known as **plasticisers** makes the plastic much softer, and increases its range of uses considerably. PTFE, or polytetrafluoroethene, is better known by the trade name 'Teflon'. This polymer of fluoroethene forms a safe, non-stick coating widely used on saucepans and frying pans to prevent food sticking to the utensil as it cooks.

The properties of polymers

The properties of polymers depend on the chains of monomers of which they are made. Polymers vary greatly, from very soft and flexible solids with very low melting points to hard, brittle materials with very high melting points. What features of the polymer chain have most effect on the properties of the polymers?

- The average length of the polymer chain – tensile strength and melting point both increase with the length of the chain until there are about 500 monomer units in the chain. Increase in chain length beyond this has relatively little effect on the properties of the polymer.
- Branching of the chain – branched chains cannot pack together so regularly as straight chains so the polymers with very branched chains tend to have low tensile strengths, low melting points and low densities.
- The presence of intermolecular forces between chains – these are of immense importance in natural polymers, and also affect synthetics. If there are strong intermolecular forces between the chains, the polymer will be strong and tend to have a high melting point. The strength of the intermolecular forces is largely determined by the side groups on the polymer chain.
- Crosslinking between chains – rigid fixed bonds between the chains make the polymer rigid, hard and brittle, usually with a very high melting point.

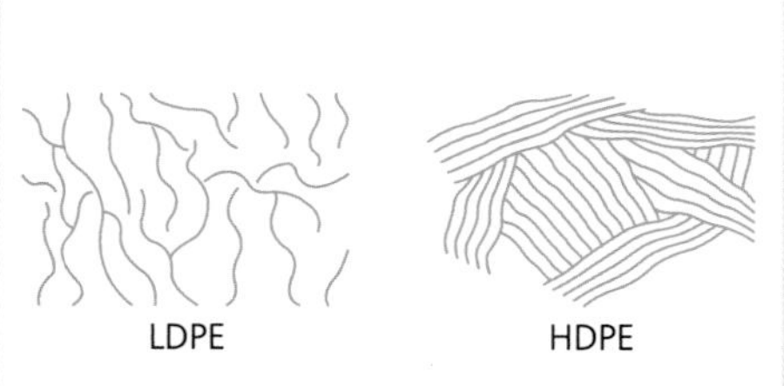

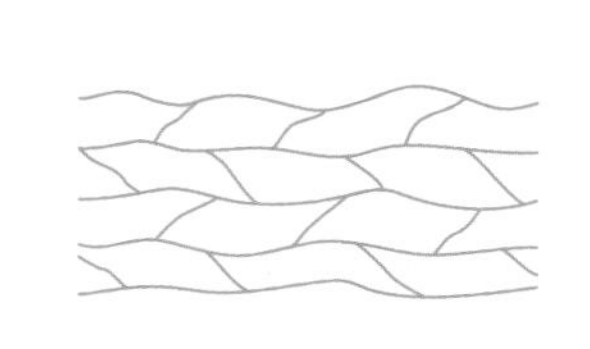

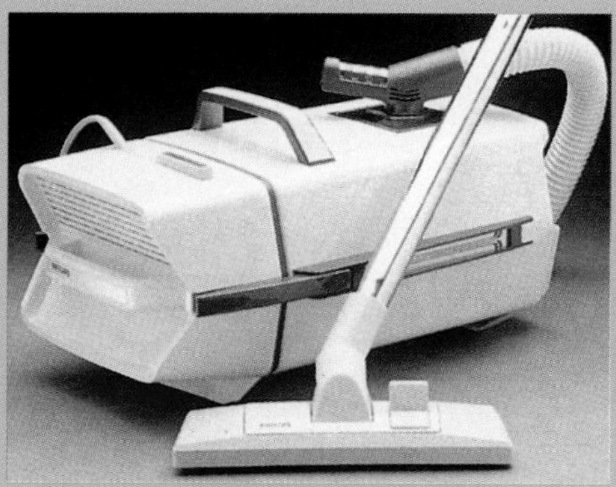

The highly crystalline structure of high-density poly(ethene) results in considerable strength in the finished polymer.

Covalent bonds linking the chains in synthetic rubber give it **resilience** – the ability to be deformed and to recover over and over again – a necessary property for a motor vehicle tyre.

Figure 4.3.21 The physical properties of a polymer are greatly influenced by the structure of the polymer chains and the forces between them.

If we take a monomer like propene or chloroethene, as the monomer molecules polymerise they produce one chiral carbon atom in the polymer molecule for each molecule of monomer that polymerises. These chiral carbon atoms can be arranged in three different ways, as figure 4.3.22 shows.

(a) An atactic polymer. The side groups (Cl in the case of poly(chloroethene) and CH_3 groups in the case of poly(propene)) are arranged randomly on either side of the polymer chain.

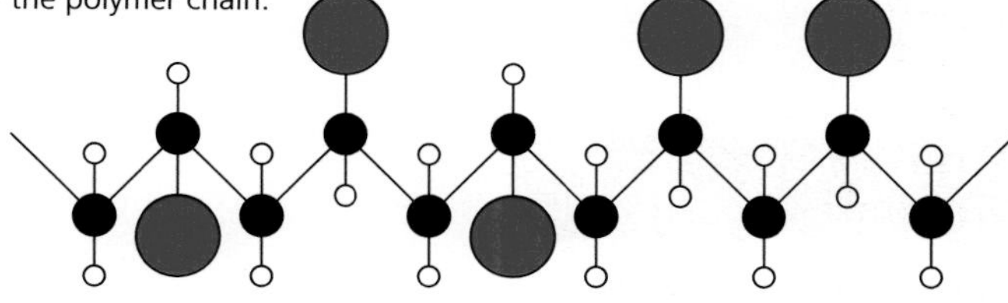

(b) A syndiotactic polymer. The side groups are arranged alternately on each side of the polymer chain.

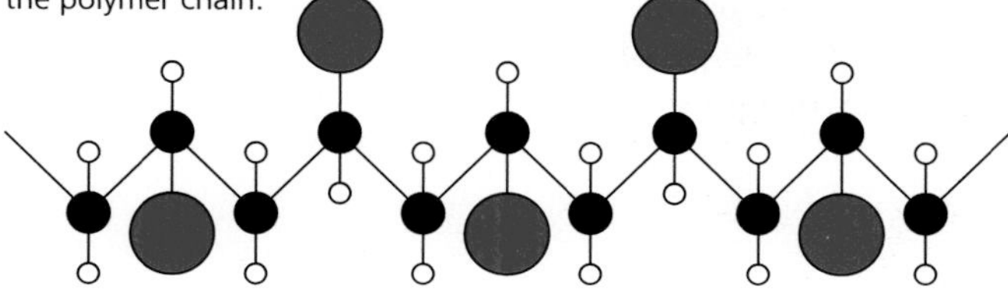

(c) An isotactic polymer. The side groups all lie on one side of the polymer chain.

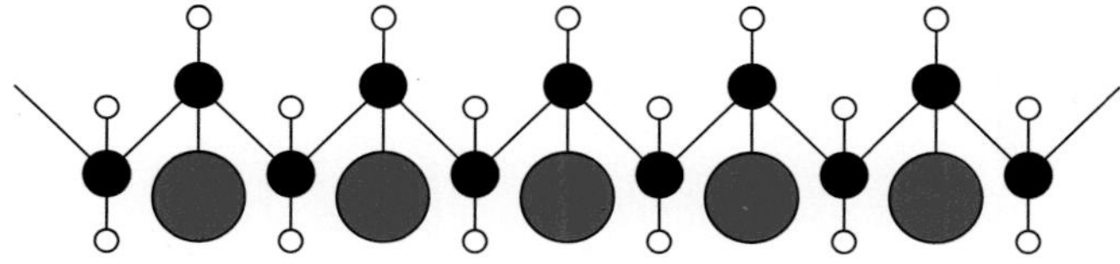

Figure 4.3.22 The different arrangements of side chains in the finished polymer greatly influences its behaviour. An atactic polymer has little ordered overall structure due to the random arrangement of the side groups – it is said to be **amorphous**. The polymer chains in isotactic polymers on the other hand can pack more easily together, producing a polymer with a crystalline structure. The formation of the different kinds of polymers can be influenced by reaction conditions.

The development of synthetic polymers has led to a great change in the world we live in, with many natural materials having been replaced by synthetic equivalents. On first sight it might seem that the development of plastics has saved the use of much natural material, and prevented the destruction of natural resources. However, the ecological effects are not as straightforward as this. The production of plastics depends on chemicals from fossil fuels, supplies of which are dwindling, and the disposal of plastics causes further difficulties.

The alkynes

The **alkynes** are a further family of hydrocarbons, containing one or more triple bonds between carbon atoms in the molecule. The monoalkynes contain one such triple bond, and have the general formula C_nH_{2n-2}. The alkynes follow the standard IUPAC nomenclature of the alkanes and alkenes, but with the ending -yne. Some examples are given in figure 4.3.23 (overleaf).

The triple bond of the alkynes makes them more reactive than the alkanes. Surprisingly, the alkynes are somewhat less reactive than the alkenes on the whole, because they are more kinetically stable than the alkenes. Apart from this difference, the chemistry of the two groups is generally very similar indeed.

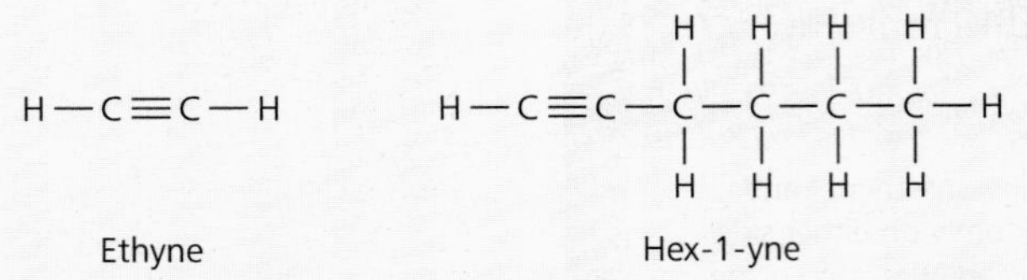

Figure 4.3.23

H—C≡C—H

Ethyne

H—C≡C—CH_2—CH_2—CH_2—CH_3 (each of the four carbons after the triple bond drawn with H above and H below; terminal C—H)

Hex-1-yne

The shape of the ethyne molecule

In ethyne, the two carbon atoms are sp hybridised, leaving two p orbitals with a single electron in each. These p orbitals overlap to form two π bonds at right-angles to each other, with one sp orbital of each carbon atom overlapping to form the σ bond. Figure 4.3.24 shows the shape of the ethyne molecule – notice that it is linear, with the two hydrogens situated at each end of the molecule.

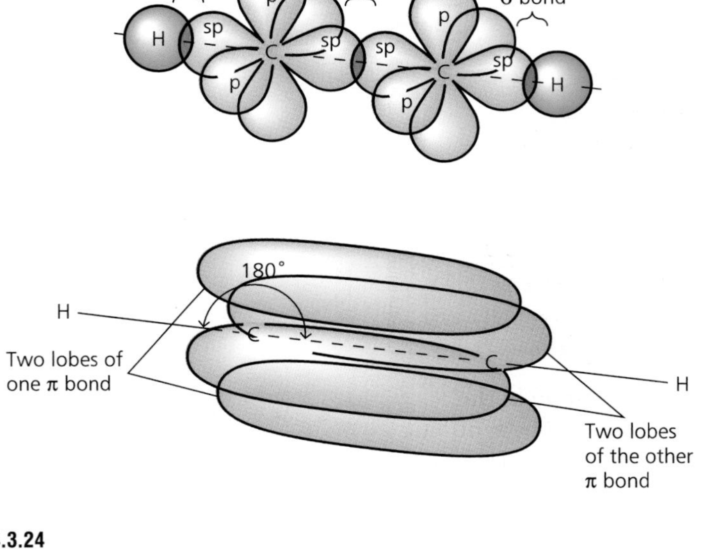

Figure 4.3.24

Reactions of the alkynes

The high electron density associated with a triple bond means that the alkynes are attacked by both electrophiles and oxidising agents. As the alkynes have a triple bond, there are frequently two alternative products in their addition reactions. Addition across one of the C—C bonds results in an alkene derivative, while addition across two of the C—C bonds results in an alkane derivative. This can be seen more clearly as we look at the reactions of the alkynes in more detail, although because their reactions are so similar to those of the alkenes we shall not look at so many examples.

Reactions with oxygen

Just like the alkenes, the alkynes can react explosively in air to form carbon dioxide and water and a large amount of energy. A controlled combustion of ethyne in air produces a flame with temperatures in excess of 2000 °C, and this is used industrially to 'cut' through metals, as shown in figure 4.3.25. The metals are oxidised rapidly in the flame.

Figure 4.3.25 The **oxyacetylene torch** (the non-systematic name for ethyne is acetylene) produces a flame so hot that it can be used for cutting through metal under water in rescue and salvage operations.

Reactions with halogens

With the halogens, the alkynes undergo addition reactions across the triple bond, as shown in figure 4.3.26. These can be very violent – ethyne and chlorine will explode if they are mixed, in a rapid reaction involving radicals. However, if the same mixing takes place in the presence of an inert material that will 'soak up' the radicals, then a much more controlled addition reaction occurs. Kieselguhr, made of silicon(IV) oxide, is a typical inert substance used to control these reactions.

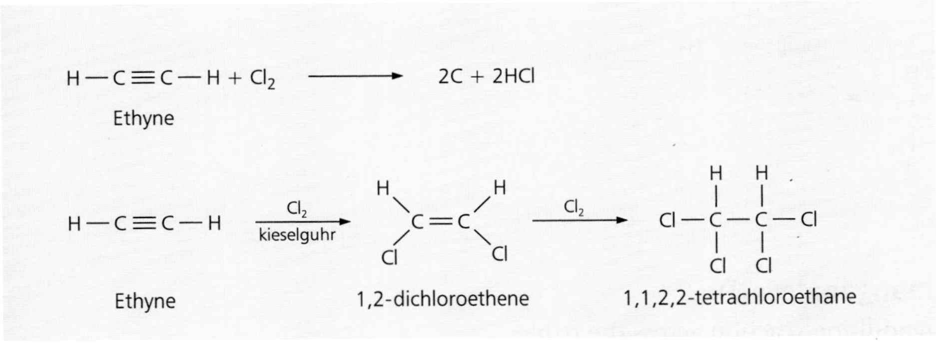

$$\underset{\text{Ethyne}}{H-C\equiv C-H} + Cl_2 \longrightarrow 2C + 2HCl$$

$$\underset{\text{Ethyne}}{H-C\equiv C-H} \xrightarrow[\text{kieselguhr}]{Cl_2} \underset{\text{1,2-dichloroethene}}{HClC=CClH} \xrightarrow{Cl_2} \underset{\text{1,1,2,2-tetrachloroethane}}{Cl_2HC-CHCl_2}$$

Figure 4.3.26

The alkynes react slowly with bromine water to decolorise it, and this reaction can be used to distinguish between alkenes and alkynes – alkenes decolorise bromine water very rapidly, but alkynes take several minutes. Figure 4.3.27 shows the reaction.

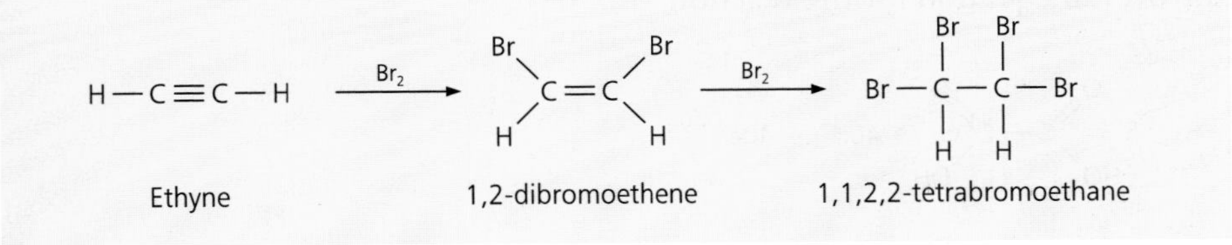

$$\underset{\text{Ethyne}}{H-C\equiv C-H} \xrightarrow{Br_2} \underset{\text{1,2-dibromoethene}}{BrHC=CHBr} \xrightarrow{Br_2} \underset{\text{1,1,2,2-tetrabromoethane}}{Br_2HC-CHBr_2}$$

Figure 4.3.27

Reactions with hydrogen halides

Addition reactions occur across the triple bond with hydrogen halides. In the first stage of the reaction, a halogenoalkene is produced as shown in figure 4.3.28 (overleaf). The second stage gives a halogenoalkane, and the addition follows Markovnikov's rule. This means that one of the carbon atoms carries both of the halogen atoms, and the other carries all of the hydrogen. The hydrogen adds to the atom already carrying the largest number of hydrogen atoms.

Reactions with hydrogen

Alkynes react with hydrogen in the presence of a finely divided nickel catalyst at 200 °C to form first the equivalent alkene, and then the alkane, as shown in figure 4.3.29 (overleaf).

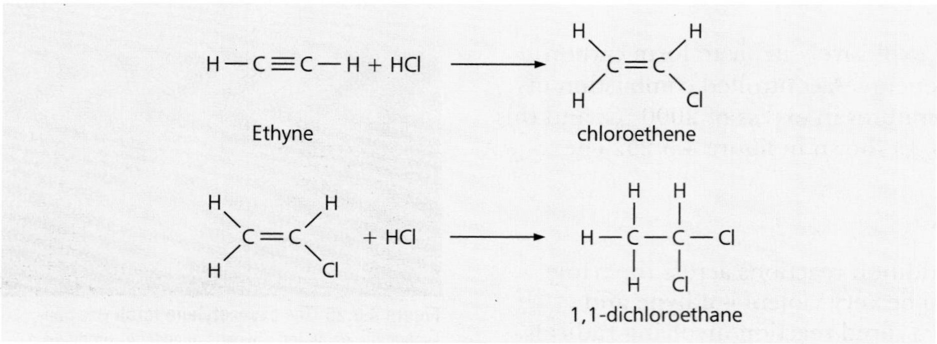

$$H-C\equiv C-H + HCl \longrightarrow H_2C=CHCl$$

Ethyne chloroethene

$$H_2C=CHCl + HCl \longrightarrow H-CH_2-CHCl-Cl$$

1,1-dichloroethane

Figure 4.3.28 Reactions with hydrogen halides.

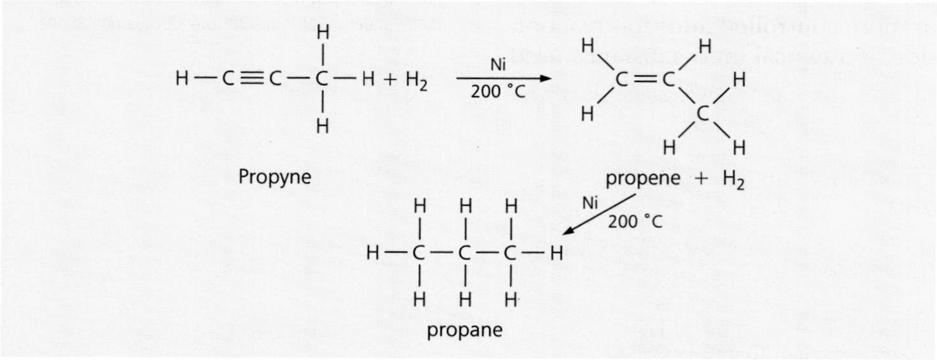

$$H-C\equiv C-CH_3 + H_2 \xrightarrow[200\ ^\circ C]{Ni} H_2C=CH-CH_3$$

Propyne propene

$$\text{propene} + H_2 \xrightarrow[200\ ^\circ C]{Ni} CH_3-CH_2-CH_3$$

propane

Figure 4.3.29 Reactions with hydrogen.

Reactions with acidified potassium manganate(VII)

The alkynes, like the alkenes, show both an addition reaction across the triple bond and an oxidation reaction to form acids. For example, ethyne reacts with acidified potassium manganate(VII) to form ethanedioic acid, as shown in figure 4.3.30. Again, the manganate(VII) solution is reduced to manganate(II) and this is shown by the change in the colour of the solution from purple to colourless. Like any redox reaction, this reaction can be written as two half-equations which can be combined to give the overall equation for the reaction.

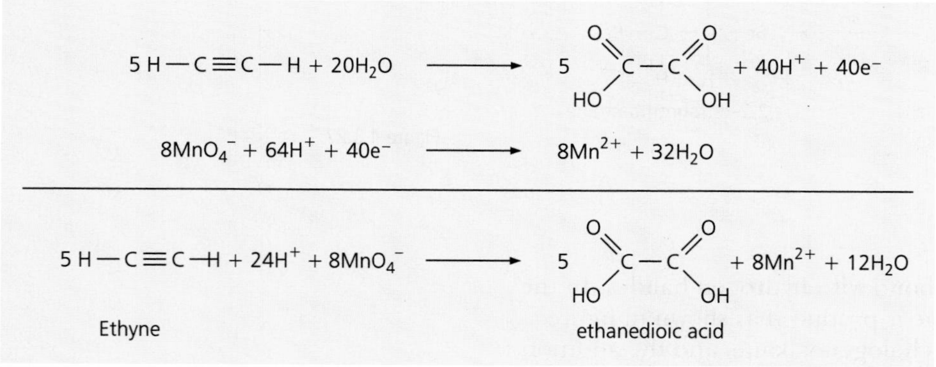

$$5\,H-C\equiv C-H + 20H_2O \longrightarrow 5\,HOOC-COOH + 40H^+ + 40e^-$$

$$8MnO_4^- + 64H^+ + 40e^- \longrightarrow 8Mn^{2+} + 32H_2O$$

$$5\,H-C\equiv C-H + 24H^+ + 8MnO_4^- \longrightarrow 5\,HOOC-COOH + 8Mn^{2+} + 12H_2O$$

Ethyne ethanedioic acid

Figure 4.3.30

The hydrocarbons are a very large and important group of organic chemicals. By studying them we can learn much about the principles of organic reactions in general. We shall now move on to look at some organic families containing elements other than the simple carbon and hydrogen of the hydrocarbons.

SUMMARY

- The **alkenes** are a family of unsaturated hydrocarbons – they contain a carbon–carbon double bond. They have a general formula of C_nH_{2n}.
- The straight-chain alkenes show regular trends in physical properties and reactivities.
- Alkenes are not found widely in nature. They are produced by the cracking of crude oil.
- The double bond has a major effect on the chemistry of the alkenes. Lack of rotation around the double bond means that *cis* and *trans* isomers occur.
- **Addition reactions** occur across the double bond more readily than the substitution reactions seen in the alkanes. Less energy is required to break the double bond than is needed to break two carbon–carbon single bonds.
- The main reactions of the alkenes are **electrophilic addition reactions** across the double bond. They are summarised in table 4.3.1.

Alkene +	Reaction
Air/oxygen	Alkenes burn readily in air to form carbon dioxide and water.
Halogens	Alkenes react rapidly with halogens to form halogenoalkanes.
Hydrogen halides	Alkenes react to give halogenoalkanes.
Hydrogen	Alkenes do not react with hydrogen at room temperature and pressure, but with a finely divided nickel catalyst at about 200 °C an addition reaction occurs, forming the equivalent alkane.
Conc sulphuric acid	Alkenes react to form alkyl hydrogensulphates.
Acidified $KMnO_4$	Alkenes react to form alkanediols. The reaction involves both addition and oxidation.

Table 4.3.1

- Alkene **monomers** can be polymerised to form long-chain **polymers**.
- **Condensation polymers** are formed by the elimination of a small molecule (often water) as bonds form between two monomer units.
- **Addition polymerisation** involves monomer units joining in an addition reaction. The polymer is the only product. Poly(ethene) is an addition polymer.
- The properties of polymers are affected by:
  the average length of the polymer chain
  the degree of branching in the chain
  the presence of intermolecular forces between regions of the chains
  cross-linking between the chains.
- The **alkynes** are a family of hydrocarbons which have a triple bond between two of the carbon atoms in the molecule. They have the general formula C_nH_{2n-2}.
- The triple bond means alkynes are generally more reactive than alkanes, but they are less reactive than the alkenes because they are kinetically more stable.
- The high electron density of the triple bond means that alkynes are attacked by both electrophiles and oxidising agents. The reactions of the alkynes are summarised in table 4.3.2.

Alkyne +	Reaction
Air/oxygen	Alkynes may react explosively in air to form carbon dioxide and water. In controlled combustion they may be used to create very hot flames.
Halogens	Alkynes react with halogens to form halogenoalkenes and then halogenoalkanes.
Hydrogen halides	Alkynes react to give halogenoalkenes and then halogenoalkanes.
Hydrogen	With a finely divided nickel catalyst at about 200 °C, alkynes will undergo an addition reaction forming first alkenes and then alkanes.
Acidified $KMnO_4$	Alkynes react to form alkanedioic acids. The reaction involves both addition and oxidation.

Table 4.3.2

QUESTIONS

1 a Give structural formulae for the following compounds:
i pent-2-ene
ii propyne
iii 2-methylbut-2-ene.

b Name the following compounds:
i $CH_3CH_2CH{=}CHCH_2CH_3$ (what are the two possible isomers?)
ii $CH_2{=}CH{-}CH{=}CH_2$
iii 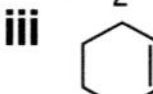
iv $CH_3CH_2C{\equiv}CH$

2 a Bond formation in the carbon atom involves the hybridisation of s and p orbitals to give sp hybridised orbitals. Explain in terms of orbitals what happens in:
i a carbon–carbon single bond
ii a carbon–carbon double bond.

b Explain the role of the carbon–carbon double bond in *cis-trans* isomerism.

3 a Describe, giving chemical equations, the reaction of ethene burning in air.

b If ethene reacts with oxygen at 180 °C in the presence of a finely divided silver catalyst, how does the reaction differ from that you have described in **a**?
The product of this reaction is very useful in several industrial syntheses. What feature of its chemistry makes it useful?

c What is the main use of the reactions of the alkynes with oxygen?

4 The reactions of the alkenes and alkynes with the halogens are very distinctive and may be used to identify them from the alkanes.

a Propene is bubbled through bromine in the dark. Describe using chemical equations the reaction you would expect.

b What is the mechanism of this reaction?

c How does the reaction with the halogens differ between the alkenes and the alkanes?

5 Predict, giving chemical equations, what would happen in the following reactions:

a Propene is mixed with hydrogen in the presence of a finely divided nickel catalyst at around 200 °C.

b Ethyne is passed through acidified potassium manganate(VII).

c Ethyne is reacted with chlorine in the presence of kieselguhr.

d Ethene is passed through cold concentrated sulphuric acid.

6 Alkenes and alkynes play an extremely important role in society as the monomer units of many polymers. Describe, using examples from the alkenes and alkynes, the process of polymerisation and the different types of plastics that may result. Give examples of some of the advantages and disadvantages of plastics.

Developing Key Skills

You have been invited to take part in a radio debate.
The proposal is that 'The use of all plastics which are not biodegradable should be banned'. Record a two minute speech either supporting or opposing this idea, backing up your view with facts and evidence as far as possible.

[Key Skills opportunities: C]

FOCUS ANTISEPTIC SAVES LIVES

Beating bacteria – antiseptics and disinfectants

The idea that diseases are caused by microorganisms is deeply entrenched in our modern philosophy of health and treatment, yet people have only been aware of the existence of bacteria, viruses and other microorganisms since the middle of the nineteenth century. Once the role of these organisms in causing disease was recognised, steps began to be taken to prevent their spread. The use of **antiseptic** techniques and **disinfectant** solutions began to reduce the numbers of deaths from infectious diseases. Some of the earliest steps were unbelievably fundamental to us today – in 1847 Ignaz Semmelweiss reduced the number of cases of childbed fever (infections that frequently killed women in the few days after giving birth) by the simple procedure of encouraging doctors to wash their hands between deliveries!

Surgery, particularly the amputation of gangrenous parts of the body, was crude and frequently ineffective in stopping the spread of infection. It had been noticed that the likelihood of a surgical wound becoming infected was reduced if it was covered with coal tar. Not only did coal tar help to keep out the air, and thus any infecting microorganisms, it also contained an unrecognised antiseptic chemical – **phenol**, also called carbolic acid.

In the mid-eighteenth century the Scottish doctor Joseph Lister read some of the work of Louis Pasteur in which he described his germ theory of disease. Lister recognised that the infections which killed so many of his patients were caused by these germs, and set out to drive them from his operating theatres. He noticed the antiseptic properties of phenol and began using it in his surgery. It seems incredible to us in the twenty-first century, but in those days a doctor would use a surgical instrument on a patient with an infected wound and then move on to amputate someone else's leg without washing the instrument between patients! But in Lister's wards the instruments, sponges and all the dressings were soaked in a strong solution of carbolic acid (phenol) before they were used. The hands of the surgeon and his assistants were also washed in phenol before they operated. During surgery a vaporiser created a carbolic spray around the wound and after surgery the site of the operation was washed in phenol before it was dressed. As a result of all these measures most of Lister's patients survived their surgery, whereas until his introduction of antiseptic conditions the great majority of all surgical patients died from post-operative infections.

The arrival of antiseptic techniques was a major step forward in the history of medicine, although phenol had some unfortunate side effects of its own. This benzene derivative is acidic, and so as well as preventing wounds from becoming infected it could also, at times, prevent them from healing as a result of its corrosive effect on the skin. Modern antiseptics are often phenol derivatives, much more effective at destroying bacteria and also much kinder to human tissues, but still with much in common with the molecule that first began to win the battle against infection.

Figure 1 The use of phenol by Lister and his team saved many of their own patients, and many thousands of other lives as the use of antiseptic techniques spread.

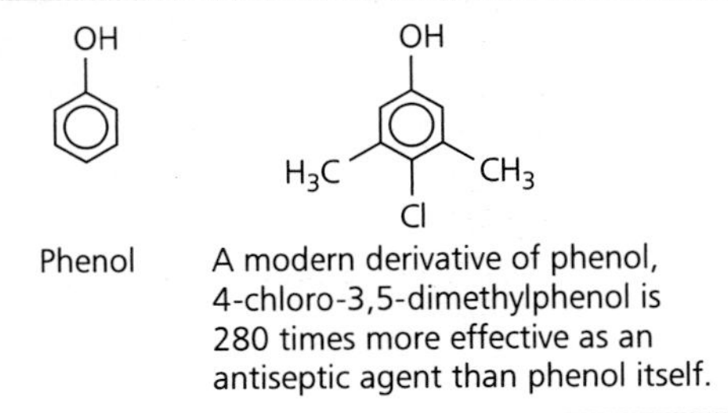

OH
OH
H_3C
CH_3
Cl
Phenol
A modern derivative of phenol, 4-chloro-3,5-dimethylphenol is 280 times more effective as an antiseptic agent than phenol itself.

Figure 2 From phenol, the first compound to be widely used to prevent infection, we now have a wide range of antiseptics and disinfectants designed for use in a variety of situations in the home, in hospitals and in industry.

4

4.4 The unsaturated hydrocarbons: The arenes

In section 4.3 we looked at two homologous series of unsaturated hydrocarbons, the alkenes and the alkynes. These are both families of straight-chain or aliphatic hydrocarbons. The group of hydrocarbons we are now going on to consider do not have straight-chain or branched molecules – they have a ring structure. They are called the **aromatic hydrocarbons**. The name 'aromatic' suggests a group of sweetly perfumed chemicals, and it originates from a time when many of the known members of the family had pleasant smells. As chemists discovered more and more 'aromatic' hydrocarbons, they found that many other members of the group smelled very unpleasant indeed. Our modern understanding of the family tells us that many of these compounds are toxic or carcinogenic (cancer causing), so inhaling the vapour regardless of the smell is not to be recommended. However, the historical name has been retained for the group, and today we are aware that the name refers to a family of organic compounds linked by common structure rather than by common smell. The systematic name for the aromatic hydrocarbons is the **arenes**.

Benzene – the founder molecule

The development of the Kekulé structure

Benzene is the compound that forms the basis of the chemistry of all the aromatic hydrocarbons. It was first isolated in 1825, and has a molecular formula of C_6H_6. (The box 'Unravelling the benzene conundrum' gives more details about the history of the structure of the benzene molecule.) Benzene is a colourless liquid that is immiscible with water and has a characteristic odour. Its boiling point is 80 °C and its melting point is 6 °C. The benzene molecule has a symmetrical six-membered ring structure, which accounts for its relatively high melting and boiling points, as the rings pack tightly into a crystal lattice. Many important compounds contain benzene rings with, for example, an alkyl group substituted for one or more of the hydrogen atoms around the carbon ring. The melting point of methylbenzene, –95 °C, illustrates the effect of these alkyl side chains on the packing of the benzene molecules.

The very high ratio of hydrogen to carbon in the C_6H_6 molecule suggests an extremely unsaturated molecule, and chemists in the past have suggested a number of aliphatic structures for it, such as those shown in figure 4.4.1.

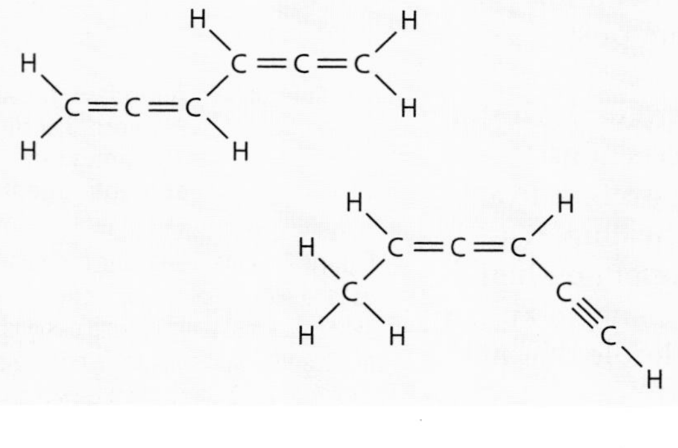

Figure 4.4.1 Two possible alternative structures for aliphatic benzene

These structures are not very good at explaining the properties of benzene, however. In the first place, benzene is very unreactive, failing to react with bromine even when heated. If either of these suggested structures for benzene were correct, the substance would be very reactive as a result of its double and/or triple bonds. Secondly, such structures would also lead to many isomers of monosubstituted compounds due to the lack of rotation around the multiple bonds, as figure 4.4.2 shows. However, only one form of bromobenzene and one form of chlorobenzene have ever been isolated.

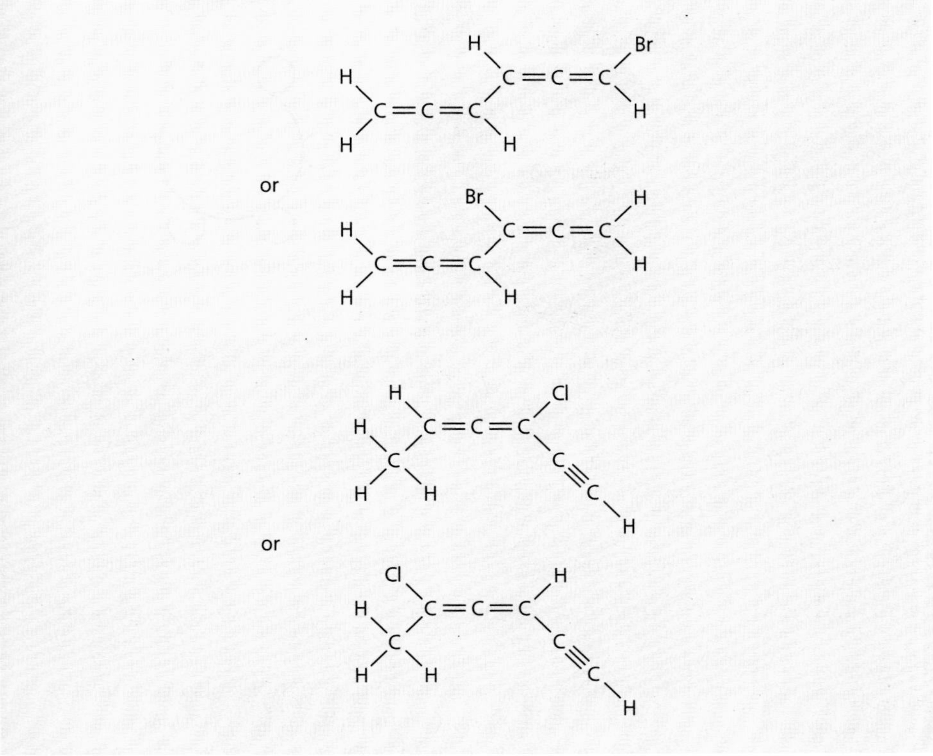

Figure 4.4.2 Isomers of monosubstituted aliphatic benzene

The aliphatic structures for benzene do not, therefore, explain its observed properties. The structure of benzene puzzled chemists for some years, until in 1865 the ring structure of benzene was recognised. In the model for benzene generally attributed to Friedrich August Kekulé, benzene consists of a six-membered carbon ring with alternate double and single bonds between the carbon atoms, as shown in figure 4.4.3. The Kekulé model of benzene was used for many years, and was capable of explaining many of the observed properties and reactions of this important compound.

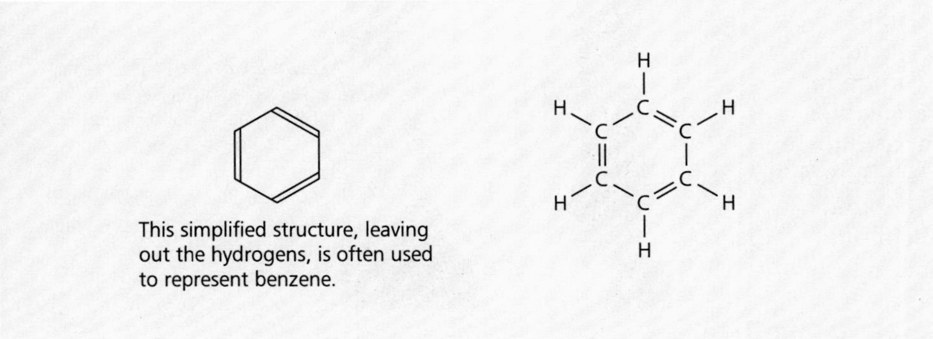

Figure 4.4.3 Kekulé's benzene ring structure

Unravelling the benzene conundrum

Figure 4.4.4 Benzene is an enormously useful chemical, usually obtained by catalytic reforming from fractions of crude oil. In the past it was obtained by the distillation of coal tar. It was first isolated by the distillation of whale oil, in the days when whale oil was a common commodity used for lighting homes.

- In 1825 Michael Faraday (also known for his work on electricity and magnetism) isolated benzene during a fractional distillation of whale oil. Because the chemical came from an oil used as a source of light, Faraday suggested the name 'phene' which, roughly translated from the Greek, means 'light-bearer'. The name 'benzene' was later taken up instead. This was the era of the general scientist, and so for Faraday to make advances in different fields of science was far from unusual.
- In 1834 the molecular formula of benzene, C_6H_6, was worked out.
- In 1882 the first pure sample of benzene was produced by Victor Meyer. The test that was commonly used at the time to show the presence of benzene failed on his sample. Careful research showed that all previous samples of benzene had been contaminated with thiophen, and that the so-called test for benzene was actually a test for thiophen! Meyer's synthesis of benzene had yielded pure product – hence the false test result.
- The years between 1834 and 1865 saw much speculation about the structure of the benzene molecule. In 1862 Friedrich August Kekulé, a German chemist who had spent some time on the problem of the structure of benzene, had what might be called a revelation about the chemistry of the molecule as he dozed by the fire. Gazing into the flickering flames, more than half asleep, he 'saw' snakelike molecules dancing and writhing in the flames, until one of the snakes grasped its tail in its mouth, making a ring which whirled round. At this point Kekulé startled awake, but held the dream snakes in his mind and in 1865 proposed the now famous benzene structure – a six-membered carbon ring with alternating single and double bonds.

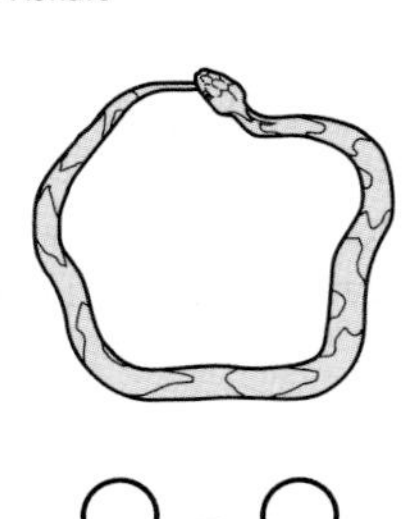

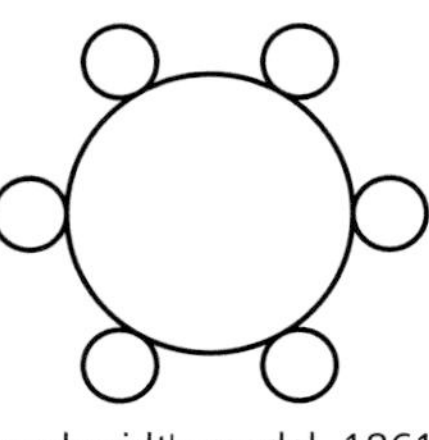

Figure 4.4.5 The story of Kekulé's dream is part of the folk history of organic chemistry. There is no doubt that Kekulé played an important role in developing chemists' understanding of the benzene molecule. However, the first mention of a cyclic structure for benzene was made in a book published in 1861 by Josef Loschmidt, an Austrian school teacher, in which he suggested a six-carbon ring as the most feasible structure for benzene. Kekulé had heard of Loschmidt and his theories, for he refers briefly to them in his paper on the structure of benzene. How much influence Loschmidt's book had on Kekulé's unconscious mind we shall never know.

- Kekulé's model of the benzene molecule accounts for much of its observed chemistry. However, as knowledge of physical chemistry grew, certain aspects of the bond lengths, energy changes and reactivity of benzene could no longer be convincingly explained using the Kekulé model. It was Linus Pauling who wrote the final chapter in the story of the structure of benzene, proposing in 1931 a model in which the benzene molecule consists of a combination of two possible electronic structures.

Figure 4.4.6 As well as proposing a radically new structure for benzene, the American chemist Linus Pauling also won a Nobel prize for his work on chemical bonds and structures. Pauling also proposed an accurate model of the way in which enzymes work, and along with Elias Corey showed the α- helical structure of proteins.

Problems with the Kekulé model

The model for benzene proposed by Kekulé held good for many years. However, as chemistry became an increasingly precise and quantified science, evidence began to build up suggesting that the accepted structure did not wholly explain the observed facts. There were three major strands to this evidence:

1. If the structure of benzene included three double bonds as proposed, it would be expected that these double bonds would show a similar level of reactivity and tendency to undergo addition reactions as the double bonds in the molecules of the alkenes. This is not the case. Most reactions involving benzene are substitution reactions, and addition reactions do not occur easily.
2. X-ray diffraction studies (see section 1.6) can be used to measure bond length. A carbon–carbon single bond in a cyclohexane ring has a bond length of 0.154 nm. A carbon–carbon double bond in cyclohexene has a bond length of 0.133 nm. If the Kekulé model of benzene structure held true, we would therefore expect there to be two different bond lengths measured in the molecule, one for the single and one for the double bonds. In fact, X-ray diffraction gives only a single value for the carbon–carbon bonds in the benzene ring, and the bond length is 0.139 nm. This suggests that all the bonds are the same, and that the bond length is somewhere between that of a single and a double bond.
3. The thermochemical evidence is also against the Kekulé model of benzene. The calculated enthalpy change for the formation of gaseous benzene (with the Kekulé structure) from its elements is $+252$ kJ mol^{-1}.
   In practice, the measured enthalpy change is only $+82$ kJ mol^{-1}. This suggests that the structure of benzene is considerably more energetically stable than the Kekulé model. Further calculations based on the hydrogenation of benzene (see box 'The hydrogenation of benzene – evidence for a new model') gave irrefutable evidence for a different, more energetically stable model of benzene.

The hydrogenation of benzene – evidence for a new model

In section 1.8 we saw how the calculation of bond dissociation enthalpies, and the concept of enthalpy changes (both theoretical and real) during the course of reactions, can help us to predict the outcome of particular reactions, and also to understand the internal arrangement of different molecules. Calculations of this sort have been extremely valuable in considering the structure of benzene.

For most reactions, the enthalpy change ΔH calculated on the basis of bond enthalpies is extremely close to the actual observed enthalpy change during the reaction. However, when we consider the enthalpy change during the hydrogenation of benzene, this simply is not the case. The theoretical enthalpy change for the hydrogenation of benzene is calculated on the basis of the enthalpy change for the hydrogenation of cyclohexene. There is one carbon–carbon double bond in cyclohexene, and when hydrogen is added across it as shown in figure 4.4.7, the enthalpy change is -120 kJ mol^{-1}.

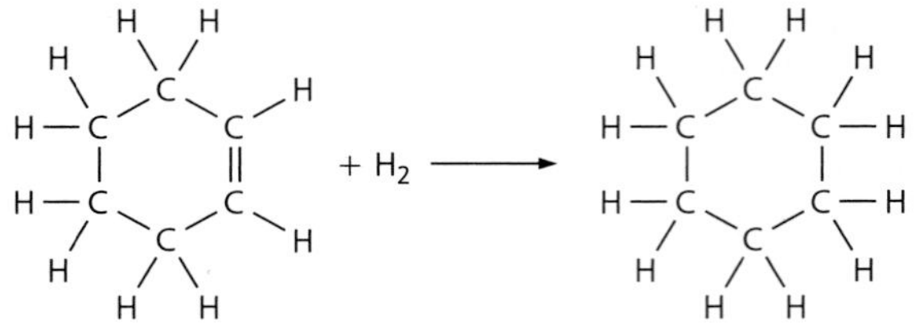

Figure 4.4.7 The hydrogenation of cyclohexene

The Kekulé model of benzene shows three such double bonds in the structure of the molecule, and so it seems reasonable to calculate that the enthalpy change when these three carbon–carbon double bonds react with hydrogen will be three times that of the hydrogenation of cyclohexene. Thus the theoretical value of ΔH for the hydrogenation of benzene based on the Kekulé model is -360 kJ mol^{-1}. When the hydrogenation is carried out practically, the measured enthalpy change is only -208 kJ mol^{-1}. This is substantially less than the theoretical value, and suggests that the addition reaction with hydrogen is not occurring across a normal double bond, as demonstrated by figure 4.4.8. This in turn leads us to a much more stable structure for the benzene molecule than had been previously proposed.

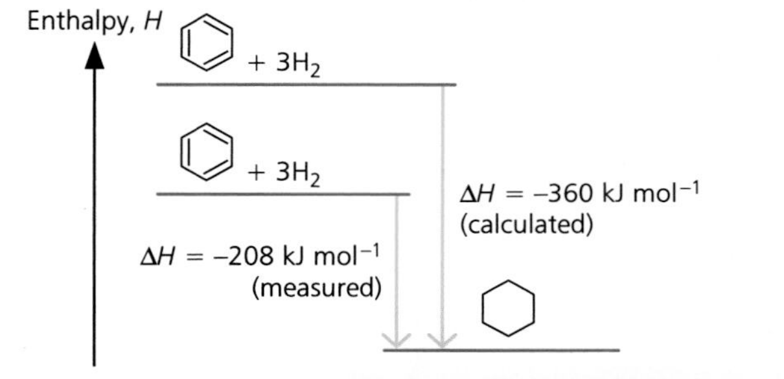

Figure 4.4.8 The difference between the theoretical enthalpy change and the actual enthalpy change for the hydrogenation of benzene can be seen clearly here. Data such as these made it imperative to find a new model for the benzene molecule to explain these discrepancies in its behaviour.

Refining the model

It was the accumulation of this kind of evidence that led Linus Pauling to propose a new way of looking at the structure of the benzene molecule in 1931. Pauling's idea was to treat the benzene molecule as if it was half-way between two possible Kekulé structures, as shown in figure 4.4.9. In this model, the real structure of benzene is called a **resonance hybrid**. The model explains the observed bond lengths in benzene, as well as its lack of reactivity, since each bond in the ring is somewhere between a single and a double bond. According to Pauling, this idea also explains the extra thermodynamic stability of benzene, since the resonance hybrid is lower in energy than either of the two possible Kekulé structures.

We need to represent this model of the structure of benzene simply on paper. The Kekulé model was easy – alternate single and double bonds around

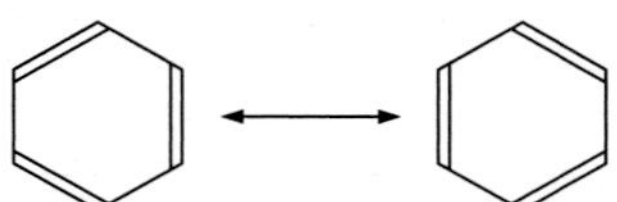

Figure 4.4.9 Pauling proposed representing the structure of 'real' benzene as being a mixture of equal parts of the two possible Kekulé structures for benzene. The double-headed arrow indicates that the real structure is half-way between the two structures (it is a resonance hybrid), *not* that there is a dynamic equilibrium between the two.

the ring. How can we represent bonds which are half-way between single and double bonds all the way round the ring? The solution is to show the benzene molecule as in figure 4.4.10.

This planar structure, in which all the bonds between the carbon atoms in the ring are identical, represents benzene as a non-polar molecule. The electrons that form the carbon–carbon bonds, rather than being closely associated with particular pairs of carbon atoms, are 'shared' around the ring, forming what is known as a **delocalised system**. This arrangement of electrons is very stable, which explains the lack of reactivity.

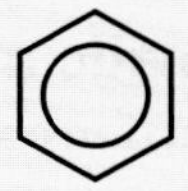

Figure 4.4.10

The arrangement of the electrons in benzene

Although Pauling's model for benzene provides one way of understanding its properties, a fuller insight comes from examining the arrangement of the orbitals of the six carbon atoms in the ring. Each carbon atom is sp^2 hybridised, and forms a σ bond with the hydrogen and with each of the two carbon atoms joined to it. This forms the skeleton of the molecule, giving a planar hexagonal ring. (Note that the angle between two sp^2 hybrid orbitals is 120°, meaning that there is no strain in this arrangement, another factor contributing to the stability of the molecule.)

Each sp^2 hybridised carbon atom has one unused p orbital containing a single electron, and the lobes of these p orbitals then fuse to form a single ring of charge above and below the σ bonded skeleton, as shown in figure 4.4.11. Within these π bonds, the electrons are free to move around the entire ring, so that these electrons are referred to as being **delocalised**. It is this delocalisation that causes all the bonds in benzene to be similar, and that also makes the benzene molecule so unexpectedly stable.

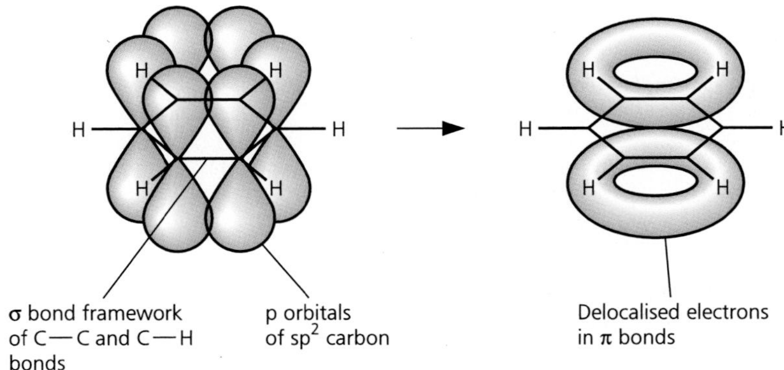

Figure 4.4.11 The delocalisation of the electrons in the π bonds of benzene explains why the chemical and physical properties of benzene are so different from those of straight-chain compounds containing double bonds.

At the beginning of this section, we considered the name 'aromatic' for the hydrocarbon compounds related to benzene, and saw that the name is not an accurate description of the family. A much more accurate definition of the aromatic hydrocarbons would be:

> **Any hydrocarbon system which has a stabilised ring of delocalised π electrons**

Reactions of benzene

Naming aromatic molecules

Most of the reactions of benzene produce other aromatic molecules, and before looking at the reactions themselves it is useful to look at the way these other aromatic hydrocarbons are organised and named.

When benzene itself is regarded as a side chain, in the same way as alkyl groups such as the methyl group, it is known as the **phenyl** group, C_6H_5-. The name 'phenyl' relates back to Michael Faraday's original suggestion of 'phene' for the newly discovered benzene.

The hydrogen atoms in the benzene ring may be substituted by other atoms or groups to give compounds such as phenol, chlorobenzene or methylbenzene, shown in figure 4.4.12(a).

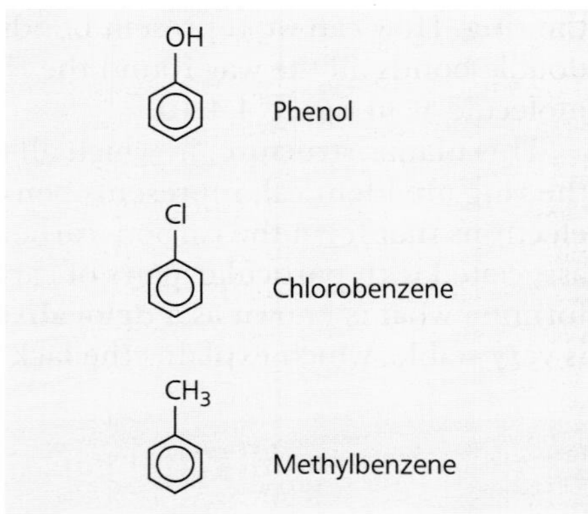

Figure 4.4.12

Because the benzene molecule is symmetrical, it does not matter where in the ring a single substitution like this is shown. It is simply convention that they are placed at the top of the ring. When two or more atoms or groups are substituted onto the benzene molecule, it becomes more critical where they are positioned. In the examples in figure 4.4.13(a), two are the same compound, but one is a structural isomer.

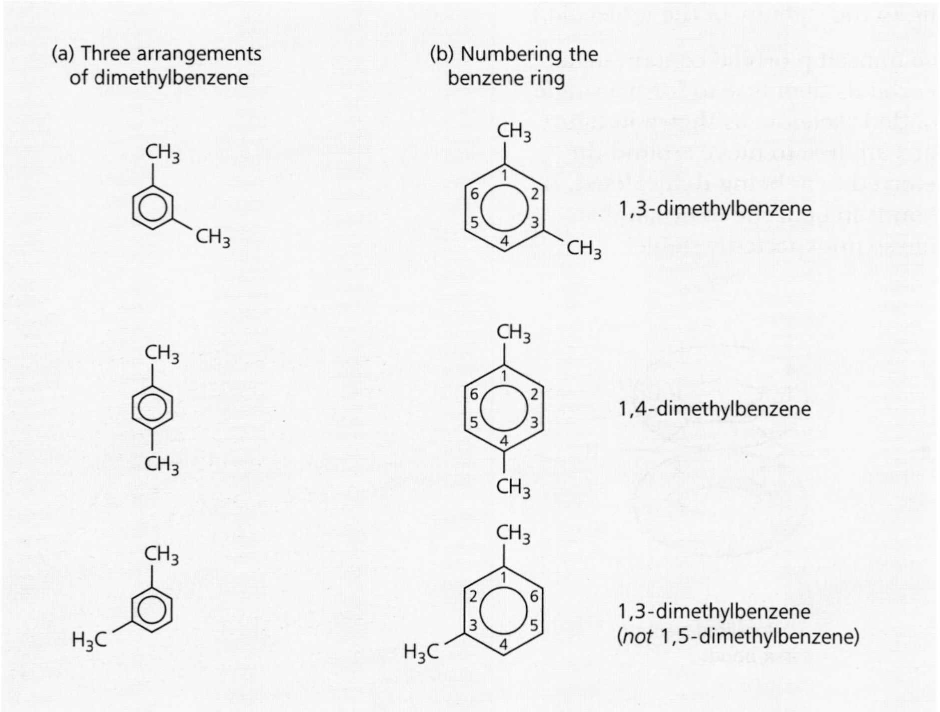

Figure 4.4.13

In order to give compounds unambiguous names, the carbon atoms are numbered. One particular group is assumed to be attached to carbon atom 1 in each of the isomers, and the others to different carbon atoms. As you can see from figure 4.4.13(b), the carbon atoms can be numbered in either direction, because the symmetry of the benzene molecule means that it does not matter which way round it we count. The important factor is to keep the same group or atom in position 1. It is conventional, however, always to count in the direction that gives the *lowest* number, so that we do not mistake 1,3-dimethylbenzene and 1,5-dimethylbenzene for different isomers – we use the numbering that gives 1,3-dimethylbenzene.

The electrophilic substitution reactions of benzene

The electron-rich regions of the benzene structure mean that benzene is attacked by electrophilic groups. The addition reactions that would be expected if benzene had the three double bonds described in the Kekulé model do not take place because of the stability of the molecule due to the delocalised electrons. Thus most of the reactions of benzene involve **electrophilic substitution**, the substitution of a hydrogen by an electrophile. Any addition reactions that do occur are the result of free radical reactions, not addition across a double bond.

Reactions with oxygen

Like all the hydrocarbons, benzene will burn in air. It produces a very sooty, smoky flame with a high percentage of carbon. This is a result of the high ratio of carbon to hydrogen in the benzene molecule.

Reactions with halogens

In the dark

Benzene does not react with the halogens unless there is a catalyst (iron(III) bromide, aluminium chloride or iron filings) present. The catalyst is known as a 'halogen carrier'. With a suitable catalyst, benzene undergoes a substitution reaction with chlorine or bromine. This forms chloro- or bromobenzene, as shown in figure 427–8.

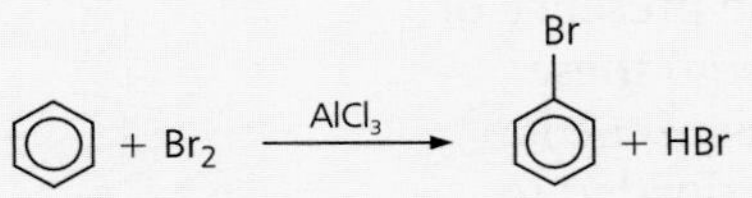

$$C_6H_6 + Br_2 \xrightarrow{AlCl_3} C_6H_5Br + HBr$$

Figure 4.4.14

The effect of the catalyst is to polarise the halogen molecule by accepting a pair of electrons from one of the halogen atoms. This leaves one end of the halogen molecule with a partial positive charge, and it is this that attacks the benzene ring. The mechanism of these reactions is discussed in the box on pages 427–8.

Phenol cannot be made directly from benzene, because the electron density of the benzene ring means that negatively charged groups are repulsed and cannot approach the molecule, so substitution of an OH^- group onto the benzene ring cannot take place directly. However, in the chlorobenzene molecule the chlorine atom tends to draw an electron towards itself, leaving the associated carbon atom with a slight positive charge and open to attack by a strong electrophile like the OH^- group. It still takes fairly extreme conditions for the reaction to occur – chlorobenzene must be heated with molten sodium hydroxide under pressure for the substitution to occur to form phenol! The reaction is shown in figure 4.4.15. As well as having antiseptic properties, phenol is also of great importance as a feedstock in the nylon industry (see the

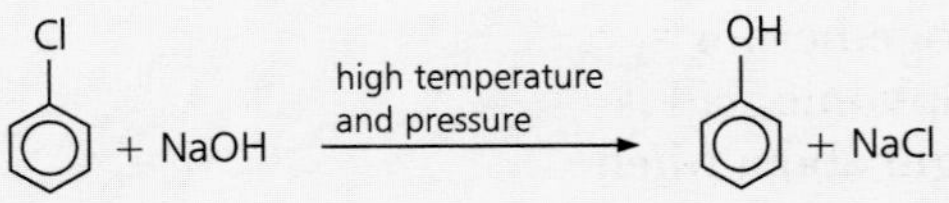

$$C_6H_5Cl + NaOH \xrightarrow{\text{high temperature and pressure}} C_6H_5OH + NaCl$$

Figure 4.4.15

Bromine and cyclohexene

If cyclohexene is mixed with bromine in the dark, an addition reaction takes place. 1,2-dibromocyclohexane is formed and the bromine is decolorised. This is in marked contrast to the reaction of benzene with bromine described here, a substitution reaction which takes place only in the presence of a halogen carrier catalyst. This is another example of the remarkable stability produced by the ring of delocalised π electrons.

4

box 'Benzene derivatives and the production of nylon' on pages 429–30). The commercial production of phenol from benzene does not use chlorobenzene, but is carried out via an intermediate compound called cumene.

In the light

Benzene will undergo addition reactions with chlorine or bromine in the presence of ultraviolet light or bright sunlight. The addition with chlorine occurs rapidly to form 1,2,3,4,5,6-hexachlorocyclohexane, as shown in figure 4.4.16. The displayed formula of this molecule looks simple enough. However, there is no rotation around the carbon–carbon bonds in a ring structure, and so there are actually eight possible geometric isomers. One of these, known commercially as Gammexane, is an effective insecticide, but it is persistent and therefore harmful to the environment. The rapidity of the reaction and the requirement for light to provide the activation energy suggest a radical reaction here, which is indeed the case.

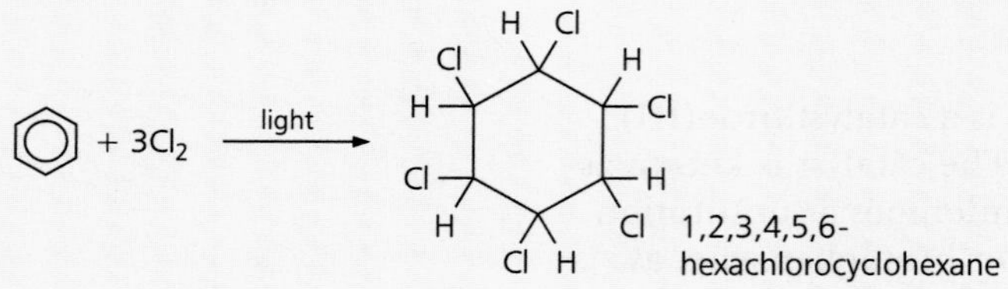

Figure 4.4.16

Reactions with hydrogen

The alkenes undergo rapid addition reactions with hydrogen in the presence of a nickel catalyst. It was the fact that benzene does not readily undergo these sort of addition reactions that pointed to its true structure. When benzene is mixed with hydrogen it needs a Raney nickel catalyst (this is finely divided to give it a very large surface area and thus very high catalytic activity) as well as a temperature of around 150 °C before it will undergo addition reactions. Benzene then reacts with hydrogen to form cyclohexane – there are no unsaturated intermediates. If one mole of hydrogen reacts with one mole of benzene, one-third of the benzene forms cyclohexane and the rest remains as benzene. Two moles of hydrogen convert two-thirds of the benzene to cyclohexane, and it takes three moles of hydrogen to give a complete hydrogenation of one mole of benzene to cyclohexane. The formation of cyclohexane from benzene, shown in figure 4.4.17, is an important industrial process as it leads to the production of nylon.

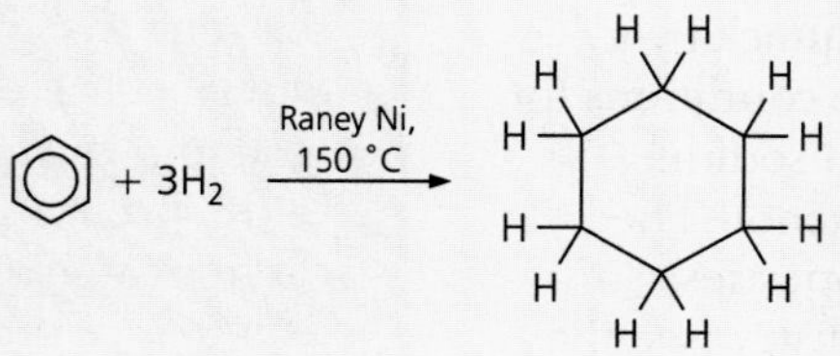

Figure 4.4.17

Reactions with sulphuric acid – sulphonation

Benzene will react with sulphuric acid, but only under fairly severe conditions. When benzene and concentrated sulphuric acid are refluxed together for several hours, electrophilic substitution occurs and benzenesulphonic acid is formed. This **sulphonation**, shown in figure 4.4.18(a), is brought about when the benzene ring is attacked by sulphur trioxide, SO_3, formed as the concentrated sulphuric acid is heated over a prolonged period of time. The sulphur atom in the SO_3 molecule has a large partial positive charge, and attacks the electron-rich benzene ring. If benzene is reacted with a solution of sulphur trioxide in concentrated sulphuric acid (known as **fuming sulphuric acid**), then sulphonation occurs in the cold, which suggests that it is indeed the sulphur trioxide rather than the sulphuric acid that is involved in the substitution reaction.

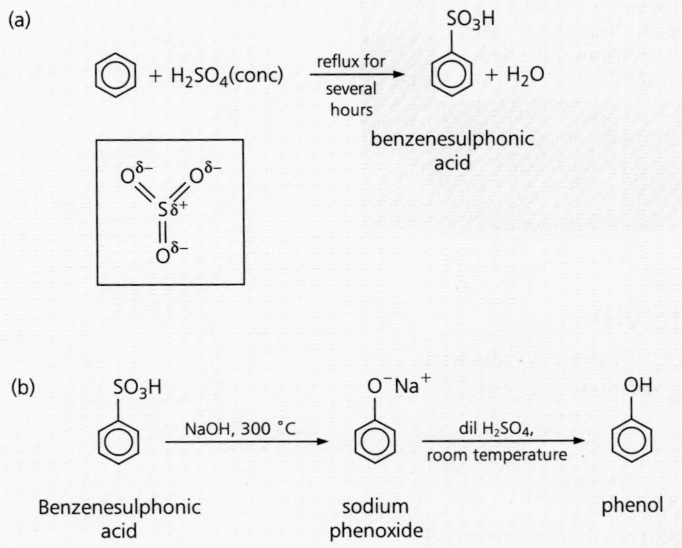

(a)

$C_6H_6 + H_2SO_4(conc) \xrightarrow{\text{reflux for several hours}} C_6H_5SO_3H + H_2O$

benzenesulphonic acid

$O^{\delta-}$ $O^{\delta-}$ $S^{\delta+}$ $O^{\delta-}$

(b)

$C_6H_5SO_3H \xrightarrow{\text{NaOH, 300 °C}} C_6H_5O^-Na^+ \xrightarrow{\text{dil } H_2SO_4\text{, room temperature}} C_6H_5OH$

Benzenesulphonic acid — sodium phenoxide — phenol

Figure 4.4.18

Like chlorobenzene, benzenesulphonic acid is important both industrially and in the laboratory. The electron-withdrawing effect of the sulphonate group activates the benzene ring to attack by nucleophiles. Once benzene has been sulphonated, the benzenesulphonic acid will react to form phenol when heated with molten sodium hydroxide, as shown in figure 4.4.18(b).

Refluxing – heating volatile reactants

When organic chemicals need to be heated for a reaction to occur, particularly when they need to be heated for some considerable time as in the sulphonation of benzene, special apparatus is needed. This is because many organic substances are volatile, and many are also either inflammable or produce toxic or carcinogenic vapours. This means that they cannot simply be heated in the laboratory in a test tube or open flask like many inorganic chemicals. A **reflux apparatus** allows organic reacting mixtures to be heated for long periods of time, by condensing the vapour that forms so that it falls back into the reacting mixture again.

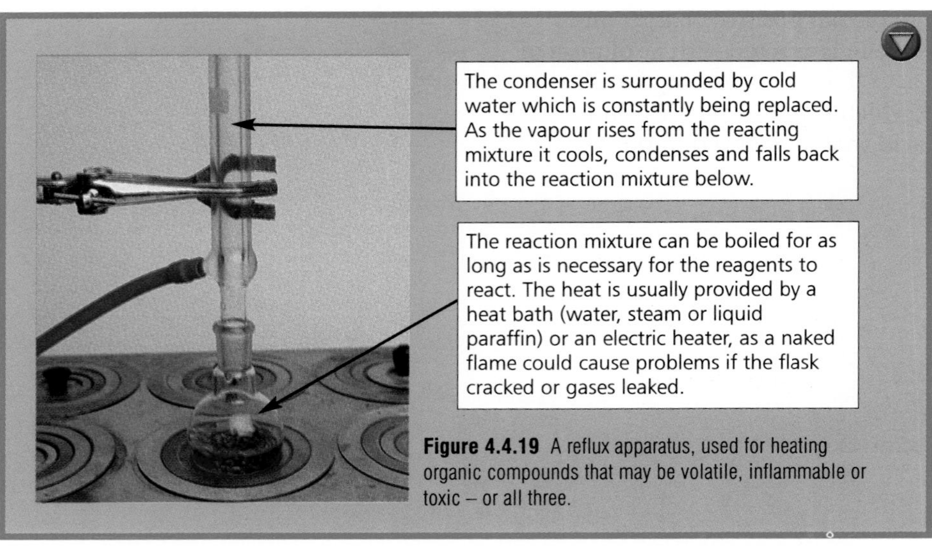

The condenser is surrounded by cold water which is constantly being replaced. As the vapour rises from the reacting mixture it cools, condenses and falls back into the reaction mixture below.
The reaction mixture can be boiled for as long as is necessary for the reagents to react. The heat is usually provided by a heat bath (water, steam or liquid paraffin) or an electric heater, as a naked flame could cause problems if the flask cracked or gases leaked.

Figure 4.4.19 A reflux apparatus, used for heating organic compounds that may be volatile, inflammable or toxic – or all three.

Reactions with acidified potassium manganate(VII)

In section 4.3 we saw the reactions of both the alkenes and the alkynes with acidified potassium manganate(VII), which is a strong oxidising agent. Benzene, however, does not react at all.

Reactions with nitric acid – nitration

Benzene will react with nitric acid, but only with a mixture of concentrated nitric acid and concentrated sulphuric acid (known as **nitrating mixture**) at 50 °C. The NO_2 group is substituted for one of the hydrogen atoms of the benzene ring, as shown in figure 4.4.20, forming a yellow oily substance which is nitrobenzene. This is another electrophilic substitution reaction of the benzene ring. Concentrated nitric acid alone reacts only very slowly with benzene at 50 °C, and at that temperature sulphuric acid does not react at all. When mixed, the two acids appear to react to form an electrophile that can attack the benzene. It is thought that this electrophile is the nitronium cation, NO_2^+ (for details of the mechanism of electrophilic substitution reactions in benzene, see the box on pages 427–9).

Nitrobenzene is used to make phenylamine, which in turn is important in the dye industry. These reactions will be considered in more detail in section 5.4.

$$C_6H_6 + HNO_3(conc) \xrightarrow[50\ °C]{conc\ H_2SO_4} C_6H_5NO_2 + H_2O$$

nitrobenzene

Figure 4.4.20

Reactions with other organic molecules

Many compounds of benzene that are used in industrial processes are produced by the reaction of benzene with other organic molecules. Any reaction between an aromatic hydrocarbon and groups such as the alkenes,

Figure 4.4.21

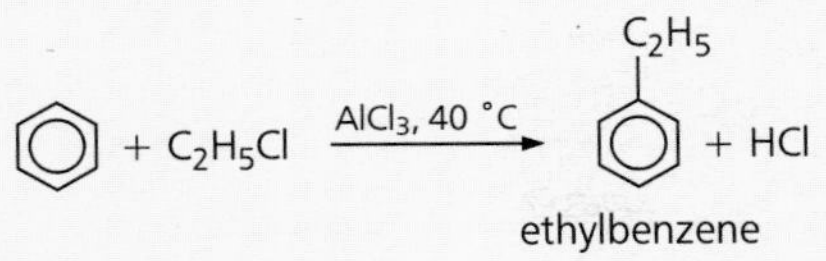

$$C_6H_6 + C_2H_5Cl \xrightarrow{AlCl_3,\ 40\ ^\circ C} C_6H_5C_2H_5 + HCl$$

ethylbenzene

alcohols or halogenoalkanes requires the formation of a carbon atom in the attacking species that is at least partially positively charged. We have already seen that a catalyst like iron(III) chloride can induce a dipole in a halogen molecule, and this process can also be used with organic molecules. For example, warming benzene with a halogenoalkane such as chloroethane and aluminium(III) chloride as a catalyst gives rise to ethylbenzene and hydrogen chloride, as shown in figure 4.4.21. The mechanism of this reaction is discussed in the box on pages 427–8.

The alkylation (substitution of an alkyl group) of benzene is important in a number of industrial processes. For example, benzene reacts with ethene in the presence of hydrogen chloride and aluminium(III) chloride to form ethylbenzene. This can then be catalytically dehydrogenated to form phenylethene. This is the monomer unit of the polymer poly(phenylethene), better known to most of us as polystyrene. Alkylation reactions also produce some of the constituents of detergents, and a variety of other chemicals.

Figure 4.4.22 Polystyrene is widely used in the building trade, for packaging and in medicine. The insulating properties of the air-filled polymer are useful either for keeping heat in, as in the wall and roof space of a house, or for keeping heat out, as in the transport of whole blood and organs for transplantation.

Reactions of alkylbenzenes

Once an aromatic hydrocarbon has another organic molecule substituted into its ring we have a molecule that is part aromatic in character (the benzene ring) and part aliphatic (the substituted group). Thus the chemistry of compounds such as methyl- and ethylbenzene and phenylethanone is a mixture of reactions of the aromatic ring and those of the aliphatic group. It will depend on the reaction conditions whether the aromatic ring or the aliphatic group will react with other chemicals.

Neither benzene nor methane can be oxidised by potassium manganate(VII). However, when methylbenzene is reacted with alkaline potassium manganate(VII), while the benzene ring remains intact (indicating its great stability), the methyl group is oxidised and benzoic acid is produced (figure 4.4.23).

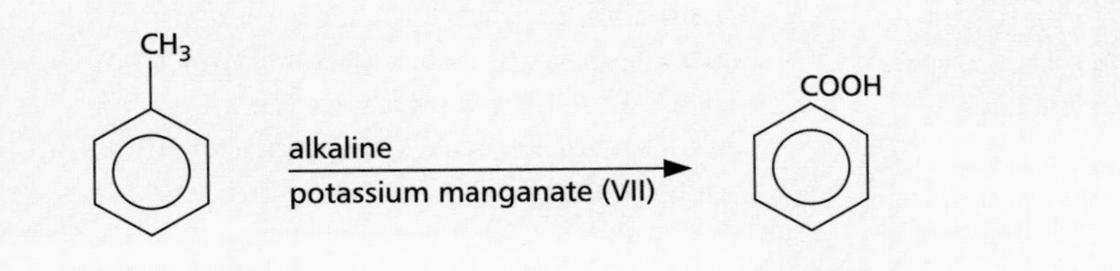

Figure 4.4.23

When chlorine is bubbled into boiling methylbenzene in the presence of ultraviolet light or strong sunlight, the methyl group undergoes a substitution reaction to form (chloromethyl)benzene. This reaction proceeds by a free radical mechanism. However, if chlorine is bubbled through methylbenzene at room temperature in the dark, with aluminium(III) chloride present, it is the benzene ring rather than the methyl group that undergoes a substitution reaction. The product is a mixture of two geometric isomers, 2-chloromethylbenzene and 4-chloromethylbenzene – the methyl group directs the position of these substitutions so that 3-chloromethylbenzene does not occur. The mechanism for

Figure 4.4.24

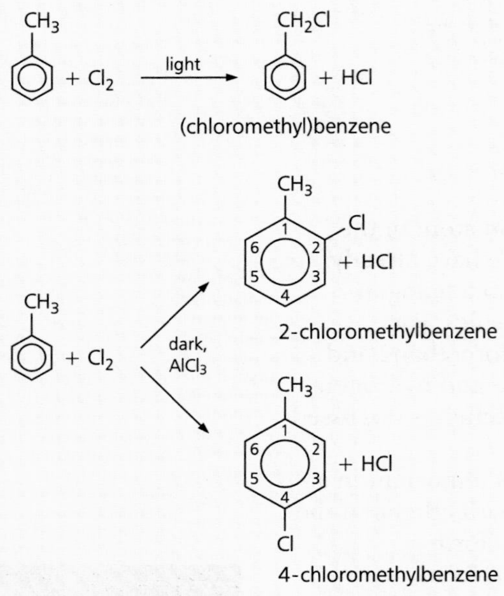

$CH_3C_6H_5 + Cl_2 \xrightarrow{\text{light}} CH_2ClC_6H_5 + HCl$

(chloromethyl)benzene

$CH_3C_6H_5 + Cl_2 \xrightarrow{\text{dark, } AlCl_3}$ 2-chloromethylbenzene $+ HCl$

$CH_3C_6H_5 + Cl_2 \xrightarrow{\text{dark, } AlCl_3}$ 4-chloromethylbenzene $+ HCl$

this reaction is electrophilic substitution, and it occurs more rapidly than the equivalent reaction between benzene and chlorine.

This reaction of methylbenzene highlights two effects that a substituent group may have on the reactions of the benzene ring:

- it directs the position in which further substitutions occur, and
- it affects the reactivity of the ring.

The effect of substituent groups on the reactions of the benzene ring

When a group replaces one of the hydrogens on the benzene ring, it may affect the electron density of the ring, so we should not be surprised that substituent groups affect the chemistry of benzene. Some groups allow subsequent substitutions to occur faster than they would on benzene itself. Other groups slow down the rate of any further substitutions. The main points can be summarised as follows:

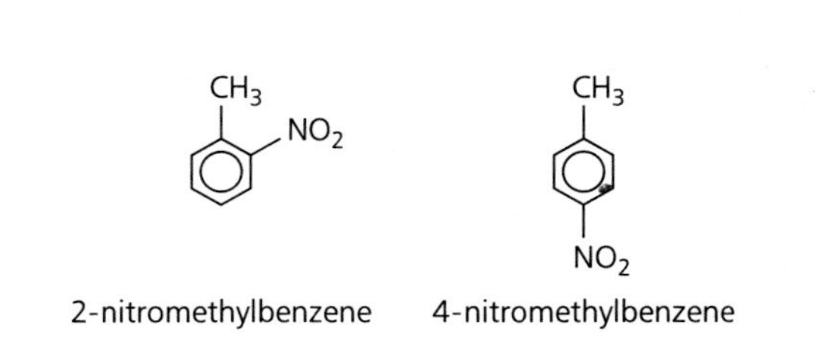

Figure 4.4.25 When one NO_2 group is substituted into methylbenzene, it tends to be directed to either the 2- or 4-position. The nitration of methylbenzene occurs at room temperature, while benzene requires heating at 60 °C for one hour for nitration to occur.

- Some functional groups donate electron density to the ring, and so cause electrophilic substitutions to occur faster than they would in benzene itself. These groups direct the majority of further substitutions to the 2- and 4-positions on the ring, and include all the alkyl groups (methyl, ethyl, etc.) as well as -OH, $-NH_2$ and $-OCH_3$, figure 4.4.25.
- Some functional groups withdraw electron density from the ring, and so cause electrophilic substitutions to occur more slowly than in benzene itself. These groups direct the majority of further substitutions to the 3-position on the ring, and include the -COOH, $-NO_2$ and $-SO_3H$ groups.

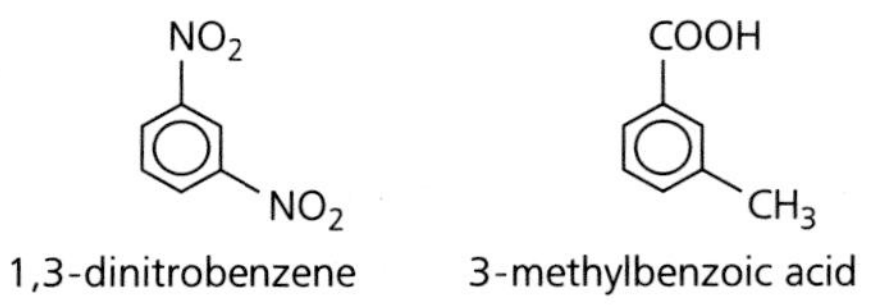

1,3-dinitrobenzene   3-methylbenzoic acid

Figure 4.4.26 The $-NO_2$ and -COOH groups tend to direct further substitution to the 3-position in the benzene ring. In this case the reactions occur considerably more slowly than for benzene.

The mechanism of electrophilic substitution in the benzene ring

All the substitution reactions of the benzene ring involve attack on the ring by electrophiles. The general mechanism of this attack is common to all reactions – here we shall be concerned with the way nitration and alkylation occur. Both these reactions take place as a result of the production in the reaction mixture of species with a positive charge (either a whole charge or a partial one), although this species is produced in different ways in the two cases. When writing reaction mechanisms for aromatic compounds it is often useful to represent them using Kekulé structures, although we should always remember that this is not a true picture of the real benzene molecule.

Nitration

The first step of nitration involves the production of the NO_2^+ (nitronium) cation from the reaction of concentrated nitric acid and concentrated sulphuric acid:

$$\mathbf{HNO_3 + 2H_2SO_4 \rightarrow NO_2^+ + 2HSO_4^- + H_3O^+}$$

Notice how nitric acid is acting as a base in this reaction.

The nitronium ion then attacks one of the carbon atoms in the benzene ring, forming the intermediate shown in figure 4.4.27. This then loses a proton to give nitrobenzene.

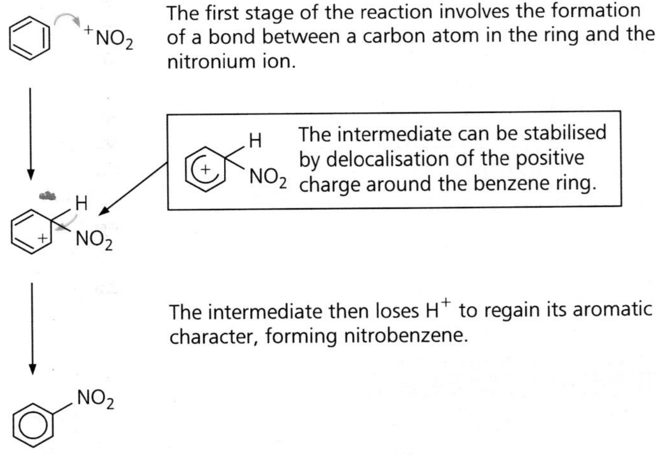
$^+NO_2$
The first stage of the reaction involves the formation of a bond between a carbon atom in the ring and the nitronium ion.
H
NO_2
The intermediate can be stabilised by delocalisation of the positive charge around the benzene ring.
H
NO_2
The intermediate then loses H^+ to regain its aromatic character, forming nitrobenzene.
NO_2

Figure 4.4.27 The reaction mechanism for the nitration of benzene

Alkyation

In 1877, 12 years after Kekulé devised his molecular structure for benzene, the American James Mason Crafts and the Frenchman Charles Friedel discovered that they could join two benzene rings together if they used an aluminium chloride catalyst. Since then, catalysts such as aluminium chloride and iron(III) bromide have been used to facilitate substitution reactions on the benzene ring.

4

The catalyst works by accepting a lone pair of electrons from an unpolarised molecule. The molecule that has lost electrons therefore becomes polarised, and the positively charged end then attacks the benzene ring. These catalysts are known as Friedel–Crafts catalysts, and are most commonly used to add alkenes, alcohols and halogenoalkanes to a benzene ring in reactions known as **Friedel–Crafts reactions.** An example of Friedel–Crafts alkylation is shown in figure 4.4.28.

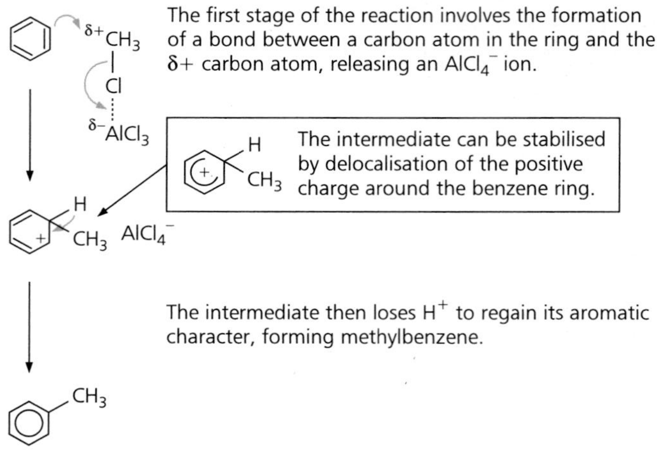

The first stage of the reaction involves the formation of a bond between a carbon atom in the ring and the δ+ carbon atom, releasing an $AlCl_4^-$ ion.

The intermediate can be stabilised by delocalisation of the positive charge around the benzene ring.

The intermediate then loses H^+ to regain its aromatic character, forming methylbenzene.

Figure 4.4.28 The reaction mechanism for the Friedel–Crafts alkylation of benzene. The mechanism for the attack of a halogen molecule on a benzene ring in the presence of a catalyst like iron(III) bromide (page 421) involves the polarisation of the Hal—Hal bond, in the same way as the C—Cl bond here.

The mechanism of both these reactions is very similar to the attack of an electrophile on a double bond. The difference lies in the second stage of the reaction, which in the aromatic compound involves the loss of a proton to regain the aromatic character. Alkenes have nothing to gain by losing a proton to reform a double bond, since they have no extra stability due to delocalisation, and so the intermediate cation is attacked by a nucleophile, forming an addition product.

Industrial importance of benzene

Benzene is of great importance industrially, because it is the starting point of a variety of processes including the manufacture of detergents and insecticides, as well as the polymers nylon and polystyrene. It is also, as a non-polar liquid, a very useful solvent for other organic molecules that will not dissolve in polar solvents such as water. It is produced industrially by the catalytic reforming of C_6–C_8 fractions of crude oil, along with a variety of other aromatic compounds. This has replaced the older method whereby benzene was produced by the destructive distillation of coal tar.

Benzene derivatives and the production of nylon

In 1934 Wallace Hume Carothers, working for the massive US chemical company Du Pont, invented nylon. He patented it the following year, but as a result of a series of tragedies and misunderstandings in both his professional and personal lives, he had died by his own hand in 1937 before his revolutionary new fibre went on the market in 1938.

Nylon is a totally synthetic fibre, a condensation polymer like many natural fibres, but made up of two monomers synthesised from benzene derivatives, and thus originating from coal or oil. Nylon has a multitude of uses in modern daily life. Apart from the obvious use in nylon stockings and tights, nylon is used in combination with many other fibres to add durability. For example, many carpets are a mixture of wool and nylon, the nylon enabling them to withstand far better the wear and tear of family or office life. Nylon ropes do not rot, nylon bearings do not wear – the list of uses is almost endless.

The monomer units of one of the most common forms of nylon are 1,6-diaminohexane and hexanedioic acid. They are obtained from benzene via phenol or cyclohexane as shown in figure 4.4.30.

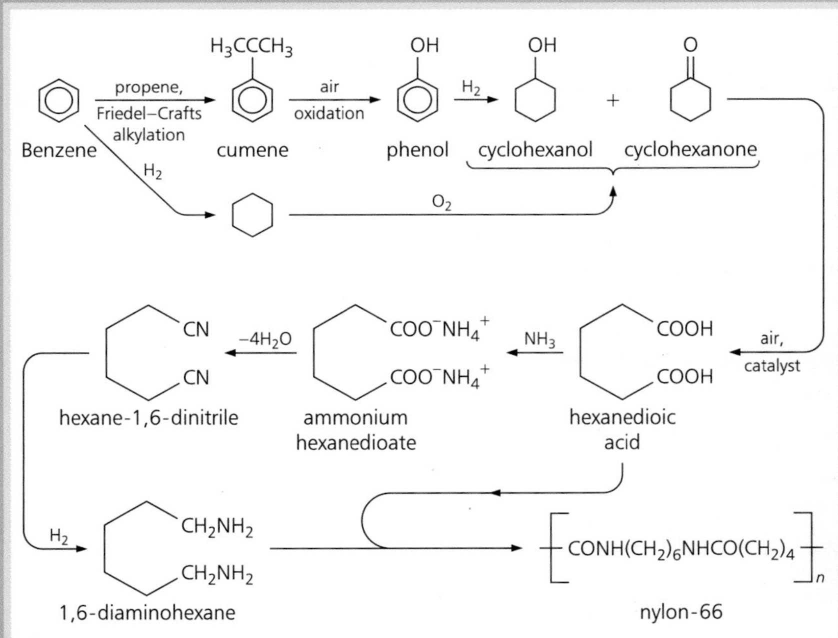

Figure 4.4.30 Nylon is produced from this complex series of reactions, resulting in strong fibres which may be put to a wide variety of uses.

Figure 4.4.29 Wallace Carothers invented nylon, a totally synthetic fibre which has been of enormous value ever since. Unfortunately he did not live long enough to reap the benefits of his own invention.

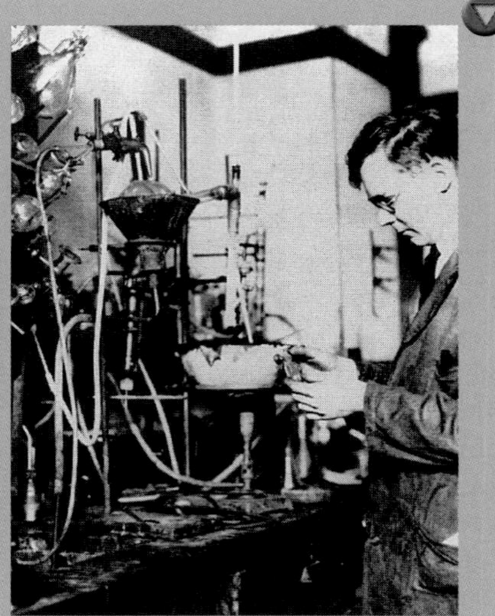

SUMMARY

- Aromatic hydrocarbons or **arenes** generally have molecules that contain a benzene ring structure.
- Benzene is the best known aromatic hydrocarbon. Benzene has a symmetrical six-membered ring structure with delocalised electrons above and below the ring.
- Benzene is produced by the catalytic reforming of crude oil fractions.
- C_6H_5- is known as the **phenyl group**.
- Most of the reactions of benzene are **electrophilic substitution reactions**. The main reactions are summarised in table 4.4.1.

Benzene +	Reaction
Air/oxygen	Benzene burns in air with a smoky, sooty flame.
Halogens	*In the dark:* Benzene does not react without a catalyst. When a suitable catalyst is present, a halogenobenzene is formed. *In the light:* Benzene reacts with chlorine or bromine to form 1,2,3,4,5,6-hexahalogenocyclohexane.
Hydrogen	With a Raney nickel catalyst at about 150 °C, benzene will undergo an addition reaction to form cyclohexane.
Nitric acid	A mixture of conc nitric and sulphuric acids (nitrating mixture) at 50 °C results in substitution to give nitrobenzene.

Table 4.4.1

- Substituted groups affect the chemistry of the benzene ring and alter the rate and position at which subsequent substitutions occur.
- The benzene ring has a marked effect on the properties of any functional group attached to it.

QUESTIONS

1 Derivatives of benzene containing more than one substituent are named by numbering the ring with the position of the substituent groups. Hence A in figure 4.4.31 is methylbenzene, B is 1,3-dimethylbenzene and C is 1-butyl-3,4-dimethylbenzene.

CH3
CH3
CH3
CH3
CH3
CH2CH2CH2CH3

A B C

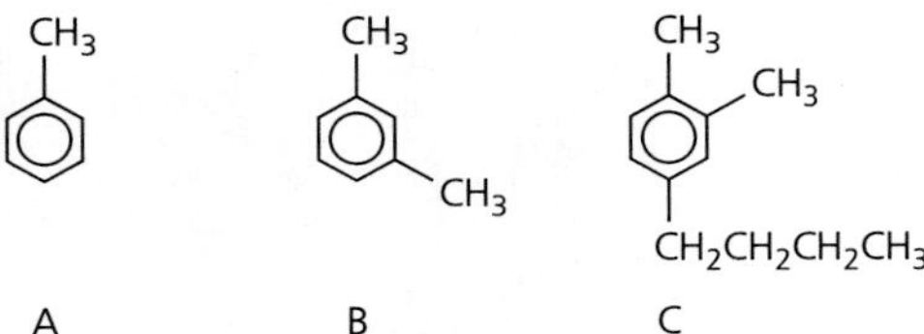

Figure 4.4.31

a Name the compounds in figure 4.4.32.

i
CH3
Cl
ii
OH
H3C
Br
iii
I
CH2CH3
iv
Cl
Cl
Cl
Cl
Cl
Cl

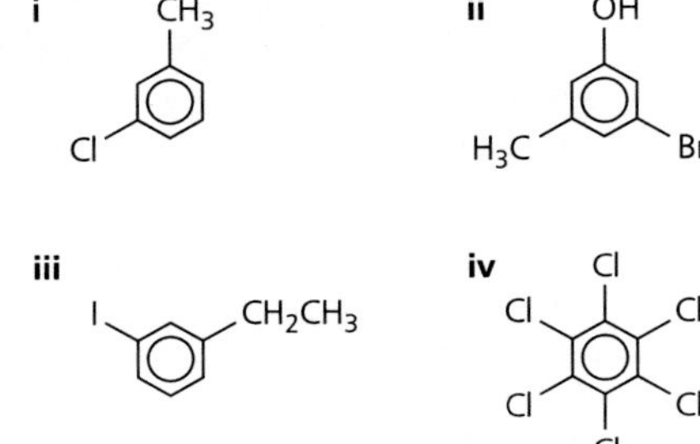

Figure 4.4.32

b Draw displayed formulae for:
i butylbenzene
ii 3-chlorophenol
iii 1,2-dimethyl-4-iodobenzene
iv 2-phenylphenol.

2 Look at figure 4.4.33.

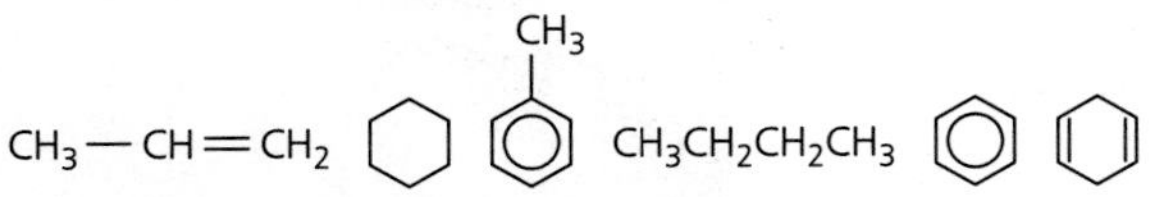

CH3 — CH = CH2
CH3
CH3CH2CH2CH3

A B C D E F

Figure 4.4.33

Which of these:
a would react with bromine in the dark
b would produce HBr when reacted with iron and bromine
c would decolorise acidified potassium manganate(VII) solution
d is a non-polar molecule
e would undergo an addition reaction with chlorine
f would react with hydrogen in the presence of a catalyst such that:
1 mole compound X + 3 moles $H_2 \rightarrow$ 1 mole of product

3 When benzene is reacted with 1-bromopropane in the presence of gallium bromide, the product shown in figure 4.4.34 is formed.

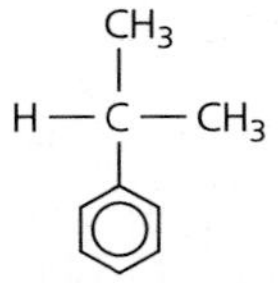

CH3
H — C — CH3

Figure 4.4.34

a What type of reaction is this?
b What is the function of the gallium bromide?
c What alternative product might be formed in this reaction?
d Suggest why this product is not formed.

4 When chlorobenzene is reacted with a strong base, the compound shown in figure 4.4.35 may be formed.

Cl
−HCl

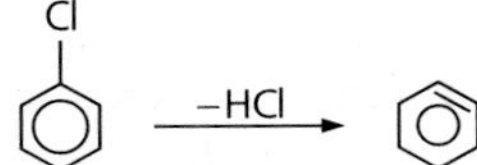

'Benzyne' – this molecule reacts very quickly with other molecules present in the reaction mixture, forming a range of products.

Figure 4.4.35

a Show how the presence of a base may lead to the loss of a molecule of HCl from a molecule of chlorobenzene.
b Why is the molecule which is formed very reactive?
c The starting molecule may be labelled as shown in figure 4.4.36, where * indicates an atom of ^{13}C. After some time the reaction mixture is found to contain some G. Suggest why this is.

Cl
*
Cl
*

Labelled chlorobenzene G

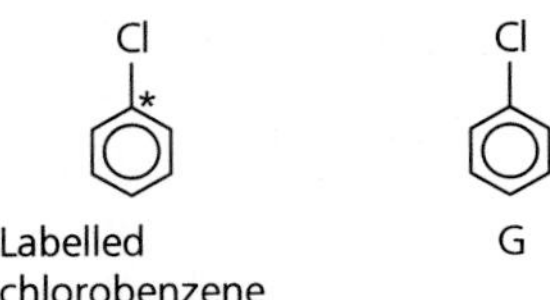

Figure 4.4.36

FOCUS A CHEMICAL SLEEP

Deep sleep

Surgery – physically removing diseased parts of the body or mending damaged internal regions – has long been attempted by the human race. However, developing finesse of technique is not possible when working on a conscious patient anaesthetised only by alcohol and held down by straps and other helpers. It was with the development of **anaesthetics** that surgery could move on from the glorified butchery of the barber-surgeons to the keyhole techniques increasingly used today.

The history of modern anaesthesia can be said to have begun in 1841, when the American Charles Jackson discovered that ethoxyethane ('ether') acts as an anaesthetic. Crawford Long used it to perform surgery in 1842, but because he did not publish his work immediately, the credit usually goes to William Moreton, who used ethoxyethane for dental surgery in 1846. Meanwhile, in 1844 Horace Wells used dinitrogen oxide ('laughing gas') as a dental anaesthetic for the first time. Ethoxyethane is a very effective anaesthetic, but it is highly flammable. Dinitrogen oxide is not flammable or toxic, but it only produces a very light anaesthesia. The final discovery of this revolutionary decade in the history of anaesthesia was by the Scot Sir James Simpson, who showed that trichloromethane ('chloroform') was an anaesthetic superior to both ethoxyethane and dinitrogen oxide. Trichloromethane received the royal stamp of approval when used by Queen Victoria during the deliveries of several of her children. Trichloromethane produces a deep sleep, but it can cause liver damage. Interestingly, although trichloromethane is no longer used for pain relief during childbirth as it is too toxic, dinitrogen oxide is still very much present as part of the 'gas and air' mixture used by many women during labour.

It was not until the middle of the twentieth century that further strides were made in the development of anaesthetics. Demand grew for compounds that could readily be inhaled and that produced deep sleep, yet were non-toxic and non-flammable. By this time, organic chemists had recognised two important features of organic compounds that helped them in their search:

- The substitution of a chlorine atom into a molecule of the organic family known as the alkanes results in a compound with anaesthetic properties – trichloromethane was a clear example. Increasing the number of chlorine atoms in the compound increases the depth of anaesthesia given, but unfortunately also increases the toxicity of the compound.
- Carbon–fluorine bonds are very stable and so their presence in a compound leads to non-flammable, non-toxic and unreactive properties.

Given this information, 30 years ago organic chemists came up with halothane, 2-bromo-2-chloro-1,1,1-trifluoroethane. With this effective compound giving deep yet safe anaesthesia, along with the similar compounds enflurane and isoflurane that followed, modern surgery developed in leaps and bounds.

Figure 1 We have come a long way since the primitive days of medicine when calling a doctor and undergoing surgery was a last resort, the alternative being certain death. Whilst even today few people relish a trip to hospital, surgery is no longer such a life-threatening or unduly painful experience thanks to the advance of organic chemistry.

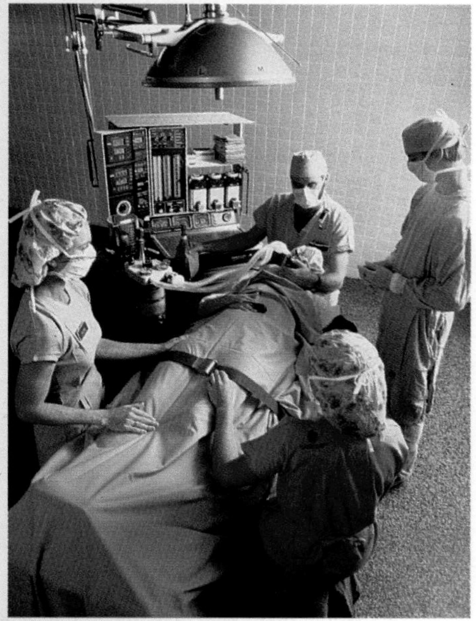

Figure 2 Halothane, a deceptively simple molecule, has allowed surgery to progress to make operations such as this safer for the patient. It has also enabled an enormous range of other surgical procedures, ranging from operations carried out through minute openings made in the body wall to massive transplant operations involving many hours in the operating theatre.

```
     F   H
     |   |
 F — C — C — Cl
     |   |
     F   Br
```

Halothane
(2-bromo-2-chloro-1,1,1-trifluoroethane)

4.5 The organohalogen compounds

The organohalogen compounds are a family of chemicals relatively rarely found in the natural world. Most of them are synthesised in the laboratory or on a larger scale in chemical plants. Organohalogens play a wide variety of roles in the modern world, many of them for the benefit of the human race. For example, as we can see from pages 405 and 432, they are important in medicine, agriculture and the production of plastics. However, they are also implicated in much environmental damage, both at the surface of the planet and in the atmosphere around it.

The organohalogen families

An organohalogen has a hydrocarbon skeleton with a halogen functional group. The hydrocarbon skeleton may be aliphatic or aromatic, and the halogen may be fluorine, chlorine, bromine or iodine. The properties of an organohalogen compound are therefore affected by three things:

- the type of hydrocarbon skeleton
- the halogen or halogens attached
- the position of the halogen in the molecule.

As we consider the different organohalogen families, we should bear these factors in mind as we predict the behaviour of each compound from our knowledge of its structure.

There are three main types of organohalogen molecules – the **halogenoalkanes**, the **halogenoarenes** (aromatic halogens) and the **acid halides**. The names tell us the basic structure of the molecules in the different families.

The halogenoalkanes

The **halogenoalkanes** are molecules in which one or more of the hydrogen atoms within an alkane molecule has been replaced by a halogen atom. Some examples are shown in figure 4.5.1.

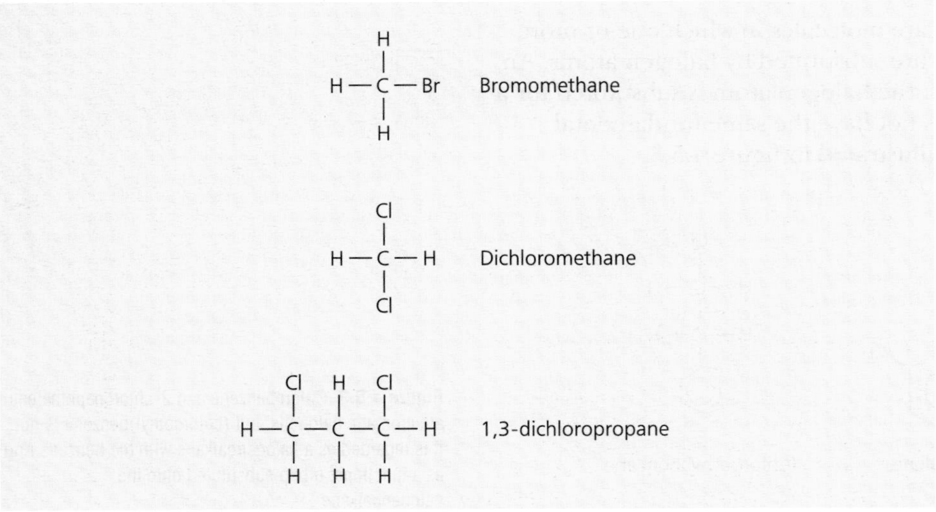

```
      H
      |
  H — C — Br      Bromomethane
      |
      H

      Cl
      |
  H — C — H       Dichloromethane
      |
      Cl

      Cl  H   Cl
      |   |   |
  H — C — C — C — H      1,3-dichloropropane
      |   |   |
      H   H   H
```

Figure 4.5.1 Members of the halogenoalkane family

Structural isomers are a very common feature of the halogenoalkanes, and so both the drawing and the naming of the compounds needs particular care. Changing the position of a halogen atom within a halogenoalkane makes a great difference to the properties of the molecule. The halogenoalkanes may be classified into **primary**, **secondary** and **tertiary** compounds.

- Primary halogenoalkanes, for example 1-chloropentane, have two hydrogens bonded to the carbon atom carrying the halogen, as shown in figure 4.5.2(a).
- Secondary halogenoalkanes, for example 2-chloropentane, have one hydrogen bonded to the carbon atom carrying the halogen, as shown in figure 4.5.2(b). The halogen-carrying carbon is bonded to two other carbons.
- Tertiary halogenoalkanes, for example 2-chloro-2-methylbutane, have no hydrogens bonded to the carbon carrying the halogen – it is bonded to three other carbons, as shown in figure 4.5.2(c).

(a) A primary halogenoalkane

```
     H   H   H   H   H
     |   |   |   |   |
 H — C — C — C — C — C — Cl     1-chloropentane
     |   |   |   |   |
     H   H   H   H   H
```

(b) A secondary halogenoalkane

```
     H   H   H   Cl  H
     |   |   |   |   |
 H — C — C — C — C — C — H      2-chloropentane
     |   |   |   |   |
     H   H   H   H   H
```

(c) A tertiary halogenoalkane

```
     H   Cl  H   H
     |   |   |   |
 H — C — C — C — C — H          2-chloro-2-methylbutane
     |   |   |   |
     H   |   H   H
         |
     H — C — H
         |
         H
```

Figure 4.5.2

The halogenoarenes

The aromatic halogens or **halogenoarenes** are molecules in which one or more of the hydrogen atoms of the benzene ring are substituted by halogen atoms. An aromatic molecule is not a halogenoarene if the halogen atom is substituted for a hydrogen atom on a side chain, as this does not have the same fundamental effect on the chemistry of the molecule, as illustrated in figure 4.5.3.

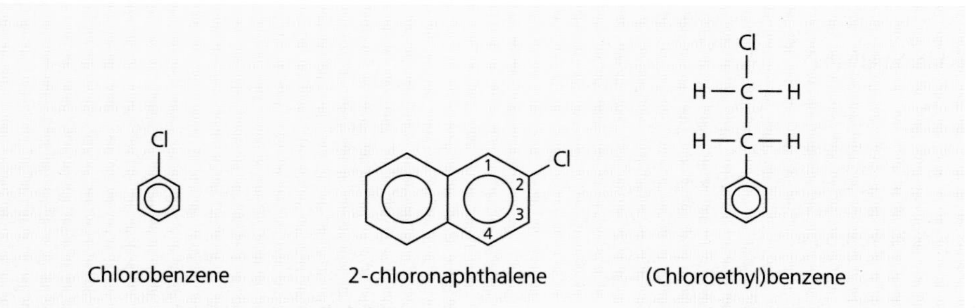

Figure 4.5.3 Chlorobenzene and 2-chloronaphthalene are aromatic halogens, but (chloroethyl)benzene is not. It is regarded as a halogenoalkane with the benzene ring as a functional group substituted onto the halogenoalkane.

As we saw in section 4.4, the position of the halogen group around an already substituted benzene ring allows for structural isomers to occur, and so the positions of the groups must be stated once more than one substitution has taken place, as in figure 4.5.4.

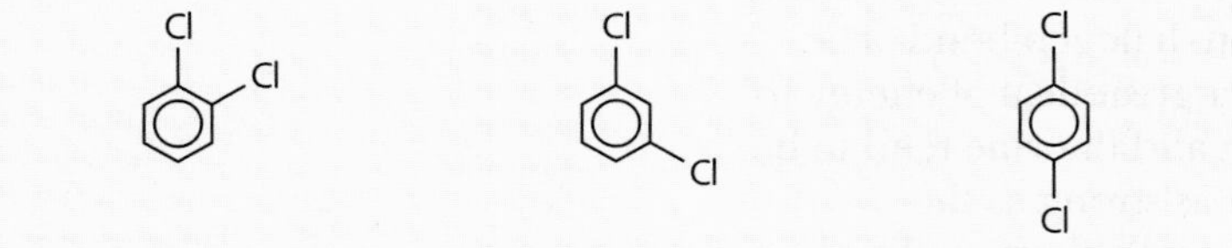

1,2-dichlorobenzene 1,3-dichlorobenzene 1,4-dichlorobenzene

Figure 4.5.4 Structural isomers of dichlorobenzene

The acid halides

The **acid halides** are the halides of the carboxylic acids. The carboxylic acids are a family of organic compounds which we will be looking at in more detail in section 5.3. They are organic acids typified by the possession of a **carboxyl** group (-COOH), and the acid halides are formed when the -OH group is replaced by a halogen atom, as shown in figure 4.5.5. Naming the acid chlorides is relatively simple. The name of the original carboxylic acid is used, with the '-oic acid' being replaced by '-oyl halide'. For example, propanoic acid becomes propanoyl chloride, and benzoic acid becomes benzoyl bromide.

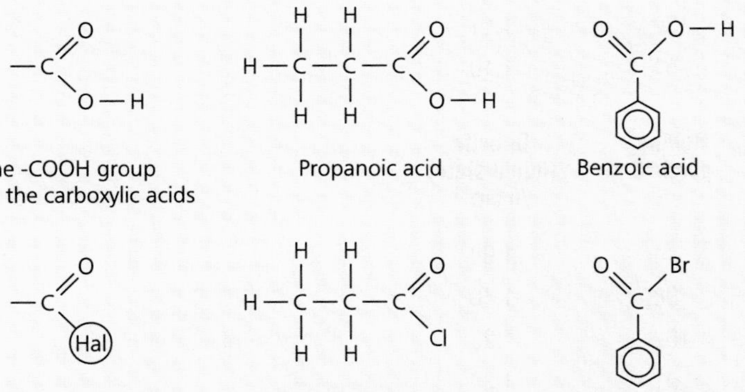

The -COOH group of the carboxylic acids

Propanoic acid

Benzoic acid

The functional group of the acid halides

Propanoyl chloride

Benzoyl bromide

Figure 4.5.5 Examples of acid halides

Physical properties of the organohalogens

There is great variety in the physical properties of the organohalogen compounds, but they are almost all either gases or liquids at room temperature. The different families – halogenoalkanes, halogenoarenes and acid halides – have different trends in their properties, and also the different halogens affect the properties of the molecules. Some of these properties for different compounds are shown in table 4.5.1 overleaf.

Large tables of data presented like this can seem very daunting, and the information they contain obscured by the sheer force of numbers. However, if smaller elements of the data are extracted and compared, as shown in figure 4.5.6, then some of the trends become clear.

The effects of the size of the molecules and the length of the longest carbon chain are both important factors in the physical properties of these compounds. They influence the number and strength of the intermolecular forces, as we saw in section 4.2 when we considered the alkanes.

The other crucial factor in the chemistry of the organohalogen compounds is the nature of the halogen atom substituted into the molecule. There is a distinct effect on both physical and chemical properties if chlorine is involved rather than bromine, and the difference between fluorine and iodine is the most marked of all. The differences depend on the nature of the carbon–halogen bond. The carbon–fluorine bond is extremely strong – it requires a great deal of energy to break it. This means that substances containing carbon and fluorine tend to be very stable and inert. The carbon–chlorine bond is not as strong as the carbon–fluorine bond, but it is nevertheless very stable and chlorinated carbon compounds tend to be fairly inert. In contrast, the carbon–bromine bond is less stable and so bromine-substituted hydrocarbons tend to undergo reactions more readily than fluoro- or chloro- substitutions. (See box 'The carbon–halogen bond' for additional information on bond dissociation enthalpies.)

Organo-chloro compound	State at 25 °C	Melting point/°C	Boiling point/°C	Density (liquid state) /g cm^{-3}
Chloromethane	Gas	–97	–24	0.92
Chloroethane	Gas	–136	12	0.90
1-chlorobutane	Liquid	–123	79	0.89
2-chlorobutane	Liquid	–121	68	0.87
2-chloro-2-methylpropane	Liquid	–25	51	0.84
Chlorobenzene	Liquid	–45	132	1.11
Ethanoyl chloride	Liquid	–112	51	1.10

Organo-bromo compound	State at 25 °C	Melting point/°C	Boiling point/°C	Density (liquid state) /g cm^{-3}
Bromomethane	Gas	–93	4	1.68
Bromoethane	Liquid	–118	38	1.46
1-bromobutane	Liquid	–112	102	1.28
2-bromobutane	Liquid	–112	91	1.26
2-bromo-2-methylpropane	Liquid	–16	73	1.22
Bromobenzene	Liquid	–32	156	1.49
Ethanoyl bromide	Liquid	–96	77	1.66

Organoiodo compound	State at 25 °C	Melting point/°C	Boiling point/°C	Density (liquid state) /g cm^{-3}
Iodomethane	Liquid	–66	33	2.28
Iodoethane	Liquid	–111	72	1.94
1-iodobutane	Liquid	–130	131	1.61
2-iodobutane	Liquid	–104	120	1.59
2-iodo-2-methylpropane	Liquid	–38	103	1.57
Iodobenzene	Liquid	–30	189	0.90
Ethanoyl iodide	Liquid	0	108	1.98

Table 4.5.1 A careful look at these physical properties allows us to draw some conclusions about the effects of various aspects of the molecular structures of organohalogen compounds on their physical properties.

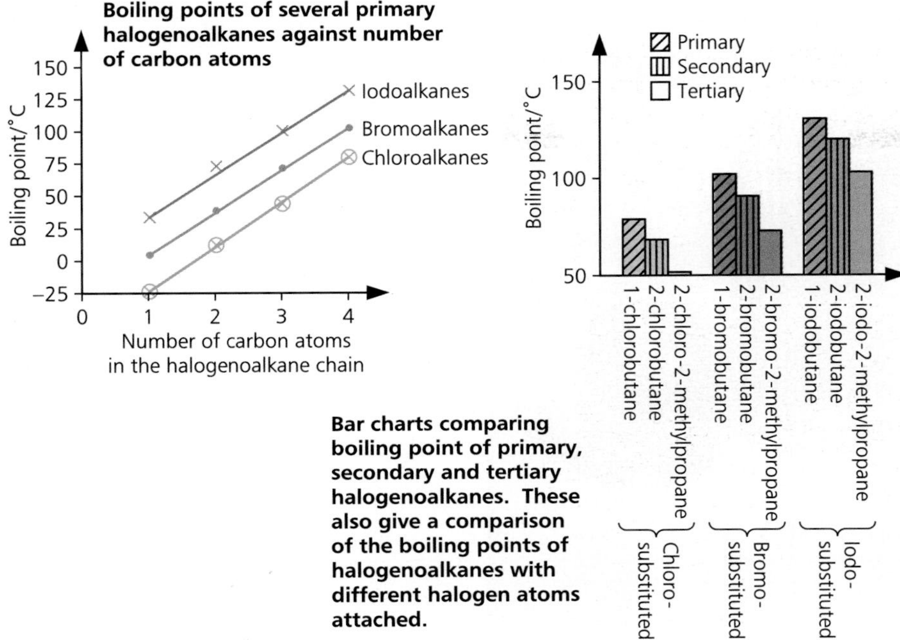

Bar charts comparing boiling point of primary, secondary and tertiary halogenoalkanes. These also give a comparison of the boiling points of halogenoalkanes with different halogen atoms attached.

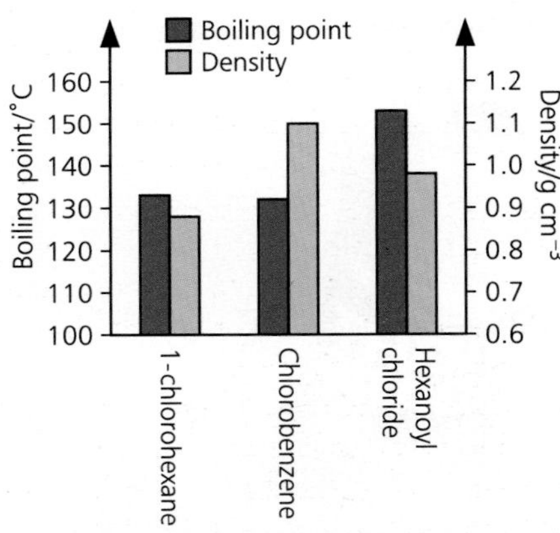

Comparison between boiling points and densities of primary halogenoalkane, halogenoarene and acid halide, with the same number of carbon atoms and the same halogen

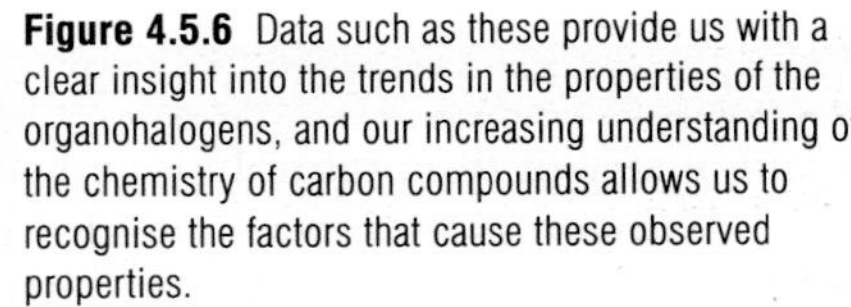

Figure 4.5.6 Data such as these provide us with a clear insight into the trends in the properties of the organohalogens, and our increasing understanding of the chemistry of carbon compounds allows us to recognise the factors that cause these observed properties.

The carbon–halogen bond

The nature of the bonds between carbon and the various halogens is seen clearly in the bond dissociation enthalpies shown in table 4.5.2.

Bond	Bond dissociation enthalpy /kJ mol^{-1}
C—F	467
C—Cl	346
C—Br	290
C—I	228

Table 4.5.2

Both the C—F bond and the C—Cl bond are very stable, with the C—F bond having exceptional strength. This observation prompted the development of a group of compounds known as the freons or chlorofluorocarbons in the 1920s and 30s. This group of compounds is exceptionally inert. They are stable and non-flammable, they are largely non-toxic and odourless and they are gases. This made them extremely valuable in a number of applications – as refrigerants, as aerosol propellants and as solvents. Unfortunately it is this same inert nature that has caused the environmental damage for which CFCs are now infamous, outlined in section 2.6. In the 1970s scientists began to realise that the CFCs which had been increasingly used over the previous 40–50 years were in fact accumulating in the upper atmosphere. CFCs were interfering with the natural ozone cycle and causing a depletion in the ozone layer.

Figure 4.5.7 Low toxicity, non-flammability, no odour, non-corrosive – chlorofluorocarbons must have seemed ideal chemicals when they were first developed. The environmental problems they were going to cause were unimagined in the 1940s, 50s and 60s. Nowadays it is a positive feature *not* to use CFCs in products!

Reactions of the organohalogens

The chemistry of the organohalogens is largely based on two factors:

- which halogen atom is present – for example, organic fluorine compounds are so stable that we shall ignore their reactions and concentrate on the chloro-, bromo- and iodo-compounds
- the position of the carbon–halogen bond within the molecule.

We shall concentrate on two types of reactions of the organohalogen compounds – **nucleophilic substitution reactions** and **elimination reactions**. Our survey will focus on the reactions of the halogenoalkanes. Where the halogenoarenes or the acid halides undergo reaction under similar conditions, we shall examine these too.

The mechanisms of the reactions of organohalogen compounds are discussed in the box on pages 442–4.

The chemistry of the halogenoalkanes – nucleophilic substitution reactions

Much of the chemistry of the halogenoalkanes involves nucleophilic substitution, in which the halogen atom is replaced by another functional group. The halogenoalkanes are an organic family in their own right, with the general formula RHal, where R is an alkyl group (methyl, ethyl, propyl, etc.) and Hal is the halogen (F, Cl, Br or I). Some examples of nucleophilic substitution reactions are described here.

Reactions with the hydroxyl (-OH) group

The halogenoalkanes have a slightly polarised C—Hal bond. Water acts as a nucleophile towards the carbon atom in this bond. As a result, the -OH group substitutes for the halogen, giving an **alcohol** and a hydrogen halide – a reaction that is sometimes called **hydrolysis** ('splitting with water'). The reaction with water is slow at room temperature – a more rapid reaction occurs with OH^- ions.

$$RHal + H_2O \rightarrow ROH + HHal$$

$$RHal + OH^- \rightarrow ROH + Hal^-$$

For example:

$$\underset{\text{Chloromethane}}{CH_3Cl} + H_2O \rightarrow \underset{\text{methanol}}{CH_3OH} + HCl$$

$$\underset{\text{1-bromopropane}}{CH_3CH_2CH_2Br} + H_2O \rightarrow \underset{\text{propan-1-ol}}{CH_3CH_2CH_2OH} + HBr$$

The C—Cl bond is more polarised than the C—Br and C—I bonds as a result of the greater electronegativity of the chlorine atom. It would therefore seem reasonable to expect the hydrolysis of the chloroalkanes to occur faster than that of the bromo- or iodoalkanes. In fact, the opposite is the case (see the box below). This is because of the greater strength of the C—Cl bond compared with the C—Br and C—I bonds, making the chloroalkanes *less* reactive than the bromo- and iodoalkanes. This order of reactivity holds good for most of the other reactions of the group, and shows that it is the bond strength that is the determining factor in the rate of reaction of the halogenoalkanes rather than the bond polarity.

Grignard reagents

If a halogenoalkane is reacted with magnesium metal in dry ethoxyethane, a **Grignard reagent** is the product – these act as nucleophiles, and are commonly used to form new carbon–carbon bonds (see page 485).

$$\underset{\text{Bromoethane}}{CH_3CH_2Br} \xrightarrow[\text{dry ethoxyethane}]{\text{magnesium in}} \underset{\text{ethyl grignard}}{CH_3CH_2MgBr}$$

Investigating the reactions of the halogenoalkanes with water

The reaction of a halogenoalkane with water involves a colourless liquid reacting to form another colourless liquid. The reaction is therefore impossible to follow without something that will detect either the formation of one of the products or the disappearance of one of the reactants. For this purpose we can use silver nitrate, $AgNO_3$, to observe the appearance of the halide ions formed by the reaction, and also to get some idea of the reaction rate.

Silver nitrate is an ionic salt that dissolves in water to form Ag^+ ions and NO_3^- ions. Neither of these ions reacts with halogenoalkanes. However, the Ag^+ ion does react with halide ions to form insoluble products which appear as precipitates, as we saw in the box 'Which halide?' in section 2.4, page 181. The reaction of a halogenoalkane with water can therefore be followed by watching the formation of a silver halide precipitate. If the time taken for the precipitate to appear is measured, then an idea of the rate of the reaction can be obtained.

The result of the investigation confirms that the rate of hydrolysis of the halogenobutanes occurs in the order:

1-chlorobutane < 1-bromobutane < 1-iodobutane

Figure 4.5.8 The rate of reaction of the different halogenoalkanes can be shown clearly using silver nitrate. In this example, 1 cm³ of silver nitrate solution was placed in each of test tubes A, B and C, together with 1 cm³ of ethanol as a solvent for both the silver nitrate and the halogenoalkane. All three tubes were placed in a water bath at 50 °C. After 10 minutes, 5 drops of 1-chlorobutane, 5 drops of 1-bromobutane and 5 drops of 1- iodobutane were added to A, B and C respectively, and the tubes were replaced in the water bath. The precipitate formed most rapidly in tube C, more slowly in tube B and slowest of all in tube A.

Reactions of the halogenoarenes and acid halides with the hydroxyl group

Halogenoarenes do not undergo nucleophilic substitution reactions under normal laboratory conditions. This is because the delocalised electron system of the benzene ring (see section 4.4, page 419) gives the C—Hal bond a certain amount of π character and so makes it stronger, whilst at the same time decreasing the polarity of the bond and so reducing the likelihood of nucleophilic attack. Thus there is no reaction with water and with aqueous sodium hydroxide when the halogen is attached to a benzene ring (although there is a reaction when chlorobenzene is reacted with molten sodium hydroxide under pressure).

Acid halides undergo nucleophilic reactions much more readily than halogenoalkanes. This is due to the polarity of both the C═O and the C—Hal bonds, producing a relatively large $\delta+$ charge on the carbon atom and so leaving it open to nucleophilic attack. This is demonstrated very clearly in the reactions of the acid halides with water – they readily undergo substitution at room temperature, forming carboxylic acids:

$$CH_3COCl + H_2O \rightarrow CH_3COOH + HCl$$

Reaction with the cyanide (CN^-) ion

Reactions with the cyanide ion might seem somewhat obscure. However, there is a very good reason for looking at this process – it gives us a controlled way of lengthening the carbon chain by just one atom at a time. The reaction shown in figure 4.5.9 takes place by a nucleophilic substitution reaction, just like the reaction with the OH^- ion, the cyanide ion acting as the nucleophile. The product is called a **nitrile**, and we shall return to the nitriles in section 5.4. The reaction does not take place in aqueous solution – the halogenoalkane must be refluxed with a solution of potassium cyanide in ethanol for the substitution to occur. Once the nitrile is formed, further refluxing with dilute sulphuric acid followed by a reaction with a reducing agent will yield an alcohol with a carbon chain one greater in length than the original halogenoalkane.

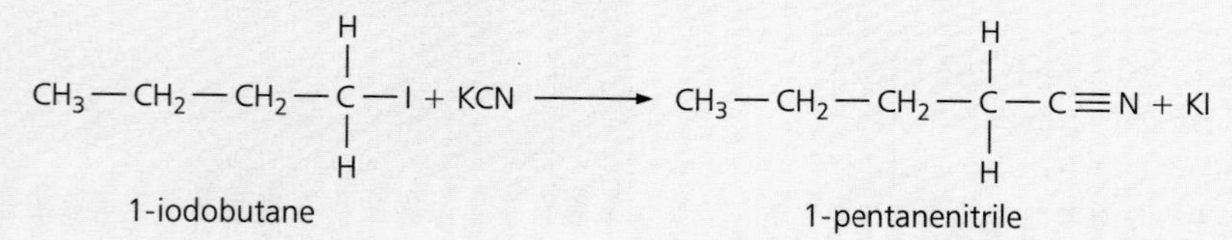

$$CH_3-CH_2-CH_2-CH_2-I + KCN \longrightarrow CH_3-CH_2-CH_2-CH_2-C{\equiv}N + KI$$

1-iodobutane → 1-pentanenitrile

Figure 4.5.9

Reaction with ammonia

Ammonia acts as a nucleophile due to the lone pair of electrons on its nitrogen atom. This means that it can easily replace a halogen atom in a halogenoalkane to form an **amine**:

$$CH_3CH_2I + NH_3 \rightarrow CH_3CH_2NH_2 + HI$$

Iodoethane **ethylamine**

This is not the end of the story, however. Ethylamine possesses a lone pair of electrons, just like ammonia. It is a stronger nucleophile than ammonia due to the electron donating effect of the alkyl group attached to its nitrogen atom, so it reacts further with iodoethane:

$$CH_3CH_2NH_2 + CH_3CH_2I \rightarrow (CH_3CH_2)_2NH + HI$$

Ethylamine **iodoethane** **diethylamine**

In turn, diethylamine is an even stronger nucleophile than ethylamine, and the reaction goes through two further stages:

$(CH_3CH_2)_2NH + CH_3CH_2I \rightarrow (CH_3CH_2)_3N + HI$
Diethylamine iodoethane triethylamine

$(CH_3CH_2)_3N + CH_3CH_2I \rightarrow (CH_3CH_2)_4N^+ I^-$
Triethylamine iodoethane tetraethylammonium iodide

The result of such reactions is a mixture of primary, secondary and tertiary amines, together with the ammonium salt – in short, a dreadful mess! For this reason, this kind of reaction is not used to make amines in the laboratory, although it is sometimes used in industry. We shall return to the amines in section 5.4.

Reactions of the halogenoarenes and acid halides with ammonia

The halogenoarenes do not react with ammonia. In contrast, the acid halides react violently with ammonia at room temperature and pressure, forming **acid amides**, as shown in figure 4.5.10. The nitrogen in amides is not particularly nucleophilic, due to the electron-withdrawing effect of the adjacent C=O group (see section 5.4, page 512). There is therefore only one product formed in these reactions, compared with the many products formed in the case of the amines.

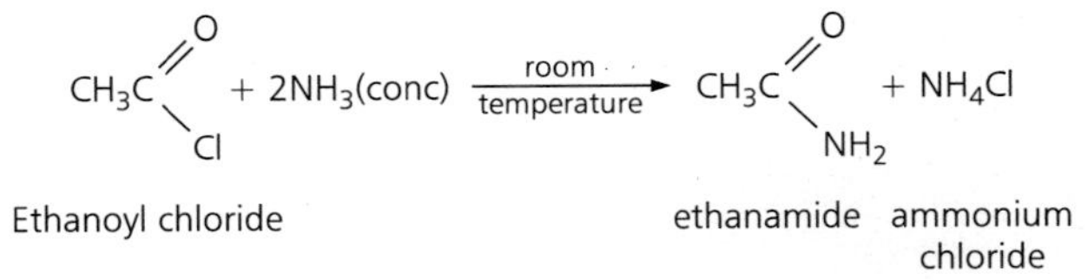

$$CH_3C(=O)Cl + 2NH_3(conc) \xrightarrow[\text{temperature}]{\text{room}} CH_3C(=O)NH_2 + NH_4Cl$$

Ethanoyl chloride ethanamide ammonium chloride

Figure 4.5.10

Reaction with the methoxide (CH_3O^-) ion

If sodium methoxide, CH_3ONa, is dissolved in methanol and refluxed with a halogenoalkane, it reacts to form an **ether**. This is yet another example of a nucleophilic substitution reaction. Instead of the methoxide ion, methanol alone will also take part in this same reaction, but just as water acts as a nucleophile far more slowly than the OH^- ion, so the substitution with methanol is much slower than with the methoxide ion.

$CH_3CH_2Br + CH_3O^- \rightarrow CH_3CH_2OCH_3 + Br^-$
Bromoethane methoxide ion methoxyethane

$CH_3CH_2Br + CH_3OH \rightarrow CH_3CH_2OCH_3 + HBr$
Bromoethane methanol methoxyethane

The methoxide ion can also act as a base under these same conditions, and an elimination reaction may take place to form ethene. This reduces the yield of the ether, and so this is not a good method of preparation of ethers, although the degree to which elimination reactions occur can be reduced by lowering the temperature.

Reactions of the halogenoarenes and acid halides with the methoxide ion

Halogenoarenes do not react with methoxide ions.

Acid halides will react vigorously with methoxide ions, even at room temperature, to form **esters**:

CH_3COCl	+	CH_3O^-	→	$CH_3COOCH_3 + Cl^-$
Ethanoyl chloride		**methoxide ion**		**methyl ethanoate**

We shall return to esters and their uses in section 5.3.

The chemistry of the halogenoalkanes – elimination reactions

In all the reactions we have considered so far, the halogenoalkanes are seen undergoing nucleophilic substitution reactions. However, there is an alternative type of reaction. This is an **elimination reaction**, in which a molecule of hydrogen halide is eliminated from the halogenoalkane, thus forming an **alkene** as a product. This kind of reaction may take place especially when species like OH^- or RO^- are involved, since these may act as a base rather than as a nucleophile:

$$CH_3CHBrCH_3 + OH^- \rightarrow CH_3CH{=}CH_2 + H_2O + Br^-$$

Elimination and substitution reactions may both take place together, and the conditions of the reaction have a substantial effect on which type of reaction predominates.

Elimination reactions and the halogenoarenes and acid halides

Simple elimination reactions such as those of the halogenoalkanes do *not* take place with either the halogenoarenes or the acid halides.

Mechanisms for the reactions of the halogenoalkanes

In this box, we shall consider briefly the mechanisms for some of the reactions we have seen in this section. The reactions fall into two classes, substitutions and eliminations, and each of these has two possible mechanisms. We shall look at substitutions first.

Substitutions

When the hydrolysis of bromoalkanes is investigated, the kinetics are found to be essentially of two types – one in which:

$$\text{Rate} \propto [\text{RHal}][\text{OH}^-]$$

and another in which:

$$\text{Rate} \propto [\text{RHal}]$$

It is found that primary halogenoalkanes such as 1-bromobutane have the first type of kinetics, and that tertiary halogenoalkanes such as 2-bromo-2-methylpropane have the second type. Chemists interpret these observations in the following way.

The hydrolysis of primary halogenoalkanes proceeds via a one-stage process, in which the attack of OH^- occurs simultaneously with the leaving of Br^-. (Br^- is called the **leaving group**.) This can be represented as shown in figure 4.5.11, and is known as an $\mathbf{S_N2}$ mechanism, standing for **second order nucleophilic substitution**.

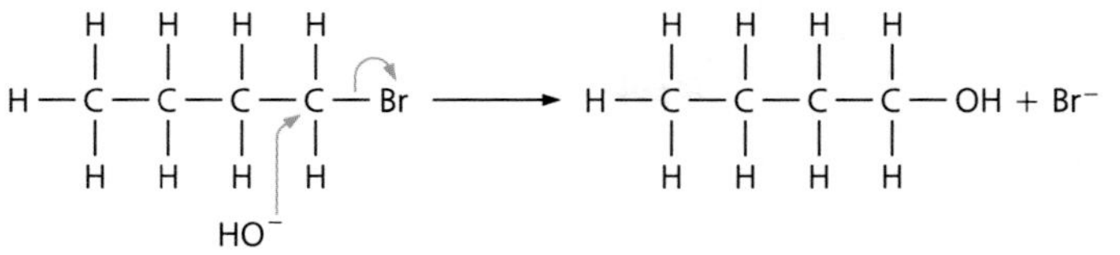

Figure 4.5.11 The hydrolysis of a primary halogenoalkane is a one-stage process, explaining why the rate expression involves terms for both the halogenoalkane concentration and the hydroxide ion concentration.

In contrast, the hydrolysis of tertiary halogenoalkanes occurs in a two-stage process, the first step being the formation of a carbocation, followed by nucleophilic attack of OH^-, as shown in figure 4.5.12. This is an $\mathbf{S_N1}$ mechanism, standing for first order nucleophilic substitution.

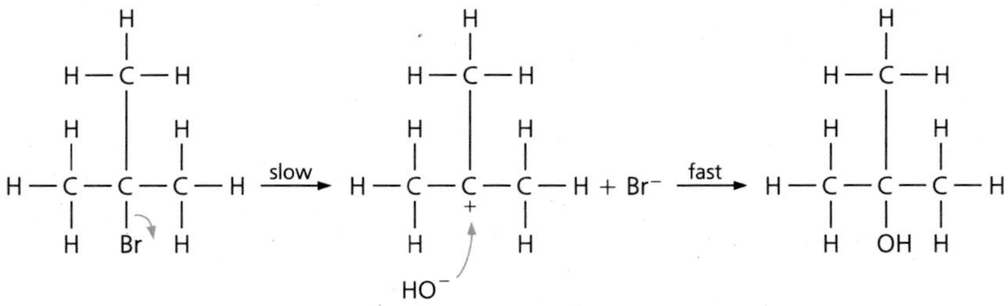

Figure 4.5.12 The hydrolysis of a tertiary halogenoalkane is a two-stage process. The formation of the carbocation is the **rate determining** (slower) step, so the rate expression involves only the concentration of the halogenoalkane. Attack by OH^- on the carbocation, once formed, is rapid.

This difference in behaviour is due to **steric differences** between the molecules – differences in their shapes. The carbon bonded to the bromine in the tertiary compound is 'hidden away' in the middle of three methyl groups, making attack by a nucleophile like OH^- very difficult. (Chemists describe atoms in a situation like this as **sterically hindered**.) On the other hand, this molecular arrangement favours the formation of the carbocation, which can be stabilised by electrons from the three methyl groups attached to it. In the case of the primary compound, the opposite is true. The formation of a carbocation is not favourable in such molecules, but the carbon attached to the bromine is much less sterically hindered, making attack by the OH^- possible.

Elimination

The two mechanisms for elimination are very similar to those for substitution, involving first and second order kinetics. These are shown in figure 4.5.13.

Once again, steric factors determine the mechanism, with carbocation formation being favoured in the case of tertiary compounds. The one-step mechanism is referred to as **E2** (second order elimination), while the two-step mechanism, in which formation of the carbocation is the rate determining step, is **E1** (first order elimination).

Formation of carbocations can be greatly influenced by the effects of solvents, as polar solvents solvate the ionic intermediate and so lower

E1

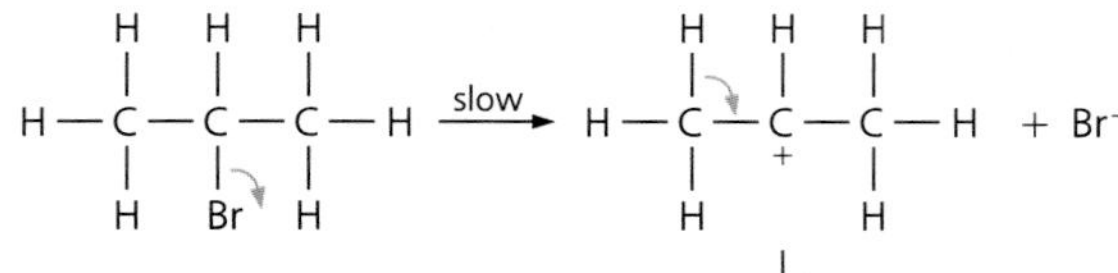

E2

Figure 4.5.13 The elimination of HBr from a bromoalkane can proceed either via a carbocation (E1) or as a concerted process (E2).

the activation energy for the reaction. Therefore the effect of the solvent on the reaction mechanism must also be taken into account – reactions in water are much more likely to involve S_N1 and E1 mechanisms than those that take place in ethanol.

Hydrolysis or elimination?

There are a number of factors which affect whether a halogenoalkane undergoes addition or elimination in the reaction with OH^-. One important factor is the solvent in which the reaction occurs. Thus, when 2-bromopropane is refluxed with aqueous sodium hydroxide, the product of the reaction is propan-2-ol.

$$CH_3CHBrCH_3 + OH^- \rightarrow CH_3CHOHCH_3 + Br^-$$

In contrast, refluxing with ethanolic sodium hydroxide produces propene:

$$CH_3CHBrCH_3 + OH^- \rightarrow CH_2{=}CHCH_3 + H_2O + Br^-$$

Perfluorides – medical molecules for the future?

Organic chemists have 'created' almost all of the organohalogens, and they are constantly working to discover new, useful molecules. One of the newest groups of organic compounds to find a major role in medicine is the perfluorocarbons (PFCs). These are linear or cyclic organic compounds in which hydrogen atoms have been substituted by fluorine atoms. The strong carbon–fluorine bond gives PFCs outstanding chemical and thermal stability. So far the most common and best known medical use of these compounds is as 'blood substitutes'. One of the most striking properties of the PFCs is that they can dissolve very large volumes of the respiratory gases oxygen and carbon dioxide. Although in a pure form they are immiscible with blood and aqueous solutions, they can be emulsified and then will mix with blood.

Experimental use of PFCs, such as the perfluorodecalin/perfluorotripropylamine mixture of the commercial product Fluosol, took place in America and Japan in the 1980s. Doctors have explored the possibility of using these compounds as blood supplements during surgery, particularly when patients could not have normal blood transfusion either because they had religious objections to the procedure or because they had rare blood groups. The first use of PFCs in Britain took place in 1991, when they were used in an operation known as 'balloon angioplasty', shown in figure 4.5.14.

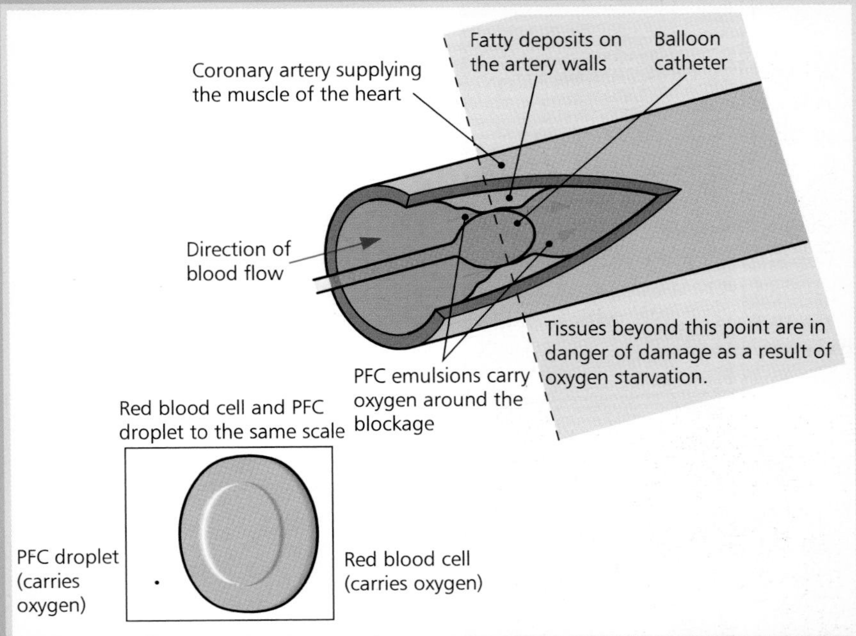

Figure 4.5.14 When fat deposits build up in the arteries supplying blood to the muscles of the heart, these vital muscles are gradually starved of oxygen. This can lead to pain (angina), heart attacks or, in severe cases, death. One technique for solving this problem involves inserting a balloon catheter and inflating the balloon in the blood vessel, widening the vessel and shifting the blockage. However, while the balloon is inflated, virtually no blood can get past it, and so the heart muscle is almost completely starved of oxygen. This is risky and limits the time available for the procedure. If PFCs are introduced into the blood vessel, they can pass the blockage due to the small size of their molecules, and carry oxygen to the tissues beyond the balloon, giving longer for the surgeon to work and increasing the effectiveness of the operation as a result.

The properties of the PFC molecules suggest a far wider range of potential applications, many of which are already at an experimental stage. For example, PFCs may have a role in cancer treatment. In a fast-growing tumour, there is often a poor blood supply, and so many of the cells are short of oxygen. Because of this, they do not respond readily to anticancer drugs. Recent research indicates that if the oxygen supply to tumours is increased using PFCs, the effectiveness of both radiation therapy and chemotherapy in destroying the cancer cells is improved. The drugs are taken up more rapidly, whilst the extra oxygen enables the radiation to produce more 'cell-killing' oxygen radicals.

Another possible role of PFCs is in diagnostic imaging techniques used increasingly to see inside the body without the invasion of surgery.

4

PFCs are also being used to maintain organs such as hearts, livers and kidneys that have been removed prior to transplantation. These organs need to be kept in perfect condition if they are to work when transplanted into the patient who needs them, and the rich supply of oxygen that PFCs can provide is one way of helping to ensure that any damage after removal is kept to a minimum. Cell cultures are also increasingly being used in medicine, both to produce compounds useful for drugs or cosmetics, as an alternative to animal testing of new pharmaceuticals and as a source of tissue for transplanting into needy patients. PFCs can be used in the oxygenation of such cultures, and thus make them much more productive.

The final area of potential use of PFCs we shall consider is their role in the survival of preterm babies. When a fetus is developing in the uterus, its lungs are very small and do not function, as all the oxygen needs are supplied via the placenta from the mother. At the moment of birth, the lungs need to inflate as breathing commences, and this is extremely difficult unless the inner surfaces are coated with a chemical known as **lung surfactant**. This is only formed in any quantity during the final weeks of pregnancy, and so if a baby is born prematurely it will probably lack this vital chemical. Artificial surfactants are now available, which help make it possible to inflate the lungs. However, the physical ventilation of tiny lungs not yet ready to be used can cause severe problems later on. Some very recent work suggests that PFCs may be able to help.

If PFCs are introduced into the immature respiratory system, then sufficient oxygen for the infant's needs can be extracted from the air with much less ventilation of the lungs. This greatly reduces the risk of permanent lung damage and so improves the chances of survival for tiny preterm babies, and also their chances of a normal, healthy life to follow. The list of possible uses for PFCs in medicine seems almost endless – if they fulfil their initial promise they will have a major impact on the health care of the future.

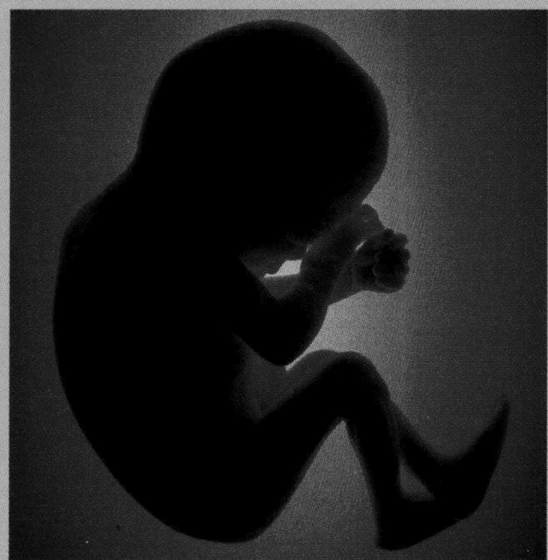

Figure 4.5.15 The use of PFCs to aid the respiration of babies born too early is an exciting development. But could PFCs lead us to the edge where science and science fiction merge? Both dogs and mice have been kept alive for some time taking all their oxygen needs from PFCs. Will the day come when developing humans will grow in artificial uteruses, with oxygen provided by the PFCs in which they float?

SUMMARY

- Organohalogen compounds have a hydrocarbon skeleton that may be aliphatic, alicyclic or aromatic, with a halogen functional group.
- The properties of organohalogen compounds are affected by:
  the type of hydrocarbon skeleton
  the halogen or halogens attached
  the position of the halogen in the molecule.
- Halogenoalkanes are alkanes in which one or more of the hydrogen atoms has been replaced by a halogen.
- Halogenoarenes have halogen atoms substituted for one or more of the hydrogen atoms of the benzene ring.
- Acid halides are the halides of carboxylic acids and may be represented as RCOHal.
- The carbon–halogen bond has a major effect on the properties of the molecule. The C—F bond in particular is very strong so compounds containing carbon and fluorine tend to be very stable.
- The C—Hal bond is polar and the partial positive charge on the carbon atom allows nucleophilic attack to occur.
- The main nucleophilic substitution reactions of the organohalogens are summarised in table 4.5.3. Generally the halogenoarenes are much less reactive than the halogenoalkanes, while the acid halides are more reactive.
- **Elimination reactions** occur when a molecule of hydrogen halide is eliminated from the halogenoalkane, leaving an alkene. Neither halogenoarenes nor acid halides undergo elimination reactions.

Reaction with	Reaction of the halogenoalkanes	Reaction of the halogenoarenes	Reaction of the acid halides
-OH group (e.g. water, NaOH)	Hydrolysis occurs to give an alcohol and a hydrogen halide.	No reaction except under extreme conditions	React very readily, due to polar C=O and C—Hal bonds, to give a carboxylic acid and hydrogen halide.
Cyanide ion	The cyanide ion acts as a nucleophile when the halogenoalkane is refluxed with a solution of KCN in ethanol and a nitrile is formed.	No reaction	React to form cyanohydrin derivatives (see page 479).
Ammonia	Forms an amine and then continues reacting to give a mixture of substituted products up to the ammonium salt.	No reaction	Vigorous reaction to give acid amides
Ethoxide ion	Sodium ethoxide dissolved in ethanol and refluxed with a halogenoalkane forms an ether.	No reaction	Vigorous reaction even at room temperature to give esters

Table 4.5.3

QUESTIONS

1 a What is a halogenoalkane? Give the structural formulae of examples of primary, secondary and tertiary halogenoalkanes.
 b What is a halogenoarene? Give the structural formula of a halogenoarene.
 c What is an acid halide? Give the structural formula of an acid chloride.

2

Halogenoalkane	Boiling point/°C
Chloromethane	–24
Bromomethane	4
Iodomethane	43
Chloroethane	12
Bromoethane	39
Iodoethane	72
1-chlorobutane	79
1-bromobutane	102
1-iodobutane	131
2-chlorobutane	68
2-bromobutane	91
2-iodobutane	120
2-chloro-2-methylpropane	51
2-bromo-2-methylpropane	73
2-iodo-2-methylpropane	100

Table 4.5.4

 a Plot the data in table 4.5.4 on a graph to show as clearly as possible the trends in boiling point.
 b What is the effect of the lengthening carbon chain on the boiling point?
 c What is the effect of the halogen atom on the boiling point?
 d What is the effect of the position of the halogen atom on the boiling point?

3 The reactivity of the organohalogen compounds is affected by the halogen atom involved.
 a Which organohalogen compounds are the most stable? Explain this stability in terms of bond enthalpies.
 b When 1-bromopropane reacts with a base, two possible products may be formed. Explain how these two products are formed, setting out the mechanism of each reaction.

4 Predict the product(s) of the following reactions:
 a bromoethane with sodium hydroxide solution
 b chlorobenzene with water
 c ethanoyl chloride and ammonia at room temperature and pressure
 d 1-iodopropane refluxed with KCN in ethanol
 e bromoethane and $LiAlH_4$
 f 2-iodopropane and sodamide ($Na^+NH_2^-$).

Developing Key Skills

Each year cases are reported in the media where people have died because their religious beliefs do not allow them to accept blood transfusions. Write an article for a popular paper heralding the arrival of perfluorocarbons as life-saving chemicals, including an explanation of how they can be used to overcome the need for blood transfusions when necessary.

[Key Skills opportunities: C]

Questions

1 a The diagram below represents the industrial fractional distillation of crude oil.

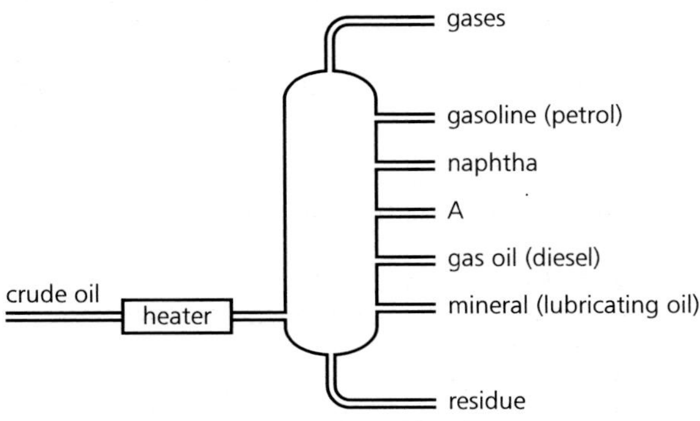

i Identify fraction **A**.

ii What property of the fractions allows them to be separated in the column? **(2 marks)**

b A gas oil fraction from the distillation of crude oil contains hydrocarbons in the C_{15} to C_{19} range. These hydrocarbons can be cracked by strong heating.

i Write the molecular formula for the alkane with 19 carbon atoms.

ii Name the type of reaction involved in cracking.

iii Write an equation for one possible cracking reaction of the alkane $C_{16}H_{34}$ when the products include ethene and propene in the molar ratio 2:1 and only one other compound.

(4 marks)

(Total 6 marks)

(AQA specimen)

2 a i Name the alkene $CH_3CH_2CH{=}CH_2$

ii Explain why $CH_3CH_2CH{=}CH_2$ does not show geometrical isomerism.

iii Draw an isomer of $CH_3CH_2CH{=}CH_2$ which does show geometrical isomerism.

iv Draw another isomer of $CH_3CH_2CH{=}CH_2$ which does not show geometrical isomerism. **(4 marks)**

b i Name the type of mechanism for the reaction shown by alkenes with concentrated sulphuric acid.

ii Write a mechanism showing the formation of the major product in the reaction of concentrated sulphuric acid with $CH_3CH_2CH{=}CH_2$

iii Explain why this compound rather than one of its isomers is the major product. **(6 marks)**

(Total 10 marks)

(AQA specimen)

3 a Give the structural formula of 2-bromo-3-methylbutane. **(1 mark)**

b Write an equation for the reaction between 2-bromo-3-methylbutane and dilute aqueous sodium hydroxide. Name the type of reaction taking place and outline a mechanism. **(4 marks)**

c Two isomeric alkenes are formed when 2-bromo-3-methylbutane reacts with ethanolic potassium hydroxide. Name the type of reaction occurring and state the role of the reagent. Give the structural formulae of the two alkenes. **(4 marks)**

(Total 9 marks)

(AQA specimen)

4 You are required to plan an experiment to determine the percentage by mass of bromine in a bromoalkane. The bromoalkane, which boils at 75°C, can be hydrolysed completely by heating with an appropriate amount of boiling, aqueous sodium hydroxide for about 40 minutes. The bromide ion released can be estimated by converting it into a silver bromide precipitate which is subsequently weighed.

a Write equations for the reactions which occur.

b Describe how you would carry out the hydrolysis, giving details of the apparatus and the conditions which you would use.

c Describe, giving details of the apparatus and reagents, how you would obtain a silver bromide precipitate from the hydrolysis solution and how you would determine the mass of the silver bromide.

d Show how the percentage by mass of bromide ion, in the original haloalkane, can be calculated. **(15 marks)**

(AQA specimen)

5 A hydrocarbon **E** has the following composition by mass: C, 90.56%; H, 9.44%. $M_r = 106$.

a i Use the data above to show that the empirical formula is C_4H_5.

ii Deduce the molecular formula. **(3 marks)**

b Compound **E** contains a benzene ring. Draw structures for all **four** possible isomers of **E**. **(4 marks)**

c The n.m.r. spectrum of **E** is shown below.

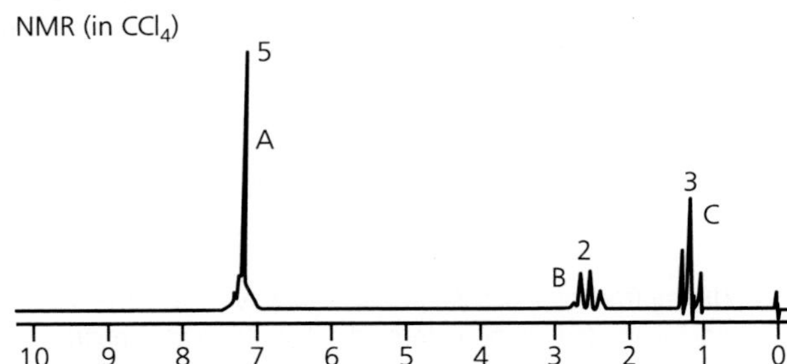

d Suggest the identity of the protons responsible for the groups of peaks **A**, **B** and **C**. For each group of peaks, explain your reasoning carefully in terms of both the chemical shift value and the splitting pattern. **(9 marks)**

e Using the evidence from the peaks in **c**, identify and draw hydrocarbon **E**. **(1 mark)**

(Total 17 marks)

(OCR specimen)

6 Carbon is able to form an enormous number of chemical compounds because of its ability to bond to itself to form chains and rings.

a Petrol is a mixture of alkanes containing between 6 and 10 carbon atoms. Some of these alkanes are structural isomers of one another.

i Explain the term *structural isomers*.

ii The alkanes are an example of a homologous series. Explain what is meant by this term.

iii State the molecular formula of an alkane that could be present in petrol. **(5 marks)**

b But-2-ene is an isomer of C_4H_8.

i Draw diagrams to show the *cis* and *trans* isomers of but-2-ene.

ii Draw diagrams of **two** isomers of C_4H_8 each of which are structural isomers of but-2-ene. Name each isomer. **(6 marks)**

c Alkenes such as but-2-ene, C_4H_8, are used by the petrochemical industry to produce many useful materials. Draw structures to represent possible compounds **A–D** in the reactions of but-2-ene shown below. **(4 marks)**

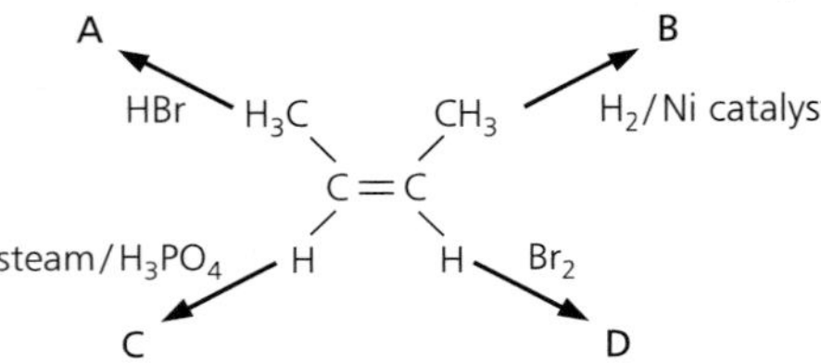

d But-2-ene is used to make a commercially important polymer.

i What type of polymerisation takes place?

ii Suggest a section of this polymer by drawing **two** repeat units. **(2 marks)**

e But-2-ene can be converted into buta-1,3-diene by a process called dehydrogenation. Buta-1,3-diene is used to make synthetic rubber.

i Suggest the structure of buta-1,3-diene.

ii Construct an equation for the dehydrogenation of but-1-ene to form buta-1,3-diene. **(2 marks)**

(Total 19 marks)

(OCR specimen)

7 Crude oil is an important source of chemicals that can be obtained by fractional distillation and subsequent processing involving cracking, isomerisation and reforming.

a During fractional distillation, explain why hydrocarbons containing few carbon atoms distil at different temperatures from hydrocarbons with many carbon atoms. **(2 marks)**

b i What is meant by *cracking*?

ii Suggest an equation which illustrates the cracking of decane, $C_{10}H_{22}$.

iii Although heat alone can be used to crack hydrocarbons, it is far more common for oil companies also to use catalysts. Suggest **two** reasons why oil companies use catalysts. **(5 marks)**

c Isomerisation produces branched hydrocarbons.

i Why should oil companies want to make branched hydrocarbons from straight-chain hydrocarbons?

ii Show the structure of a compound that could be obtained from the isomerisation of hexane. Name the compound.

iii One of the important hydrocarbons produced during reforming is benzene. Construct a balanced equation for its formation when hexane is reformed. **(4 marks)**

(Total 11 marks)

(OCR specimen)

8 Myrcene is a naturally occurring oil present in bay leaves. The structure of myrcene is shown below.

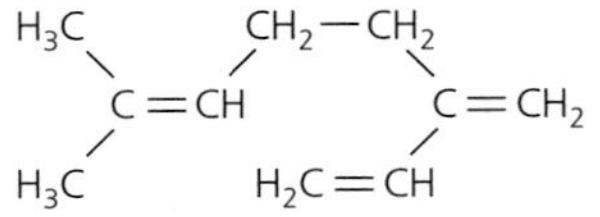

H_3C / H_3C C=CH–CH_2–CH_2–C(=CH_2)–CH=CH_2

(printed labels: H_3C, CH_2-CH_2, C=CH, C=CH_2, H_3C, H_2C=CH)

a State the molecular formula of myrcene. **(1 mark)**

b Reaction of a 0.100 mol sample of myrcene with hydrogen produced a saturated alkane **A**.

i Explain what is meant by the term *saturated* alkane.

ii Determine the molecular formula of the saturated alkane **A**.

iii Construct a balanced equation for this reaction.

iv Calculate the volume of hydrogen, measured at room temperature and pressure (r.t.p.), that reacted with the sample of myrcene. [1 mole of gas molecules occupy 24.0 dm^3 at r.t.p.] **(5 marks)**

c Squalene is a naturally occurring oil present in shark liver oil. A 0.100 mol sample of squalene reacted with 14.4 dm^3 of hydrogen, measured at r.t.p., to form a saturated hydrocarbon $C_{30}H_{62}$.

i Calculate how many double bonds there are in each molecule of squalene.

ii Suggest the molecular formula of squalene. **(3 marks)**

(Total 9 marks)

(OCR specimen)

9 a Ethane reacts with chlorine in the presence of ultraviolet light.

i What is the function of the u.v. light? **(1 mark)**

ii The reaction gives more than one chlorine-containing organic product. Explain why this is so. **(1 mark)**

iii State the name of one other product formed in this reaction which does not contain chlorine and account for its formation. **(3 marks)**

b Methylbenzene reacts with chlorine in the presence of u.v. light in a similar way to methane:

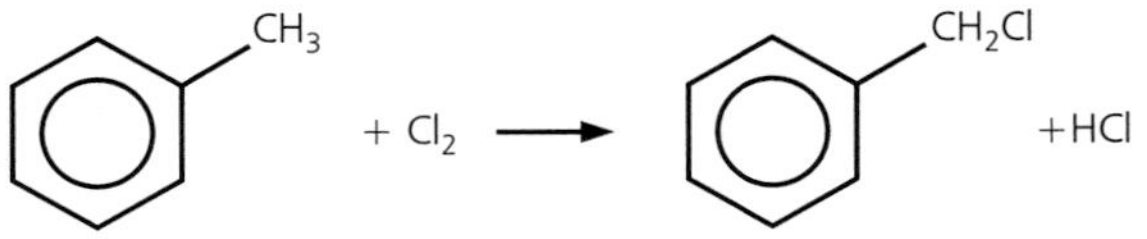

$C_6H_5CH_3 + Cl_2 \longrightarrow C_6H_5CH_2Cl + HCl$

In order for the reaction to take place, chlorine must be passed into boiling methylbenzene in the presence of u.v. light over a period of time.

i Draw an apparatus which could be used for this reaction. **(3 marks)**

ii What are the hazards **specific** to this experiment? **(2 marks)**

(Total 10 marks)

(Edexcel 1999, part)

10 This question concerns the compound, **Q**, pent-3-en-2-ol.

```
CH3CHCH=CHCH3
    |
    OH
```

a Give the structural formula of a compound isomeric with **Q** which does not contain a carbon–carbon double bond. **(1 mark)**

b **Q** shows two types of stereoisomerism.

i Draw the **geometric** isomers of **Q**, and state why **Q** shows this type of isomerism. **(2 marks)**

ii Draw the **optical isomers** of **Q** and state what effect such isomers have on the plane of plane polarised monochromatic light. **(3 marks)**

c Give the mechanism for the reaction of bromine with **Q**; you may represent the compound as >C=C< for the purposes of this question. **(3 marks)**

d i Give the structural formula of the product obtained when **Q** is reacted with potassium dichromate(VI) solution in the presence of dilute sulphuric acid. **(1 mark)**

ii Describe a test to show that the product formed in **i** is not an aldehyde. Suggest a positive test which would give the identity of the new functional group which is present in the molecule, and state what you would see. **(4 marks)**

e The **iodine number** is a measure of the degree of unsaturation of a molecule. It relates to the number of double bonds it contains. It is defined as the number of grams of iodine, I_2, that react with 1g of the compound.

Show that the iodine number of **Q** is approximately 3 by calculating the number of grams of iodine, I_2, which react with 1g of **Q**. **(3 marks)**

(Total 17 marks)

(Edexcel 1999)

11 a Define the term *structural isomerism*. **(2 marks)**

b The diagram shows part of the mass spectrum of an organic compound, **A**, which has the molecular formula C_4H_{10}.

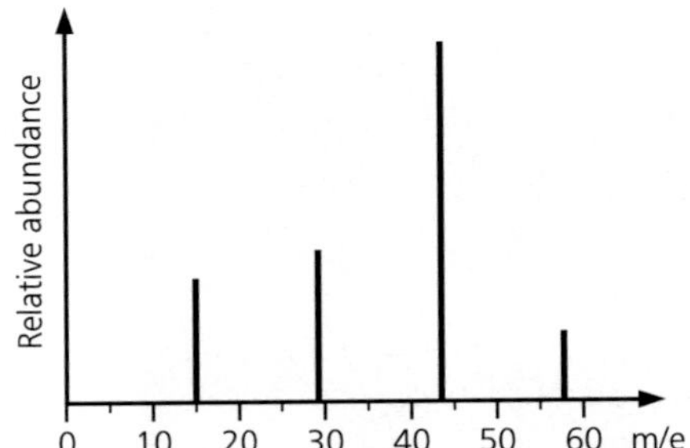

i Draw graphical formulae for the structural isomers of C_4H_{10}. **(2 marks)**

ii Suggest the formula of the fragment that corresponds to each of the following m/e values shown in the mass spectrum of compound **A**: 43, 29, 15. **(3 marks)**

iii Deduce which of the isomers drawn in **i** corresponds to compound **A**. Give a reason for your answer. **(2 marks)**

(Total 9 marks)

(AQA 1999)

12 a Explain why *cis–trans* isomerism

i does exist for 1,2-dibromoethene,

ii does **not** exist for 1,2-dibromoethane,

iii does **not** exist for 1,1-dibromoethene. **(3 marks)**

b i Draw displayed formulae for the following compounds:
hex-3-ene,
butenedioic acid.

ii State and explain which, if either, of hex-3-ene and butenedioic acid can display *cis–trans* isomerism. **(3 marks)**

c i State what is meant by a *chiral centre*.

ii Explain how such a centre gives rise to optical isomerism.

iii Which of the compounds below has optical isomers?

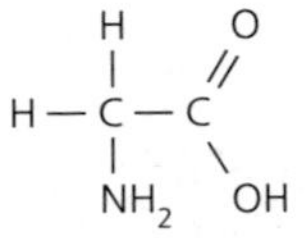

```
     H   O             H  H   O             H  H   O
     |  //             |  |  //             |  |  //
 H - C - C         H - C - C - C        H - C - C - C
     |   \             |  |   \             |  |   \
    NH2   OH          OH  H    OH           H  OH   OH

       L                    M                    N
```

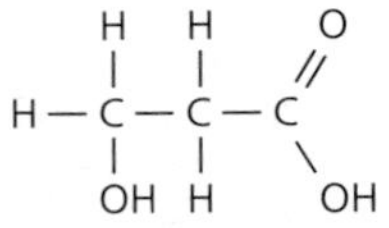

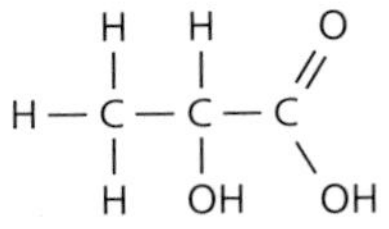

iv Epinephrine (adrenalin), shown below, can exist as optical isomers. Draw these two isomers and indicate with an asterisk (*) the chiral carbon atom.

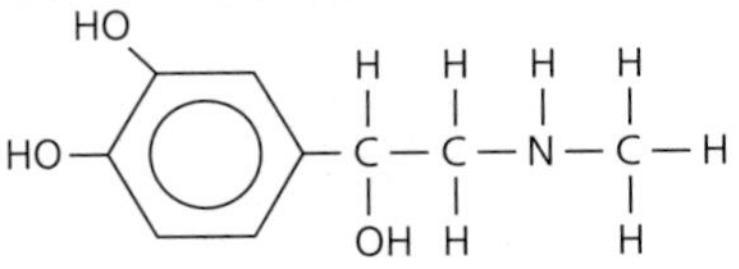

```
     HO
       \              H  H  H   H
        ____          |  |  |   |
  HO - /    \ ------- C -C -N - C - H
       \____/         |  |      |
                     OH  H      H
```

(5 marks)

(Total 11 marks)

(OCR 1998)

13 Write an equation for the sulphonation of benzene in concentrated sulphuric acid. Identify the reactive inorganic species involved in the reaction and explain how this species is able to act as an electrophile. Outline a mechanism for the substitution reaction. **(8 marks)**

(AQA 1999)

5 ORGANIC CHEMISTRY 2

Introduction

Among the 7 million different organic chemicals are some very simple ones which are a familiar part of everyday life and some very complex ones on which life itself depends. In this section of the book we will be looking at organic families which fall into both these categories, along with some compounds which are of enormous importance in the synthesis of both everyday materials and the drugs we use to help us overcome disease.

The **alcohols** have played a part in human society since the earliest days because of the effect one member of the group, ethanol, has on the human brain. But alcohols also have important uses as fuels and solvents, and they are commonly found in the natural world. **Phenols**, aromatic alcohols, are the basis of many antiseptics which help prevent the spread of disease, whilst **ethers**, the final member of this group of families, originally became noticed as anaesthetics although their main role now is as organic solvents. Another pair of closely related families of organic chemicals is the **aldehydes** and **ketones**. These carbonyl compounds have many similarities but their differences can clearly be detected by the way they react with other compounds. By understanding the basis for these differences we can learn to predict the reactions of the members of these groups in different situations.

Just as the mineral acids play an important role in inorganic chemistry, so the families of organic acids are important in organic chemistry. Some of the most common and familiar belong to the **carboxylic acids** – vinegar and lemon juice are two everyday examples. The functional group of the carboxylic acids has much in common with the alcohols, the aldehydes and the ketones, so the chemistry of the family is an interesting mixture.

The **organic nitrogen compounds** are a very complex group which includes the **nitro-compounds**, the **nitriles**, the **amides** and the **amines**. Some of these compounds are of great value to industry, particularly in the manufacture of dyes, food colourings and flavourings, sulphonamide drugs and nylon. Relatively small changes in the makeup of the molecules leads to important differences in the ways they react.

Much organic chemistry concerns naturally occuring molecules, and in the final chapter we look at some of the most important macromolecule found in living cells. These massive natural polymers – **carbohydrates**, **proteins** and **fats** – are the fundamental molecules from which cells are constructed. By applying what we have learnt about other organic families we can understand how these vital molecules are built up and how they react. The more we understand about how the chemistry of our bodies works, the better our chance of keeping everything working as it should.

Figure 1 Polyesters and other synthetic fabrics are used to make sails like these – the synthetic fibres are light and strong and do not rot.

FOCUS ETHANOL – THE DEMON DRINK?

Probably the best known use of alcohols by the human race is in alcoholic drinks containing ethanol. The use of ethanol as a pleasure-inducing drug has a long history – from earliest times the natural fermentation of fruits was used to produce alcoholic drinks. Indeed humans are not the only species to have discovered the effects. Many animals have been observed gorging on fermented wild fruits and suffering the consequences, and like their human counterparts they usually return the following day for more! Whilst many people around the world enjoy a limited intake of ethanol, with its effect of relaxing the muscles and loosening the inhibitions, rather fewer are aware of the toxicity of their chosen drink.

Ethanol is a very dangerous chemical in excess. Acute cases of ethanol poisoning are most commonly seen in young people who drink a few measures of spirits and under the influence of the alcohol decide they can manage the whole bottle. In such circumstances ethanol can cause unconsciousness and vomiting (a potentially lethal combination, as this can lead to death by drowning in vomit) along with total liver failure. Ethanol intoxication leads to greatly increased reaction times, compounded by an increased sense of being very much in control – again, frequently a lethal combination when the intoxicated person is behind the wheel of a car. Ethanol is also an addictive drug which can lead to long term heavy drinking. This causes chronic liver disease and brain damage, eventually leading to death.

Figure 1 The use of alcohol is widely accepted in many societies, although the dangers of its abuse have long been recognised as *Gin Lane*, engraved by Hogarth in 1751, clearly shows.

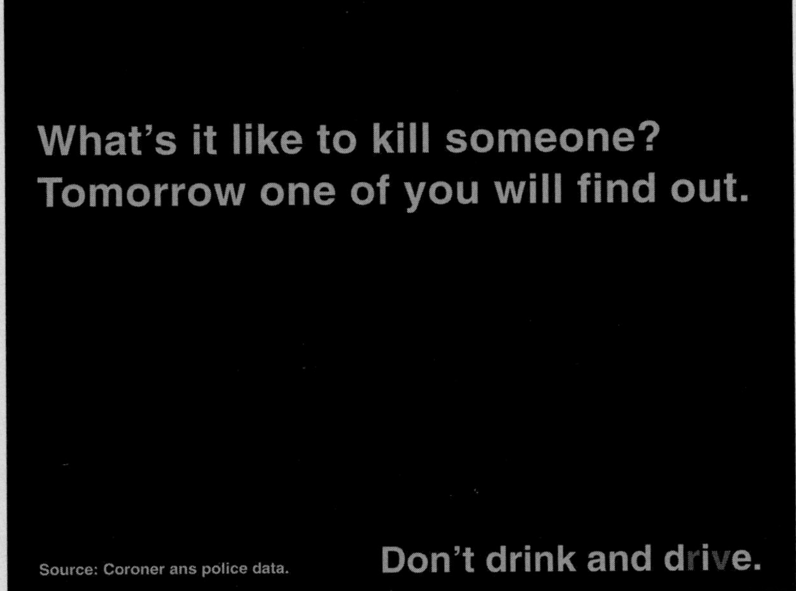

What's it like to kill someone?
Tomorrow one of you will find out.
Source: Coroner ans police data.
Don't drink and drive.

Figure 2 The numbers of young people arrested for drink-driving offences is relatively low – it is middle-aged men who are particularly likely to get behind the wheel when over the limit. But the numbers of people killed and injured by drunken motorists is still so high that regular campaigns (like this one) are set up to try and make *everyone* realise that alcohol and driving don't mix.

Drinking ethanol in various flavoured solutions is part of the culture of many societies. However, there can be little doubt that if ethanol were a newly discovered form of drug, it would be unlikely to be made legal, as it is so toxic and potentially damaging.

Nevertheless, although ethanol remains a source of death and human misery – as well as a source of considerable pleasure – through its use as a recreational drug, it may yet have a role to play in saving the Earth from damage due to greenhouse gases. Increasingly in countries like Brazil, crops such as sugar cane are being grown to be fermented on an industrial scale to produce ethanol, not for drinking, but as a renewable fuel source for vehicles (see page 459). The use of ethanol in this way replaces highly polluting fuels derived from dwindling supplies of fossil fuels. It may yet herald the time when ethanol is seen as a solution to, rather than a cause of, human problems.

5.1 Alcohols, phenols and ethers

Using ethanol in alcoholic drinks has a long history. Its use as a fuel for cars in countries such as Brazil is a more recent development. Our familiarity with ethanol stems from the fact that it is readily produced in nature and has long been part of human society. Ethanol is part of a family of similar organic compounds called the **alcohols**.

Figure 5.1.1 Sugar cane is fermented to form alcohol, a cheap alternative to petrol in Brazil.

The organic hydroxy compounds

In this section we shall look at the organic hydroxy compounds (those containing an -OH group), which include the **alcohols** and the **phenols**. We shall also look at the **ethers**. All these compounds can be regarded as organic derivatives of water, with one or both of the hydrogen atoms substituted by an alkyl or an aryl group.

- **Alcohols** are aliphatic organic molecules with an alkyl group (indicated in the general formula simply by the letter R or [R]) containing a carbon atom bonded to an -OH group, as shown in figure 5.1.2(a).

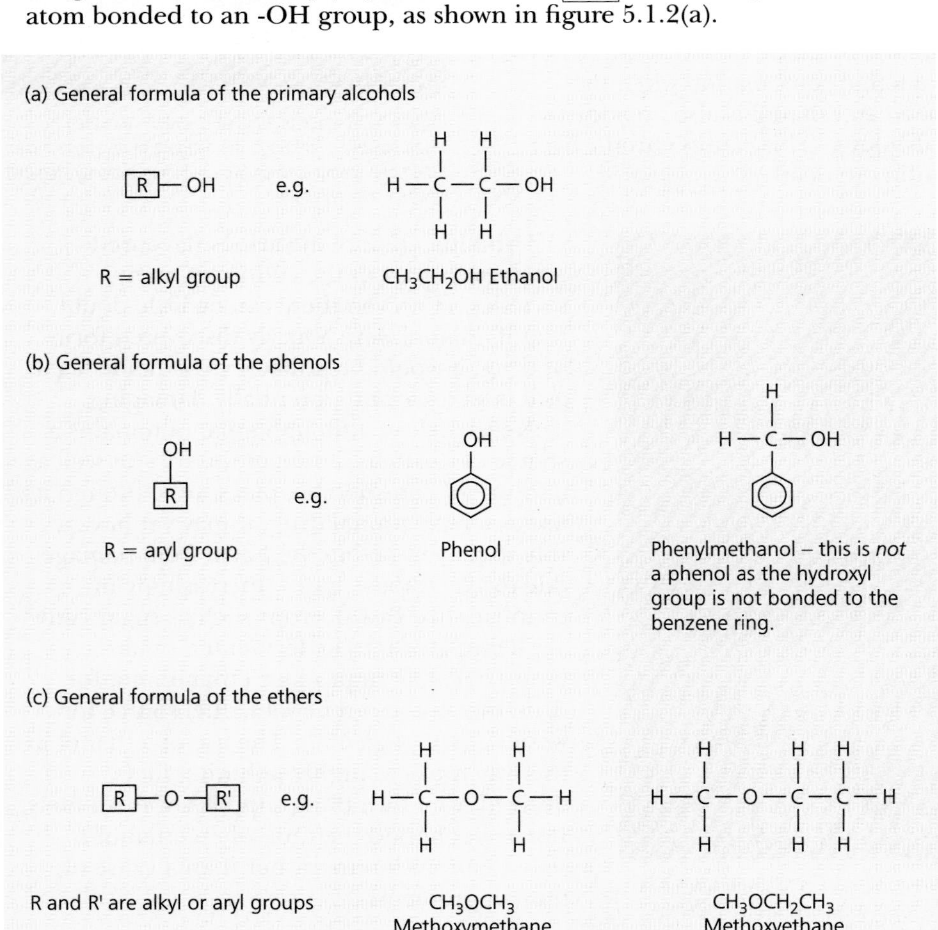

Figure 5.1.2 Alcohols, phenols and ethers

- **Phenols** are compounds with one or more -OH groups attached directly to a benzene ring, as shown in figure 5.1.2(b). Their properties are so different from those of the other alcohols that they are treated as a separate family.
- **Ethers** can be considered as water molecules in which both of the hydrogen atoms have been substituted by alkyl groups, as in figure 5.1.2(c). They have the general formula ROR′, where R and R′ may be either the same or different.

Alcohols and phenols

Naming the alcohols

The naming of the alcohols is relatively simple – the name of the alkyl group containing the longest chain of carbon atoms is used, along with the suffix '-ol'. Depending on the alcohol, a number may be used in front of the '-ol' to indicate which carbon atom the -OH group is attached to, as in figure 5.1.3. Side chains are also indicated in the usual way.

```
            H
            |
        H — C — OH              Methanol
            |
            H

        H   H
        |   |
    H — C — C — OH              Ethanol
        |   |
        H   H

    H   H   H
    |   |   |
H — C — C — C — OH              Propan-1-ol
    |   |   |
    H   H   H

    H   H   H   H
    |   |   |   |
H — C — C — C — C — OH          Butan-1-ol
    |   |   |   |
    H   H   H   H

    H   H   H   H
    |   |   |   |
H — C — C — C — C — H           Butan-2-ol
    |   |   |   |
    H   H   OH  H

            H
            |
        H — C — H
            |
    H   H   |   H
    |   |   |   |
H — C — C — C — C — H           2-methylbutan-2-ol
    |   |   |   |
    H   H   OH  H
```

Figure 5.1.3

In the same way that halogenoalkanes occur with primary, secondary and tertiary structures, so do the alcohols. Whether an alcohol is primary, secondary or tertiary depends on the number of hydrogens attached to the carbon atom bearing the -OH group, as shown in figure 5.1.4. These differences in structure affect the way the alcohols react.

- **Primary alcohols** have two hydrogens on the OH-bearing carbon – this carbon is bonded to one other carbon, as shown in figure 5.1.4(a).
- **Secondary alcohols** have two carbons and only one hydrogen substituted onto the OH-bearing carbon, as shown in figure 5.1.4(b).
- **Tertiary alcohols** have three carbons and no hydrogens on the OH-bearing carbon, as shown in figure 5.1.4(c).

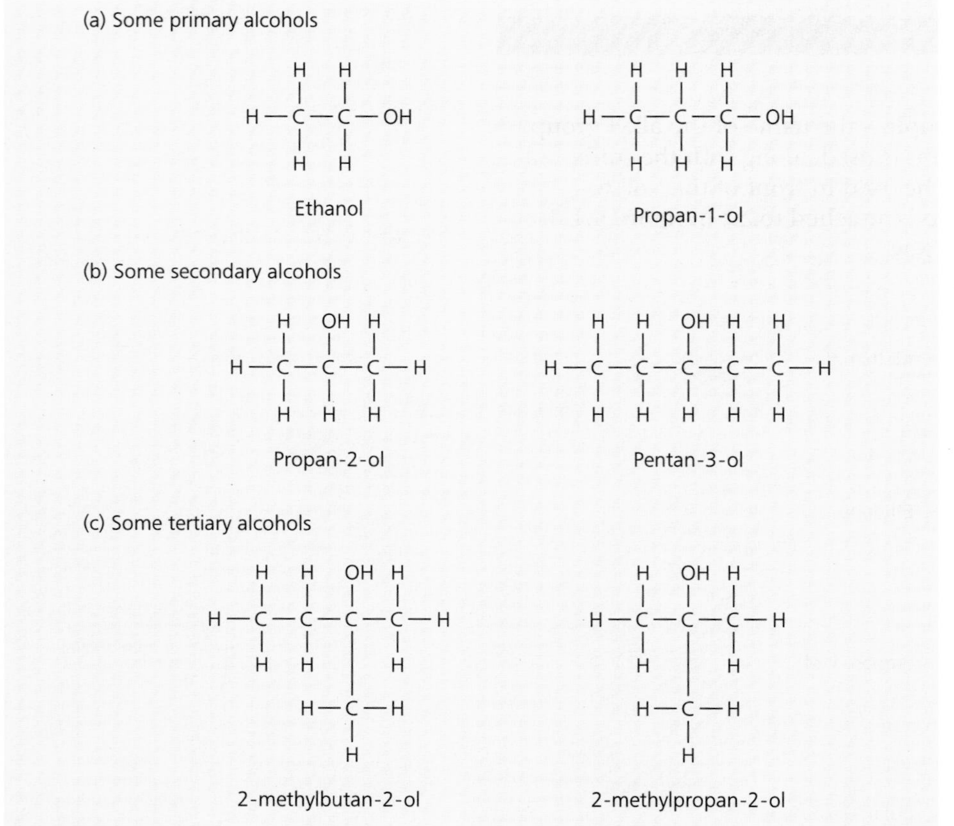

Figure 5.1.4

As well as having varying numbers of alkyl or aryl groups within the molecules, alcohols may have varying numbers of -OH groups. Alcohols with more than one -OH group are known as **polyhydric alcohols** or polyols and they are of particular importance in biological systems. Molecules with two -OH groups are known as **diols** and those with three -OH groups as **triols**. Examples include ethane-1,2-diol and butane-1,2,3-triol.

Naming the phenols

When naming the phenols, it is important to remember to number the carbon atoms and to name them very precisely, since there are often considerable differences in behaviour between compounds with -OH groups in different positions. Figure 5.1.5 (opposite) gives some examples of substituted phenols.

Physical properties of the alcohols

The physical properties of the primary alcohols change as we would predict with increasing carbon chain length. However, the properties of the family as a whole are somewhat different from those of similar sized organic molecules in other organic families. Tables 5.1.1 and 5.1.2 (page 458) and figure 5.1.6 show some physical properties of some of the alcohols, revealing the trends within

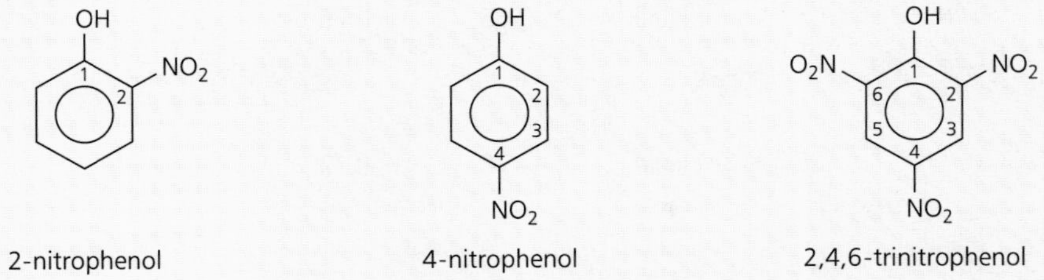

2-nitrophenol 4-nitrophenol 2,4,6-trinitrophenol

Figure 5.1.5 Some substituted phenols.

the organic family and the effect of primary, secondary and tertiary structure on the physical properties. They also allow us to compare the smaller alcohol molecules with the equivalent molecules in other organic families.

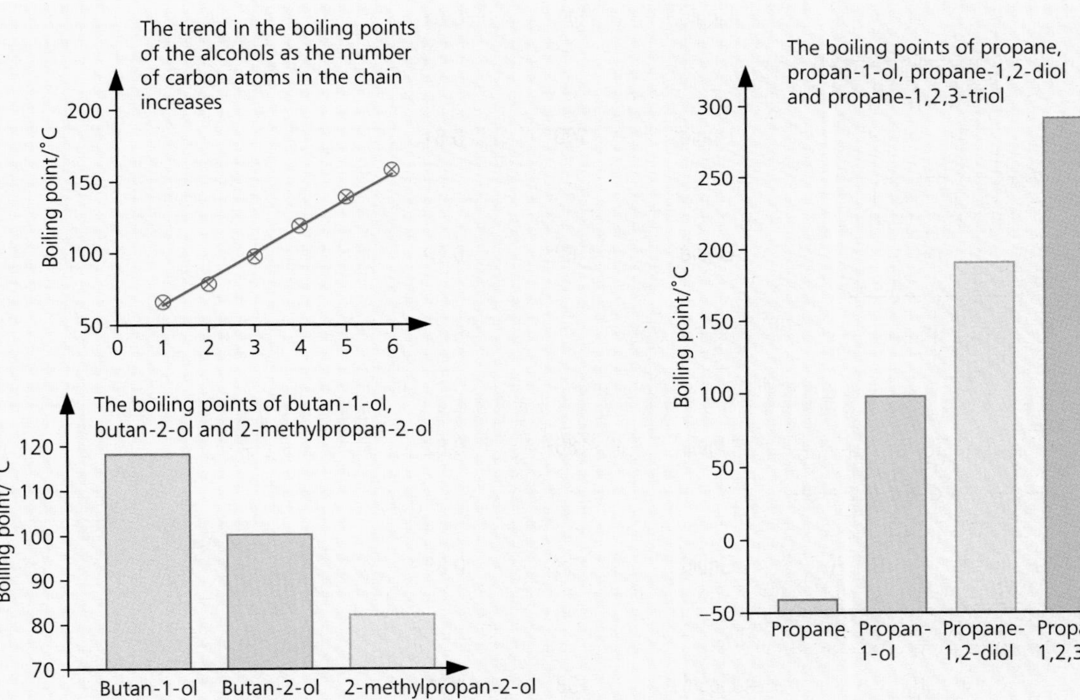

Figure 5.1.6 Displaying the data from the tables in graphical form helps us to see the trends present.

The most immediate point to be made about the physical properties of the alcohols shown in table 5.1.2 is that they are all liquids at room temperature. The short-chain alkanes, alkenes and alkynes are all gases. The boiling points and densities of the alcohols increase markedly as increasing numbers of -OH groups are added. What does this tell us about the chemistry of the alcohols?

The hydroxyl (-OH) group is polar, with a small negative charge (δ–) associated with the oxygen atom and a small positive charge (δ+) associated with the hydrogen atom. This results in the formation of hydrogen bonds between the alcohol molecules as shown in figure 5.1.7, and it is these hydrogen bonds that are responsible for the high melting and boiling points observed. With an increase in the number of -OH groups in the diols and triols, the degree of hydrogen bonding increases even more, leading to the

Compound	Boiling point/°C
Propane	–42
Propan-1-ol	97
Propane-1,2-diol	189
Propane-1,2,3-triol	290

Table 5.1.1

Alcohol	Displayed formula	State at 25 °C	Boiling point/°C	Density/ g cm^{-3}
Methanol	H H—C—OH H	Liquid	65	0.79
Ethanol	H H H—C—C—OH H H	Liquid	78	0.79
Propan-1-ol	H H H H—C—C—C—OH H H H	Liquid	98	0.80
Propan-2-ol	H OH H H—C—C—C—H H H H	Liquid	83	0.79
Butan-1-ol	H H H H H—C—C—C—C—OH H H H H	Liquid	117	0.81
Butan-2-ol	H H OH H H—C—C—C—C—H H H H H	Liquid	100	0.81
2-methylpropan-2-ol	H OH H H—C—C—C—H H H H—C—H H	Liquid	82	0.79
Pentan-1-ol	H H H H H H—C—C—C—C—C—OH H H H H H	Liquid	138	0.82
Hexan-1-ol	H H H H H H H—C—C—C—C—C—C—OH H H H H H H	Liquid	158	0.82
Ethane-1,2-diol	H H HO—C—C—OH H H	Liquid	198	1.11
Propane-1,2,3-triol	H OH H HO—C—C—C—OH H H H	Liquid	290	1.26
Phenol	OH	Solid	182	1.08

Table 5.1.2

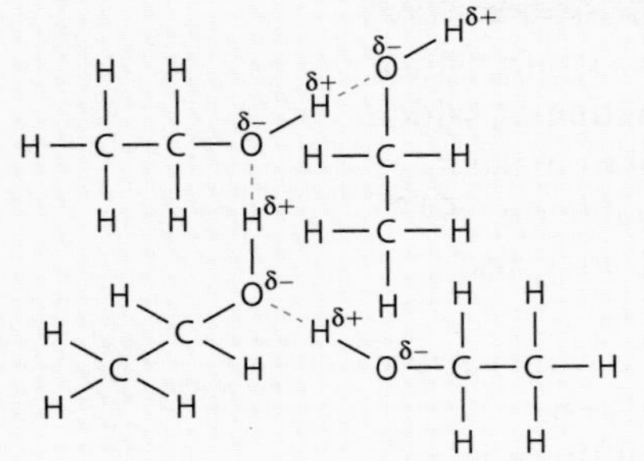

Figure 5.1.7 The hydrogen bonding between the molecules of ethanol is responsible for many of its physical properties.

observed increase in boiling point. (Hydrogen bonding was discussed in more depth in section 1.5, pages 67–70.)

Notice in figure 5.1.7 that the structure of the alkyl chain influences the boiling point of an alcohol too. Comparing the boiling points of butan-1-ol, butan-2-ol and 2-methylpropan-2-ol shows the decreasing intermolecular forces with increased branching of the alkyl chain.

The ability to form hydrogen bonds means that the shorter-chain alcohols are freely soluble in water, which is rather unusual for organic liquids. Interestingly, as the length of the carbon chain increases, the solubility of the alcohol in water decreases. The reason for this is that as the carbon chain gets longer it increasingly dominates the chemistry of the compound, and the -OH group has relatively less effect. The majority of the molecule is unable to participate in hydrogen bonding, and so the compounds are decreasingly soluble in a polar solvent such as water. Alcohols with very long carbon chains are completely insoluble in water – the non-polar character of the majority of the molecule outweighs the ability of one part of the molecule to form hydrogen bonds. The effect of the increasing hydrogen bonding with additional -OH groups in the diols and thiols is reflected in the viscosity of the alcohols. Propan-1-ol has a very low viscocity, propan-1, 2-diol is more viscous and propan-1,2,3-triol is a very viscous liquid.

Industrial ethanol from plants

As reserves of fossil fuels are used, and countries such as the USA become increasingly concerned about dependence on oil supplied by other countries, the production of ethanol by fermentation of plant material on an industrial scale is becoming more commonplace. In countries such as Brazil, sugar cane is the main source plant, whilst in America it is corn. The process involves a number of stages:

- The plant material is ground up to expose the starch.
- The material is mixed with water to form a mash and then heated. Enzymes may be used to help separate the fermentable sugars.
- The fermentation stage takes place at relatively low temperatures – above about 40 °C the yeast will be killed – and the action of the yeast cells on carbohydrates from the plant produces a mixture which contains about 10% ethanol.
- This mixture is boiled in a distillation column to produce fuel grade ethanol which is 85–95% pure.

Improvements in the process mean that ethanol can be produced in this way relatively cheaply. The conditions needed are much less severe than those needed to produce ethanol by cracking ethene, but on the other hand the process is much longer and has a number of steps. Less pollution is produced by the burning of ethanol than of conventional petrol, but at the moment industrial fuel fermentation is possible only in areas of the world where plenty of 'fuel plants' can be grown easily and cheaply. However, as ways are developed of using the cellulose portion of plants (straw and similar plant wastes) and of making the processing even more efficient, this method of obtaining ethanol looks set to become increasingly popular.

Production of the alcohols

The alcohols include ethanol, the intoxicating element of alcoholic drinks, which is toxic in high concentrations. The first member of the series is methanol, which is highly toxic. Methanol is added to industrially produced ethanol to make it poisonous and unpalatable, thus avoiding the excise duty (tax) payable on alcohol produced for human consumption. This 'methylated spirit' may also have a purple dye added to it to distinguish it from pure ethanol.

Ethanol for drinking purposes is made by the fermentation of a wide variety of plant materials. Some of the ethanol needed industrially can also be produced in this way, but this is only economically advantageous if there is a very abundant supply of plants such as sugar cane, and petroleum is in short supply. Most alcohols for industrial use, including ethanol, are manufactured from alkenes produced by the catalytic cracking of petroleum fractions, as shown in figure 5.1.9.

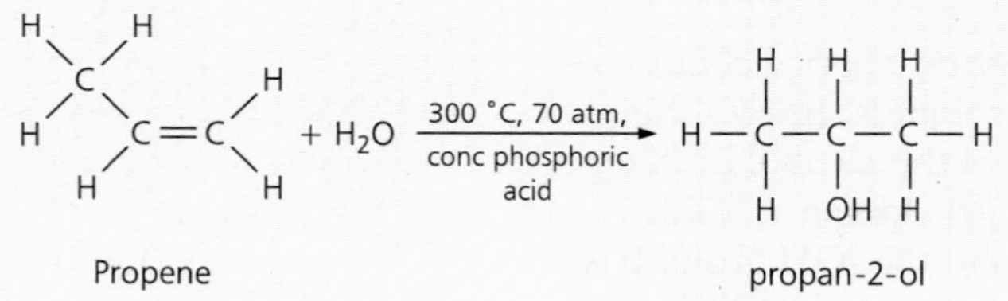

$$CH_3CH{=}CH_2 + H_2O \xrightarrow[\text{conc phosphoric acid}]{300\ ^\circ C,\ 70\ \text{atm},} CH_3CH(OH)CH_3$$

Propene → propan-2-ol

Figure 5.1.9

Preparation of alcohols

The laboratory synthesis of organic chemicals can be as much an art as a science – both reactants and products may be extremely volatile, poisonous or likely to decompose. Providing the right conditions for a particular reaction to take place with a reasonable yield of products can be a painstaking task. Throughout section 5 we shall look at a few of the organic syntheses that can be carried out in the laboratory, at some of the problems associated with them and how they are overcome.

In the laboratory we cannot easily reproduce the conditions used to produce alcohols in industrial plants. Using an alkene as the starter molecule in just the same way, we need a two-stage reaction to complete the synthesis. The alkene is reacted with concentrated sulphuric acid to form the alkyl hydrogensulphate, which is then refluxed with water to produce an alcohol.

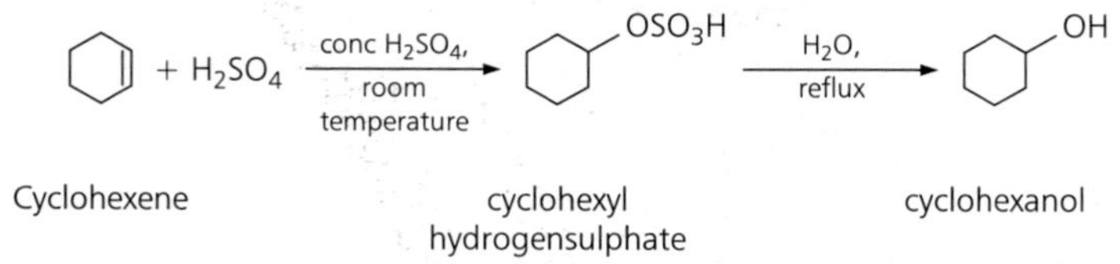

$$C_6H_{10} + H_2SO_4 \xrightarrow[\text{room temperature}]{\text{conc } H_2SO_4,} C_6H_{11}OSO_3H \xrightarrow[\text{reflux}]{H_2O,} C_6H_{11}OH$$

Cyclohexene → cyclohexyl hydrogensulphate → cyclohexanol

Figure 5.1.10 Making an alcohol from an alkene in the laboratory

The main problem when ethanol is produced in this way is that it is in aqueous solution. Ethanol is so soluble in water that it is impossible to separate the two simply by distillation, since the purest product that can be produced this way consists of 96% ethanol and 4% water. To obtain pure ethanol this distillate must be refluxed for several hours with calcium oxide, which absorbs the water to produce pure alcohol, sometimes called **absolute alcohol**. Another reason why this preparation is not used in industry is the cost of the concentrated sulphuric acid needed.

Reactions of the alcohols

When we consider the reactions of the alcohols, we are looking at two major types of reaction. One of these involves the functional group, while the other involves the carbon skeleton.

Like water, the alcohols are **amphoteric** – they can act as acids by donating protons, or as bases by accepting protons. The alcohol reacts as an acid or a base depending on the conditions, as will become clearer when we look at individual reactions.

Acidic behaviour (proton donor):

$$\mathbf{R{-}OH \rightarrow R{-}O^- + H^+}$$

Basic behaviour (proton acceptor):

$$\mathbf{R{-}OH + H^+ \rightarrow R{-}OH_2^+}$$

Reactions involving the functional group

The reactions of the functional group of the alcohols fall into two categories – those involving the fission of the O—H bond, and those involving the fission of the C—O bond. We shall look at these types of reactions in ethanol and also in phenol.

Fission of the O—H bond – reaction with sodium

Sodium and ethanol

When a small piece of sodium is added to water, it rushes about on the surface. Sodium floats on water, as its density is marginally less than that of water ($\rho_{water} = 1.0\ g\ cm^{-3}$, $\rho_{sodium} = 0.97\ g\ cm^{-3}$). The vigorous reaction between water and sodium produces hydrogen gas and sodium hydroxide, with the evolution of much heat. In this reaction, water acts as an acid, giving up a proton, H^+:

$$\mathbf{2Na(s) + 2H_2O(l) \rightarrow 2NaOH(aq) + H_2(g)}$$

When sodium is added to ethanol, it sinks – sodium is denser than ethanol ($\rho_{ethanol} = 0.79\ g\ cm^{-3}$) – and then a steady stream of hydrogen is produced. As well as hydrogen, this reaction also produces a solid called sodium ethoxide, with the evolution of rather less heat than in the reaction with water.

$$\mathbf{2Na(s) + 2CH_3CH_2OH(l) \rightarrow \underset{\text{sodium ethoxide}}{2CH_3CH_2O^-Na^+(s)} + H_2(g)}$$

The reaction between sodium and water and that between sodium and ethanol are obviously very similar. As water is acting as an acid in this reaction, this suggests that ethanol must also be acting as an acid, although since the

The combustion of alcohols

The picture of a traditional Christmas pudding, topped with holly and swimming in flaming brandy, is a familiar one to most people. It is made possible because of the flammable nature of the alcohols. When heated in the presence of oxygen in the air, all the alcohols burn. They undergo complete combustion to give carbon dioxide and water, making them a very clean fuel.

$$\mathbf{C_2H_5OH + 3O_2 \rightarrow 2CO_2 + 3H_2O}$$

5

reaction between sodium and ethanol is the less vigorous of the two, it is reasonable to assume that ethanol is a weaker acid than water. This is confirmed when the values of K_a for water and ethanol are measured, when we find that the K_a of water is 1.0×10^{-14} mol dm^{-3}, while the K_a of ethanol is 1.0×10^{-18} mol dm^{-3} (remember that weak acids have low K_a values, as we saw in section 3.2).

Sodium and phenol

If a small piece of sodium is added to a solution of phenol in an organic solvent like ethoxyethane, the sodium again sinks (it is more dense than the phenol solution), but the production of hydrogen is much more vigorous and the reaction is highly exothermic. This implies that phenol is a stronger acid than either ethanol or water, which is confirmed by its K_a of 1.3×10^{-10} mol dm^{-3}. This reaction is shown in figure 5.1.11.

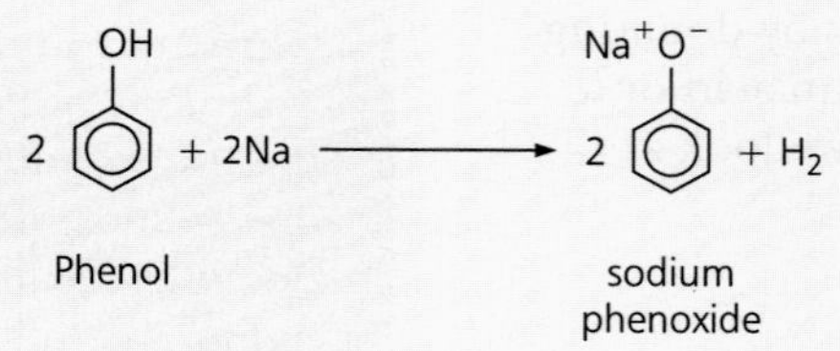

$$2\,C_6H_5OH + 2Na \longrightarrow 2\,C_6H_5O^-Na^+ + H_2$$

Phenol → sodium phenoxide

Figure 5.1.11

The relative strengths of acids

Phenol is a stronger acid than water, which in turn is a stronger acid than ethanol, because of the nature of the negative ion produced in the equilibrium:

$$\mathbf{HB \rightleftharpoons H^+ + B^-}$$

(Remember from section 3.2 that the ion B^- is referred to as the **conjugate base** of HB.) The acid–base equilibria involved are shown in figure 5.1.12.

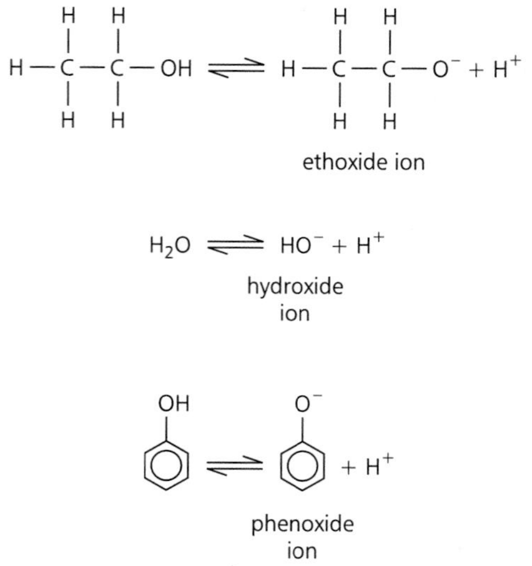

$$CH_3CH_2OH \rightleftharpoons CH_3CH_2O^- + H^+$$

ethoxide ion

$$H_2O \rightleftharpoons HO^- + H^+$$

hydroxide ion

$$C_6H_5OH \rightleftharpoons C_6H_5O^- + H^+$$

phenoxide ion

Figure 5.1.12

The negative charge on the oxygen atom in the ethoxide ion is greater than the negative charge on the oxygen atom in the hydroxide ion. This is due to the electron-donating effect of the alkyl group, a property noted on page 401. As a result of this, the ethoxide ion is less stable than the hydroxide ion, which means that the acid–base equilibrium lies further to the left in the case of ethanol than it does in the case of water. This makes ethanol a weaker acid than water.

In the case of phenol, the phenoxide ion may delocalise at least part of the negative charge on the oxygen atom round the benzene ring. This reduces the amount of negative charge on the oxygen atom, increasing the stability of the phenoxide ion so that the acid–base equilibrium lies further to the right in the case of phenol than in the case of water. The increased stability of the phenoxide ion means that it is a weaker base than the hydroxide ion, and that phenol is a stronger acid than water. (It is this acidic behaviour of phenol that reduces its effectiveness when used as an antiseptic on the body – the phenol reacts with and damages the tissues of the body, so that although they do not become infected, they do not heal well.)

It is important to remember that although we are talking here about the relative strengths of water, ethanol and phenol as acids, even phenol is a very weak acid compared with carboxylic acids such as benzoic acid, and certainly compared with the well known mineral acids such as sulphuric and hydrochloric acids.

Fission of the O—H bond – esterification

Alcohols will react with carboxylic acids in the presence of an acid catalyst to form **esters**, in a process known as **esterification**. However, the equilibrium constant for such reactions does not usually favour a very high proportion of the desired products, and so in practice esters are usually made in the laboratory by a different reaction – that between alcohols and acid halides, shown in figure 5.1.13(a).

In this reaction the alcohol, with its lone pairs of electrons around the oxygen atom, acts as a nucleophile. The same delocalisation of charge that made phenol a stronger acid than ethanol makes it a weaker nucleophile. Thus

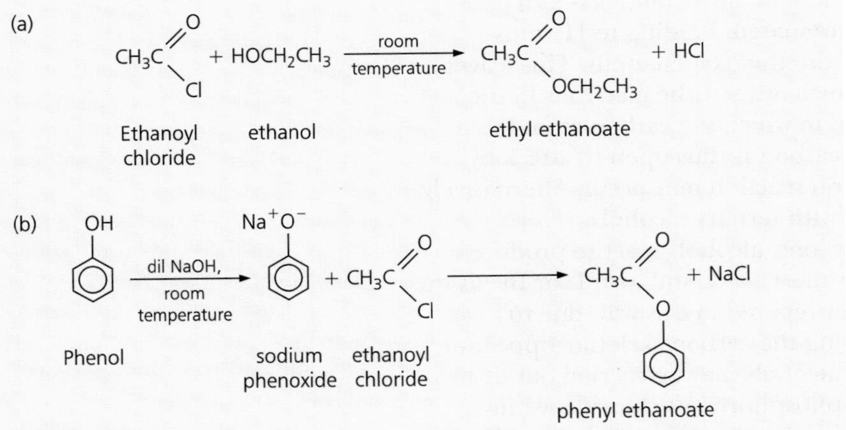

(a)

$$CH_3COCl + HOCH_2CH_3 \xrightarrow{\text{room temperature}} CH_3COOCH_2CH_3 + HCl$$

Ethanoyl chloride, ethanol, ethyl ethanoate

(b)

$$C_6H_5OH \xrightarrow{\text{dil NaOH, room temperature}} C_6H_5O^-Na^+ ; \quad C_6H_5O^-Na^+ + CH_3COCl \longrightarrow CH_3COOC_6H_5 + NaCl$$

Phenol, sodium phenoxide, ethanoyl chloride, phenyl ethanoate

Figure 5.1.13

in the case of esterification, phenols react much less readily than aliphatic alcohols, and are usually first converted into the phenoxide ion (a much stronger nucleophile) by reaction with sodium hydroxide, as shown in figure 5.1.13(b).

The mechanism of ester formation

The formation of an ester from a carboxylic acid and an alcohol under acidic conditions proceeds via a series of reversible reactions. At a given temperature the concentrations of carboxylic acid, alcohol and ester are related by the equilibrium law (see section 3.1). In accordance with this law, the amount of ester formed in the reaction mixture can be maximised by carrying out the reaction in the presence of concentrated sulphuric acid, which removes water and so pulls the reaction to the right. The reverse reaction, the acid-catalysed hydrolysis of esters, follows a similar path to the acid-catalysed formation of an ester from the corresponding alcohol and carboxylic acid, as figure 5.1.14 shows.

This equilibrium can be catalysed by a base as well as by an acid.

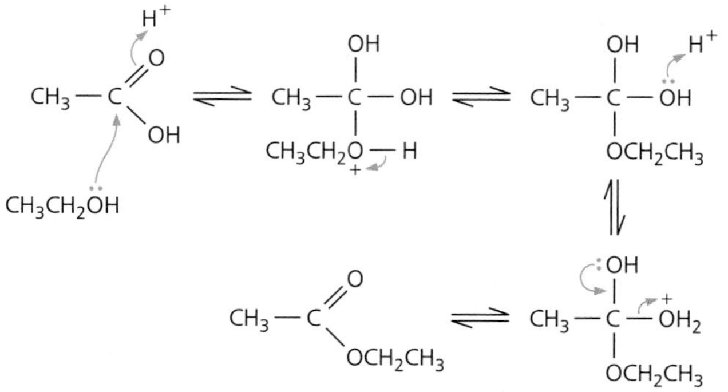

Figure 5.1.14 The mechanism of ester formation and ester hydrolysis. One piece of evidence supporting this mechanism is given in the box 'The bridging oxygen' on page 500.

Fission of the C—O bond – with halide ions

Preparation of chloroalkanes

The fission of the C—O bond only occurs under very specific conditions, in the presence of concentrated sulphuric acid. Before the C—O bond can be split, the electron clouds need to be rearranged so that the carbon atom has a large $\delta+$ charge and so becomes very attractive to nucleophiles such as halide ions.

In the presence of concentrated sulphuric acid, an alcohol acts as a base. The -OH group of the alcohol becomes **protonated**, binding to H^+ ions through the donation of an electron pair from the oxygen atom. This leaves the oxygen atom with a positive charge, which attracts the electrons in the C—O bond, forming a strongly polar bond in which the carbon atom has a large fractional positive charge ($\delta+$). This carbon is thus open to attack by nucleophiles, and a nucleophilic substitution reaction may occur. Alternatively, a carbocation may be formed, particularly with tertiary alcohols.

In strong acids in the presence of halide ions, alcohols react to produce a halogenoalkane and water. This reaction is most successful with I^- or Br^- as the nucleophile, but chloroalkanes are seldom prepared in this way, due to competing reactions (see 'Reactions involving the carbon skeleton' opposite).

The preparation of chloroalkanes from alcohols may be carried out by two methods. One of these involves the use of phosphorus(V) chloride, while a second method uses sulphur dichloride oxide (old name thionyl chloride),

$SOCl_2$. The second method is particularly useful for this preparation, since the only by-products of the reaction are gases, which simply bubble off from the reaction mixture. However, an excess of sulphur dichloride oxide is usually used, which must be distilled off from the chloroalkane.

$$\mathbf{R{-}OH(l) + PCl_5(l) \rightarrow R{-}Cl(l) + POCl_3(l) + HCl(g)}$$
$$\mathbf{R{-}OH(l) + SOCl_2(l) \rightarrow R{-}Cl(l) + SO_2(g) + HCl(g)}$$

Phenol and the halogens

The benzene ring is not susceptible to attack by nucleophiles due to its large electron density (section 4.4, page 421). In addition, the -OH attached to the benzene ring is a very weak base due to the delocalisation of the oxygen atom's electron density around the ring, as we have seen. The combination of these two factors means that phenols do not react with halide ions, even in the presence of strong acids. This is illustrated in figure 5.1.15.

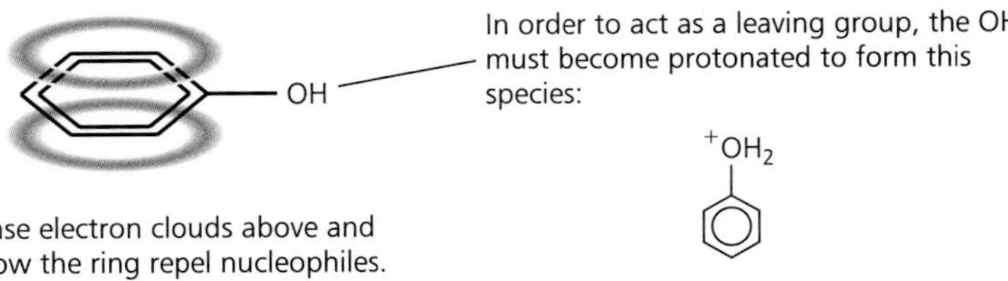

OH
In order to act as a leaving group, the OH must become protonated to form this species:
$^+OH_2$
Dense electron clouds above and below the ring repel nucleophiles.
Phenol is a weak base, due to the delocalisation of the charge on the O atom, so protonation of the OH effectively does not occur under normal reaction conditions.

Figure 5.1.15

Phenol reacts readily with halogen molecules at room temperature, though this particular reaction does not involve fission of the C—O bond. The -OH group activates the ring to attack by electrophiles in the way described on page 426. Addition of bromine water to ethanolic phenol rapidly produces a white precipitate of 2,4,6-tribromophenol, and the bromine water is decolorised. Similarly, phenol is readily nitrated at room temperature, in comparison to the elevated temperatures required by benzene.

Reactions involving the carbon skeleton

We shall consider two aspects of the reactions of the carbon skeletons of the alcohols and phenols. One is that the carbon chains may undergo typical reactions quite independently of the alcohol functional group, for example, radical substitutions will occur where there are C—H bonds in the chain. The second type of reactions involving the carbon skeleton are those where there is interaction between the carbon skeleton and the functional group during reactions with other substances. We shall consider these reactions in turn.

(1) Radical substitutions

Reactions such as the photocatalysed reactions between hydrocarbons and the halogens will take place in the presence of light, regardless of the presence of the -OH group in the molecule:

$$\mathbf{CH_3OH(l) + 3Cl_2(g) \xrightarrow{light} CCl_3OH(l) + 3HCl(g)}$$

The reaction proceeds in a series of steps, as shown for the alkanes on page 388.

5

(2) Interaction between the carbon skeleton and the -OH group – nucleophilic substitution and elimination

Substituted alkanes may be formed when the -OH group of an alcohol is protonated, as in the reaction with halide ions in the presence of a strong acid described above. If there is a nucleophile present, the carbon atom joined to the -OH group is open to attack, and a substitution reaction may occur. Alternatively, if there is no nucleophile available (for example, in a large excess of acid), the C—H bond next to the C—O bond may break so that the alcohol loses water and forms an alkene in an elimination reaction. This is **dehydration**, the effective removal of a molecule of water.

These reactions are shown in figure 5.1.16.

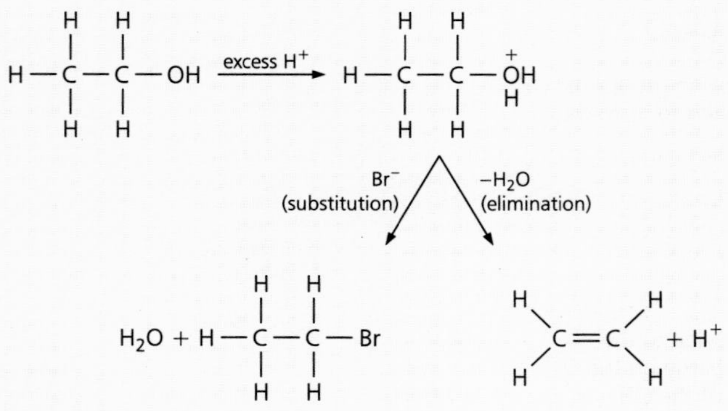

$$\mathrm{H{-}CH_2{-}CH_2{-}OH} \xrightarrow{\text{excess } H^+} \mathrm{H{-}CH_2{-}CH_2{-}\overset{+}{O}H_2}$$

$$\xrightarrow[\text{(substitution)}]{Br^-} \mathrm{H_2O + H{-}CH_2{-}CH_2{-}Br}$$

$$\xrightarrow[\text{(elimination)}]{-H_2O} \mathrm{H_2C{=}CH_2 + H^+}$$

Figure 5.1.16

Dehydrating alcohols – practical details

The dehydration of alcohols to form alkenes appears relatively straightforward on paper. In practice, the procedures need to be carried out with some care, but are nevertheless appropriate for laboratory practicals. There are two basic methods of carrying out the catalytic dehydration of an alcohol in the laboratory:

- Ethanol is heated under reflux at 170 °C with excess concentrated sulphuric acid, which acts effectively as a dehydrating agent. The ethene produced can be collected over water.
- Ethanol vapour is passed over a heated catalyst such as aluminium oxide or pumice stone. Again the ethene produced is collected over water, as shown in figure 5.1.17.

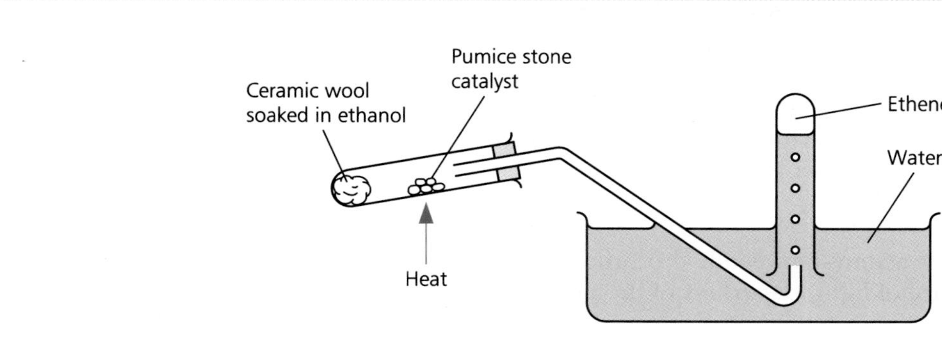

Figure 5.1.17 The apparatus used in the catalytic dehydration of ethanol

Preparation of ethers

The formation of ethers from alcohols can also be viewed as a dehydration reaction. It occurs under very similar circumstances to the reaction producing alkenes – when an alcohol reacts with concentrated sulphuric acid. However, the formation of the ether is favoured when the alcohol is in excess rather than the acid.

For example, if an excess of ethanol is refluxed with concentrated sulphuric acid for several hours at a temperature of 140 °C, the main product is the ether ethoxyethane. The reaction occurs because the ethanol molecule acts as a nucleophile, attacking the δ+ carbon atom in a protonated molecule of ethanol. Overall, a molecule of water is eliminated and an ether forms, as shown in figure 5.1.18.

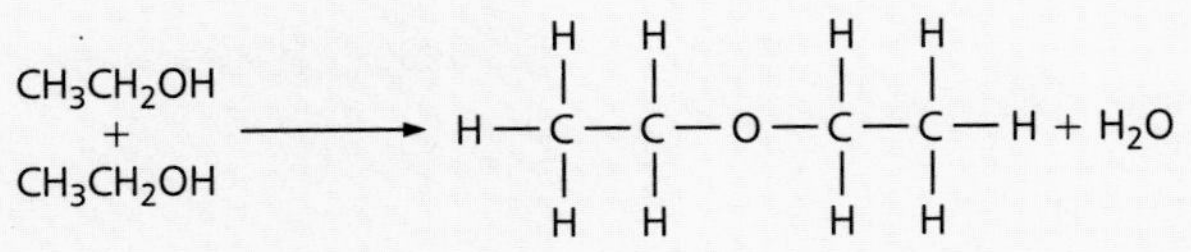

$$CH_3CH_2OH + CH_3CH_2OH \longrightarrow CH_3CH_2{-}O{-}CH_2CH_3 + H_2O$$

Figure 5.1.18

The acid-catalysed dehydration reactions of ethanol described here are examples of a reaction with two possible products, where the end result of the reaction can be influenced by the conditions under which it takes place. The reaction conditions determine whether largely alkene or largely ether is produced.

Electrophilic substitution in phenol

Dehydration reactions such as those of ethanol do not occur in the phenols. When phenol or its derivatives are reacted with concentrated sulphuric acid, electrophilic substitution reactions take place on the benzene ring to give a mixture of products, as in figure 5.1.19.

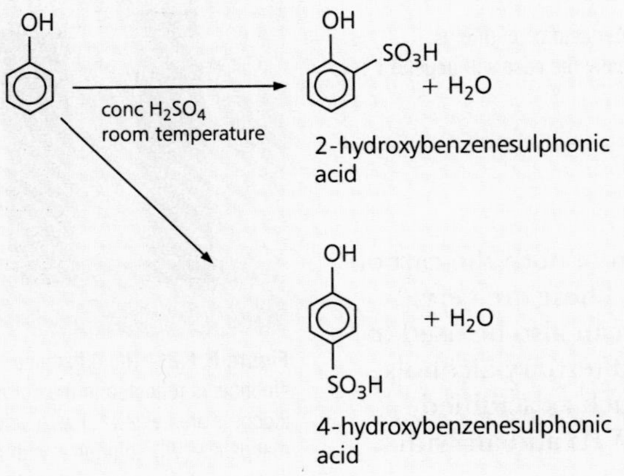

phenol (OH) → conc H_2SO_4, room temperature → 2-hydroxybenzenesulphonic acid (SO_3H ortho to OH) + H_2O

phenol (OH) → 4-hydroxybenzenesulphonic acid (SO_3H para to OH) + H_2O

Figure 5.1.19

Reaction mechanisms involving C–O bond fission

Figure 5.1.20 gives the reaction mechanism responsible for the behaviour of alcohols in acidic conditions. This explains why such a range of products is possible, due to attack on the $\delta+$ carbon atom by competing nucleophiles, together with possible elimination reactions. The major product formed will depend on several factors, including reaction conditions (temperature, pH, polarity of solvent, etc.) and the structure of the alcohol itself.

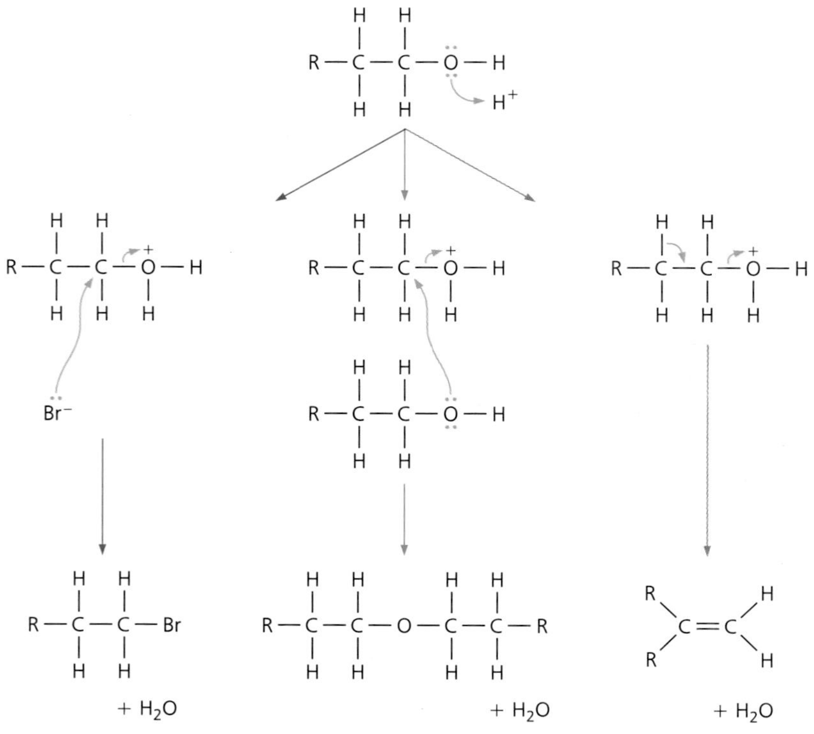

Figure 5.1.20 The competing reactions that may occur when an alcohol reacts under acidic conditions. Note that the two substitution reactions have been shown as S_N2 – whether this is really the case will depend on the considerations discussed in section 4.5, pages 442–4.

Figure 5.1.21 The difference in the structure of the alcohols is reflected in the colour changes when alcohols are heated (a) with acidified potassium manganate(VII) solution – with primary and secondary alcohols, the purple solution turns colourless as MnO_4^- is reduced to Mn^{2+}, while with tertiary alcohols there is no reaction and the purple colour remains; and (b) with acidified potassium dichromate(VI) solution – a colour change from orange to blue-green shows that the primary and secondary alcohols have undergone oxidation reactions. The $Cr_2O_7^{2-}$ ion is reduced to Cr^{3+}.

Oxidation of the alcohols

The other reactions of the alcohols involving the interaction of both the carbon skeleton and the functional group are **oxidation reactions**. These are very useful – as well as being used to make new substances, they can also be used to indicate the type of alcohol present. Primary, secondary and tertiary alcohols all give different reactions with common oxidising agents such as acidified potassium dichromate(VI), acidified potassium manganate(VII) and air with a suitable catalyst.

- **Primary alcohols** are easily oxidised to form **aldehydes**, and these aldehydes are themselves rapidly oxidised to **carboxylic acids**, figure 5.1.22(a). This means that the most common product of the oxidation of a primary alcohol is a carboxylic acid, unless the aldehyde is distilled out of the reaction mixture as it forms.

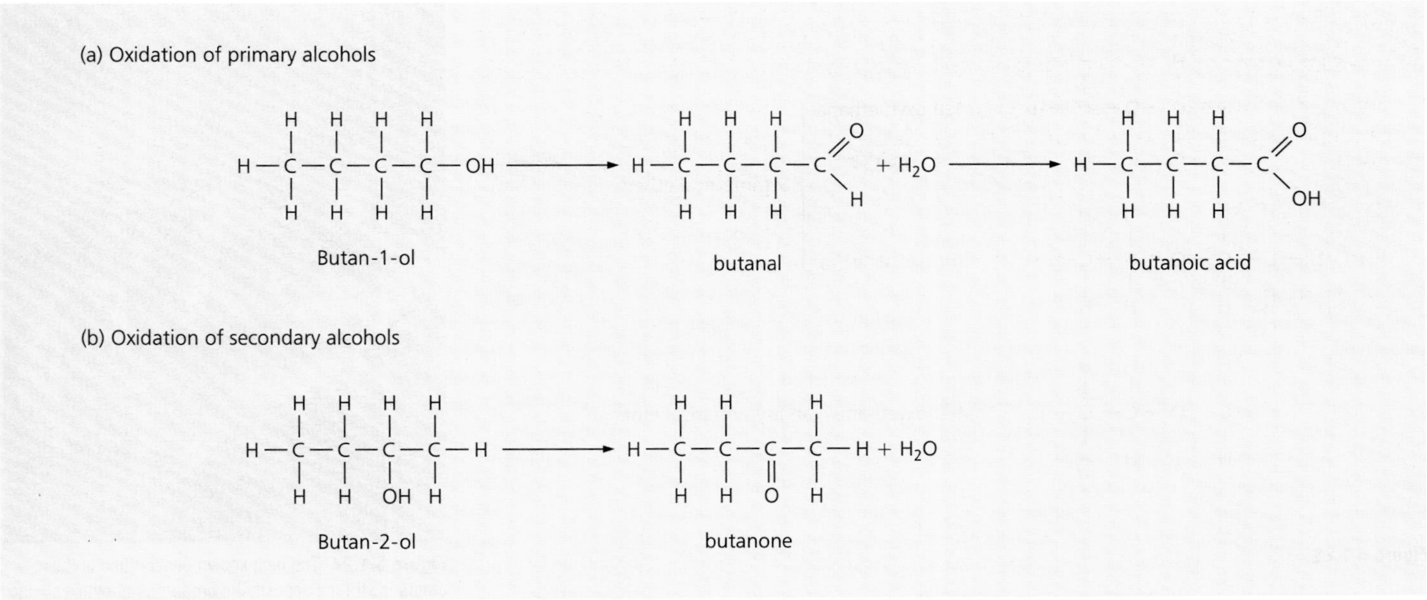

(a) Oxidation of primary alcohols

Butan-1-ol → butanal + H_2O → butanoic acid

(b) Oxidation of secondary alcohols

Butan-2-ol → butanone + H_2O

Figure 5.1.22

- **Secondary alcohols** are also readily oxidised to form **ketones**, figure 5.1.22(b), but these do not in general undergo any further oxidation and so they may be isolated as the final product of oxidation.
- **Tertiary alcohols** are not readily oxidised by any of the common oxidising agents. A careful look at the previous reactions shows why. In the case of both the primary and the secondary alcohols, the carbon atom bearing the -OH group also carries at least one other hydrogen atom. In reactions with oxidising agents, this hydrogen atom is removed in a molecule of water. Tertiary alcohols do not have a 'spare' hydrogen atom to react in this way, and thus are not readily oxidised.

  Figure 5.1.21 (opposite) shows how these different oxidation reactions can reveal whether an alcohol is a tertiary alcohol or not.
- Under normal conditions **phenol** cannot be readily oxidised, since like the tertiary alcohols, there is no 'spare' hydrogen atom.

Ethers

As we saw at the beginning of this section, ethers have the general formula ROR′ where R and R′ are alkyl or aryl groups which can be the same, as in methoxymethane, or different, as in methoxyethane. Ethers in which R and R′ are the same are known as the **symmetrical ethers**, whilst those where R and R′ are different are the **asymmetrical ethers**. Examples are shown in figure 5.1.23.

Physical properties

Perhaps the most striking difference between the ethers and the alcohols is that there is no tendency in the ethers as a group to take part in hydrogen bonding. This has a distinct effect on their physical properties, making them much more similar to the corresponding alkanes than to the alcohols. The ethers burn readily in air to form carbon dioxide and water. Ethers are therefore volatile liquids which produce vapour that is denser than air and highly flammable, so they need to be handled with great care to avoid fires and explosions.

The ethers are only very sparingly soluble in water because they form only weak hydrogen bonds with water molecules. Ethers are good solvents because

```
     H       H
     |       |
 H — C — O — C — H          Methoxymethane  ⎫
     |       |                              ⎪
     H       H                              ⎬ Symmetrical ethers
                                            ⎪
 H   H       H   H                          ⎪
 |   |       |   |                          ⎪
 H — C — C — O — C — C — H  Ethoxyethane    ⎭
 |   |       |   |
 H   H       H   H

     H       H   H
     |       |   |
 H — C — O — C — C — H      Methoxyethane – an asymmetrical ether
     |       |   |
     H       H   H
```

Figure 5.1.23

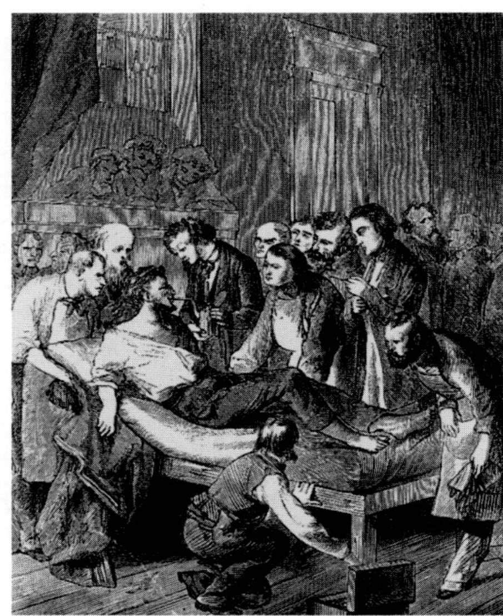

Figure 5.1.24 The best known general use of the ethers is still the anaesthetic properties of ethoxyethane (commonly referred to simply as 'ether' or as 'diethyl ether') discussed on page 432. The use of ethers as anaesthetics has now been superseded by new compounds that are far safer as a drug, and also far less likely to explode. The most common modern use of ethers is as solvents in the chemical industry.

they have both polar and non-polar parts to their molecules and they are widely used in industrial organic chemistry. Epoxyethane, the cyclic ether we met in section 4.3, is an important ether in industry, as the three-membered ring contains bonds under considerable strain, and therefore epoxyethane is more reactive than most ethers, forming a useful starting material for many organic syntheses.

Reactions

The reactions of the ethers resemble those of the alkanes – they are a very unreactive group of chemicals. However, they will dissolve in solutions of strong acids, becoming protonated and thus open to subsequent attack by nucleophiles. If they are then reacted with concentrated hydriodic acid, an iodoalkane and an alcohol result. The alcohol then reacts with the hydriodic acid, and two iodoalkanes result. Water is eliminated from the reaction, which is shown in figure 5.1.25.

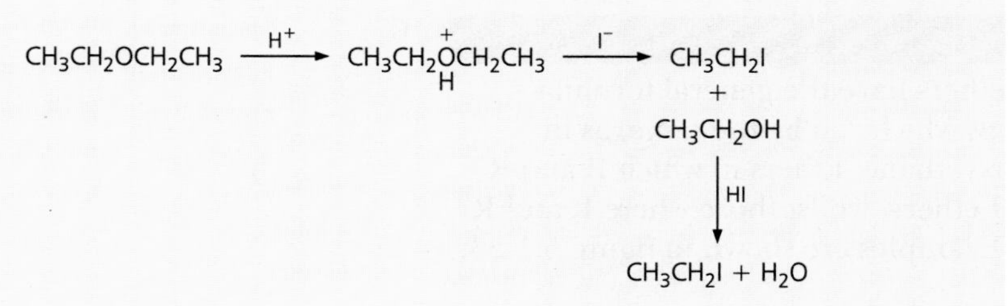

Figure 5.1.25

SUMMARY

- **Alcohols** are aliphatic molecules containing an -OH group.
- **Phenols** are aromatic molecules containing an -OH group bonded directly to the benzene ring.

- **Ethers** are molecules with two alkyl or aryl groups joined by an oxygen atom.
- **Primary alcohols** have two hydrogen atoms and one carbon atom bonded to the carbon atom bearing the hydroxyl group. **Secondary alcohols** have one hydrogen atom and two carbon atoms bonded to the carbon atom bearing the hydroxyl group. **Tertiary alcohols** have three carbon atoms and no hydrogen atoms bonded to the carbon atom bearing the hydroxyl group.
- The physical properties of the alcohols are strongly influenced by hydrogen bonding, especially alcohols with a small alkyl group.
- Alcohols are amphoteric. The -OH group may donate a proton (acidic behaviour) or may become protonated (basic behaviour).
- Phenols are stronger acids than water, which is in turn a stronger acid than aliphatic alcohols.
- Some typical reactions of ethanol are summarised in table 5.1.3.

Ethanol +	**Product**
Sodium	Sodium ethoxide, $CH_3CH_2O^-Na^+(s)$
Acid halide, RCOCl	Ester, $RCOOCH_2CH_3$
Carboxylic acid, RCOOH, with acid catalyst	Ester, $RCOOCH_2CH_3$
$HCl/PCl_5/SOCl_2$	Chloroethane, CH_3CH_2Cl
Conc H_2SO_4 at 170 °C	Ethene, $CH_2{=}CH_2$

Table 5.1.3

- Some typical reactions of phenol are summarised in table 5.1.4.

Phenol +	**Product**
Sodium or sodium hydroxide	Sodium phenoxide, 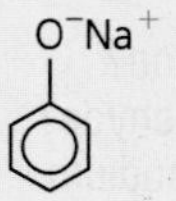
Acid halide, RCOCl	Ester, 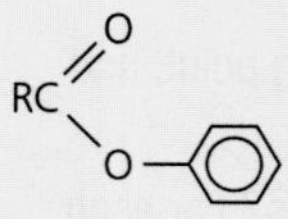
Aqueous bromine	2,4,6-tribromophenol, 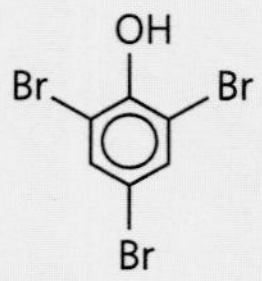
Conc sulphuric acid/conc nitric acid ('nitrating mixture')	2,4,6-trinitrophenol, 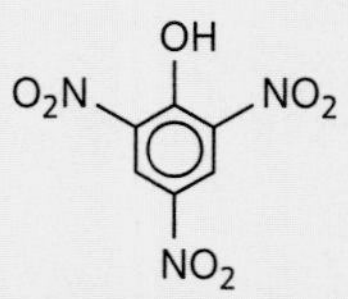

Table 5.1.4

- Primary alcohols may be oxidised to aldehydes and then to carboxylic acids. Secondary alcohols may be oxidised to ketones. Tertiary alcohols cannot readily be oxidised.
- Ethers are chemically rather unreactive.

QUESTIONS

1 The compounds A to E in figure 5.1.26 are alcohols, phenols or ethers.

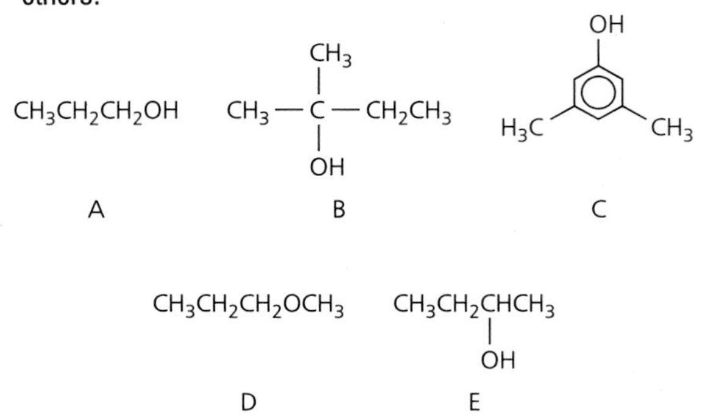

A: $CH_3CH_2CH_2OH$

B: $CH_3-C(CH_3)(OH)-CH_2CH_3$

C: OH, H_3C, CH_3

D: $CH_3CH_2CH_2OCH_3$

E: $CH_3CH_2CH(OH)CH_3$

Figure 5.1.26

a Which of these is: **i** a primary alcohol **ii** a secondary alcohol **iii** a tertiary alcohol?
b Which is a phenol?
c Which is an ether?
d Name each compound.
e Which compound may be oxidised to form a ketone?
f Which compound may be oxidised to form an aldehyde?
g Which compound would react most rapidly with sodium metal?
h What possible products might B form when reacted with excess concentrated sulphuric acid?
i Which compound would have the lowest boiling point, and why?

2 How would you carry out the following conversions? For each answer describe as fully as you can the reagents and conditions needed.
a Propan-2-ol $\rightarrow$ propanone
b Propan-1-ol $\rightarrow$ propene
c Phenol $\rightarrow$ 4-methylphenol
d Ethanol $\rightarrow$ ethoxyethane (using ethanol as the only organic starting material)

3 Phenol reacts with chlorine in solution at room temperature to form the product 2,4,6-trichlorophenol, a commonly used antiseptic.
a How would benzene react under these conditions?
b Draw a diagram showing the structural formula of 2,4,6-trichlorophenol.
c Suggest a mechanism for this reaction.
d 2,4,6-trichlorophenol is more acidic than phenol (K_a[TCP] $= 2.5 \times 10^{-8}$ mol dm^{-3}, K_a[phenol] $=$ 10^{-10} mol dm^{-3}). Suggest a reason for this.

4 Concentrated sulphuric acid is a powerful dehydrating agent, removing water from a wide range of organic molecules, including alcohols. Phenol, however, cannot be dehydrated with concentrated sulphuric acid. Why not?

5 W, C_6H_6O, is an acidic compound which reacts with hydrogen under pressure in the presence of a nickel catalyst to give X, $C_6H_{12}O$. X reacts with concentrated sulphuric acid to give Y, C_6H_{10}. Y rapidly decolorises bromine water to give Z, $C_6H_{10}Br_2$.
a Suggest structures for W, X, Y and Z.
b Write equations for the reactions involved.

Developing Key Skills

Using the Internet and any other resources available to you, collect data on the effect of ethanol on human reaction times, and the link between ethanol concentration in the blood and car accidents. Put the information together in a 'PowerPoint' presentation to be part of an evening course for sixth form students to be run by the school with the local police on 'Learning to drive – a skill for life'.

[Key Skills opportunities: C, A, IT]

FOCUS AROMA CHEMISTRY

The sense of smell is tremendously important in the appreciation of wines, but what are we actually smelling?

Whilst some of the aromas which an expert can pick up in the bouquet of a wine sound positively delicious – raspberry, almond, butterscotch and orange blossom – others, such as diesel, skunk, wet dog and hydrogen sulphide are not what you would want to meet in your glass. The chemistry of wines is enormously complex, with a wide range of organic chemicals playing their part. Carbonyl compounds – the aldehydes and ketones – have a particularly important role in the development of the aroma, both the good and the bad. Many short-chain carbonyl compounds smell quite unpleasant in the pure, concentrated form but in minute amounts in wines they contribute a great deal to the flavour without, in general, being the dominant note. However, if wine is over-exposed to air either during production or once in the bottle, some of the ethanol will be oxidised to ethanal, and the distinctive sherry-like smell which this produces serves as a warning that all is not as it should be. To avoid ethanal production, many white wines in particular are made under virtually anaerobic conditions.

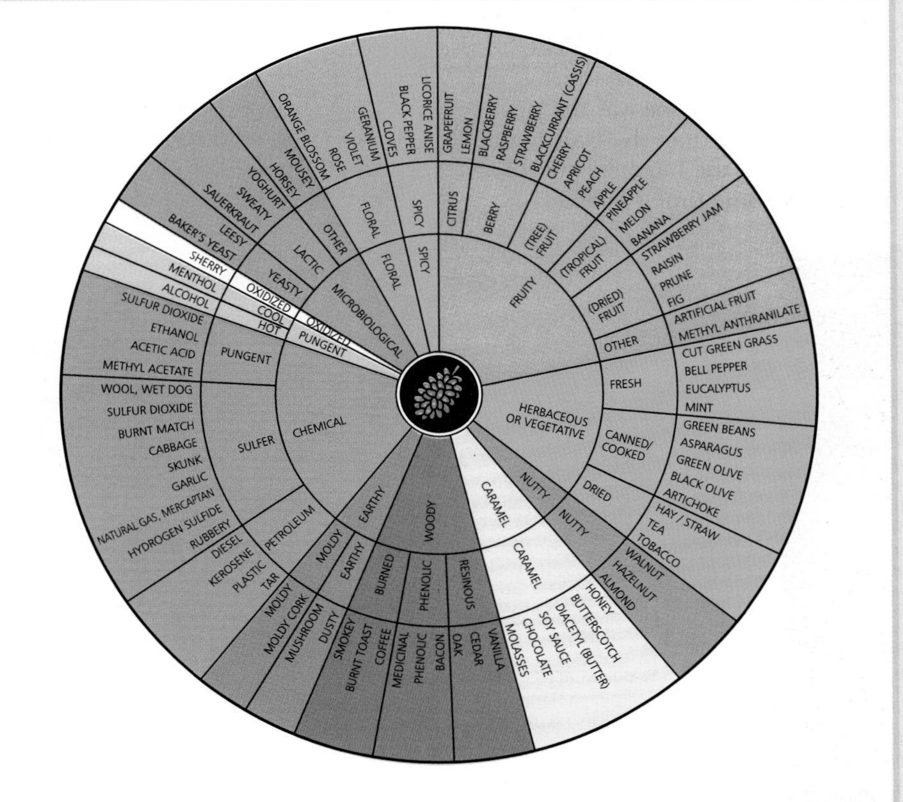

Figure 1 Wine judges are expected to be able to recognise all these different aromas in wine just as a starting point – and to use them to identify the wine, the types of grapes involved and where the wine has come from!

AROMA CHEMISTRY (*continued*)

In contrast, the oxidation of ethanol to ethanal is vital to the process of sherry making. The sherry barrels are stacked high and the wine is gradually trickled down from one barrel to another, mixing with the air. In each barrel there is always a layer of air and the *flor* yeast on the surface of the wine uses the oxygen to oxidise alcohol to form aldehydes.

Another major influence on the aroma of wine is exposure to oak barrels when it is stored and matured. Oak wood contains many aldehydes and ketones – vanillin (an aldehyde giving a vanilla flavour),oak lactones and phenolic ketones. If the wood is toasted – a fire lit in the barrel for 80–90 minutes after the barrel is made – then the balance of carbonyl compounds is altered again. So wines which are stored or matured in oak barrels take up complex flavours from the wood itself, and careful management ensures that these will enhance the flavour of the wine.

About 10 years ago a group of French oenologists (wine chemists) analysed a medium priced white burgundy wine, using GLC (gas–liquid chromatography) to separate the components of the aroma and mass spectrometry to identify them. They discovered around 120 different compounds, many of them carbonyl compounds. They then made up a synthetic version of the wine, mixing up the exact proportions of all the ingredients. A panel of wine judges undertook a blind tasting of the synthetic wine and the original wine. The experts could tell the two apart – but they could not tell which was the traditional wine and which the synthetic. The biggest difference was the price tag – it cost 30 times more to produce the synthetic bottle than the traditional one! The development of wine flavour and aroma using the biochemistry of grapes and the skill of the winemaker seems safe for some time to come!

Figure 2 The aromas which make wines like this oak-aged Bacchus from Horton Vineyard in Dorset so appealing are a result of the type of grapes used to make the wine, the soil they are grown in, the weather in the growing season, the way the wine is made and the flavours from the oak casks in which it is stored. The process is a marriage of chemistry and creativity.

5.2 Carbonyl compounds: Aldehydes and ketones

What is a carbonyl compound?

In this section we shall be considering organic molecules belonging to the group known as the **carbonyl compounds**. The carbonyl compounds include the **aldehydes** and **ketones**, which both have the **carbonyl** functional group (C═O). The carbonyl group is strongly polar ($C^{\delta+}{=}O^{\delta-}$), and this has a marked effect on the physical and chemical properties of the carbonyl compounds.

Aldehydes and ketones

What's the difference?

Aldehydes and ketones are both homologous series containing the carbonyl group. Both organic families have the general formula $C_nH_{2n}O$. What is the difference between them? The reason why there are two organic families rather than one is to do with the position of the C═O functional group in the carbon chain. The **aldehydes** have the carbonyl group at the end of the carbon chain, and have the general structure RCHO, as shown in figure 5.2.2(a) overleaf. Aldehydes are named using the stem showing the number of carbon atoms in the chain (methan-, ethan-, etc.), with the suffix '-al', for example methanal, ethanal, pentanal.

The carbonyl group of the **ketones** is not at the end of the carbon chain, figure 5.2.2(b). Its position within the molecule varies – the ketones have positional isomers – and is indicated where necessary by the numbering of the carbon atoms. The ketones are named using the suffix '-one', for example propanone, butanone, pentan-2-one. It is easier to visualise these differences by looking at the examples shown in figure 5.2.2.

Figure 5.2.1 Carbonyl compounds are commonplace in our everyday lives, though most people probably do not realise it. The poisonous aldehyde-based compound used in slug pellets and the ketones used as solvents for nail varnish and for a wide variety of industrial processes have physical and chemical properties determined by the carbonyl group.

Physical properties of the aldehydes and ketones

The physical properties of the aldehydes and the ketones are influenced by the presence of the polar carbonyl group in the molecule. The carbonyl group does not take part in hydrogen bonding, but there are permanent dipole–dipole interactions between the molecules of both aldehydes and ketones. This gives them boiling points higher than the corresponding alkanes (very volatile with only weak van der Waals interactions), but lower than the corresponding alcohols (low volatility with both van der Waals forces and hydrogen bonds between the molecules), as shown in figure 5.2.3. The influence of the carbonyl group is particularly marked in the molecules with relatively short carbon chains, because as the chains get longer they have an increasing effect on the properties of the compounds.

A further point about the boiling points of the aldehydes and ketones is that the molecules of the straight-chain isomers can approach each other more closely than the branched-chain molecules. This means that the forces between the molecules of the straight-chain isomers are stronger and more numerous than those between the molecules of the branched-chain isomers, and so the boiling points of the straight-chain isomers are higher. For example, butanal boils at 76 °C whilst its isomer 2-methylpropanal boils at 64 °C, and

5

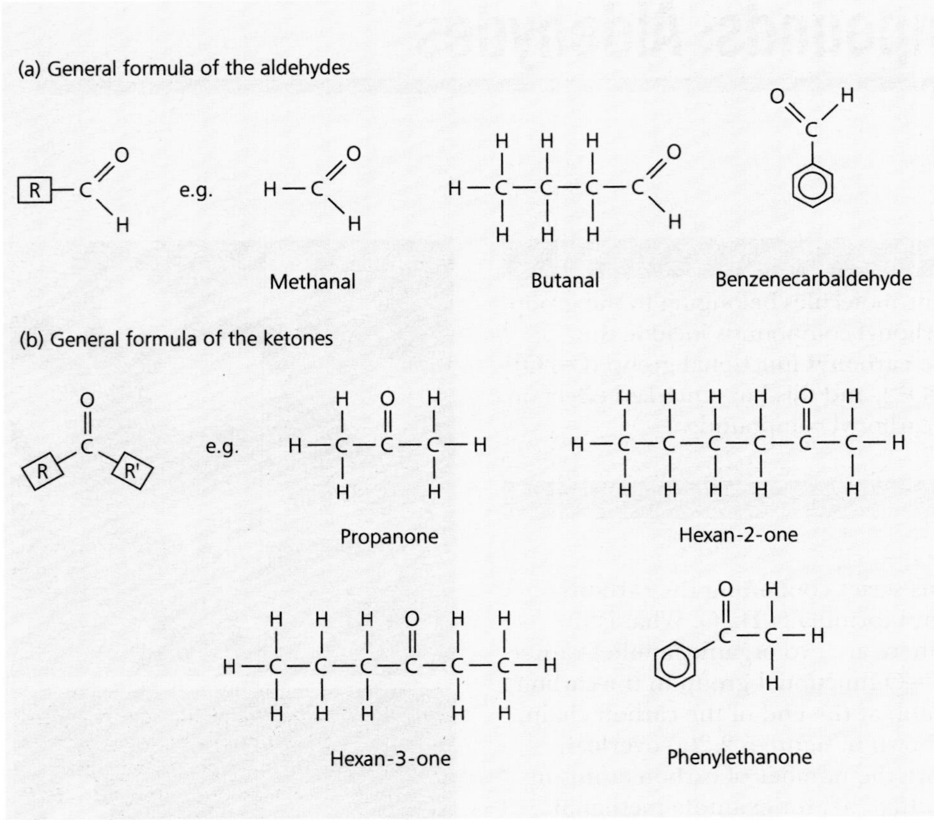

Figure 5.2.2 Some aldehydes and ketones

pentan-2-one has a boiling point of 102 °C compared with 94 °C for its isomer 3-methylbutan-2-one. (These trends were also observed in the structural isomers of the alkanes in section 4.2.)

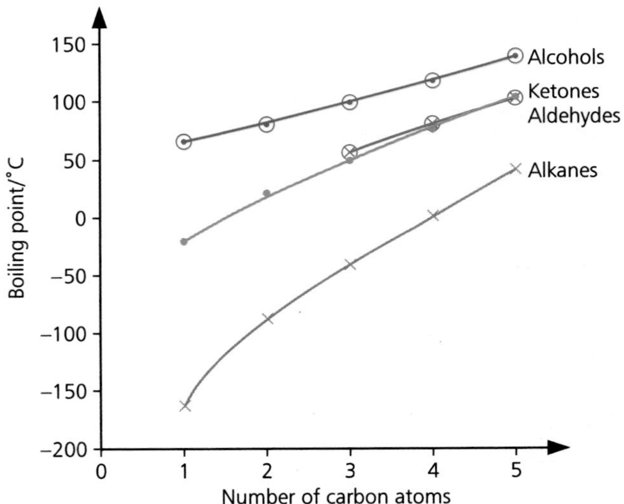

Figure 5.2.3 This graph shows the boiling points of several alkanes, aldehydes, ketones and alcohols. The effect of the intermolecular forces on the boiling points can be seen clearly, with dipole–dipole attractions resulting from the carbonyl groups of the aldehydes and ketones giving boiling points higher than the equivalent alkanes but lower than the alcohols with their hydrogen bonding.

Figure 5.2.3 also shows that the aldehydes are more volatile than the ketones. Both groups contain many compounds with strong odours, but the aldehydes in particular are responsible for imparting the characteristic smells and flavours to a variety of foods and are also one of the groups of molecules responsible for the bouquet of quality wines.

The polarity of the C=O bond leads to the short-chain aldehydes and ketones being relatively soluble in water. As organic molecules with a polar region, they can act as solvents to both polar and non-polar solutes, which gives them a wide range of uses as solvents. Propanone and butanone are just two carbonyl compounds widely used as industrial solvents.

Figure 5.2.4 Carbonyl compounds are frequently used as solvents in adhesives. When the adhesive is applied, the solvent evaporates, leaving the adhesive agent behind, as in the process of coating paper to make self-adhesive labels.

Preparation of aldehydes and ketones

Aldehydes and ketones may be made by oxidation of the corresponding alcohol.

- **Aldehydes** are made by oxidising **primary alcohols**.
- **Ketones** are made by oxidising **secondary alcohols**.

Aldehydes are formed as the initial product when a primary alcohol is refluxed with acidified potassium dichromate(VI) solution. However, as we saw in section 5.1, the aldehyde is itself oxidised to a carboxylic acid if the reaction is allowed to continue. In order to prepare an aldehyde, it must be distilled off as soon as it is formed. This requires the apparatus shown in figure 5.2.5.

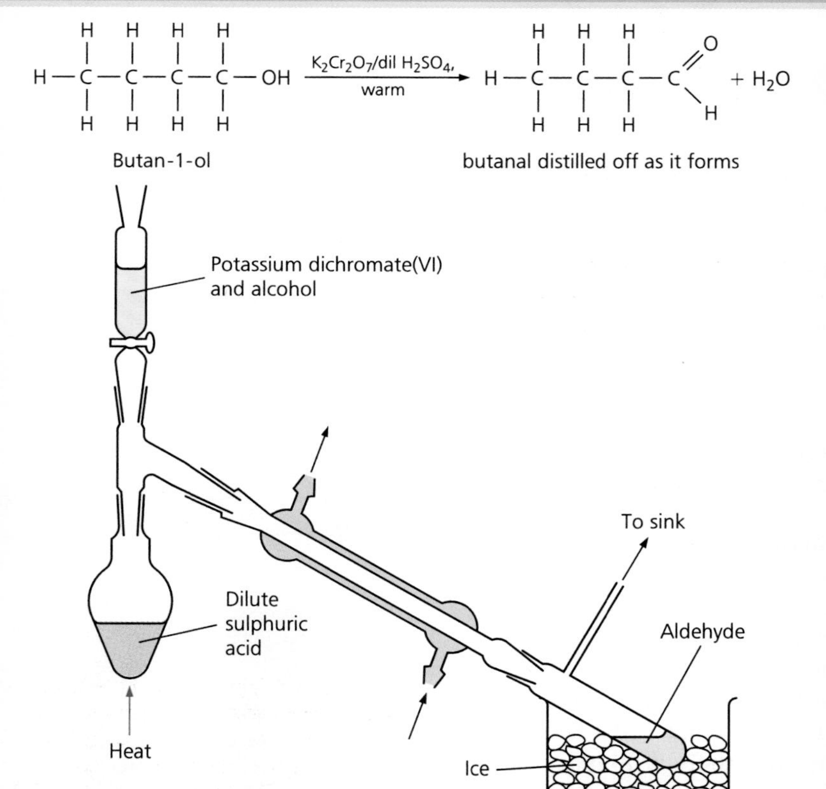

Figure 5.2.5 The apparatus used to prepare an aldehyde such as butanal

Ketones are made when a secondary alcohol is refluxed with acidified potassium dichromate(VI) solution. The oxidation does not proceed any further as ketones are resistant to oxidation under these conditions, so the preparation is somewhat more straightforward than that of the aldehydes. The equation is shown in figure 5.2.6.

5

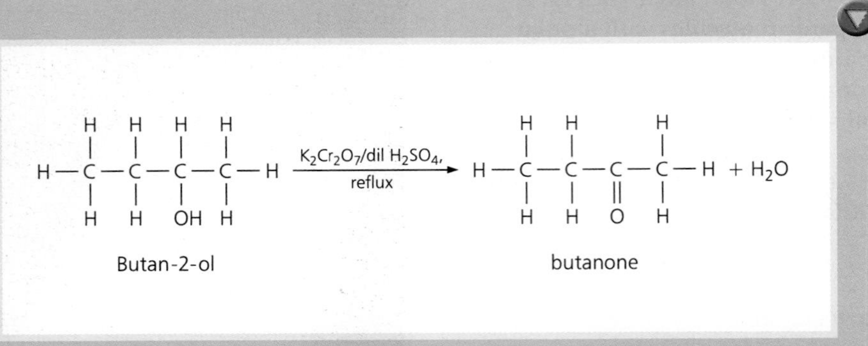

Figure 5.2.6

The industrial preparations of aldehydes and ketones are similar to the laboratory methods. However, to avoid the expense of oxidising agents such as potassium dichromate(VI), oxygen from the air is used for the oxidation along with a suitable catalyst. For example, propanone may be manufactured by oxidising the secondary alcohol propan-2-ol.

Reactions of the aldehydes and ketones

Redox reactions – oxidation of carbonyl compounds

There are many similarities in the reactions of the aldehydes and the ketones with a wide variety of reagents. However, they can be readily distinguished by the way they react with oxidising agents.

Aldehydes have a hydrogen atom that forms part of their functional group. This hydrogen is **activated** by the carbonyl group – it carries a $\delta+$ charge – and as a result aldehydes are readily oxidised to carboxylic acids by even quite mild oxidising agents. In contrast, the oxidation of ketones requires the breaking of a C—C bond. This group is therefore very resistant to oxidation, and will only react after prolonged treatment with a very strong oxidising agent.

There are two reactions with mild oxidising agents that are regularly used as qualitative tests to determine whether an unknown compound is an aldehyde or a ketone:

- **Fehling's solution** contains an alkaline solution of Cu^{2+} ions. Normally these would give a precipitate of copper(II) hydroxide under these conditions, but in Fehling's solution the copper ions are prevented from precipitating in this way as they are complexed with 2,3-dihydroxybutanedioate ions. If an aldehyde is warmed with Fehling's solution, then the aldehyde is oxidised to the salt of a carboxylic acid and the Fehling's solution is reduced, giving a red-brown precipitate of copper(I) oxide, as shown in figure 5.2.7. If a ketone is warmed with Fehling's solution in the same way, then there is no reaction. The two half-equations for the reaction with an aldehyde are:

$$2Cu^{2+}(aq) + OH^-(aq) + 2e^- \rightarrow Cu_2O(s) + H^+(aq)$$

and:

$$RCHO(aq) + H_2O(l) \rightarrow RCOOH(aq) + 2H^+(aq) + 2e^-$$

so overall:

$$2Cu^{2+}(aq) + RCHO(aq) + OH^-(aq) + H_2O(l) \rightarrow Cu_2O(s) + RCOOH(aq) + 3H^+(aq)$$

Figure 5.2.7 Fehling's solution before and after reaction with an aldehyde

- **Tollens reagent** is produced by dissolving silver nitrate in aqueous ammonia, resulting in complex diamminesilver(I) ions, $[Ag(NH_3)_2]^+$. When Tollens reagent is warmed gently with an aldehyde, the aldehyde is oxidised and the complex silver(I) ions are reduced to form silver metal. With careful preparation, this appears on the inside of the test tube as a **silver mirror** as in figure 5.2.8, or less successfully as a black precipitate. For this reason the reaction with Tollens reagent is often known as the **silver mirror test**. When ketones are warmed with the same reagent, there is no reaction. The half-equations for the reaction with an aldehyde are:

$$2Ag^+(aq) + 2e^- \rightarrow 2Ag(s)$$

and:

$$RCHO(aq) + H_2O(l) \rightarrow RCOOH(aq) + 2H^+(aq) + 2e^-$$

so overall:

$$2Ag^+(aq) + RCHO(aq) + H_2O(l) \rightarrow 2Ag(s) + RCOOH(aq) + 2H^+(aq)$$

Figure 5.2.8 The silver mirror produced when an aldehyde reacts with Tollens reagent

Redox reactions – reduction of carbonyl compounds

As we have already seen, aldehydes and ketones are frequently prepared in the laboratory by the oxidation of alcohols. These reactions can be reversed as aldehydes are readily reduced to primary alcohols and ketones to secondary alcohols. Suitable reducing agents include lithium tetrahydridoaluminate(III) ($LiAlH_4$) dissolved in ethoxyethane, sodium tetrahydridoborate(III) ($NaBH_4$) dissolved in aqueous ethanol, and zinc in dilute ethanoic acid. The reactions are shown in figure 5.2.9(a).

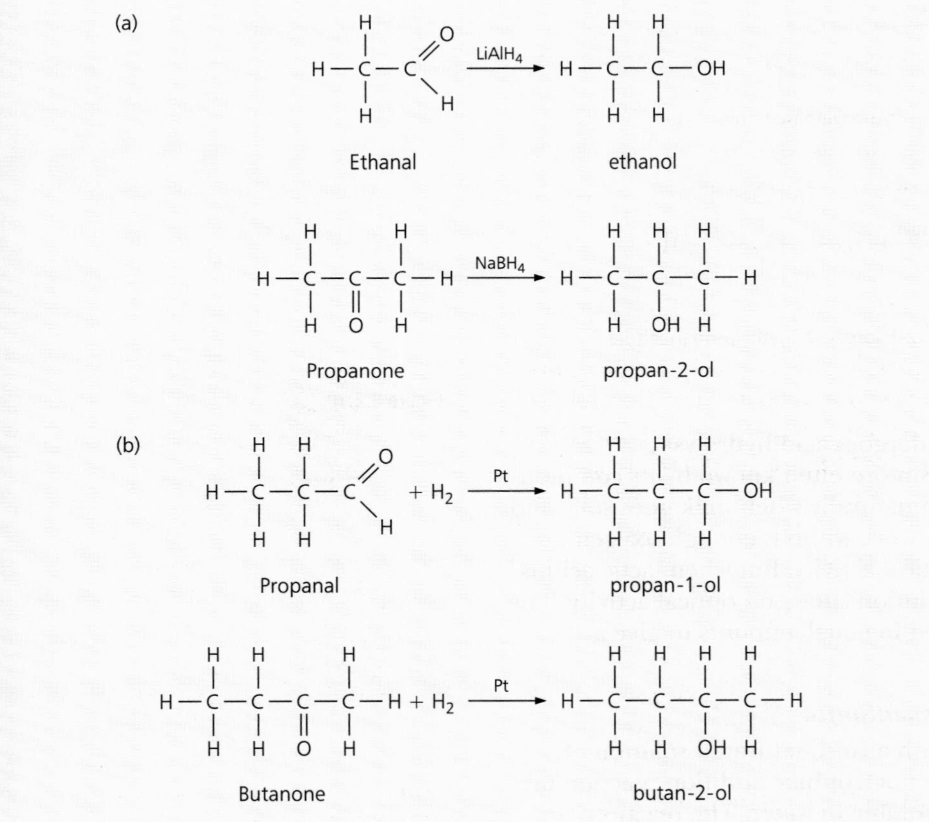

Figure 5.2.9

Aldehydes and ketones can also be reduced by the addition of hydrogen in the presence of a platinum or nickel catalyst. Once again, aldehydes give primary alcohols and ketones give secondary alcohols, as in figure 5.2.9(b).

Nucleophilic attack

One of the most important types of reactions of compounds containing the carbonyl group is **nucleophilic attack** on the δ+ carbon of the C=O bond, due to the greater electronegativity of oxygen compared with carbon. Such reactions take place with electron-rich species such as the cyanide ion ($\ddot{C}N^-$), amines ($R\ddot{N}H_2$) and even δ– carbon atoms in other organic molecules.

In general, the aldehydes are more susceptible to nucleophilic attack than the ketones, and aldehydes of low relative molecular mass are more reactive than those with long carbon chains. The aromatic aldehydes and ketones tend to react similarly to the aliphatic molecules, although generally less readily as the effect of the benzene ring is to delocalise the δ+ charge on the carbon atom of the carbonyl group.

Addition reactions with hydrogen cyanide

Aldehydes and ketones undergo a nucleophilic addition reaction with hydrogen cyanide, HCN, in cold, alkaline solution to produce hydroxynitriles (or cyanohydrins). The HCN molecule is added across the C=O bond, as shown in figure 5.2.10. The reaction can be used for the controlled lengthening of the carbon chain – each addition reaction with HCN adds one carbon atom to the chain.

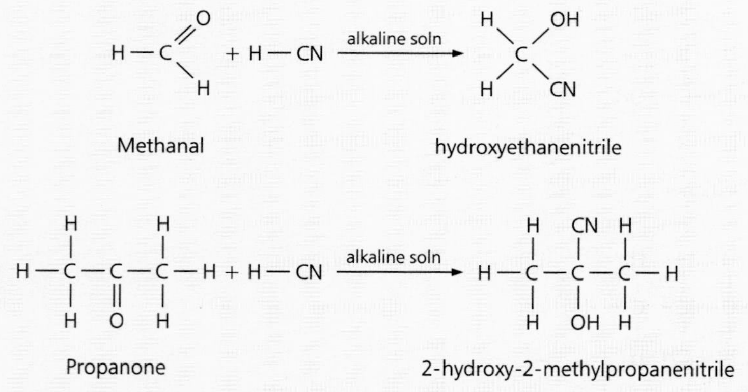

Figure 5.2.10

If 2-hydroxy-2-methylpropanenitrile undergoes acid hydrolysis, 2-hydroxypropanoic acid is formed. This is more often known by its common name of lactic acid. Lactic acid is produced naturally when milk goes sour and when the muscles of the body are forced to work without enough oxygen. This naturally formed lactic acid shows optical activity. But when lactic acid is prepared in the laboratory the resulting solution shows no optical activity. The L- and D- isomers (see page 374) are formed in equal amounts to give a **racemic mixture** (see page 481).

Addition reactions with sodium hydrogensulphite

When aldehydes and ketones are shaken with a cold, saturated solution of sodium hydrogensulphite, $NaHSO_3$, a slow nucleophilic addition reaction takes place to form ionic compounds which are soluble in water. The reactions are shown in figure 5.2.12.

Racemic mixtures

When hydrogen cyanide reacts with ethanal to form a hydroxynitrile, the molecule formed contains an asymmetrical carbon atom, and is therefore optically active. Despite this, the products of the reaction appear to show no optical activity.

Figure 5.2.11 shows why this is. The reaction produces equal quantities of the two stereoisomers of the hydroxynitrile, which rotate the plane of plane-polarised light in opposite directions. Each isomer cancels out the effect of the other on plane-polarised light, so no overall rotation of the plane of polarisation is seen. Mixtures of this type are called **racemic mixtures**.

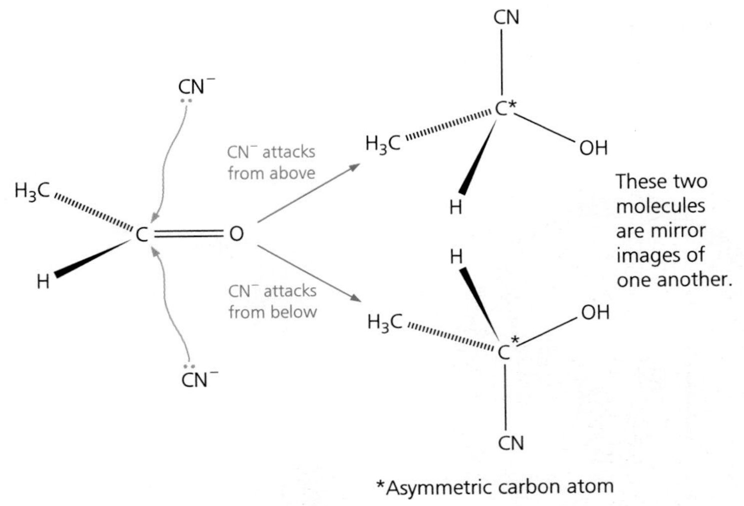

Figure 5.2.11

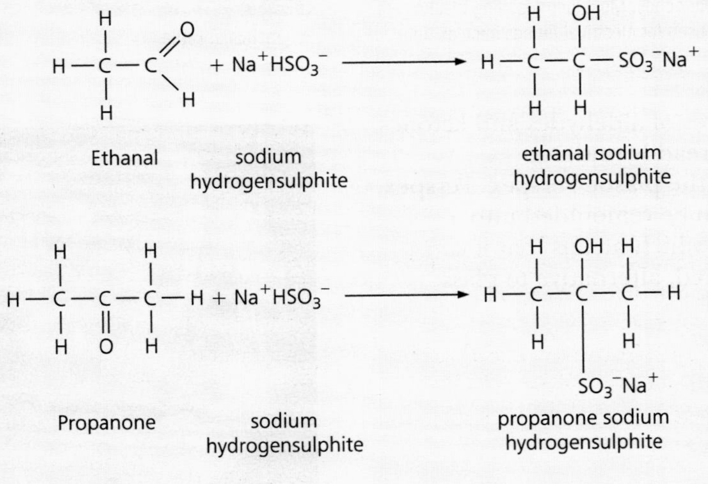

Figure 5.2.12

The reaction occurs slowly because the HSO_3^- ion is a bulky nucleophile – ketones with two large hydrocarbon groups may not react at all. The value of this addition reaction with aldehydes and ketones is that it produces a derivative which is very soluble in water. This allows aldehydes and ketones to be extracted from mixtures containing other impurities that are not soluble in water. The reaction of the aldehydes and ketones with sodium hydrogensulphite can then readily be reversed by treating the ionic compounds with dilute acids, leaving the pure organic compounds again.

Addition reactions with themselves (addition polymerisation)

Carbonyl compounds will form polymers by nucleophilic addition across the C=O bond. Ketones do not form polymers easily as they are not particularly reactive, but aldehydes readily form a variety of polymers. In some cases the polymers are made by simple addition of one monomer to the other – polymethanal, a strong plastic, is one example of this. Many such polymers contain extensive covalent links between their chains, which means that as well as having great strength they do not generally soften on heating. Such polymers are known as **thermosetting plastics** – probably the best known example of these polymers is Bakelite, an addition polymer of phenol and methanal which was patented by Leo Baekeland in 1909. Figure 5.2.13 shows the structures of polymethanal and Bakelite.

$$—O—CH_2—O—CH_2—O—CH_2—O—$$

Polymethanal

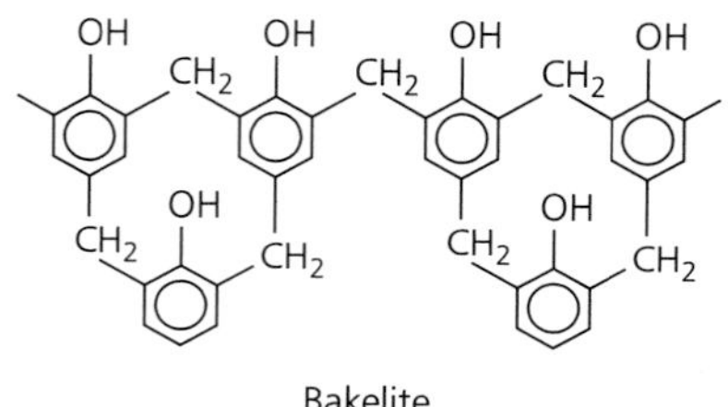

Bakelite

1930s Bakelite calendar

Polymethanal kettle

Figure 5.2.13 These are both examples of thermosetting plastics – the bonds form structures of considerable strength, although Bakelite is quite brittle, due to its rigid network of covalent bonds. Most thermosetting plastics are extremely heat resistant and do not conduct electricity, and they are widely used for electrical fittings such as light switches and plug sockets.

In other cases polymers are made with derivatives of aldehydes and ketones. For example, although ketones do not themselves readily undergo polymerisation, ester derivatives are used to make the plastic Perspex. Perspex is a **thermoplastic** – when heated it softens and can be remoulded into a different shape. It is light, strong and transparent, which means that it can frequently be used as a convenient, if easily scratched, alternative to glass.

$$\begin{array}{cccccccccccc} & H & & CH_3 & & H & & CH_3 & & H & & CH_3 \\ & | & & | & & | & & | & & | & & | \\ \cdots & C & — & C & — & C & — & C & — & C & — & C \cdots \\ & | & & | & & | & & | & & | & & | \\ & H & & CO_2CH_3 & & H & & CO_2CH_3 & & H & & CO_2CH_3 \end{array}$$

Perspex

Figure 5.2.14 A Perspex glove box allows the chemist to carry out experiments that need an inert, sterile, dry or dust-free atmosphere.

Reactions with the halogens

The aldehydes and ketones react more readily with the halogens than might be expected, with one or more of the hydrogen atoms on the carbon atom next to the carbonyl group being replaced by halogen atoms in a nucleophilic substitution reaction, as shown in figure 5.2.15. The relative ease of reaction is due to the effect of the carbonyl group. The partial positive charge on the carbon atom of the carbonyl group tends to attract electrons from neighbouring carbon atoms, making the C—H bonds polar, and thus making substitution by halogen atoms easier.

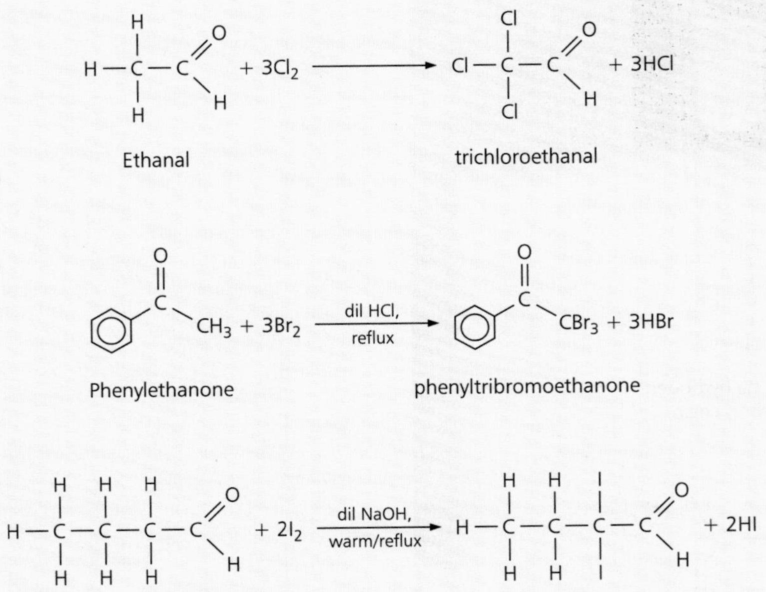

$$CH_3CHO + 3Cl_2 \longrightarrow CCl_3CHO + 3HCl$$

Ethanal → trichloroethanal

$$C_6H_5COCH_3 + 3Br_2 \xrightarrow[\text{reflux}]{\text{dil HCl,}} C_6H_5COCBr_3 + 3HBr$$

Phenylethanone → phenyltribromoethanone

$$CH_3CH_2CH_2CHO + 2I_2 \xrightarrow[\text{warm/reflux}]{\text{dil NaOH,}} CH_3CH_2CI_2CHO + 2HI$$

Figure 5.2.15

The iodoform reaction

When ethanal is warmed with iodine in alkaline solution, a two-stage reaction takes place. First the aldehyde reacts with the iodine to form triiodoethanal, and this then reacts with the alkali to form triiodomethane. Triiodomethane (or iodoform, to give its old non-systematic name) is an insoluble yellow solid with a characteristic smell. It appears as a yellow, crystalline precipitate (figure 5.2.16).

This reaction, with the formation of the characteristic yellow precipitate, is known as the **iodoform reaction**. It is not specific to the reaction of ethanal with iodine in alkaline solution, but it *is* specific to the reaction of organic compounds containing a methyl group next to a carbonyl group with iodine in the presence of a base, or a secondary alcohol that will oxidise under these conditions to form a methyl group next to a carbonyl group, as shown in figure 5.2.17.

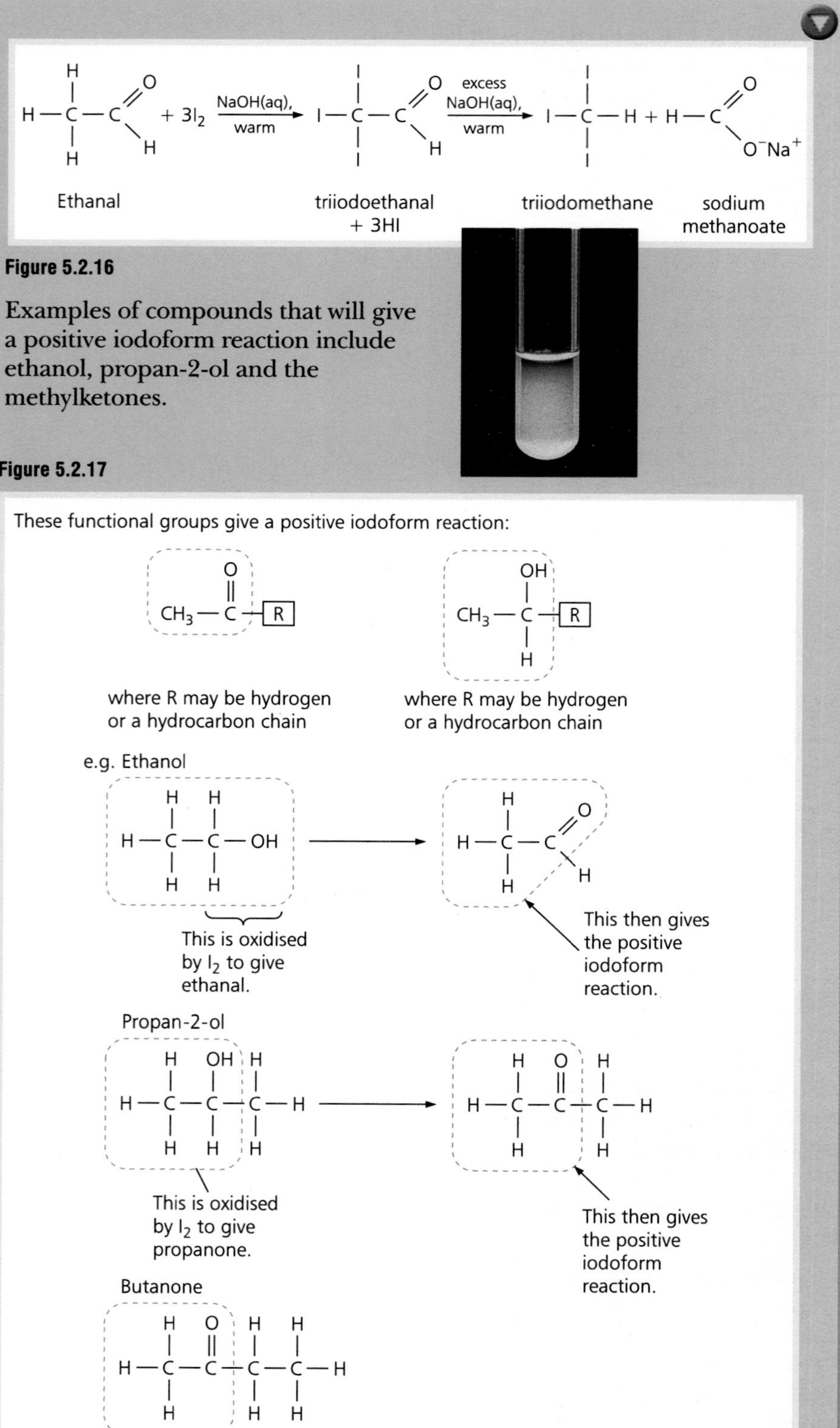

$CH_3CHO + 3I_2 \xrightarrow{NaOH(aq),\ warm} CI_3CHO + 3HI \xrightarrow{excess\ NaOH(aq),\ warm} CHI_3 + HCOO^-Na^+$
Ethanal
triiodoethanal
+ 3HI
triiodomethane
sodium
methanoate

Figure 5.2.16

Examples of compounds that will give a positive iodoform reaction include ethanol, propan-2-ol and the methylketones.

Figure 5.2.17

These functional groups give a positive iodoform reaction:
CH_3-CO-R
where R may be hydrogen or a hydrocarbon chain
$CH_3-CH(OH)-R$
where R may be hydrogen or a hydrocarbon chain
e.g. Ethanol
This is oxidised by I_2 to give ethanal.
This then gives the positive iodoform reaction.
Propan-2-ol
This is oxidised by I_2 to give propanone.
This then gives the positive iodoform reaction.
Butanone
This gives the positive iodoform reaction.

The iodoform reaction can be useful in laboratory analysis of organic molecules, but in all but the simplest laboratories chemical methods like this have been superseded by techniques such as infra-red and NMR spectroscopy and mass spectrometry.

Addition/elimination reactions

In a number of the addition reactions of the aldehydes and ketones, there is an unexpected extra element. When the reacting molecules, which may be two carbonyl monomers or a carbonyl compound and an ammonia derivative, undergo an addition reaction, a molecule of water is removed. Reactions of this type are known as **addition/elimination** or **condensation reactions**.

The reaction between phenol and methanal to make the Bakelite shown in figure 5.2.13 is an example of an addition/elimination reaction, and the general principle is shown in figure 5.2.18.

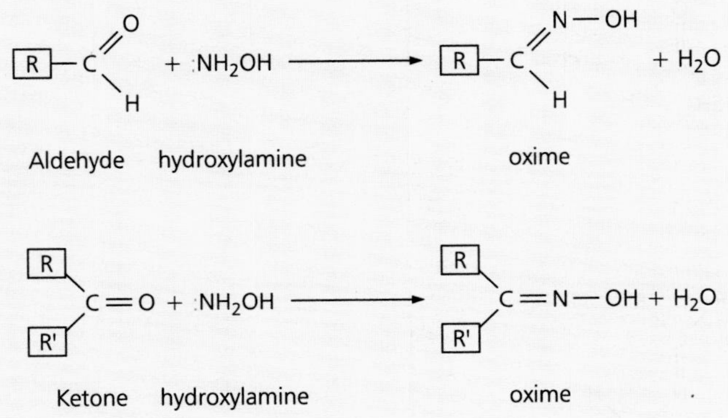

$$R\text{–}CHO + :NH_2OH \longrightarrow R\text{–}CH\text{=}N\text{–}OH + H_2O$$

Aldehyde hydroxylamine oxime

$$RR'C\text{=}O + :NH_2OH \longrightarrow RR'C\text{=}N\text{–}OH + H_2O$$

Ketone hydroxylamine oxime

Figure 5.2.18

Aldehydes and ketones undergo such addition/elimination reactions with 2,4-dinitrophenylhydrazine (2,4-DNPH), as shown in figure 5.2.19. They all form yellow/orange/red crystalline solids that have well defined and documented melting points. Before the advent of modern spectroscopy, unknown aldehydes and ketones could be identified by reaction with 2,4-dinitrophenylhydrazine followed by careful measurement of the melting point of the recrystallised product. However, like the iodoform reaction, this technique has now been superseded in all but the simplest laboratories.

$$NH_2\text{–}NH\text{–}C_6H_3(NO_2)_2 + CH_3\text{–}CHO \longrightarrow CH_3(H)C\text{=}N\text{–}NH\text{–}C_6H_3(NO_2)_2 + H_2O$$

2,4-dinitrophenylhydrazine ethanal ethanal 2,4-dinitrophenylhydrazone

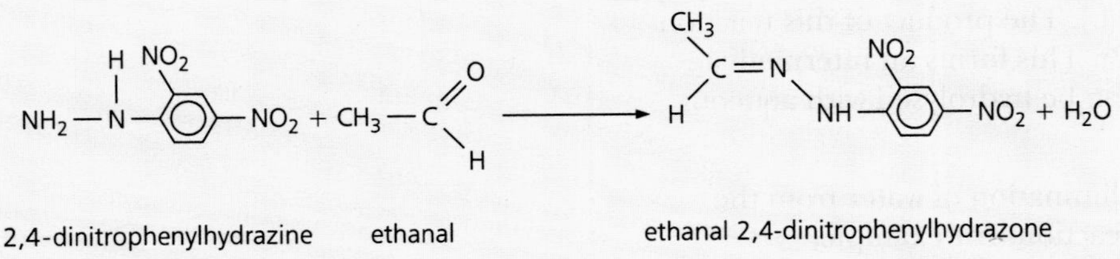

Figure 5.2.19

Nucleophilic attack on carbonyl compounds – addition, substitution or elimination?

The nucleophilic attack of an electron-rich species on the carbon atom of the C=O group results initially in the formation of an addition product. In some cases, this addition product may be the end of the reaction – for example, the attack of cyanide leading to the hydroxynitrile, as in figure 5.2.20(a).

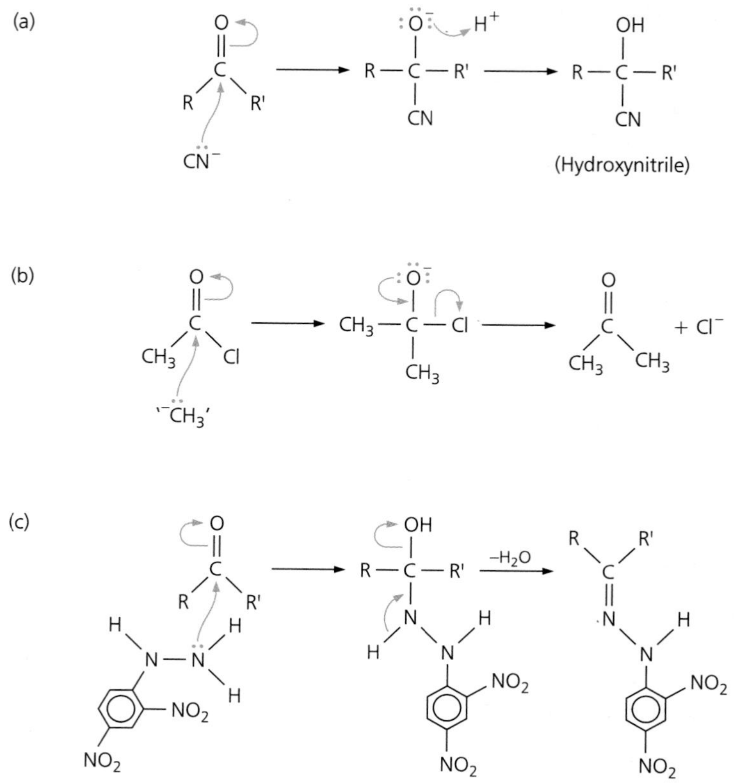

(a)

(Hydroxynitrile)

(b)

(c)

Figure 5.2.20

In other cases, if the intermediate compound can reform the C=O bond by eliminating a molecule of water or a stable ion (for example, Cl^- in an acid halide), elimination of this species from the intermediate may occur, leading to a substitution product. An example of this is the reaction of CH_3MgBr (a **Grignard reagent**, effectively a source of CH_3^- ions, (see page 439) with ethanoyl chloride, figure 5.2.20(b). The product of this reaction (propanone) can itself react with CH_3MgBr. This forms an intermediate compound $(CH_3)_2(COMgBr)CH_3$ which can be hydrolysed with aqueous acid to produce 2-methylpropan-2-ol.

Finally, some reactions may result in the elimination of water from the intermediate, leading to a condensation reaction. For example, 2,4-DNPH was mentioned on page 485. Figure 5.2.20(c) shows what happens when hydrazine and its derivatives react with a carbonyl compound.

Addition/elimination reactions of the aldehydes and ketones are of great importance in living organisms, as well as to organic chemists. Aldehydes and ketones play a major role in the structure of sugars, and addition/elimination reactions are used to build up the complex carbohydrates vital as both structural and storage molecules in living things. We will be looking at these large molecules in greater detail in section 5.5.

SUMMARY

- Carbonyl compounds, which include the aldehydes and ketones, contain the carbonyl group >C=O. Aldehydes have at least one hydrogen attached to the carbonyl carbon, while ketones have two carbons attached to the carbonyl carbon.
- Aldehydes are generally more volatile than the corresponding ketone having the same number of carbon atoms.
- The polarity of the carbonyl group leads to short-chain aldehydes and ketones being soluble in water.
- Aldehydes can be prepared by oxidising primary alcohols. Ketones can be prepared by oxidising secondary alcohols.
- Aldehydes give a positive reaction with Fehling's solution and Tollens reagent; ketones do not.
- Aldehydes and ketones may be reduced to their parent alcohols using lithium tetrahydridoaluminate(III), $NaAlH_4$, or sodium tetrahydridoborate(III), $NaBH_4$.
- The carbonyl group readily undergoes nucleophilic attack resulting in addition, addition/elimination or substitution reactions.
- Aldehydes and ketones undergo an addition reaction with hydrogen cyanide, HCN, to produce hydroxynitriles.
- Aldehydes and ketones undergo an addition reaction with sodium hydrogen-sulphite, $NaHSO_3$, to form ionic compounds that are soluble in water.
- Aldehydes readily form polymers by addition across the C=O bond under acidic conditions.
- The carbonyl group activates the hydrogen atoms on neighbouring carbon atoms, making them more reactive. An example of this is the iodoform reaction.

QUESTIONS

1 The compounds A to D in figure 5.2.21 are either aldehydes or ketones.

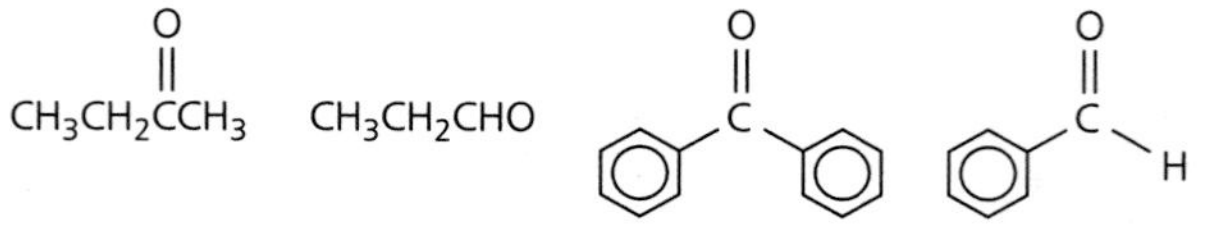

A: $CH_3CH_2COCH_3$ (CH₃CH₂C(=O)CH₃) B: CH_3CH_2CHO C: $C_6H_5COC_6H_5$ D: C_6H_5CHO

Figure 5.2.21

- **a** Which are aldehydes?
- **b** Which are ketones?
- **c** Which may result from the oxidation of a primary alcohol?
- **d** Which may be reduced to a secondary alcohol?
- **e** Name each compound.
- **f** Which would decolorise acidified potassium manganate(VII) solution?
- **g** Which would give a positive result with Tollens reagent?

2 The C═O bond is polar, due to the high electronegativity of oxygen. Discuss how the polarity of the bond is affected by the groups attached to the carbonyl group, and suggest why it is that:

- **a** aldehydes are more volatile than ketones with the same number of carbon atoms
- **b** ketones are less reactive to nucleophiles than aldehydes.

3 Predict the products formed by the following reactions:

- **a** CH_3CH_2CHO + acidified $KMnO_4$ solution
- **b** $CH_3CH_2CH_2COCH_2CH_2CH_2CH_3$ + H_2/Ni catalyst
- **c** C_6H_5CHO + HCN
- **d** $CH_3COCH_2CH_3$ + hydroxylamine
- **e** CH_3CHO + dilute acid
- **f** $C_6H_5COCH_3$ + Br_2/base

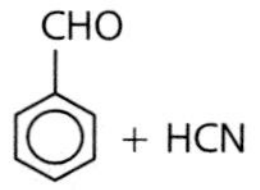

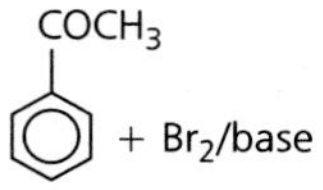

4 A compound X has the molecular formula $C_5H_{10}O$. It does not react with Fehling's solution, and may be reduced to an alcohol by lithium tetrahydridoaluminate(III). What are the possible structures of X?

FOCUS A STICKY PROBLEM

Carboxylic acids and their derivatives are widely used in the chemical industry by chemists as they try to solve various problems by the synthesis of new molecules. For example, everyone knows how uncomfortable pulling a plaster off the skin can be. But while for many of us it is simply a case of discomfort, for some people removing a plaster can cause real pain and skin damage. As people get elderly their skin becomes more fragile and loses its elasticity, whilst very young children have delicate skin which is easily damaged. So a group of chemists working in the research labs of Smith and Nephew set out to develop a new PSA (**pressure sensitive adhesive**) which could be used to keep plasters in place through the rough and tumble of life but which would peel easily when the time came to remove it – a task not as easy as it sounds!

They reasoned that to produce an adhesive that allowed the plaster to do its job but was then easy to peel off they needed a chemical with some sort of switch, to change the way it bonded to the skin. If they could introduce a mechanism that would switch the 'stickiness' off before the plaster was removed, they would have cracked the problem. First the team developed a complex polymer containing 2-ethylhexyl acrylate, *n*-butyl acrylate and acrylic acid They esterified the carboxylic acid group on the acrylic acid within the polymer with yet more complex organic compounds to form the basic adhesive which would bond effectively to skin to keep a plaster in place. Then came the tricky bit – the switch. This was done by incorporating a new and highly sensitive photoinitiator – a chemical which was sensitive to light, and which, in the presence of light, caused the formation of cross linkages within the adhesive polymer. These cross-linkages made the adhesive much less 'sticky' so it would peel more easily. This at least was the theory. It took a considerable time to work from the first switchable prototype to a chemical which was both effective and economically viable. The first molecules worked – tests on high density polyethylene surfaces showed a large reduction in the force needed to peel the adhesive once the light was switched on – but would have been prohibitively expensive: no-one could have afforded to buy them!

Once the right polymer and switch were developed the team needed to test the new adhesives on real human skin. First they had to design a 'plaster' using the new adhesive, and incorporate into the design a way of making sure that the adhesive was not exposed to light until it was time to remove the plaster. This was done by making a plaster with several layers, including a light-blocking layer which could be peeled off when needed, but which stayed in place until then! Finally came the moment of truth, when the new plasters were tried on the arms of 16 healthy volunteers. They had plasters attached to both arms for two hours, after which one of the tapes was exposed to light. The plasters were then peeled off at a constant rate and angle using a tensile testing machine which measured the force needed to remove them. Perhaps more importantly, the pain felt by the human guinea pigs, the level of reddening and soreness of the skin were also assessed. The new switchable adhesive was a success – there was a reduction of about 60% in the strength needed to peel the plaster off, with much reduced pain and skin redness experienced by the volunteers. The work continues to make current wound dressings and plasters light-switchable – the early years of the twenty-first century should see their introduction to the high-street chemists.

Figure 1 Dr Iain Webster and some of the team who set out to develop new switchable adhesives for plasters

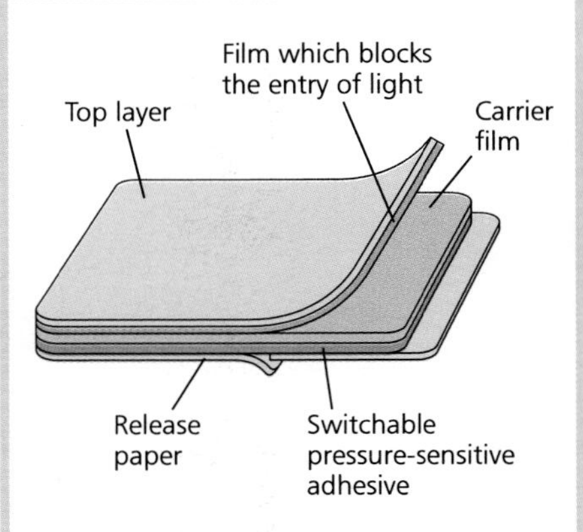

Figure 2 The benefits of light-switchable adhesives will range from a major step forward in the treatment of fragile and damaged skins to a general reduction in the 'ouch' factor of plasters for the rest of us!

5

5.3 Carboxylic acids and their derivatives

Carboxylic acids

The carboxyl group

Carboxylic acids are a family of organic chemicals containing the **carboxyl group**, -COOH, shown in figure 5.3.2. This is a complex functional group made up of a hydroxyl group, -OH (like the alcohols), and a carbonyl group, C=O (like the aldehydes and ketones). These two groups are joined together in the carboxyl group so that they have a distinct effect on each other's chemistry, and the reactions of the carboxylic acids tend to differ from those of the corresponding alcohols, aldehydes and ketones. The general formula of the monocarboxylic acids is RCOOH, where R may be a hydrogen atom, an alkyl group or an aryl group.

Naming the carboxylic acids

Naming the carboxylic acids is relatively simple. The stem (such as methan-, propan-, octan-) indicates the number of carbon atoms including the carbon of the carboxyl group. This is followed by the suffix '-oic acid'. The carboxyl carbon is taken as number 1. Examples include methanoic acid (the chemical weapon of the ant pictured in figure 5.3.1), propanoic acid and benzoic acid. Some acids contain two carboxylic groups and these are known as the **dicarboxylic acids**, for example ethanedioic acid. The structures of some of these carboxylic acids are shown in figure 5.3.2.

Figure 5.3.1 Carboxylic acids are responsible for the sourness of lemons and vinegar, and the discomfort that follows an ant bite. Their derivatives are used in a wide range of compounds.

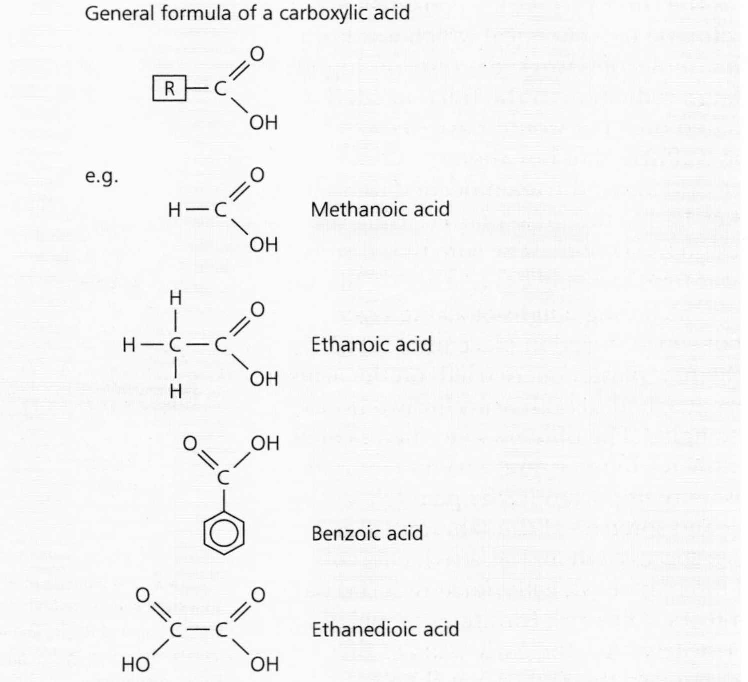

Figure 5.3.2

Properties of the carboxyl group

The properties of the carboxylic acids, both physical and chemical, are largely governed by two factors – the presence of the carboxyl group and the length of the carbon chain. As in all the organic families, as the carbon chain length increases, the influence of the functional group decreases. The carboxyl group has a great effect on the properties of the family as a whole. The presence in the molecule of two polarised groups ($C^{\delta+}{=}O^{\delta-}$ and $O^{\delta-}{-}H^{\delta+}$) means that there is a considerable degree of hydrogen bonding. Also, a carboxylic acid dissociates in water to form hydrogen ions and carboxylate ions, thus behaving as an acid, as shown in figure 5.3.3(a).

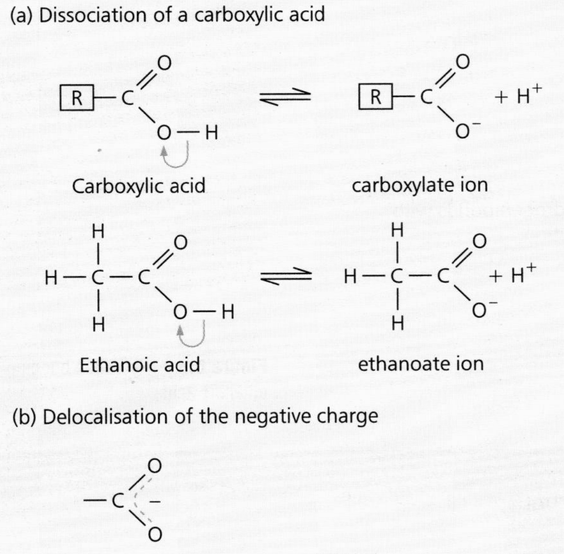

(a) Dissociation of a carboxylic acid

$$R{-}COOH \rightleftharpoons R{-}COO^- + H^+$$

Carboxylic acid ⇌ carboxylate ion

$$CH_3COOH \rightleftharpoons CH_3COO^- + H^+$$

Ethanoic acid ⇌ ethanoate ion

(b) Delocalisation of the negative charge

Figure 5.3.3

Although the carboxylate ion is drawn with one double and one single bond, X-ray diffraction work and observations of its reactions suggest that the two carbon–oxygen bonds are actually identical. This implies that the negative charge is delocalised over the whole of the carboxylate group as shown in figure 5.3.3(b). This means that the ion is stabilised, and so the equilibrium for the reaction shown in figure 5.3.3(a) is further over to the right than would be predicted. The carboxylic acids are thus quite strong acids, and the carboxylate ion (the conjugate base) is quite a weak base.

Although the carboxylic acids are much stronger acids than water and the alcohols considered in section 5.1, they are nevertheless described as weak acids. As table 5.3.1 shows, the carboxylic acids are not highly dissociated in water (remember that a high K_a indicates a strong acid with much dissociation). By contrast, the strong acid nitric acid has a K_a value of 40.

Carboxylic acid	K_a at 25 °C/ mol dm^{-3}
Methanoic	1.6×10^{-4}
Ethanoic	1.7×10^{-5}
Propanoic	1.3×10^{-5}
Butanoic	1.5×10^{-5}
Benzoic	6.3×10^{-5}
Ethanedioic:	
first K_a	5.9×10^{-2}
second K_a	5.2×10^{-5}

Table 5.3.1 The dissociation constants of some carboxylic acids

Physical properties of the carboxylic acids

The effects of hydrogen bonding

The feature of the carboxylic acids that has the greatest effect on their physical properties is their ability to form hydrogen bonds. This gives rise not only to relatively high melting and boiling points, but also to high solubilities of the smaller carboxylic acid molecules in water and in other hydrogen-bonding solvents. The larger molecules become less soluble as the organic nature of the carbon chain becomes dominant over the polar nature of the functional groups.

The polar nature of the molecules is such that they can exist as hydrogen-bonded dimers in the pure acid in the vapour, liquid or solid state. If the carboxylic acid is dissolved in a non-polar organic solvent, it will also form dimers. However, dimers do not form in a solvent that can form hydrogen bonds, and the carboxylic acid molecules form hydrogen bonds with molecules of the solvent instead, as shown in figure 5.3.4.

Dimers are formed between molecules of pure ethanoic acid in organic solvents and in the solid, liquid and vapour states. The dimers are held together by hydrogen bonding.

Ethanoic acid forms hydrogen bonds with water molecules in solution.

Figure 5.3.4 Hydrogen bonding in pure and solvated ethanoic acid

Glacial ethanoic acid

Pure ethanoic acid is a liquid at normal room temperature. However, in an unheated room in winter, it is likely to turn into a solid, as its melting point is 17 °C. Solid ethanoic acid looks remarkably like ice, and as a result of this tendency to turn into an ice-like solid, pure ethanoic acid is often referred to as **glacial ethanoic acid**. It owes this high melting point to the hydrogen bonding between the molecules – this effect can be seen very clearly when the melting points of corresponding members of other organic families are considered.

Figure 5.3.5

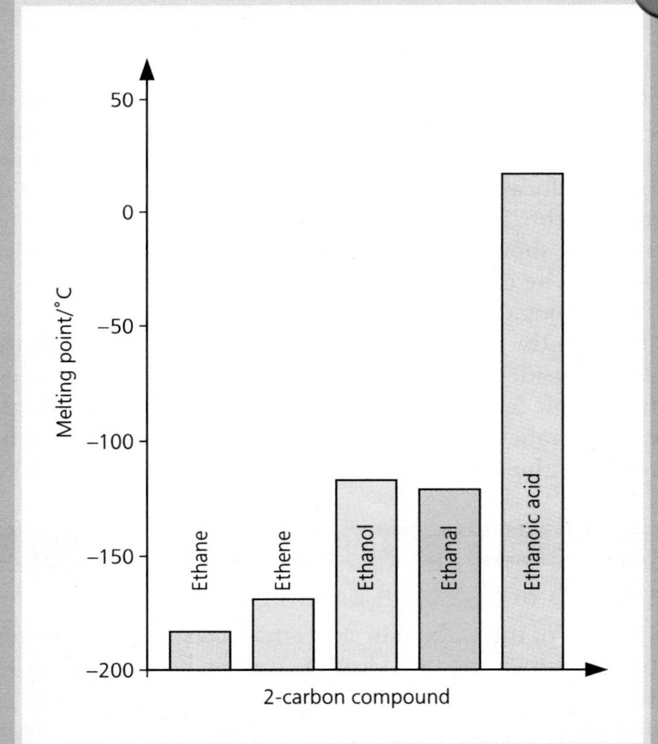

Characteristic odours

The carboxylic acids often have characteristic odours. The vinegary smell of ethanoic acid is quite unmistakable and is typical of the sharp, acidic smells associated with the short-chain carboxylic acids. Many of the carboxylic acids with slightly longer carbon chains have much stronger and often very unpleasant odours. The smell of rancid butter and unwashed sweaty socks are produced by butanoic acid. People produce their own cocktail of carboxylic acids in their sweat, and the mixture is as individual as a fingerprint.

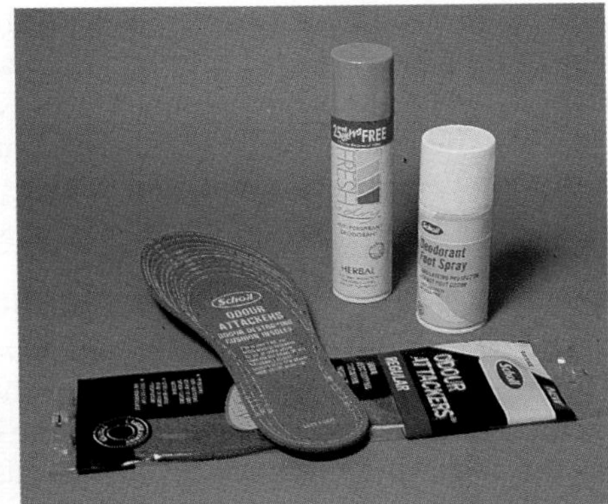

Figure 5.3.6 Our noses are very sensitive to the smells of butanoic and other carboxylic acids – their unpleasant odour is detected at concentrations of around 10^{-11} mol dm^{-3}, which means we can detect those unconcerned with their personal hygiene. In fact, this sensitivity of the human nose to unpleasant odours is the basis of an enormous chemical industry making products to reduce the production of these compounds and to mask their smell. Dogs are even more sensitive to smells than we are – they can detect a concentration of 10^{-17} mol dm^{-3} of butanoic acid. This is why tracker dogs, given a scent of a person's clothing and thus their carboxylic acid mixture, can then recognise and follow that scent to track the individual concerned.

Reactions of the carboxylic acids

Making carboxylic acids

The most common laboratory preparation of a carboxylic acid is by the oxidation of a primary alcohol or aldehyde. This is carried out by refluxing with acidified potassium dichromate(VI) or potassium manganate(VII) solution. The other general method of preparing carboxylic acids is by the hydrolysis of nitriles by refluxing with a strong acid or base. These reactions are shown in figure 5.3.7. Benzoic acid is prepared in the laboratory by the oxidation of methylbenzene with hot, acidified potassium manganate(VII). Benzoic acid is prepared industrially by the oxidation of methylbenzene using air as the oxidising agent, with a suitable catalyst.

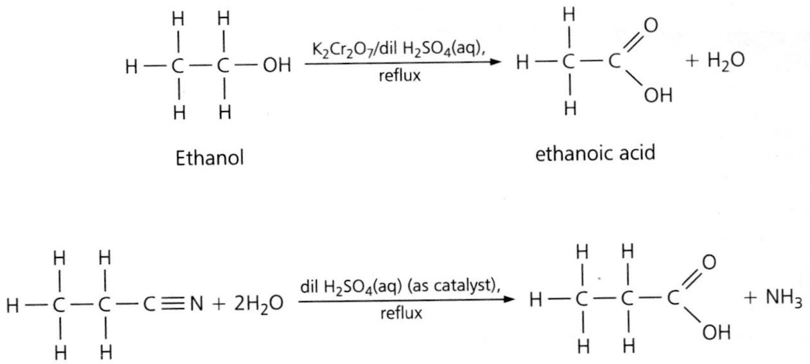

$$\mathrm{H{-}\overset{H}{\underset{H}{C}}{-}\overset{H}{\underset{H}{C}}{-}OH \xrightarrow[\text{reflux}]{K_2Cr_2O_7/\text{dil } H_2SO_4(aq),} H{-}\overset{H}{\underset{H}{C}}{-}C{\overset{\displaystyle /\!/ O}{\underset{\displaystyle \backslash OH}{}}} + H_2O}$$

Ethanol → ethanoic acid

$$\mathrm{H{-}\overset{H}{\underset{H}{C}}{-}\overset{H}{\underset{H}{C}}{-}C{\equiv}N + 2H_2O \xrightarrow[\text{reflux}]{\text{dil } H_2SO_4(aq)\ \text{(as catalyst)},} H{-}\overset{H}{\underset{H}{C}}{-}\overset{H}{\underset{H}{C}}{-}C{\overset{\displaystyle /\!/ O}{\underset{\displaystyle \backslash OH}{}}} + NH_3}$$

Propanenitrile → propanoic acid

Figure 5.3.7

Industrially, the most important carboxylic acid is ethanoic acid. Some, particularly that for human consumption, is made by the oxidation of wine. Most is made from distillation products of petrochemicals. A fraction containing alkanes with carbon chains varying from 5 to 7 carbon atoms in length is oxidised by passing pressurised air through it at 180 °C in the presence of a catalyst. The main product is ethanoic acid. This ethanoic acid is largely used in the production of ethanoic anhydride, which in turn is used in the production of the semisynthetic fibre rayon.

The chemistry of the carboxylic acids, like their physical properties, is largely a function of the carboxyl group. There are three main sites of reaction:

- the O—H bond, which splits to yield a hydrogen ion and carboxylate ion, giving the compounds their acidic character
- the C=O carbonyl bond, which gives nucleophilic addition reactions
- the C—O bond, which may break in substitution reactions.

Fission of the O—H bond – reactions with bases

Carboxylic acids react with a base to form a salt, as shown in figure 5.3.8(a). These reactions involve the fission of the O—H bond. The carboxylic acid will dissociate much more freely in alkaline solution, as the alkali 'mops up' the proton from the -OH group, and the dissociation equilibrium shifts well over to the right.

The salt formed by the reaction between a carboxylic acid and an alkali is very different in character from its organic acid parent. Salts with short carbon chains show mainly ionic character, and those with very long chains show mainly organic character. The ionic salts have crystalline structures, most of which dissolve in water. The salts of the carboxylic acids of medium chain length have a combination of organic and ionic character, and this is put to use in the manufacture and use of detergents (see box 'Soap' overleaf).

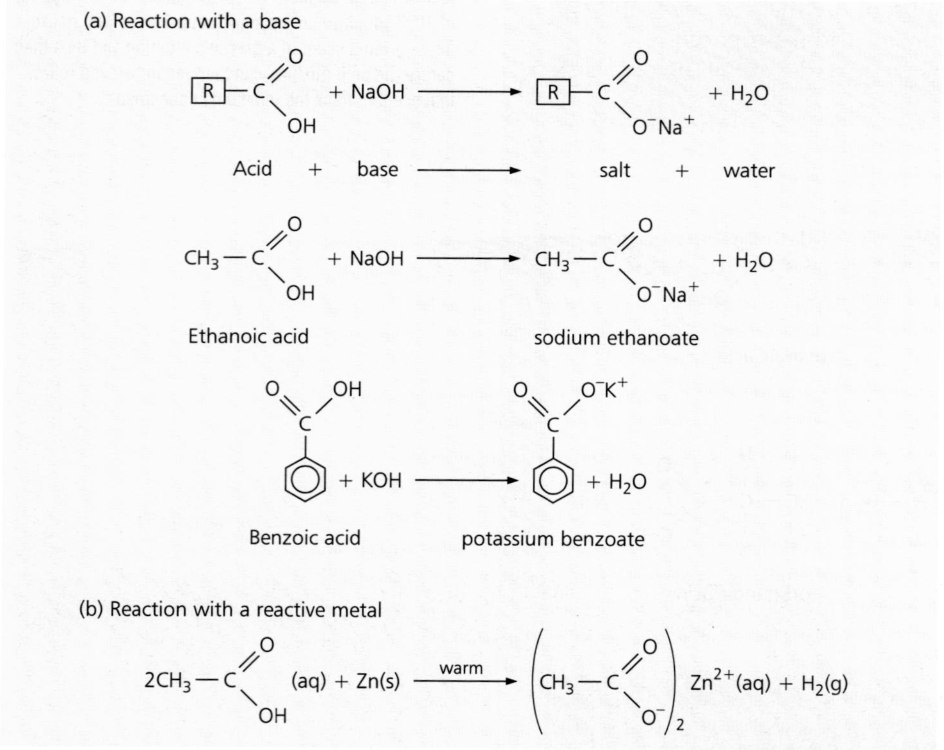

(a) Reaction with a base

$$\text{R—COOH} + \text{NaOH} \longrightarrow \text{R—COO}^-\text{Na}^+ + \text{H}_2\text{O}$$

Acid + base ⟶ salt + water

$$\text{CH}_3\text{—COOH} + \text{NaOH} \longrightarrow \text{CH}_3\text{—COO}^-\text{Na}^+ + \text{H}_2\text{O}$$

Ethanoic acid ⟶ sodium ethanoate

$$\text{C}_6\text{H}_5\text{COOH} + \text{KOH} \longrightarrow \text{C}_6\text{H}_5\text{COO}^-\text{K}^+ + \text{H}_2\text{O}$$

Benzoic acid ⟶ potassium benzoate

(b) Reaction with a reactive metal

$$2\text{CH}_3\text{—COOH(aq)} + \text{Zn(s)} \xrightarrow{\text{warm}} (\text{CH}_3\text{—COO}^-)_2\text{Zn}^{2+}\text{(aq)} + \text{H}_2\text{(g)}$$

Figure 5.3.8

Carboxylic acid salts can also be produced by the reaction of the acid with a reactive metal, for example zinc, as in figure 5.3.8(b).

Apart from their value as detergents, the salts of carboxylic acids are extremely useful in the preparation of other carboxylic derivatives. For example, acid anhydrides are formed when the anhydrous sodium salt of a carboxylic acid is heated with an acid chloride. Lead(IV) ethanoate, made by reacting Pb_3O_4 with warm glacial ethanoic acid, may be used to oxidise diols to aldehydes and ketones.

Another set of reactions involving fission of the O—H bond are the esterification reactions, considered on page 464.

Soap

Washing our clothes and ourselves has been part of human civilisation for a very long time. Human dirt includes oils and fats from our skin and food, carboxylic acids and salt from sweat, mud, dead skin cells and a wide variety of other substances. Water has been our washing medium, and though very effective at removing water-soluble dirt, it is poor at removing material that does not dissolve in water. A **detergent** is a substance that acts as a cleaning agent, improving the ability of water to wash things clean. The earliest, and for a long time the only, detergent known to the human race was **soap**.

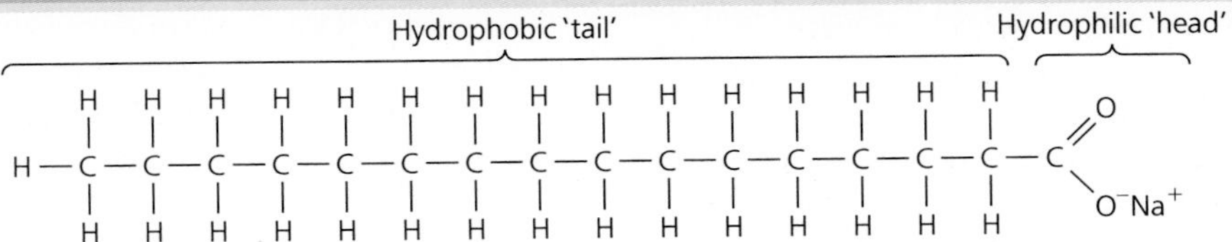

Structure of sodium hexadecanoate, $CH_3(CH_2)_{14}COONa$ (the salt of palmitic acid)

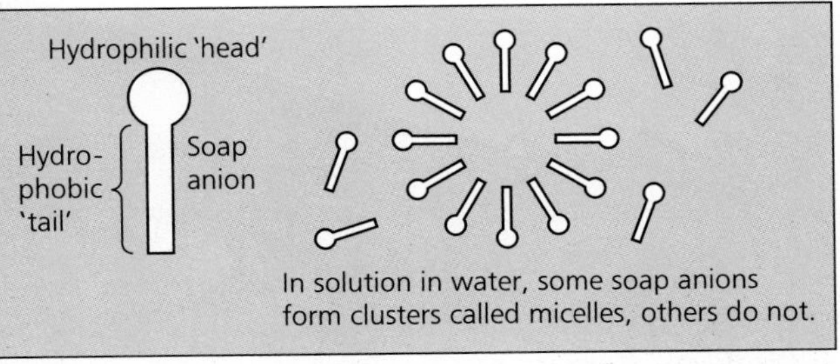

In solution in water, some soap anions form clusters called micelles, others do not.

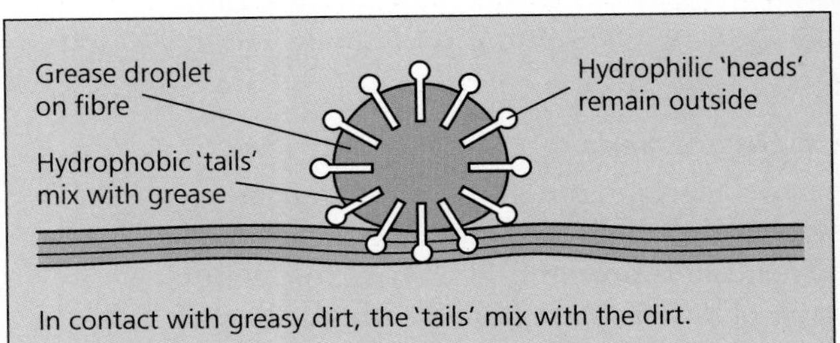

In contact with greasy dirt, the 'tails' mix with the dirt.

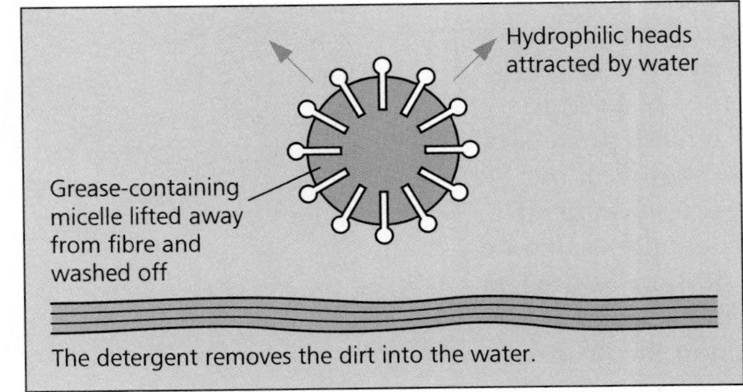

The detergent removes the dirt into the water.

Scanning electron micrograph of a dirty cotton shirt

The same shirt after washing with a modern washing powder

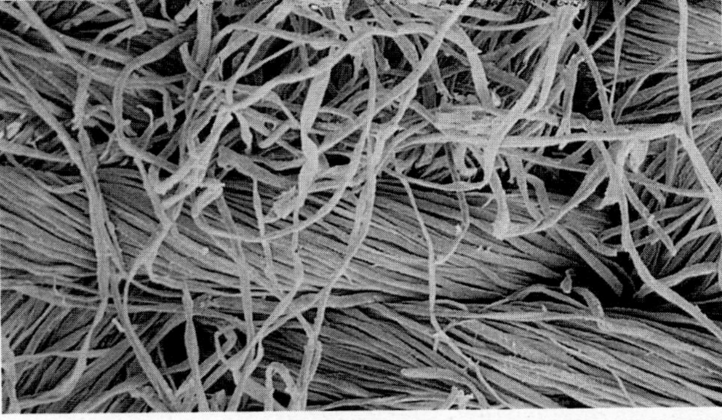

Figure 5.3.9 Soap was first manufactured in Roman times, and it is still one of the most commonly used detergents for personal cleanliness.

How soaps work

Soap is a mixture of the salts of medium-chain-length carboxylic acids. As we have seen, these are molecules with a mixture of ionic and organic characteristics – polar 'heads' and non-polar 'tails'. The polar part of the molecule is the carboxylate group, which is **hydrophilic** – it dissolves in water. The non-polar part is the carbon chain, which is **hydrophobic** – it does not dissolve in water. When soap is in water, the non-polar parts arrange themselves together as far from the water as possible, with the polar ends pointing outwards. The anions thus form clusters known as **micelles**. When soap is in contact with fat or greasy dirt, the non-polar tails mix with the non-polar dirt, so that the dirt becomes surrounded by polar 'heads'. These 'loaded' micelles are then carried away by the water and the material or skin left clean.

How soaps are made

Soaps are largely made from animal fats and vegetable oils, which are the esters of carboxylic (fatty) acids and propane-1,2,3-triol (old name glycerol). These fats are boiled with alkalis such as sodium hydroxide to form soap in a reaction known as **saponification**. Brine is pumped into the soap kettles, causing the insoluble soap to precipitate out as a curd on the top. The brine is then removed along with the glycerol and any other by-products, leaving pure soap behind. This is milled, processed and perfumed before reaching the shelves of our shops. The same saponification reaction occurs if sodium hydroxide solution is accidentally spilled on the skin – the tissue starts to feel slippery as the fats in the skin are converted to soap by the alkali, destroying the structure of the skin in the process.

Modern detergents

Soaps have their limitations – there are some types of dirt which they do not remove well. More importantly, in hard water they do not clean very efficiently as they react with the calcium and magnesium ions present to form an insoluble greasy curd-like precipitate known as **scum**, the calcium or magnesium salts of the carboxylic acids. As a result of efforts to overcome these problems, soaps are no longer our only cleansing agents. A wide range of other detergents has been manufactured from oil products, most working on the same principle as soap, with a polar and a non-polar end to the molecules. Detergents can be tailor-made to cope with a particular type of cleaning problem.

For example, some quaternary ammonium salts, with their positively charged head, act as cationic surfactants. Not only do they remove dirt effectively, they do not react with Ca^{2+} and Mg^{2+} ions in the water to form a scum. Enzymes are used to digest proteins, and organic chemicals such as the aldehydes and ketones are used in industry as solvents for specific types of dirt. However soap, manufactured from renewable resources in the plant and animal world, has been with us for at least 2000 years and will doubtless continue to be used for many years to come.

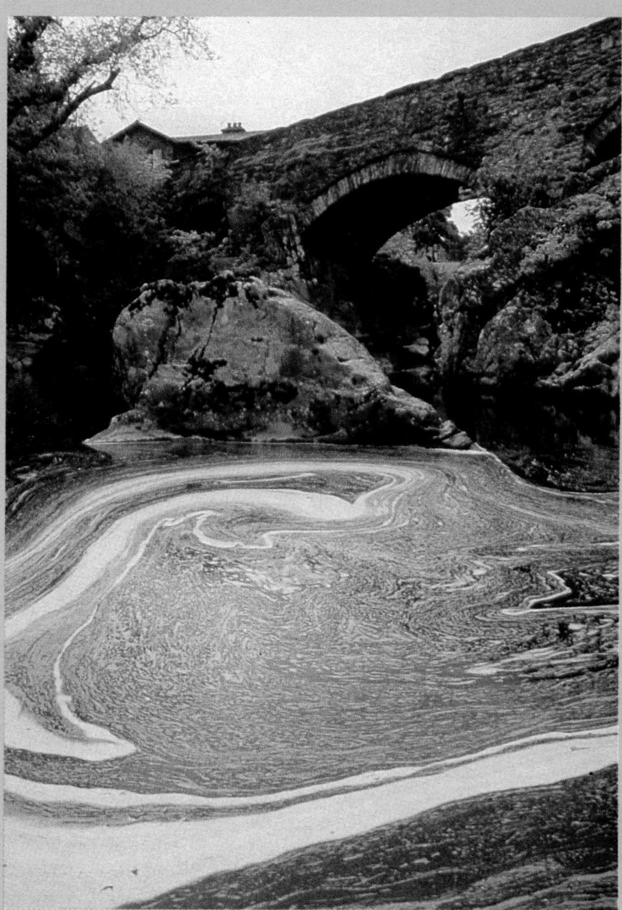

Figure 5.3.10 New oil-based detergents were developed in the 1960s and 70s to overcome the problems of scum formed by soap in hard water. However, these detergents brought their own problems. Soap is readily denatured by the action of bacteria, but many of the newer detergents were not, leading to pollution of the sort seen here in North Wales. Synthetic detergents that can be broken down by bacteria (that are **biodegradable**) are now being used in increasing quantities.

Fission of the C—O bond – halogenation

Carboxylic acids react with sulphur dichloride oxide and other halogenating reagents such as PCl_5 to form the acid chlorides, as shown in figure 5.3.11.

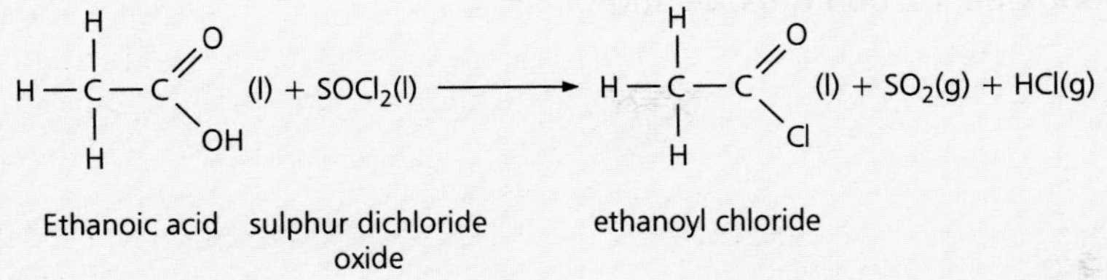

$$CH_3COOH(l) + SOCl_2(l) \longrightarrow CH_3COCl(l) + SO_2(g) + HCl(g)$$

Ethanoic acid sulphur dichloride oxide ethanoyl chloride

Figure 5.3.11

Oxidation of the carboxylic acids

The carboxylic acids are not readily oxidised – they are effectively the end product of the oxidation of a primary alcohol, as we saw in section 5.1. There are only two carboxylic acids that undergo oxidation reactions – both methanoic acid and ethanedioic acid are oxidised by warm acidified potassium manganate(VII) to form carbon dioxide and water. These reactions are shown in figure 5.3.12.

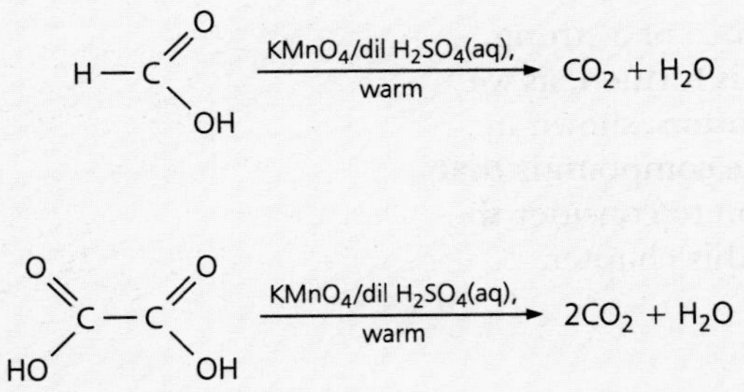

$$HCOOH \xrightarrow[\text{warm}]{KMnO_4/\text{dil } H_2SO_4(aq),} CO_2 + H_2O$$

$$HOOC{-}COOH \xrightarrow[\text{warm}]{KMnO_4/\text{dil } H_2SO_4(aq),} 2CO_2 + H_2O$$

Figure 5.3.12

Reduction of the carboxylic acids

Although the carboxylic acids are the product of the oxidation of primary alcohols, they cannot easily be reduced back to their corresponding alcohol. To bring about the reduction of a carboxylic acid a powerful reducing agent such as lithium tetrahydridoaluminate(III) ($LiAlH_4$) is needed, figure 5.3.13.

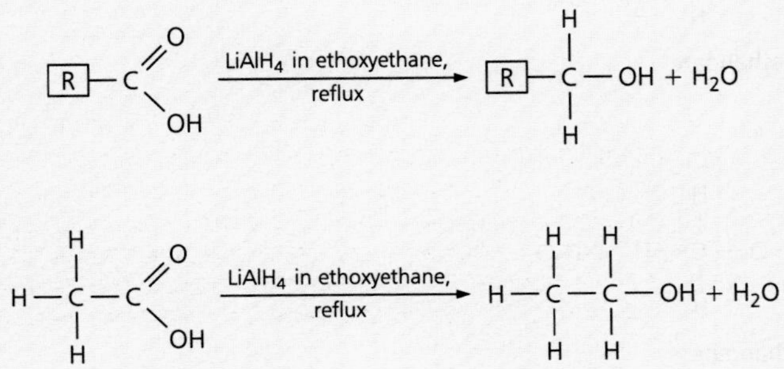

$$R{-}COOH \xrightarrow[\text{reflux}]{LiAlH_4 \text{ in ethoxyethane,}} R{-}CH_2{-}OH + H_2O$$

$$CH_3COOH \xrightarrow[\text{reflux}]{LiAlH_4 \text{ in ethoxyethane,}} CH_3CH_2OH + H_2O$$

Figure 5.3.13

5

Dehydration

In general, the carboxylic acids do not undergo dehydration reactions. However, using concentrated sulphuric acid as a dehydrating agent, the following carboxylic acids are dehydrated:

- methanoic acid is dehydrated to give carbon monoxide and water
- ethanedioic acid is dehydrated to give carbon monoxide, carbon dioxide and water.

Figure 5.3.14 shows these reactions.

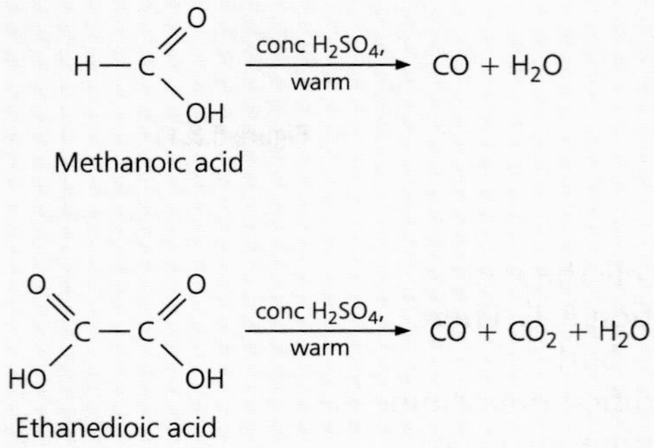

$$HCOOH \xrightarrow[\text{warm}]{\text{conc } H_2SO_4,} CO + H_2O$$

Methanoic acid

$$HOOC{-}COOH \xrightarrow[\text{warm}]{\text{conc } H_2SO_4,} CO + CO_2 + H_2O$$

Ethanedioic acid

Figure 5.3.14

Esterification

When a carboxylic acid is mixed with an alcohol in the presence of a strong acid catalyst such as sulphuric or hydrochloric acid, an **ester** is formed, as we saw in section 5.1. This type of reaction is known as **esterification**, shown in figure 5.3.15. The esters are one of several families of organic compounds that are direct derivatives of the carboxylic acids. We shall move on to consider the properties of some of these compounds in the remainder of this chapter.

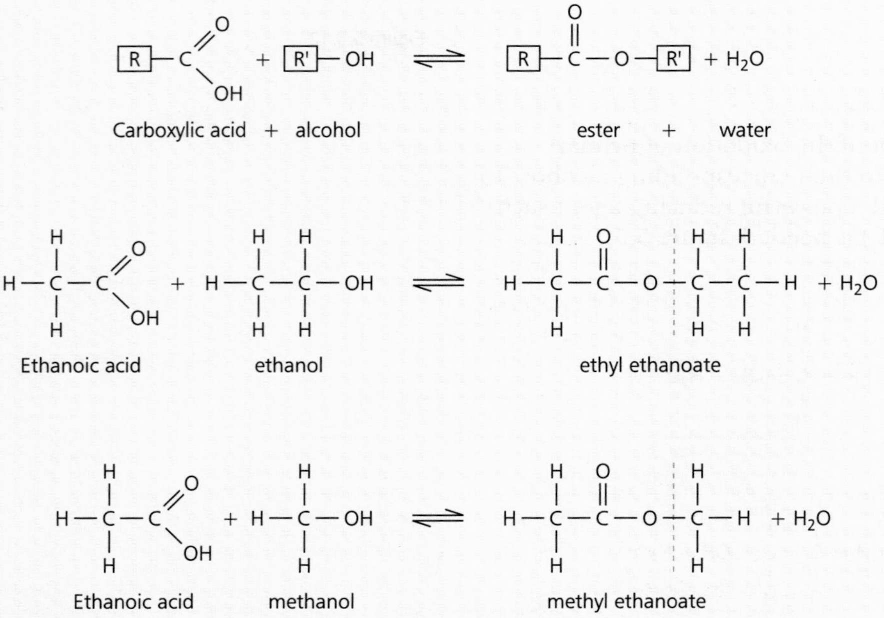

$$R{-}COOH + R'{-}OH \rightleftharpoons R{-}COO{-}R' + H_2O$$

Carboxylic acid + alcohol ⇌ ester + water

$$CH_3COOH + CH_3CH_2OH \rightleftharpoons CH_3COOCH_2CH_3 + H_2O$$

Ethanoic acid, ethanol, ethyl ethanoate

$$CH_3COOH + CH_3OH \rightleftharpoons CH_3COOCH_3 + H_2O$$

Ethanoic acid, methanol, methyl ethanoate

Figure 5.3.15

Carboxylic acid derivatives

The carboxylic acid derivatives are compounds in which the hydroxyl group of the parent carboxylic acid has effectively been replaced by another functional group, which alters the chemistry of the molecule. There are four main families of carboxylic acid derivatives – the **esters**, the **acid halides**, the **acid anhydrides** and the **acid amides**. These are shown in figure 5.3.16. We have already seen the reactions of the acid halides in section 4.5, and will meet the acid amides in section 5.4. Our main emphasis here will therefore be on the esters and the acid anhydrides.

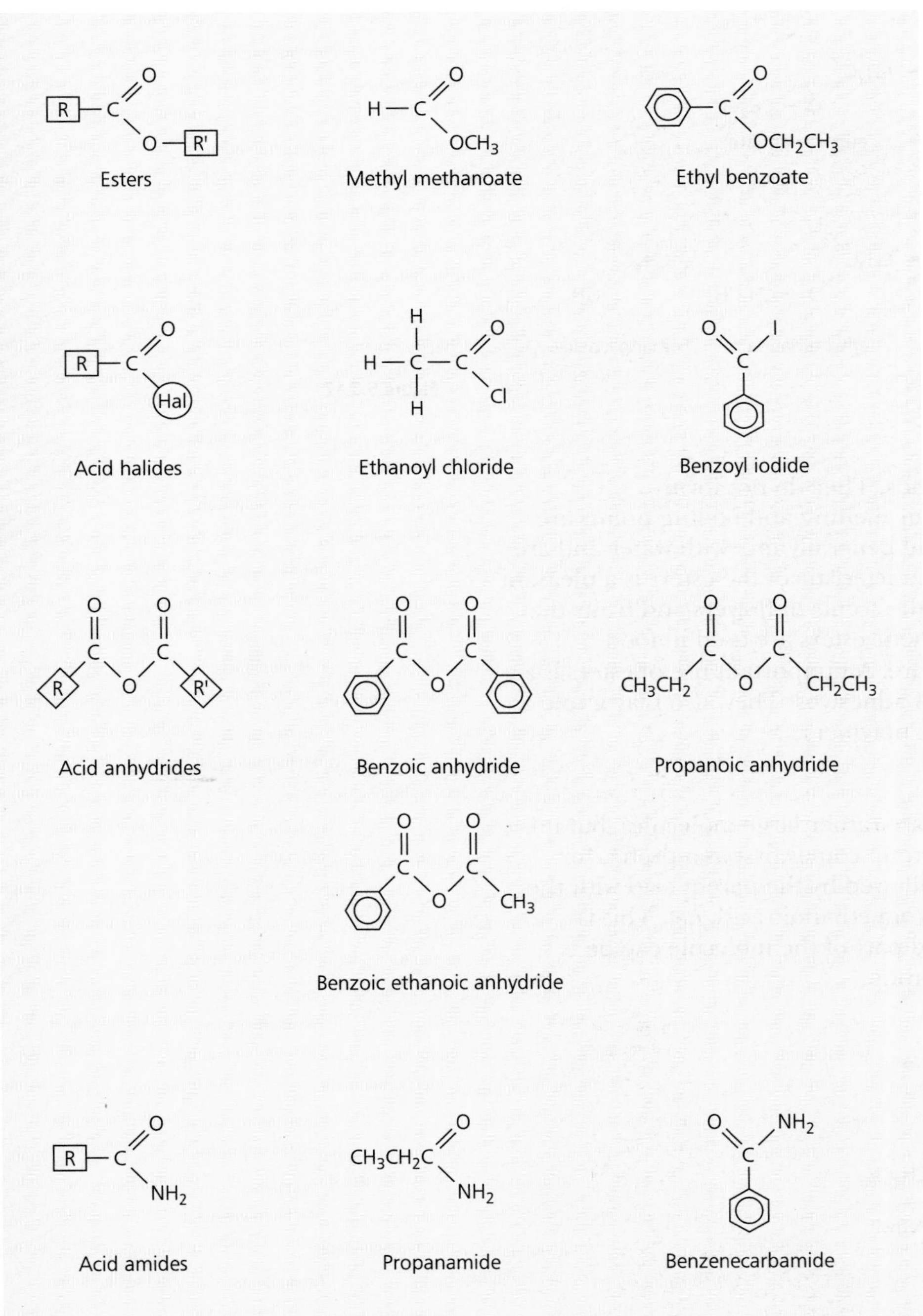

Figure 5.3.16 Some derivatives of the carboxylic acids

Esters

Preparation of esters

The formation of esters from carboxylic acids and alcohols has already been described. However, this is an equilibrium reaction that does not under normal conditions go to completion. The most common laboratory preparation of esters is by the reactions of the acid chlorides or acid anhydrides with alcohols, as shown in figure 5.3.17. These reactions take place readily and go to completion under normal laboratory conditions.

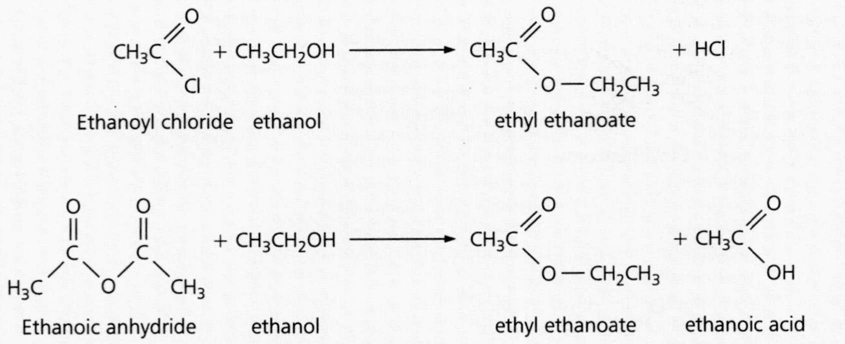

$$CH_3COCl + CH_3CH_2OH \longrightarrow CH_3COO{-}CH_2CH_3 + HCl$$

Ethanoyl chloride ethanol ethyl ethanoate

$$(CH_3CO)_2O + CH_3CH_2OH \longrightarrow CH_3COO{-}CH_2CH_3 + CH_3COOH$$

Ethanoic anhydride ethanol ethyl ethanoate ethanoic acid

Figure 5.3.17

Properties of the esters

The esters are a group of volatile compounds. They do not form intermolecular hydrogen bonds and so their melting and boiling points are relatively low. They are polar molecules and generally mix with water and are soluble in organic solvents. One typical characteristic of the esters is a pleasant smell – esters are responsible for many of the scents of flowers and fruits that we and other animals find attractive. Synthetic esters are used in food flavourings and also in perfume manufacture. An important use of esters is as solvents, particularly in the manufacture of adhesives. They also play a role as plasticisers in the manufacture of synthetic polymers.

Naming the esters

Naming esters can seem daunting as they are rather large molecules, but it is really quite straightforward. The alcohol group comes first as a prefix, for example 'propyl' from propanol. This is followed by the parent acid with the '-oic acid' replaced by '-oate' – ethanoate from ethanoic acid, etc. This is illustrated in figure 5.3.18. The parent acid part of the molecule can be identified as the part carrying the -COO group.

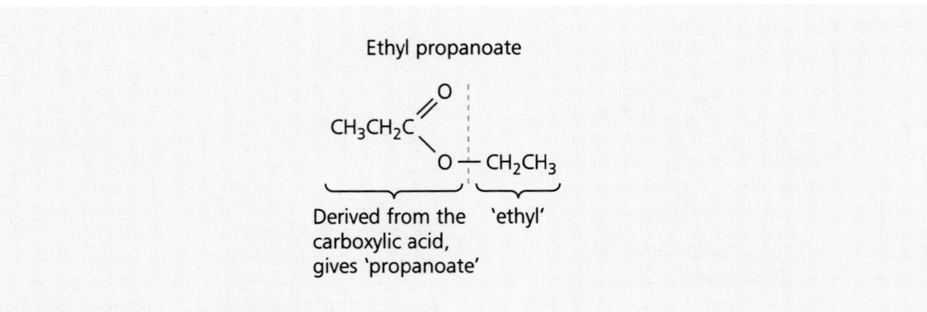

Figure 5.3.18

The bridging oxygen

A distinctive feature of ester molecules is the oxygen atom that forms a bridge between the two parts. When the carboxylic acid and alcohol join, this oxygen bridge is formed and a molecule of water is lost. Where does the oxygen atom come from – is it from the alcohol or the carboxylic acid?

The answer to this question was first discovered by two Americans, John Roberts and Harold Urey. Urey won a Nobel prize for his work in discovering isotopes of hydrogen, and it was by using oxygen isotopes that the riddle of the esters was solved. Using a technique similar to that outlined in figure 5.3.19, it was shown that the oxygen in the ester comes from the alcohol rather than the carboxylic acid, or the derivative used in the preparation of the ester.

'Labelled' ethanol is prepared, using the oxygen isotope ^{18}O in the hydroxyl group.

$$CH_3CH_2{}^{18}OH$$

The labelled ethanol is reacted with propanoic acid with a concentrated sulphuric acid catalyst.

$$CH_3CH_2{}^{18}OH + CH_3CH_2C(=O)OH$$

There are two possible sets of products: (a) if the bridging oxygen comes from the alcohol, (b) if the bridging oxygen comes from the carboxylic acid.

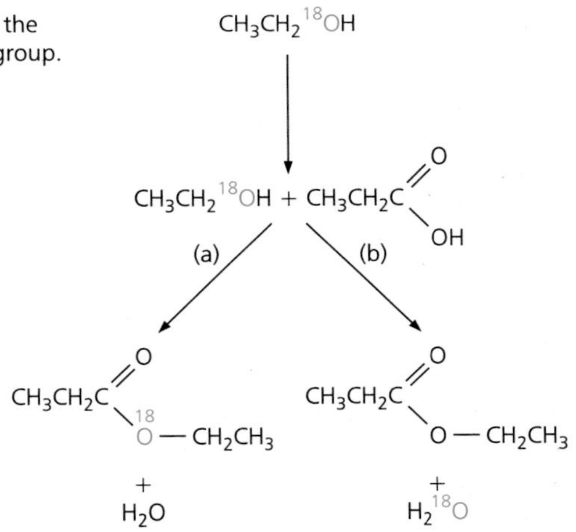

(a) $$CH_3CH_2C(=O){-}{}^{18}O{-}CH_2CH_3 + H_2O$$

(b) $$CH_3CH_2C(=O){-}O{-}CH_2CH_3 + H_2{}^{18}O$$

The ester is separated and the relative molecular mass measured using a mass spectrometer. The mass of the ester is shown to be 104, which is what would be expected if the 'heavy' oxygen was incorporated in the molecule. This demonstrates that the bridging oxygen has come from the alcohol rather than the carboxylic acid.

Figure 5.3.19

Importance of the esters

The esters are very important both in the living world and in industry. The lipids that have a vital role in the structure and metabolism of cells, as well as being used for storage and energy in the body, are esters of long-chain organic acids. We shall look at these in more detail in section 5.5. As we saw earlier, esters are also important in the production of soaps and other detergents. Their other major role is in the production of synthetic polymers – polyester is an obvious example. The polyesters are made by polymerising diols and dicarboxylic acids to produce long-chain molecules that are unreactive and that have great tensile strength. They are used widely as fibres in the production of synthetic fabrics, and because of their great tensile strength, they are also used as a bonding resin in glass fibre products to give flexible strength to the material. Polyester withstands higher temperatures than most plastics, and so it

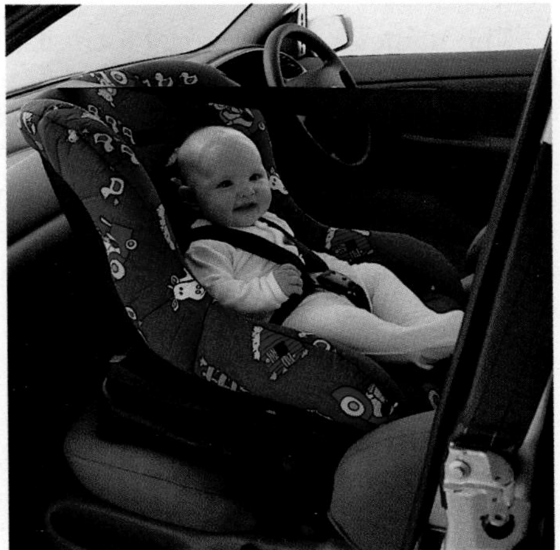

Figure 5.3.20 With their high tensile strength, polyesters like Terylene find many uses. These include the webbing used for seat belts and car safety harnesses.

is used in food roasting bags. Polyester fabrics such as Terylene have also had a big impact on the production of clothing. Many polyesters can be broken down by hydrolysis and so they are usually biodegradable, an important feature in an increasingly environmentally conscious world.

Acid anhydrides

General properties of the acid anhydrides

The acid anhydrides are simply two carboxylic acid molecules that have been joined together with the elimination of a molecule of water. They may be formed from two molecules of the same carboxylic acid (simple anhydrides), or molecules from two different carboxylic acids (mixed anhydrides). The acid anhydrides are named very simply by putting the name of the acid or acids involved in the molecule in front of the word 'anhydride'. Examples include ethanoic anhydride, propanoic anhydride and benzoic ethanoic anhydride, shown in figure 5.3.21.

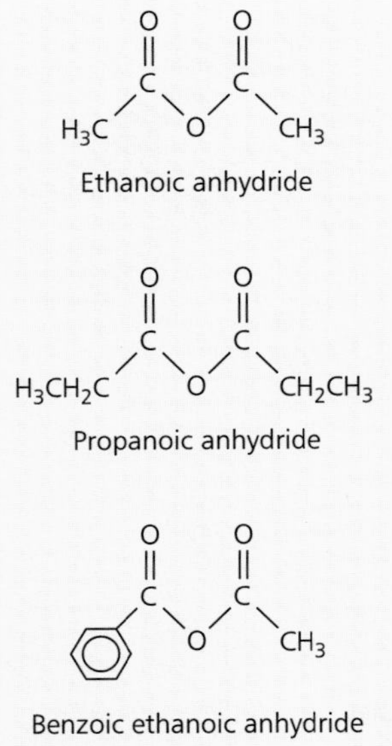

Ethanoic anhydride

Propanoic anhydride

Benzoic ethanoic anhydride

Figure 5.3.21

The names and structural formulae of the acid anhydrides imply that they can be made in a very simple dehydration reaction. This is not usually the case – the preparation of acid anhydrides is generally quite complex. However, they are extremely useful compounds. With several carbon–oxygen bonds closely associated, there is considerable shifting of the electrons from carbon atoms to oxygen atoms and therefore several areas of partial charge within the acid anhydride molecules. Figure 5.3.22 shows this effect clearly, using the molecule of ethanoic anhydride.

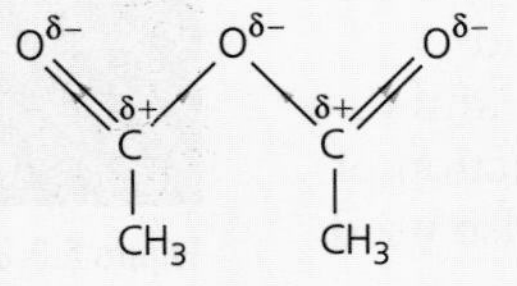

Figure 5.3.22

Reactions of acid anhydrides

The presence of these dipoles leads to very δ+ carbons within the acid anhydride molecules. This means that they are readily attacked by nucleophiles, and nucleophilic substitution reactions take place. The most commonly used nucleophiles are water, ammonia, alcohols and amines. If we show the general nucleophile as HNu, then the general reaction of an acid anhydride with a nucleophile can be represented as shown in figure 5.3.23(a).

Some of the most important reactions of the acid anhydrides are summarised below.

- Acid anhydrides react with water to form the corresponding carboxylic acids, figure 5.3.23(b).
- Acid anhydrides react with alcohols to give an ester and a carboxylic acid, figure 5.3.23(c).
- Acid anhydrides react with ammonia to give a primary amide and a carboxylic acid, figure 5.3.23(d).

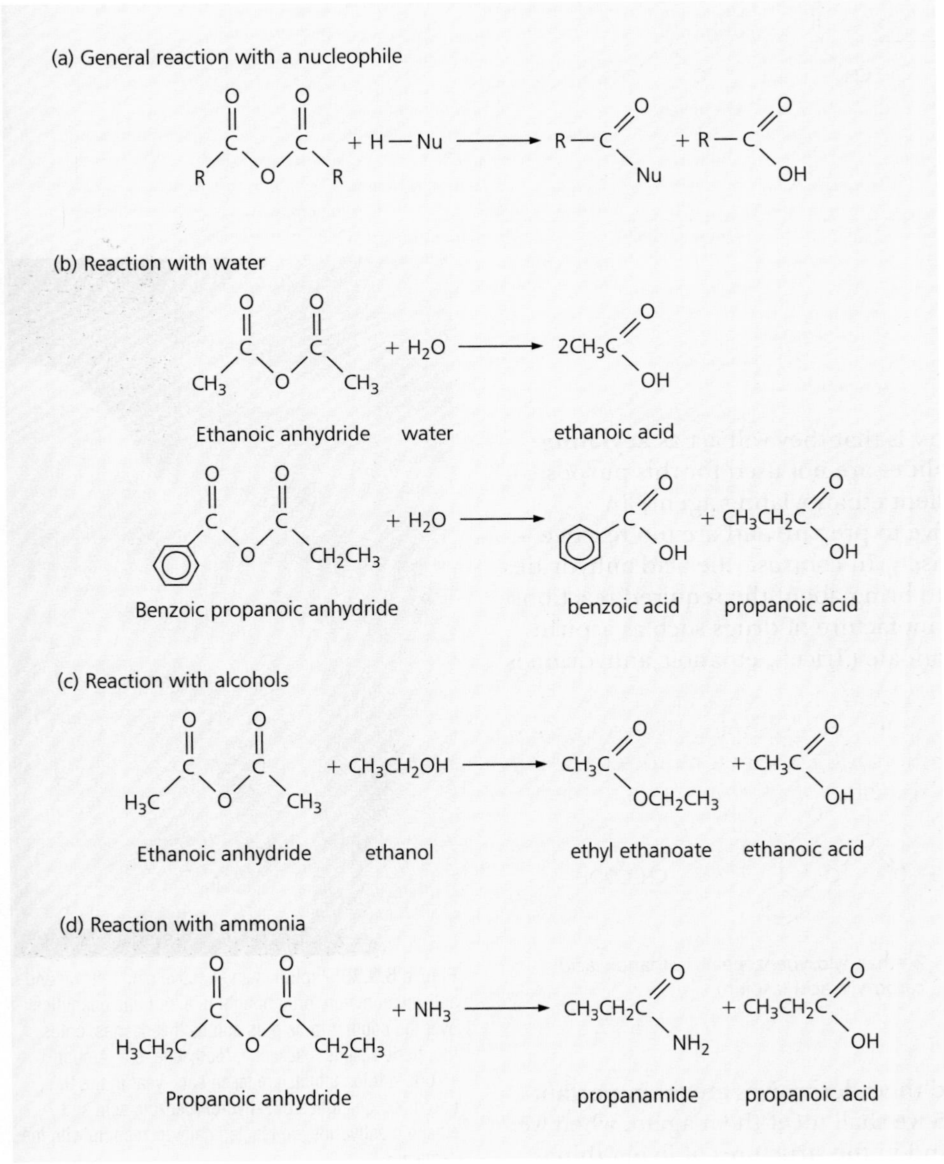

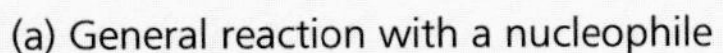

Figure 5.3.23 Reactions of acid anhydrides

5

The reaction of an acid anhydride with a nucleophile

The reaction of an acid anhydride with a nucleophile involves addition followed by elimination, as is the case with many reactions involving the carbonyl group.

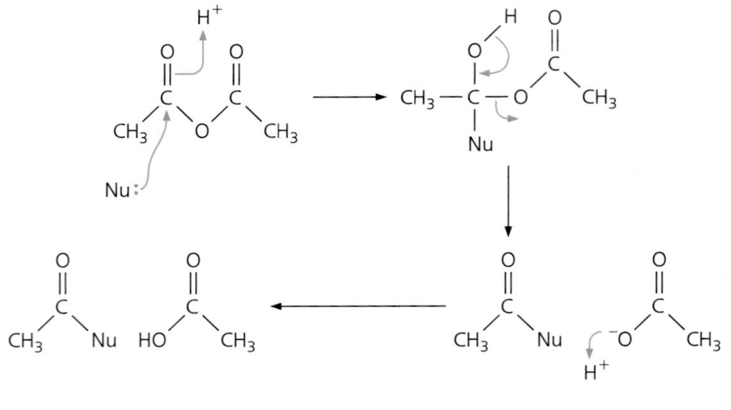

Figure 5.3.24

Importance of the acid anhydrides

The value of the acid anhydrides in industry is that they will act as **acylating agents**. It might seem odd that the acid halides are not used for this purpose – surely ethanoyl chloride would be an excellent ethanoylating agent? In practice, the acid halides are more expensive to prepare and are too reactive – it is difficult to control their reactions precisely. In contrast, the acid anhydrides are cheaper to make and reactive enough to bring about the required reactions without being too vigorous. Thus in the manufacture of drugs such as aspirin and artificial fibres such as cellulose triethanoate (Tricel), ethanoic anhydride is the ethanoylating agent used.

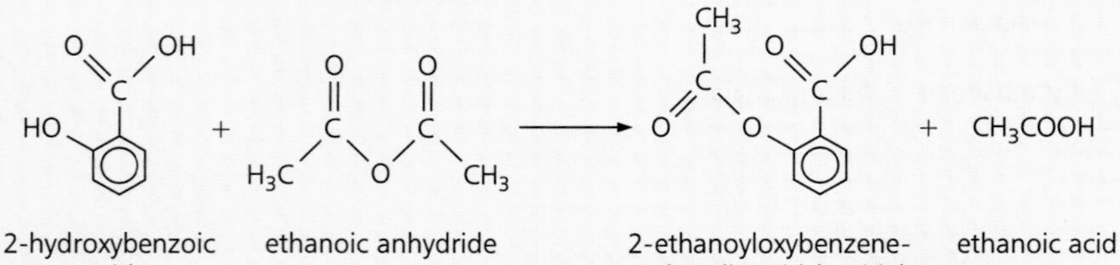

2-hydroxybenzoic acid + ethanoic anhydride → 2-ethanoyloxybenzene-carboxylic acid (aspirin) + CH_3COOH (ethanoic acid)

As we have seen, the carboxylic acids and their derivatives are an important group of organic compounds. In section 5.5 we shall meet them again, when we look at their role in the large molecules found in the structures of living things.

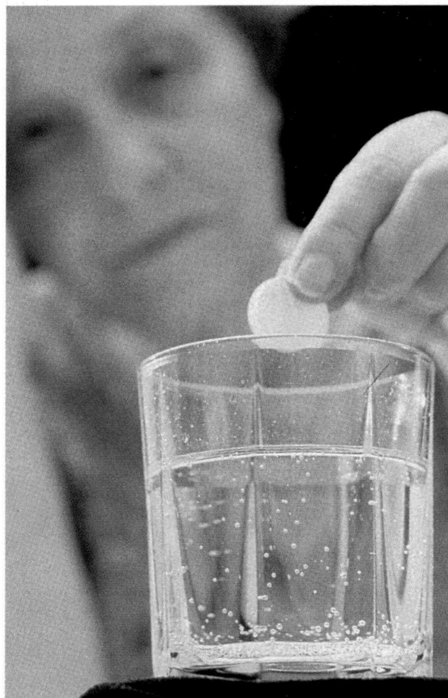

Figure 5.3.25 Aspirin is an effective pain reliever and anti-inflammatory drug bought in enormous quantities over the counter to help us combat headaches, colds, 'flu, period pains, backache, toothache, etc. Around 15 000 million tablets are taken each year in the UK. Enormous quantities of 2-hydroxybenzoic acid and ethanoic anhydride are reacted daily to keep up with the demand.

SUMMARY

- Carboxylic acids contain the **carboxyl** functional group, which is made up of the carbonyl and hydroxyl groups.
- Carboxylic acids can be prepared by oxidising primary alcohols or aldehydes.
- The carboxyl group is strongly acidic compared with water, due to the stabilisation of the carboxylate ion by the delocalisation of negative charge.
- The physical properties of carboxylic acids are strongly influenced by hydrogen bonding.
- Some typical reactions of ethanoic acid are summarised in table 5.3.2.

Ethanoic acid +	Product
NaOH	Sodium ethanoate, $CH_3COO^-Na^+$
$PCl_5/SOCl_2$	Ethanoyl chloride, CH_3COOCl
$LiAlH_4$	Ethanol, CH_3CH_2OH
Alcohol, RCH_2OH, + acid catalyst	Ester, CH_3COOCH_2R

Table 5.3.2

- Soap is a mixture of salts of medium-chain-length carboxylic acids, formed by reacting the acids with a base in a reaction called **saponification**.
- The derivatives of carboxylic acids include the **esters**, RCOOR', the **acid anhydrides**, RCOOCOR', the **acid chlorides**, RCOCl, and the **amides**, $RCONH_2$.
- The esters, anhydrides, acid chlorides and amides show similar reactions with nucleophiles, the order of reactivity being acid chloride > anhydride > carboxylic acid > ester > amide.
- The formation of esters from carboxylic acids and alcohols is an equilibrium which does not go to completion under normal conditions.
- Esters are generally volatile compounds, many of which have a pleasant smell.
- Anhydrides are important as acylating agents.

QUESTIONS

1 Consider the compounds A to E in figure 5.3.26.

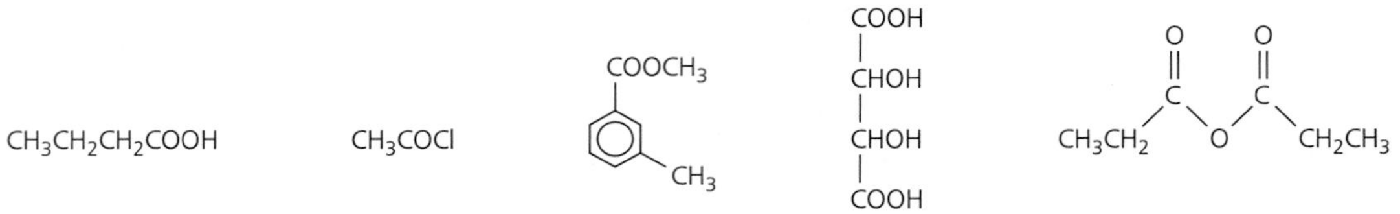

A: $CH_3CH_2CH_2COOH$

B: CH_3COCl

C: benzene ring with $COOCH_3$ and CH_3 substituents

D: $COOH$–$CHOH$–$CHOH$–$COOH$

E: CH_3CH_2–C(=O)–O–C(=O)–CH_2CH_3

Figure 5.3.26

- **a** Which are carboxylic acids?
- **b** Which is an ester?
- **c** Which is an acid anhydride?
- **d** Which is an acid chloride?
- **e** Which is a dibasic acid (an acid that contains two acidic protons)?
- **f** Which would react with an alcohol to form an ester?
- **g** Which would react vigorously with water?

2 Predict the products of the following reactions:

- **a** $CH_3CH_2CH_2COOH + LiAlH_4 \rightarrow$
- **b** $CH_3COOH + NaOH \rightarrow$
- **c** $CH_3CH_2COOH + SOCl_2 \rightarrow$
- **d** (benzene ring with COCl) $+ NH_3 \longrightarrow$

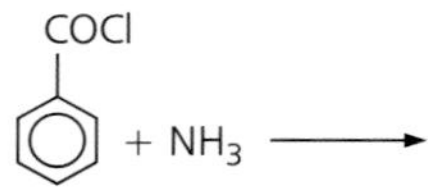

3 How would you carry out the following syntheses? Show the reactants and conditions you would need for each step in the conversion process. Two or more steps may be needed in each case.

- **a** CH_3CH_2OH from CH_3COCl
- **b** CH_3COBr from $CH_2{=}CH_2$
- **c** $(CH_3CH_2CH_2CO)_2O$ from $CH_3CH_2CH_2COOH$

4 Suggest explanations for the observations that follow.

- **a** Ethanoic acid and its chlorine-substituted derivatives have the following dissociation constants:

CH_3COOH	$CH_2ClCOOH$
1.7×10^{-5} mol dm^{-3}	1.4×10^{-3} mol dm^{-3}
$CHCl_2COOH$	CCl_3COOH
5.1×10^{-2} mol dm^{-3}	2.2×10^{-1} mol dm^{-3}

- **b** Ethanedioic acid, HOOCCOOH, has two dissociation constants, $K_{a,1}$ and $K_{a,2}$, with the values 5.9×10^{-2} mol dm^{-3} and 5.2×10^{-5} mol dm^{-3} respectively.

5 Propan-1-ol reacts with ethanoyl chloride and with ethanoic anhydride, forming propyl ethanoate as one of the products in each case.

- **a** Write the structural formula for each of the named compounds.
- **b** Write balanced equations for each of these reactions.
- **c** Suggest possible mechanisms for these reactions.

Developing Key Skills

Carry out a survey of the chemicals in your bathroom – shampoos, shower gels, hand creams, etc. – and identify as many organic chemicals as you can. Share the data with the other members of your group and produce a bar chart to show which chemicals are most widely used in personal care products. Investigate and report on the role of the three most commonly used compounds.

[Key Skills opportunities: A, IT]

FOCUS A RISKY BUSINESS

The role of the chemist in the last century or more has been one of an enabler. As a result of advances in all branches of chemistry, but most particularly in organic chemistry, the whole way society operates in the developed world has changed. So much of what we take for granted – the fabric of our everyday life, the materials we wear and use, our vehicles, the plentiful supply of food, the paints, the adhesives, the medicines which cure our illnesses and relieve our symptoms – are the result of innovations and developments which started off in the chemistry lab. The whole world economy is affected by what is going on in the petrochemical and pharmaceutical industries, to mention but two of the big chemical players. We are very happy to reap the benefits of all this chemical bounty, but it is important occasionally to pause for thought.

Figure 1 The twenty-first century in the developed world – an edifice dependent at least in part on the ingenuity and skill of chemists.

The risks

One aspect of chemical development and chemical industries that cannot be ignored is the risk factor. To produce the chemicals we need involves scales of production which can only be met by large factories (chemical plants). Many chemical processes in such factories produce highly toxic by-products that have to be disposed of safely. For these two reasons, the risk of something going wrong cannot be ignored, and chemical engineers who design such plant are careful to ensure that it can withstand the pressures and temperatures needed to carry out the specific sequence of reactions it is designed for. Large safety margins are built into the specifications so that all foreseeable problems are catered for, but in spite of this, accidents can still happen.

The disposal of toxic waste is subject to stringent regulations. Some toxins have to be incinerated at very high temperatures, while others must be buried in sealed containers in landfill sites. The public perception of toxic waste disposal is one of great suspicion, and there is often difficulty in obtaining new sites. We want the chemicals but not the waste – yet the two are integrally linked.

However carefully risks are calculated and safety precautions taken, occasionally accidents happen. Sometimes human error is the problem, in other cases the engineering of the plant is at fault and in some cases the actual cause of the accident is never discovered. It is easy to gloss over the risks involved in chemical manufacturing, for accidents are very rare indeed. But by looking at one such incident in more detail it is apparent why we should never take the safety of our chemical industries too much for granted.

The disaster at Seveso

Seveso is a town in Italy not far from Milan. On 10 July 1976, the ICMESA chemical plant was producing the antiseptic 2,4,6-trichlorophenol as usual. In the alkaline hydrolysis reactor, trichlorobenzene and sodium hydroxide were mixed in organic solution and heated to 135–150 °C at atmospheric pressure, in order to form sodium trichlorophenate. This was then fed on into the next stage of the process. No one is completely sure why, but on that day the temperature in part of the reactor rose to over 200 °C. Two slow exothermic reactions took place, which eventually set up an explosive runaway exothermic reaction yielding gaseous products and the highly toxic compound

Figure 2 People trying to leave Seveso after the tragic accident.

5

FOCUS A RISKY BUSINESS (continued)

2,3,7,8-tetrachlorodibenzo-*p*-dioxin (often simply called 'dioxin'). The enormous increase in pressure inside the reactor caused a safety device to open and a cloud of vapour was released over Seveso and the surrounding countryside.

Initially the full extent of the tragedy was unrecognised, as it was assumed that the cloud released from the plant had contained mainly 2,4,6-trichlorophenol. It was only over the following days, as table 1 shows, that the true extent of the pollution and the dioxin contained in the cloud was realised. Dioxin is toxic in parts per billion, and so stable that it does not break down or react once it is present in the environment. Although no one died as a result of the explosion, the area around the factory site is unlikely to be lived in again for the foreseeable future – the soil and the plants growing on it are permanently tainted with dioxin.

Cl, O, Cl, Cl, O, Cl

2,3,7,8-tetrachlorodibenzo-*p*-dioxin ('dioxin')

Figure 3 The dioxin molecule – and its effects. The dioxin-contaminated soil of Seveso will remain behind barriers like this for the foreseeable future.

Time	Events
Day 1	At 12.40 p.m. an exothermic explosion of unknown cause at the ICMESA TCP plant caused a safety device to rupture and release the reactor contents in a plume of liquid and vapour.
Day 2	Samples of leaves went yellow after contact with toxic cloud. ICMESA representatives issued a statement saying that a cloud of TCP had been released (with no mention of dioxins).
Day 3	People were advised not to eat vegetables due to contamination.
Day 5	First skin eruptions occurred in some children exposed to the cloud. Close to the plant some cats, rabbits and chickens died.
Day 6	More eye and skin irritations. A bigger area was sealed off and the use of local fruit and vegetables stopped.
Day 7	13 children were admitted to hospital with toxic dermatitis.
Day 8	Local authorities ordered all plants, vegetables, crops and dead animals from the areas to be burnt.
Day 10	It was admitted that dioxins had been found in samples of soil analysed.
Day 13	80 children were relocated from the most contaminated areas.
Day 16	Evacuation of the population began from the worst contaminated areas.
Day 24	Orders were issued that children under one year old and women up to three months pregnant should be kept away from the area.
Day 32	Therapeutic abortions were offered to all women exposed to the dioxin within the first three months of their pregnancy.

Table 1 Diary of a disaster – the events in the small Italian town of Seveso serve as a grim warning of the risks we run to obtain the chemicals we want. This disaster has had a lasting effect on the safety of chemical plants throughout Europe.

Chemists – for good or ill?

All of us, as members of a global society, are part of the process that determines how a chemical will be used. We give planning permission for a chemical plant to be built. We decide that the financial gain to a company and the benefit to people through the product outweigh the risks of an accident occurring. Indirectly each of us determines how far a new compound is to be tested for safety before it is allowed on the market, by the amount we are prepared to pay for the product.

We all owe an enormous debt of gratitude to chemists for many aspects of our comfortable lives in the developed world. Before apportioning blame for environmental damage we should look at the benefits we have demanded and the price we have been prepared to pay – the responsibility lies with us all.

5.4 Organic nitrogen compounds

Probably the best known and most widely occurring organic nitrogen compounds are the amino acids, fundamental building blocks of the proteins that make up a considerable part of every living organism. The amino acids and proteins will be considered in section 5.5, along with several other groups of large biochemical molecules. Here we shall concentrate on four other organic families containing nitrogen atoms – the **nitro-compounds**, the **nitriles**, the **amides** and most importantly the **amines**.

Figure 5.4.1 The short-chain aliphatic nitro-compounds are colourless liquids, often used as solvents. Nitroalkanes are used as fuel for the high-speed dragsters raced with enthusiasm in America.

The nitro-compounds

The **nitro-compounds** are defined by the presence of an NO_2 group in the molecule. The only nitro-compounds of any major industrial importance are the aromatic nitro-compounds, and we shall concentrate on these here, although aliphatic nitro-compounds do exist, as illustrated in figure 5.4.1.

Nitropropane

The polar nitro group

Physical properties of the nitro-compounds

The simplest aromatic nitro-compound is nitrobenzene. This is an oily yellow liquid, but most of the other aromatic nitro-compounds are yellow solids. They are not very soluble in water due to the aromatic influence on the molecule. The NO_2 group is very polar, and this polarity leads to strong dipole–dipole intermolecular forces. As a result, all the nitro-compounds have higher melting and boiling points than might otherwise be expected. We have already met the mono- and disubstituted nitrobenzenes in section 4.4.

Reactions of the nitro-compounds

The most important general reaction of the nitro-compounds is their reduction in acidic solution to form the amines. For example, the laboratory preparation of phenylamine involves the reduction of nitrobenzene, using tin in concentrated hydrochloric acid as the reducing agent, as shown in figure 5.4.2. The same reaction is also used in the industrial preparation of phenylamine, only the reducing agent is the cheaper iron in concentrated hydrochloric acid. Phenylamine plays an important role in the production of a variety of other organic chemicals, including the antibacterial sulphonamide drugs and many dyes, as we shall see on pages 519 and 521.

NO_2 (benzene ring) $\xrightarrow{\text{Sn/conc HCl}}$ NH_2 (benzene ring) $+ 2H_2O$

Figure 5.4.2

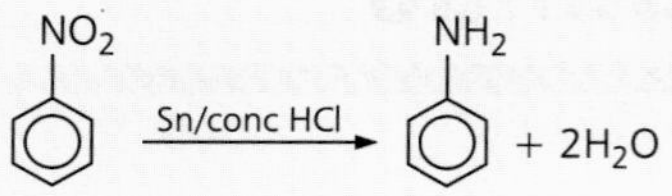

The nitriles (cyano-compounds)

The **nitriles** contain the cyano group joined to a hydrocarbon skeleton. The compounds are named using the name of the hydrocarbon with the same number of carbon atoms (including the carbon in the -CN group) but with the suffix 'nitrile', as in ethanenitrile or butanenitrile. In the aromatic nitriles, the carbon in the -CN group is not part of the benzene ring, so the monosubstituted benzene is called benzenecarbonitrile or benzonitrile. Some nitriles are shown in figure 5.4.3.

Hydrogen cyanide, HCN, and other inorganic compounds containing the -CN group are extremely poisonous as they interfere with the biochemistry of energy production in the cells of the body, but the organic nitriles are only mildly toxic.

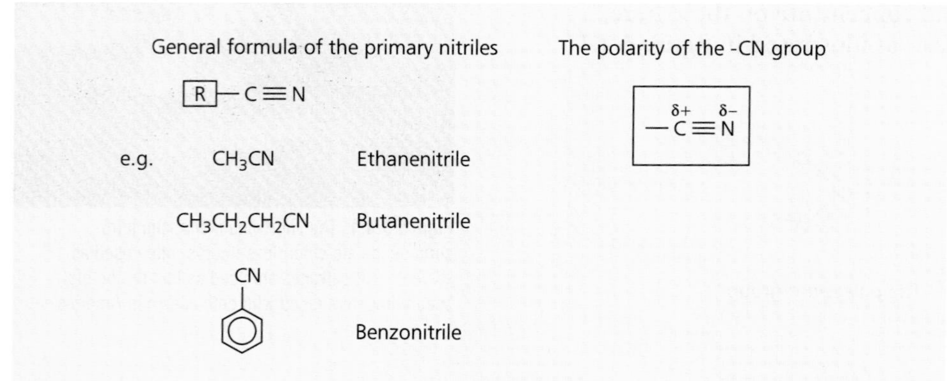

Figure 5.4.3

Physical properties of the nitriles

Figure 5.4.3 shows that the -CN group is polarised. This means that the shorter-chain aliphatic nitriles are fairly soluble in water, and also explains why all the nitriles have higher melting and boiling points than those of corresponding hydrocarbons.

Reactions of the nitriles

The main value of the nitriles to the organic chemist is their role in organic syntheses. As we have seen in earlier sections, the production of a nitrile is often a step on the way to producing a compound with a longer carbon chain. For example, if nitriles are boiled with aqueous acid, an amide is formed which hydrolyses to give a carboxylic acid, whilst if they are reduced by refluxing with lithium tetrahydridoaluminate(III) dissolved in ethoxyethane then an amine is the result:

$$\underset{\text{Ethanenitrile}}{CH_3CN} + 2H_2O \rightarrow \underset{\text{ethanoic acid}}{CH_3COOH} + NH_3$$

$$\underset{\text{Ethanenitrile}}{CH_3CN} \xrightarrow[\text{in ethoxyethane}]{LiAlH_4} \underset{\text{ethylamine}}{CH_3CH_2NH_2}$$

The amides

The **acid amides** are derivatives of the carboxylic acids in which the -OH part of the -COOH group is substituted by $-NH_2$. They have the general formula shown in figure 5.4.4.

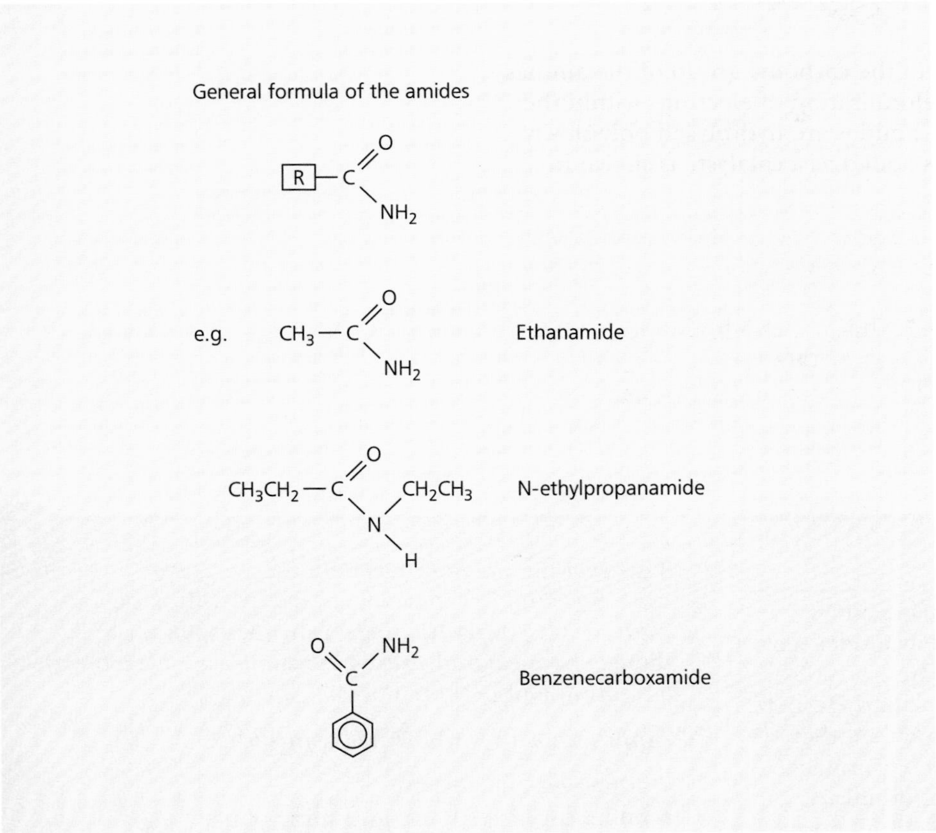

General formula of the amides

e.g. Ethanamide

N-ethylpropanamide

Benzenecarboxamide

Figure 5.4.4

Physical properties of the amides

Figure 5.4.5 shows that the amides form extensive intermolecular hydrogen bonds as a result of the polar nature of the N—H bond, and these hydrogen bonds result in relatively high melting and boiling points. Methanamide is a liquid, but all the other amides are white crystalline solids with well defined melting points, which can be used if necessary to identify them. This ability to form extensive hydrogen bonds also means that the amides are very soluble in water.

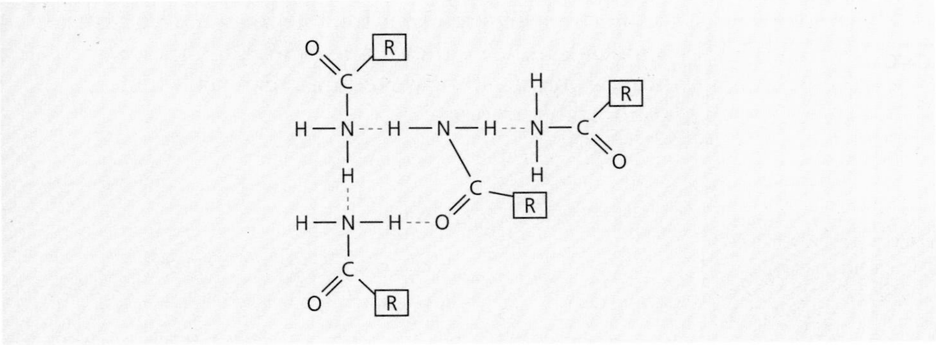

Figure 5.4.5

Reactions of the amides

The main reactions of the amides can be summarised as follows:

- Amides are reduced to amines with lithium tetrahydridoaluminate(III) dissolved in ethoxyethane as the reducing agent, for example:

$$CH_3CH_2CONH_2 \xrightarrow[\text{in ethoxyethane}]{LiAlH_4} CH_3CH_2CH_2NH_2 + H_2O$$

- Nucleophilic attack on the carbon atom of the carbonyl group of the amides occurs relatively slowly because of the delocalisation of electrons within the molecules (see box below). For example, amides are hydrolysed only slowly by boiling water unless an acid or base is added as a catalyst, as shown in figure 5.4.6.

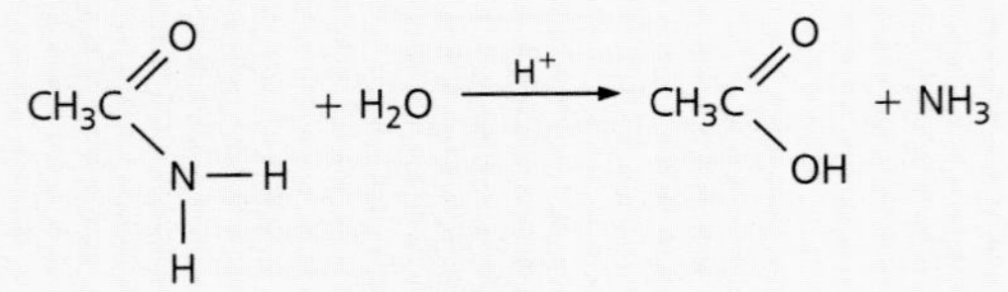

$$CH_3CONH_2 + H_2O \xrightarrow{H^+} CH_3COOH + NH_3$$

Figure 5.4.6

Kevlar®

Since the development of nylon by the Du Pont company in the 1940s (see page 429) polyamides have played a very important role in society. In the 1960s two research scientists (Stephanie Kwolek and Herbert Blades) also working at Du Pont developed another new polyamide which has had an enormous impact. Kevlar® consists of long polymer chains of linked poly-paraphenylene terephthalamide – often known as para-aramid fibres.

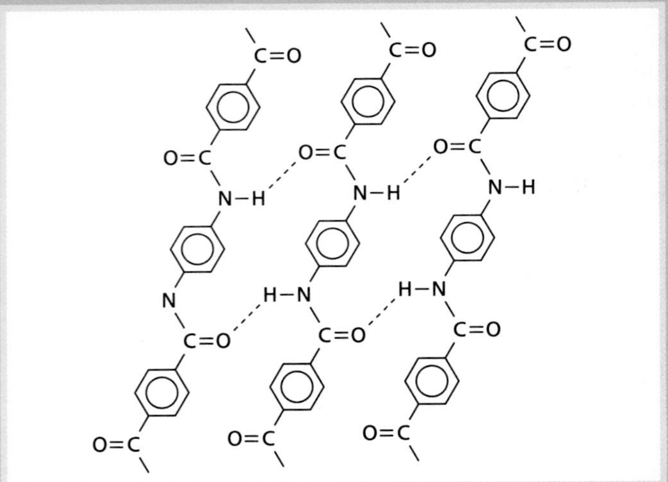

Figure 5.4.7 The regular orientation of the amide units in Kevlar® along with the very strong interchain hydrogen bonding gives the material its unique combination of properties.

This exciting polymer has high temperature resistance and low thermal conductivity. It has a very high tensile strength , doesn't shrink in the wash, is flame, chemical and cutting resistant – a material with a multitude of potential uses.

Applications of Kevlar® so far include:

- adhesives and sealants
- bullet proof vests and anti-mine boots
- transmission belts and hoses for vehicles
- structural parts of aircraft, space shuttles and boats
- communication and data transmission cables
- as a replacement for asbestos in brake pads and clutch linings
- protective clothing (bike helmets, gloves, etc)
- tyres for aircraft, cars and trucks
- ropes and cables.

This polymer really demonstrates the value of chemical ingenuity. The new material has so many properties which make it suitable for tough industrial applications, yet either alone or in combination with other fibres it can be woven into material which is comfortable to wear.

Delocalisation in the amide functional group

The lone pair of electrons on the nitrogen atom of the amide group forms part of a delocalised system of π bonds. This reduces the electron density on the nitrogen atom, making it less basic and less nucleophilic.

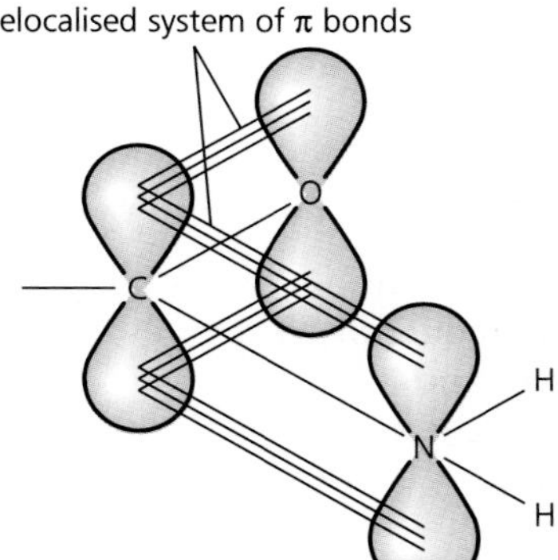

Figure 5.4.8

The fact that the nitrogen atom is not such a strong nucleophile as we might expect provides a good route for the synthesis of amines, as we shall see shortly.

- Amides are dehydrated on heating with phosphorus(V) oxide to give nitriles:

$$\underset{\text{Ethanamide}}{CH_3CONH_2} \xrightarrow[\text{reflux}]{P_4O_{10},} \underset{\text{ethanenitrile}}{CH_3CN} + H_2O$$

- Amides react with freshly made nitrous acid to give the parent acid:

$$\underset{\text{Ethanamide}}{CH_3CONH_2} + HNO_2 \xrightarrow{\text{NaNO}_2\text{/HCl mixed at 5 °C}} \underset{\text{ethanoic acid}}{CH_3COOH} + N_2 + H_2O$$

Nitrous acid is a very unstable compound which exists only in aqueous solution, and decomposes at room temperature. This means that any reactions with nitrous acid have to be carried out under very carefully controlled conditions, with the nitrous acid prepared *in situ* by mixing ice-cold solutions of sodium nitrite and dilute hydrochloric acid. All the reactions then need to be carried out at 5 °C or below, and so the reaction vessel is kept in an ice/water mixture.

- With bromine and concentrated potassium hydroxide, amides undergo a degradation reaction to form an amine with one less carbon atom in the chain. This reaction is known as the **Hofmann degradation**, and it is very valuable in synthetic processes because it enables the carbon chain to be shortened in a controlled way. An amide is mixed with bromine and then aqueous sodium or potassium hydroxide is added. The mixture is shaken gently until it turns yellow, when potassium hydroxide pellets are added and the whole mixture warmed to around 70 °C:

$$\underset{\text{Propanamide}}{CH_3CH_2CONH_2} + Br_2 + 4NaOH \rightarrow \underset{\text{ethylamine}}{CH_3CH_2NH_2} + Na_2CO_3 + 2NaBr + 2H_2O$$

The amines

The **amines** can be regarded as compounds based on ammonia, NH_3, with one, two or three of the hydrogen atoms substituted by an alkyl or aryl group, R.

- **Primary amines** have only one hydrogen atom substituted, figure 5.4.9(a).
- **Secondary amines** have two of the hydrogen atoms substituted with alkyl or aryl groups which may be the same or different, figure 5.4.9(b).
- **Tertiary amines** have all three of the original hydrogen atoms of the ammonia substituted by alkyl or aryl groups. Again, these may be all the same, figure 5.4.9(c), or they may be different.

The amines do not occur in the free state very widely in the living world except in putrefying and decaying flesh. They are formed by the action of bacteria on amino acids when proteins break down. Amines also account for some of the body odours detected in humans who do not wash very frequently, as bacteria on the skin digest amino acids from the sweat of the groin and armpits. Di- and trimethylamine occur in rotting fish and account for the very distinctive smell.

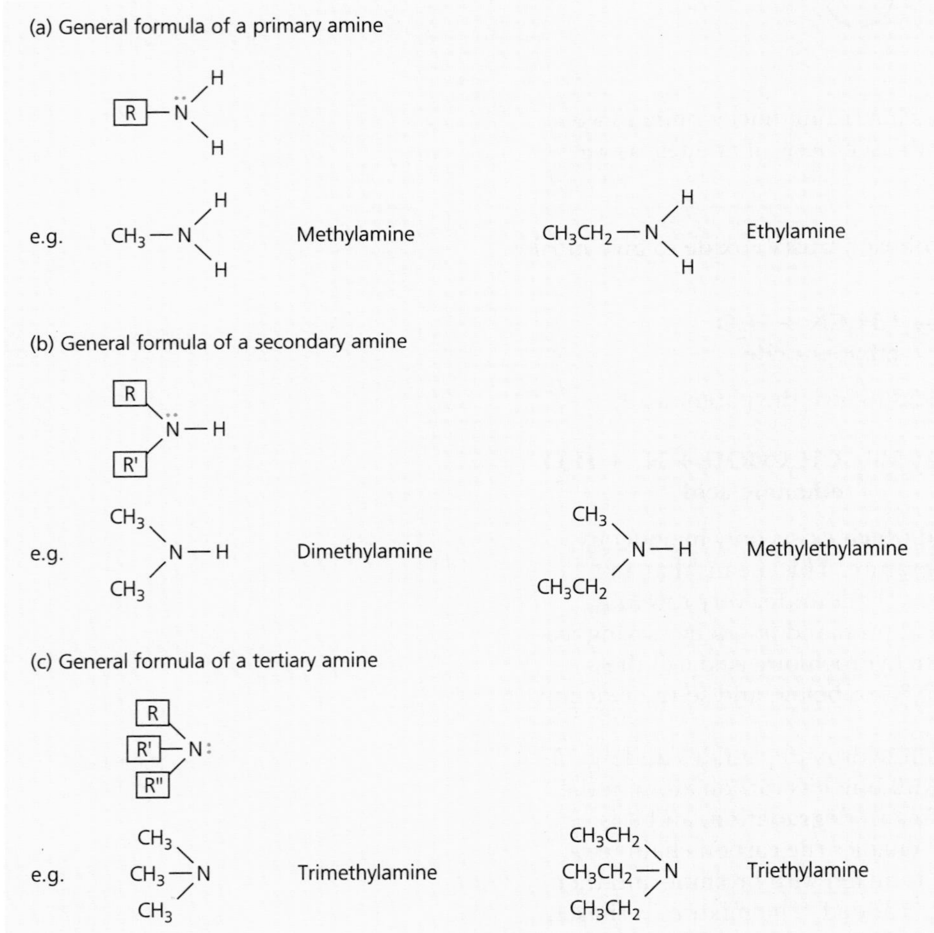

Figure 5.4.9

Physical properties of the amines

The lower aliphatic amines are all volatile liquids and the primary, secondary and tertiary amines each form an homologous series in their own right. Hydrogen bonds are important in determining the physical properties such as

Amine	Formula	Melting point/°C	Boiling point/°C	Density/ g cm^{-3}
Methylamine	$H-\overset{H}{\underset{H}{C}}-NH_2$	–93	–6	0.66
Ethylamine	$H-\overset{H}{\underset{H}{C}}-\overset{H}{\underset{H}{C}}-NH_2$	–81	17	0.68
Propylamine	$H-\overset{H}{\underset{H}{C}}-\overset{H}{\underset{H}{C}}-\overset{H}{\underset{H}{C}}-NH_2$	–83	48	0.72
Butylamine	$H-\overset{H}{\underset{H}{C}}-\overset{H}{\underset{H}{C}}-\overset{H}{\underset{H}{C}}-\overset{H}{\underset{H}{C}}-NH_2$	–49	78	0.74
Phenylamine	$C_6H_5-NH_2$	–6	184	1.02
Dimethylamine	$(CH_3)_2N-H$	–93	8	0.66
Diethylamine	$(CH_3CH_2)_2N-H$	–48	56	0.71
Trimethylamine	$(CH_3)_3N$	–117	3	0.63
Triethylamine	$(CH_3CH_2)_3N$	–115	89	0.73

Table 5.4.1 The physical properties of the primary, secondary and tertiary amines are a clear reflection of the intermolecular forces in the compounds.

melting and boiling points of the primary and secondary amines. In the tertiary amines, there is no N—H bond and so the only intermolecular forces are weak van der Waals attractions. This difference is reflected in the melting and boiling points shown in table 5.4.1 and figure 5.4.10. Like ammonia, the short-chain amines are very soluble in water, as hydrogen bonding occurs between the amine molecules and the water molecules. Phenylamine is only sparingly soluble in water because the aromatic nature of the benzene ring outweighs the tendency of the NH_2 group to form hydrogen bonds. However, it is very soluble in organic solvents. As it is also highly toxic, it must be handled with extreme caution, because it will dissolve readily in the fats in the skin and thus gain entrance to the body.

Preparation of the amines

Amines can be prepared in a variety of ways, some of which are more suitable for one type of amine than for another. We have met several of these methods

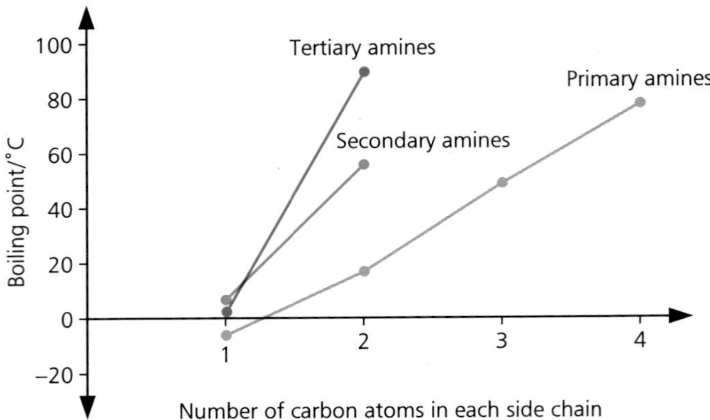

Figure 5.4.10

in earlier sections, and so here we shall simply summarise several of the most common methods of amine preparation.

Laboratory preparations of the amines

In the laboratory, **primary amines** may be prepared by:

- the reduction of **nitriles** by reducing agents such as lithium tetrahydridoaluminate(III) in ethoxyethane under reflux conditions:

$$\underset{\textbf{Ethanenitrile}}{\mathbf{CH_3CN}} \xrightarrow[\textbf{reflux}]{\textbf{LiAlH}_4\textbf{/ethoxyethane,}} \underset{\textbf{ethylamine}}{\mathbf{CH_3CH_2NH_2}}$$

- the reduction of **amides** by reducing agents such as lithium tetrahydridoaluminate(III) in ethoxyethane under reflux conditions:

$$\underset{\textbf{Propanamide}}{\mathbf{CH_3CH_2CONH_2}} \xrightarrow[\textbf{reflux}]{\textbf{LiAlH}_4\textbf{/ethoxyethane,}} \underset{\textbf{propylamine}}{\mathbf{CH_3CH_2CH_2NH_2}} + \mathbf{H_2O}$$

- the **Hofmann degradation** of amides:

$$\underset{\textbf{Propanamide}}{\mathbf{CH_3CH_2CONH_2}} + \mathbf{Br_2} + \mathbf{4NaOH} \xrightarrow{\textbf{reflux}} \underset{\textbf{ethylamine}}{\mathbf{CH_3CH_2NH_2}} + \mathbf{Na_2CO_3} + \mathbf{2NaBr} + \mathbf{2H_2O}$$

Aromatic amines may be prepared by the reduction of **nitro-compounds** with tin and concentrated hydrochloric acid, figure 5.4.11.

Figure 5.4.11

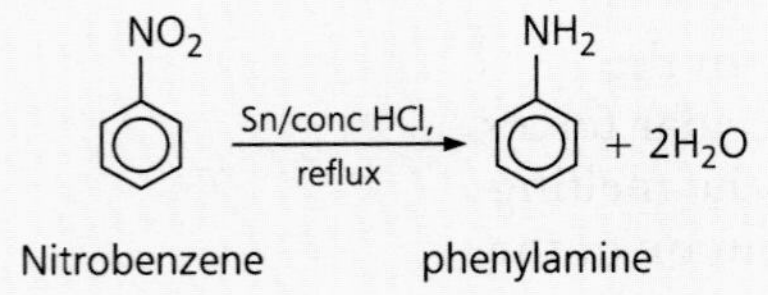

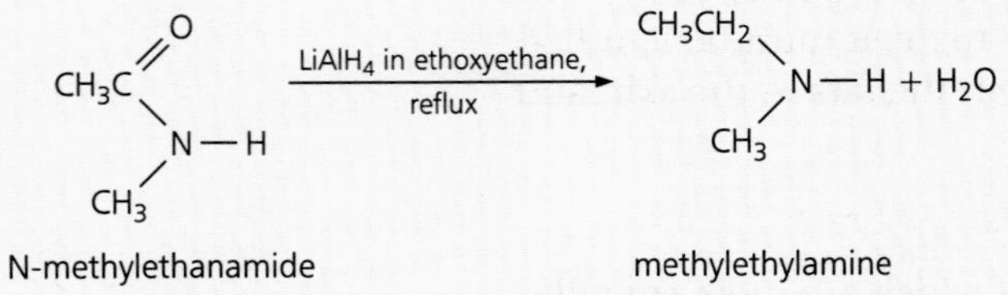

$$\underset{\text{Nitrobenzene}}{C_6H_5NO_2} \xrightarrow[\text{reflux}]{\text{Sn/conc HCl,}} \underset{\text{phenylamine}}{C_6H_5NH_2} + 2H_2O$$

$$\underset{\text{N-methylethanamide}}{CH_3C(=O)N(H)CH_3} \xrightarrow[\text{reflux}]{\text{LiAlH}_4\text{ in ethoxyethane,}} \underset{\text{methylethylamine}}{(CH_3CH_2)(CH_3)N{-}H} + H_2O$$

Figure 5.4.12

Secondary and **tertiary amines** are prepared in the laboratory by the reduction of N-substituted amides – that is, amides in which one of the hydrogen atoms of the -NH_2 group has been substituted by an alkyl or aryl group. As in the previous preparations, a strong reducing agent such as lithium tetrahydridoaluminate(III) in ethoxyethane is needed to bring about the reduction, as shown in figure 5.4.12.

Industrial preparations of the amines

Industrial preparations of the amines are not always the same as the laboratory preparations because of the difference in scale and economic factors. The most common industrial methods of preparing amines are:

- The nucleophilic substitution reactions of **halogenoalkanes** with **ammonia**, as shown in figure 5.4.13. This process gives a mixture of primary, secondary and tertiary amines which must then be separated by distillation. This method is not usually used in the laboratory but is important industrially because although the products need separation, the starting materials are plentiful and relatively cheap.

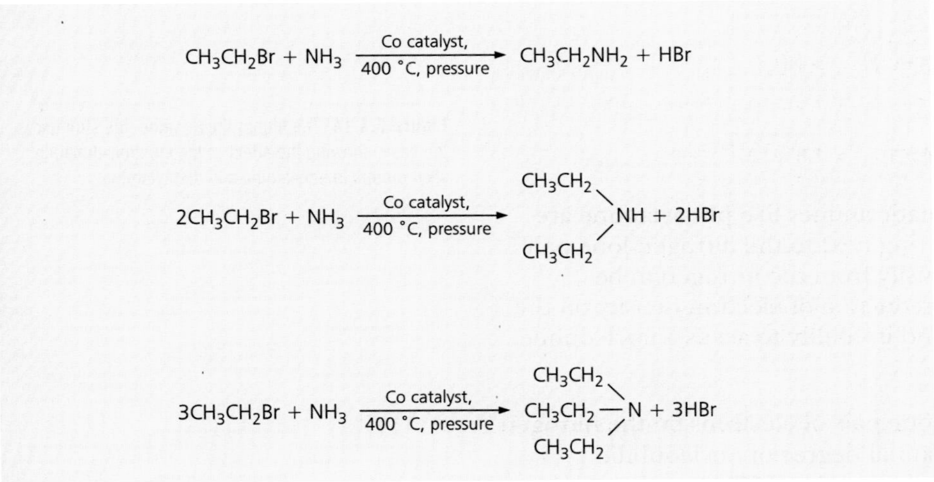

$$CH_3CH_2Br + NH_3 \xrightarrow[400\ ^\circ C,\ pressure]{Co\ catalyst,} CH_3CH_2NH_2 + HBr$$

$$2CH_3CH_2Br + NH_3 \xrightarrow[400\ ^\circ C,\ pressure]{Co\ catalyst,} (CH_3CH_2)_2NH + 2HBr$$

$$3CH_3CH_2Br + NH_3 \xrightarrow[400\ ^\circ C,\ pressure]{Co\ catalyst,} (CH_3CH_2)_3N + 3HBr$$

Figure 5.4.13

- The reduction of **nitro-compounds** as in the laboratory reaction in figure 5.4.11, but with iron rather than tin as the reducing agent as it is cheaper.

Amines are produced industrially for two main purposes. They are used in the manufacture of dyes and related products, and also in the production of nylon.

Reactions of the amines

The chemical reactions of the amines, particularly those with the shorter carbon chains, are dominated by the nitrogen atom with its lone pair of electrons. This means that these amines act as bases, and also as nucleophiles. Reactions do occur in the hydrocarbon skeleton, but in the smaller amines these are of considerably less importance.

The amines as bases

The lone pair of electrons on the nitrogen atom of the amines means that they behave as proton acceptors – in other words, as bases. When amines such as ethylamine dissolve in water, an alkaline solution is formed:

$$\mathbf{CH_3CH_2NH_2 + H_2O \rightleftharpoons CH_3CH_2NH_3^+ + OH^-}$$

Amines react with acids to form a salt and water. An example of this is the reaction of ethylamine with hydrochloric acid. Ethylamine is a short-chain

5

aliphatic amine with the characteristic unpleasant fishy smell of these compounds. When a dilute mineral acid such as hydrochloric acid is added to an aqueous solution of ethylamine, there is no visible change in the colourless liquid, but the temperature rises and the distinctive fishy amine smell disappears. The ethylammonium ion has been formed, which is not volatile and therefore has no smell.

$$\mathbf{CH_3CH_2NH_2 + HCl \rightarrow CH_3CH_2NH_3{}^+Cl^-}$$

If a strong mineral base such as sodium hydroxide is added to a solution of ethylammonium chloride in water, the amine is displaced from its salt. This reaction is detected by the return of the amine's odour.

$$\mathbf{CH_3CH_2NH_3{}^+Cl^- + NaOH \rightarrow CH_3CH_2NH_2 + NaCl + H_2O}$$

Ethylamine and diethylamine are stronger bases than ammonia. This is due to the electron-donating effect of the alkyl group which we saw in section 4.3. Diethylamine, with two electron-donating groups, is a stronger base than ethylamine, with only one, as illustrated in figure 5.4.14.

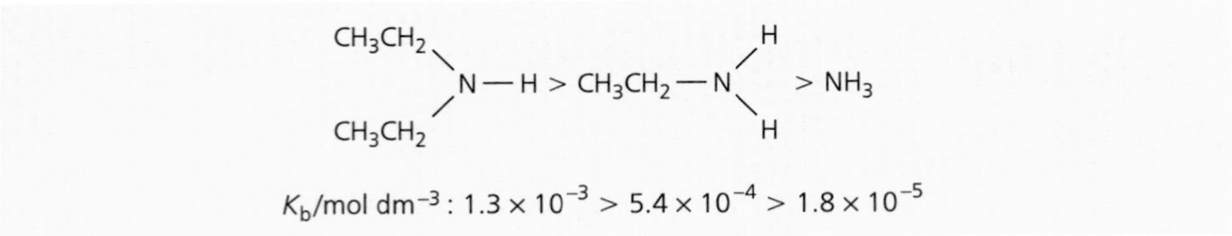

$$(CH_3CH_2)_2N{-}H > CH_3CH_2{-}NH_2 > NH_3$$

K_b/mol dm^{-3} : $1.3 \times 10^{-3} > 5.4 \times 10^{-4} > 1.8 \times 10^{-5}$

Figure 5.4.14 The higher the K_b value, the stronger is the base, showing the effect of the electron-donating alkyl groups in ethylamine and diethylamine.

In contrast to the aliphatic amines, aromatic amines like phenylamine are weaker bases than ammonia. The benzene ring next to the nitrogen lone pair of electrons causes some of the electron density from the nitrogen to be delocalised around the benzene ring. The reduction of electron density on the nitrogen atom reduces its basic character and its ability to act as a nucleophile.

The amines as nucleophiles

Besides allowing them to act as bases, the lone pair of electrons on the nitrogen atom of the amines also gives them a substantial degree of nucleophilic character. There are several reactions of the amines in which they clearly act as nucleophiles, and these are summarised below.

Amines react vigorously with **acid chlorides** to produce amides. As well as being useful as a synthesis in its own right, this reaction enables identification of unknown amines. Amides have very sharply defined melting points. By reacting an amine with an acid chloride and carefully measuring the melting point of the resulting amide, the amide can be identified, and so the amine from which it was formed can be deduced. An example of this type of reaction is shown in figure 5.4.15. Similar reactions take place between amines and **acid anhydrides**, although these occur less readily than with the acid chlorides.

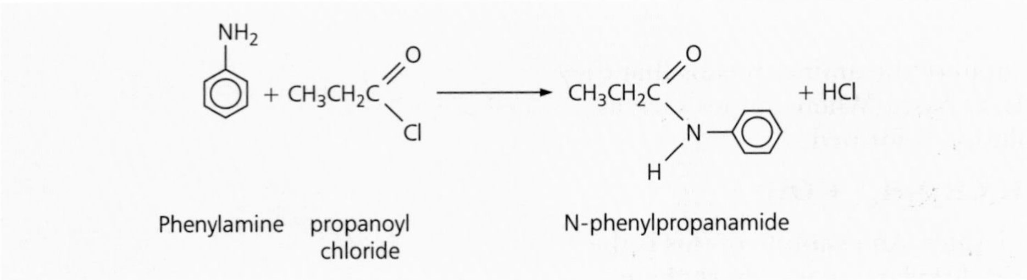

$$C_6H_5NH_2 + CH_3CH_2COCl \longrightarrow CH_3CH_2CONHC_6H_5 + HCl$$

Phenylamine propanoyl chloride N-phenylpropanamide

Figure 5.4.15

Phenylamine and the fight against bacteria

The lone pair of electrons on the nitrogen atom of phenylamine is partly delocalised around the benzene ring. This gives a high electron density around the ring, which in turn means that phenylamine is easily oxidised and readily undergoes electrophilic substitution of the benzene ring. What has this to do with the fight against disease-causing bacteria?

4-aminobenzenesulphonamide is a derivative of phenylamine which can be synthesised as a result of this ease of substitution. This compound and other similar ones derived from it form a class of drugs known as the **sulphonamides**. These were some of the earliest drugs to be really effective in destroying bacteria. They have played a major part in reducing the numbers of deaths from pneumonia, tuberculosis and infection after childbirth, and they are still in common use today.

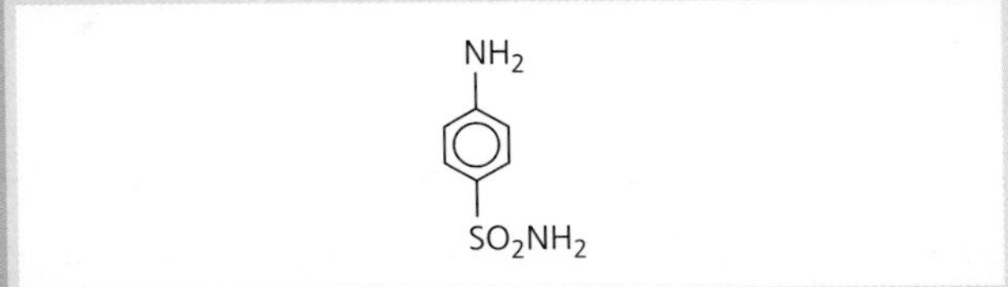

Figure 5.4.16 4-aminobenzenesulphonamide has given rise to a whole family of antibacterial drugs.

The amines react with **nitrous acid**, HNO_2, to form a variety of useful products. The nitrous acid is prepared *in situ* as described on page 513.

The reactions of the amines with this unstable acid provide us with some very interesting and valuable chemistry. When an amine reacts with nitrous acid, **diazonium compounds** may be formed. These are the starting point for a wide range of organic syntheses, and we shall come to their structure and reactions shortly. In the reactions between primary aliphatic amines such as methylamine and nitrous acid at 5 °C, the solution remains colourless and a diazonium salt is formed. This rapidly decomposes to form bubbles of nitrogen and an alcohol. Primary aromatic amines such as phenylamine also react with nitrous acid at 5 °C to form a colourless solution containing the diazonium salt. This is reasonably stable, and so nitrogen gas is not seen unless the solution is gently heated. The aromatic diazonium salts are of great value, because of the role of diazonium salts in organic synthesis.

The diazonium salts

The diazonium salts contain the highly unstable diazonium group shown in figure 5.4.17(a). As we have seen, aliphatic diazonium salts are so unstable that they decompose almost immediately they are formed, even at very low temperatures. Aromatic diazonium salts are somewhat more stable, at least up to temperatures of 5–10 °C. This additional stability is a result of the delocalised electrons of the benzene ring. An example of an aromatic diazonium salt is the benzenediazonium chloride shown in figure 5.4.17(b), a

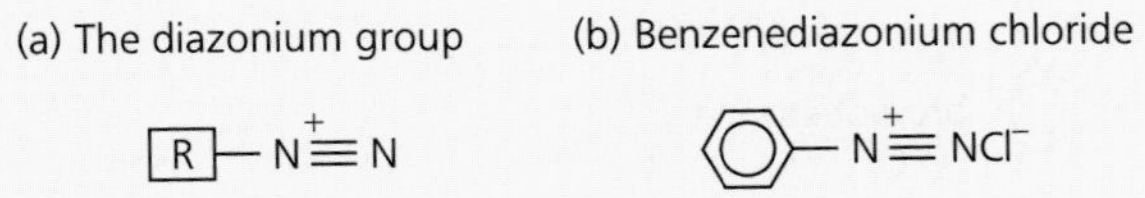

(a) The diazonium group

$R-\overset{+}{N}\equiv N$

(b) Benzenediazonium chloride

$C_6H_5-\overset{+}{N}\equiv N\,Cl^-$

Figure 5.4.17

The stability of the benzenediazonium ion

The presence of the delocalised electrons of the benzene ring stabilises the benzenediazonium ion by donation of electron density, as shown in figure 5.4.18.

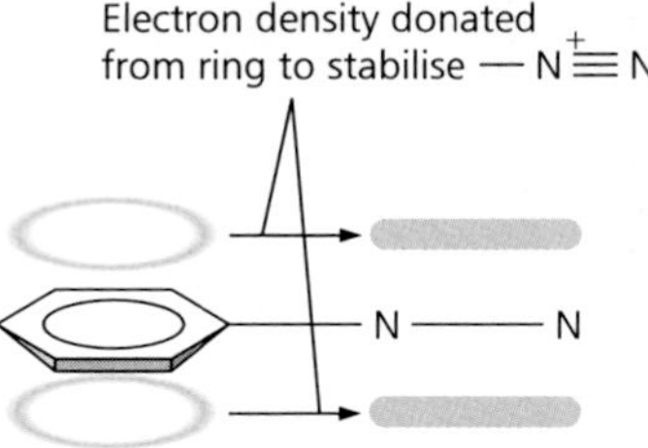

Figure 5.4.18

compound that decomposes well below room temperature and is explosive in the solid form.

Benzenediazonium chloride is prepared by adding cold sodium nitrite solution to phenylamine in concentrated hydrochloric acid at between 0 and 5 °C. Because the solid is explosive, it is always used in aqueous solution. The great reactivity of diazonium salts such as this can be used in a wide variety of situations to aid organic synthesis. The use of diazonium ions is based on two aspects of their chemistry – attack of the diazonium benzene ring by nucleophiles, and the ability of diazonium salts to couple with other molecules. When nucleophiles react with the benzene ring, nitrogen gas is evolved. When diazonium ions couple with other molecules, this does not happen.

Diazonium reactions with nucleophiles

The diazonium group makes the benzene ring open to attack by a nucleophile. Normally, benzene reacts with electrophiles, so a reaction that enables nucleophilic species such as halide ions to attack the benzene ring is extremely useful. Some of the more common syntheses in which benzenediazonium chloride or sulphate is used are summarised in figure 5.4.19.

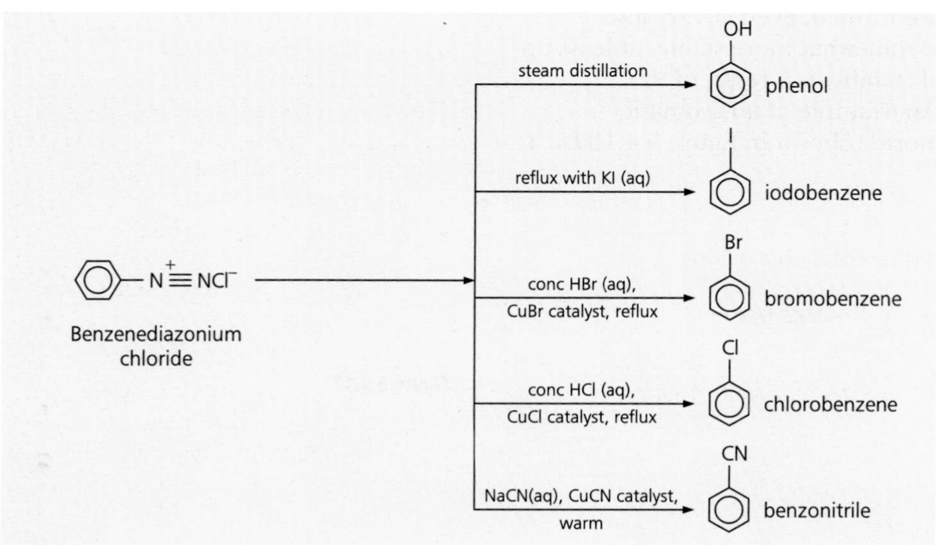

Figure 5.4.19 In all these reactions, nitrogen gas is evolved.

Coupling reactions of diazonium salts

The benzenediazonium ion carries a positive charge. It reacts readily in cold alkaline solutions both with aromatic amines and with phenols to give brightly coloured **azo-compounds**, as shown in figure 5.4.20. Nitrogen gas is *not* evolved in these reactions. These azo-compounds are complex molecules involving a minimum of two aromatic rings joined by N=N couplings. Unlike the diazonium compounds from which they arise, azo-compounds are extremely stable and unreactive. The bright colours are the result of the extensive delocalised electron systems that spread across the entire molecule through the N=N couplings. The main commercial use of these compounds is as very stable and colourfast dyes – some examples are given in figure 5.4.21.

$$C_6H_5-\overset{+}{N}{\equiv}N\,Cl^- + C_6H_5-OH \longrightarrow C_6H_5-N{=}N-C_6H_4-OH + HCl$$

Benzenediazonium chloride | phenol | (4-hydroxyphenyl)azobenzene

Figure 5.4.20

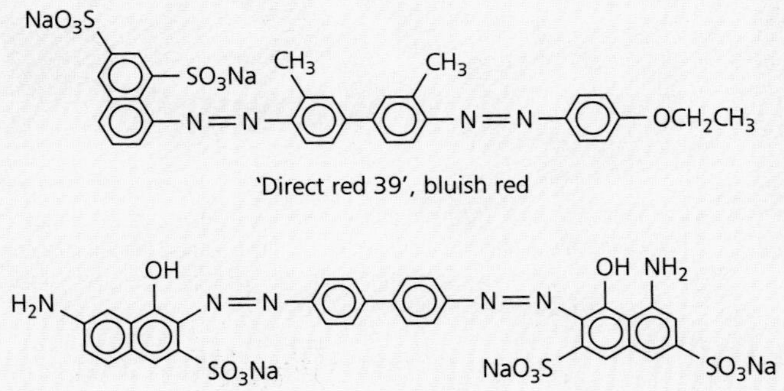

NaO_3S SO_3Na CH_3 CH_3 N=N N=N OCH_2CH_3

'Direct red 39', bluish red

H_2N OH N=N N=N OH NH_2 SO_3Na NaO_3S SO_3Na

'Direct blue 2'

Figure 5.4.21

Organic synthesis – molecules for the future

The synthesis of organic chemicals is big business. Throughout the world, the demand for commodities such as plastics, drugs, dyes, cosmetics and pesticides grows all the time. The majority of the compounds needed to fulfil this demand are organic, and organic chemists have become adept at designing molecules to make compounds with properties both desirable now and with potential uses in the future.

The multimillion pound, international industry of organic synthesis had its beginnings in 1856 with the young English student called William Perkin who produced (by accident) the first synthetic dye, subsequently known as Perkin's mauve. It made his fortune, and subsequently many other (mainly German) scientists synthesised dyes to revolutionise the textile industry. In 1875 William Perkin founded another industry by creating the first synthetic ingredient for perfume. From these beginnings in the nineteenth century, industries based on organic synthesis have simply grown and grown.

Throughout sections 4 and 5 of this book, we have looked at the chemistry of the carbon compounds and seen similarities and differences between the organic families of molecules. Most organic families take part in reactions that convert them either directly or indirectly into members of another family. The most important aspects of organic chemistry in

Figure 5.4.22 William Perkin, who founded the synthetic organic chemistry industry with the dye Perkin's mauve

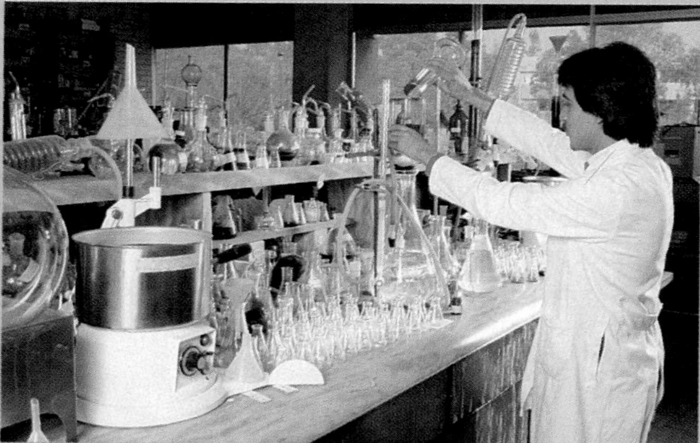

Figure 5.4.23 The main difference between the synthesis of organic chemicals in the laboratory and that in the chemical plant is one of scale. Another is that the chemical industry has to make a profit, whilst the research chemist is seen as an investment.

industry are both the ability to synthesise well known compounds efficiently, and the development of brand new ones.

Some compounds are made in large quantities, others in rather smaller amounts, but industrial preparations are always on a far larger scale than anything carried out in a chemical laboratory. Thus reactions discovered in a research laboratory may have to be modified before they can be used economically by industry. If the only catalyst that will work is extremely expensive, or the conditions needed for a reaction to go to completion are very extreme, then the price of the final product will have to reflect this expense.

Organic synthesis in the laboratory

When carrying out an organic synthesis in the laboratory, there are several key points that need to be addressed.

- The path of the reaction needs to be planned carefully. Most organic syntheses involve several stages, and the various reagents needed to carry out the oxidations, reductions, additions or substitutions must be carefully prepared before the synthesis begins.
- The conditions of the reacting mixtures need to be carefully controlled. This may involve heating or cooling the mixture, varying the pressure or adding different catalysts at different stages of the process.
- The products of the reaction need to be separated from the reaction mixture and then purified. A variety of techniques may be used for this, including distillation, fractional distillation, solvent extraction, recrystallisation and chromatography.
- The purity of the product needs to be checked. In all but the simplest laboratories this is done using infra-red, ultraviolet and mass spectra along with NMR. If these are not available, analysis of the melting or boiling point may be carried out.
- The percentage yield of the product may need to be worked out. When we look at the equation for a

given synthetic reaction, it always appears as though all the reactants are converted into products. In the real laboratory, this is far from the case, and often the percentage yield of pure product:

$$\text{Percentage yield} = \frac{\text{actual yield of pure product}}{\text{theoretical yield}} \times 100\%$$

is remarkably low. The fact that many syntheses are multi-stage reactions, each stage with a less than perfect percentage yield, can mean that the final yield is very small compared with the starting material. Carrying out the synthesis under different conditions may improve the yield.

- Throughout any organic synthesis, *safety* must be an important consideration. Not only are many of the carbon compounds used in common syntheses volatile, flammable and sometimes explosive, but a number are also very poisonous. Therefore any organic synthesis needs to be carried out observing strict safety precautions.

Industrial organic synthesis

Apart from the scale of the operation and attention to the economics of the situation, industrial syntheses are remarkably similar to those carried out on the laboratory bench.

The chemical industries are currently entering a phase when the synthesis of 'designer molecules' has moved from science fiction to science fact. This process involves generating computer images of theoretical molecules which it is thought will perform a particular function, and then setting out to synthesise the molecule. Two areas of organic chemistry where techniques such as this are proving very useful are in the polymer industry, designing new plastics and polymers to fit an enormous range specifications and tasks, and developing new drugs in the pharmaceutical industry.

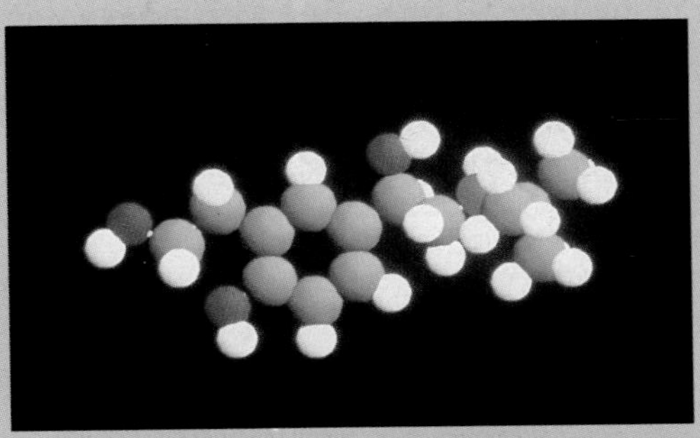

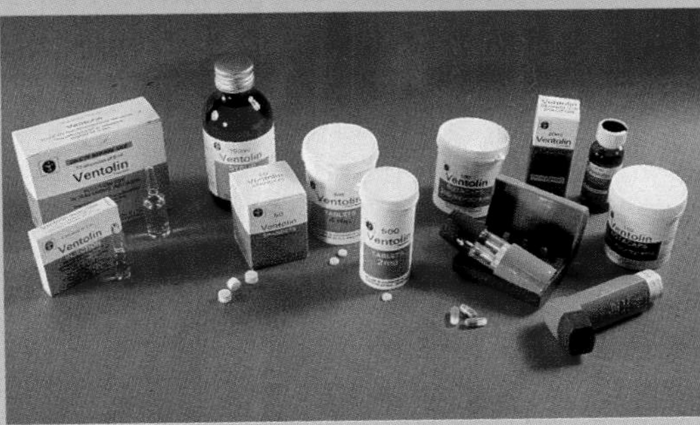

Figure 5.4.24 A computer-generated model of a molecule such as Ventolin, a life-saving molecule that gives relief to asthma sufferers, gives organic chemists a pattern to work to in their synthesis, and has helped to develop a variety of new drugs.

An understanding of the organic families of compounds we have studied and the reactions that they undergo allows organic chemists to build up 'route maps' of organic pathways, and by consulting these, possible synthetic routes for a particular desired compound can be worked out. A process of trial and error is still needed to decide which synthetic route is the best in terms of both cost and yield, but the orderly arrangement of carbon compounds into families and their very predictable reactions allows us to travel ever further along the path towards 'off-the-peg' chemicals for almost every need.

SUMMARY

- Homologous groups containing nitrogen include the nitro-compounds, the nitriles, the amides and the amines.
- The **nitro-compounds** contain an -NO_2 group. The main nitro-compounds that are important industrially are the aromatic nitro-compounds such as nitrobenzene.
- The **nitriles** contain the cyano group, -C≡N. Their main use in organic chemistry is in the controlled lengthening of carbon chains.

5

- Nitriles may be hydrolysed by boiling with aqueous acid, forming a carboxylic acid. Reduction with $LiAlH_4$ produces an amine.
- **Amides** are carboxylic acid derivatives. Some typical reactions of propanamide are given in table 5.4.2.

Propanamide +	Product
$LiAlH_4$	Propylamine, $CH_3CH_2CH_2NH_2$
P_2O_5	Propanenitrile, CH_3CH_2CN, a dehydration reaction
HNO_3	Propanoic acid, CH_3CH_2COOH
Br_2/conc KOH	Ethylamine, $CH_3CH_2NH_2$, the Hofmann degradation

Table 5.4.2

- The **amines** are compounds based on ammonia, NH_3. In **primary amines**, one hydrogen has been replaced by an alkyl or aryl group, RNH_2. In **secondary amines**, two hydrogens have been replaced by alkyl or aryl groups, RR'NH. In **tertiary amines**, three hydrogens have been replaced by alkyl or aryl groups, RR'R"N.
- Amines may be prepared by:

  the reduction of nitriles by $LiAlH_4$

  the reduction of amides by $LiAlH_4$

  the Hofmann degradation of amides

  the reduction of nitro-compounds using Sn/HCl.
- Amines are basic.
- Primary aromatic amines form **diazonium salts** when treated with nitrous acid, HNO_2, at low temperatures. These salts form a useful starting point in organic syntheses. They produce azo-compounds, in which benzene rings are coupled together. The diazonium group also opens the benzene ring up to attack by nucleophilic species such as halide ions.

QUESTIONS

1 Consider the compounds A to E in figure 5.4.25.

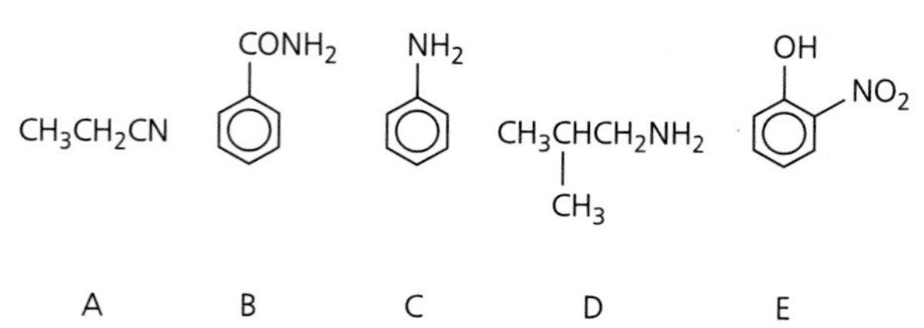

Figure 5.4.25

a Which is a nitrile?
b Which is an aliphatic amine?
c Which is an amide?
d Which is a nitro-compound?
e Which is an aromatic amine?
f Which is the strongest base?
g Which could be hydrolysed to give an acidic compound?
h Which would react with P_2O_5 to form a nitrile?

QUESTIONS

2 Methyl α-cyanoacrylate has the structure shown in figure 5.4.26.

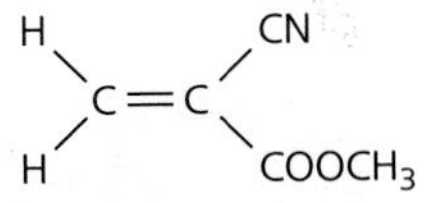

$$\begin{array}{c} H \quad\quad\quad CN \\ \quad C=C \quad \\ H \quad\quad\quad COOCH_3 \end{array}$$

Figure 5.4.26

This substance is unusual, in that it undergoes addition polymerisation which is promoted by water. A tube of 'superglue' is simply a tube of this monomer. When applied to a surface, the monomer polymerises to form a substance that is a very strong adhesive.

- **a** Suggest a polymerisation mechanism for poly(methyl α-cyanoacrylate).
- **b** Draw a representative section of the structure of poly(methyl α-cyanoacrylate).
- **c** Why can superglue stick your fingers together extremely quickly?

3 Suggest ways of carrying out the following changes. State clearly the reagents and conditions needed in each case. More than one step may be needed.

- **a** $CH_3CH_2CONH_2 \rightarrow CH_3CH_2CN$
- **b** $C_6H_5COOH \longrightarrow C_6H_5CH_2COOH$
- **c** $CH_3CH(CH_3)CH_2CH_2NH_2 \rightarrow CH_3CH(CH_3)COOH$
- **d** $C_6H_5NH_2 \longrightarrow C_6H_5COOCH_3$

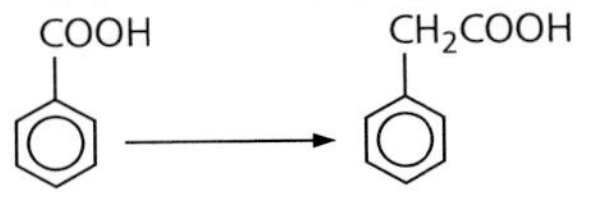

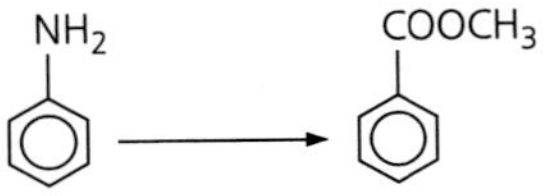

4 The methyl and dimethyl derivatives of ammonia are basic, with strength as bases decreasing in the following order:

$(CH_3)_2NH$ dimethylamine > CH_3NH_2 methylamine > NH_3 ammonia

- **a** What particular feature of these molecules makes them act as bases?
- **b** Draw the displayed formula of each of these molecules, showing clearly the feature identified in **a**.
- **c** Write down equations showing how each of these molecules reacts with water.
- **d** Using your answers to **a** and **b**, explain the factors likely to influence the strength of an amine as a base.
- **e** Explain the trend in basic strength seen above.
- **f** Trimethylamine is a stronger base than ammonia, but a weaker base than methylamine. Why? (*Hint*: Remember that these reactions are taking place in aqueous solution, and that all the substances need to be solvated.)

5 Suggest explanations for the following observations:

- **a** Ethanamide is less basic than ethylamine.
- **b** Phenylamine is more soluble in dilute hydrochloric acid than in water.
- **c** Treatment of propylamine, $CH_3CH_2CH_2NH_2$, with a mixture of concentrated sulphuric and nitric acids leads to the production of a range of products which include propene, propan-1-ol and propan-2-ol.
- **d** 2,4,6-trinitrophenol ('picric acid') is a powerful explosive.

Developing Key Skills

There are plans for the siting of a new chemical factory in the area where you live. Both the reactants and the products of the process are liquids which will arrive at the factory and be transported away by tankers. The reactants are harmless and if spilled will simply degrade. The products are vitally important in the production of many everyday substances including some commonly used pharmaceuticals. However, if the liquid comes into contact with the skin it can cause burning and the vapour it produces can result in breathing difficulties if it is inhaled.

Acting as a representative of the chemical company, choose a site in your local area which you feel would be suitable for such a factory. Plan a presentation which you will give to the planning meeting and at the public meeting which will inevitably follow. You need to justify your choice of site and convince others that your risk analysis is accurate and that the benefits from the site will outweigh the possible problems.

[Key Skills opportunities: C]

FOCUS THE CHEMISTRY OF LIFE

In the millions of years which have passed since life first appeared on Earth an enormous range of living organisms, both animals and plants, has evolved. Although these organisms display great variety, when we examine their chemistry we find much that is common to all of them. The basic biochemistry of a living cell is based around carbohydrates, proteins and fats, but there are other molecules of enormous biochemical importance too. Perhaps the most important of all is DNA.

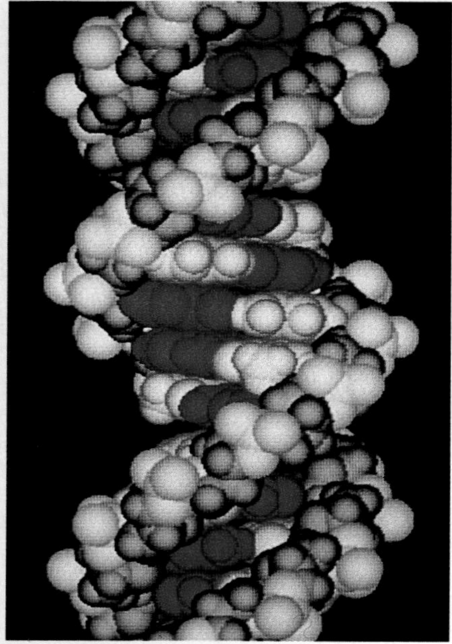

Figure 1 The DNA double helix holds the blueprint of life itself.

DNA – the molecule of life

All living organisms, from bacteria to elephants, can reproduce themselves. This amazing feat is made possible by one molecule – **deoxyribonucleic acid** or DNA. This huge molecule is a polymer made up of repeating units of 5-carbon sugars, nitrogen-containing bases and phosphate groups. Two strands of DNA are held together by hydrogen bonds to form a double helix, and the arrangement of the repeating units within those strands carries the instructions for the making of a new cell or a whole new organism. The genetic code holds the instructions for making new proteins – enzymes – which will then make possible all the other reactions needed to synthesise new cell material, to respire and to carry out the biochemistry of life.

Familiar families

As soon as we begin to dip into biochemistry – the chemistry of living organisms – we find members of familiar organic families playing a variety of roles. When our cells break down sugar to release energy in the presence of oxygen they produce carbon dioxide and water, but if we start exercising hard we cannot fully utilise the sugar and instead form **lactic acid** (a carboxylic acid) as a waste product. Plants and many yeasts form **ethanol** in the same circumstances, a reaction which we make use of in many ways.

Most people produce a protein called insulin which allows glucose in the cells of their body to be used. People who do not produce insulin suffer from diabetes. In the absence of glucose as a fuel the body cells begin to break down protein from the muscles, but this is a somewhat inefficient process and results in the formation of lots of **ketones**. These are circulated in the blood and cause the person concerned to feel unwell, whilst their breath has a characteristic odour of 'peardrops' from the ketones present in their system.

Smelly feet and sweaty armpits are a fact of life for everyone once they are past puberty. The unpleasant odours are largely the result of **carboxylic acids** produced by the body itself and also of the waste products produced by the bacteria which feed on and break down the sweat.

Hormones, the chemical messages which control much of the activity of living organisms, are organic chemicals which range in complexity from ethene which causes fruit to ripen in plants, through to the large steroid sex hormones which control the maturation of our bodies and the production of new life.

On the boundaries where organic chemistry and biochemistry merge, chemists work to produce new drugs to keep us healthy, new fertilisers and pesticides to ensure a better food supply for the world and new compounds to enhance the quality of our lives. As a society we remain deeply in their debt.

Figure 2 The enormous variety of organic molecules found in living organisms such as these mean that biochemistry is a subject of almost infinite scope and potential for helping both people and the whole of the natural world.

5.5 Naturally occurring macromolecules

Building blocks of life

Polymers – synthetic and natural

Throughout sections 4 and 5 of this book, we have concerned ourselves with organic families and the compounds within them. We have also noted the ability of carbon atoms to form long chain molecules, though most of the molecules in the reactions we have looked at have had relatively short carbon chains of between 1 and 10 carbon atoms. The main exception to this has been the synthetic polymers, formed when certain organic compounds such as ethene are reacted in carefully controlled conditions, resulting in extremely long molecules made up of repeating monomer units. As we saw in section 4.3, these monomers may be joined by a simple addition reaction, or by an addition/elimination reaction in which a small molecule (often water) is produced each time another monomer joins the polymer chain. These two types of polymer are represented in figure 5.5.1. The physical and chemical properties of these polymers vary depending on the monomer units of which they are made up.

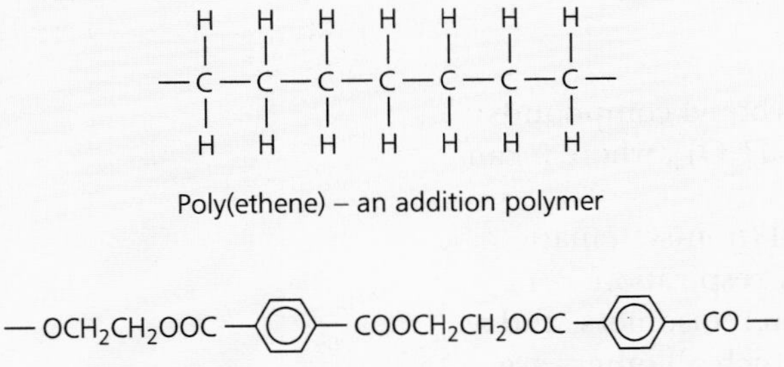

$$-\left(CH_2\right)_7- \quad \text{(seven C atoms each bearing two H: } -C-C-C-C-C-C-C-\text{)}$$

Poly(ethene) – an addition polymer

$$-OCH_2CH_2OOC-C_6H_4-COOCH_2CH_2OOC-C_6H_4-CO-$$

Poly(ester) – an addition/elimination polymer

Figure 5.5.1 The structure of two types of synthetic polymer

However, synthetic polymers comprise a very small proportion of the vast numbers of large organic molecules. Most of these organic macromolecules are naturally occurring in living organisms. Many of them are polymers, some of them are not, but between them they make up the bulk of the biological material on the Earth. In this section we shall look at some of these biological macromolecules in greater depth, using our knowledge of the nature of the different organic families to help us understand these more complex structures.

Cells – the units of life

To a chemist, a substance such as a blue crystalline solid in a jar, a yellow oil in a test tube or a colourless solution needs to be understood not simply by what can be observed – does it dissolve in water? what happens when we heat it? – but in terms of the molecules within it, the atoms that make up the molecules and the relationships of the subatomic particles within those atoms and molecules. Together these give us a model of the structure of the compound and why it reacts as it does. In a similar way, to a biologist, a living organism needs to be studied not only in terms of the behaviour of the whole organism, but also in terms of its **cells**, which can be regarded as the fundamental units of life. But cells themselves are made up of smaller units called **organelles**, and these in turn are constructed from a vast array of molecules, some very simple, many extraordinarily complex. To understand how cells function, and so in turn to understand how whole living organisms function, it is vital to have a grasp of the building blocks that make up the units of life.

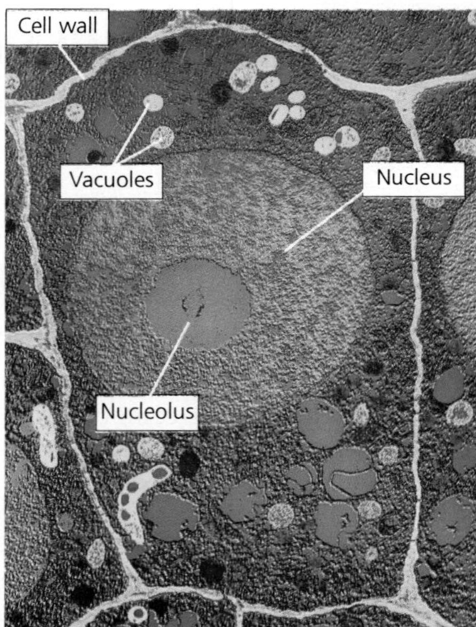

Figure 5.5.2 Whilst the cell may be regarded as the basic unit of living organisms, as this electron micrograph shows, in magnifying a single maize root tip cell a complex infrastructure is revealed within the cytoplasm of the cell, and this is constructed largely from organic macromolecules.

What are cells made of?

Around 65% of a living organism is water. The remainder is made up of other very important groups of molecules, many of which are organic compounds. There are three main types of organic macromolecules found in living cells – **carbohydrates**, **proteins** and **fats**. We shall look at each in turn.

Carbohydrates

The basic structure of all carbohydrates is the same. They are made up of carbon, hydrogen and oxygen. Carbohydrates provide energy for an organism – they are **respired** in cells and as a result drive many other reactions such as muscle contraction, hormone secretion, etc. They are also involved in storing energy, and in plants they form an important part of the cell wall.

The most commonly known carbohydrates are the sugars and starches. We have all come across **sucrose** in our sugar bowls and in cooking. **Glucose**, the energy supplier in sports and health drinks, may well be familiar to you, and we meet **starch** in flour and potatoes. But the range of chemicals known to the biochemist as carbohydrates is much wider than this.

The simple sugars – monosaccharides

Monosaccharides are naturally occurring, sweet-tasting carbonyl compounds (aldehydes and ketones). They have the general formula $(CH_2O)_n$, where n can be any number, but is usually less than 10.

The **triose** ($n = 3$) sugars such as 2,3-dihydroxypropanal (non-systematic name glyceraldehyde) are important in the biochemistry of respiration. 2,3-dihydroxypropanal, shown in figure 5.5.3(a), exists as stereoisomers, and as we have seen in section 4.1 it is the standard molecule to which all others are compared in order to determine whether they are the D- or L-form. The triose sugars have the formula $C_3H_6O_3$.

The **pentose** ($n = 5$) sugars such as ribose (2,3,4,5-tetrahydroxypentanal) and deoxyribose are most important in the structure of the nucleic acids that make up the genetic material of the cell. Except for the deoxy compounds, they have the formula $C_5H_{10}O_5$. Ribose is shown in figure 5.5.3(b).

The best known monosaccharides, including glucose, are the **hexose** sugars, with $n = 6$. Thus they have the general formula $C_6H_{12}O_6$. The molecules of many of these sugars form rings, and the molecules exhibit both geometrical and optical stereoisomerism, as shown in figure 5.5.3(c). Note that each of the carbons 1 to 5 is a chiral centre.

Looking at the ring molecules of the monosaccharide sugars shown in figure 5.5.3(c), it may be difficult to see why they are described as carbonyl compounds. However, the ring molecules can also exist as straight-chain molecules. There is a ready conversion between the two, as shown in figure 5.5.4, with the equilibrium position strongly over to the left. The ring form predominates over the straight-chain form due to the tendency of the carbonyl group to undergo an internal nucleophilic addition reaction with one of the -OH groups attached to another carbon atom.

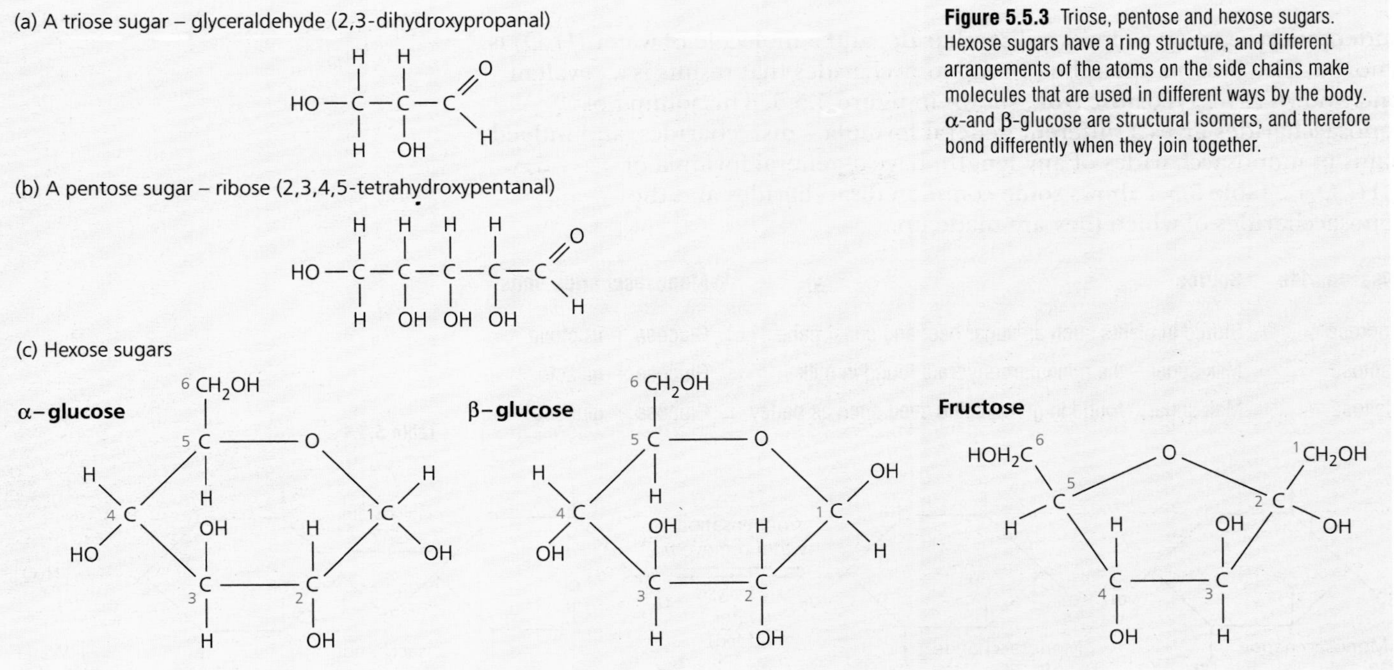

(a) A triose sugar – glyceraldehyde (2,3-dihydroxypropanal)

(b) A pentose sugar – ribose (2,3,4,5-tetrahydroxypentanal)

(c) Hexose sugars

α-glucose

β-glucose

Fructose

Figure 5.5.3 Triose, pentose and hexose sugars. Hexose sugars have a ring structure, and different arrangements of the atoms on the side chains make molecules that are used in different ways by the body. α-and β-glucose are structural isomers, and therefore bond differently when they join together.

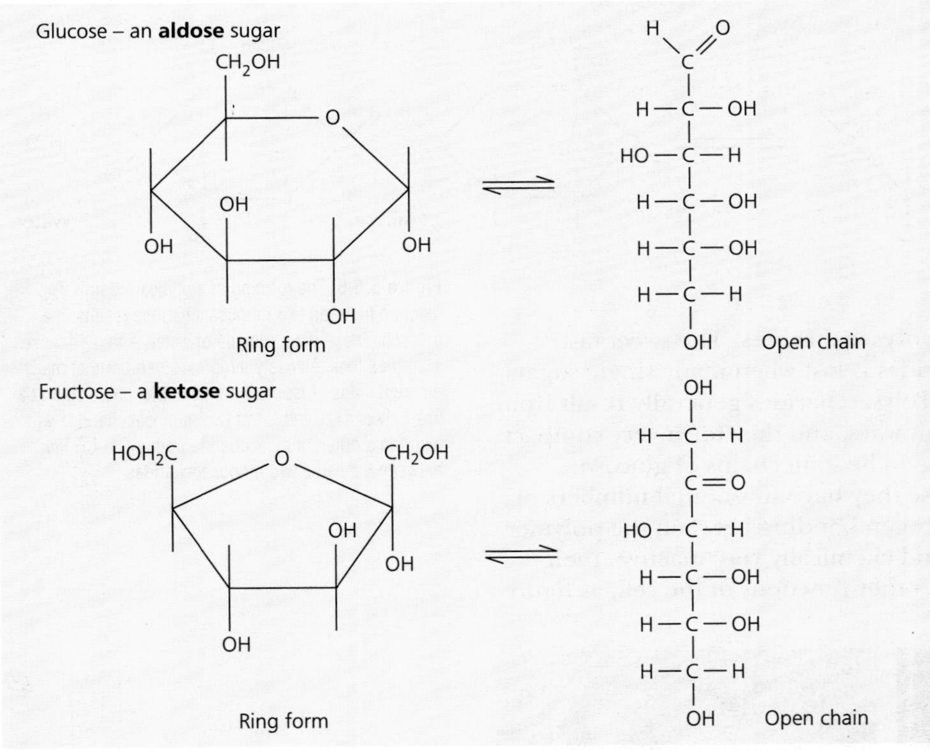

Glucose – an **aldose** sugar

Ring form

Open chain

Fructose – a **ketose** sugar

Ring form

Open chain

Figure 5.5.4 Sugars such as glucose which are aldehydes may be referred to as **aldose sugars**, whilst those such as fructose which are ketones are the **ketose sugars**.

Aldose and ketose sugars can be identified using the Fehling's solution and Tollens reagent (see section 5.2) as they give typical aldehyde or ketone results.

The double sugars – disaccharides

Disaccharides are molecules made up of two monosaccharides joined together. Sucrose (the substance you know as 'sugar') is the result of a molecule of α-glucose joining with a molecule of β-fructose. Two monosaccharides join in a condensation reaction to form a disaccharide, and a molecule of water (H_2O) is removed. The link between the two monosaccharides that results is a covalent bond known as a **glycosidic link**, shown in figure 5.5.5. This joining of monosaccharides gives a different general formula – disaccharides, and indeed chains of monosaccharides of any length, have a general formula of $(C_6H_{10}O_5)_n$. Table 5.5.1 shows some common disaccharides and the monosaccharides of which they are made up.

Disaccharide	Source	Monosaccharide units
Sucrose	Stored in plants such as sugar beet and sugar cane	Glucose + fructose
Lactose	Milk sugar – the main carbohydrate found in milk	Glucose + galactose
Maltose	Malt sugar – found in germinating seed such as barley	Glucose + glucose

Table 5.5.1

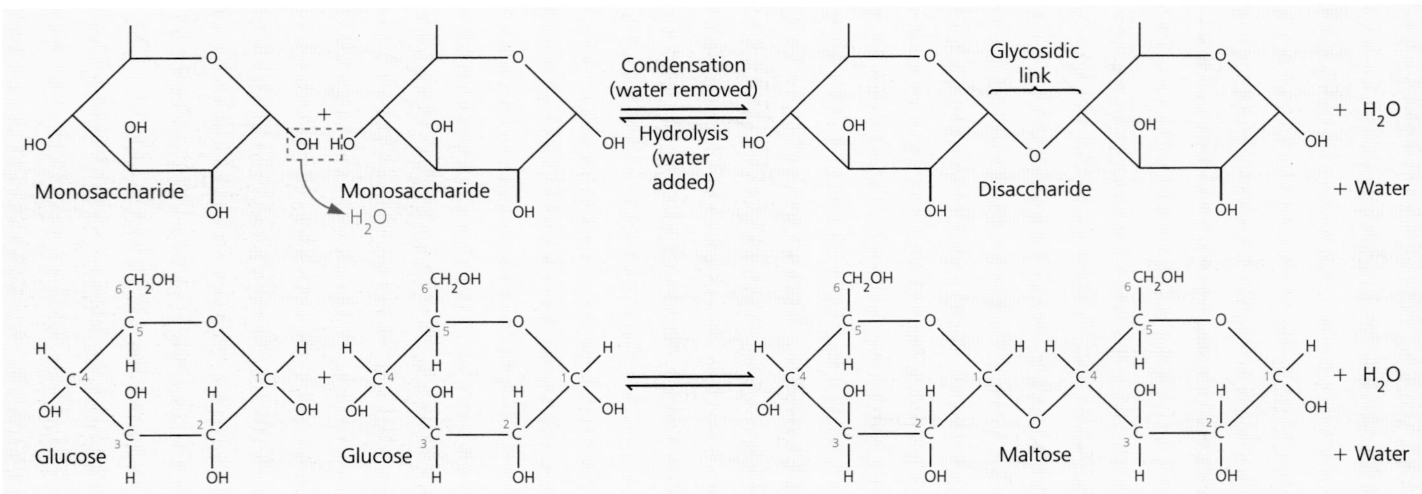

Figure 5.5.5 The forming of a glycosidic link. The reaction between two monosaccharides results in a disaccharide and a molecule of water. Although these examples look relatively simple, remember that many different kinds of isomers are possible, and also that a link between C1 and C4 of two monosaccharides will result in a different disaccharide from a C1–C6 link between the same two monosaccharides.

Polysaccharides

The most complex carbohydrates are the **polysaccharides**. The sweet taste characteristic of both mono- and disaccharides is lost when many single sugar units are joined to form a polysaccharide. Polysaccharides generally result from the linking of glucose molecules in different ways, and they form very compact molecules which are ideal for storing energy. The long chains of glucose monomers are held together tightly because they have substantial numbers of -OH groups, and so there is extensive hydrogen bonding between the polymer chains. As polysaccharides are physically and chemically very inactive, their presence in the cell does not interfere with other functions of the cell, as figure 5.5.6 illustrates.

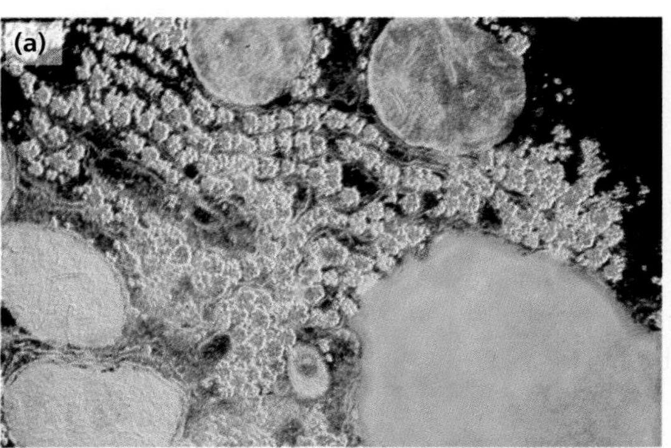

Figure 5.5.6 Storage carbohydrates play important roles in both plant and animal cells.
(a) Glycogen granules (pink) stored in the cells of the liver provide us with an energy reserve. (b) Starch grains in plant cells supply energy when the Sun doesn't shine.

Starch is one of the best known polysaccharides. It is particularly important as an energy store in plants. The sugars produced by photosynthesis are rapidly converted into starch. Particularly rich starch sources for us are plant storage organs such as potatoes. Starch is made up of long coiled chains of α-glucose, joined with 1–4 glycosidic links. The chains have branches in places – the more branches there are, the less easily the starch forms a colloidal solution in water.

Glycogen is sometimes referred to as 'animal starch'. It is the only carbohydrate energy store found in animals. Chemically it is very similar to starch, also being made up of many α-glucose units. The difference is that in glycogen, there are some 1–6 glycosidic links as well as 1–4 links, as shown in figure 5.5.7. Glycogen is found in muscle tissue and the brain as well as the liver, where the glycogen store is used as a reserve for maintaining the glucose level of the blood.

Cellulose is an important structural material in plants. It is the main constituent in plant cell walls. Like starch and glycogen, it consists of long chains of glucose molecules, but in this case β-glucose held together by 1–4 linkages, also shown in figure 5.5.7. This is an important difference because mammals, and indeed most animals, do not possess the enzymes needed to break β-1–4 linkages and so they cannot digest cellulose. (Some mammals such as cows have bacteria living in their guts that can digest cellulose for them.) The way that the monomers are joined means that most of the -OH groups stick out from the molecule, giving rise to many hydrogen bonds holding neighbouring chains together. All this makes cellulose a material with considerable strength. In the cell wall, groups of about 2000 chains form interweaving microfibrils which can be seen by the electron microscope. The structure of cellulose shown in figure 5.5.7, combined with its indigestibility, make it one of the most used polysaccharides in human civilisation. It makes up the structure of cotton, wood and many other plant products that we use. Cellulose obtained from wood is used to make the fibre rayon, and also sheets of cellophane.

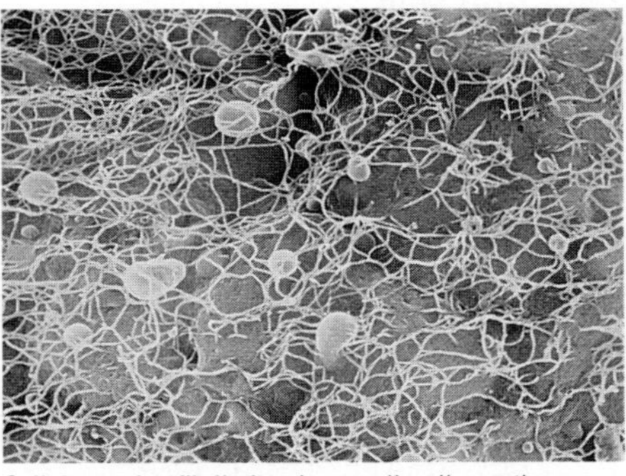

Cellulose microfibrils forming a cell wall over the surface of the cell membrane

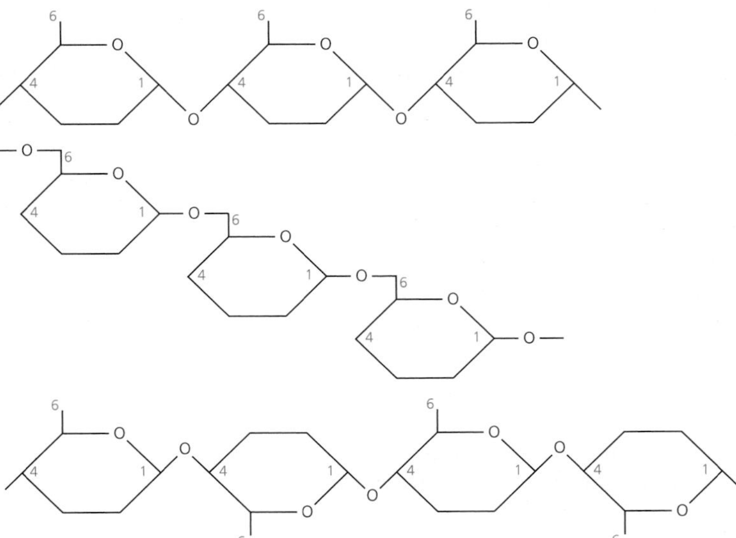

Starch : α-1–4 linkages

Glycogen : α-1–6 linkages

Cellulose : β-1–4 linkages

Figure 5.5.7 These seemingly small differences in the molecules of starch, glycogen and cellulose make all the difference in the world to where they are found and how they behave.

Lipids

Another group of organic chemicals that go to make up the structure of cells is the **lipids**. Lipids include some chemicals with a high profile in public health issues at present – **cholesterol** and **fat**. The media constantly remind us of the importance of a low-fat diet and the dangers of high cholesterol levels. But are fats really harmful, and what *is* cholesterol? To achieve answers to questions such as these we need a clear understanding of the chemistry of the molecules concerned.

Lipids are an extremely important group of chemicals which play major roles in living systems. They are an important source of energy in the diet of many animals and the most effective form for living things to store energy – they contain more energy per gram than carbohydrates or proteins. Many plants and animals convert spare food into fats or oils for use at a later date. Combined with other molecules, lipids also play vital roles in cell membranes and in the nervous system.

Lipids are naturally occurring esters of carboxylic acids. They dissolve in organic solvents, but are insoluble in water. This is important because it means they do not interfere with the many reactions which go on in aqueous solution in the cytoplasm of a cell.

As in carbohydrates, the chemical elements that make up lipids are carbon, hydrogen and oxygen. In lipids, however, there is a considerably lower proportion of oxygen than in carbohydrates.

Propane-1,2,3-triol (glycerol)

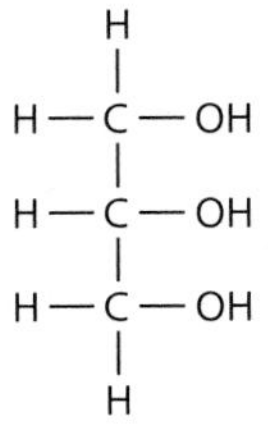

General formula of a fat or oil

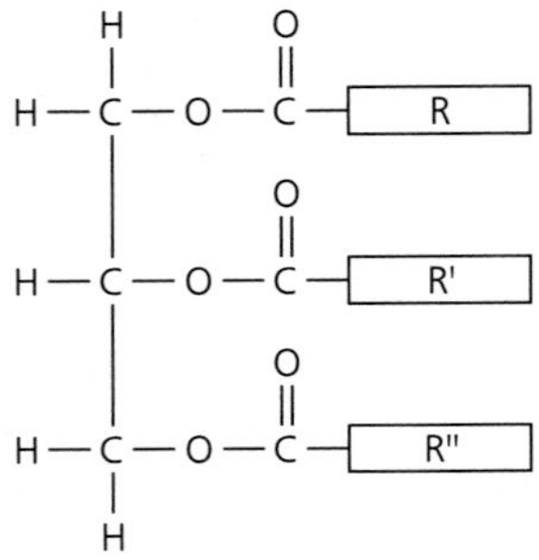

R, R' and R'' are long-chain hydrocarbons which may be the same or different.

Figure 5.5.8

Fats and oils

The main groups of lipids are the **fats** and **oils**. They are chemically extremely similar, but fats (for example, butter) are solids at room temperature and oils (for example, olive oil) are liquids. Fats and oils are esters of **fatty acids** (long-chain carboxylic acids) and **glycerol** (propane-1,2,3-triol). Propane-1,2,3-triol has the formula $C_3H_8O_3$. Its structure is shown in figure 5.5.8.

There is a wide range of long-chain carboxylic or fatty acids. Over 70 different ones have been extracted from living tissues, and the nature of a fat or oil depends mainly on the parent carboxylic acids. The carboxylic acids vary in two main ways. The length of the carbon chain can differ, although in living organisms it is frequently between 15 and 17 carbon atoms long. More importantly, the carboxylic acid may be **saturated** or **unsaturated**.

In a saturated carboxylic acid, each carbon atom is joined to the next by a *single* covalent bond. The example shown in figure 5.5.9 is stearic acid, systematic name octadecanoic acid.

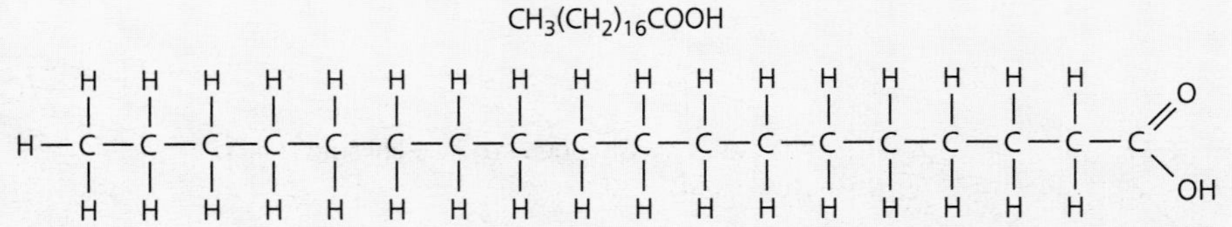

$CH_3(CH_2)_{16}COOH$

Figure 5.5.9 Stearic acid – a saturated acid

In an unsaturated carboxylic acid, the carbon chains have one or more carbon–carbon *double* bonds in them. An example shown in figure 5.5.10 is linoleic acid, an essential fatty acid in the diet of mammals, including ourselves, as it cannot be synthesised within mammalian metabolic pathways.

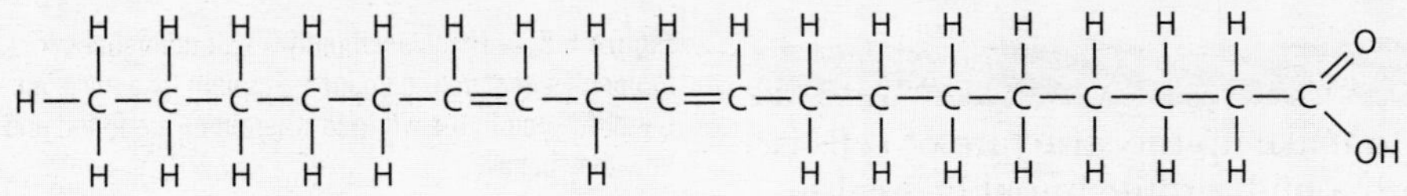

Figure 5.5.10 Linoleic acid – an unsaturated acid

A fat or oil results when propane-1,2,3-triol combines with one, two or three carboxylic acids to form a **mono-**, **di-** or **triester**. A condensation reaction takes place, involving the removal of a molecule of water, and the resulting bond is an **ester link**, as we saw in section 5.3. The process is shown in figure 5.5.11. Most fats are triesters. The carboxylic acid molecules that combine with the propane-1,2,3-triol may all be the same, or they may be different. The physical and chemical properties of a triester are greatly affected by the parent carboxylic acids – saturated acids result in fats, that are solid at room temperature, while unsaturated acids result in oils, that are liquid at room temperature.

```
                                                   Ester link
      H                                         H        O
      |            O                            |        ||
H — C — OH   +      \C — (CH2)nCH3  →  H — C — O — C — (CH2)nCH3  +  H2O
      |            OH/                          |
H — C — OH               H2O             H — C — OH
      |                                         |
H — C — OH                               H — C — OH
      |                                         |
      H                                         H
```

Propane-1,2,3-triol + carboxylic acid → monoester (monoglyceride) + water

The process is repeated to give a diester (diglyceride)... ...and finally a triester (triglyceride).

```
      H       O                               H       O
      |       ||                              |       ||
H — C — O — C — (CH2)nCH3              H — C — O — C — (CH2)nCH3
      |       O                               |       O
      |       ||                              |       ||
H — C — O — C — (CH2)nCH3              H — C — O — C — (CH2)nCH3
      |                                       |       O
      |                                       |       ||
H — C — OH                             H — C — O — C — (CH2)nCH3
      |                                       |
      H                                       H
```

Figure 5.5.11 Esterification of propane-1,2,3-triol and the molecules that may result

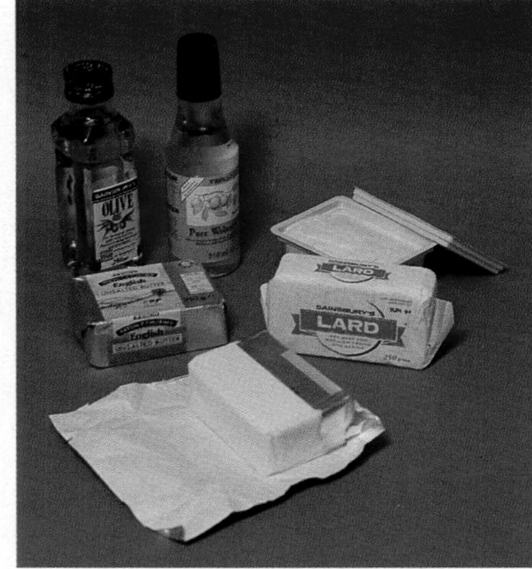

Figure 5.5.12 It is the combination of parent carboxylic acids in a triester that determines what it is like. Saturated carboxylic acids give solid fats like butter and lard, whereas unsaturates give a liquid like olive oil.

Biochemistry and affairs of the heart

Recent medical research seems to indicate that high levels of fat, particularly saturated fat, in our diet are not good for our long term health. Fatty foods are very high in energy, and so a diet high in fats or oils is likely to result in obesity. Worse than this, however, is the implication that saturated fats (saturates) – found particularly in animal products such as dairy produce, hard margarine and meat – can affect the metabolism, leading to fatty deposits in the arteries. In the long term this can lead to all sorts of problems, including heart disease and death. Unsaturated fats (unsaturates), found mainly in plants, do not seem to have this effect and so people are being encouraged to replace saturates in their diets with unsaturates whenever possible.

But there is a further twist to this tale. Unsaturates themselves can be further divided into **polyunsaturates** and **monounsaturates**. Most of the parent carboxylic acids in polyunsaturates have two or more double bonds in the carbon chain. It seems that these do not have the damaging effects of saturated fats. The majority of the parent carboxylic acids in monounsaturates have only one double bond in the carbon chain – and these seem to have a positively beneficial effect, helping the body to cope better with saturated fats.

Healthy margarines?

The origin of margarine goes back to Napoleon III in 1869. He proposed a competition with the aim of discovering 'for the use of the working class and, incidentally, the Navy, a clean fat, cheap and with good keeping qualities, suitable to replace butter'. A modern definition of margarine might be a 'mixture of edible oils, fats and water prepared in the form of a solid or semi-solid emulsion (water in oil).' It includes all substances made in imitation or semblance of butter.

For many years after margarine first became available, Napoleon's words must have haunted the margarine manufacturers. Their product was seen as a 'cheap and nasty' alternative to butter, used widely only by those who could not afford to buy butter, and the most persuasive advertising did not really help this image. However, over the last 20 years or so, margarine has come into its own. An increasing awareness of the role of fats – particularly the saturated animal fats found in dairy products such as butter – in the development of heart disease has led many people to radically rethink their eating habits.

Polyunsaturated vegetable oils such as sunflower oil, palm oil and corn oil are the esters of long-chain unsaturated carboxylic acids. Whilst being perceived as far less damaging to the health than the saturated fats of butter, as liquids they are of little use for spreading on bread. By adding hydrogen across some of the double bonds in the presence of a nickel catalyst, the vegetable oils can be given a rather more saturated nature. This **hydrogenation** makes them into soft, low-melting-point solids which have better spreading properties than butter, and yet less of the saturated fats that people want to avoid. The degree of hydrogenation controls the consistency of the margarine – increased hydrogenation gives a harder product, and also results in a saturation level and health risk approaching that of an animal fat. An added bonus in the use of vegetable margarines is that the long-chain carboxylic acid molecules of the vegetable oils are less prone to oxidation than the shorter-chain molecules of butter, and therefore margarine does not so readily turn rancid.

Margarines are now perceived as a 'healthy' part of the diet. Whilst this is not strictly true, as too much fat of most types does not seem good for the health, margarine has become a well established part of many diets, and butter consumption has fallen.

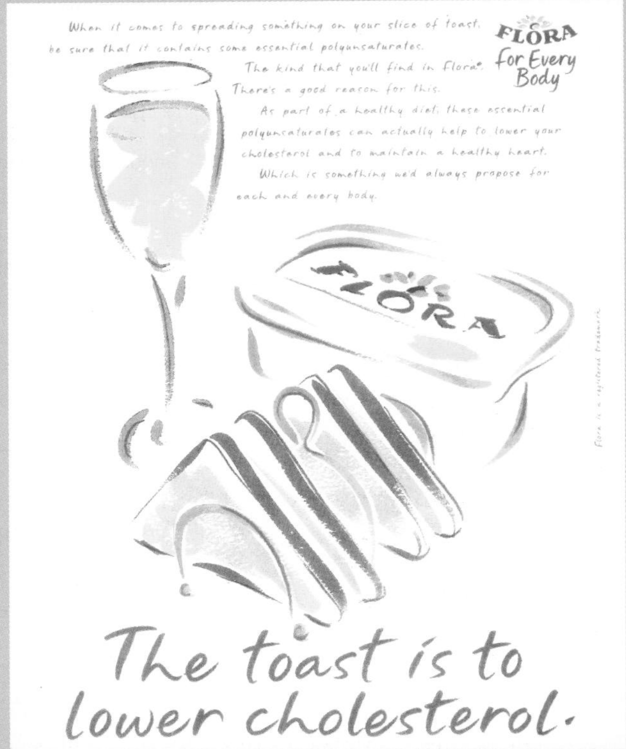

When it comes to spreading something on your slice of toast, be sure that it contains some essential polyunsaturates.
The kind that you'll find in Flora®. There's a good reason for this.
As part of a healthy diet, these essential polyunsaturates can actually help to lower your cholesterol and to maintain a healthy heart.
Which is something we'd always propose for each and every body.
FLORA
For Every Body
FLORA
The toast is to lower cholesterol.

Figure 5.5.13

Other types of lipid

Fats and oils are not the only types of lipids. Some of the others that also play a major role in living organisms are summarised below.

Phospholipids

Sometimes one of the hydroxyl groups of propane-1,2,3-triol undergoes a reaction with phosphate instead of with a carboxylic acid, and a simple **phospholipid** is formed, as shown in figure 5.5.14. Fats and oils are quite insoluble in water – they are not polar molecules. This makes them useful as inert storage materials, but limits their use elsewhere. Phospholipids are very important because the lipid and phosphate parts of the molecule give it very different properties. The lipid part is neutral and insoluble in water – it is

hydrophobic – whilst the phosphate part of the molecule is highly polar and dissolves readily in water – it is **hydrophilic**. This means part of the molecule can be dissolved in a fatty material such as the cell membrane, whilst the other part interacts with substances dissolved in water. Phospholipids are a vital component of cell membranes, and as such are present in almost all living things.

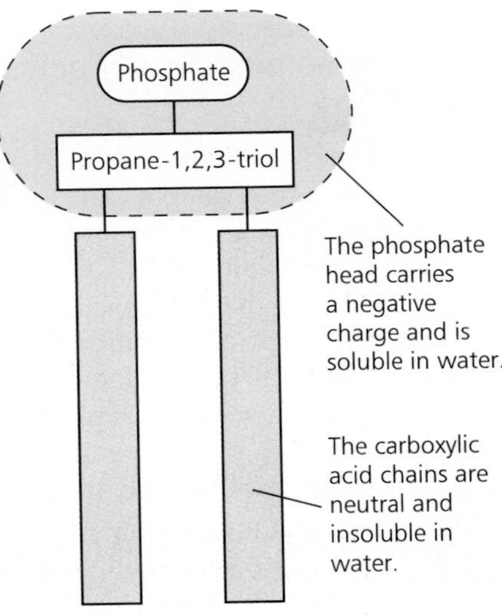

Figure 5.5.14 The structure of a phospholipid

Waxes

Waxes are lipids which are esters of very long-chain carboxylic acids and alcohols. The difference between fats and waxes is that in a wax, the ester is made of only one fatty acid with one alcohol, as the alcohols have only one hydroxyl group, not three as in propane-1,2,3-triol. Waxes are very insoluble in water and are largely used by both plants and animals for waterproofing. Some waxes produced by insects to protect their cuticles from water loss can withstand extremely high temperatures without melting.

Steroids

Steroids are sometimes classed as lipids, though apart from being insoluble in water, they have little in common with the others. But steroids are of great biological importance, particularly as hormones in both animals and plants. Steroids are made up of very large numbers of carbon atoms arranged in complex ring structures. Figure 5.5.15 shows an example.

Proteins

About 18% of the human body is made up of protein – a high proportion second only to water. Proteins are involved in a great variety of functions in living systems – muscle contraction, the clotting of the blood, enzymes that bring about respiration and photosynthesis, and structural tissues in almost all animal life. An understanding of the structure of proteins and hence the way they function will help us develop an insight into the biology of all living things.

Proteins form very large molecules, polymers of **amino acids**. They are organic nitrogen compounds – amino acids are amino-substituted derivatives of carboxylic acids. Amino acids combine in long chains in particular combinations to produce proteins. There are about 20 different naturally occurring amino acids that can polymerise in almost any order directed by the cell. This means the potential variety of proteins is vast – more than 10^{6000} different possible combinations for a protein containing 5000 amino acid units.

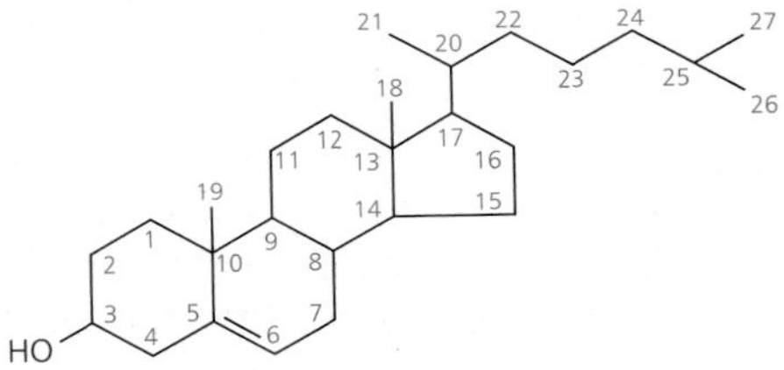

Figure 5.5.15 One of the best known steroids – cholesterol. A high blood cholesterol level can mean heart trouble ahead.

Amino acids

All naturally occurring amino acids have the same basic structure, containing an amino ($-NH_2$) group and a carboxyl ($-COOH$) group attached to a carbon atom, as shown in figure 5.5.16.

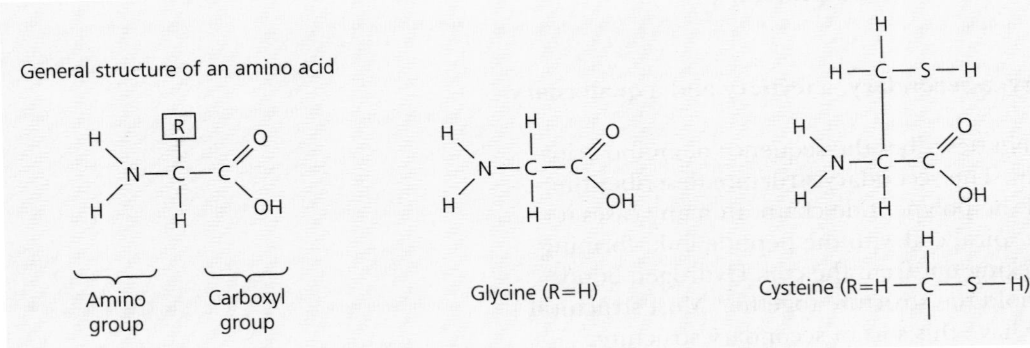

Figure 5.5.16 The R group varies from one amino acid to another. In the simplest amino acid, glycine, R is a single hydrogen atom. In a larger amino acid such as cysteine, R is a more complex group.

Are amino acids really acids?

The carboxyl (-COOH) group of an amino acid is indeed acidic in nature. It will ionise in water to give hydrogen ions. However, the amino ($-NH_2$) group is basic in nature. It attracts hydrogen ions in aqueous solution. In acidic solutions an amino acid acts like a base, while in alkaline solutions it acts like an acid. In the mainly neutral conditions found in the cytoplasm of most living organisms, it can and does act as both. This ability means it is **amphoteric**. The R group also has an effect on how the amino acid behaves – some R groups are more acidic in nature than others.

Pure amino acids are high-melting-point solids which are very soluble in water, but only very sparingly soluble in organic solvents. These properties are those of an ionic salt, and amino acids are regarded as existing as **internal ionic salts** or **zwitterions**. They can exist in three different states depending on the pH of the solution, as shown in figure 5.5.17.

The combination of all these factors means that different amino acids can be separated by a sophisticated form of chromatography that takes place in silica gel or on paper, called **electrophoresis**. There is more about this in the box 'How to unravel a protein' on pages 539–40.

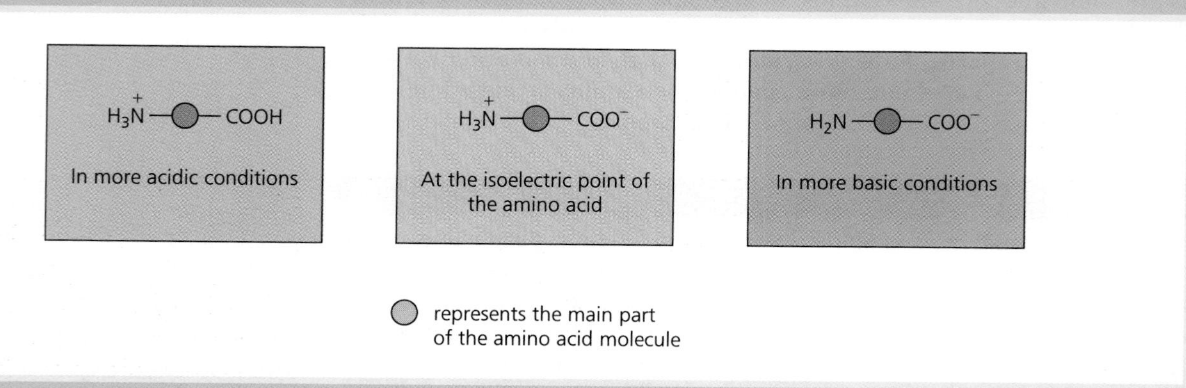

Figure 5.5.17 At a pH called the isoelectric point, an amino acid is neutral. At a lower pH it is basic, and at a higher pH it is acidic.

Proteins from amino acids

Amino acids join together by a reaction between the amino group of one amino acid and the carboxyl group of another. A condensation reaction takes place and a molecule of water is lost. The bond formed is known as a **peptide link**, the same linkage as in the amides in section 5.4. When two amino acids join, a **dipeptide** is the result, as in figure 5.5.18. More and more amino acids can join together to form **polypeptide chains**, which may be from around ten to many thousands of amino acids long. A polypeptide can fold or coil or become associated with other polypeptide chains to form a **protein**.

Protein structure

Proteins are described by a **primary**, a **secondary**, a **tertiary** and a **quaternary structure**, shown in figure 5.5.19.

The primary structure of a protein describes the sequence of amino acids that make up the polypeptide chain. The secondary structure describes the three-dimensional arrangement of the polypeptide chain. In many cases it forms a right-handed (α-) **helix** or spiral coil with the peptide links forming the backbone and the R groups sticking out from the coil. Hydrogen bonds between N—H and O═C groups hold the structure together. Most structural proteins such as the keratin in hair have this sort of secondary structure.

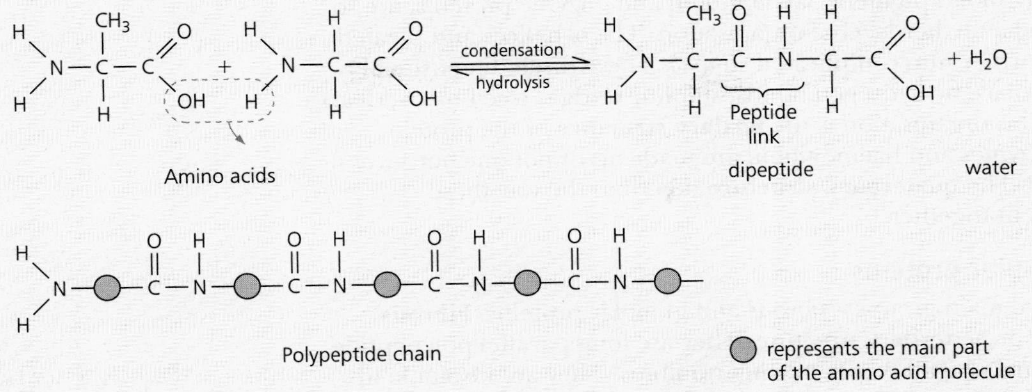

Amino acids

dipeptide

water

Peptide link

Polypeptide chain

represents the main part of the amino acid molecule

Figure 5.5.18 Amino acids can be joined in a seemingly endless variety of orders to produce an almost infinite variety of polypeptides.

In other proteins, the polypeptide chains fold up into pleated sheets, again with the pleats held together by hydrogen bonds. Sometimes there is no regular secondary structure, and the polypeptide forms a random coil.

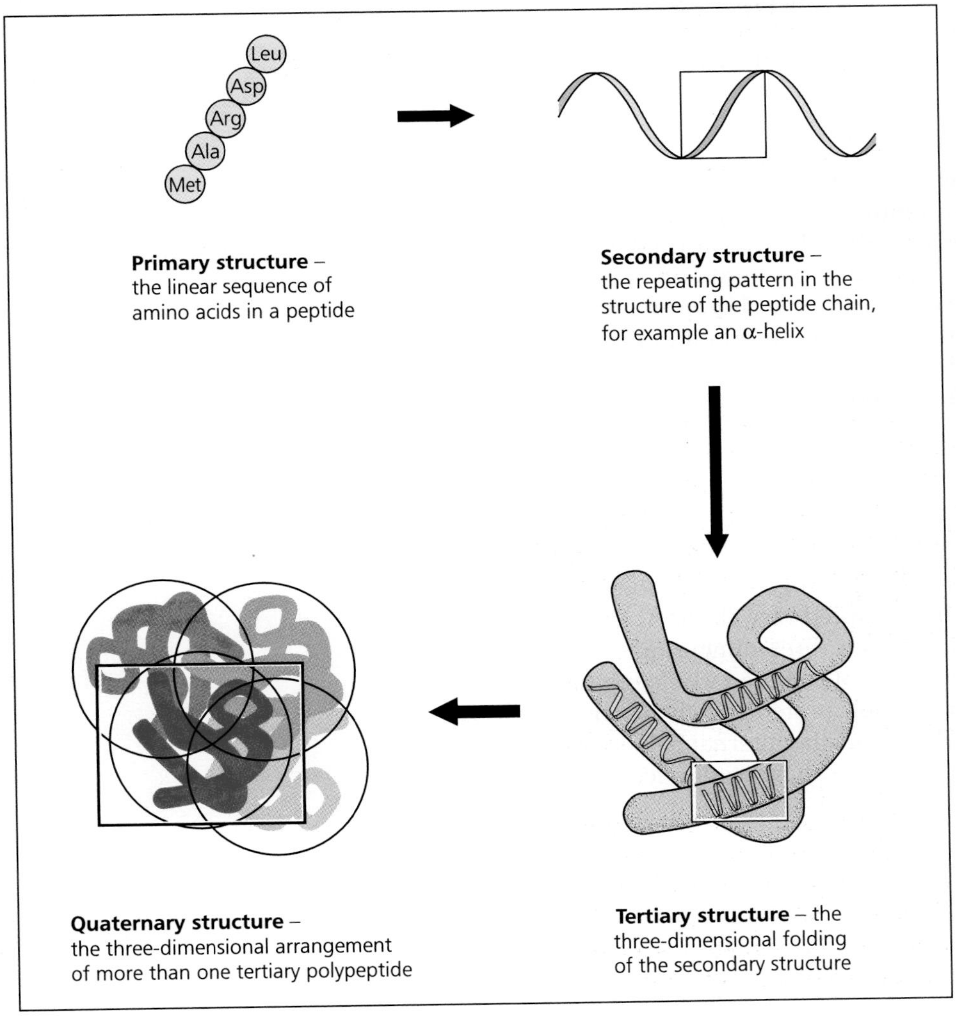

Primary structure – the linear sequence of amino acids in a peptide

Secondary structure – the repeating pattern in the structure of the peptide chain, for example an α-helix

Tertiary structure – the three-dimensional folding of the secondary structure

Quaternary structure – the three-dimensional arrangement of more than one tertiary polypeptide

Figure 5.5.19 It is not only the sequence of amino acids but also the arrangement of the polypeptide chains that determines the characteristics of a protein.

Acid hydrolysis of proteins

Proteins and peptide chains can be broken down by acid hydrolysis – in the presence of hot HCl the peptide links are broken and free amino acids are the result. This reaction takes place regularly in the human stomach. The stomach lining produces HCl which is kept at body temperature (37 °C). When protein food such as meat is eaten it enters the stomach and hydrolysis begins in the acid environment. The protein molecules are broken down to give amino acids which can be used by the body. This process is continued further along the gut by the action of specific protein digesting enzymes.

Some proteins are very large molecules indeed, consisting of thousands of amino acids joined together. The globular proteins (described below), such as the oxygen-carrying blood pigment haemoglobin and enzyme proteins, are so large that they need a further level of organisation. The α-helices and pleated sheets are folded further into complicated shapes. These three-dimensional shapes are held in place by hydrogen bonds, sulphur bridges (see box overleaf) and ionic bonds. This organisation is the **tertiary structure** of the protein.

Finally, some enzymes and haemoglobin are made up of not one but several polypeptide chains. The **quaternary structure** describes the way these polypeptide chains fit together.

Fibrous and globular proteins

Proteins fall into two main groups – fibrous and globular proteins. **Fibrous proteins** have little or no tertiary structure. They are long parallel polypeptide chains with occasional cross-linkages making up fibres. They are insoluble in water and are very tough, which makes them ideally suited to their mainly structural functions within living things. They are found in connective tissue, in tendons and the matrix of bones (collagen), in the structure of muscles, in the silk of spiders' webs and silkworm cocoons and as the keratin making up hair, nails, horns and feathers.

Globular proteins have complex tertiary and sometimes quaternary structures. They are folded into spherical (globular) shapes. Globular proteins make up the immunoglobulins (antibodies) in the blood. They form enzymes and some hormones and are important for maintaining the structure of the cytoplasm in the cell.

Conjugated proteins

Sometimes a protein molecule is joined with or **conjugated** to another molecule, called a **prosthetic group**. For example, the **glycoproteins** are proteins with a carbohydrate prosthetic group. Many lubricants used by the human body, such as mucus and the synovial fluid in the joints, are glycoproteins, as are some proteins in the cell membrane. **Lipoproteins** are proteins conjugated with lipids and they too are most important in the structure of cell membranes. Haemoglobin, the complex oxygen-carrying molecule in the blood, is a conjugated protein too, with an inorganic iron prosthetic group.

The properties of proteins

The secondary, tertiary and quaternary structures of proteins can be relatively easily damaged or **denatured**. Although the peptide links in the polypeptide chains are not readily broken, the relatively weak hydrogen bonds holding the different parts of the chains together can be disrupted very easily. As the functions of most proteins rely very heavily on their three-dimensional structure, this means that the entire biochemistry of cells and whole organisms is very sensitive to changes that might disrupt their proteins, such as a rise in temperature or a change in pH which would distort the internal balance of charges. These changes to the structure of proteins can be readily seen on boiling an egg or adding an acid to milk.

Sulphur bridges

The secondary, tertiary and quaternary structures of proteins depend on intermolecular forces to maintain them. The majority of these forces are hydrogen bonds, but sometimes much stronger covalent bonds are formed. These are formed when two cysteine molecules, which contain sulphur atoms, are close together in the structure of a polypeptide. An oxidation reaction takes place involving the two sulphur-containing groups resulting in a covalent bond known as a **sulphur bridge** or **disulphide link**. These sulphur bridges are much stronger than hydrogen bonds.

A simple demonstration of the difference in strength between these two types of bond is shown by blow-drying and perming. When you wash your hair, you break the hydrogen bonds in the protein and reform them with your hair curled in a different way as you blow-dry it. Next time you wash your hair it returns to its natural style as the original hydrogen bonds reform. If you have a perm, the chemicals of the perm solutions break the sulphur bridges between the polypeptide coils and reform them in a different place. Perming is an involved procedure and the effect on that piece of hair is permanent.

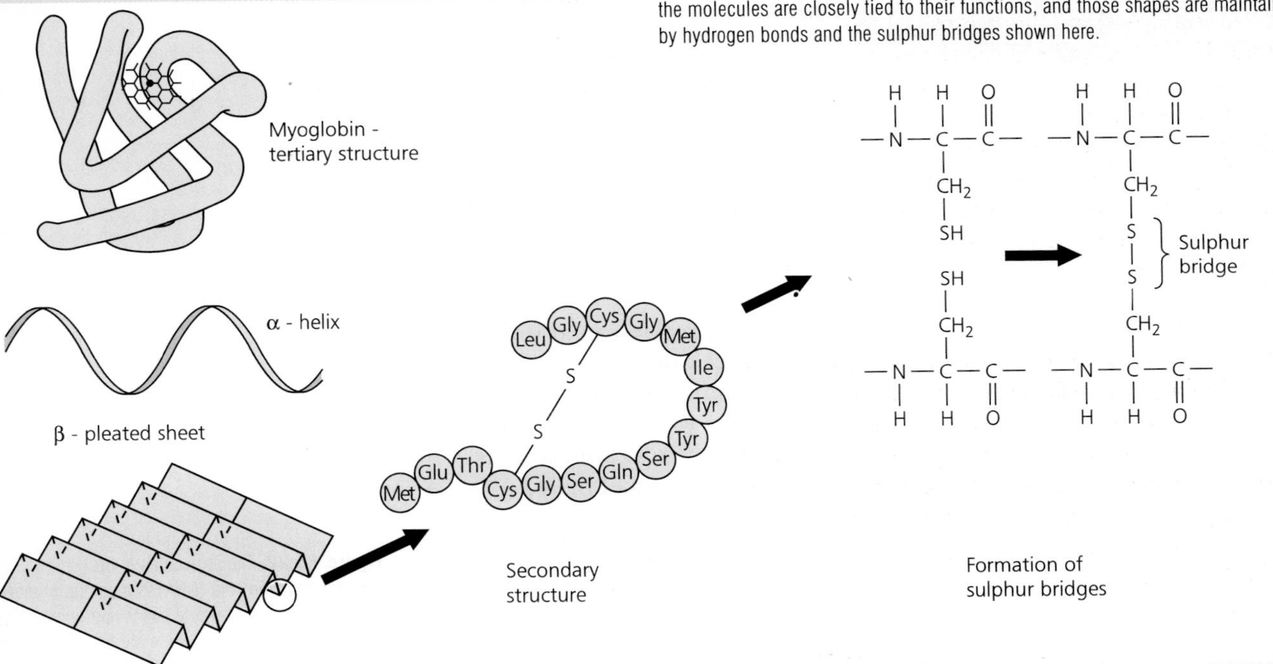

Figure 5.5.20 Some very complex molecule shapes result in our hair, the myoglobin in our muscles and the enzymes vital for all forms of life. The shapes of the molecules are closely tied to their functions, and those shapes are maintained by hydrogen bonds and the sulphur bridges shown here.

How to unravel a protein

There are many different tools used by scientists to discover the molecular structure of proteins, but two techniques in particular have played a large part in helping us to understand proteins.

The first technique is called **electrophoresis**. Chromatography can be used to separate amino acids quite successfully, but electrophoresis gives even better results. Known amino acids are first placed on a special support medium in a buffering solution. This is to keep the pH constant. Depending on their isoelectric points, some amino acids will be positively charged, some negatively charged and some neutral at this pH. The isoelectric point is affected by the nature of the R group. A measured

5

electric current is passed through the medium, and the amino acids then move on the medium at different rates according to whether they are positively or negatively charged, and to what degree. Once the medium has been removed from the solution and dried, the amino acids can be revealed using a ninhydrin spray. This reacts with the amino acids to form a coloured product. The distance each known amino acid has travelled under these known conditions of pH and electric field can then be measured and recorded.

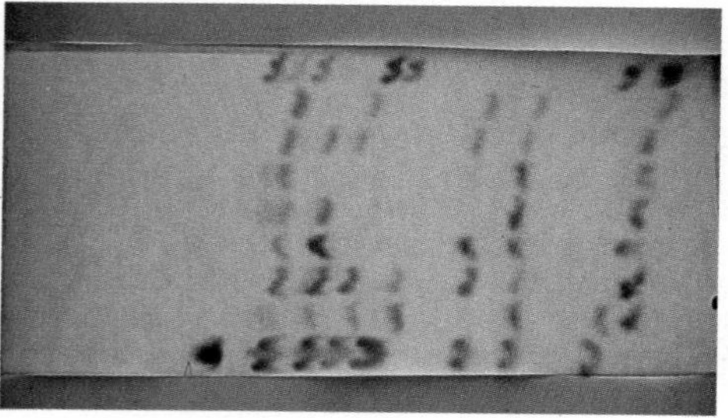

Figure 5.5.21 Electrophoresis enables us to work out which amino acids are present in a protein.

This technique can be used to find out exactly which amino acids make up a particular protein. First the protein is broken down into its component amino acids. This is done using enzymes that break the peptide links in the protein. After electrophoresis, the distance the component amino acids have travelled can be measured and compared with how far the known amino acids have travelled under the same conditions.

The precise order of the amino acids in a particular protein can also be worked out. This involves special enzymes which break peptide links one at a time, starting from either the amino or the carboxyl end of the molecule. It is a very time-consuming process – when Frederick Sanger did it for the first time in 1958, working out the sequence of amino acids in the relatively small protein insulin, he was awarded a Nobel prize.

The second technique that has been vital in helping us to build up a picture of what protein molecules are like – and so, very often, how they work – is X-ray crystallography, described in section 1.6.

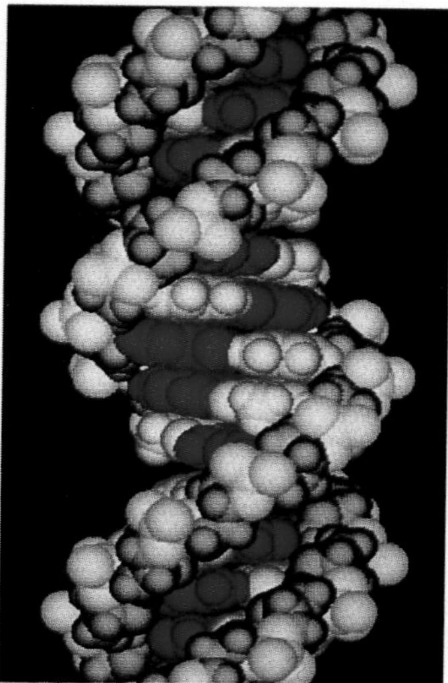

Figure 5.5.22 The molecules of life – deoxyribonucleic acid (DNA) is made up of repeating units of sugars, nucleotide bases and phosphates. Within this complex molecule are the instructions to synthesise the enzyme proteins that will in turn bring about the formation of the enormous variety of organic molecules represented in the life forms seen here. The scope of organic chemistry and biochemistry is infinitely greater than a text such as this could hope to cover.

SUMMARY

- The three main groups of organic macromolecules in living cells are carbohydrates, proteins and fats.
- **Carbohydrates** are made up of carbon, hydrogen and oxygen, with the general formula $C_n(H_2O)_n$.
- **Monosaccharides** (general formula $(CH_2O)_n$, where $n < 10$) are naturally occurring aldehydes and ketones.
- **Disaccharides** (general formula $(C_6H_{10}O_5)_n$) consist of two monosaccharide units joined together in a condensation reaction.
- **Polysaccharides** are polymers of many monosaccharide units. There may be branching and cross-linking of the chains. Polysaccharides are generally chemically and physically inert.
- **Lipids** are naturally occurring esters of carboxylic acids.
- **Fats** and **oils** are esters of long-chain carboxylic acids and propane-1,2,3-triol (glycerol).
- Long-chain carboxylic acids are also known as **fatty acids**. They may be **unsaturated** (containing one or more carbon–carbon double bonds) or **saturated**.
- **Proteins** are polymers of amino acids.
- **Amino acids** contain an amino group and a carboxyl group attached to the same carbon atom, $RR'C(NH_2)COOH$. Amino acids are amphoteric.
- Amino acids join in a condensation reaction between the amino group of one molecule and the carboxyl group of another, producing a **peptide link**.
- Proteins are described by their primary, secondary, tertiary and quaternary structure.
- Intermolecular forces are important in maintaining the structure of proteins. Such forces may arise from hydrogen bonds and through ionic interactions. Covalent linkages (S—S bridges) also play a role in protein structure.

QUESTIONS

1 Consider the compounds A to F in figure 5.5.23.

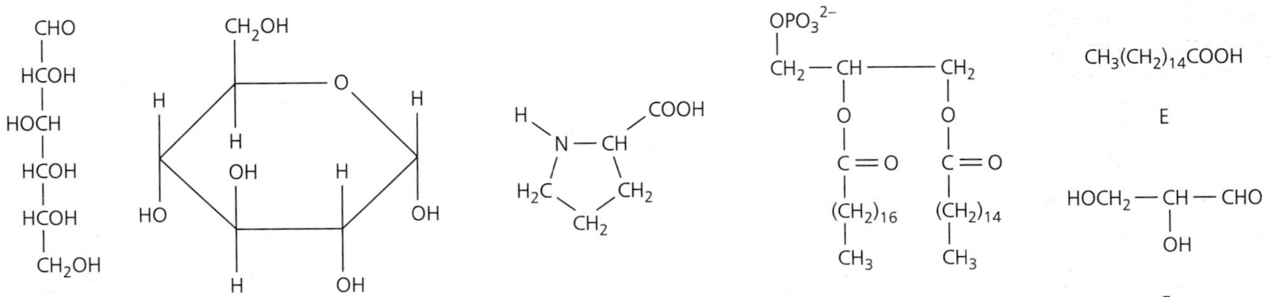

A: CHO—HCOH—HOCH—HCOH—HCOH—CH_2OH

B: ring structure with CH_2OH, O, H, OH, HO substituents

C: H—N—CH(COOH) ring with H_2C, CH_2, CH_2

D: OPO_3^{2-}—CH_2—CH—CH_2 with O—C=O—$(CH_2)_{16}$—CH_3 and O—C=O—$(CH_2)_{14}$—CH_3

E: $CH_3(CH_2)_{14}COOH$

F: $HOCH_2$—CH(OH)—CHO

Figure 5.5.23

5

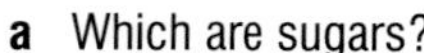

QUESTIONS

a Which are sugars?
b Which is an amino acid?
c Which is a fatty acid?
d Which is a lipid?
e Which might form part of a protein molecule?
f Which are carbohydrates?

2 Ribose and fructose have the structures shown in figure 5.5.24.

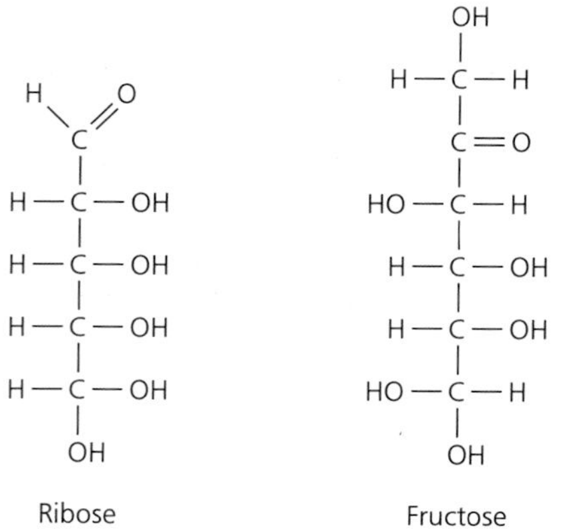

H—C=O
H—C—OH
H—C—OH
H—C—OH
H—C—OH
OH

Ribose

OH
H—C—H
C=O
HO—C—H
H—C—OH
H—C—OH
HO—C—H
OH

Fructose

Figure 5.5.24

a Which would give a positive reaction with Fehling's solution? Why?
b In biological systems, the ring form of sugars is more important than the straight-chain form shown above. Suggest a possible ring structure for ribose.
c Fructose and ribose are optically active – why?

3 a Why are fatty acids such as palmitic acid, $CH_3(CH_2)_5CH{=}CH(CH_2)_7COOH$, insoluble in water, while ethanoic acid, CH_3COOH, is soluble?
b Write an equation to show how palmitic acid reacts with glycerol (propane-1,2,3-triol) to form a monoglyceride.
c Fatty acids, triglycerides and lipids may form micelles in aqueous solutions. How do they do this?
d Why is butter hard to spread when it is taken straight out of the refrigerator, yet many margarines are easy to spread when they are cold?

4 The amino acids serine, leucine and cysteine have the structures shown in figure 5.5.25.
a What part of the serine molecule makes it behave as a base?
b What part of the serine molecule makes it behave as an acid?
c Electrophoresis of a solution containing serine under acidic conditions causes the amino acid to move towards the cathode, while electrophoresis under alkaline conditions causes it to move towards the anode. Explain these observations.

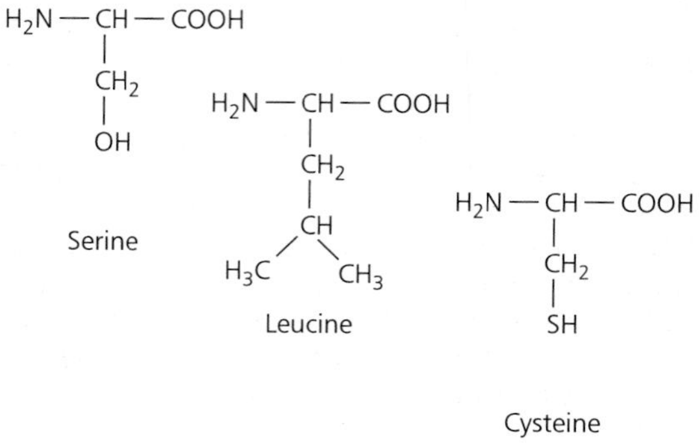

H_2N—CH—COOH
CH_2
OH

Serine

H_2N—CH—COOH
CH_2
CH
H_3C CH_3

Leucine

H_2N—CH—COOH
CH_2
SH

Cysteine

Figure 5.5.25

d Show how a serine molecule may join with a molecule of leucine and one of cysteine to form a tripeptide.
e How many different peptides are possible which contain two molecules of serine, one of leucine and one of cysteine?

5 From your knowledge of the chemistry of amino acids, suggest why the structure of proteins is very sensitive to pH.

Developing Key Skills

A greater understanding of the chemistry of life and sophisticated computer modelling techniques mean that increasing numbers of drugs are being produced by the pharmaceutical companies to target very specific health problems. These drugs are of enormous benefit to many people. At the same time increasing numbers of people are becoming dissatisfied with conventional medicine, particularly for chronic but not life-threatening conditions. Many are turning to complementary therapies to keep them healthy. Very popular alternatives to prescribed drugs include nutritional supplements (such as gamma linoleic acid, peptide supplements, vitamins and minerals) and herbal remedies (such as St John's Wort for depression and Oil of Evening primrose for premenstrual tension). You have been commissioned by a popular magazine to look at the differences and similarities between prescribed medicines and herbal and nutritional remedies, and to discuss some of the advantages and disadvantages attached to each type of therapy.

[Key Skills opportunities: C]

5 Questions

1 The two functional groups in compound **A**, (cyclohexene ring)–CHO, behave independently.

a State what would be observed if a few drops of compound **A** were added to Fehling's solution and heated. Give the structure of the organic reaction product. **(2 marks)**

b Using RCHO to represent compound **A**, write an equation for the reaction between RCHO and hydrogen cyanide. State the type of reaction taking place and outline a mechanism. **(5 marks)**

c State the type of isomerism shown by 2-hydroxypropanoic (*lactic*) acid, $CH_3CH(OH)COOH$, and point out the structural feature of the molecule which causes the existence of two isomers. With the aid of diagrams, show the structural relationship between the two isomers and state how these isomers can be distinguished. **(5 marks)**

(Total 12 marks)

(AQA specimen)

2 a Write an equation for the formation of ethyl ethanoate from ethanoyl chloride and ethanol. Name and outline the mechanism for the reaction taking place. **(6 marks)**

b Explain why dilute sodium hydroxide will cause holes to appear in clothing made from polymers such as Terylene but a poly(phenylethene) container can be used to store sodium hydroxide. **(2 marks)**

3 Ethane can be cracked at high temperatures to yield ethene and hydrogen, according to the equation:

$$C_2H_6(g) \rightleftharpoons C_2H_4(g) + H_2(g)$$

The standard enthalpy of formation of ethene is positive whereas that of ethane is negative.

a Discuss the effect on the equilibrium constant, K_p, of changes to **i** the temperature **ii** the pressure. **(3 marks)**

b Calculate the value of the equilibrium constant, K_p, for this cracking reaction, given that 1.00 mol of ethane under an equilibrium pressure of 180 kPa at 1000 K can be cracked to produce an equilibrium yield of 0.36 mol of ethene. **(7 marks)**

(Total 10 marks)

(AQA specimen)

4 Compound **A**, $C_5H_{10}O$, reacts with $NaBH_4$ to give **B**, $C_5H_{12}O$. Treatment of **B** with concentrated sulphuric acid yields compound **C**, C_5H_{10}. Acid-catalysed hydration of **C** gives a mixture of isomers, **B** and **D**.

Fragmentation of the molecular ion of **A**, $[C_5H_{10}O]^{+\cdot}$, leads to a mass spectrum with a major peak at *m*/*z* 57. The infra-red spectrum of compound **A** has a strong band at 1715 cm^{-1} and the infra-red spectrum of compound **B** has a broad absorption at 3350 cm^{-1}. (see table).

Table of infra-red absorption data

Bond	Wavenumber/cm^{-1}
C—H	2850–3300
C—C	750–1100
C=C	1620–1680
C=O	1680–1750
C—O	1000–1300
O—H (alcohols)	3230–3550
O—H (acids)	2500–3000

The proton n.m.r. spectrum of **A** has two signals at δ 1.06 (triplet) and 2.32 (quartet), respectively (see Spectrum).

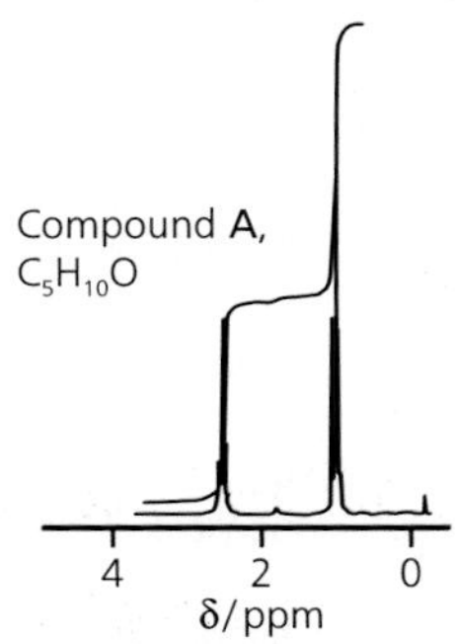

Use the analytical and chemical information provided to deduce structures for compounds **A**, **B**, **C** and **D**, respectively. Include in your answer an equation for the fragmentation of the molecular ion of **A** and account for the appearance of the proton n.m.r. spectrum of **A**. Explain why isomers **B** and **D** are formed from compound **C**.

(Total 20 marks)

(AQA specimen)

5 a Butanone has the infra-red spectrum below.

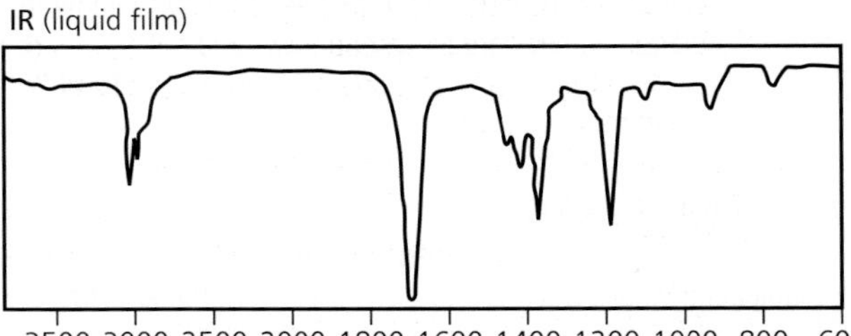

How does this infra-red spectrum confirm the presence of the functional group present in butanone? **(2 marks)**

b Butanone can be reduced with $NaBH_4$ to form compound **G**.

i Show the structure of **G**.

ii The product of this reaction can exist as two optical isomers. Draw the isomers and identify the chiral carbon with an asterisk, (*). **(3 marks)**

c Butanone reacts with hydrogen cyanide in the presence of potassium cyanide.
  i Describe, with the aid of curly arrows, the mechanism of this reaction.
  ii What type of reaction is this? **(4 marks)**

(Total 10 marks)

(OCR specimen)

6 A student prepared benzoic acid, C_6H_5COOH by hydrolysing methyl benzoate, $C_6H_5COOCH_3$ using the following method.

- Dissolve 4.0 g of sodium hydroxide in water to make 50 cm³ of an alkaline solution.
- Add the aqueous sodium hydroxide to 2.70 g of methyl benzoate in a 100 cm³ flask and set up the apparatus for reflux.
- Reflux this mixture for 30 minutes.
- Distil the mixture and collect the first 2 cm³ of distillate.
- Pour the residue from the flask into a beaker and add dilute sulphuric acid until the solution is acidic.
- Filter the crystals obtained and re-crystallise from hot water to obtain the benzoic acid.

The overall equation for this hydrolysis is:

$$C_6H_5COOCH_3 + H_2O \rightarrow C_6H_5COOH + CH_3OH$$

The student obtained 1.50 g of benzoic acid, C_6H_5COOH.

a i Name the functional group that reacts during this hydrolysis.
  ii Draw the structure of methyl benzoate and clearly show the bond that breaks during the hydrolysis. **(3 marks)**

b i Calculate how many moles of methyl benzoate were used.
  ii What was the concentration, in mol dm^{-3}, of the aqueous sodium hydroxide used?
  iii Calculate the percentage yield of the C_6H_5COOH obtained by the student.
  iv Suggest why the percentage yield was substantially below 100%. **(9 marks)**

c i Why was the residue from the flask acidified before recrystallising?
  ii Why were the crystals recrystallised? **(2 marks)**

d Infra-red spectroscopy can be used to monitor the progress of a chemical reaction.
  i Predict the key identifying features of the infra-red spectra of methyl benzoate and its hydrolysis products, benzoic acid and methanol.
  ii How could you use infra-red spectroscopy to show that the ethanol did **not** contain any benzoic acid? **(6 marks)**

(Total 20 marks)

(OCR specimen)

7 The spectra shown below were obtained from an organic compound **G**. Using data from the three spectra, suggest a structure for **G**, indicating what evidence you have used from the spectra.

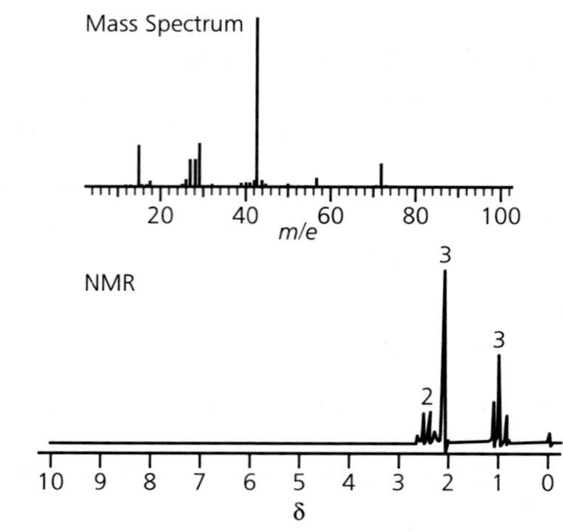

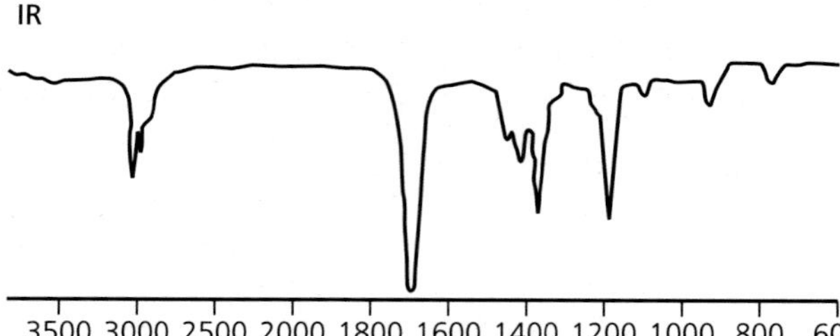

(Total 9 marks)

(Quality of Written Communication 2 marks)

(OCR specimen)

8 Compounds **J** and **K** contribute to the 'leafy' odour of violet oil.

CH_3CH_2 CH_2OH

C=C

H   H

J

a Name the functional groups present in compound **J**. **(2 marks)**

b What is the molecular formula of compound **J**? **(1 mark)**

c Draw the structure of the organic product formed by the reaction of compound **J** with
  i Br_2
  ii CH_3COOH in the presence of an acid catalyst. **(2 marks)**

d A chemist reacted compound **J** with HBr. He separated 2 structural isomers **K** and **L** with the molecular formula $C_5H_{10}Br_2$. Draw structures for **K** and **L**. **(2 marks)**

e Compound **M** below can be prepared from compound **J**.

H

CH_3CH_2 C=O

C=C

H   H

M

  i Suggest reagent(s) for the conversion of **J** into **M**.
  ii Draw the structure of a possible organic impurity (other than **J**) which might contaminate the product. Explain your choice. **(3 marks)**

(Total 10 marks)

(OCR specimen)

9 The kinetics of a reaction are used to clarify reaction mechanisms. An experiment to determine the kinetics of the substitution reaction between 2-chloro-2methylpropane and sodium hydroxide uses **equal initial** concentrations of these substances in aqueous ethanol solvent. A mixture was maintained at 25°C, and samples taken at intervals. The samples are quenched in about twice their volume of cold propanone, and the concentration of sodium hydroxide is found.

Time/min:	0	7	15	27	44	60
Conc. mol dm^{-3}	0.080	0.065	0.054	0.041	0.028	0.020

a i Show by means of a suitable graph that the reaction is first order. **(4 marks)**

ii As performed, the results cannot distinguish between the rate laws

rate = k[OH$^-$] and rate = k[halogenoalkane]

Outline a further experiment which must be performed to enable the distinction to be made, showing how the new data would be used to establish the rate law. **(3 marks)**

iii The reaction is in fact first order with respect to the halogenoalkane. Write the mechanism for the substitution reaction, identifying the rate-determining step. **(4 marks)**

iv Nucleophilic substitution is usually accompanied by elimination as a competing reaction. Write the name and structural formula of the product of the elimination reaction with 2-chloro-2-methylpropane and state the conditions which favour elimination over substitution. **(3 marks)**

b The method of preparation of 2-chloro-2-methylpropane reverses the above substitution reaction. A modified version of part of the method is given below.

- Place 25 g of 2-methylpropan-2-ol and 85 cm^3 of concentrated hydrochloric acid in a stoppered separating funnel.
- Shake the mixture periodically over 20 minutes, releasing the pressure from time to time.
- Allow the layers to separate, draw off and discard the lower acid layer.
- Wash the organic layer with 20 cm^3 of 5% $NaHCO_3$ solution, and then with water; discard the aqueous layer in each case.
- Add 5 g of anhydrous calcium sulphate to the organic layer.
- Filter the organic layer into a dry distillation apparatus, and distil; collect the fraction between 49 °C and 51 °C.
- The yield is 28 g (90%).

The first step in the mechanism for this reaction involves protonation of the hydroxyl group of the alcohol.

i Representing the alcohol as ROH, use dot-and-cross diagrams to show the structure of the protonated alcohol, and suggest why this species is substituted by a chloride ion whereas in the absence of acid the reaction does not work. **(4 marks)**

ii Explain why the organic product is treated with sodium hydrogencarbonate solution, then with water, and then with anhydrous calcium sulphate. Show, on the basis of the quantity of alcohol used, that the yield of the product is 90% as claimed, and give reasons why the yield is less than 100%. **(7 marks)**

(Total 25 marks)

(Edexcel 1999)

10 There are four structurally isomeric alcohols of molecular formula $C_4H_{10}O$. Graphical formulae of these isomers, labelled **A**, **B**, **C** and **D**, are shown below.

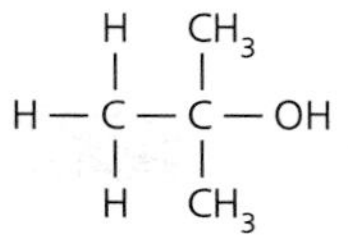

```
    H   CH3
    |   |
H — C — C — OH
    |   |
    H   CH3
      A

    H   H   CH3
    |   |   |
H — C — C — C — OH
    |   |   |
    H   H   H
        B

    H   H   H   H
    |   |   |   |
H — C — C — C — C — OH
    |   |   |   |
    H   H   H   H
          C

    H   CH3 H
    |   |   |
H — C — C — C — OH
    |   |   |
    H   H   H
        D
```

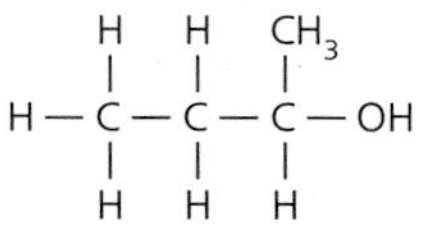

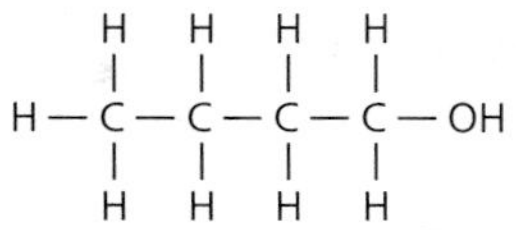

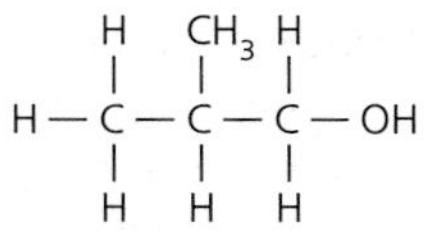

a Identify the type of alcohol represented by **A** and by **B**. **(2 marks)**

b Give the name of alcohol **A**. **(1 mark)**

c Select one of the alcohols **A**, **B**, **C** or **D** which will, on oxidation, produce an aldehyde.

i Give the structural formula of the aldehyde produced by this reaction. **(1 mark)**

ii State the reagents and conditions required for the aldehyde to be the main product of the oxidation reaction. **(3 marks)**

d All the alcohols **A**, **B**, **C** and **D** may be readily dehydrated.

i Explain what is meant by the term *dehydration*. **(1 mark)**

ii State the type of compound formed by dehydration of alcohols. **(1 mark)**

iii Suggest suitable reagent(s) and condition(s) for the dehydration of alcohols. **(2 marks)**

iv Select one of the alcohols **A**, **B**, **C** or **D** which, on dehydration, would give a single product. Draw the structural formula of this product. **(1 mark)**

v Select one of the alcohols **A**, **B**, **C** or **D** which, on dehydration, would give two products which are **structurally** isomeric. Draw structural formulae for these two structural isomers and explain why the formation of two structural isomers is possible in this case. **(4 marks)**

(Total 16 marks)

(AQA 1999)

11 The molecular formulae of some compounds that can be prepared from ethanoic acid are given in the scheme below.

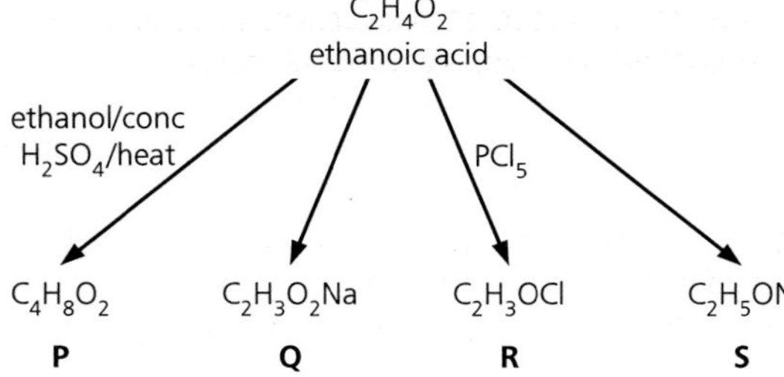

a i Give the name and graphical formula of **P**. **(2 marks)**

ii Give the name of the type of reaction which occurs when **P** is formed from ethanoic acid. **(1 mark)**

b Ethanoic acid can be obtained from **P**.

i Give the name of the reagent(s) and state the conditions required. **(2 marks)**

ii Write a balanced equation for the reaction. **(1 mark)**

5

c **i** State the reagent and reaction conditions that could be used for converting ethanoic acid into **Q**. **(2 marks)**

ii Give the name and graphical formula of the organic product of the reaction between anhydrous samples of **Q** and **R**. **(2 marks)**

iii State how the product formed in **c ii** could be converted into ethanoic acid and write an equation for the reaction. **(2 marks)**

d **i** Give the name and graphical formula of the amide, **S**. **(2 marks)**

ii State the reagent(s) and reaction conditions that could be used for converting ethanoic acid into **S**. **(2 marks)**

iii Write a balanced equation for the reaction between **S** and aqueous hydrochloric acid. **(1 mark)**

(Total 17 marks)

(AQA 1999)

12 Below is a reaction scheme for some aromatic compounds.

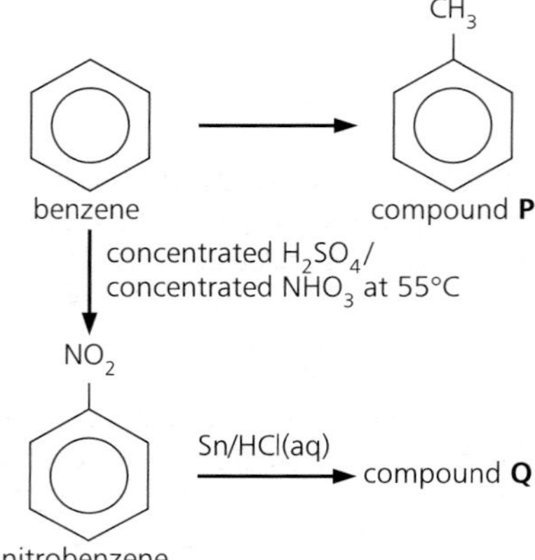

a **i** Give the reagents and conditions for the conversion of benzene into compound **P**. **(3 marks)**

ii Give the name of the mechanism of this reaction. **(2 marks)**

b Draw the graphical formulae of the possible organic products when excess chlorine is passed through boiling compound **P** in strong sunlight. **(3 marks)**

c **i** Classify the type of reaction occurring when nitrobenzene is converted into compound **Q**. **(1 mark)**

ii Draw the graphical formula of compound **Q**. **(2 marks)**

d Classify the types of reaction and draw the graphical formulae of the organic products of the reaction of propanal with:

i sodium tetrahydridoborate(III), $NaBH_4$; **(2 marks)**

ii Fehling's solution; **(2 marks)**

iii hydrogen cyanide. **(2 marks)**

(Total 17 marks)

(AQA 1999)

13 **a** Starch is a naturally occurring polymer that can be converted into glucose. The equation for this conversion is:

$$(C_6H_{10}O_5)_n + nH_2O \rightarrow nC_6H_{12}O_6$$

i Give the name of the type of reaction occurring. **(1 mark)**

ii Give the name of a substance, other than an enzyme, that will catalyse the reaction. **(1 mark)**

b **i** Draw the graphical formula of the chain structure of glucose. **(1 mark)**

ii A glucose molecule contains four asymmetric carbon atoms. Identify any **two** of these by placing a * next to the relevant carbon atoms on the graphical formula drawn in **b i**. **(1 mark)**

iii Give the type of isomerism shown by glucose because of the chirality in the molecule. State the property shown by glucose as a result of the chirality. **(2 marks)**

c **i** Give the names of the **two** products of the reaction of sucrose with water. **(1 mark)**

ii Explain why this reaction is known as the *inversion of sucrose*. **(3 marks)**

(Total 10 marks)

(AQA 1999)

14 Terylene has the following repeat unit.

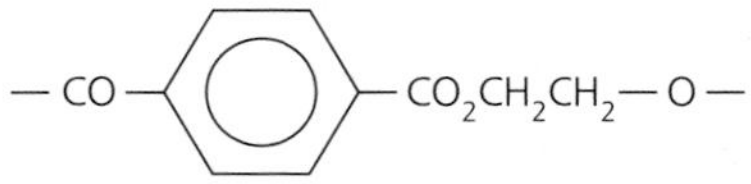

— CO —⟨benzene ring⟩— $CO_2CH_2CH_2$ — O —

a Draw the displayed formulae of the two monomers used to make Terylene.

b State the type of polymerisation which occurs when Terylene is made.

c Explain why holes form when aqueous sodium hydroxide is spilled on a Terylene shirt. **(4 marks)**

(OCR 1998)

15 Ferulic acid is a plant derivative. It is a natural antioxidant because it terminates free-radical chain reactions. It is used commercially to give photo-protection in skin lotions and sunscreens as well as in a range of medical applications. Ferulic acid is an active ingredient in many ancient Chinese herbal remedies.

Ferulic acid occurs in onions with related acids having the structures given below.

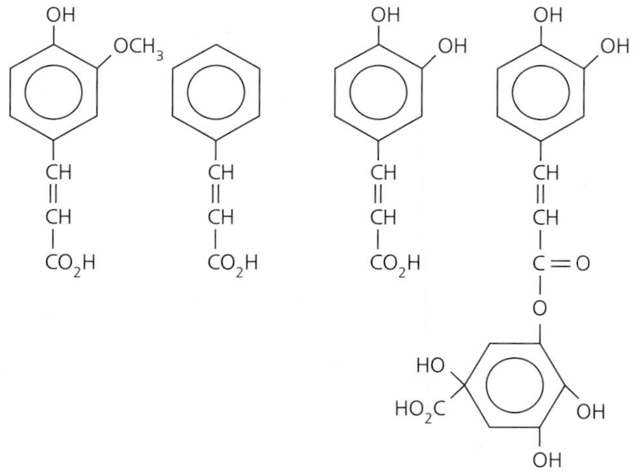

ferulic acid	cinnamic acid	caffeic acid	chlorogenic acid
A	**B**	**C**	**D**

a Use the letters to rank the four acids in order of *decreasing* water-solubility, and explain your answer. **(4 marks)**

An extract from onion is subjected to chromatographic analysis. The mobile phase is water and the stationary phase consists of small beads made from inert silane macromolecules to which $C_{18}H_{37}$ alkyl groups are attached.

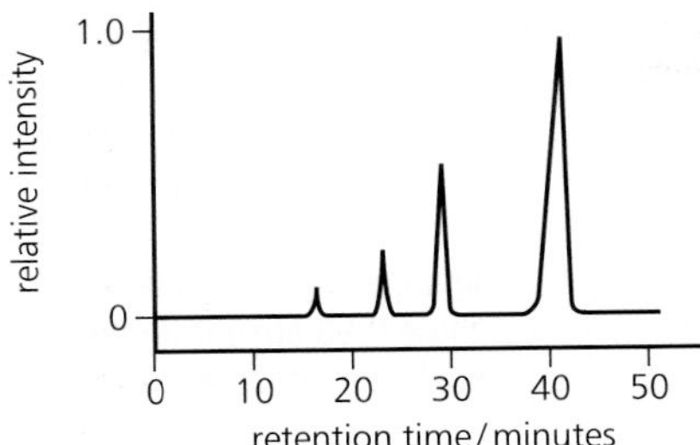

b The water is kept at a pH of 2.0 for reproducibility of retention times. Suggest a chemical reason why a higher pH is not used. **(2 marks)**

c The flow rate of the water is 0.5 $cm^3\ min^{-1}$. Calculate the volume of water which flowed through the chromatograph before the acid with the longest retention time was detected by the recorder. **(1 mark)**

d i Explain why the acid which is least water-soluble should have the longest retention time in the chromatogram.

ii Hence copy and complete the table below.

acid	approximate retention time
A	
B	
C	
D	

(3 marks)
(Total 10 marks)
(OCR 1998)

16 a The diagram shows the high resolution proton n.m.r. spectrum of a sample of laboratory ethanol.

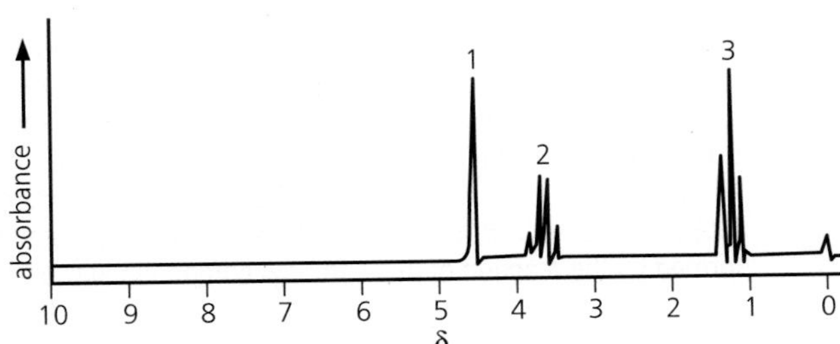

i Explain why the different protons present absorb at different δ values.

ii Explain why some of the peaks are split.

iii Suggest why the peak produced by the proton in the –OH group is a singlet. **(6 marks)**

b The two spectra shown below were obtained from isomeric compounds **J** and **K** with the formula C_7H_8O.

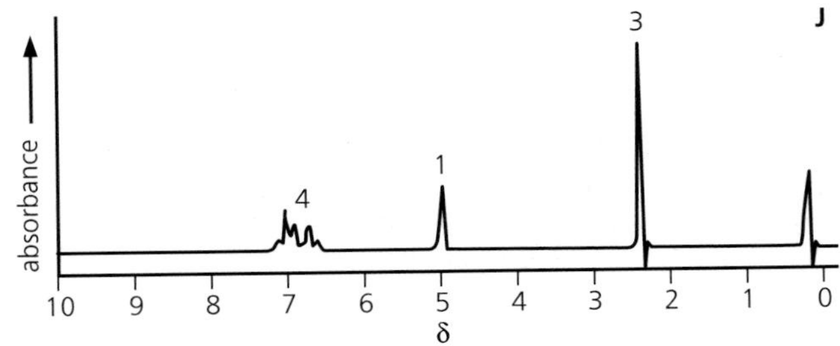

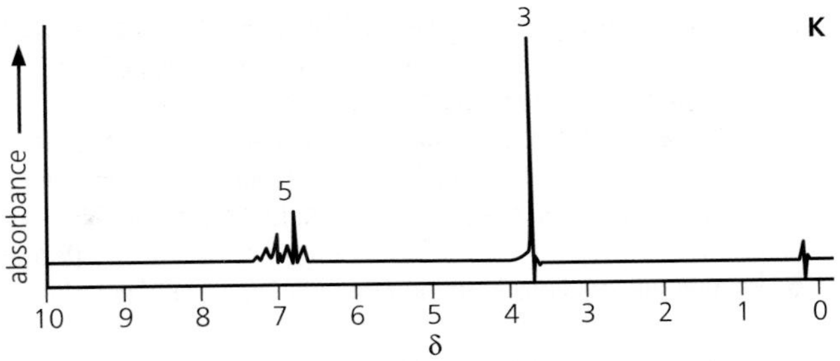

Deduce possible structures for compounds **J** and **K**, explaining in detail how you arrived at your answers. **(7 marks)**

c One of the compounds forms a deep violet complex with Fe^{3+} ions.

i Outline how you could use colorimetry to determine the formula of this complex.

ii The compound which forms this complex is a known industrial pollutant of streams and rivers. Suggest briefly how you might monitor concentrations of this pollutant in water samples. **(8 marks)**
(Total 21 marks)
(OCR 1998)

17 Esters are often described as having "fruity smells". An ester can be prepared in the laboratory by the reaction of an alcohol and a carboxylic acid in the presence of an acid catalyst.

a Ethanol reacts with 2-methylbutanoic acid to produce an ester which is found in ripe apples.

i Draw the displayed formula of 2-methylbutanoic acid.

ii When ethanol reacts with 2-methylbutanoic acid, the ester produced has the formula $CH_3CH_2CH(CH_3)CO_2CH_2CH_3$. Write a balanced equation for the formation of this ester. **(2 marks)**

b The following experiment was carried out by a student.

A 9.2 g sample of ethanol and 20.4 g of 2-methylbutanoic acid were mixed in a flask and 2.0 g of concentrated sulphuric acid was added. The mixture was refluxed for four hours and then fractionally distilled to give 17.4 g of the crude ester. The ester was washed repeatedly with aqueous sodium carbonate until there was no more effervescence. After further washing with distilled water and drying, 15.6 g of pure ester were obtained.

By referring to the experimental procedure above,

i explain the meaning of *refluxed*;

ii explain why the crude ester was *washed repeatedly with aqueous sodium carbonate*;

iii state which gas was responsible for the *effervescence*.

iv Calculate how many moles of each reactant were used [ethanol (M_r: 46), 2-methylbutanoic acid (M_r : 102)]

v Use your answers to **a ii** and to **b iv** to calculate the percentage yield of pure ester obtained in the above experiment. **(9 marks)**

c The ester produced in the above experiment smelt of apples. A similar ester, $CH_3CO_2CH_2CH_2CH(CH_3)_2$, contributes to the flavour of ripe pears.
Suggest structures of a carboxylic acid and an alcohol that could be used for the preparation of the ester $CH_3CO_2CH_2CH_2CH(CH_3)_2$. **(2 marks)**
(Total 13 marks)
(OCR 1998)

5

18 2-Aminopropanoic acid (alanine), $CH_3CH(NH_2)CO_2H$, has a chiral centre and hence can exist as two optical isomers.

a **i** State what is meant by a *chiral centre*.

ii Explain how a chiral centre gives rise to optical isomerism.

iii Draw diagrams to show the relationship between the two optical isomers. State the bond angle around the chiral centre. **(5 marks)**

b In aqueous solution, 2-aminopropanoic acid exists in different forms at different pH values. The zwitterion predominates between pH values of 2.3 and 9.7.
Draw the displayed formula of the predominant form of 2-aminopropanoic acid at pH values of 2.0, 6.0 and 10.0. **(3 marks)**

c 2-Aminopropanoic acid can react with an amino acid, **K**, to form the dipeptide below.

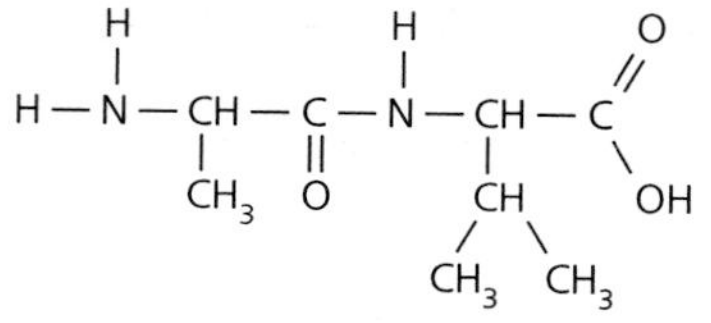

```
     H              H           O
     |              |          //
H — N — CH — C — N — CH — C
         |    ||        |      \
        CH3   O        CH       OH
                      /  \
                    CH3   CH3
```

i Copy the structural formula and draw a circle around the peptide linkage.

ii Draw the displayed formula of the amino acid, **K**.

iii 2-Aminopropanoic acid can react with the amino acid, **K**, to form a different dipeptide from that shown above. Draw the structural formula of this other dipeptide. **(3 marks)**

(Total 11 marks)

(OCR 1998)

19 The proton n.m.r. spectrum of an alcohol, **A**, $C_5H_{12}O$, is shown below.

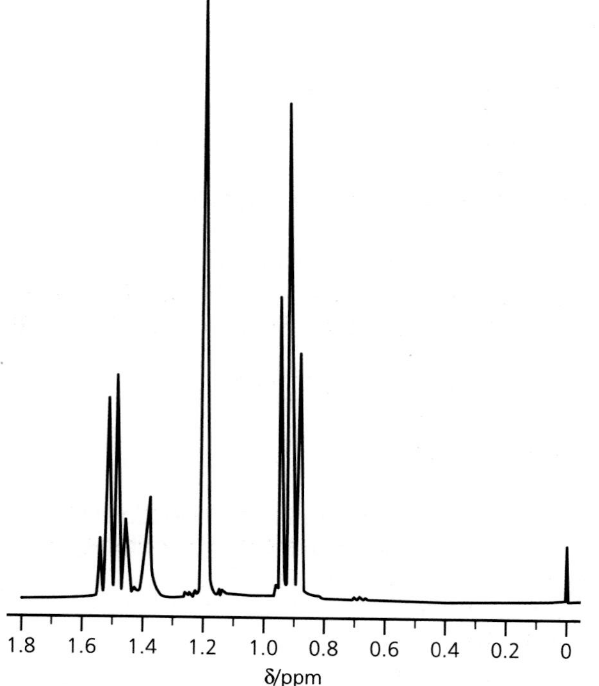

The measured integration trace gives the ratio 0.90 to 0.45 to 2.70 to 1.35 for the peaks at δ 1.52, 1.39, 1.21 and 0.93, respectively.

a What compound is responsible for the signal at δ 0? **(1 mark)**

b How many different types of proton are present in compound **A**? **(1 mark)**

c What is the ratio of the numbers of each type of proton? **(1 mark)**

d The peaks at δ 1.52 and δ 0.93 arise from the presence of a single alkyl group. Identify this group and explain the splitting pattern. **(3 marks)**

e What can be deduced from the single peak at δ 1.21 and its integration value? **(1 mark)**

f Give the structure of compound **A**. **(1 mark)**

(Total 8 marks)

(AQA 1999)

20 Draw a Maxwell–Boltzman curve for a sample of gas at a temperature, T_1. On the same axes draw a second curve for the same sample of gas at a higher temperature, T_2. Label the curves and the axes clearly. Explain, using these curves, why an increase in temperature increases the rate of a gas phase reaction.

Suggest how and why the first curve would change if a larger mass of gas were used at temperature T_1. **(10 marks)**

(AQA 1999)

21 **a** Draw the structure of each of the three ketones which have the molecular formula $C_5H_{10}O$. For each compound give the ratio of the areas under each peak in its low-resolution proton n.m.r. spectrum. **(6 marks)**

b Draw the structure of each of the four aldehydes which also have the molecular formula $C_5H_{10}O$. Label with the letter **X** the compound which has only two peaks in its low-resolution proton n.m.r. spectrum. Label with the letter **Y** the compound which has five peaks with the ratios of the areas under each peak 3:3:2:1:1 in its low-resolution proton n.m.r. spectrum. Label with the letter **Z** the compound which shows optical isomerism. **(7 marks)**

c When carbonyl compounds react with HCN, racemic mixtures are often produced. Name the type of mechanism involved and explain what is meant by the term *racemic mixture*. Choose any carbonyl compound which does **not** form a racemic mixture when it reacts with HCN and draw the structure of the product formed by the reaction of this carbonyl compound with HCN. **(4 marks)**

d Explain why aldehydes react with Tollen's (or Fehling's) reagent but ketones do not. **(3 marks)**

(Total 20 marks)

(AQA 1999)

Answers

Note that numerical and organic identification answers only are given.

Answers to end-of-chapter questions

1.1 Atoms: The basis of matter (pg 17)

1 **a** 79.990
b 107.974
c 52.056
2 **a** 1.806×10^{24} **c** 2.408×10^{24}
b 1.003×10^{23} **d** 1.650×10^{23}
3 **a** 198 g **c** 164 g
b 291 g **d** 396 g
4 **a** 3 mol **c** 408 g
b 1 mol **d** Al_2O_3 12.75 g, HI 96 g
5 **a** C 0.03211 mol, H 0.05178 mol, N 0.01605 mol, P 0.009616 mol, O 0.04170 mol
b 10:16:5:3:13
c $C_{10}H_{16}O_{13}N_5P_3$
d $C_{10}H_{16}O_{13}N_5P_3$
6 **a** 3.5 mol
b Old volume = 0.03 m^3, new volume = 1312.5 m^3
7 **b** 0.08 mol dm^{-3}
c 0.596 g

1.2 The structure of the atom (pg 28)

3 **a** 7, 8, 7 **d** 92, 143, 92
b 46, 59, 46 **e** 50, 70, 50
c 19, 20, 19 **f** 9, 10, 9
4 **a** 18.0105 u **d** 28.0654 u
b 43.9898 u **e** 101.9477 u
c 20.0231 u
5 **a** 8.146×10^{-15} m
b 2.264×10^{-42} m^3
c 1.73×10^{-32} m^3
6 **b** 143 km^3
7 **a** 7.33×10^{17} g m^{-3}
b 174 km

1.3 The arrangement of electrons in atoms (pg 46)

1 **c** 4.8×10^{-19} J

1.4 Chemical bonding (pg 63)

5 **a** –2 **e** N +3, O –2
b P –3, H +1 **f** K +1, I +5, O –2
c H +1, S +6, O –2 **g** Na +1, S +2, O –2
d I +7, F –1

1.5 Intermolecular forces (pg 72)

5 **b** 17.4 kJ mol^{-1}
c 23.7 kJ mol^{-1}

1.6 Solids (pg 86)

5 **a** 4 **c** $4.743\,8 \times 10^{-29}$ m^3
b $4.220\,8 \times 10^{-25}$ kg **d** 8900 kg m^{-3}

1.7 Liquids and gases (pg 100)

4 99.997%

1.8 Energy changes (pg 121)

1 762 g
2 **a** 2194.5 J
b –433.5 kJ mol^{-1}
3 66 kJ mol^{-1}
4 –510.7 kJ mol^{-1}
5 ΔH_f° [CaCl] = –178.1 kJ mol^{-1}
ΔH_f° [$CaCl_2$] = –759.4 kJ mol^{-1}
ΔH_f° [$CaCl_3$] = +1493.3 kJ mol^{-1}

2.1 Patterns of matter (pg 141)

2 **a** Mg^{2+}, K^+, Al^{3+}, Sr^{2+}, S^{2-}, Br^-, N^{3-}, Cl^-, O^{2-}
b i KBr
ii $AlCl_3$
iii Mg_3N_2
iv MgS
v SrO
vi K_2O
3 **a** CH_4, NH_3, HCl, H_2S, PH_3, H_2O

2.2 The periodicity of compounds (pg 152)

3 **b** Li 1.1, Be 0.6, B 0.1, C –0.4, N –0.9, O –1.4, F –1.9, Na 1.2, Mg 0.9, Al 0.6, Si 0.3, P 0, S –0.4, Cl –0.9

2.3 The s-block elements (pg 167)

4 **a** 738 kJ mol^{-1}
b 2189 kJ mol^{-1}

2.4 The p-block elements: The halogens (pg 183)

3 **a** HOCl +1, $HClO_2$ +3, $HClO_3$ +5, $HClO_4$ +7
d $HClO_2$ +3, ClO_2 +4, HCl –1
4 **c** 9.2×10^{-4} mol
d 4.6×10^{-4} mol
e 9.2×10^{-4} mol
f 4.6×10^{-3} mol
g 34%

2.5 The p-block elements: Aluminium (pg 192)

3 **a** −1
4 **a** Al^{3+}, C^{4-}

2.7 The p-block elements: Carbon and group IV (pg 205)

2.8 The d-block elements: The transition metals (pg 227)

3 **a** $[TiF_6]^{2-}$, $[VO(H_2O)_5]^{2+}$, $[Cr(CO)_6]$, $[Mn(H_2O)_6]^{2+}$, $[Fe(H_2O)_5NO]^{2+}$, $[Ni(H_2O)_2(NH_3)_4]^{2+}$
c 0
d $[Ni(H_2O)_2(NH_3)_4]^{2+}$
4 **d ii** Cr(0) → Cr(+3) → Cr(+2)
6 **c ii** 7.15×10^{-4} mol
iii 3.57×10^{-3} mol
iv 3.57×10^{-2} mol
v 2.00 g

3.1 Chemical equilibria: How far? (pg 253)

3 **a** kPa^{-2}
5 **a** $K_c = \dfrac{[CH_3COOCH_2CH_3(aq)][H_2O(l)]}{[CH_3COOH(aq)][CH_3CH_2OH(aq)]}$
b 4
c $\frac{1}{4}$
d 0.1066 mol
6 **a** $K_p = \dfrac{p_{PCl_5}}{p_{PCl_3} p_{Cl_2}}$
c $p_{PCl_3} = 0.777$ kPa $p_{Cl_2} = 0.247$ kPa
$p_{PCl_5} = 3.843$ kPa
7 **a** $mol^{-1}\ dm^3$
f i $0.0185\ mol\ dm^{-3}$
ii $0.0315\ mol\ dm^{-3}$

3.2 Acid–base equilibria (pg 268)

2 **a** 0.01 mol
b 0.005 mol
c 0.53 g
d 3%
4 **a ii** $K_a = \dfrac{[H^+(aq)][PABA^-(aq)]}{[HPABA(aq)]}$
b $6.026 \times 10^{-4}\ mol\ dm^{-3}$
c $6.026 \times 10^{-4}\ mol\ dm^{-3}$
d 7.2×10^{-4} mol
e $0.03\ mol\ dm^{-3}$
f $K_a = 1.21 \times 10^{-5}\ mol\ dm^{-3}$
6 **a** $2.98 \times 10^{-8}\ mol\ dm^{-3}$
b $2.98 \times 10^{-8}\ mol\ dm^{-3}$
c 7.52
7 **a** $1.10 \times 10^{-5}\ mol\ dm^{-3}$
b i $9.54 \times 10^{-12}\ mol\ dm^{-3}$
c ii 11.02

3.3 Buffers and indicators (pg 279)

2 **a i** $4.04 \times 10^{-3}\ mol\ dm^{-3}$
ii 2.40
b $1 \times 10^{-4}\ mol\ dm^{-3}$
c $0.1635\ mol\ dm^{-3}$
d New pH = 4.52

3.4 Precipitation and complex ion formation (pg 290)

1 $2.0 \times 10^{-5}\ mol^2\ dm^{-6}$
2 $2.33 \times 10^{-3}\ g\ dm^{-3}$
3 $2 \times 10^{-9}\ mol\ dm^{-3}$
5 4×10^{-34}

3.5 Electrochemistry (pg 312)

3 **c** Copper
4 **a** 2 V
b 0.59 V
c 0.79 V
d 2.53 V
5 **b i** $Mn^{3+}(aq)$
ii $Ce^{3+}(aq)$
c $Mn^{3+}(aq)$
d $V^{2+}(aq)$, $Ce^{3+}(aq)$
e $Hg^{2+}(aq)$, $Mn^{3+}(aq)$
6 **a, d**
7 **a** 0.12 V, $Pb^{2+}(aq) + Ni(s) \rightarrow Pb(s) + Ni^{2+}(aq)$
b 0.03 V, $Ag^+(aq) + Fe^{2+}(aq) \rightarrow Ag(s) + Fe^{3+}(aq)$
c 1.40 V, $\frac{3}{2}Cl_2(g) + Fe(s) \rightarrow 3Cl^-(aq) + Fe^{3+}(aq)$

3.6 Reaction kinetics: How fast? (pg 339)

2 **a** Negative
b Positive
c Double the rate
d Reduce the rate by $1/\sqrt{2}$
3 **a** $\frac{1}{2}$
b $\frac{1}{4}$
c $\frac{1}{1024}$
4 **b** 2 years ago
5 **a** First order
b Second order
c $47.62 \times 10^{-3}\ mol^{-2}\ dm^6\ s^{-1}$
6 **a** $-d[A]/dt = k[A]$
c $0.010\,65\ s^{-1}$

3.7 Thermodynamics (pg 354)

4 334.8 K
5 **i a**
ii b
iii c
iv b
v a
vi b
6 **a** $2Fe^{2+}(aq) + Ni^{2+}(aq) \rightarrow 2Fe^{3+}(aq) + Ni(s)$
c $197\ kJ\ mol^{-1}$

4.1 Organising the range of organic molecules (pg 377)

2 **a** $C_{21}H_{22}N_2O_2$
  b $C_{21}H_{22}N_2O_2$

4 **c i** Identical
  ii Structural isomers
  iii Structural isomers
  iv Geometric isomers
  v Optical isomers

4.2 The saturated hydrocarbons (pg 392)

4.3 The unsaturated hydrocarbons: The alkenes and alkynes (pg 412)

1 **b i** Hex-3-ene
  ii But-1,3-diene
  iii Cyclohexene
  iv But-1-yne

4.4 The unsaturated hydrocarbons: The arenes (pg 431)

1 **a i** 1-chloro-3-methylbenzene
  ii 2-bromo-4-methylphenol
  iii 1-iodo-3-ethylbenzene
  iv Hexachlorobenzene

2 **a** A, F
  b E
  c A, C, F
  d B, D, E
  e A, F
  f C, E

4.5 The organohalogen compounds (pg 448)

3 **a** Organofluorides

5.1 Alcohols, phenols and ethers (pg 472)

1 **a i** A **ii** E **iii** B
  b C
  c D
  d A propan-1-ol, B 2-methylbutan-2-ol, C 3,5-dimethylphenol, D methoxypropane, E butan-2-ol
  e E
  f A
  g C
  h 2-methylbut-1-ene and 2-methylbut-2-ene
  i D

5 **a** W
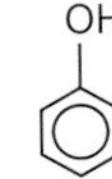

X
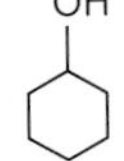

Y

Z
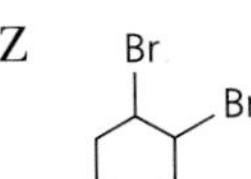

5.2 Carbonyl compounds: Aldehydes and ketones (pg 488)

1 **a** B, D
  b A, C
  c B, D
  d A, C
  e A butanone, B propanal, C diphenylmethanone, D benzenecarbaldehyde
  f B, D
  g B, D

4 $CH_3COCH_2CH_2CH_3$ $CH_3CH_2COCH_2CH_3$
  $CH_3COCH(CH_3)_2$

5.3 Carboxylic acids and their derivatives (pg 506)

1 **a** A, D
  b C
  c E
  d B
  e D
  f B, C, E
  g B

5.4 Organic nitrogen compounds (pg 524)

1 **a** A
  b D
  c B
  d E
  e C
  f D
  g A, B
  h B

5.5 Naturally occurring macromolecules (pg 541)

1 **a** A, B, F
  b C
  c E
  d D
  e C
  f A, B, F

2 **a** Ribose

Answers to end-of-section questions

The author takes sole responsibility for these answers, which have not been provided by the Examination Boards.

Section 1

1 a ii A is HCN, B is CNH_5 (or CH_3NH_2)
b i $C_3H_7{}^{37}Cl^+$
ii $C_2H_5{}^{81}Br^+$
2 c ii 1197 kJ mol^{-1}
3 b i –124 kJ mol^{-1}
4 b 4817 kJ mol^{-1}
5 a –692 kJ mol^{-1}
6 a i

Isotope	Percentage composition	Protons	Neutrons	Electrons
^{39}K	92	19	20	19
^{41}K	8	19	22	19

iii 39.16
7 a $PbCO_3$
b ii 685.0 g **iii** 68.5 g
9 b i 1.64×10^{-4} mol
ii 1.64×10^{-4} mol
iii 3.28×10^{-3} mol dm^{-3}, 0.21 g dm^{-3}
11 b ii –1561 kJ mol^{-1} **c** 46.4 K
12 b 6.021×10^{23} mol^{-1}
c i 197 atm **ii** 246 atm
14 d Kr, 83.93 **e** $C_3H_2O_2N$, $C_6H_4O_4N_2$
15 b i IF_5 **c ii** 108°
16 b –145 kJ mol^{-1}
c –247 kJ mol^{-1}
e +44 kJ mol^{-1}
18 a $^{34}_{16}S$
b $1s^2\,2s^2\,2p^6$
e 1.65×10^{-24} g, 12.99, 12.01
19 b ii 0.015 mol dm^{-3}

Section 2

1 a ii +6
c i Cr^{3+}
d i MnO_3Cl
ii +7
3 d i 0.002 mol **ii** 0.01 mol **iv** 2
6 a A is $FeCl_2$, B is $FeCl_3$
b Fe: $1s^2\,2s^2\,2p^6\,3s^2\,3p^6\,3d^6\,4s^2$
A: $1s^2\,2s^2\,2p^6\,3s^2\,3p^6\,3d^6$
B: $1s^2\,2s^2\,2p^6\,3s^2\,3p^6\,3d^5$
9 b i 0.05 mol^{-1} **ii** 50 cm^3 **iii** 1.2 dm^3
11 a [Ar] $3d^7$
b i +2 **iii** 6
12 c $[Fe(H_2O)_5OH]^{2+}$
13 d 26% Br, 56% I

Section 3

1 a i NO – 2nd order, CO – zero order, O_2 – 1st order
ii $k = 4400\ dm^6\ mol^{-2}\ s^{-1}$
c $NaNO_3$, 0.48 dm^3
2 d i +0.46 V
6 c iii 35%
10 a 1200 cm^3
11 a iii 10^{-14} $mol^2\ dm^{-6}$
b i 0.06 mol dm^{-3}, pH1.2 **ii** pH12.9
c i 0.32 mol dm^{-3} **ii** 14.6 cm^3
14 a ii 54%

Section 4

1 a i kerosene/paraffin **b i** $C_{19}H_{40}$
2 a i but-1-ene
iii $CH_3CH{=}CHCH_3$
3 a $(CH_3)_2CHCHBrCH_3$
c $(CH_3)_2C{=}CHCH_3$, $(CH_3)_2CH{-}CH{=}CH_2$
5 a ii C_8H_{10}
e

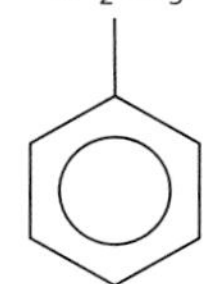

6 b i

H_3C, CH_3 / C=C / H, H — *cis*-isomer

H, CH_3 / C=C / H_3C, H — *trans*-isomer

ii

CH_3CH_2, H / C=C / H, H — but-1-ene

H_3C, H / C=C / H_3C, H — methylpropene

c

A: $H-C(CH_3)(H)-C(CH_3)(H)-Br$

B: $H-C(CH_3)(H)-C(CH_3)(H)-H$

C: $H-C(CH_3)(H)-C(CH_3)(H)-OH$

D: $H-C(CH_3)(Br)-C(CH_3)(Br)-H$

d ii $-C(CH_3)(H)-C(CH_3)(H)-C(CH_3)(H)-C(CH_3)(H)-$

e i $CH_2{=}CHCH{=}CH_2$

8 **a** $C_{10}H_{16}$
b ii $C_{10}H_{22}$ **iv** 7.2 dm^3
c i 6 double bonds **ii** $C_{30}H_{50}$

10 **a** $CH_3CH_2CH_2CH_2CHO$
b i

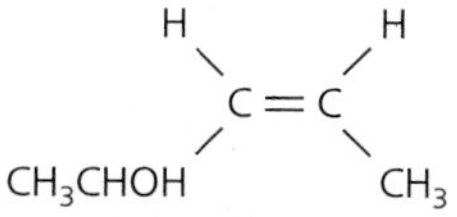

H H
C=C
CH_3CHOH CH_3

ii

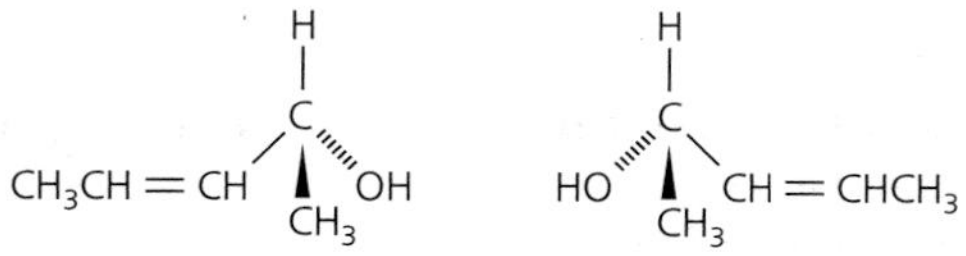

H H
C C
$CH_3CH=CH$ OH HO $CH=CHCH_3$
CH_3 CH_3

d i $CH_3COCH{=}CHCH_3$

11 **b i**

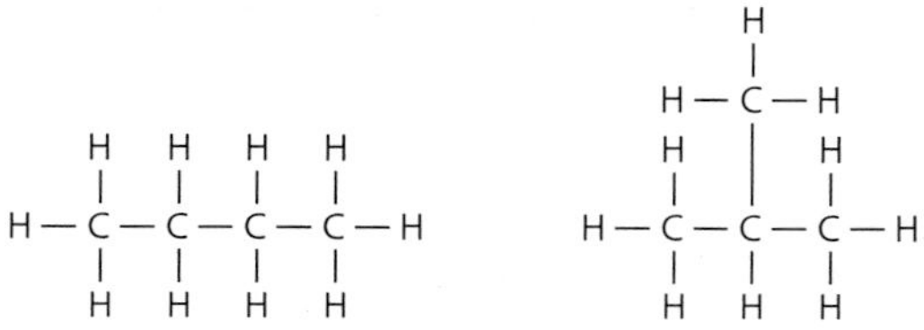

H H H H
H—C—C—C—C—H
H H H H

H
H—C—H
H H
H—C—C—C—H
H H H

ii 43 – C_3H_7, 29 – C_2H_5, 15 – CH_3

12 **b i**

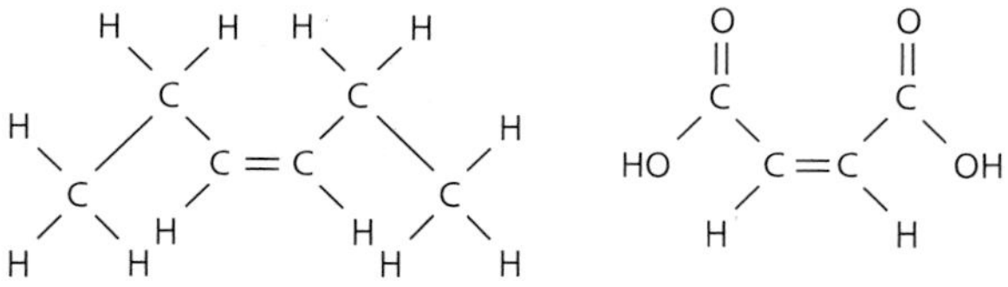

H H H H
C C
H C=C H
C C
H H H H
H H

O O
C C
HO C=C OH
H H

c iv

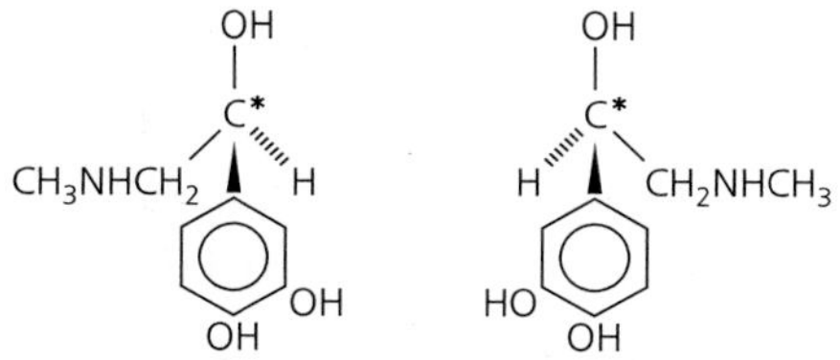

OH OH
C* C*
CH_3NHCH_2 H H CH_2NHCH_3
OH HO
OH OH

Section 5

1 **a**

COOH

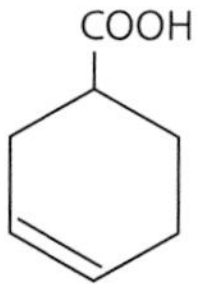

3 **b** 26.8 kPa

4 A – $CH_3CH_2COCH_2CH_3$
B – $CH_3CH_2CHOHCH_2CH_3$
C – $CH_3CH_2CH{=}CHCH_3$
D – $CH_3CHOHCH_2CH_3$

5 **b i**

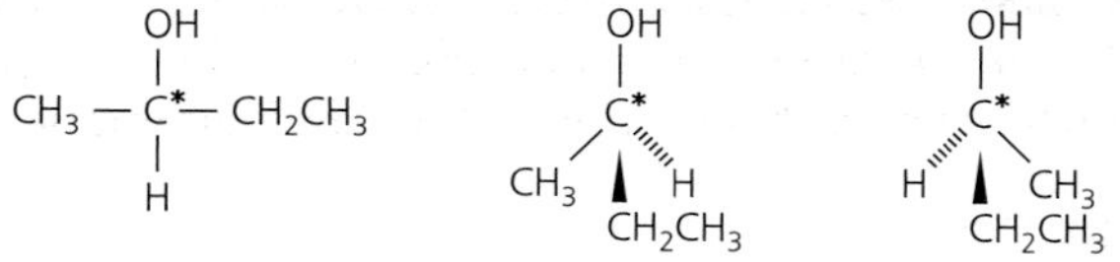

OH
CH_3 — C* — CH_2CH_3
H

OH
C*
CH_3 H
CH_2CH_3

OH
C*
H CH_3
CH_2CH_3

6 **a ii**

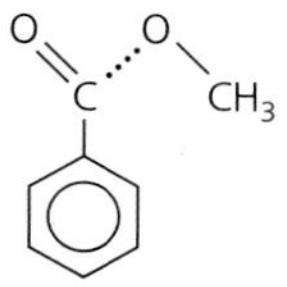

O O
C CH_3

b i 0.020 mol
ii 2.0 mol dm^{-3}
iii 61.9%

7 $CH_3CH_2COCH_3$

8 **b** $C_5H_{10}O$
c i $CH_3CH_2CHBrCHBrCH_2OH$
ii $CH_3CH_2CH{=}CHCH_2OCOCH_3$
d $CH_3CH_2CH_2CHBrCH_2Br$
$CH_3CH_2CHBrCH_2CH_2Br$

9 **a iv** 2-methylpropene $CH_2{=}C(CH_3)_2$

10 **b** 2-methylpropan-2-ol
c i C gives $CH_3CH_2CH_2CHO$
D gives $CH_3CH(CH_3)CHO$
d iv A gives $(CH_3)_2C{=}CH_2$
C gives $CH_3CH_2CH{=}CH_2$
D gives $(CH_3)_2C{=}CH_2$
v B gives $CH_3CH{=}CHCH_3$ and $CH_3CH_2CH{=}CH_2$

11 **a i** ethyl ethanoate

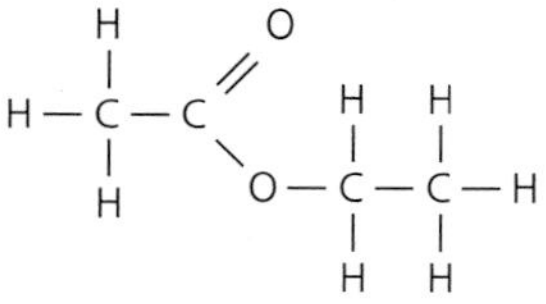

H O
H—C—C H H
H O—C—C—H
H H

c ii ethanoyl ethanoate

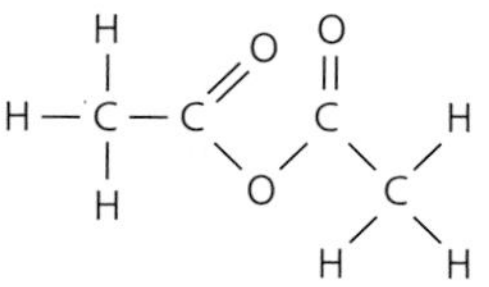

H O O
H—C—C C H
H O C
H H

d i ethanamide

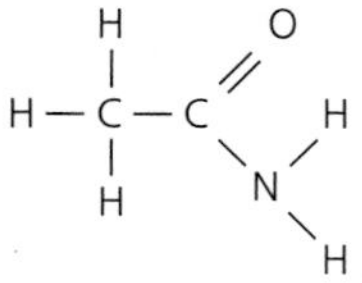

H O
H—C—C H
H N
H

12 a i CH_3Cl, $AlCl_3$, heat

c ii

```
    NH2
     |
  (benzene ring)
```

d i

```
    H   H   H
    |   |   |
H — C — C — C — O — H
    |   |   |
    H   H   H
```

ii

```
    H   H      O
    |   |     //
H — C — C — C
    |   |     \
    H   H      O — H
```

iii

```
             H
            /
    H   H   O
    |   |   |
H — C — C — C — C≡N
    |   |   |
    H   H   H
```

13 b i, ii

```
    H       H       H
   /       /       /
   O   H   O   H   O     O
   |   |   |   |   |    //
H — C — C* — C — C — C* — C
   |   |   |   |   |     \
   H   O   H   O   H      H
        \       \
         H       H
```

14 a

```
 COOH              H   H
  |                |   |
(benzene ring)  HO — C — C — OH
  |                |   |
 COOH              H   H
```

15 a C > D > A > B

c 20 cm^3

d ii

acid	approximate retention time
A	29
B	40
C	17
D	23

16 b

```
        OH                     CH3
        |                     /
J – (benzene ring)     K –   O
        |                    |
        CH3            (benzene ring)
```

17 a i

```
                H
                |
            H — C — H
    H   H       |        O
    |   |       |       //
H — C — C — C — C
    |   |   |           \
    H   H   H            O — H
```

b iv 0.2 mol ethanol, 0.2 mol 2-methylbutanoic acid

v 60%

c CH_3COOH $(CH_3)_2CHCH_2CH_2OH$

18 a iii

```
      H                      H
      |                      |
      C                      C                108°
     / ⋮ \                  / ⋮ \
  CH3  ▲  COOH        HOOC   ▲   CH3
      NH2                   NH2
```

b

```
pH 2.0                 pH 6.0                 pH 10.0
     H                      H                      H
     |                      |                      |
CH3 — C — COOH        CH3 — C — COO⊖         CH3 — C — COO⊖
    ⊕NH3                    |                      |
                          ⊕NH3                    NH2
```

c ii

```
H        H        O
 \       |       //
   N  —  C  —  C
 / H     |       \
H  |     |        O — H
H — C — C — H
   |     |
   H     |
     H — C — H
         |
         H
```

iii

```
                                  H        O
H        H        O               |       //
 \       |       //               C  —  C
   N  —  C  —  C                 /|       \
 / H     |       \              / |        O — H
H  |     |         NH — C — H
H — C — C — H      |    |
   |     |          H    H
   H     |
     H — C — H
         |
         H
```

19 b 4 **c** 2:1:6:3 **d** methyl (CH_3) group

f

```
             H
             |
         H — C — H
    H   H    |
    |   |    |
H — C — C — C — O
    |   |    |   \
    H   H    |    H
         H — C — H
             |
             H
```

21 a $CH_3COCH_2CH_2CH_3$ 3:2:2:3 $CH_3CH_2COCH_2CH_3$ 3:2

$CH_3COCH(CH_3)_2$ 3:1:6

b $CH_3CH_2CH_2CH_2CHO$ $(CH_3)_3CCHO$[X]

$(CH_3)_2CHCH_2CHO$[Y] $CH_3CH_2CH(CH_3)CHO$[Z]

Useful data

Physical constants

Quantity	Symbol	Value and units	
Speed of light in a vacuum	c	$299\,792\,458$ m s^{-1}	(exact)
Plank constant	h	6.626×10^{-34} J s	
Avogadro constant	N_A	6.022×10^{23} mol^{-1}	
Molar gas constant	R	8.314 J mol^{-1} K^{-1}	
Boltzmann constant	k	1.381×10^{-23} J K^{-1}	
Molar volume of ideal gas, RT/p (T = 273.15 K, p = 100 kPa)	V_m	22.4 dm^3 mol^{-1}	
Elementary charge	e	1.602×10^{-19} C	
Electron specific charge	$-e/m_e$	-1.759×10^{-11} C kg^{-1}	
Electron mass	m_e	9.109×10^{-31} kg	
Proton mass	m_p	1.673×10^{-27} kg	
Neutron mass	m_n	1.675×10^{-27} kg	

Relative atomic masses

The relative atomic masses of selected elements are given correct to one decimal place.

Element	Symbol	A_r	Element	Symbol	A_r
Aluminium	Al	27.0	Molybdenum	Mo	95.9
Antimony	Sb	121.8	Neodymium	Nd	144.2
Argon	Ar	39.9	Neon	Ne	20.2
Arsenic	As	74.9	Nickel	Ni	58.7
Barium	Ba	137.3	Niobium	Nb	92.9
Beryllium	Be	9.0	Nitrogen	N	14.0
Bismuth	Bi	209.0	Osmium	Os	190.2
Boron	B	10.8	Oxygen	O	16.0
Bromine	Br	79.9	Palladium	Pd	106.4
Cadmium	Cd	112.4	Phosphorus	P	31.0
Caesium	Cs	132.9	Platinum	Pt	195.1
Calcium	Ca	40.1	Potassium	K	39.1
Carbon	C	12.0	Praseodymium	Pr	140.9
Cerium	Ce	140.1	Rhenium	Re	186.2
Chlorine	Cl	35.5	Rhodium	Rh	102.9
Chromium	Cr	52.0	Rubidium	Rb	85.5
Cobalt	Co	58.9	Ruthenium	Ru	101.1
Copper	Cu	63.5	Samarium	Sm	150.4
Dysprosium	Dy	162.5	Scandium	Se	45.0
Erbium	Er	167.3	Selenium	Se	79.0
Europium	Eu	152.0	Silicon	Si	28.1
Fluorine	F	19.0	Silver	Ag	107.9
Gadolinium	Gd	157.3	Sodium	Na	23.0
Gallium	Ga	69.7	Strontium	Sr	87.6
Germanium	Ge	72.6	Sulphur	S	32.1
Gold	Au	197.0	Tantalum	Ta	180.9
Hafnium	Hf	178.5	Tellurium	Te	127.6
Helium	He	4.0	Terbium	Tb	158.9
Holmium	Ho	164.9	Thallium	Tl	204.4
Hydrogen	H	1.0	Thorium	Th	232.0
Indium	In	114.8	Thulium	Tm	168.9
Iodine	I	126.9	Tin	Sn	118.7
Iridium	Ir	192.2	Titanium	Ti	47.9
Iron	Fe	55.8	Tungsten	W	183.9
Krypton	Kr	83.8	Uranium	U	238.0
Lanthanum	La	138.9	Vanadium	V	50.9
Lead	Pb	207.2	Xenon	Xe	131.3
Lithium	Li	6.9	Ytterbium	Yb	173.0
Lutetium	Lu	175.0	Yttrium	Y	88.91
Magnesium	Mg	24.3	Zinc	Zn	65.4
Manganese	Mn	54.9	Zirconium	Zr	91.2
Mercury	Hg	200.6			

Electronic configurations

Atomic number	Symbol	Electronic configuration
1	H	$1s^1$
2	He	$1s^2$
3	Li	[He] $2s^1$
4	Be	[He] $2s^2$
5	B	[He] $2s^22p^1$
6	C	[He] $2s^22p^2$
7	N	[He] $2s^22p^3$
8	O	[He] $2s^22p^4$
9	F	[He] $2s^22p^5$
10	Ne	[He] $2s^22p^6$
11	Na	[Ne] $3s^1$
12	Mg	[Ne] $3s^2$
13	Al	[Ne] $3s^23p^1$
14	Si	[Ne] $3s^23p^2$
15	P	[Ne] $3s^23p^3$
16	S	[Ne] $3s^23p^4$
17	Cl	[Ne] $3s^23p^5$
18	Ar	[Ne] $3s^23p^6$
19	K	[Ar] $4s^1$
20	Ca	[Ar] $4s^2$
21	Sc	[Ar] $4s^23d^1$
22	Ti	[Ar] $4s^23d^2$
23	V	[Ar] $4s^23d^3$
24	Cr	[Ar] $4s^13d^5$
25	Mn	[Ar] $4s^23d^5$
26	Fe	[Ar] $4s^23d^6$
27	Co	[Ar] $4s^23d^7$
28	Ni	[Ar] $4s^23d^8$
29	Cu	[Ar] $4s^13d^{10}$
30	Zn	[Ar] $4s^23d^{10}$
31	Ga	[Ar] $4s^23d^{10}4p^1$
32	Ge	[Ar] $4s^23d^{10}4p^2$
33	As	[Ar] $4s^23d^{10}4p^3$
34	Se	[Ar] $4s^23d^{10}4p^4$
35	Br	[Ar] $4s^23d^{10}4p^5$
36	Kr	[Ar] $4s^23d^{10}4p^6$
37	Rb	[Kr] $5s^1$
38	Sr	[Kr] $5s^2$
39	Y	[Kr] $5s^24d^1$
40	Zr	[Kr] $5s^24d^2$
41	Nb	[Kr] $5s^14d^4$
42	Mo	[Kr] $5s^14d^5$
43	Tc	[Kr] $5s^24d^5$
44	Ru	[Kr] $5s^14d^7$
45	Rh	[Kr] $5s^14d^8$
46	Pd	[Kr] $4d^{10}$
47	Ag	[Kr] $5s^14d^{10}$
48	Cd	[Kr] $5s^24d^{10}$
49	In	[Kr] $5s^24d^{10}5p^1$
50	Sn	[Kr] $5s^24d^{10}5p^2$
51	Sb	[Kr] $5s^24d^{10}5p^3$
52	Te	[Kr] $5s^24d^{10}5p^4$
53	I	[Kr] $5s^24d^{10}5p^5$
54	Xe	[Kr] $5s^24d^{10}5p^6$
55	Cs	[Xe] $6s^1$
56	Ba	[Xe] $6s^2$
57	La	[Xe] $6s^25d^1$
58	Ce	[Xe] $6s^24f^15d^1$
59	Pr	[Xe] $6s^24f^3$
60	Nd	[Xe] $6s^24f^4$
61	Pm	[Xe] $6s^24f^5$
62	Sm	[Xe] $6s^24f^6$
63	Eu	[Xe] $6s^24f^7$
64	Gd	[Xe] $6s^24f^75d^1$
65	Tb	[Xe] $6s^24f^9$
66	Dy	[Xe] $6s^24f^{10}$
67	Ho	[Xe] $6s^24f^{11}$
68	Er	[Xe] $6s^24f^{12}$
69	Tm	[Xe] $6s^24f^{13}$
70	Yb	[Xe] $6s^24f^{14}$
71	Lu	[Xe] $6s^24f^{14}5d^1$
72	Hf	[Xe] $6s^24f^{14}5d^2$
73	Ta	[Xe] $6s^24f^{14}5d^3$
74	W	[Xe] $6s^24f^{14}5d^4$
75	Re	[Xe] $6s^24f^{14}5d^5$
76	Os	[Xe] $6s^24f^{14}5d^6$
77	Ir	[Xe] $6s^24f^{14}5d^7$
78	Pt	[Xe] $6s^14f^{14}5d^9$
79	Au	[Xe] $6s^14f^{14}5d^{10}$
80	Hg	[Xe] $6s^24f^{14}5d^{10}$
81	Tl	[Xe] $6s^24f^{14}5d^{10}6p^1$
82	Pb	[Xe] $6s^24f^{14}5d^{10}6p^2$
83	Bi	[Xe] $6s^24f^{14}5d^{10}6p^3$
84	Po	[Xe] $6s^24f^{14}5d^{10}6p^4$
85	At	[Xe] $6s^24f^{14}5d^{10}6p^5$
86	Rn	[Xe] $6s^24f^{14}5d^{10}6p^6$
87	Fr	[Rn] $7s^1$
88	Ra	[Rn] $7s^2$
89	Ac	[Rn] $7s^26d^1$
90	Th	[Rn] $7s^26d^2$
91	Pa	[Rn] $7s^25f^26d^1$
92	U	[Rn] $7s^25f^36d^1$
93	Np	[Rn] $7s^25f^46d^1$
94	Pu	[Rn] $7s^25f^6$
95	Am	[Rn] $7s^25f^7$
96	Cm	[Rn] $7s^25f^76d^1$
97	Bk	[Rn] $7s^25f^9$
98	Cf	[Rn] $7s^25f^{10}$
99	Es	[Rn] $7s^25f^{11}$
100	Fm	[Rn] $7s^25f^{12}$
101	Md	[Rn] $7s^25f^{13}$
102	No	[Rn] $7s^25f^{14}$
103	Lr	[Rn] $7s^25f^{14}6d^1$
104	Rf	[Rn] $7s^25f^{14}6d^2$
105	Ha	[Rn] $7s^25f^{14}6d^3$
106		[Rn] $7s^25f^{14}6d^4$

INDEX

The Periodic table
(wide form)

	I	II																												III	IV	V	VI	VII	0
1s																								1 H											2 He
2s	3 Li	4 Be																											2p	5 B	6 C	7 N	8 O	9 F	10 Ne
3s	11 Na	12 Mg																											3p	13 Al	14 Si	15 P	16 S	17 Cl	18 Ar
4s	19 K	20 Ca	3d	21 Sc																22 Ti	23 V	24 Cr	25 Mn	26 Fe	27 Co	28 Ni	29 Cu	30 Zn	4p	31 Ga	32 Ge	33 As	34 Se	35 Br	36 Kr
5s	37 Rb	38 Sr	4d	39 Y																40 Zr	41 Nb	42 Mo	43 Tc	44 Ru	45 Rh	46 Pd	47 Ag	48 Cd	5p	49 In	50 Sn	51 Sb	52 Te	53 I	54 Xe
6s	55 Cs	56 Ba	5d	57 La	4f	58 Ce	59 Pr	60 Nd	61 Pm	62 Sm	63 Eu	64 Gd	65 Tb	66 Dy	67 Ho	68 Er	69 Tm	70 Yb	71 Lu	72 Hf	73 Ta	74 W	75 Re	76 Os	77 Ir	78 Pt	79 Au	80 Hg	6p	81 Tl	82 Pb	83 Bi	84 Po	85 At	86 Rn
7s	87 Fr	88 Ra	6d	89 Ac	5f	90 Th	91 Pa	92 U	93 Np	94 Pu	95 Am	96 Cm	97 Bk	98 Cf	99 Es	100 Fm	101 Md	102 No	103 Lr	104 Rf*	105 Ha	106													

*or Ku

Blocks
- Reactive metals
- Transition metals
- Poor metals
- Metalloids
- Non-metals
- Noble gases